T0215702

Lecture Notes in Computer Science   12047

Shivakumara Palaiahnakote ·
Gabriella Sanniti di Baja ·
Liang Wang · Wei Qi Yan (Eds.)

Pattern Recognition

5th Asian Conference, ACPR 2019
Auckland, New Zealand, November 26–29, 2019
Revised Selected Papers, Part II

 Springer

Editors
Shivakumara Palaiahnakote
University of Malaya
Kuala Lumpur, Malaysia

Gabriella Sanniti di Baja
Consiglio Nazionale delle Ricerche, ICAR
Naples, Italy

Liang Wang
Chinese Academy of Sciences
Beijing, China

Wei Qi Yan
Auckland University of Technology
Auckland, New Zealand

ISSN 0302-9743 ISSN 1611-3349 (electronic)
Lecture Notes in Computer Science
ISBN 978-3-030-41298-2 ISBN 978-3-030-41299-9 (eBook)
https://doi.org/10.1007/978-3-030-41299-9

LNCS Sublibrary: SL6 – Image Processing, Computer Vision, Pattern Recognition, and Graphics

This Springer imprint is published by the registered company Springer Nature Switzerland AG
The registered company address is: Gewerbestrasse 11, 6330 Cham, Switzerland

Preface

Our heartiest welcome to ACPR 2019, the 5th Asian Conference on Pattern Recognition, held in Auckland, New Zealand, during 26–29 November, 2019. It is truly an honor to host our premier conference in the business capital of New Zealand; the first time that this conference has been hosted in this part of the world. Auckland is New Zealand's largest city and most important economic region. It is host to some of the most picturesque seaside views in New Zealand as well as fantastic shopping and dining opportunities. Auckland is a true multicultural city – it has the largest Polynesian population of any city in the world as well as a significant percentage of Maori and Asian peoples. We did our best to prepare every aspect of this conference and hope that you enjoyed your stay in Auckland.

ACPR was initiated to promote the scientific exchanges and collaborations of pattern recognition researchers in the Asia-Pacific region, and it also welcomed participation from other regions of the world. The fifth ACPR followed the previous editions, ACPR 2011 in Beijing, China, ACPR 2013 in Okinawa, Japan, ACPR 2015 in Kuala Lumpur, Malaysia, and ACPR 2017 in Nanjing, China. For ACPR 2019, there was not only a well-presented technical program, but we also invited participants to experience the diverse culture of the city of Auckland.

ACPR 2019 received 214 full submissions from 37 countries. The program chairs invited 68 Program Committee members and 54 additional reviewers to review the submitted papers. Each paper received at least two reviews, and most papers received three reviews. Based on the reviews, the Program Committee accepted a total of 125 papers, leading to an oral paper acceptance rate of 16.82% and for posters a 41.58% acceptance rate. The technical program included nine oral sessions, three poster spotlights, three poster sessions, and three keynotes. The keynote speeches were given by three internationally renowned researchers active in pattern recognition and computer vision. They are:

- Professor Andrea Cavallaro of Queen Mary University of London, UK, "Multimodal Learning for Robust and Privacy Preserving Analytics"
- Professor Yihong Wu of the Chinese Academy of Sciences, China, "Possibility to Localize a Camera without Matching"
- Professor Dacheng Tao of the University of Sydney, Australia, "AI at Dawn - Opportunities and Challenges"

A large event like ACPR depends almost exclusively on a team of volunteers who work hard on the program, infrastructure, and facilities. We would like to thank the chairs of the different committees for their help and support. Our special thanks to our workshop chairs for arranging six workshops in this iteration of ACPR. We gratefully acknowledge the financial support of our sponsors which helped to reduce costs and provide various awards including best paper awards. We would like to thank sponsorship chairs and the AUT Professional Conference Organizer for helping us with this event. Our sincere thanks to all the researchers who have shown an interest in ACPR 2019 by sending contributed papers, workshop and tutorial proposals, and by participating. Thanks also to the Program Committee members, reviewers, and Local Organizing Committee members for their strong support and active participation. Finally, we would like to express our deep appreciation for the valuable guidance of the ACPR Advisory Board.

We hope you found your stay to be fruitful and rewarding, and that you enjoyed the exchange of technical and scientific ideas during ACPR 2019, as well as the flavor of the diverse and beautiful city of Auckland.

November 2019

Reinhard Klette
Brendan McCane
Umapada Pal
Shivakumara Palaiahnakote
Gabriella Sanniti di Baja
Liang Wang

Organization

Steering Committee

Seong-Whan Lee	Korea University, South Korea
Cheng-Lin Liu	Chinese Academy of Sciences, China
Sankar K. Pal	Indian Statistical Institute, India
Tieniu Tan	Chinese Academy of Sciences, China
Yasushi Yagi	Osaka University, Japan

General Chairs

Reinhard Klette	Auckland University of Technology, New Zealand
Brendan McCane	University of Otago, New Zealand
Umapada Pal	Indian Statistical Institute, India

Program Chairs

Gabriella Sanniti di Baja	Institute of High Performance Computing and Networking, Italy
Shivakumara Palaiahnakote	University of Malaya, Malaysia
Liang Wang	Chinese Academy of Sciences, China

Publication Chair

Wei Qi Yan	Auckland University of Technology, New Zealand

International Liaison Chairs

Chokri Ben Amar	University of Sfax, Tunisia
Wang Han	Nanyang Technology University, Singapore
Edwin Hancock	University of York, UK
Anil K. Jain	University of Michigan, USA
Domingo Mery	Pontificia Univerisdad Catolica, Chile

Workshop Chairs

Michael Cree	University of Waikato, New Zealand
Fay Huang	National Ilan University, Taiwan
Junsong Yuan	State University of New York at Buffalo, USA

Tutorial Chairs

Michael Blumenstein   University of Technology Sydney, Australia
Yukiko Kenmochi   French National Centre for Scientific Research, France
Ujjwal Maulik   Jadavpur University, India

Sponsorship Chair

Koichi Kise   Osaka University, Japan

Local Organizing Chair

Martin Stommel   Auckland University of Technology, New Zealand

Organizing Committee

Terry Brydon   Auckland University of Technology, New Zealand
Tapabrata Chakraborty   University of Otago, New Zealand
Gisela Klette   Auckland University of Technology, New Zealand
Minh Nguyen   Auckland University of Technology, New Zealand

Web Manager

Andrew Chen   The University of Auckland, New Zealand

Program Committee

Alireza Alaei   Griffith University, Australia
Fernando Alonso-Fernandez   Halmstad University, Sweden
Mahmoud Al-Sarayreh   Auckland University of Technology, New Zealand
Yasuo Ariki   Kobe University, Japan
Md Asikuzzaman   University of New South Wales, Australia
George Azzopardi   University of Groningen, The Netherlands
Donald Bailey   Massey University, New Zealand
Nilanjana Bhattacharya   Bose Institute, India
Saumik Bhattacharya   IIT Kanpur, India
Partha Bhowmick   IIT Kharagpur, India
Michael Blumenstein   University of Technology Sydney, Australia
Giuseppe Boccignone   University of Milan, Italy
Phil Bones   University of Canterbury, New Zealand
Gunilla Borgefors   Uppsala University, Sweden
Murk Bottema   Flinders University, Australia
Alfred Bruckstein   Technion, Israel
Weidong Cai   The University of Sydney, Australia
A. Campilho   University of Porto, Portugal
Virginio Cantoni   Università di Pavia, Italy

Sukalpa Chanda	Uppsala University, Sweden
Chiranjoy Chattopadhyay	IIT Jodhpur, India
Songcan Chen	Nanjing University of Aeronautics and Astronautics, China
Li Cheng	University of Alberta, Canada
Hsiang-Jen Chien	Auckland University of Technology, New Zealand
Michael Cree	University of Waikato, New Zealand
Jinshi Cui	Peking University, China
Zhen Cui	Nanjing University of Science and Technology, China
Dao-Qing Dai	Sun Yat-sen University, China
Daisuke Deguchi	Nagoya University, Japan
Andreas Dengel	German Research Center for Artificial Intelligence, Germany
Alberto del Bimbo	University delgi Studi di Firenze, Italy
Eduardo Destefanis	Universidad Tecnológica Nacional, Spain
Claudio de Stefano	Università degli studi di Cassino e del Lazio Meridionale, Italy
Changxing Ding	South China University of Technology, China
Jihad El-Sana	Ben Gurion University of the Negev, Israel
Miguel Ferrer	Universidad de Las Palmas de Gran Canaria, Spain
Robert Fisher	University of Edinburgh, UK
Gian Luca Foresti	University of Udine, Italy
Huazhu Fu	Inception Institute of Artificial Intelligence, UAE
Fei Gao	Hangzhu Dianzi University, China
Guangwei Gao	Nanjing University of Posts and Telecommunications, China
Edel Bartolo Garcia Reyes	Academy of Sciences of Cuba, Cuba
Andrew Gilman	Massey University, New Zealand
Richard Green	University of Canterbury, New Zealand
Yi Guo	Western Sydney University, Australia
Michal Haindl	Czech Academy of Sciences, Czech Republic
Takatsugu Hirayama	Nagoya University, Japan
Shinsaku Hiura	University of Hyogo, Japan
Kazuhiro Hotta	Meijo University, Japan
Du Huynh	The University of Western Australia, Australia
Ichiro Ide	Nagoya University, Japan
Masaaki Iiyama	Kyoto University, Japan
Yoshihisa Ijiri	OMRON, Japan
Atsushi Imiya	IMIT Chiba University, Japan
Koichi Ito	Tohoku University, Japan
Yumi Iwashita	Kyushu University, Japan
Motoi Iwata	Osaka Prefecture University, Japan
Yunde Jia	Beijing Institute of Technology, China
Xiaoyi Jiang	University of Münster, Germany
Xin Jin	Beijing Electronic Science and Technology Institute, China

Jianjun Qian	Nanjing University of Science and Technology, China
Ramachandra Raghavendra	Norwegian University of Science and Technology, Norway
Mokni Raouia	University of Sfax, Tunisia
Daniel Riccio	University of Naples Federico II, Italy
Paul Rosin	Cardiff University, UK
Partha Pratim Roy	IIT Roorkee, India
Kaushik Roy	West Bengal State University, India
Punam Kumar Saha	University of Iowa, USA
Fumihiko Sakaue	Nagoya Institute of Technology, Japan
Nong Sang	Huazhong University of Science and Technology, China
Carlo Sansone	University of Naples Federico II, Italy
Yoichi Sato	University of Tokyo, Japan
Atsushi Shimada	Kyushu University, Japan
Xiangbo Shu	Nanjing University of Science and Technology, China
Xiaoning Song	Jiangnan University, China
Martin Stommel	Auckland University of Technology, New Zealand
Suresh Sundaram	IIT Guwahati, India
Tomokazu Takahashi	Gifu Shotoku Gakuen University, Japan
Masayuki Tanaka	Tokyo Institute of Technology, Japan
Hiroshi Tanaka	Fujitsu Laboratories, Japan
Junli Tao	The University of Auckland, New Zealand
Wenbing Tao	Huazhong University of Science and Technology, China
Jules-Raymond Tapamo	University of KwaZulu-Natal, South Africa
Roberto Togneri	The University of Western Australia, Australia
Kar-Ann Toh	Yonsei University, Japan
Zhiyong Wang	The University of Sydney, Australia
Brendon J. Woodford	University of Otago, New Zealand
Yihong Wu	Chinese Academy of Sciences, China
Haiyuan Wu	Wakayama University, Japan
Yanwu Xu	Institute for Infocomm Research (A*STAR), Singapore
Hirotake Wamazoe	Ritsumeikan University, Japan
Wei Qi Yan	Auckland University of Technology, New Zealand
Mengjie Zhang	Victoria University of Wellington, New Zealand
Jinxia Zhang	Southeast University, China
Cairong Zhao	Tongji University, China
Quan Zhou	Nanjing University of Posts and Telecommunications, China

Contents – Part II

Contents – Part I

Adversarial Learning and Networks

Computational Photography

Learning Theory and Optimization

Applications, Medical and Robotics

Computer Vision and Robot Vision

Pattern Recognition and Machine Learning

Margin Constraint for Low-Shot Learning

Xiaotian Wu[✉] and Yizhuo Wang

Beijing Institute of Technology, Beijing 100081, China
{xiaotianwu,frankwyz}@bit.edu.cn

Abstract. Low-shot learning aims to recognize novel visual categories with limited examples, which is mimicking the human visual system and remains a challenging research problem. In this paper, we introduce the margin constraint in loss function for the low-shot learning field to enhance the model's discriminative power. Additionally, we adopt the novel categories' normalized feature vectors as the corresponding classification weight vectors directly, in order to provide an instant classification performance on the novel categories without retraining. Experiments show that our method provides a better generalization and outperforms the previous methods on the low-shot leaning benchmarks.

Keywords: Low-shot learning · Margin constraint · Normalized vectors

1 Introduction

In visual recognition field, deep learning methods [1,5,6,8,16] have achieved great results in recent years. However, these supervised learning methods rely on a large amount of labelled data, and therefore can not maintain good performance when the labeled data is scarce. In contrast, humans can identify new categories with only a small number of samples. This phenomenon motivates the development of low-shot learning [7,15,18–20] which aims to produce models that can recognize novel categories from rare samples.

However, most prior low-shot methods do not consider two important factors: (*a*) When learning to recognize novel categories, the model should not forget the previous categories that it has learned; (*b*) The model should learn to recognize novel categories quickly, which is one of the abilities of humans. Motivated by this observation, we consider a realistic scenario that the model is initially trained with a relatively large dataset containing the base categories, and then exposed to the novel categories with very few samples. The goal is to obtain a model that performs well on the combined set of categories.

To obtain such a model, firstly, we need to train a base model that can recognize the base categories precisely, which means a learner has acquired the rich knowledge already. However, due to the data insufficiency in the low-shot learning field, existing methods [10,19,20] could not learn an excellent metric space. So images may be wrongly classified by the classifier when mapped in the

© Springer Nature Switzerland AG 2020
S. Palaiahnakote et al. (Eds.): ACPR 2019, LNCS 12047, pp. 3–14, 2020.
https://doi.org/10.1007/978-3-030-41299-9_1

metric space. To alleviate this problem, we introduce the margin constraint in the loss function, *i.e.*, the margin softmax loss, which can enforce samples from the different categories to be mapped as far apart as possible. Meanwhile, the margin constraint can also enforce samples from the same categories to be mapped as compact as possible [10]. Finally, the model can learn a more discriminative metric space. As a result, the recognition accuracy can be improved.

Additionally, to expand the model's capacity of recognizing novel categories quickly, we adopt the normalized novel categories' feature vectors as the corresponding classification weight vectors directly. The reason is that the feature vectors have a tight relationship with the classification weight vectors of the same category [14]. So our method does not need any iterative optimization to learn the novel categories' classification weight vectors, which means the model can recognize novel categories quickly. Experiments show that our method performs well in the low-shot learning benchmarks.

2   Related Work

Recently, there is a series of methods for low-shot learning problems. We will briefly introduce the relevant methods in the following.

Metric Learning Based Methods. Metric learning is a commonly used method in the low-shot learning field. It aims to learn a good discriminative space so that similar samples are mapped relatively close, whereas different samples are relatively far apart. Metric based low-shot learning methods mainly differ in the choice of metric. For example, Vinyals et al. [18] propose the Matching Network which uses the cosine distance as the metric to find the query example's nearest neighbor within the support set. Prototypical Network [20] applies the average feature vectors of each category as the prototype and measures the distance between the query example and the prototype by European distance. Relation Network [10] uses the network to learn a non-linear distance.

Meta Learning Based Methods. Meta-learning is another promising way to solve the low-shot classification problem. It aims to train the model through a series of tasks to obtain the meta-knowledge, which can be transferred into new tasks quickly within just a few iterations [17]. Specifically, Finn et al. [2] propose a model-agnostic meta-learning (MAML) approach which learns how to initialize the model so that it can adapt to new tasks within a few gradient-descent update steps. Ravi and Larochelle [15] propose an LSTM-based optimizer which can provide a good parameter initialization and can also be trained effectively for fine-tuning. Under the meta-learning framework, models are trained by the episode-based strategy, which is designed to mimic the low-shot learning task so that the model can adapt the test environment and therefore enhance its generalization. However, this evaluation setup and framework can only test the model's performance on the novel categories at the inference stage. Different from that, we consider an alternative setup similar to [4] and [13], where the model is trained on the abundant samples in advance and then test its performance on a combined set of categories including samples from both the base and novel categories.

Due to data scarcity in the low-shot learning field, it is difficult to learn enough discriminative metric space. Recently, a popular research direction is to incorporate the margin constraint for softmax loss to enhance the discriminative power in the face verification field. For example, Sphereface [9] proposes the idea of angular margin penalty, leading to a better discriminative power of the trained model, but their networks cannot be trained stably. CosFace [22,24] adds the cosine margin in the loss function, which achieves a better result and is easier to implement. As a result, our approach adopts the margin constraint for the low-shot learning field. More similar to our work is [13], which uses the normalized novel feature vectors as the corresponding weight vectors. The difference is that our work introduces the margin constraint in loss function during the training phase, which can enhance the discriminative power and improve the model's generalization capability and finally boost the model's performance.

3   Method

3.1   Problem Formulation

Notice that we first want to train a base model that can recognize the base categories precisely. And then expand its capability of recognizing novel categories quickly without forgetting the base categories. So there are two natural training stages: in the first stage, we have a large labeled dataset to train a good base model, and in the second stage, we expect the model to have the ability to recognize novel categories without retraining, which means it can learn quickly.

In the first stage, the model is trained with a dataset C_{base}, which contains samples from the base categories. Feature vectors z are first extracted by the feature extractor $F(.|\theta)$ (with learnable parameters θ), i.e. $z = F(x|\theta)$ where x is the input image. Then, the classifier $L(.|W)$ takes z as inputs and returns the classification score $s = L(z|W)$ of each category. W is a set of classification weight vectors - one per category. Note that in a typical ConvNet, the feature extractor is the part of the network that starts from the first layer and ends at the last hidden layer. And the classifier is the last linear layer. The parameters θ and W_{base} (the base categories' classification weight vectors) are initialized randomly and during this training stage, the model will learn the parameters by using the dataset C_{base} with backpropagation. Finally, the model will be able to recognize the base categories by setting $W = W_{base}$.

In the second stage, the model must learn to distinguish the novel categories C_{novel} while not forgetting the base categories. For each unseen novel category, only n shots are available, such as $n \in \{1, 2, 5, 10, 20\}$. The model needs to add the novel classification weight vectors W_{novel}, which are generated from the novel samples, to the original weights vector incrementally, without changing the existed weight vectors W (i.e. $W = W_{base} \cup W_{novel}$). In the testing stage, we measure the model's performance on the previously unseen test images from the combined set of categories $C_{base} \cup C_{novel}$.

The key factor is how to generate the novel classification weight vectors W_{novel} efficiently to recognize the novel categories, meanwhile they do not affect the model's ability of recognizing the base categories. In the following sections, we will describe it in detail.

3.2   Cosine-Similarity Based Classifier

During the forward propagation, the model first extracts the feature vectors z, and then uses the classifier to get a classification score $s_c = z^T w_c$ by the inner product, where w_c is the c-th classification weight vector and is randomly initialized. After that, utilize the softmax operator to compute the classification probability scores across all categories:

$$p_i = \frac{\exp(z^T w_i)}{\sum_c \exp(z^T w_c)} \tag{1}$$

Different from the standard ConvNet classifier, our model replaces the inner product with the cosine similarity, which is also suggested by [3,11,13,14]. Specifically, we add an additional layer at the end of feature extractor to normalize the feature vectors to the unit length, i.e. $\bar{z} = \frac{z}{\|z\|_2}$. Additionally, we also normalize the weight vectors to unit length, i.e. $\bar{w} = \frac{w}{\|w\|_2}$. Notice that $\cos\theta = \frac{z^T w}{\|z\|_2 \|w\|_2}$, so the inner product $\bar{z}^T \bar{w}$ measures cosine similarity. Due to $\bar{z}^T \bar{w} \in [-1, 1]$, which means the model can be trained stably. However, consider that even if the samples are classified perfectly, i.e. value 1 for the correct categories and value -1 for incorrect categories, the final probability score $p_i = e^1/[e^1 + (C-1)e^{-1}] = 0.036$ is a very small number (assuming $C = 200$ categories). We can see that p_i is far away from the one-hot encoding of the ground truth label. So the cosine similarity may affect the optimization of cross-entropy loss. To alleviate this problem, we adapt a learnable parameter τ, which is suggested by [23] (In all of our experiments, τ is initialized to 10). Finally, the loss function is:

$$L_1 = -\frac{1}{n} \sum_{i=1}^{n} \log \frac{\exp(\tau \bar{z}^T \bar{w}_i)}{\sum_c \exp(\tau \bar{z}^T \bar{w}_c)} \tag{2}$$

3.3   Incorporating the Margin Constraint

Due to the data scarcity of novel samples, getting a more discriminative model in training stage is crucial. The main optimization direction of Eq. 2 is to separate the different categories away, but not good at making the features of the same categories compact [22]. To address this and to improve the model's generalization capacity, we add the margin penalty on the target logit as suggested by [22,24] which has been used widely in face recognition field. Finally, the objective function in first training stage is:

$$L_2 = -\frac{1}{n} \sum_{i=1}^{n} \log \frac{\exp(\tau(\bar{z}^{\mathrm{T}} \bar{w}_i - m))}{\exp(\tau(\bar{z}^{\mathrm{T}} \bar{w}_i - m)) + \sum_{j=1, j \neq i}^{c} \exp(\tau \bar{z}^{\mathrm{T}} \bar{w}_j)} \tag{3}$$

To better understand the effect of margin constraint in target logit, we can consider the decision boundaries of the models. Taking the binary classification as example, as shown in Fig. 1, the decision boundary of class 1 and class 2 is $\tau(\cos\theta_1 - \cos\theta_2) = 0$ for Eq. 2; As for Eq. 3 the decision boundary of class 1 is $\tau(\cos\theta_1 - m - \cos\theta_2) = 0$, whereas the decision boundary of class 2 is $\tau(\cos\theta_2 - m - \cos\theta_1) = 0$, which means there are two different decision boundaries. The decision margin in Eq. 3 enforces extra intra-class compactness and inter-class discrepancy simultaneously. So the samples can be mapped in a more discriminative metric space, and can be recognized more easily and accurately by the classifier.

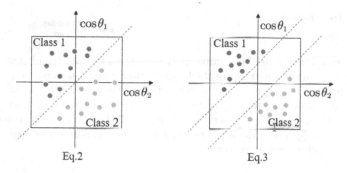

Eq.2   Eq.3

Fig. 1. Decision boundary of Eqs. 2 and 3. The dashed line represents the decision boundary. We can see that there is a margin between Eq. 3's decision boundary.

Notice that the margin softmax loss is only used in the first training stage to obtain a discriminative base model. During the second training stage, the challenge is how to generate the classification weight vectors for the novel limited samples incrementally. So in the testing stage, the model can recognize both the novel and the base categories. Since [14] has proven that the feature vectors have a tight relationship with the classification weight vectors of the same categories, we directly use the normalized feature vectors as the corresponding classification weight vectors. When $n > 1$ samples are available for a novel category, we compute the novel weight vectors by first averaging the normalized feature vectors *i.e.* $\tilde{w}_{novel} = \frac{1}{n} \sum_{i=1}^{n} \bar{z}_{novel}$, and then re-normalizing the resulting weight vectors to unit length *i.e.* $\bar{w}_{novel} = \tilde{w}_{novel} / \|\tilde{w}_{novel}\|_2$. So in the second training stage, there is no need to proceed with any iterative optimization. The overall framework is shown by Fig. 2.

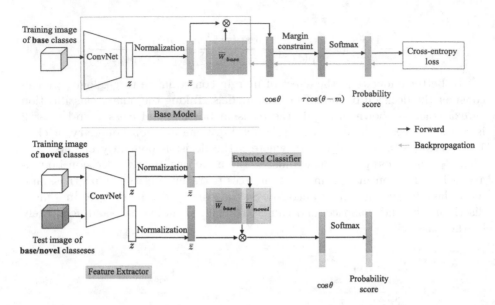

Fig. 2. The overall framework. The upper part represents the first training stage, which incorporates the margin constraint in the loss function. The bottom part represents the second stage and the testing stage. In the second stage, the model extends its' capability by adopting the normalized feature vectors as the classification weight vectors of that categories. In the testing stage, we can test the model's performance on the combined set of classes. Note $\otimes$ means the inner product between the normalized feature vectors and the corresponding weight vectors.

4   Experiments

We evaluate our method on the CUB-200-2011 dataset [21]. In the following paragraphs, we will describe the results and discuss them in detail. All the experiments are implemented based on PyTorch [12].

4.1   Dataset

The CUB-200-2011 dataset has 117888 images with 200 categories birds, which provides the training/test split. Specifically, we construct two disjoint sets of categories C_{base} and C_{novel}, each has 100 categories. In the first training stage, we use the training samples from C_{base} categories, and each category has about 30 images on average. In the second training stage, the training samples from the C_{novel} categories and each category contains only n examples, where $n \in \{1, 2, 5, 10, 20\}$ in our experiments. All the training samples belong to the original training split. During the testing stage, the test samples from the test split include both novel and base categories. We measure the final classifier's performance both on the only novel and on all categories.

4.2   Implementation Details

Based on the standard ConvNet, we add an extra linear layer to map the feature vectors' dimension to 256 at the end of the feature extractor. After that, a l_2 normalization layer is used for normalizing the feature vectors. The final classifier layer's output dimension is equal to the number of categories, *i.e.* in the first training stage is 100 while in the second training stage and the testing stage is 200. In the first training stage, the classification parameters W_{base} are initialized randomly. We resize the input images to 256×256 and crop them to 224×224, following [13]. During training, we augment inputs with random cropping and random horizontal flipping to the inputs. In the testing stage, we set up two main experiments, experiment (a) and experiment (b), for the different network architectures of base models, which aims to evaluate whether our method still works in the different architectures. In experiment (a), the base model's architecture is InceptionV1. The learning rate starts at 0.0001 for feature extractor, and 0.001 for the classifier. We use exponential decay every four epochs with a decay rate of 0.94. And we use the RMSProp optimizer with momentum 0.9, which is suggested by [13]. In experiment (b), the learning rate's setting is the same as (a), and it is divided by 10 at every ten epochs. The optimizer is stochastic gradient descent with 0.9 momentum and the weight decay is 0.0001.

In practice, we evaluate to choose the suitable margin m for different settings. Figure 3 represents the test results when the base model's architecture is InceptionV1. Notice that when $m = 0$, the model then equals to the *Imprinting* model [13]. And in Fig. 4, the base model's architecture is ResNet-50. From the two figures, we can see that the best choice of margin m is 0.10 when the base model's architecture is InceptionV1. And when the base model's architecture is ResNet-50, the best choice of margin m is 0.15.

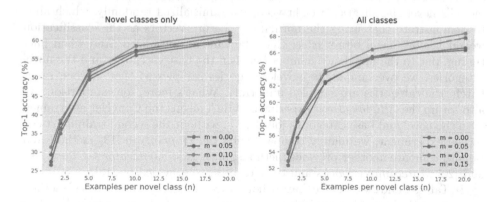

Fig. 3. Top-1 accuracy on CUB-200-2011 val with different choices of margin m for novel examples and all examples respectively. The base model's architecture is InceptionV1.

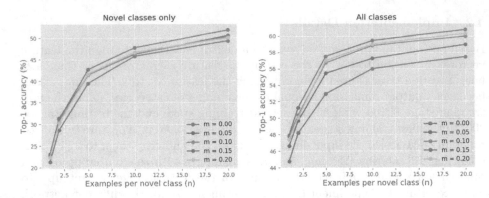

Fig. 4. Top-1 accuracy on CUB-200-2011 val with different choices of margin m for novel examples and all examples respectively. The base model's architecture is ResNet-50.

4.3 Results

Tables 1 and 2 represent the test results for novel examples and all examples in CUB-200-2011 respectively. The base model's architecture is InceptionV1, which is consistent with *Imprinting* [13]. Alternatively, we evaluate the data augmentation experiment ($+Aug$), and after augmentation, the amount of novel categories is five times than before. In Table 2, we can see that there involves zero backpropagation during the second training stage, however, when the capability of the hardware device is enough or the time is allowed, fine-tuning can be applied after the second training stage. Because our model architecture has the same form as ordinary ConvNet. So we also evaluate the fine-tuning ($+FT$) experiments. Additionally, the comparative experiment $Rand + FT$ means the model's novel classification weight vectors are initialized randomly, which aims to demonstrate that using the normalized feature vectors as the classification weight vectors has a better effect than the random initialized. And when fine-tuning, due to the imbalance number between the base categories and the novel categories, we oversample the novel categories samples so that in each mini-batch, the categories are sampled uniformly. What's more, for comparison, we also set up the *AllClassJoint* experiment which means the ConvNet are trained with the novel and base categories jointly, *i.e.* without the second training stage. In the experiment, we mainly compare with the *Imprinting* [13] method and we also compare another previous methods, such as [4] *Generator + Classifier* and [20] *MatchingNetworks* and their results are provided by [13].

In Tables 1 and 2 we can see that without fine-tuning, the margin-based method can provide the best classification performance such as 23.14% on novel and 48.01% on all categories for one-shot examples. Additionally, we can see that data augmentation does not bring a significant improvement, it may because the transformations can't change the feature vectors extracted from the base categories a lot. The second block of 3 rows shows the results of fine-tuning the network weights with backpropagation, models which is initialized with margin-based method achieve better final accuracies than *Imprinting* [13] model and the randomly initialized model *Rand + FT*, especially when the novel examples is extremely scarce, *i.e.* $n = 1, 2, 5$. The results are recorded after 200 epochs, in that case the performance is close to saturation so that fine-tuning with more epochs can lead to the accuracy drop on the novel evaluation categories.

Table 1. Top-1 accuracy (%) of 200-way classification for novel-class examples in CUB-200-2011. The base model's architecture is InceptionV1. For different settings, bold: the best low-shot accuracy, and italic: the second best.

		$n = 1$	$n = 2$	$n = 5$	$n = 10$	$n = 20$
w/o FT	Imprinting [13]	21.26	28.69	39.52	45.77	49.32
	Imprinting+Aug [13]	21.40	30.03	39.35	46.35	49.80
	ours ($m = 0.10$)	*23.14*	*31.13*	**42.73**	*47.57*	**51.80**
	ours+Aug ($m = 0.10$)	**25.05**	**32.49**	*42.67*	**49.07**	*51.33*
w/FT	AllClassJoint	3.89	10.82	33.00	50.24	64.88
	Rand+FT	5.25	13.41	34.95	54.33	65.60
	Imprinting+FT [13]	*18.67*	*30.17*	*46.08*	*59.39*	**68.77**
	ours ($m = 0.10$)	**23.34**	**32.46**	**48.78**	**59.44**	*67.32*
	Generator + Classifier [4]	18.56	19.07	20.00	20.27	20.88
	Matching networks [20]	13.45	14.57	16.65	18.81	25.77

Table 2. Top-1 accuracy of the 200-way classification across examples in all classes (100 base plus 100 novel classes) of CUB-200-2011. The Margin-based method retains similar advantage.

		$n = 1$	$n = 2$	$n = 5$	$n = 10$	$n = 20$
w/o FT	Imprinting [13]	44.75	48.21	52.92	55.59	57.47
	Imprinting+Aug [13]	44.60	48.48	52.78	56.51	57.84
	ours ($m = 0.10$)	*47.91*	**51.27**	**57.52**	**59.44**	*60.77*
	ours+Aug ($m = 0.10$)	**48.01**	*51.14*	*57.34*	*59.34*	**60.87**
w/FT	AllClassJoint	38.02	41.89	52.24	61.11	68.31
	Rand+FT	39.26	43.36	53.69	63.17	68.70
	Imprinting+FT [13]	*45.81*	*50.41*	*59.15*	*64.65*	**68.75**
	ours+FT ($m = 0.10$)	**48.31**	**53.54**	**62.01**	**66.09**	*68.73*
	Generator + Classifier [4]	45.42	46.56	47.79	47.88	48.22
	Matching networks [20]	47.71	43.15	44.46	45.65	48.63

To evaluate that our approach still has a good effect on different network architectures, we reproduce all the experiments when the base model's architecture is ResNet-50, and the results are shown in Tables 3 and 4. We can see that our method remains a similar advantage, which proves that our method still works in different architectures.

Table 3. Top-1 accuracy of the 200-way classification for novel-class examples in CUB-200-2011. Different Table 1, the base model's network architecture is ResNet-50. We can see that the margin-based method still remains advantages.

		n = 1	n = 2	n = 5	n = 10	n = 20
w/o FT	Imprinting	27.37	35.11	50.34	57.06	60.37
	Imprinting+Aug	28.39	37.20	50.51	56.38	60.81
	ours ($m = 0.15$)	**31.36**	*38.62*	*51.23*	*58.63*	*62.25*
	ours+Aug ($m = 0.15$)	*31.29*	**40.54**	**52.10**	**57.85**	**62.45**
w/FT	AllClassJoint	3.89	10.82	33.00	50.24	64.88
	Rand+FT	5.25	13.41	34.95	54.33	65.60
	Imprinting+FT	*26.41*	*34.67*	*54.71*	*64.61*	**73.17**
	ours+FT ($m = 0.15$)	**29.32**	**37.82**	**55.70**	**66.62**	*72.67*

Table 4. Top-1 accuracy of 200-way classification across examples in all classes. Different Table 2, the base model's network architecture is ResNet-50.

		n = 1	n = 2	n = 5	n = 10	n = 20
w/o FT	Imprinting	52.36	55.76	62.51	65.46	66.75
	Imprinting+Aug	51.82	56.51	62.70	64.77	67.05
	ours ($m = 0.15$)	*54.05*	**58.01**	*63.92*	**66.52**	*68.05*
	ours+Aug ($m = 0.15$)	**54.07**	*57.78*	**64.10**	*66.48*	**68.57**
w/FT	AllClassJoint	38.02	41.89	52.24	61.11	68.31
	Rand+FT	39.26	43.36	53.69	63.17	68.75
	Imprinting+FT	*53.02*	*57.82*	*67.72*	*71.73*	**76.25**
	ours+FT ($m = 0.15$)	**53.76**	**58.90**	**67.19**	**72.93**	*75.89*

5   Conclusions

This paper introduces the margin constraint in loss function for the low-shot learning field, which enhances the discriminative power of the trained model so that the accuracy of recognition can be improved. Additionally, we adopt the normalized feature vectors as the novel classification weight vectors, which means the model can quickly extend its capability of recognizing novel categories. Experiments demonstrate that our work has a good effect on the low-shot learning field.

References

1. Chollet, F.: Xception: deep learning with depthwise separable convolutions. In: Proceedings of Conference on Computer Vision and Pattern Recognition (2017)
2. Finn, C., Abbeel, P., Levine, S.: Model-agnostic meta-learning for fast adaptation of deep networks. In: Proceedings of International Conference on Machine Learning (2017)
3. Gidaris, S., Komodakis, N.: Dynamic few-shot visual learning without forgetting. In: Proceedings of Conference on Computer Vision and Pattern Recognition (2018)
4. Hariharan, B., Girshick, R.: Low-shot visual recognition by shrinking and hallucinating features. In: Proceedings of Conference on Computer Vision and Pattern Recognition (2017)
5. He, K., Zhang, X., Ren, S., Sun, J.: Deep residual learning for image recognition. In: Proceedings of Conference on Computer Vision and Pattern Recognition (2016)
6. Huang, G., Liu, Z., Van Der Maaten, L., Weinberger, K.Q.: Densely connected convolutional networks. In: Proceedings of Conference on Computer Vision and Pattern Recognition (2017)
7. Koch, G., Zemel, R., Salakhutdinov, R.: Siamese neural networks for one-shot image recognition. In: Proceedings of International Conference on Machine Learning Workshop (2015)
8. Krizhevsky, A., Sutskever, I., Hinton, G.E.: ImageNet classification with deep convolutional neural networks. In: Proceedings of Conference on Neural Information Processing Systems (2012)
9. Liu, W., Wen, Y., Yu, Z., Li, M., Raj, B., Song, L.: SphereFace: deep hypersphere embedding for face recognition. In: Proceedings of Conference on Computer Vision and Pattern Recognition (2017)
10. Liu, W., Wen, Y., Yu, Z., Yang, M.: Large-margin softmax loss for convolutional neural networks. In: Proceedings of International Conference on Machine Learning (2016)
11. Luo, C., Zhan, J., Xue, X., Wang, L., Ren, R., Yang, Q.: Cosine normalization: using cosine similarity instead of dot product in neural networks. In: Kůrková, V., Manolopoulos, Y., Hammer, B., Iliadis, L., Maglogiannis, I. (eds.) ICANN 2018. LNCS, vol. 11139, pp. 382–391. Springer, Cham (2018). https://doi.org/10.1007/978-3-030-01418-6_38
12. Paszke, A., et al.: Automatic differentiation in PyTorch. In: Proceedings of Conference on Neural Information Processing Systems Workshop (2017)
13. Qi, H., Brown, M., Lowe, D.G.: Low-shot learning with imprinted weights. In: Proceedings of Conference on Computer Vision and Pattern Recognition (2018)
14. Qiao, S., Liu, C., Shen, W., Yuille, A.L.: Few-shot image recognition by predicting parameters from activations. In: Proceedings of Conference on Computer Vision and Pattern Recognition (2018)
15. Ravi, S., Larochelle, H.: Optimization as a model for few-shot learning. In: Proceedings of International Conference on Learning Representations (2017)
16. Russakovsky, O., et al.: ImageNet large scale visual recognition challenge. Int. J. Comput. Vision 115(3), 211–252 (2015)
17. Schmidhuber, J., Zhao, J., Wiering, M.: Shifting inductive bias with success-story algorithm, adaptive Levin search, and incremental self-improvement. Mach. Learn. 28(1), 105–130 (1997)
18. Snell, J., Swersky, K., Zemel, R.: Prototypical networks for few-shot learning. In: Proceedings of Conference on Neural Information Processing Systems (2017)

19. Sung, F., Yang, Y., Zhang, L., Xiang, T., Torr, P.H., Hospedales, T.M.: Learning to compare: relation network for few-shot learning. In: Proceedings of Conference on Computer Vision and Pattern Recognition (2018)
20. Vinyals, O., Blundell, C., Lillicrap, T., Wierstra, D., et al.: Matching networks for one shot learning. In: Proceedings of Conference on Neural Information Processing Systems (2016)
21. Wah, C., Branson, S., Perona, P., Belongie, S.: Multiclass recognition and part localization with humans in the loop. In: Proceedings of International Conference on Computer Vision (2011)
22. Wang, F., Cheng, J., Liu, W., Liu, H.: Additive margin softmax for face verification. IEEE Signal Process. Lett. **25**(7), 926–930 (2018)
23. Wang, F., Xiang, X., Cheng, J., Yuille, A.L.: NormFace: L_2 hypersphere embedding for face verification. In: Proceedings of the 25th ACM International Conference on Multimedia (2017)
24. Wang, H., et al.: CosFace: large margin cosine loss for deep face recognition. In: Proceedings of Conference on Computer Vision and Pattern Recognition (2018)

Enhancing Open-Set Face Recognition by Closing It with Cluster-Inferred Gallery Augmentation

Floris De Feyter$^{(\boxtimes)}$ ⓘ, Kristof Van Beeck ⓘ, and Toon Goedemé ⓘ

KU Leuven—EAVISE,
Jan De Nayerlaan 5, 2860 Sint-Katelijne-Waver, Belgium
floris.defeyter@kuleuven.be
https://iiw.kuleuven.be/onderzoek/eavise

Abstract. In open-set face recognition—as opposed to closed-set face recognition—it is possible that the identity of a given query is not present in the gallery set. In that case, the identity of the query can only be correctly classified as "unknown" when the similarity with the gallery faces is below a threshold that was determined a priori. However, in many use-cases, the set of queries contains multiple instances of the same identity, whether or not this identity is represented in the gallery. Thus, the set of query faces lends itself to identity clustering that could yield representative instances for unknown identities. By augmenting the gallery with these instances, we can make an open-set face recognition problem more *closed*. In this paper, we show that this method of Cluster-Inferred Gallery Augmentation (CIGA) does indeed improve the quality of open-set face recognition. We evaluate the addition of CIGA for both a private dataset of images taken in a school context and the public LFW dataset, showing a significant improvement in both cases. Moreover, an implementation of the suggested approach along with our experiments are made publicly available on https://gitlab.com/florisdf/acpr2019.

Keywords: Open-set face recognition · Face clustering · Gallery augmentation

1 Introduction

Facial recognition can be categorized as face verification and face identification [23]. Both paradigms work with a set of known faces (the *gallery*), and an unknown face (the *query* or *probe*) that needs to be identified. In face verification, the query is compared in a one-to-one fashion to the gallery items. Each of these comparisons yields either a match (same identity) or a mismatch (different identity). In face identification, the query is compared to the whole gallery at once in a one-to-many fashion, yielding the identity of the best matching gallery item. For both face recognition categories, it is possible that the identity of the query face is not present in the gallery. This is referred to as *open-set recognition*.

© Springer Nature Switzerland AG 2020
S. Palaiahnakote et al. (Eds.): ACPR 2019, LNCS 12047, pp. 15–26, 2020.
https://doi.org/10.1007/978-3-030-41299-9_2

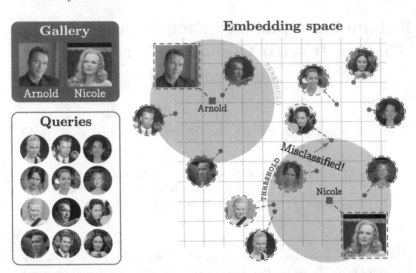

Fig. 1. Simply thresholding a similarity measure might not suffice to avoid problems with open-set recognition. Here, Angelina Jolie is *unknown* and one of her embeddings is misclassified as "Nicole" because it falls within the threshold limit. The fact that the embedding is close to other embeddings of Angelina Jolie, is ignored.

When all query identities are present in the gallery, the recognition is referred to as *closed-set*.

1.1 Closing Open-Set Recognition

The problem with open-set face recognition is that when a given query does not appear in the gallery, the system will certainly misclassify the query. This misclassification can be prevented, however, by setting a proper threshold on the similarity score of the query face and the gallery items. Whenever the similarity score between query and gallery is below such a threshold, the system could report that the query has an *unknown* identity. Still, with this approach, much emphasis is put on the threshold and any query of an unknown identity with a similarity score slightly above the threshold, will certainly be misclassified. Figure 1 illustrates this problem. Two identities are present in the gallery: "Arnold" and "Nicole". In the query set, however, a third unknown identity is present (Angelina Jolie). One of these unknown queries has a face embedding that is just close enough to the gallery embedding of "Nicole" to fall within the threshold limit and get misclassified as "Nicole".

At the same time, many identities will occur more than once in the set of all queries. For example, when searching for a person through security camera footage, most people will be visible in multiple frames and thus appear more than once as a query. Therefore, the set of queries has a valuable structure that would be ignored when we simply focus on setting a good threshold to avoid misclassifications. More specifically, one can try to exploit the structure of the query set to infer reference images of identities that are not present in the gallery set. These references of *unknown* identities can then be added to the gallery set. As such, one makes the open-set recognition problem more *closed* by adding identities to the gallery that previously were not present but that do appear in the query set. This will alleviate the stress on setting a good threshold, since a false positive matching score that was slightly above the threshold for one of the known gallery items, can now have a much higher true positive matching score with one of the unknown identities in the augmented gallery. Hence, the queries that do match with the known identities will be more reliable, leading to a more precise identification system. This solution is shown in Fig. 2. The embeddings of the query set are analyzed and a third identity is discovered, labeled "id3". A representative embedding is chosen and added to the gallery. The query that was misclassified in Fig. 1, has a better match with this new embedding and is therefore not classified as "Nicole", but as "id3".

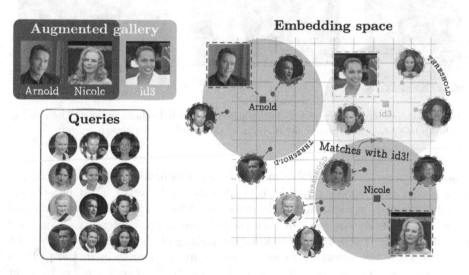

Fig. 2. By inspecting the structure of the query embeddings, we can augment the gallery with identities that were previously unknown. That way, a previously false positive match can now have a much better true positive match with the new gallery item.

We propose a proof-of-concept that is based on an initial step of clustering face embeddings to augment the original gallery with extra (unknown) identities. Section 3 discusses this approach—which we coin Cluster-Inferred Gallery Augmentation (CIGA)—in more detail.

1.2   Use-Case: Individualized School Photo Albums

As a concrete example, we frame our proposed approach in the use-case of composing individualized school photo albums. After a year of photographing the

Fig. 3. Random samples from the Jade dataset of labeled school images. The color of the bounding boxes represents the identity of the child. (Color figure online)

school activities and memorable moments, teachers have created a collection of photos that can be very valuable to many parents. However, most parents are mainly interested in pictures of their own child. A facial recognition system could be used to identify the faces in each picture, based on a reference image that parents provide. These references form the *gallery* of the identification system, since we know the identity of each reference image. The faces in the pictures that were taken throughout the year correspond to the *query* set, i.e. we want to identify each of these faces by comparing them to the reference images. This use-case is indeed an open-set recognition problem as we assume that there will be children in the query set of whom the parents did not provide a reference image.

In the evaluation of our proposed approach, we use a private real-world dataset—referred to as the *Jade dataset*—that was created for the development of such an album-creating system. The Jade dataset contains 1629 labeled images of 29 unique identities. It contains a wide variety of image types: close-up portraits, wider portraits, group photos etc. Fig. 3 shows multiple randomly selected samples of the Jade dataset. The color of the bounding boxes corresponds to the identity of the face. It is clear that the images are unconstrained in terms of pose, facial expression, face angle, blur, resolution and lighting.

2   Related Work

Much research has been done on open-set recognition, mostly in the general case of open-set image classification [1–3,12,13,16]. Scheirer et al. [16] were the first to formally define open-set recognition as a constrained minimization problem. They propose a 1-vs-set SVM that has a second plane in the feature space which delimits the reasonable support of known classes. Many other papers further built upon this idea [1,10,17,18]. Apart from the 1-vs-set-based approaches, an algorithm based on deep neural networks has also been proposed [2]. Bendale and Boult present a methodology that adds an extra layer—called OpenMax—to the network. The additional layer estimates the probability that an input belongs to an unknown class.

The more specific case of open-set face recognition is often considered and evaluated [8,20], but only rarely algorithms are specifically targeted at dealing with open-set situations [6]. Instead, most works on face recognition focus on creating algorithms that transform faces into robust embeddings [4,14,19,21,22]. They assume that a robust embedding combined with a proper threshold should prevent problems with open-set recognition.

The previous works ignore that in many practical situations of open-set recognition, the query set contains multiple examples of the same unknown classes. For example, in our use-case of individualized school album creation described in Sect. 1.2, an unknown child that is in the same class as a known child, will probably appear multiple times in the series of images. Our approach attempts to exploit this query structure to discover unknown identities such that the recognition of the identities of interest (i.e. the known identities) will improve.

3   Cluster-Inferred Gallery Augmentation

As previously motivated, discovering the structure of the query set might improve the precision of a facial recognition system that is used in an open-set context. More specifically, we argue that one can infer representative instances of identities that are not present in the gallery set but that are present in the query set. We propose an approach called Cluster-Inferred Gallery Augmentation (CIGA). It consists of the following steps:

1. Add the descriptors from the query set *and the gallery* (i.e. the *known* identities) into one large set of descriptors and cluster this set of descriptors;
2. Discard the clusters where items from the original gallery ended up;
3. For each remaining cluster, find the medoid;
4. Add the cluster medoids as new identities to the gallery;
5. Discard *newly added* gallery items that lie unusually close to another gallery item.

Note that we made the CIGA algorithm publicly available via this repository: https://gitlab.com/florisdf/acpr2019. The README file explains how you can use the CIGA algorithm yourself. The following sections explain the algorithm in more detail.

3.1 Clustering

We propose to exploit the structure of the query set by clustering the face embeddings. It is, however, unfeasible to make a realistic a priori estimation of the amount of separate identities that are present in the dataset. Therefore, a clustering algorithm like K-means clustering [7] is inapplicable for CIGA as it takes the number of clusters as an argument. In our proof-of-concept, we use the Density-Based Spatial Clustering of Applications with Noise (DBSCAN) algorithm [5] to discover the query structure. This algorithm defines clusters as areas of high density separated by areas of low density. The algorithm has two parameters: ε and `min_samples`. When at least `min_samples` samples are within a distance ε of a sample (including the sample itself), that sample is considered a *core sample*. A sample that is in the neighbourhood of a core sample can either be a core sample itself, or a *non-core sample*, which means that there are less than `min_samples` samples within a distance ε around that sample. A cluster contains at least one core sample, and hence at least `min_samples` samples. Samples that are not in the neighbourhood of a core sample, do not belong to a cluster and are considered *outliers*. Figure 4 shows the DBSCAN algorithm applied on a small set of data points.

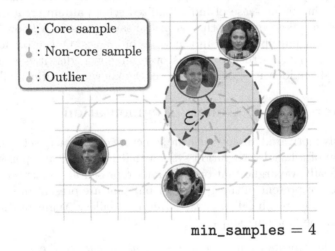

Fig. 4. The DBSCAN algorithm applied on a set of 5 points. The parameter min_samples is set to 4 and the distance ε is chosen as shown in the figure. There is one core sample (dark blue), three non-core samples (light blue) and one outlier (grey). As such, the clustering results in one cluster (the blue points) and one outlier. (Color figure online)

3.2 Selecting References for Unknown Identities

After clustering the query and gallery faces, we want to select references for unknown identities to add to the gallery. However, we must avoid augmenting the gallery with an already *known* identity, since, in that case, query faces of that known identity might later get classified as *unknown*. Therefore, we look for the clusters where the original gallery faces end up. These clusters are ignored for finding new gallery items. For the remaining clusters, we choose to add the medoid embedding to the gallery as a new identity, i.e. the embedding that has, on average, the smallest distance to the other embeddings in the same cluster. This process is illustrated in Fig. 5. The clustering resulted in 3 clusters. However, two of them contain an item from the original gallery. These two clusters are therefore ignored. For the remaining cluster, we search the medoid and add this as a reference of a new, unknown, identity to the gallery. Hence, we obtain the *augmented* gallery.

3.3 Discarding Similar Gallery Items

The final step involves calculating the mean difference between all the embeddings in the augmented gallery set and removing the descriptors that lie closer than k standard deviations from the mean difference. This is, again, to avoid that an identity that was already in the gallery would have a second representation that is marked as *unknown*. In our experiments, we chose a fixed $k = 3$. Further experiments are needed to determine the best value of this parameter and to evaluate its influence on the performance of CIGA.

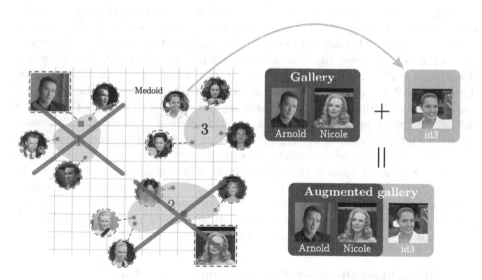

Fig. 5. After clustering the query and gallery embeddings, we ignore clusters with gallery items. For the remaining cluster, the medoid is used to add as a reference of a new identity to the gallery.

4   Experiments and Results

In the following sections, we evaluate our proposed method that involves exploiting the structure of the query set to improve the classification result of an open-set recognition problem. In Sect. 4.1, we perform a grid search to find the optimal parameters for the DBSCAN clustering algorithm that is used by CIGA. Next, in Sect. 4.2 we evaluate CIGA on both the Jade dataset and a subset of the LFW dataset. The experiments from Sect. 4.2 can be reproduced from the code on our repository: https://gitlab.com/florisdf/acpr2019. The README file contains further details on how to reproduce the results.

4.1   DBSCAN Hyperparameters

The values to choose for the parameters ε and `min_samples` of the DBSCAN clustering algorithm, depend on the density of the descriptors to cluster. We performed a grid search to discover the hyperparameters that yield truthful clusters. We used two metrics for evaluating the clustering results: the Adjusted Rand Index (ARI) and the number of clusters found. The ARI is a metric to determine the similarity of two clusterings, with chance normalization [9]. It yields a result between -1 and 1 where 1 means that the two clusterings are identical. The ground truth clusters are obtained by simply grouping the labeled faces per identity.

Our experiments are done on the Jade dataset, assuming that—when using the same facial feature extractor—the hyperparameters would generalize to other datasets as well. This assumption will be confirmed in Sect. 4.2. Figure 6(a) shows the ARI between the clusters arising from the DBSCAN clustering and the ground truth clusters for multiple combinations of ε and `min_samples`. The darker the color, the closer the ARI is to 1 and thus the better the DBSCAN clustering. While the highest ARI scores are obtained in the leftmost column of the heat map, setting `min_samples` to 1 would not be a good choice as is apparent from Fig. 6(b), which shows that the number of clusters becomes very high in that case. This makes sense, because `min_samples` $= 1$ means that every sample is a cluster by itself. Ideally, the number of clusters would be equal to 30 ($= 29$ + 1 for outliers) corresponding to the 29 identities present in the Jade dataset. The color map in Fig. 6(b) is centered around 30, such that the darkest areas are those with the correct amount of clusters. When comparing Figs. 6(a) and (b), a reasonable choice for the parameters would be $\varepsilon = 0.42$ and `min_samples` $= 5$.

4.2   CIGA in an Open-Set Face Recognition Pipeline

We evaluate the effect of adding CIGA for two different datasets: our private Jade dataset and a subset of the publicly available LFW [8] dataset. In both cases, we use the MTCNN face detector [24], a facial feature extractor from the Dlib library [11] and the DBSCAN clustering algorithm [5] from the Scikit-learn library [15]. For each dataset, we take the following steps after detecting the faces and extracting the face embeddings:

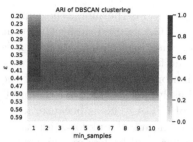

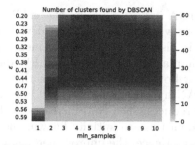

(a) ARI vs. DBSCAN hyperparameters

(b) Number of clusters vs. DBSCAN hyperparameters

Fig. 6. Two types of grid search to find the optimal parameters for the DBSCAN algorithm. (a) ARI between the DBSCAN clustering and the ground truth clusters for multiple combinations of ε and min samples; *darker is better.* (b) Number of clusters according to the DBSCAN clustering for multiple combinations of ε and min samples; *darker is better.*

1. Compose a random gallery with N_{gal} items;
2. Apply CIGA, adding N_{CIGA} extra items to the gallery;
3. Classify the query items into the $N_{gal} + N_{CIGA}$ classes defined by the new gallery using Euclidean distance as a similarity measure;
4. Calculate the average precision (AP) score of the micro-averaged PR-curve for the classification of the identities that were initially in the gallery, i.e. the *known* identities.

This experiment is repeated multiple times for different values of N_{gal}. For each of these values, we repeat the experiment 30 times, each time randomly selecting a certain combination of N_{gal} initial gallery items. For the Jade dataset, N_{gal} is varied between $3 \ldots 29$. For the LFW dataset, N_{gal} is varied between $3 \ldots 150$. It is important to note here that we only used a subset of the LFW dataset, i.e. we used the 150 identities with the most images. We do this because many identities in the LFW dataset have only a single image. Of course, these identities are of no interest to the current experiment, as we want to check if exploiting the query structure improves the classification result. Therefore, the query set must contain multiple instances of the same identity. To reproduce our results, please visit https://gitlab.com/florisdf/acpr2019.

Figures 7 and 8 show the results for the Jade and the LFW dataset, respectively. The solid lines indicate the mean of the AP score (mAP) for the 30 experiments performed with the corresponding N_{gal}. The colored bands around the solid lines show the standard deviation between the APs of those 30 experiments. We observe that in both cases, using CIGA systematically improves the classification result when the gallery contains few items. Table 1 shows the results for the extreme case when only 3 identities were initially present in the gallery, i.e. the leftmost value on the horizontal axis of Figs. 7 and 8. We see that CIGA indeed improves the mAP in both cases. Furthermore, the experiments show a smaller standard deviation when CIGA is used, indicating that CIGA is also a reliable improvement. This decrease in standard deviation is more than a factor of 10 for the LFW dataset. The fact that we see an improvement for both datasets, motivates the assumption on the generalizability of the DBSCAN hyperparameters found in Sect. 4.1. Lastly, we observe that the mAP clearly reduces when N_{gal} increases in the case of CIGA. This is because, when more identities are initially known, the relative amount of unknown identities decreases. Hence, the improvement that CIGA offers will be less pronounced.

Table 1. Results on the Jade dataset and the LFW dataset when only 3 identities were initially present in the gallery.

Dataset	CIGA (mAP)	No CIGA (mAP)
Jade	$(59.12 \pm 11.65)\%$	$(44.61 \pm 13.39)\%$
LFW	$(99.74 \pm 0.46)\%$	$(94.65 \pm 6.03)\%$

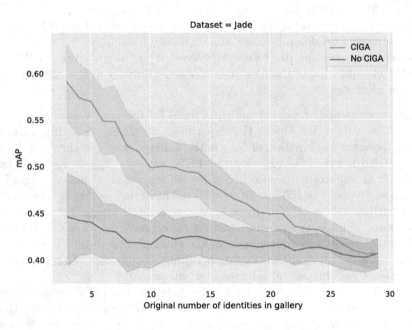

Fig. 7. Evaluation of CIGA on the Jade dataset.

Fig. 8. Evaluation of CIGA on the LFW dataset.

5 Conclusion

In this paper, we motivate that in many practical open-set face recognition applications a query might contain a valuable structure that is often ignored. We proposed a proof-of-concept that consists of a Cluster-Inferred Gallery Augmentation step and evaluated it on a private real-world dataset and a subset of the public LFW dataset. As such, we showed that exploiting the query structure can indeed significantly improve the performance of an open-set face recognition problem. Our implementation of the algorithm and the experiments can be found on https://gitlab.com/florisdf/acpr2019.

References

1. Bendale, A., Boult, T.: Towards open world recognition. In: The IEEE Conference on Computer Vision and Pattern Recognition (CVPR), pp. 1893–1902 (2015)
2. Bendale, A., Boult, T.E.: Towards open set deep networks. In: Proceedings of the IEEE Conference on Computer Vision and Pattern Recognition, pp. 1563–1572 (2016)
3. Busto, P.P., Iqbal, A., Gall, J.: Open set domain adaptation for image and action recognition. IEEE Trans. Pattern Anal. Mach. Intell. **42**(2), 413–429 (2018)
4. Deng, J., Guo, J., Niannan, X., Zafeiriou, S.: ArcFace: additive angular margin loss for deep face recognition. In: CVPR (2019)
5. Ester, M., Kriegel, H.P., Sander, J., Xu, X., et al.: A density-based algorithm for discovering clusters in large spatial databases with noise. In: KDD, vol. 96, pp. 226–231 (1996)

6. Günther, M., Cruz, S., Rudd, E.M., Boult, T.E.: Toward open-set face recognition. In: Proceedings of the IEEE Conference on Computer Vision and Pattern Recognition Workshops, pp. 71–80 (2017)
7. Hartigan, J.A., Wong, M.A.: Algorithm as 136: a k-means clustering algorithm. J. Roy. Stat. Soc. Ser. C (Appl. Stat.) **28**(1), 100–108 (1979)
8. Huang, G.B., Mattar, M., Berg, T., Learned-Miller, E.: Labeled faces in the wild: a database for studying face recognition in unconstrained environments. In: Workshop on Faces in 'Real-Life' Images: Detection, Alignment, and Recognition (2008)
9. Hubert, L., Arabie, P.: Comparing partitions. J. Classif. **2**(1), 193–218 (1985)
10. Jain, L.P., Scheirer, W.J., Boult, T.E.: Multi-class open set recognition using probability of inclusion. In: Fleet, D., Pajdla, T., Schiele, B., Tuytelaars, T. (eds.) ECCV 2014. LNCS, vol. 8691, pp. 393–409. Springer, Cham (2014). https://doi.org/10.1007/978-3-319-10578-9_26
11. King, D.E.: Dlib-ml: a machine learning toolkit. J. Mach. Learn. Res. **10**, 1755–1758 (2009)
12. Malalur, P., Jaakkola, T.: Alignment based matching networks for one-shot classification and open-set recognition. arXiv preprint arXiv:1903.06538 (2019)
13. Meyer, B.J., Drummond, T.: The importance of metric learning for robotic vision: open set recognition and active learning. arXiv preprint arXiv:1902.10363 (2019)
14. Parkhi, O.M., Vedaldi, A., Zisserman, A., et al.: Deep face recognition. In: BMVC, vol. 1, p. 6 (2015)
15. Pedregosa, F., et al.: Scikit-learn: machine learning in Python. J. Mach. Learn. Res. **12**, 2825–2830 (2011)
16. Scheirer, W.J., De Rezende Rocha, A., Sapkota, A., Boult, T.E.: Toward open set recognition. IEEE Trans. Pattern Anal. Mach. Intell. **35**(7), 1757–1772 (2013). https://doi.org/10.1109/TPAMI.2012.256
17. Scheirer, W.J., Jain, L.P., Boult, T.E.: Probability models for open set recognition. IEEE Trans. Pattern Anal. Mach. Intell. **36**(11), 2317–2324 (2014). https://doi.org/10.1109/TPAMI.2014.2321392
18. Scherreik, M.D., Rigling, B.D.: Open set recognition for automatic target classification with rejection. IEEE Trans. Aerosp. Electron. Syst. **52**(2), 632–642 (2016). https://doi.org/10.1109/TAES.2015.150027
19. Schroff, F., Kalenichenko, D., Philbin, J.: FaceNet: a unified embedding for face recognition and clustering. In: Proceedings of the IEEE Conference on Computer Vision and Pattern Recognition, pp. 815–823 (2015)
20. Sun, Y., Liang, D., Wang, X., Tang, X.: DeepID3: face recognition with very deep neural networks. arXiv preprint arXiv:1502.00873 (2015)
21. Taigman, Y., Yang, M., Ranzato, M., Wolf, L.: DeepFace: closing the gap to human-level performance in face verification. In: Proceedings of the IEEE Conference on Computer Vision and Pattern Recognition, pp. 1701–1708 (2014)
22. Wang, H., et al.: CosFace: large margin cosine loss for deep face recognition. In: Proceedings of the IEEE Conference on Computer Vision and Pattern Recognition, pp. 5265–5274 (2018)
23. Wang, M., Deng, W.: Deep face recognition: a survey. arXiv preprint arXiv:1804.06655 (2018)
24. Zhang, K., Zhang, Z., Li, Z., Qiao, Y.: Joint face detection and alignment using multitask cascaded convolutional networks. IEEE Signal Process. Lett. **23**(10), 1499–1503 (2016)

Action Recognition in Untrimmed Videos with Composite Self-attention Two-Stream Framework

Dong Cao[1,2](✉), Lisha Xu[1,2], and HaiBo Chen[2]

[1] Institute of Cognitive Intelligence, DeepBlue Academy of Sciences, Shanghai, China
doocao@gmail.com, xuls@deepblueai.com
[2] DeepBlue Technology (Shanghai) Co., Ltd., No. 369, Weining Road, Shanghai, China
chenhaibo@deepblueai.com

Abstract. With the rapid development of deep learning algorithms, action recognition in video has achieved many important research results. One issue in action recognition, Zero-Shot Action Recognition (ZSAR), has recently attracted considerable attention, which classify new categories without any positive examples. Another difficulty in action recognition is that untrimmed data may seriously affect model performance. We propose a composite two-stream framework with a pre-trained model. Our proposed framework includes a classifier branch and a composite feature branch. The graph network model is adopted in each of the two branches, which effectively improves the feature extraction and reasoning ability of the framework. In the composite feature branch, 3-channel self-attention modules are constructed to weight each frame of the video and give more attention to the key frames. Each self-attention channel outputs a set of attention weights to focus on the particular stage of the video, and a set of attention weights corresponds to a one-dimensional vector. The 3-channel self-attention modules can inference key frames from multiple aspects. The output sets of attention weight vectors form an attention matrix, which effectively enhances the attention of key frames with strong correlation of action. This model can also implement action recognition under zero-shot conditions, and has good recognition performance for untrimmed video data. Experimental results on relevant datasets confirm the validity of our model.

Keywords: Action recognition · Self-attention · Graph network

1 Introduction

1.1 Action Recognition

In computer vision, video understanding is one of the most important fields. As a brunch of video understanding, action recognition is the base of visual reasoning tasks, e.g. video captioning, video relational reasoning. Current mainstream models for action recognition are almost derived from Two-stream Convolutional Network and Convolution-3D Network.

S. Palaiahnakote et al. (Eds.): ACPR 2019, LNCS 12047, pp. 27–40, 2020.
https://doi.org/10.1007/978-3-030-41299-9_3

Some studies have found that inputting optical flow into 2D CNNs made a better performance than merely RGB frames, as optical flow represents the motion information at some extent. Thus it is grant to propose Two-stream networks, which take RGB images and optical flow images as input, providing the state of the art result [1, 11]. One of the derivation of Two-stream networks, Temporal Segment Network(TSN), uses multiple two-stream modules to learn parameters with combining video-level supervision and sparse temporal sampling strategy. This algorithm results in good performance in long-term temporal structure even given limited training samples [18].

However, since optical flow is computation-expensive and supposed to calculate offline, it cannot meet the demand of real-time application. 3D spatiotemporal CNN models which use 3D ConvNet enable model to extract spatial features and temporal features simultaneously, over 10 times faster than Two-stream models even though sacrificing some accuracy, have drawn more attention [8].

Basic 3D spatiotemporal CNNs suffer from large parameters and are not easy to train from scratch. Therefore, a model named I3D, based on the pre-trained model on ImageNet, inflates the 2D ConvNets into 3D to increase the temporal dimension, having much fewer parameters. Furthermore, I3D found that two-stream design still improve the accuracy [1]. Another $(2 + 1)$D architecture decomposes 3D convolution filters into separate spatial and temporal components, so as to renders the optimization, maintains the solid performance with 2D CNNs in single frame as well as speeds up the calculation [15].

To balance the accuracy and calculation speed, some works also adjust the input data of two stream. Like extracting key-frames for spatial stream [20], or processing kinds of optical flow information for temporal stream [2, 3, 13], relax the burden of calculation and store.

1.2 Zero-Shot Learning

Zero-shot learning aims to construct new classifiers dynamically based on the semantic descriptors provided by human beings or the existing knowledge, rather than labeled data. Due to poor scalability of exhaustive annotation for supervised learning [19] and expectation to emulate the human ability to learn semantics with few examples, zero-shot have drawn considerable attention. There have been various architectures proposed for zero-shot action recognition (ZSAR) [9, 10, 22]. In these studies, the information of unseen classes come from human-defined attributes, language in the form of text descriptions or word embedding, which are effective in trimmed videos datasets. In this work, we propose a novel architecture that perform well in untrimmed videos.

2   Related Work

Zero-Shot Action Recognition. With the explosive growth of videos and the difficulty of scalability of manual annotation, there have been approaches to recognize the un-known action in the way of zero-shot. ZSAR associates known categories and un-known categories based on semantic space, including manual attributes [23–27], text descriptions [28] and word vectors [29, 30]. However, attributes are not easy to define and the manually-specified attributes are highly subjective. Thus, it is hard to generate undefined but definite categories in real scenes. Word embeddings are more objective for addressing ZSAR. [19, 39] embedded the labels of videos as word vectors for a shared semantic space. [31] utilizes both category-level semantics and intrinsic data structures by adopting error-correcting output codes to address domain shift problem. But these methods usually ignore the temporal information of videos, which take significant advantages for visual understanding [32]. [5] proposed a zero-shot action recognition framework using both the visual clues and external knowledge to show relations between objects and actions, also applied self-attention to model the temporal information of videos.

Transfer Learning. In many real-world applications, it is expensive to re-collect training data and re-model when task changes [33]. Knowledge transfer or transfer learning is needed between task domains, research referring has drawn increasing attention. [34] used the similarity of classifier in source domain and target domain to calculate the gradient between them. [35, 36] used a max mean discrepancies (MMD) as the distance between source and target. Recently, the study about domain adaption are expected to improve the knowledge transfer [37].

3 Our Approach

In this section, we present the details of our proposed model, which consists of a composite two-stream structure, including recognition branch and composite feature branch. Inspired by literatures [5, 38], a dual sub-branch structure is constructed in the composite feature branch. Each sub-branch adopts a multi-channel self-attention module for feature extraction in untrimmed and trimmed videos respectively.

We introduce the novel model from three stage as follows. In Sect. 3.1, the first stage is the sub-branch of the multi-channel self-attention model for trimmed videos. In Sect. 3.2, the second stage is the composite feature branch for both trimmed and untrimmed videos. In Sect. 3.3, the third stage is the composite two-stream structure for zero-shot action recognition, whether it is for trimmed videos or untrimmed videos.

3.1 Sub-branch of the Multi-channel Self-attention Model

Video Data Preprocessing
This sub-branch processes trimmed video data, and implements action recognition in the dataset. Firstly, the original videos need to be preprocessed. The purpose of the preprocessing is to extract the features of the video, including the feature from the spatial dimension and the feature from the temporal dimension.

Specifically, the feature extraction from the spatial dimension is to extract RGB information. As a deep neural network, the ResNet101 network perform well in image classification task. For each frame of the original video having three-channel RGB information, we adopt the pre-trained ResNet101 network to directly perform feature extraction tasks. The output of the ResNet101 is a s-dimension vector corresponding to each video frame. Suppose there are G frames in a video, G s-dimension vectors are output, and a $s \times G$ matrix is formed sequentially, denoted as $\mathcal{S} \in \mathbb{R}^{s \times G}$. We name this extractor as a spatial feature extractor.

The latter feature extraction is to extract motion information in the temporal dimension. Rather extracting motion feature from the original video directly, we process the original video to generate flow images by the general optical flow algorithm. Then the optical flow images are inputted to a deep neural network model. Here we also adopt pre-trained ResNet101 network, where this ResNet101 network is independent of the network in the first extractor above. After the feature extraction of the ResNet101 network, the output is a t-dimension vector corresponding to each flow image. Because optical flow image is calculated by adjacent frames, we get G-1 flow images in horizontal direction and vertical direction respectively. Through ResNet101, $(G-1)$ t-dimension vectors can be obtained, which form a matrix of $t \times (G-1)$, denoted as $\mathcal{T} \in \mathbb{R}^{t \times (G-1)}$. We name this extractor as a temporal feature extractor.

Multi-channel Self-attention Model

After data preprocessing, we obtain two kinds of features map. The first is the spatial feature information matrix $\mathcal{S}$ obtained by the spatial feature extractor and video frames. The second is the temporal feature information matrix $\mathcal{T}$ obtained by the temporal feature extractor and optical flow images. Then, the spatial feature information matrix $\mathcal{S}$ and the temporal feature information matrix $\mathcal{T}$ are respectively input into independent multi-channel self-attention modules. The output result is two information matrixes weighted by attention, corresponding to the input spatial feature information matrix $\mathcal{S}$ and temporal feature information matrix $\mathcal{T}$. These two matrixes are denoted as $\mathbb{S}$ and $\mathbb{T}$, respectively [38]. The multi-channel self-attention module consists of two independent parts, which are respectively applied to the spatial feature information matrix $\mathcal{S}$ and the temporal feature information matrix $\mathcal{T}$, named as multi-channel spatial self-attention (MC-spatial Self-Attention) and multi-channel temporal self-attention (MC-temporal Self-Attention).

The structure of the MC-spatial Self-Attention is shown in the left part of Fig. 1. The 3-channel self-attention structure is used in the structure. Although the construction of each channel is the same, different activation functions are used. Specifically speaking, sigmod, tanh, and Leaky ReLU can be selected to facilitate the extraction of spatial information to form a multi-angle attention weight vector. The input data passes through the fully connected layer FC11, the activation function, the fully connected layer FC12 and softmax function. Fully connected layer FC11, correspond to parameters $W_{1j} \in \mathbb{R}^{a \times s}, j = 1, 2, 3$, where a is a configurable hyperparameter. Activation functions are sigmod, tanh and Leaky ReLU. Fully connected layer FC12 correspond to the parameter $u_{1j} \in \mathbb{R}^{1 \times a}, j = 1, 2, 3$. The softmax output results with the attention vector V_{1j}, where $V_{1j} \in \mathbb{R}^{G \times 1}, j = 1, 2, 3$. The operation relationships are as follows.

$$V_{11} = softmax[u_{11} \cdot sigmoid(W_{11}\mathcal{S})]$$
$$V_{12} = softmax[u_{12} \cdot tanh(W_{12}\mathcal{S})]$$
$$V_{13} = softmax[u_{13} \cdot LeakyReLU(W_{13}\mathcal{S})]$$

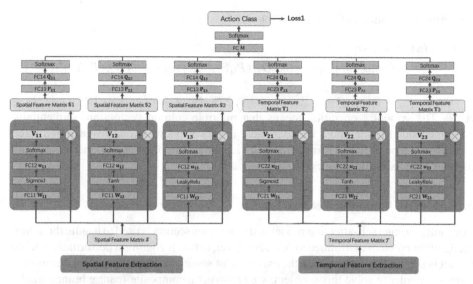

Fig. 1. The structure of multi-channel self-attention models

Further $V_{1j} \in \mathbb{R}^{G \times 1}$, $j = 1, 2, 3$ are multiplied by the matrix of the spatial feature matrix S respectively, to obtain the attention spatial feature matrix $\mathbb{S}_i$, where $\mathbb{S}_i \in \mathbb{R}^{s \times G}$, $i = 1, 2, 3$.

The structure of the MC-temporal Self-Attention is shown in the right part of Fig. 1. The 3-channel self-attention structure is also used. Like the spatial structure, the temporal self-attention structure of each channel is the same. But different activation functions are utilized in order to facilitate the extraction of temporal information from multiple angles of attention, forming a multi-angle attention weight vector. The input data passes through the fully connected layer FC21, the activation function, the fully connected of FC22 and softmax. The fully connected layer FC21 correspond parameters $W_{2j} \in \mathbb{R}^{b \times t}$, $j = 1, 2, 3$, where b is configurable hyperparameter. Activation functions are sigmod, tanh, Leaky ReLU. The fully connected FC22, correspond parameters $u_{2j} \in \mathbb{R}^{1 \times b}$, $j = 1, 2, 3$. The softmax function output result with the attention vector V_{2j}, where $V_{2j} \in \mathbb{R}^{(G-1) \times 1}$, $j = 1, 2, 3$. The operation relationships are as follows.

$$V_{21} = softmax[u_{21} \cdot sigmoid(W_{21}\mathcal{T})]$$
$$V_{22} = softmax[u_{22} \cdot tanh(W_{22}\mathcal{T})]$$
$$V_{23} = softmax[u_{23} \cdot LeakyReLU(W_{23}\mathcal{T})]$$

Further $V_{2j} \in \mathbb{R}^{(G-1) \times 1}$, $j = 1, 2, 3$ are multiplied by the temporal feature matrix $\mathcal{T}$, respectively, to obtain the attention temporal feature matrix $\mathbb{T}_i$, where $\mathbb{T}_i \in \mathbb{R}^{t \times (G-1)}$, $i = 1, 2, 3$.

After the attention spatial feature matrix $\mathbb{S}_i$ and the temporal attention feature matrix $\mathbb{T}_i$ are obtained by the above operations, the following operations are performed. This part is to calculate the loss function in the training phase, expressed as $loss1$, and used for action recognition in the inference phase.

$$loss1, inference \Leftrightarrow$$
$$softmax\left\{M\left\{\left[softmax\left[Q_{1i}(P_{1i}\mathbb{S}_i)^T\right]\right]\|\left[softmax\left[Q_{2i}(P_{2i}\mathbb{T}_i)^T\right]\right]\right\}^T\right\}$$

Where $P_{1i} \in \mathbb{R}^{1\times s}$, $Q_{1i} \in \mathbb{R}^{k\times G}$, $i = 1, 2, 3$, $P_{2i} \in \mathbb{R}^{1\times t}$, $Q_{2i} \in \mathbb{R}^{k\times(G-1)}$, $M \in \mathbb{R}^{1\times 6}$. The symbol $\|$ operation means that multiple column vectors are sequentially combined side by side into a matrix, where three vectors of spatial features and three vectors of temporal features are combined side by side to obtain $k \times 6$ matrix. k is the number of classifications for action recognition.

3.2 Model of Composite Feature Branch

The multi-channel self-attention model in the previous section is used to handle the action recognition problem in trimmed videos. However, action recognition performance of the model is significantly affected by the existance of some interfering frames in untrimmed videos. In order to solve this problem, we construct a composite feature branch model [38], as shown in Fig. 2 below.

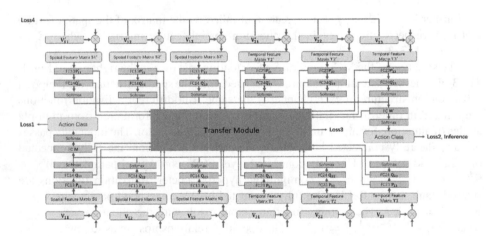

Fig. 2. Composite feature branch model

We designed the two-stream parallel composite architecture with multi-channel self-attention model. The upper stream is to process untrimmed videos, and the lower stream is to process trimmed videos.

The upper part deals with the model of untrimmed videos, where the loss function loss2 is defined in the output part of the action classification. The loss2 is calculated in the form of standard multi-class cross entropy, which is related to the following representation.

$$loss2, inference \Leftrightarrow$$
$$softmax\left\{M'\left\{\left[softmax\left[Q'_{1i}(P'_{1i}\mathbb{S}'_i)^T\right]\right]\|\left[softmax\left[Q'_{2i}(P'_{2i}\mathbb{T}'_i)^T\right]\right]\right\}^T\right\}$$

Where $P'_{1i} \in \mathbb{R}^{1 \times s}$, $Q'_{1i} \in \mathbb{R}^{k \times G}$, $i = 1, 2, 3$, $P'_{2i} \in \mathbb{R}^{1 \times t}$, $Q'_{2i} \in \mathbb{R}^{k \times (G-1)}$, $M' \in \mathbb{R}^{1 \times 6}$. Here are three column vectors of temporal features and spatial features. The three column vectors are combined side by side to obtain a $k \times 6$ matrix. k is the number of classifications for action recognition.

The key point to make good use of knowledge of the trimmed videos dataset to improve the performance of the model on the untrimmed videos is through transfer learning.

First, the model learning from the trimmed videos dataset, is to train the lower stream model shown in Fig. 2. The lower stream applies the loss function loss1, the specific form of which is standard classification cross entropy (see expression "loss1, inference"). After the training of the lower stream model is completed, the parameters of the model learn the quasi-optimal distribution values to infer the trimmed videos. With the multi-channel self-attention model used in both space and time, the classifier parameters in the total 6 channels are transmitted to the transfer module, including: P11, P12, P13 in the fully connected layer FC13; Q11, Q12, Q13 in the fully connected layer FC14; P23, P22, P23 in the fully connected layer FC23; Q21, Q22, Q23 in the fully connected layer FC24; M in the fully connected layer FC.

Then, with the multi-channel self-attention spatial model and temporal model of the upper stream, the corresponding classifier parameters of the total 6 channels are transmitted to the transfer module in the meantime, including: P'_{11}, P'_{12}, P'_{13} in the fully connected layer FC13; Q'_{11}, Q'_{12}, Q'_{13} in the fully connected layer FC14; P'_{21}, P'_{22}, P'_{23} in the fully connected layer FC23; Q'_{21}, Q'_{22}, Q'_{23} in the fully connected layer FC24; M' in the fully connected layer FC. The loss function loss3 of the transfer module is constructed by calculating the generalized maximum mean discrepancy (MMD) of the upper and lower models by the literature [35].

In the third step, during the training process of the upper stream, the attention vectors $V'_{ij}, i = 1, 2, j = 1, 2, 3$, need to be regularized to optimize the structure of the loss function loss4.

Given $V'_{1j} \in \mathbb{R}^{G \times 1}$, $V'_{2j} \in \mathbb{R}^{(G-1) \times 1}$, $j = 1, 2, 3$, let $V'_{1j} = \left(v'_{1j1}, v'_{1j2}, \cdots, v'_{1jG} \right)^{T}$, $V'_{2j} = \left(v'_{2j1}, v'_{2j2}, \cdots, v'_{2j(G-1)} \right)^{T}$, where T represents a transposition operation. Thus $loss4$ is calculated as follows:

$$
\begin{aligned}
loss4 = & \sum_{j=1}^{3} \sum_{n=1}^{G-1} \left(v'_{1jn} - v'_{1j(n+1)} \right)^{4} + \sum_{j=1}^{3} \sum_{n=1}^{G-2} \left(v'_{2jn} - v'_{2j(n+1)} \right)^{4} \\
& + \sum_{j=1}^{3} \left(\left\| V'_{1j} \right\|_{1} + \left\| V'_{2j} \right\|_{1} \right) \\
= & \sum_{j=1}^{3} \{ 2((V'_{1j})^{T} \circ (V'_{1j})^{T})(V'_{1j} \circ V'_{1j}) - ((V'_{1j})^{T} \circ (V'_{1j})^{T})\mathcal{A}(V'_{1j} \circ V'_{1j}) \\
& - ((V'_{1j})^{T} \circ (V'_{1j})^{T})\mathcal{B}(V'_{1j} \circ V'_{1j}) + 6((V'_{1j})^{T} \circ (V'_{1j})^{T})\mathcal{C}(V'_{1j} \circ V'_{1j}) \\
& - 4((V'_{1j})^{T} \circ (V'_{1j})^{T} \circ (V'_{1j})^{T})\mathcal{C}V'_{1j} - 4(V'_{1j})^{T} \left(\left(\mathcal{C}V'_{1j} \right) \circ \left(\mathcal{C}V'_{1j} \right) \circ \left(\mathcal{C}V'_{1j} \right) \right) \} \\
& + \sum_{j=1}^{3} \{ 2((V'_{2j})^{T} \circ (V'_{2j})^{T})(V'_{2j} \circ V'_{2j}) - ((V'_{2j})^{T} \circ (V'_{2j})^{T})\mathcal{D}(V'_{2j} \circ V'_{2j})
\end{aligned}
$$

$$- ((V'_{2j})^T \circ (V'_{2j})^T) \mathcal{E}(V'_{2j} \circ V'_{2j}) + 6((V'_{2j})^T \circ (V'_{2j})^T) \mathcal{F}(V'_{2j} \circ V'_{2j})$$

$$- 4((V'_{2j})^T \circ (V'_{2j})^T \circ (V'_{2j})^T) \mathcal{F} V'_{2j} - 4(V'_{2j})^T \left(\left(\mathcal{F} V'_{2j} \right) \circ \left(\mathcal{F} V'_{2j} \right) \circ \left(\mathcal{F} V'_{2j} \right) \right) \}$$

$$+ \sum_{j=1}^{3} \left(\left\| V'_{1j} \right\|_1 + \left\| V'_{2j} \right\|_1 \right)$$

Where $\circ$ represents Hadamard product, $\| \|_1$ represents 1-norm, matrixes $\mathcal{A}, \mathcal{B}, \mathcal{C}, \mathcal{D}, \mathcal{E}, \mathcal{F}$ are as follows:

$$\mathcal{A} = \begin{bmatrix} 1\,0 \\ 0\,0 \\ \ddots \\ 0\,0 \\ 0\,0 \end{bmatrix}_{G \times G}, \mathcal{B} = \begin{bmatrix} 0\,0 \\ 0\,0 \\ \ddots \\ 0\,0 \\ 0\,1 \end{bmatrix}_{G \times G}, \mathcal{C} = \begin{bmatrix} 0\,1 \\ 0\,0 \\ \ddots \\ 0\,1 \\ 0\,0 \end{bmatrix}_{G \times G},$$

$$\mathcal{D} = \begin{bmatrix} 1\,0 \\ 0\,0 \\ \ddots \\ 0\,0 \\ 0\,0 \end{bmatrix}_{(G-1) \times (G-1)}, \mathcal{E} = \begin{bmatrix} 0\,0 \\ 0\,0 \\ \ddots \\ 0\,0 \\ 0\,1 \end{bmatrix}_{(G-1) \times (G-1)}, \mathcal{F} = \begin{bmatrix} 0\,1 \\ 0\,0 \\ \ddots \\ 0\,1 \\ 0\,0 \end{bmatrix}_{(G-1) \times (G-1)}$$

Finally, the complete loss function to train model on untrimmed videos is used on the upper stream of Fig. 2, which is set as follows:

$$loss = loss2 + loss3 + loss4$$

3.3 Model of Composite Feature Branch

We use the multi-channel self-attention composite model in Sect. 3.2 as a pre-training model, and then replace the attention module of Two-Stream Graph Convolutional Network in the [5], to implement a multi-channel self-attention composite two-stream model. This model can achieve zero-shot action recognition through sharing knowledge space with corresponding objects and action between these objects.

4 Experiment

In this section, we evaluate the performance of our novel architecture for zero-shot untrimmed action recognition on THOUMOS14, which contains a large number of trimmed and untrimmed videos.

4.1 Experimental Setup

Dataset. The training dataset of THUMOS14 is UCF101, including 101-class trimmed videos. The validation and test dataset of THUMOS14 contain 2500 and 1574 untrimmed videos respectively. Furthermore, the validation is the same as the video category in the training set, while the test data contains only part of 101 categories.

Following the transductive and generalized configuration, we implement our experiments. The former is to access the unseen data in training, while the latter is to combine part of the seen data and the unseen data as test data [12, 39]. Here we design three data splits (TD, G and TD+G), to evaluate the performance of zero-shot action recognition in untrimmed videos.

Since most zero-shot methods currently evaluate their performance on UCF101, we also follow the data split proposed by (Xu, Hospedales, and Gong 2017), i.e. to evaluate our model.

Among the data splitting ways, the setup of TD+G is shown in Table 3. During training, on one hand, we set the 30-class trimmed videos as part of the seen data, responsible for explicit knowledge, which is the source of transfer module. On the other hand, for THUMOS14, we select 80% of other 20-class untrimmed videos as seen data and 20% of remaining 51-class untrimmed videos as unseen data for the target of transfer module. During testing, for THUMOS14, we combine the 20% of those 20-class untrimmed videos and 80% of those 51-class untrimmed videos.

Table 1. The split way to evaluate our model in transductive setting.

THUMOS14				
TD	Train		Test	
	Seen	Unseen	Seen	Unseen
Trimmed	30 classes	0	0	0
Untrimmed	20 classes	51 classes * 20%	0	51 classes * 80%

Table 2. The split way to evaluate our model in generalized setting

THUMOS14				
G	Train		Test	
	Seen	Unseen	Seen	Unseen
Trimmed	30 classes	0	0	0
Untrimmed	20 classes * 80%	0	20 classes * 20%	51 classes

Table 3. The split way to evaluate our model in both transductive and generalized setting.

THUMOS14				
TD+G	Train		Test	
	Seen	Unseen	Seen	Unseen
Trimmed	30 classes	0	0	0
Untrimmed	20 classes * 80%	51 classes * 20%	20 classes * 20%	51 classes * 80%

4.2 Evaluation

We evaluate the model in three phases below, and adopt classification accuracy as metric.

1. The trimmed action recognition (TAR) task is to predict the label of trimmed videos. Here we use the training data of THUMOS, and 3 splits of UCF101 to examine the performance of MC-spatial Self-Attention module.
2. The untrimmed action recognition (UTAR) task is to predict the primary action label of untrimmed videos. This task examines the performance of transfer module that learns the explicit knowledge from trimmed videos.
3. The zero-shot action recognition (ZSAR) is to predict the primary action label of untrimmed videos, the categories of which are absent during training.

4.3 Implement Details

The novel model consists of a composite two-stream structure, and including recognition branch and composite feature branch. For the composite feature branch, we process the optical information by generic end-to-end way, and extract the spatial and temporal features of the input RGB and optical flow information respectively by using the residual network pre-trained in kinetics. In order to reduce the influence caused by different inputs, we uniformly limit the input size to $16 \times 224 \times 224$.

For the recognition branch, we encode action categories through Glove, and construct knowledge graph by ConceptNet. To train the whole model, we optimize the parameter with SGD where initial learning rate is 0.0001 and weight decay 0.0005 every 1000 iterations.

4.4 Result

In TAR, we compare our sub-model with some baselines: (1) Two-stream CNN (Simonyan et al. 2014) (2) Two-stream+LSTM (Ng et al. 2015) (3) Two-stream fusion (Feichtenhofer et al. 2016.) (4) C3D (Tran et al. 2015) (5) Two-stream+I3D (Carreira and Zisserman 2017) (6) TSN (Wang et al. 2016).

Overall, according Table 1, compared with the Two-stream I3D and its pre-trained model, our 3 channels sub-model outperforms 0.5% and 0.7%, respectively.

In UTAR, Table 2 demonstrates that our architecture improves the untrimmed videos recognition about 3.5% on TUMOS14 thanks to the transfer learning, that combine the explicit knowledge from trimmed videos. As for ZSAR, following the set of [5], our results have not improved significantly. We hope to find better algorithm in future works (Tables 4 and 5).

Table 4. TAR accuracy that is the average accuracy over 3 splits of UCF101.

Method	UCF101
Two-stream CNN [11]	88.0
Two-stream+LSTM [6]	88.6
Two-stream Fusion [4]	92.5
C3D 3nets [14]	90.4
Two-stream+I3D [1]	93.4
Two-stream+I3D (kinetics pre-trained) [1]	98.0
TSN [18]	94.0
Two-stream+multi-channels self-attention	93.9
Two-stream (kinetics pre-trained)+multi-channels self-attention	98.1

Table 5. UTAR accuracy on THUMOS14.

Method	THUMOS14
IDT+FV [16]	63.1
Two-stream [39]	66.1
Objects+Motion [7]	66.8
Depth2Action [21]	–
C3D [14]	–
TSN 3seg [18]	78.5
UntrimmedNet [17]	82.2
OUR	85.7

5 Conclusion

In this paper, based on a pre-training multi-channels self-attention model. We propose a composite two-stream framework. The novel framework includes a classifier stream and a composite instance stream. The graph network model is adopted in each of the two streams, the performance of feature extraction is significantly improved, and reasoning ability of the framework becomes better. In the composite instance stream, a multi-channel self-attention models are constructed to weight each frame in the video and give more attention to the key frames. Each self-attention models channel outputs a set of attention weights to focus on a particular aspect of the video. The multi-channels self-attention models can evaluate key-frames from multiple aspects, and the output sets of attention weight vectors form an attention matrix. This model can implement action recognition under zero-shot conditions, and has good recognition performance for untrimmed video data. Experimental results on relevant data sets confirm the validity of our model.

Acknowledgments. We are very grateful to DeepBlue Technology (Shanghai) Co., Ltd. and DeepBlue Academy of Sciences for their support. Thanks to the support of the Equipment pre-research project (No. 31511060502). Thanks to Dr. Dongdong Zhang of the DeepBlue Academy of Sciences.

References

1. Carreira, J., Zisserman, A.: Quo vadis, action recognition? A new model and the kinetics dataset. In: Proceedings of the IEEE Conference on Computer Vision and Pattern Recognition, CVPR (2017)
2. Dosovitskiy, A., et al.: Flownet: learning optical flow with convolutional networks. In: Proceedings of the IEEE International Conference on Computer Vision, ICCV (2015)
3. Fan, L., Huang, W., Chuang G., Ermon, S., Gong, B., Huang, J.: End-to-end learning of motion representation for video understanding. In: Proceedings of the IEEE Conference on Computer Vision and Pattern Recognition, CVPR (2018)
4. Feichtenhofer, C., Pinz, A., Zisserman, A.: Convolutional two-stream network fusion for video action recognition. In: Proceedings of the IEEE Conference on Computer Vision and Pattern Recognition, CVPR (2016)
5. Gao, J., Zhu, T., Xu, C.: I know the relationships: zero-shot action recognition via two-stream graph convolutional networks and knowledge graphs. In: AAAI (2019)
6. Ng, J.Y.-H., Hausknecht, M., Vijayanarasimhan, S., Vinyals, O., Monga, R., Toderici, G.: Beyond short snippets: deepnetworks for video classification. In: Proceedings of the IEEE Conference on Computer Vision and Pattern Recognition, CVPR (2015)
7. Jain, M., Gemert, J.C., Snoek, C.G.M: What do 15000 object categories tell us about classifying and localizing actions? In: Proceedings of the IEEE Conference on Computer Vision and Pattern Recognition, CVPR (2015)
8. Ji, S., Xu, W., Yang, M., Yu, K.: 3D convolutional neural networks for human action recognition. IEEE Trans. Pattern Anal. Mach. Intell. **35**(1), 221–231 (2013)
9. Mishra, A., Verma, V.K., Reddy, M.S.K., Arulkumar, S., Rai, P., Mittal, A.: A generative approach to zero-shot and few-shot action recognition. In: WACV (2018)
10. Norouzi, M., et al.: Zero-shot learning by convex combination of semantic embeddings. In: ICLR (2014)
11. Simonyan, K., Zisserman A.: Two-stream convolutional networks for action recognition in videos. In: Advances in Neural Information Processing Systems (NIPS), pp. 568–576 (2014)
12. Song, J., Shen, C., Yang, Y., Liu, Y., Song, M.: Transductive unbiased embedding for zero-shot learning. In: Proceedings of the IEEE Conference on Computer Vision and Pattern Recognition, CVPR (2018)
13. Sun, S., Kuang, Z., Sheng, L., Ouyang, W., Zhang, W.: Optical flow guided feature: a fast and robust motion representation for video action recognition. In: Proceedings of the IEEE Conference on Computer Vision and Pattern Recognition, CVPR (2018)
14. Tran, D., Bourdev, L., Fergus, R., Torresani, L., Paluri, M.: Learning spatiotemporal features with 3D convolutional networks. In: Proceedings of the IEEE Conference on Computer Vision and Pattern Recognition, CVPR (2015)
15. Tran, D., Wang, H., Torresani, L., Ray, J., LeCun, Y., Paluri, M.: A closer look at spatiotemporal convolutions for action recognition. In: Proceedings of the IEEE Conference on Computer Vision and Pattern Recognition, CVPR (2018)
16. Wang, H., Schmid, C.: Action recognition with improved trajectories. In: ICCV, pp. 3551–3558 (2013)

17. Wang, L., Xiong, Y., Lin, D., Gool, L.: UntrimmedNets for weakly supervised action recognition and detection. In: Proceedings of the IEEE Conference on Computer Vision and Pattern Recognition, CVPR (2017)
18. Wang, L., et al.: Temporal segment networks: towards good practices for deep action recognition. In: Leibe, B., Matas, J., Sebe, N., Welling, M. (eds.) ECCV 2016. LNCS, vol. 9912, pp. 20–36. Springer, Cham (2016). https://doi.org/10.1007/978-3-319-46484-8_2
19. Xu, X., Hospedales, T.M., Gong, S.: Multi-task zero-shot action recognition with prioritised data augmentation. In: Leibe, B., Matas, J., Sebe, N., Welling, M. (eds.) ECCV 2016. LNCS, vol. 9906, pp. 343–359. Springer, Cham (2016). https://doi.org/10.1007/978-3-319-46475-6_22
20. Zhu, W., Hu, J., Sun, G., Cao, X., Qiao, Y.: A key volume mining deep framework for action recognition. In: Proceedings of the IEEE Conference on Computer Vision and Pattern Recognition, CVPR (2016)
21. Zhu, Y., Newsam, S.: Depth2Action: exploring embedded depth for large-scale action recognition. In: Hua, G., Jégou, H. (eds.) ECCV 2016. LNCS, vol. 9913, pp. 668–684. Springer, Cham (2016). https://doi.org/10.1007/978-3-319-46604-0_47
22. Zhu, Y., Long, Y., Guan, Y., Newsam, S., Shao, L.: Towards universal representation for unseen action recognition. In: Proceedings of the IEEE Conference on Computer Vision and Pattern Recognition, CVPR (2018)
23. Farhadi, A., Endres, I., Hoiem, D., Forsyth, D.: Describing objects by their attributes. In: Proceedings of the IEEE Conference on Computer Vision and Pattern Recognition, CVPR (2009)
24. Lampert, C.H., Nickisch, H., Harmeling, S.: Learning to detect unseen object classes by between-class attribute transfer. In: Proceedings of the IEEE Conference on Computer Vision and Pattern Recognition, CVPR (2009)
25. Akata, Z., Perronnin, F., Harchaoui, Z., Schmid, C.: Label embedding for attribute-based classification. In: Proceedings of the IEEE Conference on Computer Vision and Pattern Recognition, CVPR (2013)
26. Morgado, P., Vasconcelos, N.: Semantically consistent regularization for zero-shot recognition. In: Proceedings of the IEEE Conference on Computer Vision and Pattern Recognition, CVPR (2017)
27. Xian, Y., Lampert, C.H., Schiele, B., Akata, Z.: Zero-shot learning-a comprehensive evaluation of the good, the bad and the ugly. In: IEEE Transactions on Pattern Analysis and Machine Intelligence (2017)
28. Frome, A., Corrado, G.S., Shlens, J., Bengio, S., Dean, J., Mikolov, T., et al.: Devise: a deep visual-semantic embedding model. In: Advances in Neural Information Processing Systems, pp. 2121–2129 (2013)
29. Reed, S., Akata, Z., Lee, H., Schiele, B.: Learning deep representations of fine-grained visual descriptions. In: Proceedings of the IEEE Conference on Computer Vision and Pattern Recognition, CVPR (2016)
30. Zhang, L., Xiang, T., Gong, S.: Learning a deep embedding model for zero-shot learning. In: Proceedings of the IEEE Conference on Computer Vision and Pattern Recognition, CVPR (2017)
31. Qin, J., et al.: Zero-shot action recognition with error-correcting output codes. In: Proceedings of the IEEE Conference on Computer Vision and Pattern Recognition, CVPR (2017)
32. Niebles, J.C., Chen, C.-W., Fei-Fei, L.: Modeling temporal structure of decomposable motion segments for activity classification. In: Daniilidis, K., Maragos, P., Paragios, N. (eds.) ECCV 2010. LNCS, vol. 6312, pp. 392–405. Springer, Heidelberg (2010). https://doi.org/10.1007/978-3-642-15552-9_29
33. Pan, S.J., Yang, Q.: A survey on transfer learning. IEEE Trans. Knowl. Data Eng. 22(10), 1345–1359 (2010)

34. Ganin, Y., Lempitsky, V.: Unsupervised domain adaptation by backpropagation. In: International Conference on Machine Learning (2015)
35. Long, M., Cao, Y., Wang, J., Jordan, M.I.: Learning transferable features with deep adaptation networks. In: International Conference on Machine Learning (2015)
36. Gretton, A., Borgwardt, K.M., Rasch, M.J., Scholkopf, B., Smola, A.J.: A kernel two-sample test. JMLR **13**, 723–773 (2012)
37. Tzeng, E., Hoffman, J., Saenko, K., Darrell, T.: Adversarial discriminative domain adaptation. In: Proceedings of the IEEE Conference on Computer Vision and Pattern Recognition, CVPR (2017)
38. Zhang, X., Shi, H., Li, C., Zheng, K., Zhu, X., Duan, L.: Learning transferable self-attentive representations for action recognition in untrimmed videos with weak supervision. In: AAAI (2019)
39. Xu, X., Hospedales, T., Gong, S.: Transductive zero-shot action recognition by word-vector embedding. Int. J. Comput. Vis. **123**(3), 309–333 (2017)

Representation Learning for Style and Content Disentanglement with Autoencoders

Jaemin Na$^{(\boxtimes)}$ and Wonjun Hwang

Ajou University, Suwon, Korea
{osial46,wjhwang}@ajou.ac.kr

Abstract. Many approaches have been proposed to disentangle the style and content from image representations. Most existing methods aim to create new images from the combinations of separated style and content. These approaches have shown impressive results in creating new images in such a way as to swap or mix features, but they have not shown the results of disentanglement respectively. In this paper, to show up the results of disentanglement on each, we propose a two-branch autoencoder framework. Our proposed framework, which incorporates content branch and style branch, enables complementary and effective end-to-end learning. To demonstrate our framework in a qualitative assessment, we show the disentanglement results of style and content respectively in handwritten digit datasets: MNIST-M and EMNIST-M. Furthermore, for the quantitative assessment, we examined our method to the classification task. In this experiment, our disentangled content images from digit datasets can achieve competitive performance in classification tasks. Our code is available at https://github. com/najaemin92/disentanglement-pytorch.

Keywords: Disentanglement · Representation learning · Auto-encoders

1 Introduction

The studies to understand embedded features in the latent space of neural network has great attention among the research community. In this interest, many approaches have emerged to disentangle the features from image representation. Approaches to disentangling factors can be widely used to solve various problems. In image-to-image translation, [4] have proposed the approaches to separate domain-specific features and domain-invariant features. In digit recognition, there are many methods to separate style and content [5–9]. While impressive, these studies aim to create new images obtained by combining content and style features. They focus only on swapping or mixing features to create new images but not confirm the disentanglement.

In this paper, we focus on separating style and content from the images not generating new images. To do this, we propose a two-branch autoencoder network that integrates style and content networks. The style branch performs style disentanglement from the input image whilst the content branch is trained to separate the content from the input image. We demonstrate successful results of our framework on digit datasets, where we obtain style and content representations respectively.

© Springer Nature Switzerland AG 2020
S. Palaiahnakote et al. (Eds.): ACPR 2019, LNCS 12047, pp. 41–51, 2020.
https://doi.org/10.1007/978-3-030-41299-9_4

The main contributions of our work are summarized as follows: *(1)* We propose a novel framework to disentangle the style and content feature representations. The proposed framework has a two-branch autoencoder that incorporates style and content networks. We place a shared encoder and a classifier between style and content branches. These architectures enable us to End-to-End learning and complementary training between two-branch autoencoders. *(2)* We demonstrate the disentanglement results of style and content respectively. Many Existing methods are only aimed to create new images from the results of disentanglement. Unlike previous works, we focus on the consequences of each one and each one. *(3)* To demonstrate our method, we use the four-digit datasets: MNIST [16], MNIST-M [17], EMNIST [18], and EMNIST-M. We evaluate our framework for both quantitative and qualitative experiments. In these experiments, we achieved successful results. The details are described in Sect. 3.

2 Related Work

Representation Learning. In deep learning, representation learning or feature learning is discovering representations of input data, which makes it easier to perform a task like a feature detection or classification. As the importance of finding and understanding representations increases day by day, many studies have been conducted [11, 12]. One of the major interests in image representation learning is disentangling the specific representations that are relevant to the task. Disentangling representations are intended to exclude unnecessary representations and to use useful representations for the specific task. However, it is difficult to strike a balance between the two goals of disentangling as many features as possible at the sacrificing of as little data as possible. With this in mind, we designed an effective and compact network for the disentanglement. Our network finds the class informative representations from lower layers and guides to forget this information. On the other hand, class uninformative representations from higher layers are guided to be maintained.

Autoencoders. An autoencoder [1–3] is a kind of feedforward neural network that is constituted by an encoder and a decoder. The main goal of the autoencoder is to compress the input into lower-dimensional representations and reconstruct the same output as the input. An encoder extracts the features from the input images and the decoder reconstructs the input using these extracted features. With these properties, autoencoders are widely used to feature representation learning as well as dimensionality reduction. Basically, our framework is based on autoencoders. We build two-branch of autoencoders to disentangle the content representations and style representations. The content autoencoder is trained to separate out the content features from the input images and generate content images except for style information. On the other hand, the style autoencoder separates the style features form the input images and construct the style images not included content information.

Disentanglement. There are many approaches to disentangle the factors from the images. In a supervised manner, [11] proposed a variational autoencoder that can separate the class label information from the image style information. To incorporate the class label information, they imposed a one-hot label vector to the decoder. Similarly, [12] proposed an autoencoder model, in which the decoder is conditioned on a categorical variable indicating the class label. However, In the case of small-sized label vectors, they are too small for the size of the inputs to the encoder and decoder model, which can be ignored. Unlike [11, 12], one-hot label vectors are not suitable for our work, because we must consider the content feature as well as the style features. Instead of one-hot vectors, we imposed the class label information on the features by complementary learning of two-branch autoencoders.

Many recent disentanglement studies are based on autoencoders [5, 6, 8, 13]. [5] introduced a method to disentangle factors of variation in an unsupervised manner. By mixing features from two images, they keep the invariant attribute to changes in other attributes. [6] also proposed the network to generate the image by mixing the style and content representations. [8] proposed a simple autoencoder network with an adversarial classifier which has a two-step training process. [13] proposed a network composed of four components: an encoder, a generator, a content discriminator, and a style discriminator. Since their training process is based on generative adversarial networks (GANs) [14], Wasserstein GAN [15] algorithm is used to deal with the unstable optimization problems in GAN. And, they exploit style labels to disentangle the feature representations. Contrary to [13], our framework does not require any style labels and additional algorithms. Furthermore, our framework is designed to enable end-to-end learning through complementary training between two-branch autoencoders. Also, in contrast to [6, 8, 13], our framework has two decoders to deal with the style and content representations respectively.

3 Proposed Method

In this section, we introduce our two-branch autoencoder framework that incorporates style and content networks. As shown in Fig. 1, our proposed framework contains four main parts: the shared encoder, two autoencoders, two classifiers, and a reversal classifier. The shared encoder $Enc_\theta(x)$ is composed of a stack of convolutional layers followed by the content branch and the style branch. The content branch is composed of content encoder Enc_C and content decoder Dec_C. Meanwhile, the style branch is constructed with style encoder Enc_S and style decoder Dec_S. These architectures enable us to end-to-end learning and complementary training between two-branch autoencoders.

3.1 Style Disentanglement

As shown in Fig. 1, let us denote $f_S \in R^{N_S}$ as the style features, written as $f_S = Enc_S(Enc_\theta(x))$. To disentangle f_S, we assume that the representations at lower layers are often low-level features such as edges and corners. Under this assumption, we introduce a reversal classifier Cls_{rev} which guides the shared encoder to "forget" the class information in low-level features. The reversal classifier is being trained to minimize the negative cross-entropy loss. The objective for learning reversal classifier can be written as follows:

$$L_{rev} = \frac{1}{N} \sum_{i=1}^{N} y_i \log \hat{y}_i = \frac{1}{N} \sum_{i=1}^{N} y_i \log(Cls_{rev}(Enc_\theta(x_i))), \qquad (1)$$

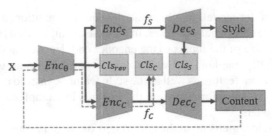

Fig. 1. Illustration of our network structure. The green dashed lines show how decoded content images are passed back through the encoder to obtain predicted label, $\hat{y}$. (Color figure online)

where N is the size of data, y_i is ground truth of labels, and $\hat{y}_i$ is predicted labels. Note that the gradients for L_{rev} are back-propagated only to the reversal classifier but do not update the shared encoder.

On the other hand, if all the class information is removed, we cannot disentangle the style feature properly. To keep the class information, we introduce another classifier Cls_S. Besides, we define z^s as the output feature from the second layer of $Dec_S(f_S)$ and concatenate the $Enc_\theta(x)$ and z^s features in the channel dimension to preserve $Enc_\theta(x)$. Let us denote $Enc_\theta(x) \oplus z^s$ as the concatenated features. Then the Cls_S is trained to minimize the cross-entropy loss on $Enc_\theta(x) \oplus z^s$. In contrast to Cls_{rev}, the gradients are back-propagated to the Enc_S and Enc_θ. The objective for learning Cls_S can be written as,

$$L_{class}^S = -\frac{1}{N}\sum_{i=1}^N y_i \log \hat{y}_i = -\frac{1}{N}\sum_{i=1}^N y_i \log\big(Cls_S\big(Enc_\theta(x_i) \oplus z_i^s\big)\big). \quad (2)$$

The last layer of Dec_S concludes the disentanglement of style with $Enc_\theta(x) \oplus z^s$. We apply the squared L_2 norm as the reconstruction loss. The objective for style disentanglement can be defined as,

$$L_{recon}^S = \frac{1}{N}\sum_{i=1}^N \big\| Dec_S^{last}\big(Enc_\theta(x_i) \oplus z_i^s\big) - x_i \big\|_2^2, \quad (3)$$

where Dec_S^{last} indicates the last layer of Dec_S.

Leveraging the balance between two classification loss and reconstruction loss allows us to disentangle the style features. The parameters for style-disentanglement are updated with gradients from the following loss function:

$$L_{style} = \lambda_\alpha L_{rev} + \lambda_\beta L_{class}^S + \lambda_\gamma L_{recon}^S, \quad (4)$$

where λ_α, λ_β, and λ_γ are regularization coefficients.

3.2 Content Disentanglement

The content branch extracts the $f_C \in R^{N_C}$ and generates the content image as illustrated in Fig. 1. Under the same assumption in the style branch, we exploit the reversal classifier Cls_{rev} in the same position. However, in contrast to the style branch, a classifier Cls_C

is applied to f_C directly to capture the relatively high-level features. The objective for learning Cls_C can be written as,

$$L_{class}^C = -\frac{1}{N}\sum_{i=1}^N y_i \log \hat{y}_i = -\frac{1}{N}\sum_{i=1}^N y_i \log(Cls_C(f_C)). \tag{5}$$

Moreover, we introduce an auxiliary classification loss to help end-to-end learning. Let us denote $\hat{y}$ to indicate the predicted label of a disentangled content image. Then the auxiliary classification loss can be written as follows:

$$L_{aux} = -\frac{1}{N}\sum_{i=1}^N y_i \log \hat{y}_i = -\frac{1}{N}\sum_{i=1}^N y_i \log(Cls_C(Enc_C(Enc_\theta(\tilde{x}_i)))), \tag{6}$$

where $\tilde{x}$ is the disentangled content image from input image x.

In the content branch, we define z^c as the output feature from the second layer of $Dec_C(f_C)$ and concatenate the $Enc_\theta(x)$ and z^c features in the same way as the style branch. Then the objective for content disentanglement as follows:

$$L_{recon}^C = \frac{1}{N}\sum_{i=1}^N \left\| Dec_C^{last}\left(Enc_\theta(x_i) \oplus z_i^c\right) - x_i \right\|_2^2, \tag{7}$$

where Dec_C^{last} indicates the last layer of Dec_C.

The parameters of the content branch are updated with gradients from the following loss function:

$$L_{content} = \lambda_\delta L_{rev} + \lambda_\varepsilon L_{class}^C + \lambda_\epsilon L_{recon}^C + \lambda_\zeta L_{aux}, \tag{8}$$

where λ_δ, λ_ε, λ_ϵ, and λ_ζ are regularization coefficients.

3.3 Overall Framework

To disentangle the style features and content features, our framework uses many different losses. Nevertheless, since the shared encoder $Enc_\theta(x)$ integrates our two branches, our training process can become end-to-end and complementary learning. Note that since L_{rev} is used in both two branches, the gradients of L_{rev} are back-propagated twice. Our overall objective is a weighted combination:

$$L_{final} = \lambda_\alpha L_{rev} + \lambda_\beta L_{class}^S + \lambda_\gamma L_{recon}^S + \lambda_\delta L_{class}^C + \lambda_\epsilon L_{recon}^C + \lambda_\epsilon L_{aux}. \tag{9}$$

4   Experiments

In this section, we evaluate our disentanglement method on digit datasets. For the qualitative results, we show up the disentangled style images and content images. And, we demonstrate our disentanglement model with an assessment of the classification task to show quantitative results.

4.1 Datasets

We tested our model on four-digit datasets: MNIST [16], MNIST-M [17], EMNIST [18] and EMNIST-M. The MNIST dataset is handwritten digits which have 10 different classes with the size of 28×28 pixels. The MNIST contains 60,000 examples for the training set and 10,000 examples for the test set. The MNIST-M dataset was first introduced in [18]. The MNIST-M dataset consists of MNIST digits blended with random color patches from the BSDS500 [10] dataset.

The EMNIST dataset is a set of handwritten character digits derived from the NIST datasets [19]. These images have a size of 28×28 pixels. There are six different splits provided in this dataset, we use only "EMNIST Letters" split set. This dataset merges a balanced set of the uppercase and lowercase letters into a single 26 classes. It contains 124,800 examples for the training and 20,800 examples for the test. According to the procedure proposed in [17], we created EMNIST-M samples from EMNIST dataset.

4.2 Implementation Details

Encoders. Our proposed network contains three encoding parts: the shared encoder Enc_θ, style encoder Enc_S, and content encoder Enc_C. We constructed Enc_θ to have only one convolutional layer followed by a ReLU. Enc_S and Enc_C have the same architecture, consisting of two convolutional layers with 5×5 kernels and stride 1.

Decoders. The style decoder Dec_S and content decoder Dec_C have the same network, involving three convolutional layers with 5×5 kernels and stride 1. All convolutional layers are followed by ReLU and there is no Batch Normalization [22] performed on our models.

Classifiers. Our network is composed of three classifiers: the reversal classifier Cls_{rev}, style classifier Cls_S, and content classifier Cls_C. All these classifiers have the same architecture, consisting of a 1×1 convolutional layer followed by a global average pooling layer. We used a softmax function for the classification.

Hyperparameters. The batch size is set to be 64. We apply the Adam optimizer [20] with a fixed learning rate 10^{-3} for the content autoencoder and reversal classifier. For the style autoencoder, we used Adam optimizer with a learning rate of 0.0005. For the MNIST-M dataset, weight parameters in the final objective are as follows: $\{\lambda_\alpha = 0.4, \lambda_\beta = 0.4, \lambda_\gamma = 0.1, \lambda_\delta = 0.1, \lambda_\varepsilon = 0.1, \lambda_\epsilon = 0.1\}$. For the EMNIST-M dataset, we set as follows: $\{\lambda_\alpha = 0.3, \lambda_\beta = 0.5, \lambda_\gamma = 0.1, \lambda_\delta = 0.1, \lambda_\varepsilon = 0.1, \lambda_\epsilon = 0.1\}$.

4.3 Experimental Results

Qualitative Results. As depicted in Fig. 2, we demonstrate our method on MNIST-M and EMNIST-M datasets. The first row of this figure presents input images, and the second row shows disentangled content images from the input images. The last row shows the disentangled style images. We show up the images that have various

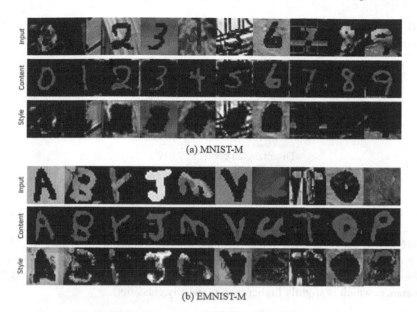

(a) MNIST-M

(b) EMNIST-M

Fig. 2. Disentanglement results on MNIST-M and EMNIST-M datasets.

backgrounds to demonstrate our method obviously. When the background of input is simple, we can get clearly disentangled content images. However, when the input has a complex background, our disentangled content images have some noise in edges. This is because the content encoder Enc_C has been guided to work in low-level layers. In some images, even digits that are hard to identify with a person's eye can be seen more clearly in disentangled content images. Meanwhile, the disentangled style images have masking on the position of the digits. The size of the masking area can be adjusted with our parameters in final objective. The experiments on this are covered in Sect. 5.

Quantitative Results. To demonstrate our method in quantitative measurement, we evaluate our method on the classification task. As well known, the CNN network trained on MNIST has huge performance degradation on the MNIST-M dataset because the MNIST-M have no longer constant background and strokes. If our disentanglement method is correctly performed, a high classification performance will be obtained when the content disentangled from MNIST-M is applied to the MNIST pre-trained model. On the other hand, if the disentangled style images are applied, it would give us near-random results because the style images have no class information. From this point of view, we evaluate the classification performance on MNIST, EMNIST, MNIST-M and EMNIST-M datasets. In this experiment, we adopted VGG16 [21] model as a baseline classification model.

First, we pre-trained MNIST on VGG16 and got 97.34% accuracy. As shown in Table 1, when the MNIST-M data is applied to this pre-trained VGG16, classification performance has a huge drop to 48.51%. On the other hand, our disentangled content image achieved 93.39%, improving performance by 44.88% compared to the direct use of MNIST-M. And, our disentangled style image attained 13.69% performance, similar to that of random classifications.

Table 1. Classification performance on MNIST(-M).

Input	Accuracy
MNIST	97.34%
MNIST-M	48.51%
Disentangled content	93.39%
Disentangled style	13.69%

We tested our method on EMNIST and EMNIST-M dataset in the same way as above. First, we pre-trained VGG16 with EMNIST dataset. As shown in Table 2, when the EMNIST-M data is applied to this model, classification accuracy is 36.77%. On the other hand, with our disentangled content, we achieved 83.65%, an improvement of 46.88% from 36.77%. And with our disentangled style images, we achieved an 8.94% performance, which is slightly higher than randomly classified.

Table 2. Classification performance on EMNIST(-M).

Input	Accuracy
EMNIST	94.15%
EMNIST-M	36.77%
Disentangled content	83.65%
Disentangled style	8.94%

Furthermore, we compared our method with other methods in terms of unsupervised domain adaptation. Since there are no comparable models in the EMNIST(-M) dataset, we only compared in MNIST(-M) dataset. As shown in Table 3, our method outperforms other unsupervised domain adaptation methods. From this, we observed that our disentanglement method can be utilized to solve the domain adaptation problem.

Table 3. Comparison with domain adaptation methods.

Method	MNIST → MNIST-M
Source-only	48.51%
CycleGAN [23]	74.50%
DANN [17]	76.66%
MMD [24]	76.90%
DSN [25]	83.20%
AssocDA [26]	89.50%
Ours (Content)	93.39%
Ours (Style)	13.69%
Target-only	97.34%

These experiments show that our method disentangles the style and the content representations from the image properly.

4.4 Ablation Study

In this section, we take the MNIST-M dataset as an example to analysis our framework. Figure 3 shows the ablation study on how parameters λ_α and λ_β of the final objective affect the style-disentanglement results. We observed the results by setting λ_α and λ_β at a ratio of $\{1:1, 1:2, 1:3, 1:5\}$. In this experiment, other parameters are all fixed. As shown in Fig. 3, we observed that as λ_β increased, the area of masking grows. Too large λ_β causes large masking beyond the area in which the number exists. On the other hand, too small λ_β produces results that are not significantly different from the input image.

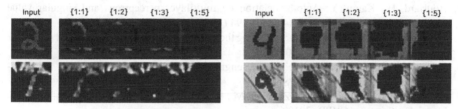

Fig. 3. Variation of the disentangled style image with the change of parameters λ_α and λ_δ.

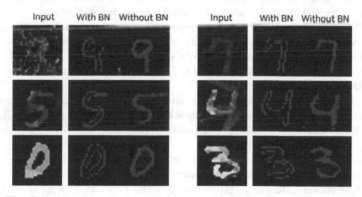

Fig. 4. Effect of Batch Normalization on the disentangled content image.

Figure 4 shows the effect of Batch Normalization (BN) [22] on our framework. When the Batch Normalization is applied to entire autoencoder branches, only the edges of the content are drawn. The content images disentangled with BN could not produce good results in the classification task.

5 Conclusion

We have proposed a two-branch autoencoder framework that integrates style and content branches. Our framework enables us to end-to-end and complementary learning between two-branch autoencoders. The overall network can be trained stably with a balanced weight of final objectives. Qualitative and quantitative evaluations on digit datasets demonstrate the superiority of our proposed framework.

Acknowledgments. This work was supported by the Technology Innovation Program (20000316, Scene Understanding and Threat Assessment based on Deep Learning for Automatic Emergency Steering) funded By the Ministry of Trade, Industry & Energy (MOTIE, Korea).

References

1. Bengio, Y., Courville, A., Vincent, P.: Representation learning: a review and new perspectives. IEEE Trans. Pattern Anal. Mach. Intell. **35**(8), 1798–1828 (2013)
2. Bourlard, H., Kamp, Y.: Auto-association by multilayer perceptrons and singular value decomposition. Biol. Cybern. **59**(4), 291–294 (1988)
3. Hinton, G.E., Salakhutdinov, R.R.: Reducing the dimensionality of data with neural networks. Science **313**(5786), 504–507 (2006)
4. Gonzalez-Garcia, A., van de Weijer, J., Bengio, Y.: Image-to-image translation for cross-domain disentanglement. In: NIPS (2018)
5. Hu, Q., Szabó, A., Portenier, T., Zwicker, M., Favaro, P.: Disentangling factors of variation by mixing them. In: CVPR (2018)
6. Zhang, Y., Zhang, Y., Cai, W.: Separating style and content for generalized style transfer. In: CVPR (2018)
7. Mathieu, M., Zhao, J., Sprechmann, P., Ramesh, A., LeCun, Y.: Disentangling factors of variation in deep representation using adversarial training. In: NIPS (2016)
8. Hadad, N., Wolf, L., Shahar, M.: A two-step disentanglement method. In: CVPR (2018)
9. Szabó, A., Hu, Q., Portenier, T., Zwicker, M., Favaro, P.: Challenges in disentangling independent factors of variation. In: ICLRW (2018)
10. Arbelaez, P., Maire, M., Fowlkes, C., Malik, J.: Contour detection and hierarchical image segmentation. PAMI **33**(5), 898–916 (2011)
11. Kingma, D.P., Welling, M.: Auto-encoding variational Bayes. In: ICLR (2014)
12. Makhzani, A., Shlens, J., Jaitly, N., Goodfellow, I.: Adversarial autoencoders. In: ICLR (2016)
13. Liu, Y., Wang, Z., Jin, H., Wassell, I.: Multi-task adversarial network for disentangled feature learning. In: CVPR (2018)
14. Goodfellow, I., et al.: Generative adversarial nets. In: NIPS (2014)
15. Arjovsky, M., Chintala, S., Bottou, L.: Wasserstein GAN. In: ICML (2017)
16. Lecun, Y., Bottou, L., Bengio, Y., Haffner, P.: Gradient-based learning applied to document recognition. Proc. IEEE **86**, 2278–2324 (1998)
17. Ganin, Y., et al.: Domain-adversarial training of neural networks. JMLR **17**, 1–35 (2016)
18. Cohen, G., Afshar, S., Tapson, J., van Schaik, A.: EMNIST: an extension of MNIST to handwritten letters. arXiv preprint arXiv:1702.05373 (2017)
19. Grother, P., Hanaoka, K.: NIST special database 19 handprinted forms and characters database. National Institute of Standards and Technology (2016)
20. Kinga, D., Adam, J.B.: Adam: a method for stochastic optimization. In: ICLR (2015)
21. Simonyan, K., Zisserman, A.: Very deep convolutional networks for large-scale image recognition. arXiv preprint arXiv:1409.1556 (2014)
22. Ioffe, S., Szegedy, C.: Batch normalization: accelerating deep network training by reducing internal covariate shift. In: ICML (2015)

23. Zhu, J., Park, T., Isola, P., Efros, A.: Unpaired image-to-image translation using cycle-consistent adversarial networks. In: ICCV (2017)
24. Tzeng, E., Hoffman, J., Zhang, N., Saenko, K., Darrell, T.: Deep domain confusion: Maximizing for domain invariance. arXiv preprint arXiv:1412.3474 (2014)
25. Bousmalis, K., Trigeorgis, G., Silberman, N., Krishnan, D., Erhan, D.: Domain separation networks. In: NIPS (2016)
26. Haeusser, P., Frerix, T., Mordvintsev, A., Cremers, D.: Associative domain adaptation. In: ICCV (2017)

Residual Attention Encoding Neural Network for Terrain Texture Classification

Xulin Song[1,2]⬤, Jingyu Yang[1,2], and Zhong Jin[1,2](✉)

[1] School of Computer Science and Engineering,
Nanjing University of Science and Technology, Nanjing, China
Jessicasxl@163.com, yangjy@mail.njust.edu.cn, zhongjin@njust.edu.cn
[2] Key Laboratory of Intelligent Perception and Systems for High-Dimensional
Information of Ministry of Education, Nanjing University of Science and Technology,
Nanjing, China

Abstract. Terrain texture classification plays an important role in computer vision applications, such as robot navigation, autonomous driving, etc. Traditional methods based on hand-crafted features often have suboptimal performances due to the inefficiency in modeling the complex terrain variations. In this paper, we propose a residual attention encoding network (RAENet) for terrain texture classification. Specifically, RAENet incorporates a stack of residual attention blocks (RABs) and an encoding block (EB). By generating attention feature maps jointly with residual learning, RAB is different from the usually used which only combine feature from the current layer with the former one layer. RAB combines all the preceding layers to the current layer and is not only minimize the information loss in the convolution process, but also enhance the weights of the features that are conducive to distinguish between different classes. Then EB further adopts orderless encoder to keep the invariance to spatial layout in order to extract feature details before classification. The effectiveness of RAENet is evaluated on two terrain texture datasets. Experimental results show that RAENet achieves state-of-the-art performance.

Keywords: Residual learning · Attention learning · Encoding layer · Terrain texture classification

1 Introduction

Terrain texture classification aims to design algorithms for declaring which category that a given image or texture patch belongs to. It is a challenging but important task for computer vision application in robot navigation and autonomous

This work is partially supported by National Natural Science Foundation of China under Grant Nos 61872188, U1713208, 61602244, 61672287, 61702262, 61773215.

© Springer Nature Switzerland AG 2020
S. Palaiahnakote et al. (Eds.): ACPR 2019, LNCS 12047, pp. 52–63, 2020.
https://doi.org/10.1007/978-3-030-41299-9_5

driving, etc. Solving such terrain texture classification problems can improve driving safety and comfort passengers.

Hand-crafted features, such as Scale Invariant Feature Transform (SIFT) [32], Local Binary Pattern and its variants (LBPs) [3,7,12,35,37] have dominated many computer vision fields for a long time until deep convolutional neural networks (CNNs) [26] achieved record-breaking performances for image classification tasks. Conventional approaches generally employ hand-crafted filters to extract the histograms [29,42] or bag-of-words [11,23,24,30,38] as feature of the images. With the rapid development of deep learning, CNNs becomes the de facto approach in many object recognition problem. Since the goals of texture algorithms and object algorithms are similar which are both to capture orderless representation encompassing some spatial repetition. Gatys et al. [13] first utilize CNN to synthesis texture feature. Fisher Vector CNN (FV-CNN) [10] has been proposed to replace handcrafted filter for extracting feature. These models mentioned above are the early attempts which use CNN for texture learning. Many state-of-the-art approaches design models usually based on popular generic CNN structures, including VGGNet [39], InceptionNet [41], GoogleNet [41], ResNet [16], DenseNet [20], etc.A full review of these networks is beyond the scope of this paper, we suggest readers to the original papers [16,20,39,41] or to excellent surveys [5,7,15,28] for additional details. Among these classical networks, they are too deep and complex for terrain texture classification since the terrain texture image we adopt is usually a local patch of a complete image. The image patch contains some details of the terrain texture which is an important cue for classification. In order to perform classification on image patches we have to design a CNN structure that works for terrain texture classification.

Generally, attention is used as a tool to balance the allocation of available processing resources of the inputs [22,27,33,34]. Since attention mechanisms selectively magnify valuable feature maps and suppress valueless ones, and they have been successfully applied in many areas such as image localisation and understanding [6], image captioning [8,43], human pose estimation [9], visual tracking [36], etc. [43] introduces both 'soft' and 'hard' attention into image captioning. [9] fuses multi-content attention mechanism and hourglass residual units to predict human pose. A residual attentional siamese network [36] presents different kinds of attention mechanisms for high performance object tracking. The above attention mechanism generates attention weights from the input (or features extracted from the CNNs) and then apply them on some feature maps that generated from the input. Stimulated by the advantage of attention mechanism which adds weight on feature maps and reduces information redundancies, we first introduce attention that specialised to model channel-wise relationships into CNN model for terrain texture classification. In contrast, our attention integrated with residual learning not only keep the lightweight design but also minimize the information loss in the convolutional process.

Traditional computer vision methods often utilize SIFT or filters to capture the texture feature, then learn a dictionary to encode global feature details. However, texture recognition use orderless representation, so the global spatial information is not of great importance. It has been realized that texture recognition can adopt orderless representation to preserve invariance to spatial layout [10, 31, 44, 46]. Recent CNN structures have incorporated the conventional encoder in their CNN models [4, 44, 45]. [4] utilizes a generalized vector of locally aggregated descriptors (VLAD) to place recognition. Zhang et al. [45, 46] propose the encoding layer to capture orderless representation for texture classification and semantic segmentation, respectively. Therefore, encoding for the orderless representation can be effective utilized to terrain texture classification. In this paper, we extend the encoding layer to capture feature details for terrain classification.

To efficiently utilize the information in each convolution layer, we propose residual attention encoding neural network (RAENet) which combines residual learning, attention mechanism and encoding layer for terrain texture classification. Our method follows the above basically attention, but differs from these studies in that we consider the attention and the residual learning together. The conventional combination [19] with attention and residual learning neglect to fully use information of each convolutional layer because each convolutional layer only have access to its former layer. In the RAENet, each convolutional layer has direct access to the preceding layers differs from the generic residual connection [16, 17]. Based on the residual attention block, we further extend the encoding layer to capture texture details for terrain texture classification.

The paper is organized as follows. Section 2 will present the architecture of the proposed RAENet followed by the details of the residual attention as well as the encoding layer. Sections 3 and 4 will introduce the experimental results and some conclusions, respectively.

2   Residual Attention Encoding Neural Network

In this section, we first introduce the architecture of the proposed RAENet, and then show details of the residual attention block and the encoding block which compose the core components of the proposed model.

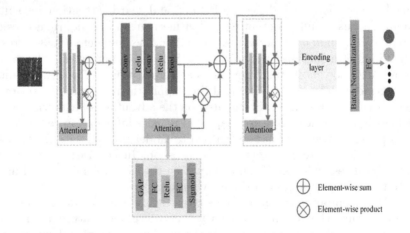

Fig. 1. The architecture of our proposed RAENet. (GAP means global average pooling and FC is fully connection)

2.1 Network Structure

The framework and its components of residual attention encoding neural network are illustrated in Fig. 1. RAENet is mainly constructed by stacking multiple residual attention blocks (RABs) followed by an encoding block (EB). In Fig. 1, we omit the softmax layer for simplicity. Specially, we first use two convolutional layers to extract shallow features from the original input image

$$F_1 = H_{RAB,1}(F_0) \tag{1}$$

where $H_{RAB,1}$ is the composite function of operations with convolution, rectified linear units (ReLU) [14], pooling, residual learning and attention learning. Note, we denote the input image as $F_0 = IMG_{input}$, F_1 is the output of the first RAB. Supposing we have N RABs, the output F_N of the N-th RAB can be further obtained by

$$F_N = H_{RAB,N}(F_{N-1}) = H_{RAB,N}(H_{RAB,N-1}(\cdots(H_{RAB,1}(F_0)))) \tag{2}$$

where F_N denotes the operations of the N-th RAB, $H_{RAB,N}$ can be a composite function of operations and we take $N = 3$. F_N is produced by the N-th RAB and can make full use of features from all the preceding layers. More details of RAB will be given in Sect. 2.2.

After extracting features from a set of RABs, we further conduct an encoding block (EB) to extract the texture details. The EB is formulated as

$$F_{EB} = H_{EB}(F_N) \tag{3}$$

F_{EB} is the output feature map of the encoding layer by using the composite function H_{EB}. More details of EB can be found in Sect. 2.3. Finally, F_{EB} is regarded as input feature to the batch normalization layer following with fully connected layer.

2.2 Residual Attention Block

In CNN, a convolution kernel can filter a certain type of visual pattern, each channel can be treated as a visual pattern from one convolutional kernel. Therefore, different channel represents different visual pattern. Meanwhile, with the idea of residual, we emphasize that the identity mapping [17] need information not only from the former layer but also need the current layer. Motivated by this note that not every channel has equal importance [47] and the convolution operation may lose information, we construct a new block, named residual attention block. By integrating the residual learning and the attention learning, the information loss in the convolution process can be minimized. Now we present details about our proposed residual attention block in Fig. 2.

The input to the RAB is a 3-D tensor $X_n^{in} \in \mathbb{R}^{W \times H \times C}$, where W, H, C denote the numbers of pixels in the width, height and the channel dimensions, respectively (n indicates the level of the block). Letting $X_n^{out} \in \mathbb{R}^{W_1 \times H_1 \times C_1}$

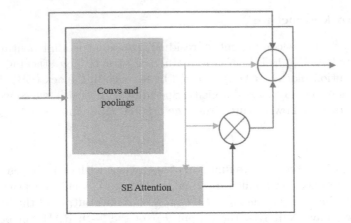

Fig. 2. The proposed residual attention block. The red arrow contains information from the former RAB, the yellow arrow contains information which after some convolutions and pooling, the green arrow include information with attention mask and the blue arrows are the information combination. (Color figure online)

denotes the output of the n-th RAB, the proposed residual learning jointly with attention learning can be formulated as

$$X_{n+1}^{in} = X_n^{out} + (1 + A) X_{n+1}^{out} \qquad (4)$$

where A denotes the mask matrix of the attention mechanism as in [19]. As shown in Fig. 2, the red arrow from previous RAB represents the residual information, and the yellow arrows from the convs and poolings indicate the feature maps after some convolution and pooling operation in the current RAB and the green arrow from the attention is attention mask. The rest two blue arrows represent the attention fusion in the form of the element-wise product and the residual fusion in the form of element-wise sum.

Here, we model the channel attention with a squeeze-and-excitation (SE) [19] layer. As described in [19], SE layer is a computational unit that can be constructed in any given transformation. Next, we briefly introduce the SE layer. Firstly, we perform the squeeze operation via a global average pooling to aggregate the spatial information in channel space

$$sq^c = \frac{1}{W \times H} \sum_{i=1}^{W} \sum_{j=1}^{H} X_{i,j}^c \ (c = 1, \cdots, C) \qquad (5)$$

where X is the input of the channel attention layer, i.e. the output of the convs and poolings with the shape of $W \times H \times C$ as displayed in Fig. 2. The result of squeeze is a vector $Sq = [sq^1, sq^2, \cdots, sq^C] \in \mathbb{R}^C$ that conveys the per-channel filter response from the whole feature map. Sq is a collection of those local descriptors whose statistics are essentially for the input image. Then the excitation operation is formulated as

$$Ex = \sigma \left(W_2^{ca} \times \varepsilon \left(W_1^{ca} Sq \right) \right) \qquad (6)$$

where $\sigma \left(\cdot \right)$ is the sigmoid activation function and $\varepsilon \left(\cdot \right)$ denotes the ReLU activation function. For simplicity, all the bias are omitted. $W_1^{ca} \in \mathbb{R}^{\frac{C}{r} \times C}$ and $W_2^{ca} \in \mathbb{R}^{C \times \frac{C}{r}}$ (r is the ratio of dimensionality reduction) denote the weight matrix. This RAB reduces the parameter number since the residual operation do not add parameter and the channel attention reduces the parameter number from C^2 (a single convolution layer) to $\frac{2C^2}{r}$. By applying residual learning jointly with attention learning, RABs minimize the information loss in the convolution process.

2.3 Encoding Block

The texture encoding layer [46] integrates dictionary learning and visual encoding as a single layer into CNN pipeline, which provides an orderless representation for texture modeling. Encoding layer can be considered as a global feature pooling on top of the convolutional layers. After extracting local feature details with a set of RABs, we further extend an encoding layer to exploit global texture details. We briefly describe the encoding layer [46] as follows.

The input feature map from the last RAB is with the shape of $W \times H \times C$. Let $F = \{f_1, f_2, \cdots, f_N\}$ denotes the N visual descriptors, where N is the feature number and equals to $W \times H$. The dictionary D that will be learned can be expressed as $D = \{d_1, d_2, \cdots, d_K\}$ with each codeword $d_k \in \Re^C, k = 1, \cdots, K$ and K is the number of codewords. Similar to these classic dictionary learning and encoders, encoding layer can encode more texture details by increasing the number of learnable codewords K. So the larger K is, the more information encoding layer learns. Then each visual descriptor f_i is assigned with a weight $w_{i,j}$ to each codeword d_k. The residual vector is computed by $r_{i,j} = f_i - d_j$, where $i = 1, \cdots, N$ and $j = 1, \cdots, K$ After aggregating the residual vector with the soft weight, the output of encoding layer is expressed as

$$w_{i,k} = \frac{\exp \left(-s_k \| r_{i,k} \|^2 \right)}{\sum_{j=1}^{K} \exp \left(-s_j \| r_{i,j} \|^2 \right)} \tag{7}$$

$$e_k = \sum_{i=1}^{K} w_{i,k} r_{i,k} \tag{8}$$

where $s_1, s_2, \cdots, s_K$ is a set of learnable smoothing factors. After the encoding layer, the texture details of the feature map F will be expressed with a set of residual encoding vectors $E = e_1, e_2, \cdots, e_K$.

3 Experiments

In this section, we display the experimental results of the proposed RAENet on two datasets: Outex and NJUSTPCA. Sections 3.1 and 3.2 introduce the experimental settings and the details of datasets, respectively. The experimental results are shown in Sect. 3.3.

3.1 Implementation Details

We implemented our RAENet model in the Tensorflow [2] framework. All the terrain texture images have been resized to 64×64. The baseline is designed with 6 convolution layers which based on the framework of ResNet, we set 3×3 as filter size of all the convolutional layers. After that we add a 64-dimension fully-connected layer followed by the batch normalization [21], ReLU [18],and Dropout [40]. Therefore, the length of the terrain image features is 64. For our RAENet, we add the RABs and EB which do not change the dimension of the features. We set the reduction ratio r = 2 in attention and the codewords K = 32 in encoding layer. For model optimization, we use the ADAM [25] algorithm at the initial learning rate 8×10^{-4}. We set the batch size as 128 and epoch as 300, respectively, and Weight decay equals to 0.98 after every 20 epoches. Notely, we also used horizontal flipping and image whiten for data augmentation and the RAENet model is trained start from scratch.

3.2 Datasets

Outex Dataset: The Outex dataset [1] consists of Outex-0 and Outex-1, each of them includes 20 outdoor scene images. These images can be marked as one type of sky, tree, bush, grass, road and building. Some image samples of the Outex dataset are displayed in Fig. 3. Since the original image size in both Outex0 and Outex1 is 2272×1704, we have cut the marked area of every image into 64×64 patch and then each patch is regarded as one sample in the experiments. So the total number of image patches is 6000 and 4213 in Outex1 and Outex2. Specially, Outex0 has all the six categories mentioned above while Outex1 has only five categories as it does not have bush.

Fig. 3. Image patch samples of the Outex (Outex0 and Outex1) terrain texture dataset.

NJUSTPCA Dataset: This terrain dataset was created by Pattern computing and Application Laboratory of Nanjing University of Science and Technology (NJUSTPCA). It is composed of three classes of ground terrain, namely 54978 dirt road images, 23479 sandstone road images and 50065 grass images. Each image is color image with the size of 64 × 64. Because of the huge interclass differences, we first perform simple clustering on each class. Finally we got four subclasses of the dirt road, three subclasses of the sandstone road and two sub-classes of the grass, respectively. So, all the images were divided into nine classes. The samples of NJUSTPCA have been shown in Fig. 4.

dirt road

gravel road

grass

Fig. 4. Image patch samples of the NJUSTPCA terrain texture dataset. The initial three classes have been divided into corresponding four, three and two subclasses due to the huge interclass difference, respectively.

3.3 Experimental Results

The proposed residual attention block is not only different from the classical residual notion but also different from the general attention mechanism, we use a framework with the traditional residual as the baseline. The configuration of the baseline and our RAENet are displayed in Table 1. The difference between the baseline and RAENet is that RAENet add attention mechanism and encoding layer to the baseline. From Table 1 we can conclude that RAENet do not add more parameters except sum or product operation, so it takes lower computation resource and is time saving.

We compare the classification results in Table 2 among the baseline, the model with encoding layer based on ResNet, DEP [44] and RAENet. Our model out-performs other methods with 3.8%, 1.0%, 16.4% on Outex1, 1.1%, 0.9%, 12.0% on Outex2 and 4.1%, 1.5% on NJUSTPCA. We conclude that our RAENet has the highest performance with the best accuracy 97.5%, 99.4% and 87.6% on the two datasets, respectively.

Table 1. The configuration of the baseline and RAENet. (Left) Baseline. (Right) RAENet. conv and avgpool denote convolution and average pooling operation, pad indicates the paddings in the spatial space or the channel space and A is the attention mask. bn denotes the batch normalization and fc is the used fully connected layer in our RAENet.

	Baseline	RAENet
mlp1	[conv, 3 × 3, 16]	[conv, 3 × 3, 16]
	[conv, 3 × 3, 16]	[conv, 3 × 3, 16]
	[avgpool, (2 × 2, 2)]	[avgpool, (2 × 2, 2)]
mlp1	-	mlp1 (1+A)
mlp2	[conv, 3 × 3, 32]	[conv, 3 × 3, 32]
	[conv, 3 × 3, 32]	[conv, 3 × 3, 32]
	[avgpool, (2 × 2, 2)]	[avgpool, (2 × 2, 2)]
mlp2	pad(mlp1) + pad(mlp2)	pad(mlp1) + pad(mlp2)
	-	mlp2 + mlp1 × (1+A)
mlp3	[conv, 3 × 3, 64]	[conv, 3 × 3, 64]
	[conv, 3 × 3, 64]	[conv, 3 × 3, 64]
	[avgpool, (2 × 2, 2)]	[avgpool, (2 × 2, 2)]
mlp3	pad(mlp3) + pad(mlp2)	pad(mlp3) + pad(mlp2)
mlp3	-	mlp3 + mlp2 ×(1+A)
Encoding	-	32 codewords
bn + fc	64 × n	64 × n

Table 2. Classification result comparisons based on Outex and NJUSTPCA datasets among different methods.

	Baseline	Encoding layer only	DEP [20]	RAENet
Outex0	93.70%	96.50%	81.10%	**97.50%**
Outex1	98.30%	98.50%	87.40%	**99.40%**
NJUSTPCA	83.50%	86.10%	-	**87.60%**

4   Conclusion

In this paper, we have presented a residual attention encoding neural network (RAENet), a novel architecture for terrain texture classification. It mainly consists of a stack of residual attention blocks (RABs) and an encoding block (EB). RABs integrate the residual learning and the attention in a novel former that minimize the information loss in the convolution process and enhance the weights of significant features, which enrich the power for distinguishing between different classes. EB keeps the invariance to spatial layout and encodes the global feature for recognition to further extract texture details. Additionally, the proposed residual attention block is general. It is a plug-and-play block and would help other vision tasks.

References

1. University of Oulu texture database. http://www.outex.oulu.fi/temp/
2. Abadi, M., et al.: Tensorflow: A system for large-scale machine learning (2016)
3. Akl, A., Yaacoub, C., Donias, M., Costa, J.P.D., Germain, C.: A survey of exemplar-based texture synthesis methods. Comput. Vis. Image Underst. **172**, 12–24 (2018)
4. Arandjelovic, R., Gronat, P., Torii, A., Pajdla, T., Sivic, J.: NetVLAD: CNN architecture for weakly supervised place recognition. IEEE Trans. Pattern Anal. Mach. Intell. TPAMI **99**, 1 (2017)
5. Bengio, Y., Courville, A., Vincent, P.: Representation learning: a review and new perspectives. IEEE Trans. Pattern Anal. Mach. Intell. TPAMI **35**(8), 1798–1828 (2013)
6. Cao, C., Liu, X., Yi, Y., Yu, Y., Huang, T.S.: Look and think twice: capturing top-down visual attention with feedback convolutional neural networks. In: IEEE International Conference on Computer Vision, ICCV (2016)
7. Chatfield, K., Simonyan, K., Vedaldi, A., Zisserman, A.: Return of the devil in the details: delving deep into convolutional nets. Computer Science (2014)
8. Chen, L., et al.: SCA-CNN: Spatial and channel-wise attention in convolutional networks for image captioning. In: IEEE Computer Society Conference on Computer Vision and Pattern Recognition, CVPR (2017)
9. Chu, X., Yang, W., Ouyang, W., Ma, C., Yuille, A.L., Wang, X.: Multi-context attention for human pose estimation. In: IEEE Computer Society Conference on Computer Vision and Pattern Recognition, CVPR (2017)
10. Cimpoi, M., Maji, S., Kokkinos, I., Vedaldi, A.: Deep filter banks for texture recognition, description, and segmentation. Int. J. Comput. Vis. **118**(1), 65–94 (2016)
11. Csurka, G.: Visual categorization with bags of keypoints. Workshop Stat. Learn. Comput. Vis. **44**(247), 1–22 (2004)
12. Di, H., Shan, C., Ardabilian, M., Wang, Y., Chen, L.: Local binary patterns and its application to facial image analysis: a survey. IEEE Trans. Syst. Man Cybern. Part C Appl. Rev. **41**(6), 765–781 (2011)
13. Gatys, L.A., Ecker, A.S., Bethge, M.: Texture synthesis using convolutional neural networks. Adv. Neural Inf. Process. Syst. **70**(1), 262–270 (2015)
14. Glorot, X., Bordes, A., Bengio, Y.: Deep sparse rectifier neural networks. In: International Conference on Artificial Intelligence and Statistics (2011)
15. Gu, J., Wang, Z., Kuen, J., Ma, L., Shahroudy, A., Bing, S., Liu, T., Wang, X., Gang, W.: Recent advances in convolutional neural networks. Computer Science (2015)
16. He, K., Zhang, X., Ren, S., Sun, J.: Deep residual learning for image recognition. In: IEEE Computer Society Conference on Computer Vision and Pattern Recognition, CVPR (2016)
17. He, K., Zhang, X., Ren, S., Sun, J.: Identity mappings in deep residual networks. In: Leibe, B., Matas, J., Sebe, N., Welling, M. (eds.) ECCV 2016, Part IV. LNCS, vol. 9908, pp. 630–645. Springer, Cham (2016). https://doi.org/10.1007/978-3-319-46493-0_38

18. Hinton, G.E.: Rectified linear units improve restricted boltzmann machines Vinod Nair. In: Proceedings of the 27th International Conference on Machine Learning, ICML (2010)

19. Hu, J., Shen, L., Albanie, S., Sun, G., Wu, E.: Squeeze-and-excitation networks. IEEE Trans. Pattern Anal. Mach. Intell. TPAMI **99**, 1 (2017)

20. Huang, G., Liu, Z., Laurens, V.D.M., Weinberger, K.Q.: Densely connected convolutional networks. In: IEEE Computer Society Conference on Computer Vision and Pattern Recognition, CVPR (2017)

21. Ioffe, S., Szegedy, C.: Batch normalization: accelerating deep network training by reducing internal covariate shift. In: Proceedings of the 32nd International Conference on Machine Learning, ICML (2015)

22. Itti, L., Koch, C.: Computational modelling of visual attention. Nat. Rev. Neurosci. **2**(3), 194–203 (2001)

23. Sivic, J., Russell, B.C., Efros, A.A., Zisserman, A., Freeman, W.T.: Discovering objects and their location in images. In: IEEE International Conference on Computer Vision, ICCV (2005)

24. Joachims, T.: Text categorization with support vector machines: learning with many relevant features. In: Proceedings of the Conference on Machine Learning (1998)

25. Kingma, D.P., Ba, J.: Adam: a method for stochastic optimization. Computer Science (2014)

26. Krizhevsky, A., Sutskever, I., Hinton, G.: Imagenet classification with deep convolutional neural networks. In: International Conference on Neural Information Processing Systems, NIPS (2012)

27. Larochelle, H., Hinton, G.E.: Learning to combine foveal glimpses with a third-order Boltzmann machine. In: International Conference on Neural Information Processing Systems, NIPS (2010)

28. Lecun, Y., Bengio, Y., Hinton, G.: Deep learning. Nature **521**(7553), 436–444 (2015)

29. Leung, T., Malik, J.: Representing and recognizing the visual appearance of materials using three-dimensional textons. Int. J. Comput. Vis. IJCV **43**(1), 29–44 (2001)

30. Li, F.F., Perona, P.: A bayesian hierarchical model for learning natural scene categories. In: IEEE Computer Society Conference on Computer Vision and Pattern Recognition, CVPR (2005)

31. Lin, M., Chen, Q., Yan, S.: Network in network. Computer Science (2013)

32. Lowe, D.G.: Distinctive image features from scale-invariant keypoints. Int. J. Comput. Vis. **60**(2), 91–110 (2004)

33. Mnih, V., Heess, N., Graves, A., Kavukcuoglu, K.: Recurrent models of visual attention. In: International Conference on Neural Information Processing Systems, NIPS, vol. 3 (2014)

34. Newell, A., Yang, K., Deng, J.: Stacked hourglass networks for human pose estimation. In: Leibe, B., Matas, J., Sebe, N., Welling, M. (eds.) ECCV 2016, Part VIII. LNCS, vol. 9912, pp. 483–499. Springer, Cham (2016). https://doi.org/10.1007/978-3-319-46484-8_29

35. Ojala, T., Pietikainen, M., Maenpaa, T.: Multiresolution gray-scale and rotation invariant texture classification with local binary patterns. IEEE Trans. Pattern Anal. Mach. Intell. TPAMI **24**(7), 971–987 (2002)

36. Wang, Q., Teng, Z., Xing, J., Gao, J., Hu., W., Maybank, S.: Learning attentions: Residual attentional siamese network for high performance online visual track. In: IEEE Computer Society Conference on Computer Vision and Pattern Recognition, CVPR (2018)
37. Raad, L., Davy, A., Desolneux, A., Morel, J.M.: A survey of exemplar-based texture synthesis. Arxiv (2017)
38. Lazebnik, S., Schmid, C., Ponce, J.: Beyond bags of features: spatial pyramid matching for recognizing natural scene categories. In: IEEE Computer Society Conference on Computer Vision and Pattern Recognition, CVPR (2012)
39. Simonyan, K., Zisserman, A.: Very deep convolutional networks for large-scale image recognition. Computer Science (2014)
40. Srivastava, N., Hinton, G., Krizhevsky, A., Sutskever, I., Salakhutdinov, R.: Dropout: a simple way to prevent neural networks from overfitting. J. Mach. Learn. Res. **15**(1), 1929–1958 (2014)
41. Szegedy, C., et al.: Going deeper with convolutions. In: IEEE Computer Society Conference on Computer Vision and Pattern Recognition, CVPR (2015)
42. Varma, M., Zisserman, A.: A statistical approach to texture classification from single images. Int. J. Comput. Vis. **62**(1–2), 61–81 (2005)
43. Xu, K.: Show, attend and tell: Neural image caption generation with visual attention. In: Proceedings of the 34nd International Conference on Machine Learning, ICML (2017)
44. Xue, J., Zhang, H., Dana, K.: Deep texture manifold for ground terrain recognition. In: IEEE Computer Society Conference on Computer Vision and Pattern Recognition, CVPR (2018)
45. Zhang, H., et al.: Context encoding for semantic segmentation. In: IEEE Computer Society Conference on Computer Vision and Pattern Recognition, CVPR (2018)
46. Zhang, H., Xue, J., Dana, K.: Deep ten: texture encoding network. In: IEEE Computer Society Conference on Computer Vision and Pattern Recognition, CVPR (2017)
47. Zhang, X., Wang, T., Qi, J.: Progressive attention guided recurrent network for salient object detection. In: IEEE Computer Society Conference on Computer Vision and Pattern Recognition, CVPR (2016)

The Shape of Patterns Tells More

Using Two Dimensional Hough Transform to Detect Circles

Yuan Chang[(✉)], Donald Bailey, and Steven Le Moan

Centre for Research in Image and Signal Processing, Massey University,
Palmerston North, New Zealand
ychang2@massey.ac.nz

Abstract. For image processing applications, an initial step is usually extracting features from the target image. Those features can be lines, curves, circles, circular arcs and other shapes. The Hough transform is a reliable and widely used method for straight line and circle detection, especially when the image is noisy. However, techniques of Hough transform for detecting lines and circles are different; when detecting circles it usually requires a three-dimensional parameter space while detecting straight lines only requires two. Higher dimensional parameter transforms suffer from high storage and computational requirements. However, in the two dimensional Hough transform space, straight lines and circles yield patterns with different shapes. By analysing the shape of patterns within the Hough transform space it is possible to reconstruct the circles in image space. This paper proposes a new circle detection method based on analysing the pattern shapes within a two-dimensional line Hough transform space. This method has been evaluated by a simulation of detecting multiple circles and a group of real-world images. From the evaluation our method shows ability for detecting multiple circles in an image with mild noise.

Keywords: Hough transform · Circle detection · Shape of peaks

1 Introduction

Straight lines and circles are important features for many computer vision and image processing tasks, and detecting them becomes an initial task for many applications. Although, straight lines are more likely to be contained in a man-made environment, in natural science circular objects are more common, for example, red blood cells and pupils [1]. Circular arcs are important features in images of industrial parts or tools, and in real-time recognition systems, the most popular detection targets are straight lines and circular arcs. Many approaches have been proposed to detect straight lines and circles. The Hough transform (HT) is one of the most popular feature extraction techniques used in image analysis, computer vision, and digital image processing. It was first invented as a straight line detection method by Hough in 1959 [2]. In later research,

© Springer Nature Switzerland AG 2020
S. Palaiahnakote et al. (Eds.): ACPR 2019, LNCS 12047, pp. 64–75, 2020.
https://doi.org/10.1007/978-3-030-41299-9_6

the HT was extended to circles, ellipses, and even arbitrarily shaped objects [3]. The basic principle behind the HT is that the target is parametrised, and each detected point votes for all possible sets of parameters with which it is compatible. When multiple points in the image share the same set of parameters, the parameters of the associated target will receive many votes. Therefore targets are detected by finding peaks within parameter space. However, the parameter space for detecting straight lines and circles are quite different.

When detecting straight lines, the HT usually requires a two-dimensional parameter space. There are two famous parametrizations for detecting straight lines, the original HT and the standard HT. The original HT (OHT) was invented for machine analysis of bubble chamber photographs by Hough in 1959 [2]. The OHT uses the slope and intercept as parameters to represent a line in image space. Unfortunately, such a parameterisation is unsuitable for vertical lines (where the slope becomes infinite), which require a second parameter space for vertical lines. Now, the most commonly used parametrization is that introduced by Duda and Hart [4]. It represents lines as:

$$\rho = xcos(\theta) + ysin(\theta) \tag{1}$$

where ρ is the distance from the origin to the closest point on the straight line and θ is the angle between the x axis and the line connecting the origin with that closest point. Although the HT is a popular tool for line detection, it suffers from high computational cost. To solve this problem and make the Hough transform more suitable for real-time applications, the gradient HT was introduced. The gradient HT measures the orientation of the edge and rather than voting for all possible lines through that point, it only votes for a small range of angles [5,6].

Detecting more complex objects usually requires a higher dimensional Hough space. The HT can be used to detect circular arcs by choosing the centre and radius as parameters. The Circle Hough Transform (CHT) [4] is one of the best-known algorithms for detecting circular shapes. For a circle with unknown radius, the algorithm should vote for all possible radius to form a three-dimensional parameter space, where two dimensions represent the position of the centre and the third represents the radii. In the CHT parameter space, a local maximum indicates the circle on which the arc lies. However, the CHT is limited by its high storage and computational requirement especially when dealing with real-time tasks. Many attempts have been reported to increase the performance of the CHT. The randomized Hough transform (RHT) is an efficient approach introduced by Xu et al. [7]. This method selects three non-collinear edge pixels and votes for the circle parameters found by fitting a circle through those points. The mechanisms of the random sampling of points and many-to-one mapping from the image space to parameter space were for the first time proposed in this method. However, without decreasing the dimensionality of the Hough space, the randomized method still suffers from low speed.

There has been many attempts to use a lower dimensional Hough space to detect circular arcs. Ho and Chen [3] used the global geometrical symmetry of circles to reduce the dimension of the parameter space. In another attempt by

Kimme *et al.* [5], the circle's centre is detected by measuring the direction of the gradient of the edge points. This modification reduces computational requirements as an arc only needs to vote perpendicular to the edge orientation at a distance r from the edge point. Mintor and Sklansky [8] extended the use of edge orientation, by plotting a line in the edge direction to detect circles over a range of sizes simultaneously. Based on the symmetry of the gradient pair vectors on each circle, Rad *et al.* [9] presented a circle detection method, however, their method requires the circle to be totally brighter or darker than their background. There are other approaches to reducing the storage and computational requirement and increase the speed for the HT. Hollitt [10] present a new algorithm for calculating the CHT by recasting it as a convolution. This approach allows the transform to be calculated using the fast Fourier transform, yielding an algorithm with lower computational complexity.

However, none of the above approaches were based on analysing the peak shapes within the two dimensional line HT space. The two-dimensional line HT provides more information than just peaks. Szentandrasi et al. [11] used the pattern of peaks within the whole HT to detect grid structures and QR codes. Furukawa and Shinagawa [12] analysed the shape of the peaks to directly identify the end-points of line segments within an image. Chang et al. used the shape of peaks to detect and correct lens distortion [13].

In this paper, we propose a circle detection method based on analysing the shape of patterns within a two-dimensional Hough transform space. In the parameter space, a straight line appears with a concentrated peak, because a straight line can be represented as a single set of parameters. However, a curved line results in a blurred peak in the HT space. This results because different points along the curved line have different parameters, and the votes are no longer concentrated at a single set of parameters. A circle will further separate high energy points in HT space into a sinusoidal wave. By analysing the shape of the sinusoidal wave it is possible to get information about the circle. Figure 1 illustrates the shape of peaks in Hough space corresponding to a straight line and a curved line. Figure 2 illustrates the pattern of points in Hough space corresponding to a circle.

The rest of this paper is organised as follows. Section 2 analyses the shape of peaks in Hough space and investigates the relationship between the circle in image space and the shape of the Hough pattern. The basic steps of our method are presented in Sect. 3. Experiment results and a brief discussion are covered in Sect. 4. Section 5 provides the conclusion and future research goals.

2   Analysing the Shape of Pattern in Hough Space

In this section, the shape of patterns in the two dimensional HT space is analysed. To reduce the processing we use the gradient HT which will remove the repeated information and the clutter while keeping the necessary information. The gradient HT measures the orientation of the edge and rather than voting for all possible lines through that point, it only votes for a small range of angles

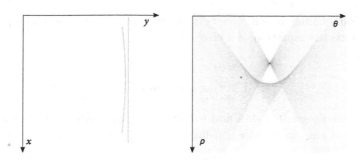

Fig. 1. The left image is the straight line and the line curved by lens distortion, the right image is the peak of the straight line and the blurred peak of the curved line in the Hough space [13].

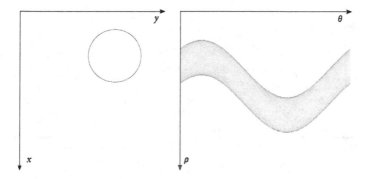

Fig. 2. The left image is a circle, the right image is the sinusoidal pattern of the circle in the Hough space.

[14]. Figure 3 compares the parameter space between the gradient HT and the full HT. In the gradient HT space a circle separates votes into a single sinusoidal wave.

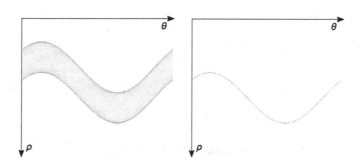

Fig. 3. The left image is the full HT shape for a circular object and the right image is the gradient HT space.

Unlike straight lines which have two parameters, a circle has three parameters. The parametric equation of a circle is:

$$\begin{cases} x(\theta) = a + r\cos\theta \\ y(\theta) = b + r\sin\theta \end{cases} \tag{2}$$

where θ is in the range $[0, 2\pi]$, a and b define the coordinates of the centre and r is the radius of the circle. Substituting (2) into (1) gives:

$$\rho = a\cos\theta + b\sin\theta + r \tag{3}$$

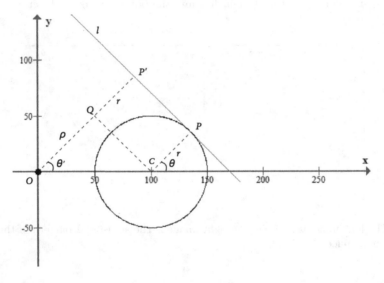

(a) a circle in image space

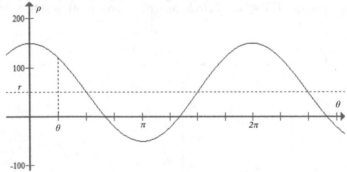

(b) the corresponding patten in the HT space

Fig. 4. A circle in image space a its corresponding sinusoidal wave in HT space.

Equation (3) can be written as:

$$\rho = \sqrt{a^2 + b^2}\,\sin(\theta + \alpha) + r \qquad (4)$$

where:

$$\cos \alpha = \frac{b}{\sqrt{a^2 + b^2}} \qquad (5)$$

therefore, in the HT space, the amplitude of the sinusoidal wave is equal to the distance from the centre of the image to the centre of the circle, the offset of the sinusoidal wave is equal to the radius of the circle.

Figure 4 shows the relationship between a circle in image space and the sinusoidal wave in HT space. In Fig. 4(a) the centre of the circle is point C, and the radius of the circle is r. For a given angle θ, the tangent line of the circle (l) will be perpendicular to CP. The perpendicular line of l through the origin, is OP'. From the definition of the Hough transform the ρ is equal to the length of line OP'. The length of OP' can be further represented as OQ plus QP' which is equal to the circle radius plus the distance from the centre of the circle to the centre of the image multiplied by $cos(\theta)$. By changing the angle θ from 0 to 2π there will be a sinusoidal wave in the HT space. The position of the circle centre is also related to the phase shift of the sinusoid and the amplitude. By measuring the sinusoidal wave in HT space, it is possible to estimate the parameters of the circle. In the next section the basic steps of this method will be explained.

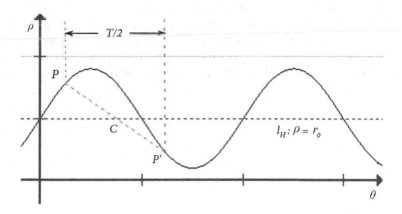

Fig. 5. Estimating the offset (r_0) of a sinusoidal wave.

3   Basic Steps

3.1   Estimating the Radius of the Circle

The radius of the circle is estimated by a one dimensional HT method. In Fig. 5 we assume the expression of the circle is:

$$\rho = A \sin(\theta + \alpha) + r. \qquad (6)$$

Estimating the offset of the sinusoidal wave in the HT space, therefore, provides an estimate of the radius. Zou and Shi [15] provide a Hough transform like method to estimate the offset of a sinusoidal wave. The period T of the sinusoidal wave in the HT space is known. Let P be an arbitrary point on the sinusoidal trace, with P' offset from P by half of the period (π). The mid point of PP' is C, the ρ axis value of C is equal to r. This feature works for every point along the sinusoidal wave. By using a Hough transform like method, every pair of points separated by $T/2$ will vote for the offset they have, and the offset will be a peak in the one-dimensional space. By finding the peak in the parameter space, the radius of the circle in HT space can be estimated. Figure 6 shows the one-dimensional space with a significant peak for the offset of the sinusoidal wave (radius of the circle).

3.2   Estimating the Centre of the Circle

From (4), the centre of the circle is related to the amplitude and the phase of the sinusoidal wave in the HT space. To estimate the centre of the circle, it is necessary to get the expression of the sinusoidal wave. Figure 7 shows a target sinusoidal wave (green) in the HT space, and another sinusoidal wave corresponding to another circle. The red dotted line is the offset of the target sinusoid which is equal to the radius of the circle. Points z_1 and z_2 are where the sinusoidal wave cross the offset line. Point p_1 is the positive peak of the sinusoid while point n_1 is the negative peak. Locating these four points provides sufficient information to estimate the expression of the sinusoidal wave. Points z_1 and z_2 have an angle π between them. (The purple sinusoidal wave also has intersections with $\rho = r$, however, the distance between them on the angle axis is not equal to π.) In Fig. 7 the angle distance between z_1 and z_2 and the two

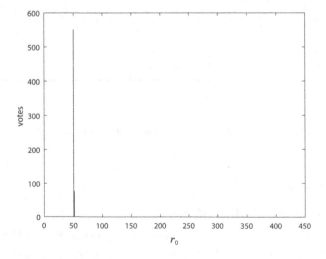

Fig. 6. One dimensional space to estimate r_0.

peak points p_1 and n_1 is $\pi/2$, while the ρ-axis distance from p_1 and n_1 to the offset is equal; using these features, the peak points can be located. Then the centre of the circle can be estimated from:

$$a = A\sin(\alpha)$$
$$b = A\cos(\alpha)$$

(7)

where a is the x-axis value and b is y-axis value of the circle, the α is phase shift which can be found by locating points z_1 and z_2, A is the amplitude which can be calculated from:

$$A = \frac{\rho(p_1) - \rho(n_1)}{2}$$
$$\alpha = \frac{\pi}{2} - \theta(p_1)$$

(8)

4   Simulation Results and Discussion

To test our method, 20 images were used. Each image contains one to three circles with different radius. The test images are from the internet or captured by a camera on an iPhone 8. The resolution of captured images is 1440×1080, and the light conditions within each image is different. Figure 8 shows some of the test images. A Canny edge detection has been applied to the image before the gradient Hough transform, and the angle range of the GHT is $\pm 0.01\pi$. Left are original images and right are images with detected circles highlighted in red. From the results of this test, our method shows reliable detection in different backgrounds and shows an ability to detect multiple circles. This method requires the target circle to have a sufficient contrast with the background to be reliably

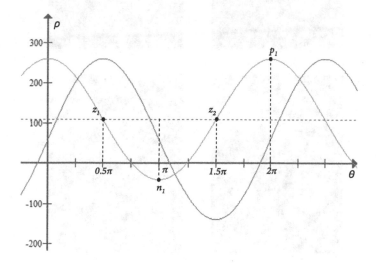

Fig. 7. Estimate the centre of the circle. (Color figure online)

Fig. 8. Example test images, left shows the original image; right shows the detected circles highlighted in red. (Color figure online)

detected by the Canny filter. This is because this method relies on the quality of the edge image to give a clear sinusoidal wave.

To better evaluate this method, an experiment against noise has been designed. In this experiment, white Gaussian noise has been added to the grey level image with a variance from 0 to 0.1. Figure 9 shows the original image without noise and the image with noise (variance in this image is 0.05). By increasing the noise level in steps of 0.001 variance, a group of images has been tested. Each image has been tested twice to reduce the random error, because the noise added to the image is random. The error rate is measured by whether the circle has been detected. Figure 10 shows the error rate of this method with an increasing noise level. The dotted line shows that the error rate is approximately proportional to the noise level. When the variance is less than 0.075 (SNR > 7.7)

(a) Images without and with Gaussian noise

(b) Edge images without and with Gaussian noise

(c) Hough space without and with Gaussian noise

Fig. 9. Influence from white Gaussian noise (variance in this image is 0.05).

the error rate is less than 50%. When the noise level in the image is low and mild, this method provides a reliable detection rate. However, the error rate increases with the noise level; even though the Hough transform itself is robust against noise, locating feature points in HT space is still influenced by the noise. Noise in the image influences the quality of the edge detection result (Fig. 9(b)), and further influences the quality of the sinusoidal wave within the HT space (Fig. 9(c)); a cluttered sinusoidal wave will bring difficulties to locating points on it. Without accurate feature points, there will be significant errors in estimated parameters of the sinusoidal wave, and further influence in the accuracy of the circle's parameters.

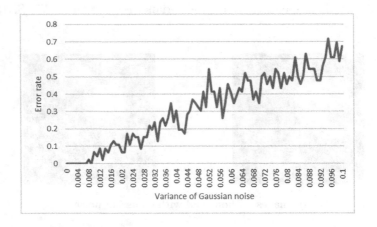

Fig. 10. Error rate in the circle detection with a increase noise level.

5   Conclusion and Future Works

This paper proposes a new circle detection method based on analysing the pattern shapes within a two-dimensional HT space. The radius of the circle is estimated by a one-dimensional Hough-like transform, and the centre of the circle is estimated by locating feature points. From the experimental result, this method can detect multiple circles in an image with mild noise.

For increasing the accuracy of this method, one future work is to find a more robust way to detect the sinusoidal wave in the HT space. This method changes the problem of detecting circle in image space into detecting the sinusoidal wave in parameter space. Using a more efficient and robust way of detecting the sinusoidal wave can provide more accurate information of the circle. We also plan to explore this method for ellipse detection. Figure 11 shows an ellipse and detected ellipse highlighted in red. This ellipse is detected by locating feature points in a two dimensional line HT space.

Fig. 11. Ellipse detection.

References

1. Smereka, M., Dulęba, I.: Circular object detection using a modified Hough transform. Int. J. Appl. Math. Comput. Sci. **18**(1), 85–91 (2008)
2. Cauchie, J., Fiolet, V., Villers, D.: Optimization of an Hough transform algorithm for the search of a center. Pattern Recogn. **41**(2), 567–574 (2008)
3. Ballard, D.H.: Generalizing the Hough transform to detect arbitrary shapes. Pattern Recogn. **13**(2), 111–122 (1981)
4. Duda, R.O., Hart, P.E.: Use of the Hough transformation to detect lines and curves in pictures. Commun. ACM **15**(1), 11–15 (1972)
5. Kimme, C., Ballard, D., Sklansky, J.: Finding circles by an array of accumulators. Commun. ACM **18**(2), 120–122 (1975)
6. O'Gorman, F., Clowes, M.: Finding picture edges through collinearity of feature points. IEEE Trans. Comput. **25**(4), 449–456 (1976)
7. Xu, L., Oja, E., Kultanen, P.: A new curve detection method: randomized Hough transform (RHT). Pattern Recogn. Lett. **11**(5), 331–338 (1990)
8. Minor, L.G., Sklansky, J.: The detection and segmentation of blobs in infrared images. IEEE Trans. Syst. Man Cybern. **11**(3), 194–201 (1981)
9. Rad, A.A., Faez, K., Qaragozlou, N.: Fast circle detection using gradient pair vectors. In: Seventh International Conference on Digital Image Computing: Technique and Applications, pp. 879–887 (2003)
10. Hollitt, C.: Reduction of computational complexity of Hough transforms using a convolution approach. In: 24th International Conference Image and Vision Computing New Zealand, pp. 373–378 (2009)
11. Szentandrási, I., Herout, A., Dubská, M.: Fast detection and recognition of QR codes in high-resolution images. In: Proceedings of the 28th Spring Conference on Computer Graphics, pp. 129–136 (2013)
12. Furukawa, Y., Shinagawa, Y.: Accurate and robust line segment extraction by analyzing distribution around peaks in Hough space. Comput. Vis. Image Underst. **92**(1), 1–25 (2003)
13. Chang, Y., Bailey, D., Le Moan, S.: Lens distortion correction by analysing peak shape in Hough transform space. In: 2017 International Conference on Image and Vision Computing New Zealand (IVCNZ), pp. 1–6 (2017)
14. Fernandes, L.A., Oliveira, M.M.: Real-time line detection through an improved Hough transform voting scheme. Pattern Recogn. **41**(1), 299–314 (2008)
15. Zou, C., Shi, G.: A fast approach to detect a kind of sinusoidal curves using Hough transform. Comput. Eng. Appl. **3**(4), 1–3 (2002)

A Spatial Density and Phase Angle Based Correlation for Multi-type Family Photo Identification

Anaica Grouver[1], Palaiahnakote Shivakumara[1], Maryam Asadzadeh Kaljahi[1], Bhaarat Chetty[2], Umapada Pal[3], Tong Lu[4(✉)], and G. Hemantha Kumar[5]

[1] Faculty of Computer Science and Information Technology, University of Malaya, Kuala Lumpur, Malaysia
anaicagrouver@gmail.com, {shiva,asadzadeh}@um.edu.my
[2] Google Developers Group, NASDAQ, Bangalore, India
bhaarat.chetty@gmail.com
[3] Computer Vision and Pattern Recognition Unit, Indian Statistical Institute, Kolkata, India
umapada@isical.ac.in
[4] National Key Lab for Novel Software Technology, Nanjing University, Nanjing, China
lutong@nju.edu.cn
[5] University of Mysore, Mysore, Karnataka, India
ghk.2007@yahoo.com ·

Abstract. Due to change in mindset and living style of humans, the numbers of diversified marriages are increasing all around the world irrespective of race, color, religion and culture. As a result, it is challenging for research community to identify multi type family photos, namely, normal family (family of the same race, religion or culture), multi-culture family (family of different culture, religion or race) from the family and non-family photos (images with friends, colleagues, etc.). In this work, we present a new method that combines spatial density information with phase angle for multi-type family photo classification. The proposed method uses three facial key points, namely, left-eye, right-eye and nose, for the features which are based on color, roughness and wrinkleless of faces, these are prominent for extracting unique cues for classification. The correlations between features of Left & Right Eyes, Left Eye & Nose and Right Eye & Nose are computed for all the faces in an image. This results in feature vectors for respective spatial density and phase angle information. Furthermore, the proposed method fuses the feature vectors and feeds them to the Convolutional Neural Network (CNN) for the classification of the above-three class problem. Experiments conducted on our database which contains three classes, namely, multi-cultural, normal and non-family images and the benchmark databases (due to Maryam et al. and Wang et al.) which contain two class-family and non-family images, show that the proposed method outperforms the existing methods in terms of classification rate for all the three databases.

Keywords: Spatial density · Fourier transform · Phase angle · Correlation coefficients · Convolutional Neural Network · Family and non-family photos

© Springer Nature Switzerland AG 2020
S. Palaiahnakote et al. (Eds.): ACPR 2019, LNCS 12047, pp. 76–89, 2020.
https://doi.org/10.1007/978-3-030-41299-9_7

1 Introduction

As generation changes and technology advances, the mindset and lifestyle of people also changes. One such change reflects in diversified families instead of families of the same community, religion, culture, etc. This makes identifying or classifying family photos more challenging. In addition, due to more than million uploads of images per day on social media, the database results in a large collection of diversified and heterogeneous images [1]. This makes identifying family photos more challenging and complex. The family photo identification can help scattered family members to bring together including refugees. In addition, it can assist immigration department in the airport to prevent human trafficking by retrieving family photos of each person using face images. For classifying and retrieving family images, there are umpteen methods that explore biometric features, such as face, gait and action. However, these methods [2] are good only when images contain uniform emotions and single actions but not as shown in Fig. 1, where one can expect multiple emotions, expressions and actions in a single image. This makes the problem of finding photos that belong to the same family more complex and challenging. Therefore, there is a need for developing a new method to classify multi-type family photo from non-family photos. This solution can play a vital role in preventing human trafficking, kinship recognition, and the problem of identifying/locating refugees [3].

    Multi-cultural Family   Normal Family   Non-family

Fig. 1. Sample images for multi-cultural, normal and non-family classes

This work presents a method for multi-type family classification including non-family photos, namely, normal, multi-culture and non-family photos based on characteristics defined below [4–6] and our observations. We define the following characteristics for multi-type family and non-family photos in this work. **For multi-cultural family**: Photos will consist of parents and their children either sitting or standing in a cascaded order with different colors of faces. Photos should contain only one family. **For normal family**: Photos will consist of parents and their children either sitting or standing in a cascaded order with the uniform color. The number of persons in an image should be more than 3, including parents and one child. **For non-family**: Photos must have persons with almost the same age, and it is expected persons of different families, for example, friends and colleagues, might be present. Images must have persons with different poses and any order with any background. Example for multi-cultural, normal and non-family photos can be seen in Fig. 1. It is observed from Fig. 1 that color of faces of parents and children in multi-cultural family photos differs, while color of faces of parents and children is almost the same for normal family photos. On the other hand, color and shape of faces in non-family are totally different. Besides, actions, emotions and expressions of persons in multi-cultural, normal and non-family are also different.

2 Related Work

It is noted from the previous section that methods developed based on biometric features may not work well for multi type family and non-family photo identification. This section presents a review on the methods that focus on kinship recognition, family, non-family photo classification and related methods.

Ng et al. [7] proposed social relationship discovery and face annotations for personal photo collections. This method explores the combination of ensemble RBFNN with pairwise social relationships as context for recognizing people. Dandekar et al. [8] proposed verification of family relationships from parents' and children's facial images. The method uses local binary pattern features and degree of similarity between faces of children and parents. Xia et al. [9] proposed face clustering in a photo album, by exploring spectral features, similarity features, the minimum cost flow, and clustering. The proposed features are extracted from cropped face images. The main objective of this method is to find images, which share the same faces. Qin et al. [10] proposed tri-subject kinship verification for understanding the core of a family. The method proposes a degree of similarity between children and parents, resulting in a triangular relationship. Dai et al. [11] proposed family member identification from photo collections. The method explores an unsupervised EM joint inference algorithm with a probabilistic CRF. The proposed model identifies role assignments for all the detected faces along with associated pairwise relationships between them. Robinson et al. [3] proposed visual kinship recognition of families in the wild. This method explores deep learning for face verification, clustering and boosted baseline scores. The method involves multimodal labeling to optimize the annotation process. Wang et al. [4, 5] proposed leveraging geometry and appearance cues for recognizing family photos. The methods identify facial points for each face in an image. Based on facial points, the method constructs polygons to study geometric features of faces. Recently, Maryam et al. [6] proposed a geometric and fractional entropy-based method for classifying family and non-family images. The method explores the cues of foreground (faces) and background (other than face) for classification of family and non-family images.

In the backdrop of the above methods, one can note that most of the methods focus on kinship and social relationship recognition. These methods use face detection and recognition. As discussed in the above, these methods may not be robust for multi-type family and non-family photo identification due to complex background and multiple emotions, expression and actions. A few methods for family and non-family image classification are proposed. However, the scope of the methods is limited to two classes, namely, family and non-family. In addition, the methods consider multi-cultural and normal family photos as one class to make the problem simple. Therefore, we can conclude that there is a demand for identification of multi-type and non-family photos.

Hence, this work focuses on developing a method for three-class classification. As pointed out in the Introduction Section by referring to sample images in Fig. 1 skin color, wrinkles and roughness are important clues for differentiating multi-type family photos from non-family photos. To extract such observations, motivated by the method [6] where it is stated that spatial information and orientation are sensitive to smoothness, roughness and color of images, we explore spatial density and phase angle for identifying multi-type and non-family photos in this work. It is valid because as age of a person varies, color and skin surface also varies in terms of smoothness, roughness and wrinkles. In the same way, as emotion and expression changes, skin structure also changes. In this context, spatial density and phase angle are good features for studying those changes.

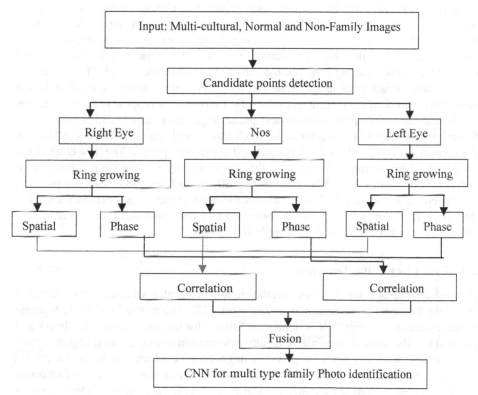

Fig. 2. Flow chart of the proposed method for multi-type and non-family photo identification

3 Proposed Method

It is noted that face, skin and actions in images are important features for identifying multi-type family and non-family photos. With this notion, we propose to extract facial information by using three key facial points, namely, left-eye, right-eye and nose. To study local and neighboring changes around key points, the proposed method constructs rings with radius of 5 pixels until the ring growing reaches any facial boundary. The radius value is determined empirically and is presented in the Experimental Section. In this work, we prefer to choose only three key facial points because nose is the center one and the ring growing covers the full face, which can help us study smoothness and roughness of skin. Similarly, left-right eye covers the exact wrinkles around those key points. The reason to choose ring constructions is that the proposed intention is to make features invariant to rotation and scaling. For each ring, as mentioned in the previous section, the proposed method counts number of white pixels in Canny of the input image, which is called Spatial Density features. Since the considered problem is complex, spatial density features alone may not be enough, hence we propose to explore phase angle given by Fourier transform, which are calculated for each ring of all the three key points. For each key point, ring growing process results in feature vectors. Therefore, for each face, the propose method gets three vectors using one procedure and finds correlations between nose with left-eye, nose with right eye and left-eye with right-eye, which are defined as triangular relationship. However, input images may have multiple faces, thus the proposed method considers the mean, median and standard deviation of corresponding correlation coefficient values of the triangle. This results in three values for each input image. The same thing is for phase angle information, which results in three correlation coefficient values. Furthermore, we propose to fuse correlation coefficient values of spatial density and phase angle information, which results in a feature vector. The feature vector is fed to Convolutional Neural Networks (CNN) for the identification of multi-cultural, normal and non-family photos. The flow chart of the proposed method is shown in Fig. 2. In this work, we prefer to use the combination of hand crafted features and CNN because it is difficult to learn and extract the features, which represent non-family photos. This is valid because there is no boundary and clear definition for the non-family photos compared to family photos. In this situation, the combination is better than sole learning based models.

3.1 Facial Key Points Detection

For each input image, we propose to use the method that explores the concept of 'locality' principle for facial point detection in this work [12]. The method follows a learning based approach. The principle is defined as follow: for locating a certain landmark at a given stage, the most discriminative texture information lies in the local region around the estimated landmark from the previous stage. Shape context, which gives locations of other landmarks and local textures of this landmark, provides sufficient information. With these observations, the method first learns intrinsic features to encode local textures for each landmark independently. It then performs joint regression to incorporate shape context. The method first learns a local feature mapping function to generate local binary features for each landmark. Here, it uses the standard regression random forest to learn each local mapping function. Next, it concatenates all the local features to obtain the mapping functions. It learns linear projections by linear regression. This learning process is repeated stage by stage in a cascaded fashion. After that, a global feature mapping function and a global linear projection and objective function are used to incorporate shape context. This process can effectively enforce the global shape constraint to reduce local errors. In the case of the testing phase, shape increment is directly predicted and applied to update the current estimated shape. More details regarding implementation can be found in [12]. The reason to choose this method is that it is said to be generic, efficient and accurate for finding facial key points. In addition, it can cope up with issues of partial occlusion and distortion. This is justifiable because the proposed work considers multi-type family and non-family images with complex backgrounds and diversified content. The sample results of the above method are illustrated in Fig. 3, where (a) gives the results of candidate point detection for the input image, while Fig. 3(b) shows samples of facial key points for family and non-family images.

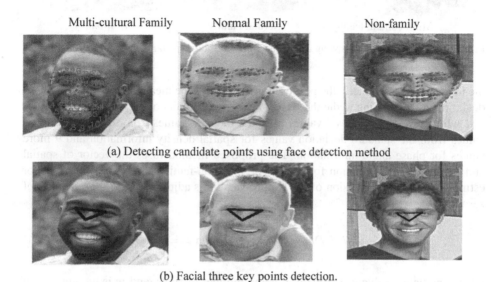

Multi-cultural Family   Normal Family   Non-family

(a) Detecting candidate points using face detection method

(b) Facial three key points detection.

Fig. 3. Detecting three key points for the faces of multi type and non-family images.

3.2 Spatial Density and Phase Angle Based Correlation Feature Extraction

For facial three key points detected by the step presented in previous section, the proposed method finds centroid using x and y coordinates of the cluster points of respective three facial points. To study the local information of skin and face, the proposed method constructs rings for each key points by considering centroid of each facial point as a the center of the ring as shown in Fig. 4 for the three images, where different colors indicate ring growing of different key points. The radius used for constructing rings is 5 pixels according to our experiments. For each ring of each key points, the proposed method counts the number of white pixels as spatial density as defined in Eq. (1) through Canny edge image of the input image. Similarly, for each ring of each key points, the proposed method calculates phase angle using Fourier transform as defined in Eqs. (2)–(4). The ring growing process continuous till it reaches any of facial boundary points as shown in Fig. 4. For each ring, the proposed method obtains spatial density (sum of white pixels) value and mean phase angle. This process results in feature vector containing either spatial density or mean phase angles with dimension of the number of rings.

To find the relationship between vectors of the three key points, the proposed method defines triangular relationship between them, which estimates the Correlation Coefficient (CC) as defined in Eq. (5) as three pairs of each triangle, namely, Nose-Left Eye, Nose-Right Eye and Left Eye-Right Eye. This step outputs three CC values for each face in

Multi-cultural Family   Normal Family   Non-family

Fig. 4. Illustrating the ring growing for the facial key of multi type family and non-family images

the image. For all the faces, the proposed method finds the mean, median, and standard deviation corresponding to the three CC of the triangle pairs of all the faces in images. The whole process includes 3 values from the mean, 3 values from the median, and 3 values from the standard deviation values for spatial density information and 9 more values for phase angle information. The proposed method fuses the vector of spatial and phase angle information for multi-type and non-family photo classification. For estimating CC, the dimension of the feature vectors is adjusted by padding number of zeros.

$$Spatial\ Density = \sum_{i=1}^{n}\sum_{i=1}^{m}(pix == 1) * 1 \tag{1}$$

where, pix is the value of the pixel, n is the number of rings, and m is the number of pixels in each ring.

$$f(x) = a_0 + \sum_{n=1}^{\infty}\left(a_n \cos\frac{n\pi x}{L} + b_n \sin\frac{n\pi x}{L}\right) \tag{2}$$

$$X = \frac{(\sum_{i=1}^{n}\cos\theta)}{n}; Y = \frac{(\sum_{i=1}^{n}\sin\theta)}{n} \tag{3}$$

where, θ is the phase angle for a particular pixel in the ring, and n is the number of pixels in the ring.

$$\Phi = \tan\frac{Y}{X} \tag{4}$$

where, ϕ is equal to the circular mean phase of a ring

$$CC(x, y) = \frac{\sum_{n=0}^{n-1}x[n] * y[n]}{\sqrt{\sum_{n=0}^{n-1}x[n]^2} * \sqrt{\sum_{n=0}^{n-1}y[n]^2}} \tag{5}$$

where, x and y are the vectors with respect to key points, and n denotes the size of the vectors.

The effect of feature vectors with respect to spatial density and phase angle can be seen in Fig. 5(a) and (b), respectively, where the vector containing 9 values and

Multi-cultural Family   Normal Family   Non-family

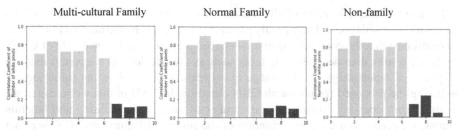

(a) The bar graphs for mean, median and standard deviation of correlation coefficients computed by spatial density features of respective three facial key points of all the faces in the respective three images. Green, pink and maroon denote mean, median and standard deviation of three key points, respectively for all the above three graphs.

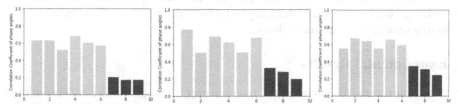

(b) The bar graphs for mean, median and standard deviation of correlation coefficients computed by Phase values of respective three facial key points of all the faces in the respective three images. Green, pink and maroon colors denote mean, median and standard deviation of three key points, respectively for all the above three graphs.

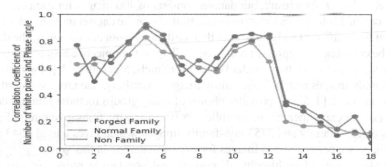

(c) The line graphs for the number of features (18 features) vs mean correlation values after fusing 9 features of spatial density and phase angle based methods.

Fig. 5. The feature extraction by fusing spatial density and phase angle based methods for three images. (Color figure online)

the corresponding CC values are drawn. It is observed from Fig. 5(a) and (b) that bar graphs of both the approaches exhibit the unique behavior for multi type and non-family photos. This shows that spatial density and phase angle based features contribute equally for identification of multi type and non-family photos. In the same way, when we look line graphs of the fusion results shown in Fig. 5(c), where the line graphs representing each class in different color exhibit distinct behavior. Therefore, one can argue that the extracted features are good enough to differentiate multi type and non-family photos.

The result of fusion is fed to a fully connected Convolutional Neural Network (CNN) for classifying multi type and non-family photos [13]. Inspired by the method [14] where it is mentioned that the combination of handcrafted features and the ensemble of CNNs give better results than deep learning tools such as GoogleNet and ResNet50 that use raw pixels of input images for bioimage classification, we explore the same idea of combining the proposed features with CNN classification [15] in this work. Since the proposed work does not provide a large number of samples for training and labeling samples, we prefer to use the combination of the proposed features and CNNs rather than raw pixels with the recent deep learning models. In addition, it is difficult to choose pre-defined samples that represent non-family photos because there is no limitation for defining non-family photos. For learning parameters of the classifier, we follow a 10-fold cross-validation procedure, which splits datasets into training and testing components. The training samples are used for learning and adjusting parameters of the classifier and the testing samples are used for evaluation.

4 Experimental Results

For evaluating the classification of multi type and non-family photos, we collect our own dataset for multi-cultural family photo and use the standard dataset for other two classes because there is no standard dataset for multi-culture class. Therefore, we collect images for multi-culture family class according to the definitions mentioned in the Introduction Section from different sources, such as Facebook, Reddit, Flickr, Instagram, internet and other websites. As a result, our dataset consists of 400 images for each class, which gives a total of 1200 images for experimentation. Since the dataset includes more images collected from different websites rather than collecting from our own way, the dataset is said to be complex and represent real images. To test the objectiveness of the proposed method, we also test it on two standard datasets, namely, Maryam et al. [6] which consists of 388 family images and 382 non-family images. Similarly, one more standard dataset called Wang et al. [4, 5, 16], provides photos of many groups including family and non-family. For our experimentation, we collect as per the instructions given in [4, 5], which consists of 1790 family and 2753 non-family images, which gives a total 4543 images. In summary, we consider 6513 images for experimentation in this work.

Sample images of multi-cultural, normal and non-family photos of our dataset, Maryam et al. [6] and Wang et al. [4, 5] datasets are shown in Fig. 6(a)–(c), respectively. Figure 6(a)–(c) show that images of all the three datasets appear different in terms of background, family arrangements and actions in case of non-family images. The sample images shown in Fig. 6 are the results of successful classification of the proposed method.

To show the usefulness of the proposed method, we implement three state-of-the-art methods, namely, Maryam et al.'s [6] which explores geometrical feature of face and fractional entropy based features for classifying family and non-family images, Wang et al.'s [4] which explores facial geometric features and facial appearance model-based features, and its extension [5] for the classification of normal family and non-family images. Note that the same idea is extended and the results are improved in [5] for the purpose of family and non-family image classification. Since these are three state-of-the-art methods for the classification of family and non-family photos, we use the same

(a) Multi-cultural   Normal   Non-family

(b)   Family   Non-family

(c)   Family   Non-family

Fig. 6. Successful classification of the proposed method for our dataset and two standard datasets.

for comparative study in this work. To measure the performance of the proposed and existing methods, we generate confusion matrices for the three classes of our dataset and two classes of the other two standard datasets to calculate Classification Rate (CR). The Classification Rate (CR) is defined as the number of images classified correctly by the proposed method divided by the total number of images in the class. Average Classification Rate (ACR) is calculated for diagonal elements of confusion matrices to evaluate the overall performances of the proposed and existing methods.

For extracting features in the proposed methodology, we mention the value of radius is 5. To validate this value, we conduct experiments on our own dataset for calculating average classification rate *vs.* different radii as shown in Fig. 7, where it is noted that the classification rate gradually increases as radii increases and at radius 5, the classification rate reaches the highest peak and then the classification rate decreases for higher values of radius. Therefore, 5 is considered as the optimal value for all the experiments.

4.1 Experiments on Three Classes Classification

It is noted from the proposed methodology, spatial and phase angle based features contribute equally for multi type and non-family photo classification. To validate the statement, we conduct experiments using only spatial density based features and only phase angle based feature as reported in Table 1, where it is observed that both scores almost the same results in term of CR for three classes of our dataset. This is because spatial density based feature is for extracting smoothness and roughness skin, while phase angle is good for extracting wrinkleless of skin. As a result, the proposed method, which fuses both spatial density and phase angle based features, achieves better results than individual features in terms of classification rate. The proposed method uses CNN for

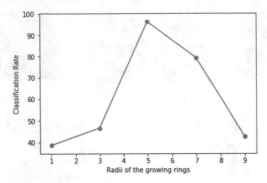

Fig. 7. Determining the optimal radius values for ring growing process.

classification of multi type family and non-family photos. To show CNN is better than SVM, we conduct experiments and the results are reported in Table 2, where it is seen that the proposed method with CNN is better than the proposed method with SVM in terms of CR. This is because the CNN can learn the complex relationship between input and output, while SVM does not as it depends on kernel size and type. Qualitative results of the proposed and existing methods for the three classes of our dataset are reported in Table 3, where we can see that the proposed method is better than the two existing methods in terms of CR. When we compare the results of two existing methods, Wang et al.'s [5] is better than Wang et al.'s [4]. This is correct because Wang et al.'s [5] is an improved version of the Wang et al.'s [4]. The reason for the poor results of the existing methods is the scope is limited to two classes, namely, family and non-family. On the other hand, the proposed method is better than the existing methods because of the contributions of spatial density and phase angle based features.

Table 1. Performances of the proposed spatial density, phase angle based features and the combined features (Spatial + Phase angle) for our databases (in %). MCF: Multi-Cultured NF: Normal and NoF: Non-family classes.

Methods	Spatial			Phase angle			Proposed method (Spatial + Phase Angle)		
Classes	MCF	NF	NoF	MCF	NF	NoF	MCF	NF	NoF
Multi-cultural	**82.3**	9.5	8.2	**82.6**	10.2	7.2	**93.5**	4.4	2.1
Normal	7.2	**88.2**	4.6	7.9	**86.5**	5.6	1.3	**98**	0.7
Non family	7.5	7.2	**85.3**	6.5	7.5	**86**	1.6	0.9	**97.5**
CR	85.3			85			96.3		

Table 2. Classification rates for the proposed method with CNN and SVM (in %)

Methods	Proposed method with CNN			Proposed method with SVM		
Classes	MCF	NF	NoF	MCF	NF	NoF
Multi-cultural	**93.5**	4.4	2.1	**85**	8	7
Normal	1.3	**98**	0.7	6.3	**87.4**	6.3
Non-family	1.6	0.9	**97.5**	8.6	6	**85.4**
CR	96.3			86.0		

Table 3. Classification rates of the proposed and existing methods for our database (in %)

Methods	Proposed method			Maryam et al. (2019)			Wang et al. (2015)			Wang et al. (2017)		
Classes	MCF	NF	NoF	MCF	NF	NoF	MCF	NF	NoF	MCF	NF	NoF
Multi-cultural	**93.5**	4.4	2.1	**83.2**	**16.3**	**0.5**	**53.5**	46.5	0	**42.8**	0	57.2
Normal	1.3	**98**	0.7	74	23.2	2.8	20.7	**78.8**	0.5	1	**45.8**	53.2
Non Family	1.6	0.9	**97.5**	0	26.3	73.7	0	78.2	**21.8**	0	11.5	**88.5**
CR	96.3			60.0			51.4			59.03		

4.2 Experiments on Two Class Classification

To assess the performance of the proposed and existing methods for two classes, we divide the three-class problem into a two-class problem for experiments, namely, Multi-Cultural *vs* Normal and Multi-Cultural *vs* Non-family. The experimental results of the proposed and existing methods for our datasets and the other two standard datasets are reported in Tables 4, 5 and 6, respectively. Tables 4, 5 and 6 show that the proposed method is the best at CR compared to the existing methods for all the three datasets, and CR is almost consistent. However, the existing methods are not consistent and score poor results for our dataset and good for the other two standard datasets. It is expected because the existing methods are not capable of handling multi-cultural since the extracted features are not robust. For the other two standard datasets, the existing methods achieve better results compared to our dataset. The reason is the methods are developed for these two classes. When we compare the results of the existing methods, Wang et al.'s method is better than Wang et al.'s in terms of classification rate. The reason is the same as discussed in the previous section. Overall, based on experimental results, we can conclude that the proposed method is good for three classes as well as two classes, and the proposed method is consistent. It is evident that the proposed method can be extended to more classes and more images per class.

Table 4. Classification rates of the proposed and existing methods for two classes on our database (Multi-Cultural vs. Normal and Multi-Culture vs. Non-family)

Methods	Proposed		Maryam et al. (2019)		Wang et al. (2015)		Wang et al. (2017)	
Classes	MCF	NF	MCF	NF	MCF	NF	MCF	NF
MCF	**95**	5	**85.2**	14.8	**88.5**	11.5	**83.2**	16.8
NF	3	**97**	11.7	**88.3**	78.8	**21.2**	52.7	**47.3**
CR	96		86.75		56.		65.3	
Classes	MCF	NoF	MCF	NoF	MCF	NoF	MCF	NoF
MCF	**96**	4	**82.5**	17.5	**88.2**	11.8	**92.8**	7.2
NoF	2	**98**	38.75	**61.25**	77.5	**22.5**	45.5	**54.5**
CR	97		71.88		55.35		73.6	

Table 5. Classification rates of the proposed and existing methods on benchmark database (Maryam et al.)

Methods	Proposed		Maryam et al. (2019)		Wang et al. (2015)		Wang et al. (2017)	
Classes	NF	NoF	NF	NoF	NF	NoF	NF	NoF
NF	**95.8**	4.2	**88.41**	11.59	**76.28**	23.72	**82.73**	17.27
NoF	3.8	**96.2**	15.71	**84.29**	37.44	**62.56**	22.51	**77.49**
CR	96		86.34		69.42		80.11	

Table 6. Classification rates of the proposed and existing methods on benchmark database (Wang et al. 2017)

Methods	Proposed		Maryam et al. [6]		Wang et al. [4]		Wang et al. [5]	
Classes	NF	NoF	NF	NoF	NF	NoF	NF	NoF
NF	**98.7**	1.3	**96.36**	3.64	**88.55**	11.45	**94.52**	5.48
NoF	1	**99**	1.16	**98.84**	24.95	**75.05**	14.38	**85.61**
CR	98.8		97.59		81.79		90.06	

5   Conclusion and Future Work

We have proposed a new work for classifying multi-type and non-family photos in this work based on the combination of spatial density and phase angle based features. The proposed method uses face detection to find facial points, and then it chooses three facial key points for feature extraction. Ring growing technique is proposed for studying local information of skin and face using spatial density and phase angle given by Fourier transformation. Correlation coefficients for feature vectors of three facial key points are estimated to find triangular relationship between three key points. Correlation coefficients given by spatial density phase angle based feature are fused to obtain the final feature vector. The feature vector is then fed to CNN for the identification of multi-cultural, normal and non-family photos. Experimental results on our own dataset of three classes and two standard datasets of two classes show that the proposed method is better than the existing methods in term classification rate for all the three datasets. Our future plan is to extend the same idea for more classes by dividing family and non-family into sub-classes.

References

1. Cai, G., Hio, C., Bermingham, L., Lee, K., Lee, I.: Sequential pattern mining of geo-tagged photos with an arbitrary regions-of-interest detection method. Expert Syst. Appl. **41**, 3514–3526 (2014)
2. Mehta, J., Ramnani, E., Singh, S.: Face detection and tagging using deep learning. In: Proceedings of ICCCCSP (2018)
3. Robinson, J.P., Shao, M., Wu, Y., Gillis, T., Fu, Y.: Visual kinship recognition of families in the wild. IEEE Trans. PAMI **40**, 2624–2837 (2018)
4. Wang, X., Guo, G., Rohith, M., Kambhamettu, C.: Leveraging geometry and appearance cues for recognizing family photos. In: Proceedings of ICWAFG, pp. 1–8 (2015)
5. Wang, X., et al.: Leveraging multiple cues for recognizing family photos. Image Vis. Comput. **58**, 61–75 (2017)
6. Maryam, A.K., et al.: A geometric and fractional entropy-based method for family photo classification. Expert Syst. Appl. (2019)
7. Ng, W.W.Y., Zheng, T.M., Chan, P.P.K., Yeung, D.S.: Social relationship discovery and face annotations in personal photo collection. In: Proceedings of ICMLC, pp. 631–637 (2011)
8. Dandekar, A.R., Nimbarte, M.S.: A survey: verification of family relationship from parents and child facial images. In: Proceedings of SCEECS (2014)
9. Xia, S., Pan, H., Qin, A.K.: Face clustering in photo album. In: Proceedings of ICPR, pp. 2844–2848 (2014)
10. Qin, X., Tan, X., Chen, S.: Tri-subject kinship verification: understanding the core of a family. IEEE Trans. Multimedia **17**, 1855–1867 (2015)
11. Dai, Q., Carr, P., Sigal, L., Hoiem, D.: Family member identification from photo collections. In: Proceedings of WCACV, pp. 982–989 (2015)
12. Ren, S., Cao, X., Wei, Y., Sun, J.: Face alignment at 3000 fps via regressing local binary features. In Proceedings of CVPR, pp. 1685–1692 (2014)
13. McAllister, P., Zheng, H., Bond, R., Moorhead, A.: Towards personalized training of machine learning algorithms for food image classification using a smartphone camera. In: Proceedings of ICUCAI, pp. 178–190 (2016)
14. Nanni, L., Chidoni, S., Brahnam, S.: Ensemble of convolutional neural networks for bioimage classification, Appl. Comput. Inf. (2018)
15. Arora, R., Suman: Comparative analysis of classification algorithms on different datasets using WEKA. Int. J. Comput. Appl. **54**, 21–25 (2012)
16. Gallagher, A.C., Chen, T.: Understanding images of groups of people. In: Proceedings of CVPR, pp. 256–263 (2009)

Does My Gait Look Nice? Human Perception-Based Gait Relative Attribute Estimation Using Dense Trajectory Analysis

Allam Shehata[1,2(✉)], Yuta Hayashi[1], Yasushi Makihara[1], Daigo Muramatsu[1], and Yasushi Yagi[1]

[1] Institute of Scientific and Industrial Research, Osaka University,
8-1 Mihogaoka Ibaraki, Osaka 567-0047, Japan
{allam,hayashi,makihara,muramatsu,yagi}@am.sanken.osaka-u.ac.jp
[2] Informatics Department, Electronics Research Institute, Cairo, Egypt
allam@eri.sci.eg

Abstract. Relative attributes play an important role in object recognition and image classification tasks. These attributes provide high-level semantic explanations for describing and relating objects to each other instead of using direct labels for each object. In the current study, we propose a new method utilizing relative attribute estimation for gait recognition. First, we propose a robust gait motion representation system based on extracted dense trajectories (DTs) from video footage of gait, which is more suitable for gait attribute estimation than existing heavily body shape-dependent appearance-based features, such as gait energy images (GEI). Specifically, we used a Fisher vector (FV) encoding framework and histograms of optical flows (HOFs) computed with individual DTs. We then compiled a novel gait dataset containing 1,200 videos of walking subjects and annotation of gait relative attributes based on subjective perception of gait pairs of subjects. To estimate relative attributes, we trained a set of ranking functions for the relative attributes using a Rank-SVM classifier method. These ranking functions estimated a score indicating the strength of the presence of each attribute for each walking subject. The experimental results revealed that the proposed method was able to represent gait attributes well, and that the proposed gait motion descriptor achieved better generalization performance than GEI for gait attribute estimation.

Keywords: Dense trajectories · Relative attribute · Gait attribute estimation · Ranking functions · Histogram of optical flow · Fisher vector

A. Shehata—On leave from Informatics Department.

S. Palaiahnakote et al. (Eds.): ACPR 2019, LNCS 12047, pp. 90–105, 2020.
https://doi.org/10.1007/978-3-030-41299-9_8

1 Introduction

Walking style (i.e., gait) of pedestrians can be considered a powerful biometric modality, and gait can be recorded remotely without human participation or any need for special sensors. Recently, several gait-based approaches have been proposed for a range of applications [20,28]. These applications include, but are not limited to: gait analysis/recognition [34], age/gender estimation [19,21], person identification, and person verification [18,20]. Existing human gait recognition systems use two main types of modeling: model-based and model-free. Model-based approaches typically model the kinematics of human joints to measure physical gait parameters such as step length and angular speed [2]. As a consequence, model-based models suffer from high computational cost due to parameter calculation. They are also influenced by the quality of gait sequences. In the model-free approach, a compacted representation of gait sequences is considered, without explicit modeling of body structure [24]. These models use features extracted from the motion or shape of the walking subject and hence require substantially less computation. They are, however, not always robust against covariates such as viewpoints, clothing, or carrying status [29]. In addition, both modeling types require pre-processing steps to extract binary silhouettes, skeletons, or body joints to encode gait motion/appearance information. This limits the performance of the methods, due to the presence of dynamic backgrounds, noisy segmentation, and inaccurate body joint localization. Instead, gait motion information can be computed directly using optical flow through the spatial-temporal domain [37]. Thus, local motion descriptors have become widely used in human action recognition research [4]. To build these descriptors, highly dense sample points are detected and tracked through video footage of the action. Motion information relevant to these points is then aggregated in the form of histograms. For instance, a recent study [4] used local motion descriptors based on dense trajectories (DTs) for gait recognition instead of binary silhouettes. The researchers proposed the extraction of local motion features for different body regions, combining the extracted descriptors into a single global descriptor using the Fisher vector (FV) encoding mechanism [26].

Although satisfactory performance can be achieved using gait analysis/recognition approaches, to the best of our knowledge, the challenge of gait attribute estimation based on human perception and relating people to each other based on their gait attributes has not been examined in depth. In an early attempt, one study [34] reported a process for encoding socially relevant information (i.e., human attributes) into biomedical motion patterns. The researchers proposed an interactive linear classifier tool to discriminate male from female walking patterns based on instantaneous changes in a set of defined human attributes, such as age, gender, and mode. Furthermore, another proposed method [11] adopted attribute-based classification to overcome the computational complexity of the traditional multi-class gait classification task. This research has primarily sought to define certain attributes based on similarities between subject classes. However, these methods have been applied only for improving gait recognition and have not examined gait attribute estimation. Another recent study [41] proposed a method using

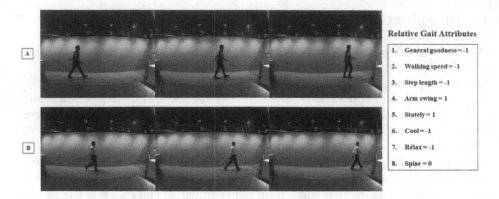

Fig. 1. The gait relative attributes for two walking subjects, A and B. Eight relative attributes are considered. A value of 1 indicates that the attribute strength for person A is greater than that for person B. A value of 1 indicates that the attribute strength for person B is greater than for person A. A value of 0 indicates that the attribute strengths for person A and B are similar.

convolutional neural networks and a multi-task learning model to identify human gait, and simultaneously estimate subjects' attributes. According to the researchers, this was the first multi-attribute gait identification system to be proposed. Nevertheless, because this method still depends on a GEI-based representation, it still suffers from the requirement of special considerations, and is influenced by body shape changes. Recent developments in deep learning models have made it easier to identify a person based on his/her gait style if a model is sufficiently trained with the walking sequence [33, 39]. However, it is difficult for such models to automatically recognize the gait of a person they have not seen before (i.e., in a zero-shot learning scenario). Moreover, several classification-based systems directly associate low-level features with absolute annotation labels. In contrast, discriminative visual descriptions cannot be easily characterized by an absolute label. Therefore, to describe action properties, it may be more appropriate to define high-level semantic concepts in terms of relative attribute labels instead of absolute labels. Recent research indicates that these attributes can be adopted in human recognition tasks as an intermediate level of description for human visual properties [42]. Using this approach, Parikh et al. [25] proposed the concept of relative attributes for learning in a rank classification model. This method utilizes the associated strengths of relative visual features in objects/scenes to learn a set of ranking functions and uses these functions to describe/relate new unseen instances to training images. In their proposed method, different images datasets of faces and natural scenes are examined, achieving excellent performance exceeding that of traditional binary attribute prediction methods. In the proposed method in the current study, we utilized the concept of relative attributes to build the first human gait relative attributes estimation framework. This framework is designed to assess the gait styles of walking subject pairs based on a set of gait relative attributes as shown in Fig. 1. Given the attribute annotation labels of the training pairs, the model learned a set of ranking functions based on a Rank-SVM classifier [25] and used these ranking functions to relate the subjects to each other and predict the gait relative attributes of previously unseen subjects. For gait motion encoding, we used the FV encoding method to build robust gait motion representation based on the extracted DTs from the gait videos. The main contributions of this work can be summarized as follows:

1. **Introducing the first gait relative attribute estimation approach**
   In gait-based approaches, the concept of relative attributes has not yet been explored. We propose a method that utilizes this concept and introduce an approach for estimating the gait relative attributes of walking subjects based on human perception.
2. **Building a new gait data set with attribute annotation**
   We describe a new annotated gait dataset based on subjective human perception. We hired seven annotators to watch videos of 1,200 walking subjects. We defined eight gait attributes and asked the annotators to assess each attribute for each pair of subjects, based on their subjective perception. This evaluation was recorded in the form of annotation labels that express the presence of an attribute for each subject pair (see Fig. 1). We used these annotations to train and evaluate our proposed gait relative attribute estimation method.
3. **Proposal of a robust dense trajectory-based gait motion representation**
   Instead of using existing and widely used appearance-based gait descriptors such as GEI [9], we used DTs to encode motion information of gait dynamics. We built a HOF for each trajectory. These HOFs were then encoded into a single robust global motion descriptor using FV encoding. This global vector carries the gait motion of the entire walking sequence.

The rest of the paper is organized as follows. Section 2 outlines closely related research. In Sect. 3, we present a detailed explanation of our proposed motion feature representation. The ranking learning process, the relative attribute annotation, and the proposed evaluation framework are described in Sect. 4. The details of our experiments are included in Sect. 5. Finally, Sect. 6 concludes the paper.

2 Related Work

Previous gait recognition systems can be classified into two main types: model-based and model-free approaches. In model-based approaches, a predefined model is introduced for model human body movement [3]. Model-free approaches directly depend on image features (i.e., appearance) for analysis and recognition. The GEI descriptor has been adopted by most gait recognition approaches as a baseline to represent a person's walking style. GEI is extracted by averaging over a binary silhouette sequence of the walking person [9]. Thus, the measure mainly depends on extracting pure foreground pixels, and its robustness may be degraded by conditions involving severe occlusion, camera movement, and or cluttered visual scenes. To partially overcome the limitations described above, new descriptors based on the local motion features of sampling points instead of binary silhouettes were considered. These descriptors have been embraced in the field of human action recognition research [5,37]. A new approach, as proposed in one previous study [37], to describe videos using DTs[1] by sampling dense points from each frame and tracking them based on displacement information from a dense optical flow field. The resulting trajectories are robust to fast irregular motion, and appropriately cover the motion information in videos. Moreover, a new motion descriptor can be designed based on differential motion scalar quantities, divergence, curl and shear features (DCS). This descriptor adequately decomposes visual motion into dominant and residual motion, in the extraction of space-time trajectories. In recent studies, local feature descriptors have been leveraged for gait-based analysis/recognition systems, and space-time interest points (STIPs) [14] have been used for representing a person's gait [12]. In addition, the gait modality has been utilized for enhancing multi-person tracking in crowded scenes based on trajectory-based clustering [32]. The key novel feature of this method is that it uses the person's individual characteristics (i.e., their gait features in the frequency domain) and the temporal consistency of local appearance to track each individual in a crowd. Previous experiments have indicated that the use of these combined features contributes to a significant performance improvement for tracking people in crowded scenes. New motion descriptors have been proposed in several previous studies [5,17,23] for improving gait recognition based on DT-based pyramidal representation. The person region is divided into sub-regions and the local features are extracted then combined into a single high-level gait descriptor using FV encoding [27]. Recently, attribute-based methods have demonstrated the usefulness of object attributes for high-level semantic description, functioning well for action recognition and zero-shot learning [1,16,25]. Another previously method [7] shifted the goal of object recognition to description, by describing objects and learning from these descriptions. The researchers claimed that if the object's attributes were adopted as the anchor representation for object recognition systems, more details about an object than just its name could be revealed. The direct attribute prediction (DAP) model was introduced in a previous study [13] to predict the presence of each attribute, for training object models. The relative attribute concept was later proposed to semantically describe the relationships between objects based on a set of defined attributes [25].

---

[1] Throughout the manuscript, we will refer to the dense trajectory as DT, gait energy image as GEI, histogram of optical flow as HOF, and Fisher vector as FV.

Fig. 2. (a) Sampled frames from the walking sequence, (b) Collected DTs from the video sequence, (c) Location probability map for trajectories distribution, and (d) Motion magnitude probability map.

3 The Proposed Approach

In the following subsections, we explain the proposed DT-based gait motion representation method in detail. In addition, we describe the learning of the rank model based on the proposed gait relative attribute annotations.

3.1 Dense Trajectory Extraction from Gait Videos

Local motion feature descriptors have recently been utilized in the field of action recognition [22, 37]. These descriptors are primarily built on extracted short fragment trajectories of the tracked sampled points. The descriptors can often be extracted directly from the original frames without pure foreground segmentation as shown in Fig. 2(b). These trajectories best describe the actual spatial-temporal dynamics of the moving objects and can represent their kinematics appropriately (see Fig. 2(c, d)). These local feature descriptors have been utilized in gait recognition systems [5, 23, 38], providing marked performance enhancements in gait analysis/recognition tasks. In our work, we extracted a set of DTs from walking videos using a state of the art DT extractor method [37]. We used the extracted DTs to build a global gait motion descriptor for the gait styles of walking subjects.

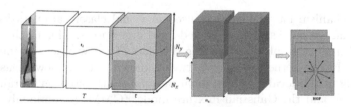

Fig. 3. HOF descriptor computation over $N_x \times N_y \times T$ spatial-temporal volume along trajectory t_i. $N_x \times N_y$ is local neighborhood pixels and T is the temporal length. The volume was divided into three main blocks ($N_x \times N_y \times t$). To capture the actual motion dynamics, every block was subdivided into four ($n_x \times n_y \times t$) sub-blocks. For each sub-block, the HOF descriptor was computed. The global HOF of the trajectory was the concatenation of the HOFs resulting from sub-blocks.

3.2 Dense Trajectory-Based HOF Descriptor

HOF descriptors have become widely used in many recognition and video classification approaches [15,35–37]. To encode the the motion information of DTs, we propose computation of the HOF descriptors from the 3D spatial-temporal volume around the DTs. As shown in Fig. 3, we assign $N_x \times N_y \times T$ volume along the trajectory where $N_x \times N_y$ is the spatial dimension and T is temporal length the of the DT. We extracted a $N_x \times N_y$ spatial window around each xy point lying on the DT, and considered this point as the window anchor. These extracted windows were stacked to form the 3D volume. It should be noted that these extraction criteria enable us to properly embed the structure information of the DT. The optical flow displacement vectors in the horizontal and vertical directions were then computed for each 3D volume using the classical Farneback optical flow method [8]. Furthermore, the average velocity was computed for each trajectory and subtracted from displacement vectors to suppress fast irregular motions. Because both of the average speed-subtracted displacement vectors were 2-dimensional vector fields per frame, we were able to use them to compute the corresponding motion magnitudes and orientations. For robust motion encoding, we divided the 3D volume into three main blocks $(N_x \times N_y \times t)$ and every main block was, in turn, subdivided into four $(n_x \times n_y \times t)$ sub-blocks. Note that $t = T/3$ and $n_x = N_x/2$, and these values were experimentally selected for best performance. Now the number of sub-blocks in the entire 3D volume was 12 sub-blocks[2], and we computed a HOF histogram for each sub-block. Given the obtained magnitude and orientation grids, the magnitude was quantized in nine-bin orientations using the full orientations and aggregated over every sub-block in both the spatial and temporal directions as shown in Fig. 3. We concatenated all of the adjacent sub-blocks responses into one global HOF descriptor. We normalize this global descriptor using its **L1**-norm followed by the signed square root [36]. For now, each DT is represented by its corresponding global 108-dimensional HOF descriptor. As a result, the motion information of the entire walking video was encoded into a group of global HOF descriptors.

3.3 Fisher Vector Encoding for Gait Motion

The FV mechanism has been widely used in visual classification tasks to represent an image/scene as a set of pooled local descriptors. This method follows the Fisher kernel principle of estimating the class of a new object by minimizing the average Fisher kernel distance between known and unknown classes [10]. FV encoding of the local D-dimensional descriptors is based on a trained generative model, and the Gaussian mixture model (GMM) is commonly used for this purpose. Let $H = (h_1, \ldots, h_N)$ be a set of D-dimensional HOF descriptors extracted from the walking video, with one HOF for each DT. Suppose that the K-mixtures GMM were considered to model the generative process of these

---

[2] Three main blocks and each one was subdivided into four sub-blocks. Each sub-block had dimensions of $8 \times 8 \times 5$. The HOF descriptor for each trajectory had a dimension of $12 \times 9 = 108$.

local descriptors. Thus, we estimated the GMM parameters by fitting them to H descriptors. Given this K-mixture GMM, where each mixture has its mixture weight, mean, and covariance parameters respectively, (π_k, μ_k, Σ_k). Let Θ denote the vector of Q parameters of the K-mixture GMM, $[\theta_1, \ldots, \theta_Q]^T \in \mathbb{R}^Q$ where $Q = (2D+1)K - 1$. Given each θ expressed as (π_k, μ_k, Σ_k), the parameter vector Θ can be represented as follows:

$$
\begin{aligned}
\Theta &= (\pi_1, \ldots, \pi_K; \mu_1, \ldots, \mu_K; \Sigma_1, \ldots, \Sigma_K) \\
&= (\theta_1, \ldots, \theta_Q).
\end{aligned}
\tag{1}
$$

Following a previous study [26], the gradient vector of D-dimensional HOFs descriptors w.r.t. the parameters of GMM can be written as:

$$
\begin{aligned}
&\nabla_\theta \log f_\theta(H|\Theta) \\
&\qquad = \left(\frac{\partial \log f(H|\Theta)}{\partial \theta_1}, \ldots, \frac{\partial \log f(H|\Theta)}{\partial \theta_Q} \right)^T.
\end{aligned}
\tag{2}
$$

Note that f is the probability density function of the GMM, which models the generative process of H, and is expressed as follows

$$
f(h|\Theta) = \sum_{k=1}^{K} \pi_k \exp \left(-\frac{1}{2}(h - \mu_k)^T \Sigma_k^{-1} (h - \mu_k) \right).
\tag{3}
$$

The function $\log f(H|\Theta)$ in Eq. (2) is the log-likelihood of the density function of the GMM and can be rewritten as $\log f(H|\Theta) = \sum_{n=1}^{N} f(h_n|\Theta)$. FV is approximately the concatenation of the gradient results from Eq. (2) for $k = 1, \ldots, K$. Finally, the FV is normalized using power normalization criteria [26]. For the detailed derivation of FV encoding, please see [26]. In the current study, we leveraged this method to encode the gait motion information from the walking video into a single global descriptor.

4   Learning Gait Relative Attributes

In the field of object recognition research, visual properties of objects have become widely used. These properties provide high-level semantic descriptions of the objects and their explicit and implicit features. To learn such attributes from a given training sample, researchers have used two approaches. The first approach involves learning the mapping of the examples to an ordered set of numerical ranks [30]. The other approach involves learning a set of ranking functions from relative ranking preferences between example pairs [25]. In our proposed method, we adopted the second approach. Given the computed global FV descriptors from the training gait videos, we sought to learn a set of ranking functions from these descriptors, then used these ranking function scores to predict the relative attributes of unseen test samples. To learn these ranking functions, we used our novel human perception-based pairwise gait relative attribute annotations to build the ordered and un-ordered pair matrices. We describe the proposed annotation method in detail in Sect. 4.1. Given the walking subject pairs $(A, B) = \{(a_1, b_1), \ldots, (a_z, b_z)\}$, $z = 1, \ldots, M$ and the corresponding relative attribute annotations. For each annotated attribute l_p, we selected all subjects' pairs with $l_p = 1$ or -1, and adopted them as ordered pairs. We then built the ordered pairs sparse matrix O with dimensions of $I \times n$, where I was the total number of ordered pairs and n was the number of subjects. This matrix has a row for each subject pair and each row only contains single 1 and single 1. For instance, if l_p strength in subject a is greater than in b (i.e., $a \succ b$), then the row i contains $O(i, a) = 1$ and $O(i, b) = -1$ respectively. Similarly, we selected all of the subject pairs that have $l_p = 0$ to build the un-ordered pairs sparse matrix $\{U_p\}$, with dimensions of $J \times n$. If both subjects a and b had similar attribute strength, then $U(j, a) = 1$ and $U(j, b) = -1$. Following the ranking process in [25], we denoted a set of ranking functions to be learned for each of the annotated attributes. This ranking function can be defined as $r_p(h_i) = \boldsymbol{w}_p^T \boldsymbol{h}_i$ under conditions satisfying the following constraints:

$$\forall (a, b) \in O_p : \boldsymbol{w}_p^T \boldsymbol{h}_a > \boldsymbol{w}_p^T \boldsymbol{h}_b$$
$$\forall (a, b) \in U_p : \boldsymbol{w}_p^T \boldsymbol{h}_a = \boldsymbol{w}_p^T \boldsymbol{h}_b. \tag{4}$$

The ranking function is typically assumed to be a linear function [6]. Because of this, it appears in the form $\boldsymbol{w}_p^T \boldsymbol{h}_i$ where $\boldsymbol{w}_p$ is the zero-shot weight vector produced by the Rank-SVM classifier [25].

Table 1. Sample from our proposed gait relative attribute annotations based on the subjective perception of the annotators. $a \succ b$ indicates that the strength of an attribute for subject a was greater than in subject b, and $a \prec b$ indicates the opposite. In addition, $a \sim b$ indicates that both subjects have the same attribute strength, while 1 indicates that $a \succ b$, -1 indicates that $a \prec b$, and 0 indicates that $a \sim b$.

Subjects pairs	Annotators	General goodness	Stately	Cool	Relaxed	Arm swing	Step length	Walking speed	Spine
28 ⇔ 30	Annotator X	−1	1	−1	−1	1	−1	−1	0
	Annotator Y	−1	−1	−1	1	1	0	0	0
	Annotator Z	−1	−1	0	0	−1	−1	0	0
1423 ⇔ 1479	Annotator X	−1	−1	0	1	−1	0	−1	−1
	Annotator Y	1	1	0	0	−1	1	−1	0
	Annotator Z	0	−1	1	0	−1	0	0	−1

4.1   Gait Attribute Annotation

To prepare the annotations of gait relative attributes, we selected 1,200 subjects' walking sequences from the publicly available OULP-Age dataset [40]. These sequences were presented to seven annotators. Each annotator was instructed to watch binary silhouette sequences for 1,200 subject pairs and reported their observations for each pair based on the annotators' subjective perception. We defined eight gait attributes using the terms {*General goodness, Stately, Cool, Relaxed, Arm swing, Walking speed, Step length, Spine*}. Each of these attributes describes a visual property of the walking subject and could receive three different labels, $\{1, 0, -1\}$. The annotator was instructed to assign 1 to the attribute if they observed that the strength of the attribute in the gait style of subject a was greater than that for subject b. Similarly, the annotator was instructed to assign -1 if b had a greater attribute strength than a. If the annotator judged that both subjects had the same strength of the attribute, they assigned the label 0. In Table 1, the relative attribute annotations for two subject pairs are shown for illustration. For each pair, the relative attribute labels were listed for three different annotators. It can be clearly seen that for the same gait attribute, different annotators assigned different/same labels according to their subjective perception. For instance, for subject pair (28, 30), annotators X and Y observed that *Arm swing* in 28 was better than in 30 (i.e., assign 1). However, annotator Z observed the opposite (i.e., assign -1). In addition, all annotators agreed that both 28 and 30 had the same *Spine* (i.e., assign 0).

4.2 Evaluation Criteria

To evaluate the predicted attribute labels against the ground truth annotation, we derived new multi-threshold evaluation criteria. It should be noted that the learned ranking functions produced real-valued ranks. We used these ranks to evaluate the decision of the classifier. Given the test descriptor h_z and the weights vector $w \in \mathbb{R}^D$, which were obtained from the R-SVM training stage, we computed the predicted scores using the inner-product, $w_p^T h_z$. Suppose a set of s thresholds $L = (L_1, \ldots, L_s)$ and $L_1 < L_2 < \cdots < L_s$. The real-valued ranks had one score for each subject (i.e., h_z). We computed the difference scores for each pair then thresholded them for mapping to predicted attribute labels; for instance, the score difference between the ranking function scores of subjects a and b is $d_{ab} = r_p(h_a) - r_p(h_b)$. Because it was necessary to predict three labels $(1, 0, -1)$, we derived the following multi-threshold evaluation criteria

$$l_p = \begin{cases} 1 & \forall \quad d_{ab} > L_i, \quad i = 1, \ldots, s \\ 0 & \forall \quad d_{ab} \quad \nexists \quad d_{ab} > L_i \quad \wedge \quad d_{ab} < -L_i \\ -1 & \forall \quad d_{ab} < -L_i \end{cases} \tag{5}$$

In Eq. (5), the first condition was used for attribute assignment 1, the second condition for 0 and the third for -1. For instance, the attribute l_p for a specific subject pair will be 1, if the score difference for subject pairs (a, b) is greater than L_i and, hence, we predict that $a \succ b$ for that attribute. Given the computed score differences, we mapped them to attribute assignments using Eq. (5) and counted the mapped assignments for each threshold. To evaluate the predicted attributes against the GT annotation, we computed the accuracy precision measure for each threshold and adopted the highest precision value.

5 Experiments

We evaluated our proposed approach using a selective gait dataset from the publicly available OULP-Age gait dataset [40]. We carefully select 1,200 subjects, all aged in their thirties. Each subject performed a complete walking sequence under a laboratory environment. We extracted the binary silhouette sequences for all subjects. We then hired seven annotators to watch the binary sequences for each pair of subjects and recorded their observations, as shown in Table 1 in terms of attribute assignment (1,0, or −1). For each gait video, we extracted DTs, and for each DT, we assigned 3D volume around it. The 3D volume had a spatial-temporal dimension $16 \times 16 \times 15$. We divided the volume into 12 sub-blocks, and each sub-block had dimensions of $8 \times 8 \times 5$. For each sub-block, we computed the mean-speed subtracted HOF (nine-bin orientation quantization) then normalized it by its L1-norm followed by a square root. All the sub-blocks HOFs were then concatenated into single DT-HOF. Thus, we obtained a DT-HOF descriptor for each DT. As the number of extracted DTs-HOFs was large for each gait video, we adopted the Fisher encoding mechanism to encode the motion information of the entire video into a single global feature vector. Following the formulation in Sect. 3.3, we selected the corresponding DTs-HOFs of 800 subjects from the dataset for training a GMM with 256 clusters. We then used the estimated parameters to compute the FVs for all 1,200 subjects, as described above. Thus, we obtained a single FV for each video. The FV only encoded the motion information for each subject (i.e., we did not yet consider the appearance information).

5.1 Zero-Shot Learning Results

For training, we used the Rank-SVM classifier, which was adapted from that used by Parikh et al. [25], to handle relative attributes instead of binary labels. We completely disjointed the dataset into 1,000 FVs for training and 200 FVs for testing. We used the training set to build the ordered and un-ordered pairs of sparse matrices, which were needed for the classifier. The output of the classifier is the zero-shot weights vector. We then used these estimated weights to compute a set of ranking functions for training and testing (totally unseen) samples using Eq. (4). Finally, we used the resulting real-valued ranks for evaluating the classifier against the GT annotation using the evaluation criteria proposed in Sect. 4.2. Note that, we assume multi-threshold evaluation and examined the performance of the proposed method under individual annotations. We used the seven gait attribute annotations individually and measured the classification accuracy of the predicted attributes for each annotator. The quantitative results are listed in Table 2. Several key observations can be made from the results. First, as the pairwise gait attribute annotations depend on human perception, different annotators may report different labels of the same attribute. This, in turn, influences the estimation process. The proposed method adapts appropriately to the subjective perception of the annotators. Second, the experimental results revealed

Table 2. The attribute estimation accuracies for seven different annotators. Both the training and testing phases accuracies are reported. The average accuracy for each annotation is shown in the rightmost column.

Subjects pairs	Annotators	General goodness	Stately	Cool	Relax	Arm swing	Step length	Walking speed	Spine	Average accuracy (%)
1,200 Training	Annotator 1	66	78	95	75	73	81	73	75	77
	Annotator 2	71	72	73	75	77	76	72	72	73
	Annotator 3	83	70	68	75	83	80	74	82	77
	Annotator 4	75	76	71	74	76	80	68	75	74
	Annotator 5	72	65	68	62	75	79	75	75	71
	Annotator 6	77	78	68	69	79	74	77	74	75
	Annotator 7	70	68	74	80	67	66	75	70	71
Testing	Annotator 1	56	73	85	68	69	82	67	77	72
	Annotator 2	55	49	64	70	62	81	72	48	63
	Annotator 3	65	57	42	71	81	74	72	64	66
	Annotator 4	54	51	52	72	51	61	60	54	57
	Annotator 5	49	44	61	52	63	64	67	48	56
	Annotator 6	73	71	49	62	64	69	66	58	64
	Annotator 7	62	57	57	74	57	54	59	52	59

that the gait style had key feature attributes that influenced the annotators' decisions. Table 2 shows higher accuracy rates for all annotators at *arm swing*, *step length*, and *walking speed* attributes. This means that although each annotator performed the annotation separately without bias, they largely agreed that these attributes were strongly present in subjects' gait styles. This observation may lead gait researchers to closely examine walking patterns and focus on the discriminative features that best describe the gait style. At a high semantic level of analysis, the social relationships between subjects could be investigated based on estimated relative attributes. Moreover, the subjective perception of the annotators can be inferred. To measure the robustness of our DT-based representation for gait attribute estimation, we compared it against the GEI-based deep feature representation under the same classifier and evaluation criteria. We used VGG16 deep architecture [31], which produced a 4,096-dimensional feature descriptor from the walking person's GEI. It should be noted that GEI encodes both the appearance and motion information of the walking subject, in contrast to our representation, which carries only motion information. The evaluation results based on the GEI-based feature representation are listed in Table 3 in the first and third columns. It can be seen that the GEI-based representation slightly outperformed our representation at the training stage. In contrast, in the testing stage, our DT-based representation outperforms the GEI-based representation by most accuracy measures, with marked performance enhancements. Sample results based on our proposed representation are shown in Table 3 in the second and fourth rows (DTs features) and the average accuracy for all attributes appears in the rightmost column. In the current experiment, we concatenated all seven annotations to produce 8,389 subject pairs after excluding the unrecognizable pairs.

Table 3. The accuracy of gait attribute estimation based on our proposed DT-based representation versus GEI-deep based representation.

Attributes		General goodness	Stately	Cool	Relax	Arm swing	Step length	Walking speed	Spine	Average accuracy (%)
Training	GEI features	**66.08**	**62.36**	**62.72**	**60.93**	**62.93**	**64.17**	**61.50**	**68.46**	**63.56**
	DTs features	63.62	60.28	60.37	58.32	61.89	63.27	60.41	65.20	61.67
Testing	GEI features	**55.89**	**48.08**	52.86	47.92	46.01	58.59	47.61	50	50.87
	DTs features	55.41	47.61	**57.96**	**56.84**	**58.75**	**61.30**	**61.78**	48.56	**56.03**

6   Conclusion

In the current study, we utilized the concept of learning with ranking for human gait attribute estimation, for the first time. Instead of using the traditional binary classification problem, the model directly learned a set of ranking functions based on the preference relationships between walking subject pairs. We considered each walking subject to have a gait style that could be represented by a set of relative attributes. We proposed novel pairwise gait attribute annotations for 1,200 walking subjects based on human perception. In the current method, we used encoding of only the gait motion information of subjects, based on DT-based representation. We built global gait motion descriptors for walking subjects based on the HOF descriptors of the extracted DTs. For gait attribute prediction, the model learned a set of ranking functions from training samples of the proposed annotations. The initial results revealed the robustness of the proposed DT-based gait motion representation compared with the GEI-deep feature representation which encodes both appearance and motion information. The experimental results indicated that the proposed method was able to represent gait attributes well, and that the proposed gait motion descriptors had better generalization capability than GEI for the gait attribute estimation task.

Acknowledgment. This work was supported by JSPS Grants-in-Aid for Scientific Research (A) JP18H04115. We thank Benjamin Knight, MSc., from Edanz Group (www.edanzediting.com/ac) for editing a draft of this manuscript.

References

1. Akae, N., Mansur, A., Makihara, Y., Yagi, Y.: Video from nearly still: an application to low frame-rate gait recognition. In: 2012 IEEE Conference on Computer Vision and Pattern Recognition (CVPR), pp. 1537–1543. IEEE (2012)
2. Bobick, A., Johnson, A.: Gait recognition using static activity-specific parameters. In: Proceedings of the 14th IEEE Conference on Computer Vision and Pattern Recognition, vol. 1, pp. 423–430 (2001)
3. Bouchrika, I., Nixon, M.S.: Model-based feature extraction for gait analysis and recognition. In: Gagalowicz, A., Philips, W. (eds.) MIRAGE 2007. LNCS, vol. 4418, pp. 150–160. Springer, Heidelberg (2007). https://doi.org/10.1007/978-3-540-71457-6_14

4. Castro, F.M., Marín-Jiménez, M.J., Mata, N.G., Muñoz-Salinas, R.: Fisher motion descriptor for multiview gait recognition. Int. J. Pattern Recognit. Artif. Intell. **12**(7), 756–763 (2017)
5. Castro, F.M., Marín-Jimenez, M.J., Medina-Carnicer, R.: Pyramidal fisher motion for multiview gait recognition. In: 2014 22nd International Conference on Pattern Recognition (ICPR), pp. 1692–1697. IEEE (2014)
6. Crammer, K., Singer, Y.: Pranking with ranking. In: Advances in Neural Information Processing Systems, pp. 641–647 (2002)
7. Farhadi, A., Endres, I., Hoiem, D., Forsyth, D.: Describing objects by their attributes. In: 2009 IEEE Conference on Computer Vision and Pattern Recognition, pp. 1778–1785. IEEE (2009)
8. Farnebäck, G.: Two-frame motion estimation based on polynomial expansion. In: Bigun, J., Gustavsson, T. (eds.) SCIA 2003. LNCS, vol. 2749, pp. 363–370. Springer, Heidelberg (2003). https://doi.org/10.1007/3-540-45103-X_50
9. Han, J., Bhanu, B.: Individual recognition using gait energy image. IEEE Trans. Pattern Anal. Mach. Intell. **28**(2), 316–322 (2006)
10. Jaakkola, T., Haussler, D.: Exploiting generative models in discriminative classifiers. In: Advances in Neural Information Processing Systems, pp. 487–493 (1999)
11. Kusakunniran, W.: Attribute-based learning for gait recognition using spatio-temporal interest points. Image Vis. Comput. **32**(12), 1117–1126 (2014)
12. Kusakunniran, W.: Recognizing gaits on spatio-temporal feature domain. IEEE Trans. Inf. Forensics Secur. **9**(9), 1416–1423 (2014)
13. Lampert, C.H., Nickisch, H., Harmeling, S.: Learning to detect unseen object classes by between-class attribute transfer. In: 2009 IEEE Conference on Computer Vision and Pattern Recognition, pp. 951–958. IEEE (2009)
14. Laptev, I.: On space-time interest points. Int. J. Comput. Vision **64**(2–3), 107–123 (2005)
15. Laptev, I., Marszałek, M., Schmid, C., Rozenfeld, B.: Learning realistic human actions from movies (2008)
16. Liu, T.Y., et al.: Learning to rank for information retrieval. Found. Trends® Inf. Retr. **3**(3), 225–331 (2009)
17. López-Fernández, D., Madrid-Cuevas, F.J., Carmona-Poyato, Á., Marín-Jiménez, M.J., Muñoz-Salinas, R.: The AVA multi-view dataset for gait recognition. In: Mazzeo, P.L., Spagnolo, P., Moeslund, T.B. (eds.) AMMDS 2014. LNCS, vol. 8703, pp. 26–39. Springer, Cham (2014). https://doi.org/10.1007/978-3-319-13323-2_3
18. Makihara, Y., Sagawa, R., Mukaigawa, Y., Echigo, T., Yagi, Y.: Gait recognition using a view transformation model in the frequency domain. In: Leonardis, A., Bischof, H., Pinz, A. (eds.) ECCV 2006. LNCS, vol. 3953, pp. 151–163. Springer, Heidelberg (2006). https://doi.org/10.1007/11744078_12
19. Makihara, Y., Mannami, H., Yagi, Y.: Gait analysis of gender and age using a large-scale multi-view gait database. In: Kimmel, R., Klette, R., Sugimoto, A. (eds.) ACCV 2010. LNCS, vol. 6493, pp. 440–451. Springer, Heidelberg (2011). https://doi.org/10.1007/978-3-642-19309-5_34
20. Makihara, Y., Matovski, D.S., Nixon, M.S., Carter, J.N., Yagi, Y.: Gait recognition: databases, representations, and applications, pp. 1–15. Wiley (1999). https://doi.org/10.1002/047134608X.W8261
21. Makihara, Y., Okumura, M., Iwama, H., Yagi, Y.: Gait-based age estimation using a whole-generation gait database. In: 2011 International Joint Conference on Biometrics (IJCB), pp. 1–6. IEEE (2011)
22. Marín-Jiménez, M.J., de la Blanca, N.P., Mendoza, M.A.: Human action recognition from simple feature pooling. Pattern Anal. Appl. **17**(1), 17–36 (2014)

23. Marín-Jiménez, M.J., Castro, F.M., Carmona-Poyato, Á., Guil, N.: On how to improve tracklet-based gait recognition systems. Pattern Recogn. Lett. **68**, 103–110 (2015)

24. Nordin, M., Saadoon, A.: A survey of gait recognition based on skeleton mode l for human identification. Res. J. Appl. Sci. Eng. Technol. (2016)

25. Parikh, D., Grauman, K.: Relative attributes. In: 2011 International Conference on Computer Vision, pp. 503–510. IEEE (2011)

26. Perronnin, F., Dance, C.: Fisher kernels on visual vocabularies for image categorization. In: 2007 IEEE Conference on Computer Vision and Pattern Recognition, pp. 1–8. IEEE (2007)

27. Perronnin, F., Sánchez, J., Mensink, T.: Improving the fisher kernel for large-scale image classification. In: Daniilidis, K., Maragos, P., Paragios, N. (eds.) ECCV 2010. LNCS, vol. 6314, pp. 143–156. Springer, Heidelberg (2010)

28. Rida, I., Almaadeed, N., Almaadeed, S.: Robust gait recognition: a comprehensive survey. IET Biom. **8**(1), 14–28 (2019). https://doi.org/10.1049/iet-bmt.2018.5063

29. Rida, I., Al Maadeed, N., Marcialis, G.L., Bouridane, A., Herault, R., Gasso, G.: Improved model-free gait recognition based on human body part. In: Jiang, R., Almaadeed, S., Bouridane, A., Crookes, D., Beghdadi, A. (eds.) Biometric Security and Privacy. SPST, pp. 141–161. Springer, Cham (2017). https://doi.org/10.1007/978-3-319-47301-7_6

30. Shashua, A., Levin, A.: Ranking with large margin principle: two approaches. In: Advances in Neural Information Processing Systems, pp. 961–968 (2003)

31. Simonyan, K., Zisserman, A.: Very deep convolutional networks for large-scale image recognition. arXiv preprint arXiv:1409.1556 (2014)

32. Sugimura, D., Kitani, K.M., Okabe, T., Sato, Y., Sugimoto, A.: Using individuality to track individuals: clustering individual trajectories in crowds using local appearance and frequency trait. In: 2009 IEEE 12th International Conference on Computer Vision, pp. 1467–1474. IEEE (2009)

33. Takemura, N., Makihara, Y., Muramatsu, D., Echigo, T., Yagi, Y.: On input/output architectures for convolutional neural network-based cross-view gait recognition. IEEE Trans. Circuits Syst. Video Technol. **PP**(99), 1 (2017). https://doi.org/10.1109/TCSVT.2017.2760835

34. Troje, N.F.: Decomposing biological motion: a framework for analysis and synthesis of human gait patterns. J. Vis. **2**(5), 2–2 (2002)

35. Uijlings, J., Duta, I.C., Sangineto, E., Sebe, N.: Video classification with densely extracted HOG/HOF/MBH features: an evaluation of the accuracy/computational efficiency trade-off. Int. J. Multimed. Inf. Retr. **4**(1), 33–44 (2015)

36. Uijlings, J.R., Duta, I.C., Rostamzadeh, N., Sebe, N.: Realtime video classification using dense HOF/HOG. In: Proceedings of International Conference on Multimedia Retrieval, p. 145. ACM (2014)

37. Wang, H., Kläser, A., Schmid, C., Cheng-Lin, L.: Action recognition by dense trajectories (2011)

38. Weng, J., Lu, W., Xu, J., et al.: Multi-gait recognition based on attribute discovery. IEEE Trans. Pattern Anal. Mach. Intell. **40**, 1697–1710 (2017)

39. Wu, Z., Huang, Y., Wang, L., Wang, X., Tan, T.: A comprehensive study on cross-view gait based human identification with deep CNNs. IEEE Trans. Pattern Anal. Mach. Intell. **PP**(99), 1 (2016). https://doi.org/10.1109/TPAMI.2016.2545669

40. Xu, C., Makihara, Y., Ogi, G., Li, X., Yagi, Y., Lu, J.: The OU-ISIR gait database comprising the large population dataset with age and performance evaluation of age estimation. IPSJ Trans. Comput. Vis. Appl. **9**(1), 24 (2017)

41. Yan, C., Zhang, B., Coenen, F.: Multi-attributes gait identification by convolutional neural networks. In: 2015 8th International Congress on Image and Signal Processing (CISP), pp. 642–647. IEEE (2015)
42. Zhang, Z., Wang, C., Xiao, B., Zhou, W., Liu, S.: Robust relative attributes for human action recognition. Pattern Anal. Appl. **18**(1), 157–171 (2015)

Scene-Adaptive Driving Area Prediction Based on Automatic Label Acquisition from Driving Information

Takuya Migishima[1(✉)], Haruya Kyutoku[2], Daisuke Deguchi[1],
Yasutomo Kawanishi[1], Ichiro Ide[1], and Hiroshi Murase[1(✉)]

[1] Nagoya University, Nagoya, Aichi, Japan
migishimat@murase.is.i.nagoya-u.ac.jp, murase@nagoya-u.jp
[2] Toyota Technological Institute, Nagoya, Aichi, Japan

Abstract. Technology for autonomous vehicles has attracted much attention for reducing traffic accidents, and the demand for its realization is increasing year-by-year. For safety driving on urban roads by an autonomous vehicle, it is indispensable to predict an appropriate driving path even if various objects exist in the environment. For predicting the appropriate driving path, it is necessary to recognize the surrounding environment. Semantic segmentation is widely studied as one of the surrounding environment recognition methods and has been utilized for drivable area prediction. However, the driver's operation, that is important for predicting the preferred drivable area (scene-adaptive driving area), is not considered in these methods. In addition, it is important to consider the movement of surrounding dynamic objects for predicting the scene-adaptive driving area. In this paper, we propose an automatic label assignment method from actual driving information, and scene-adaptive driving area prediction method using semantic segmentation and Convolutional LSTM (Long Short-Term Memory). Experiments on actual driving information demonstrate that the proposed methods could both acquire the labels automatically and predict the scene-adaptive driving area successfully.

Keywords: Semantic segmentation · Path prediction · Autonomous vehicle

1 Introduction

Technology for autonomous vehicle has attracted much attention for reducing traffic accidents, and the demand for its realization is increasing year-by-year. Although some automotive manufacturers are already providing autonomous driving functions on expressways, it is still difficult to provide such functions on urban roads due to the diversity of the surrounding environment. To drive safely on urban roads, it is indispensable to predict an appropriate driving path even if various objects exist in the environment. Here, the path can be considered

© Springer Nature Switzerland AG 2020
S. Palaiahnakote et al. (Eds.): ACPR 2019, LNCS 12047, pp. 106–117, 2020.
https://doi.org/10.1007/978-3-030-41299-9_9

(a) In-vehicle camera image.

(b) Example of drivable area.   (c) Example of preferred drivable area.

Fig. 1. Difference between ordinary path and scene-adaptive path. (Color figure online)

as a trajectory of future vehicle positions, which should be predicted appropriately since it will affect vehicle control during autonomous driving. Since vehicle driving paths heavily depend on the environment, it is necessary to recognize the surrounding objects, the situation of pedestrians, and other vehicles in the environment in detail for path prediction. Therefore, the technology to predict a future vehicle path is strongly needed especially for autonomous driving in urban environment.

On the other hand, semantic segmentation, which is a task to predict object labels pixel-by-pixel in an image, is widely studied as one of the surrounding environment recognition methods [2,7]. Barnes et al. [1] and Zhou et al. [8] tried to extend the problem of semantic segmentation to that of predicting the drivable area that cannot be observed as image features. They constructed the semantic segmentation model from the training data generated by projecting the trajectory of vehicle positions onto the road surface. For example, the green area in Fig. 1(b) indicates the drivable area label that is used as a ground-truth by Zhou's method. As shown in Fig. 1(a), which is the original image corresponding to Fig. 1(b), there is a stop sign on the path. However, as shown in Fig. 1(b), Zhou's method does not consider this sign and thus the necessity of braking. In addition, although the drivable area should be adaptively changed by considering the context of surrounding dynamic objects (e.g. pedestrians, vehicles, etc.), Zhou's method does not consider the necessity of braking against those objects. Thus, we tackle the problem of predicting the preferred drivable area considering the safety required for automated driving and the movement

Fig. 2. Example of a scene containing oncoming vehicles. The yellow rectangle indicates the oncoming vehicles. (Color figure online)

of surrounding dynamic objects. Figure 1(c) shows an example of the preferred drivable area in the context that can follow traffic rules around the intersection. Since the preferred drivable area should adapt to the environment, we call this as the "scene-adaptive driving area", and propose a method for its prediction in this paper.

To predict the scene-adaptive driving area accurately, it is necessary to solve the following two issues: (i) Generation of the ground-truth labelling of the scene-adaptive driving area, (ii) How to handle the object movement in the vehicle front. As described above, since the scene-adaptive driving area should end before stop signs or other objects, it is necessary to prepare training data satisfying this requirement for training a semantic segmentation model that can predict a scene-adaptive driving area. On the other hand, the scene-adaptive driving area should change dynamically due to the existence and the movement of oncoming vehicles, pedestrians, and other objects. Figure 2 shows an example of a scene containing an oncoming vehicles. In this situation, we cannot cross the road safely without considering the moving state of the oncoming vehicles indicated by the yellow rectangle in the image. That is, it is necessary to consider the movement of surrounding objects for predicting an appropriate scene-adaptive driving area.

To tackle the first issue, we refer to the vehicle speed to generate the training data. If a driver operates the brake pedal, we can assume that the main cause for that should exist in the vehicle front. However, there are possibilities such as the existence of blind intersections, traffic signs, or red traffic signals. From these points-of-view, we try to generate appropriate labels for scene-adaptive driving area automatically by referring to the reduction of the own vehicle speed as a key.

To tackle the second issue, we introduce feature learning from frame sequences. In a deep learning framework, LSTM (Long Short-Term Memory), which is a variant of RNN (Recursive Neural Network), is one of the popular techniques to handle frame sequences. Generally speaking, LSTM can be used to learn sequential (temporal) information, but it loses spatial information. To pre-

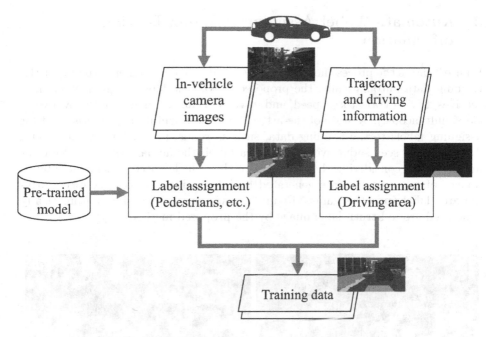

Fig. 3. Process flow of the automatic label assignment.

serve the spatial information in LSTM, we use ConvLSTM (Convolutional Long Short-Term Memory) proposed by Shi et al. [6]. The ConvLSTM is a network where fully connected layers in LSTM are replaced with convolutional layers. It can be applied to the semantic segmentation task for predicting labels of a future frame [4]. The proposed method incorporates ConvLSTM in the prediction of the scene-adaptive driving area to learn the movement of other objects.

Based on the above concept, we propose a method to automatically acquire training data from actual driving information and predict a scene-adaptive driving area. The main contributions of this paper can be summarized as follows:

- Proposal of the concept of "scene-adaptive driving area" considering the necessity of braking.
- Automatic label assignment for training a model to predict a scene-adaptive driving area referring to the own vehicle's speed.
- Prediction of a scene-adaptive driving area that considers the movement of other objects using ConvLSTM.

In the following, Sect. 2 proposes the automatic label assignment method, Sect. 3 describes the construction of the scene-adaptive driving area predictor using ConvLSTM, Sect. 4 reports the evaluation experiments, and Sect. 5 concludes this paper.

2   Automatic Label Assignment Using Driving Information

Figure 3 shows the process flow of the automatic label assignment. To output the training data automatically, the proposed automatic label assignment method receives driving trajectory, speed, and in-vehicle camera images of the own vehicle simultaneously. A state-of-the-art semantic segmentation model is used for assigning labels to the training data, such as roads, pedestrians, vehicles, etc. To assign the scene-adaptive driving area label, the actual driving trajectories are projected considering the reduction of the vehicle speed. Then, the labeled images of training data are generated by integrating the scene-adaptive driving area label and other labels. Figure 4 shows an example of an automatically generated ground-truth label image by the proposed method.

Fig. 4. Examples of scene-adaptive driving areas.

2.1   Assigning Labels: Pedestrian, Vehicle, Roads, etc.

To train the semantic segmentation model, it is necessary to prepare images and pixel-wise annotation to each of them. Since drivers usually determine their driving path from the relationship between the own vehicle and its surrounding objects (e.g. pedestrians, vehicles, and roads), it is important to recognize those objects and scene-adaptive driving area simultaneously. Here, the labels of pedestrians, vehicles, and roads can be easily and accurately extracted by applying a state-of-the-art semantic segmentation model.

2.2   Assigning Labels: Scene-Adaptive Driving Area

The proposed method assigns the ground-truth label for scene-adaptive driving area from the own vehicle's speed and the actual trajectory of the own vehicle estimated by a LIDAR-based localization method. Here, let X_t be the vehicle center position at time $t \in T$ in the world coordinate system and $F_{t'}$ be a transformation matrix that converts the vehicle position from the world coordinate system to a coordinate system whose origin is $X_{t'}$ (t'-th coordinate system).

Based on these notations, the vehicle position $\widetilde{X}_t$ in the t'-th coordinate system at time t is calculated as

$$\widetilde{X}_t = F_{t'} X_t. \tag{1}$$

By iterating the above transformation until the following conditions (2) are met, a set of vehicle positions $\widetilde{\mathcal{X}}_{t'} = \{\widetilde{X}_{t'}, \widetilde{X}_{t'+1}, ...\}$ in the t'-th coordinate system is obtained.

$$\|X_t - X_{t'}\|_2 > D \quad \text{or} \quad \alpha_t \leq -2 \quad \text{or} \quad v_t = 0, \tag{2}$$

where α_t is the acceleration [m/s^2] at time t, and v_t is the velocity [m/s] at time t.

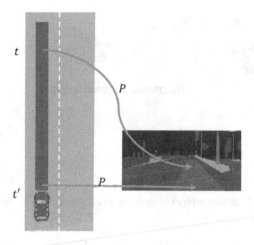

Fig. 5. Projection of vehicle positions onto an image at time t'.

Here, since a scene-adaptive driving area far from the own vehicle cannot be observed in the image plane, the proposed method calculates the distance between the vehicle position $X_{t'}$ and each point X_t, and terminates the transformation process if the distance becomes larger than a threshold D. In addition, it is necessary to consider traffic signs and traffic signals requesting the vehicle to stop for safety. Through observation of actual driving data, we found that these situations are observed with a specific driving behavior like the reduction of the vehicle speed. From these points-of-view, the proposed method terminates the transformation when the vehicle's speed decreases.

Subsequently, the vehicle trajectory is projected onto an image. Figure 5 shows the schematic diagram of the projection. The vehicle trajectory is obtained using the set of vehicle center position $\widetilde{\mathcal{X}}_{t'}$ and vehicle width. Here, the pixel position $\{x'_t, y'_t\}$ on the image is calculated as

$$\begin{bmatrix} x'_t \\ y'_t \\ 1 \end{bmatrix} = P \begin{bmatrix} \widetilde{X}_t \\ 1 \end{bmatrix}, \tag{3}$$

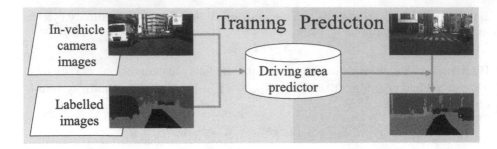

Fig. 6. Overview of the proposed scene-adaptive driving area predictor.

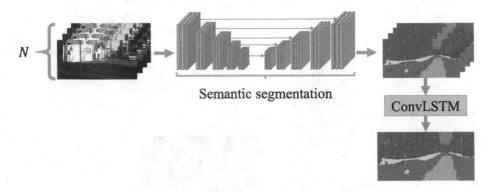

Fig. 7. Scene-adaptive driving area prediction model.

where P is a projection matrix. By filling with the corresponding class label between the line segments that indicate the trajectories of left and right tires, the scene-adaptive driving area is obtained.

3   Scene-Adaptive Driving Area Prediction Model Using Semantic Segmentation and ConvLSTM

We build a scene-adaptive driving area prediction model from the training data acquired by the above procedure. As shown in Fig. 6, the model is trained from pairs of an in-vehicle camera image and a generated label, and assigns label using only from the in-vehicle camera image for prediction. To predict the scene-adaptive driving area considering movement of the surrounding object, ConvLSTM is integrated into a scene-adaptive driving area prediction model as shown in Fig. 7. The proposed method employs a semantic segmentation model inspired by U-Net [5] that is based on an encoder-decoder architecture with skip connections; The encoder extracts the latent features from the in-vehicle camera images. Then, the pixel-wise label likelihoods are estimated by concatenating and restoring the features using the decoder. To obtain the movement of objects, we remove the logits layer in the segmentation model and join the end of the

Fig. 8. Vehicle used for the experiment.

segmentation model to the ConvLSTM model. By applying the proposed model to N in-vehicle camera images, N label likelihoods are obtained. Here, the ConvLSTM model receives N label likelihoods. By learning the movements of other objects from N label likelihoods, we can predict the scene-adaptive driving area accurately.

4   Experiment

We conducted an experiment for evaluating the proposed method. In this experiment, the accuracy of the semantic segmentation results was compared by changing the length of an input sequence N. We evaluated the proposed method based on Intersection over Union (IoU) that is a measure to calculate the overlap of the prediction results and their ground-truth. Here, IoU is calculated by

$$\text{IoU} = \frac{\mathcal{A} \cap \mathcal{B}}{\mathcal{A} \cup \mathcal{B}}, \tag{4}$$

where $\mathcal{A}$ is the ground-truth constructed from the acquired dataset, and $\mathcal{B}$ is the prediction result.

4.1   Dataset

The data were collected by driving around Nagoya Station in Nagoya, Japan, with a special vehicle as shown in Fig. 8. DeepLabv3+ [2] model trained by Cityscapes dataset [3] was utilized to assign the labels other than the driving area. Here, we merged the labels of "car", "truck", "bus" in the Cityscapes dataset into a single "vehicle" label. Finally, the proposed method merged these labels with the label of the scene-adaptive driving area that is automatically acquired from driving information. Figure 9 shows an example of the constructed dataset. In the following evaluations, 4,245 data were used for training, and 1,405 data were used for evaluation.

Table 1. Experiment results (IoU) of driving area prediction for each class.

Labels	$N = 1$	2	3	4	5	6	7
Sky	8.5	12.1	12.8	**14.7**	**14.7**	9.9	14.4
Building	76.4	**77.3**	77.2	77.1	77.1	77.0	76.7
Pole	16.3	19.3	19.9	19.3	16.9	19.2	**21.6**
Road	**76.0**	75.8	75.9	75.7	75.6	75.7	75.7
Terrain	58.9	60.4	**60.8**	60.5	59.9	60.1	60.4
Vegetation	**38.1**	36.0	34.0	33.6	34.3	32.5	35.1
Traffic sign	12.5	12.6	13.2	14.2	13.9	13.4	**14.6**
Fence	17.4	18.9	**19.7**	18.8	18.2	18.5	17.4
Car	57.8	58.9	**59.3**	58.6	58.9	58.5	57.5
Person	9.4	12.7	**13.0**	12.4	11.8	12.8	12.0
Rider	8.9	8.7	8.9	8.6	7.5	8.1	**9.1**
Drivable path	65.1	64.9	65.2	65.2	**65.3**	**65.3**	65.2
mIoU	37.1	38.1	**38.3**	38.2	37.9	37.6	**38.3**

4.2   Results and Discussions

From Fig. 9, we confirmed that the proposed method could acquire training data considering stop signs and other objects.

Table 1 shows the IoU of each class and mean IoU, and the column of $N = 1$ indicates the prediction results by a conventional semantic segmentation model that does not use ConvLSTM. On the other hand, the $N \geq 2$ columns indicate the prediction results by the proposed scene-adaptive driving area prediction model employing ConvLSTM. From Table 1, we confirmed that the IoU of the scene-adaptive driving area and the mIoU improved by increasing the length of an input sequence N. The IoU of the scene-adaptive driving area at $N = 2$ was lower than that at $N = 1$. In the case of $N = 2$, since the frame-interval between the two input images was about 0.125 s, the change of object positions in the surrounding environment was small during this short interval. Therefore, we consider that objects' movements were not trained appropriately because sufficient information was not given. In addition, IoU of scene-adaptive driving area improved from $N = 2$ to $N = 6$, while the prediction result degraded by increasing N to 7. Comparing the experimental results when $N = 6$ and $N = 7$, we can see that the IoU of the person and the vehicle degraded at $N = 7$ in addition to the scene-adaptive driving area. Since the movement of the surrounding environment objects increases as the length of an input sequence increases, as a result, we

(a) In-vehicle camera images.   (b) Generated labels.

Fig. 9. Examples from the dataset.

consider that it could not be learned correctly. This indicates that it is necessary to investigate the optimal value of N. Figure 10 shows an example of prediction results using the proposed model. We confirmed that the proposed predictor can estimate these scene-adaptive driving area with 65.3%. From these results, we confirmed that the proposed method could acquire the labels automatically and predict highly accurate scene-adaptive driving area using the proposed semantic segmentation model incorporating ConvLSTM.

(a) In-vehicle camera images. (b) Ground-truth labelling.   (c) Predicted labelling.

Fig. 10. Examples of prediction results.

5 Conclusion

In this paper, we proposed a method to acquire training labels of scene-adaptive driving area automatically from driving information of the in-vehicle camera image, own vehicle's speed, and location. Here, the training labels are acquired by using the actual trajectory of the own vehicle. According to the own vehicle speed, the proposed method generates the label of the scene-adaptive driving area reflecting the driving context. We also proposed a scene-adaptive driving area predictor based on the semantic segmentation model introducing ConvLSTM trained with the acquired training data.

To evaluate the proposed method, we created 5,650 labels and predicted the scene-adaptive driving area by the proposed method with 65.3%. We also confirmed the effectiveness of considering the movement of the surrounding object.

Future work will include an improvement of scene-adaptive driving area prediction, an enhancement of the network structure, and experiments using a larger dataset including various patterns of the scene-adaptive driving area.

Acknowledgements. Parts of this research were supported by MEXT, Grant-in-Aid for Scientific Research 17H00745, and JST-Mirai Program Grant Number JPMJMI17C6, Japan.

References

1. Barnes, D., Maddern, W., Posner, I.: Find your own way: weakly-supervised segmentation of path proposals for urban autonomy. In: Proceedings of 2017 IEEE International Conference on Robotics and Automation, pp. 203–210 (2017)
2. Chen, L.C., Zhu, Y., Papandreou, G., Schroff, F., Adam, H.: Encoder-decoder with atrous separable convolution for semantic image segmentation. In: Proceedings of 2018 European Conference on Computer Vision, pp. 833–851 (2018)
3. Cordts, M., et al.: The Cityscapes dataset for semantic urban scene understanding. In: Proceedings of 2016 IEEE Conference on Computer Vision and Pattern Recognition, pp. 3213–3223 (2016)
4. Nabavi, S., Rochan, M., Wang, Y.: Future semantic segmentation with convolutional LSTM. In: Proceedings of 2018 British Machine Vision Conference, pp. 137-1–137-12 (2018)
5. Ronneberger, O., Fischer, P., Brox, T.: U-Net: convolutional networks for biomedical image segmentation. In: Navab, N., Hornegger, J., Wells, W.M., Frangi, A.F. (eds.) MICCAI 2015. LNCS, vol. 9351, pp. 234–241. Springer, Cham (2015). https://doi.org/10.1007/978-3-319-24574-4_28
6. Shi, X., Chen, Z., Wang, H., Yeung, D., Wong, W., Woo, W.: Convolutional LSTM network: a machine learning approach for precipitation nowcasting. In: Advances in Neural Information Processing Systems, vol. 28, pp. 802–810 (2015)
7. Zhao, H., Shi, J., Qi, X., Wang, X., Jia, J.: Pyramid scene parsing network. In: Proceedings of 2017 IEEE Conference on Computer Vision and Pattern Recognition, pp. 6230–6239 (2017)
8. Zhou, W., Worrall, S., Zyner, A., Nebot, E.M.: Automated process for incorporating drivable path into real-time semantic segmentation. In: Proceedings of 2018 IEEE International Conference on Robotics and Automation, pp. 1–6 (2018)

Enhancing the Ensemble-Based Scene Character Recognition by Using Classification Likelihood

Fuma Horie[1]([✉]), Hideaki Goto[1,2], and Takuo Suganuma[1,2]

[1] Graduate School of Information Sciences, Tohoku University, Sendai, Japan
fuma.horie.s3@dc.tohoku.ac.jp
[2] Cyberscience Center, Tohoku University, Sendai, Japan
hgot@cc.tohoku.ac.jp, suganuma@tohoku.ac.jp

Abstract. Research on scene character recognition has been popular for its potential in many applications including automatic translator, signboard recognition, and reading assistant for the visually-impaired. The scene character recognition is challenging and difficult owing to various environmental factors at image capturing and complex design of characters. Current OCR systems have not gained practical accuracy for arbitrary scene characters, although some effective methods were proposed in the past. In order to enhance existing recognition systems, we propose a hierarchical recognition method utilizing the classification likelihood and image pre-processing methods. It is shown that the accuracy of our latest ensemble system has been improved from 80.7% to 82.3% by adopting the proposed methods.

Keywords: Hierarchical recognition method · Ensemble voting classifier · Synthetic Scene Character Data

1 Introduction

Recognition of text information in scenes, which is often referred to as scene character recognition, has some important applications such as automatic driving system, automatic translation, and reading assistant for the visually-impaired. Scene character recognition is more difficult in comparison with printed character recognition because there are various factors such as rotation, geometric distortion, uncontrolled lighting, blur, noise and complex design of characters in scene images. Japanese scene character recognition is a more challenging task since there are many similar characters in thousands of character classes in the language. Current Optical Character Recognition (OCR) systems have not gained practical accuracy for arbitrary scene characters, although some effective methods were proposed in the past.

Many researchers have investigated the ensemble techniques combining the multiple classifiers to improve the accuracy. Breiman proposed Random Forests which is a combination of tree predictors [1]. Bay demonstrated that multiple

© Springer Nature Switzerland AG 2020
S. Palaiahnakote et al. (Eds.): ACPR 2019, LNCS 12047, pp. 118–128, 2020.
https://doi.org/10.1007/978-3-030-41299-9_10

Nearest Neighbor Search (NNS) classifiers are able to reduce both bias and variance components of error [2]. Kim et al. demonstrated that their Support Vector Machine (SVM) ensemble scheme outperforms the single SVM in terms of the classification accuracy [3]. Hansen and Salamon investigated an ensemble scheme of Multi-Layer Perceptron (MLP) and proposed several methods for improving the performance of the ensemble scheme [4]. Also, an ensemble theory was constructed and Bootstrap aggregating (Bagging) method was proposed in [5].

Some previous researches introduced a data augmentation method using Synthetic Scene Character Data (SSCD) which is randomly generated by some particular algorithms such as filter processing, morphology operation, color change, and geometric distortion from the font sets of printed characters [6–8,10]. Jader et al. and Ren et al. have shown that the accuracy of the deep neural network model can be improved by adding SSCD to the training data. It has been demonstrated that the augmentation methods are effective for improving the accuracy of the scene character recognition. Figure 1 shows some examples of the Japanese characters in natural scenes.

In our previous work for Japanese scene character recognition [8], the ensemble scheme and the SSCD generation were used to improve the generalization ability of the classifier, and they proved that some ensemble approaches by learning SSCD are effective for Japanese scene character recognition. In this paper, we propose a hierarchical recognition method utilizing the classification likelihood calculated from the ensemble voting and some image pre-processing methods. In the proposed system, a query character image is changed by some image processings such as deformation and filtering before the recognized ensemble system in order to recognize noisy and distorted images. Experimental results show the effectiveness of the proposed methods. Although the longer recognition time is required in the proposed method, the improvement of the accuracy is prioritized in this paper.

This paper is organized as follows. Section 2 describes the hierarchical recognition method in the ensemble scheme. Section 3 shows the process of experiments and the results. Conclusions and future work are given in Sect. 4.

2   Hierarchical Recognition Method in the Ensemble Scheme

2.1   Flow of the Hierarchical Recognition Method

It is effective to introduce the ensemble scheme to the pattern recognition system for improving the generalization ability [5]. The hierarchical recognition method proposed in our study is adopted in the classification scheme which are able to calculate the classification likelihood in a system such as the ensemble one and neural network-based one. Figure 2 shows the flow of our recognition system. The proposed method is applied not to the training stage but only to the recognition stage.

Fig. 1. Examples of Japanese scene characters.

2.2 Classification Likelihood of the Ensemble Voting Method

As the fundamental ensemble system, we suppose our earlier ensemble scheme includes Random Image Feature (RI-Feature) method which applies some random image processing to the character image before extracting the character features as shown Fig. 3 [10]. The random processing is applied to the training data and the query image. The processing parameters are prepared for each classifier. According to the ensemble learning theory shown by Tumor and Ghosh [11], the smaller overall correlation leads to the smaller error rate of the ensemble system. It is expected that the RI-Feature method makes an overall correlation smaller by adding some fluctuations to the input data. The results obtained in [10] show the effectiveness of introducing RI-Feature method to the ensemble method.

We consider obtaining and utilizing classification likelihood of each class in the ensemble system. Here we assume the plural and hard voting method in the ensemble scheme. Let T be the number of classifiers and N_c be the number of classes, and let $\boldsymbol{h} = \{h_i\}$, $i \in \{x \in \mathbb{N} \mid 1 \in x \in N_c\}$ be the histogram obtained by the ensemble voting. Also, let A be the output of the ensemble classifier and η be the likelihood predicting that A is the correct answer, then the following formulas hold.

$$A = \underset{i}{\mathrm{argmax}}(h_i), \tag{1}$$

$$\eta = \frac{h_A}{T}, \ 0 \le \eta \le 1. \tag{2}$$

η is used for judging whether the recognition sequence should be terminated. Let η_t be a threshold likelihood parameter. Then if $\eta > \eta_t$, the recognition sequence finishes. A large η may lead to the better improvement of the accuracy while the recognition time becomes longer.

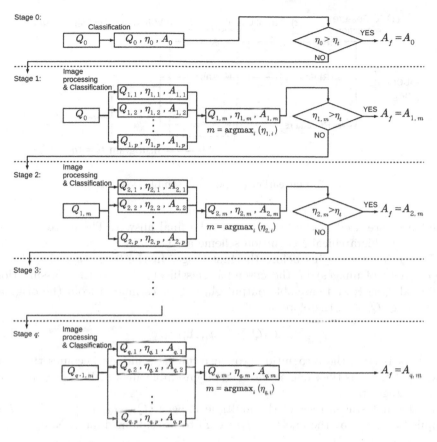

Fig. 2. Flow of the hierarchical recognition system. Q, η, and A denote a query image, a classification likelihood, and an ensemble answer, respectively. Q_o and A_f indicate the original query image and the final answer in the recognition system, respectively.

2.3 Pre-processing Method and Judgement of Final Answer

Character images in natural scenes have various noises and distortions. In order to deal with noisy and distorted scene character images, we propose the hierarchical scheme utilizing the classification likelihood in this paper. The classification likelihood is thought as the confidence value of whether the output answer is correctly recognized or not and have used in some classification systems such as neural network [12]. The hierarchical image processing generates multiple images as shown in Fig. 4 and a normalized image may be included in the images.

We utilize image preprocessings such as rotation, distortion, and filtering in the proposed hierarchical scheme. The multiple query images are generated from the original query image by the image processings. Each image is recognized by the ensemble system using the same feature extraction and classification method. Each confidence value is calculated by the ensemble classification. The

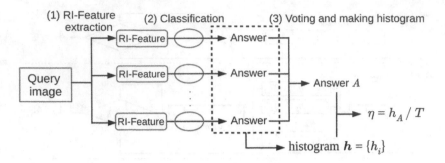

Fig. 3. Our earlier proposed ensemble system.

confidences are used in the judgement of the final answer. There are multiple stages in the hierarchical recognition scheme as shown in Fig. 2.

The 0-th stage begins before the first stage. In the 0-th stage, let $C(x \in I)$ (I is a set of images) be the ensemble classification, then the classification likelihood $\eta_0 \in \mathbb{R}$ and ensemble output $A_0 \in \mathbb{N}$ is estimated from the original query image $Q_o \in I$, therefore

$$C(Q_0) = (\eta_0, A_0). \tag{3}$$

If $\eta_0 > \eta_t$ is true, the recognition sequence finishes, and A_0 becomes the final answer. Then A_0 is the same as the original ensemble answer. If $\eta_0 > \eta_t$ is false, the first stage starts.

In the first stage, a set of the multiple images $\{Q_{1,1}, Q_{1,2}, ..., Q_{1,p}\} \subset I$ is compiled from Q_o by the injection $f_1(x \in I)$ to be explained later. Namely,

$$f_1(Q_0) = \{Q_{1,1}, Q_{1,2}, ..., Q_{1,p}\}. \tag{4}$$

Then, the classification likelihood $\{\eta_{1,1}, \eta_{1,2}, ..., \eta_{1,p}\}$ and ensemble answers $\{A_{1,1}, A_{1,2}, ..., A_{1,p}\}$ are estimated by $C(x)$ from an query image as follows:

$$C(Q_{1,j}) = (\eta_{1,j}, A_{1,j}). \tag{5}$$

If $\eta_{i,m_1} > \eta_t$ is true, the recognition sequence finishes, and A_{m_1} becomes the final answer. Otherwise, the second stage starts, where $m_1 = \underset{j}{\mathrm{argmax}}(\eta_{1,j})$.

Generally, in the i-th ($1 \le i \le q-1$) stage, regarding $\{Q_{i,1}, Q_{i,2}, ..., Q_{i,p}\}$, $\{\eta_{i,1}, \eta_{i,2}, ..., \eta_{i,p}\}$, and $\{A_{i,1}, A_{i,2}, ..., A_{i,p}\}$, then the following formulas hold.

$$f_1(Q_o) = \{Q_{1,1}, Q_{1,2}, ..., Q_{1,p}\}, \tag{6}$$
$$f_i(Q_{i-1,m_{i-1}}) = \{Q_{i,1}, Q_{i,2}, ..., Q_{i,p}\} \ (2 \le i), \tag{7}$$
$$C(Q_{i,j}) = (\eta_{i,j}, A_{i,j}), \tag{8}$$

where $m_i = \underset{j}{\mathrm{argmax}}(\eta_{i,j})$. If $\eta_{i,m_i} > \eta_t$ is true, the recognition sequence finishes, and A_{i,m_i} becomes the final answer. Otherwise, the next stage recognition starts. However, in the last (q-th) stage, A_{q,m_q} immediately becomes the final answer.

Let $g_j : I \to I$ $(1 \le j \le p)$ be the image processing injections, then the injection $f_i(x \in I)$ is defined in the following formula:

$$\forall x \in I, \ f_i(x) = \{g_j(x)\} \subset I. \tag{9}$$

Therefore, f_i generates p images from an image by the image processings g_j. In the hierarchical system, each g_j is not altered in each stage. Some pre-processing examples are shown as follows:

- Gaussian Filter $F_G(x \in I, \sigma) \in I$
  3×3 matrices are used as the Gaussian filter kernels defined by the following formula;

$$K^{GF}(x, y) = \frac{1}{2\pi\sigma^2} \exp\left(-\frac{x^2 + y^2}{2\sigma^2}\right), \quad \sigma \in \mathbb{R}. \tag{10}$$

  (x, y) is the offset from the kernel center.
  An image is convoluted with the kernel K^{GF}.
- Affine transformation $F_A(x \in I, a_1, a_2, a_3, a_4) \in I$
  The affine matrix is calculated as follows:

$$\begin{bmatrix} x' \\ y' \\ 0 \end{bmatrix} = \begin{bmatrix} 1 & 0 & C_x \\ 0 & 1 & C_y \\ 0 & 0 & 0 \end{bmatrix} \begin{bmatrix} a_1 & a_2 & 0 \\ a_3 & a_4 & 0 \\ 0 & 0 & 1 \end{bmatrix} \begin{bmatrix} 1 & 0 & -C_x \\ 0 & 1 & -C_y \\ 0 & 0 & 1 \end{bmatrix} \begin{bmatrix} x \\ y \\ 0 \end{bmatrix}, \tag{11}$$

$$a_1, a_2, a_3, a_4 \in \mathbb{R},$$

where (x, y) and (x', y') are the coordinate of the image before and after the transformation, respectively, and (C_x, C_y) is the center coordinate of the image.
- Rotation $F_R(x \in I, \phi) \in I$
  Rotating an image is realized by the affine transformation. Let $\phi \in \mathbb{R}$ be the rotation angle, then

$$(a_1, a_2, a_3, a_4) = (\cos(\phi), -\sin(\phi), \cos(\phi), \sin(\phi)). \tag{12}$$

- Horizontally/vertically scaling $F_S(x \in I, s_x, s_y) \in I$
  Horizontally/vertically scaling is realized by the affine transformation. Let $s_x \in \mathbb{R}$ be the horizontal scale, and $s_y \in \mathbb{R}$ be the vertical scale, then

$$(a_1, a_2, a_3, a_4) = (s_x, s_y, s_x, s_y). \tag{13}$$

It is expected that the hierarchical recognition system improves the accuracy of the ensemble scheme without the alternation of the feature extraction, classifier, and training data.

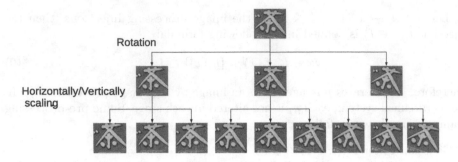

Fig. 4. An example of the hierarchical image processing.

3 Performance Evaluation of Hierarchical Recognition Method

3.1 Experimental Environment and Dataset

Experimental environment is as follows:

- CPU: Intel Core i7-3770 (3.4 GHz)
- Memory: 16 GB
- Development language: C/C++, Python

We used the same Japanese scene character dataset developed in our earlier work [8]. The dataset consists of Hiragana, Katakana and Kanji (1,400 images and 523 classes) taken in real scenes. All character images are in color and in arbitrary size. Seven Japanese fonts (3,107 classes, total 21,749 characters) are used for training. The training dataset does not include real scene characters since it is difficult to collect characters of all classes in Japanese. SSCD is used in order to compile the training dataset and to enhance the recognition system in this paper. The SSCD set is generated through some processes such as distortion, color change, morphology operation, background blending and various filters from the font image set. We adopt the generation processes used in [8].

3.2 Pre-processing Parameters in the Hierarchical Recognition Method

We have used the following preprocessing injection in the evaluation as follows:

$$f_1(x) = \{x, F_G(x, 0.5), F_G(x, 1.0)\}, \tag{14}$$

$$f_2(x) = \{x, F_R(x, -5(\deg)), F_R(x, 5(\deg))\}, \tag{15}$$

$$f_3(x) = \{x, F_S(x, 1.0, 1.1), F_S(x, 1.1, 1.0)\}, \tag{16}$$

therefore, $p = 3$. The query image is resized to 64×64 image size without saving the aspect ratio before the above processings.

3.3 Evaluation with Simple Ensemble Scheme

We have evaluated the hierarchical recognition system with the simple ensemble scheme [8]. The details are shown as follows:

- Feature extraction method: HOG (Parameters are shown in Table 1.) [9]
- Classifier: NNS (Nearest Neighbor Search) or SVM
- Voting method: plural and hard
- Image size: 64 × 64
- The number of classifiers $T = 90$
- A classifier is trained from 9,000 synthetic scene character images generated from 7 Japanese fonts.

The results are shown in Table 2. It is shown that the hierarchical ensemble method contributes to improving the ensemble scheme. The greater ϕ and q tend to make the accuracy improving. As a results, the accuracy of the NNS and SVM ensemble scheme is improved from 60.4 % to 63.4 %, 62.4 % to 63.6 %, respectively.

Table 1. Parameters of a HOG feature.

Image size	Cell size	Block size	Orientation	Dimension
16 × 16	2	16	5	320
32 × 32	4	32	5	320
64 × 64	8	64	5	320

Table 2. Evaluation results of the hierarchical recognition method with the simple ensemble scheme.

NNS	η_t					
	0	0.1	0.2	0.3	0.4	0.5
q 0	60.4 %	–	–	–	–	–
1	–	60.4 %	61.2 %	61.4 %	61.4 %	61.4 %
2	–	61.1 %	62.5 %	62.6 %	62.4 %	62.3 %
3	–	61.2 %	62.9 %	**63.4 %**	63.1 %	63.1 %

SVM	η_t					
	0	0.1	0.2	0.3	0.4	0.5
q 0	62.4 %	–	–	–	–	–
1	–	62.5 %	62.4 %	62.5 %	62.4 %	62.4 %
2	–	62.8 %	63.1 %	63.4 %	63.4 %	**63.6 %**
3	–	62.9 %	62.9 %	63.1 %	63.4 %	**63.6 %**

Table 3. Evaluation results of the hierarchical recognition method with the latest ensemble scheme.

NNS	η_t					
	0	0.1	0.2	0.3	0.4	0.5
q 0	80.7 %	–	–	–	–	–
1	–	80.9 %	80.8 %	80.6 %	80.7 %	80.7 %
2	–	81.8 %	81.8 %	81.9 %	82.0 %	82.0 %
3	–	82.0 %	82.1 %	82.2 %	82.2 %	**82.3 %**

successful

failed

Fig. 5. Some recognition examples.

3.4 Evaluation with Latest Ensemble Scheme

We have evaluated the hierarchical recognition system with the latest ensemble scheme using RI-Feature method [10]. The details of the scheme are shown as follows:

- Feature extraction method: HOG (Parameters are shown in Table 1.) with RI-Feature method
- Classifier: MLP
- Voting method: plural and hard
- The number of classifiers $T = 90$
- A classifier is trained from 200,000 synthetic scene character images generated from 7 Japanese fonts.

The results are shown in Table 3. It is shown that the hierarchical ensemble method contributes improving the latest ensemble scheme. The latest accuracy is updated from 80.7% to 82.3%.

Figure 5 shows some recognition examples. The images at the top are correctly recognized by the $(\eta_t, p) = (0.5, 3)$ scheme and incorrectly recognized by the $(\eta_t, p) = (0, 0)$ scheme, and the images at the bottom are correctly recognized by the $(\eta_t, p) = (0, 0)$ scheme and incorrectly recognized by the $(\eta_t, p) = (0.5, 3)$ scheme. The ensemble recognition system with the proposed method has become able to recognize the images including a large rotation. However, the result shows that the ensemble system may incorrectly classify characters into a different similar character class by introducing the hierarchical method.

4   Conclusion

We have proposed a hierarchical recognition method for improving the ensemble-based scene character recognition. The hierarchical recognition method is to recognize the preprocessed query image in the multiple stages. The multiple query images are generated by some image processings such as deformation and filtering in the hierarchical recognition scheme. We use the classification likelihood calculated by the ensemble scheme in order to judge the final answer.

Experimental results have shown that our latest ensemble scheme is furthermore improved by the proposed method. It is shown that the accuracy has been improved from 80.7% to 82.3% by the newly introduced hierarchical recognition method in the ensemble scheme.

Our future work includes analyzing the proposed method at other preprocessing methods for the hierarchical ensemble method and evaluating the proposed method in various ensemble methods.

References

1. Breiman, L.: Random forests. Mach. Learn. **45**(1), 5–32 (2001)
2. Bay, S.D.: Combining nearest neighbor classifiers through multiple feature subsets. In: Proceedings of ICML 1998, pp. 37–45 (1998)
3. Kim, H., Pang, S., Je, H., Kim, D., Bang, S.Y.: Constructing support vector machine ensemble. Pattern Recognit. **36**(12), 2757–2767 (2003)
4. Hansen, L.K., Salamon, P.: Neural network ensemble. IEEE Trans. Pattern Anal. Mach. Intell. **12**(10), 993–1001 (1990)
5. Breiman, L.: Bagging predictors. Mach. Learn. **24**(2), 123–140 (1996)
6. Jaderberg, M., Simonyan, K., Vedaldi, A., Zisserman, A.: Synthetic data and artificial neural networks for natural scene text recognition. In: Workshop on Deep Learning, NIPS (2014)
7. Ren, X., Chen, K., Sun, J.: A CNN based scene chinese text recognition algorithm with synthetic data engine. arXiv preprint arXiv:1604.01891 (2016)
8. Horie, F., Goto, H.: Synthetic scene character generator and multi-scale voting classifier for Japanese scene character recognition. In: Proceedings of IVCNZ 2018 (2018)
9. Yi, C., Yang, X., Tian, Y.: Feature representations for scene text character recognition. In: Proceedings of ICDAR 2013, pp. 907–911 (2013)
10. Horie, F., Goto, H.: Japanese scene character recognition using random image feature and ensemble scheme. In: Proceedings of ICPRAM 2019, vol. 1, pp. 414–420 (2019)

11. Tumor, K., Ghosh, J.: Theoretical foundations of linear and order statistics combiners for neural pattern classifiers. Technical report TR-95-02-98, Computer and Vision Research Center, University of Texas, Austin (1995)
12. Li, P., Peng, L., Wen, J.: Rejecting character recognition errors using CNN based confidence estimation. Chin. J. Electron. **25**(3), 520–526 (2016)

Continual Learning of Image Translation Networks Using Task-Dependent Weight Selection Masks

Asato Matsumoto and Keiji Yanai[✉]

The University of Electro-Communications, Tokyo, Japan
{matsumo-a,yanai}@mm.cs.uec.ac.jp

Abstract. Continual learning is training of a single identical network with multiple tasks sequentially. In general, naive continual learning brings severe catastrophic forgetting. To prevent it, several methods of continual learning for Deep Convolutional Neural Networks (CNN) have been proposed so far, most of which aim at image classification tasks. In this paper, we explore continual learning for the task of image translation. We apply Piggyback [1], which is a method of continual learning using task-dependent masks to select model weights, to an Encoder-Decoder CNN so that it can perform different kinds of image translation tasks with only a single network. By the experiments on continual learning of semantic segmentation, image coloring, and neural style transfer, we show that the performance of the continuously trained network is comparable to the networks trained on each of the tasks individually.

Keywords: Continual learning · Lifelong learning · Image translation · Catastrophic forgetting

1 Introduction

Humans and animals can learn and fine-tune their knowledge continually throughout their lives. This ability is realized by the rich neurocognitive function of the brain. As a result, humans and animals can learn new knowledge through many experiences over a long period, and never forget old knowledge. In this way, learning that adapts to new knowledge while retaining previously learned knowledge is called continual learning. Since a CNN operated in the real world is given continual information and tasks sequentially, it is important to learn knowledge over a long period and repeat fine-tuning in such a situation. In addition, continual learning is considered to contribute to the achievement of general-purpose artificial intelligence, which needs to perform various tasks given in large quantities, because it does not forget tasks learned previously. Besides, since multiple tasks can be executed by one CNN, the size of the learned model can be reduced from a practical point of view, and it is thought that CNN applications also contribute to the implementation of smartphones and devices. Although continual learning is related to general-purpose artificial intelligence,

© Springer Nature Switzerland AG 2020
S. Palaiahnakote et al. (Eds.): ACPR 2019, LNCS 12047, pp. 129–142, 2020.
https://doi.org/10.1007/978-3-030-41299-9_11

continual learning in CNNs is an unsolved problem because of the property of learning only in specific situations of machine learning. When humans do continual learning, they can learn new tasks without forgetting tasks they have learned in the past. On the other hand, the knowledge CNNs acquired depends on training datasets, and in order to adapt to changes in data distribution, it is necessary to re-train the parameters of CNN for the entire data set. As we learn about new tasks given over time, the accuracy of old tasks decreases. In this way, continual learning in CNNs brings catastrophic forgetting, which results in forgetting the learning results of old tasks while learning new tasks. Catastrophic forgetting causes CNN to suffer from performance degradation and that new knowledge overwrites old knowledge. For this reason, it is difficult for CNNs to perform continual learning like humans and animals. Until now, pseudo-rehearsal [2], EWC [3], network expansion [4], pruning [5], and selection of weights [1] have been proposed as methods to avoid fatal oblivion. However, most of these methods relate to image classification, object detection, and reinforcement learning, and continual learning on image translation tasks have hardly been conducted. Therefore, we propose a method of continual learning in the image translation tasks including semantic region segmentation, neural style transfer, and coloring using an Encoder-Decoder CNN.

This paper treats with continual learning for image translation tasks where both input and output are images. Specifically, as shown in Fig. 1, applying "Piggyback" [1], which is a continual learning method using binary masks that select the weights of the trained Encoder-Decoder CNN model. In the original paper of "Piggyback", it was applied to only image classification tasks, and its effectiveness of the other task such as image translation tasks has not confirmed so far. Then, the purpose of this paper is to explore the effectiveness of "Piggyback" on image transformation tasks with a single Conv-Deconv CNN and small task-dependent binary masks.

To summarize it, the contributions of this work are follows: (1) We applied the "Piggyback" method to continual learning of multiple image translation tasks. (2) We confirmed that continual learning of a Conv-Deconv CNN was possible and its performance of each of the trained tasks was comparable to the performance of individually trained model. (3) We analyzed the trained binary masks and found that the distributions of masked-out weights for image translation tasks were totally different from ones for image classification tasks.

2   Related Work

A common way to learn new tasks with already learned CNN is fine-tuning. Fine-tuning is a method to re-train weights of pre-trained CNN for new tasks. Because of re-training, changing the value of CNN weights causes catastrophic forgetting that the accuracy of the old task decrease. To overcome it, many works have been proposed so far. Pseudo-rehearsal [2], optimization [3], network expansion [4], pruning [5], weight selection [1], etc. are representative methods. Here, we will particularly describe optimization [3] and weight selection [1] methods.

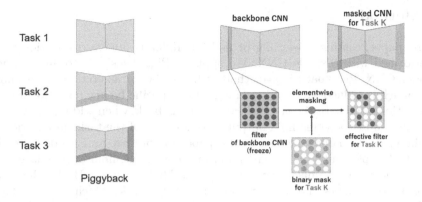

Fig. 1. Outline of the proposed method.

2.1   EWC

EWC [3], standing for Elastic Weight Consolidation, makes it suppress to change important weights for previous tasks. This method reduces the disruption of the learning result in the previous task. Thus the performance degradation of the previous task by learning a new task is expecting to be prevented. The importance of the weights is determined based on a Fisher information matrix of the trained CNN, and the learning rate of each parameter is adjusted in proportion to the importance of the weights. By adjusting learning rate on each weight adaptively, it becomes possible to learn new tasks while maintaining the important weights of the previously learned tasks. However, if the previous task and the new task are very different or if many tasks are added repeatedly, the performance of the previous task will be degraded, because the value of the weights with high importance are expected to be changed greatly even in EWC.

2.2   Piggyback

On the other hand, Piggyback [1] solved the performance degradation of the previous task which is the problem of EWC and succeeded in learning many tasks by only a single CNN with high accuracy. In Piggyback, first we train a general-purpose backbone CNN with a large data set such as ImageNet, and next train a binary mask that selects high importance weights for each of the tasks when learning additional new tasks. Since only the binary mask is learned, the weight value of the backbone CNN does not change. That is why the performance of the previous task is never degraded. While EWC trains less importance weights for additional tasks, Piggyback freezes the backbone CNN and select important weights from it for additional tasks. However, in the original paper, Piggyback was applied to only classification CNN. Then, in this paper, we explore the effectiveness of Piggyback for an Encoder-Decoder CNN.

3   Method

In this paper, continual learning of multiple different image translation tasks is performed with a single CNN. We applied "Piggyback" [1] to an Encoder-Decoder CNN. Piggyback is a method of fixing the parameters of a backbone CNN which is trained first, and learning a task-specific binary mask for each of the additional tasks. In the original paper of Piggyback, when adding a new task, we prepare a task-specific final output layer for each of the tasks in addition to the binary masks. Following this, in this paper, we prepare an Encoder-Decoder CNN, task-dependent binary masks and task-dependent final output layers for continual learning of image translation tasks. That is, "PiggyBack" shares all the backbone weights and needs to prepare binary masks which have 1 bit per backbone CNN parameter and final output layers independently for each of the additional tasks. The learning procedure of Piggyback is shown in Fig. 2.

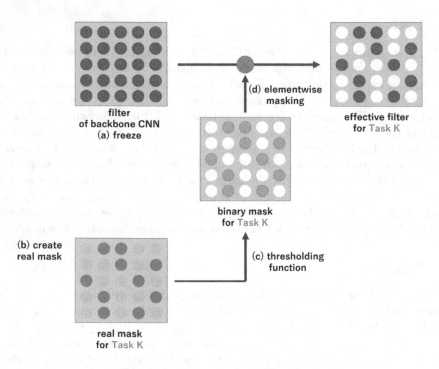

Fig. 2. Learning masks in Piggyback

In this paper, we apply PiggyBack to image translation tasks. First we train a CNN for one of the image translation tasks to build a backbone network. The task for a backbone network should be trained with a large-scale dataset. The learning of the mask is performed by the following procedure. Note that real masks are the real weights the number of which are the same as one of the backbone CNN.

1. Train and fix the weight of backbone CNN for the first task ((a) in Fig. 2)
2. Train real mask weights for the additional tasks ((b) in Fig. 2)
3. Binarize real weights with a threshold and create binary masks ((c) in Fig. 2)
4. Select weights with masking the weight of backbone CNN with the binary masks ((d) in Fig. 2)
5. Calculate forward gradients and back propagation using (4)
6. Update of real mask weights.
7. Repeat (3) to (6)

In the experiment, the value of the real mask is initialized to 1e−2 for all parameters, as in the paper of Piggyback [1]. Also, the threshold used for creating the binary mask was 5e−3 in the same way as [1].

In this experiment, we expected that a highly functional backbone CNN was necessary to be trained with the largest datasets within all the tasks to be trained with a single Conv-Deconv network. Therefore, we utilized semantic segmentation with a large dataset as the first image translation task. In particular, the first image translation task was semantic segmentation with MS COCO where approximately 200,000 images were annotated with 80 objects. The second and subsequent tasks learned binary masks that selects valid weights for the task newly added from the backbone CNN. Furthermore, in this experiment, we have used U-Net [6] which has skip connections between the Encoder and the Decoder CNN and is used for various image translation tasks. Note that instead of Batch Normalization before the activation function ReLU other than the output layer, we used Instance Normalization, which has the effect of accelerating the convergence of learning.

4   Experiments

4.1   Evaluation on Continual Learning of Image Translation Tasks

In the experiment, we performed continual learning of five different image translation tasks, and evaluated the performance of the proposed method and baselines. The tasks used in the experiments were semantic segmentation with two different dataset, gray image coloring, neural style transfer [7] with two different styles, and edge image coloring as shown in Table 1. In training, Tasks 1, 2, 3, 4, 5, and 6 were sequentially trained with a single Conv-Deconv network. Exceptionally, for Task 5 with "fine-tune" which is one of the baselines, the model was trained after not Task 4 but Task 3.

Both MS COCO and Pascal VOC are image datasets containing general images such as people, vehicles and animals. MS COCO consists of about 330,000 images including 80 categories with pixel-wise segmentation masks, and Pascal VOC consists of about 10,000 images including 20 categories included in the eighty categories of MS COCO. Edges2handbags [8] is a dataset consisting of about 137,000 handbags images and binary edges extracted from the images. The reason for performing the same semantic segmentation in Task 1 and Task 2 is to confirm whether continual learning is possible for the same task with

Table 1. Five tasks for continual learning.

	Task category	Dataset
Task 1	Semantic segmentation	MS COCO
Task 2	Semantic segmentation	Pascal VOC 2012
Task 3	Gray image coloring	MS COCO
Task 4	Style transfer (Gogh)	MS COCO
Task 5	Style transfer (Munk)	MS COCO
Task 6	Edge image coloring	edges2handbags

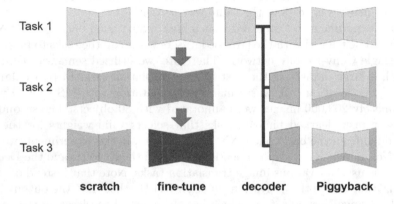

Fig. 3. The overview of three baselines and "Piggyback".

different datasets. After Task 3, experiments were performed on the tasks that differ from semantic segmentation, and we verified whether continual learning across different tasks was possible. In the original "Piggyback" paper [1], only classification tasks are performed with different datasets. On the other hand, in this experiments, we perform continual learning of heterogeneous tasks with different loss functions and different datasets. This point is completely different from the original work.

In the framework of Piggyback, the first task is importance since the network trained in the first task is used as a backbone CNN. It should be comprehensive and trained with as large dataset as possible. This is why we used MSCOCO for training of Task 1.

In this experiment, three types of baselines were prepared for comparison. The outline of the three baselines and Piggyback is shown in Fig. 3. Three types of the baselines are "scratch" in which we learn task-dependent models from scratch separately for each task, "fine-tune" in which we fine-tune a single identical model with five tasks continuously, and "decoder" in which we train both the encoder and decoder part when training of Task 1, fix the encoder part and train only task-dependent decoder parts for all the other tasks independently. "Piggyback" is the method we explore its effectiveness for image translation tasks in this paper.

Table 2. Results of continual learning (See the Table 1 for the contents of each task)

	Scratch	Fine-tune	Decoder	Piggyback
Task 1 (mIoU(%))	**21.47**			
Task 2 (mIoU(%))	58.59	**64.87**	61.63	61.45
Task 3 (MSE)	244.000	**237.92**	241.66	242.49
(SSIM)	0.9138	**0.9148**	0.9121	0.9058
Task 4 (SSIM)	**0.3678**	0.3555	0.3595	0.3501
(total loss)	413,833	**405,893**	473,723	528,587
Task 5 (loss)	**447,480**	490,490	544,348	521,476
Task 6 (MSE)	211.96	**207.76**	237.53	232.02
Task 1 after Task 2 (mIoU)	–	0.70	21.47	21.47
Task 2 after Task 3 (mIoU)	–	1.87	61.63	61.45
Task 3 after Task 4 (MSE)	–	870.18	241.66	242.49
(SSIM)	–	0.5321	0.9121	0.9058
Model size (MB)	338.4 (56.4 * 6)	338.4 (56.4 * 6)	158.9 (56.4 + 20.5 * 5)	**65.4** (56.4 + 1.8 * 5)

For training for each of the tasks, the loss functions are as follows: In Task 1 and Task 2, we use Cross-Entropy Loss, in Task 3 we use L2 loss, Tasks 4 and Task 5 we used Content Loss and Style Loss used in the fast style transfer proposed by Johnson et al. [7], and in Task 6 we used GAN Loss and L1 used in the pix2pix by Isola et al. [8].

The input image are RGB images in Task 1, Task 2, Task 4, and Task 5, gray images are used for Task 3, and binary edges in Task 6. The outputs were as follows: In Task 1 and Task 2 they are segmentation masks of 81 channels and 21 channels, respectively. In Task 3, they are the CbCr components which need to be combined with input grayscale images to obtain RGB images. In Task 4, Task 5, and Task 6, they were RGB images.

For evaluation, the test dataset is used, and the performance was evaluated as follows: For Task 1 and Task 2 we use mean intersections over union (mIoU), for Task 3 we use mean square error (MSE) and structural similarity (SSIM), for Task 4 we use SSIM and total loss values, for Task 5 we use only total loss values, and for Task 6 we use MSE. As described in Gatys's style translation paper [9] we use the total loss that combines Content Loss and Style Loss for Task 4 and 5.

Table 2 shows the evaluation values of each task, the evaluation values of the previous task after training of the next task represented as "Task n after Task $(n + 1)$", and the total size of trained models over five tasks. Note that the most important part of this table is the part of "Task n after Task $(n + 1)$", since the values of this part are expected to be degraded greatly if catastrophic forgetting happens.

This table indicates continual learning by "decoder" and "Piggyback" never brought catastrophic forgetting, while it happens and the performance was degraded greatly in case of "fine-tune". Regarding the size of the total models, the "Piggyback" model is about half of the "decoder" model. "Decoder" has independent decoder parts and shares only the encoder part across tasks, while "Piggyback" shares the backbone model which is trained with the first task and has independent binary masks for four other tasks than the first task. In general, the size of one binary masks is 1/32 of the size of the backbone model, since each element of the binary masks is represented in one bit and each element of the backbone model is represented in a 32-bit floating value. From this results, we can confirm the effectiveness of "Piggyback" for image translation tasks.

Some examples of generated images of Task 2, 3, 4, 5, and 6 are shown in Figs. 4, 5, 6, 7 and 8, respectively. Compared between the results by "Piggyback" and the results by the other methods, the quality of output images are almost comparable and not degraded at all.

4.2   Analysis on Trained Binary Masks

Here, we analyze the binary masks trained by Piggyback in the same way as Mallya et al. [1]. We examined the ratio of 0 in the trained binary masks. The ratio of 0 in the binary mask represents how many parameters are required to changed in the backbone CNN for performing each of the tasks or how effective the backbone CNN was initialized in the MS COCO semantic segmentation task for each task. The ratio of 0 in the binary mask learned for each layer of U-Net is shown from Figs. 9, 10, 11, 12 and 13 for Task 2, 3, 4, 5, and 6, respectively. In addition, for comparison on the ratio of 0, Fig. 14 shows the ratio of 0 in the binary mask for the Wiki Art classification task used in Mallya et al. [1]. The horizontal axis of the graph of Figs. 9, 10, 11, 12 and 13 represents each layer of U-Net, and "conv" and "up-conv" in the graph correspond to "conv" and "up-conv" in the U-Net.

The ratio of 0 in the binary masks of classification task for the VGG16 network tended to be low in the low layer and to increase with depth of the layer gradually as shown in Fig. 14. From this, for continual learning of classification tasks with Piggyback, it is considered that the rate of re-using backbone CNN is high in the lower layer because the lower layer is a part to extract low-level local features. Regarding the uppper layers, the ratio of 0 was higher than one of the lower layers, since task-dependent semantics information are extracted.

On the other hand, in the case of image translation with U-Net, the above-mentioned tendency is only slightly seen in the Encoder part of Task 2. Conversely, the common tendency in classification tasks are not seen in the Encoder of other tasks and in the Decoder of all tasks. In addition, the ratio of 0 in both Encoders and Decoders of all tasks was about 40% to 60%. Moreover, the ratio of 0 to the first layer of the U-Net in Task 2 is about 30 points which is much higher than the first layer of VGG16, and the weight re-use ratio of the backbone CNN becomes lower.

Furthermore, in the tasks after Task 3, other than semantic segmentation, the ratio of 0 in the lower layer also increased. The reason why the ratio of 0 is higher in all layers of U-Net with image translation compared to VGG16 with classification is expected to be attributed to the following three reasons. The first reason is that the number of U-Net layers used in the experiment is larger than VGG16. By increasing the number of parameters in the CNN, it was considered that the number of important parameters possessed by one layer decreased, and

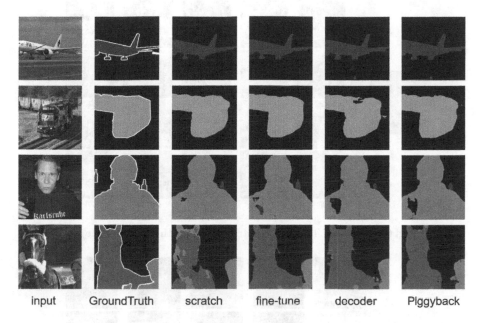

input   GroundTruth   scratch   fine-tune   decoder   Piggyback

Fig. 4. Results of Task 2 (semantic segmentation for Pascal VOC)

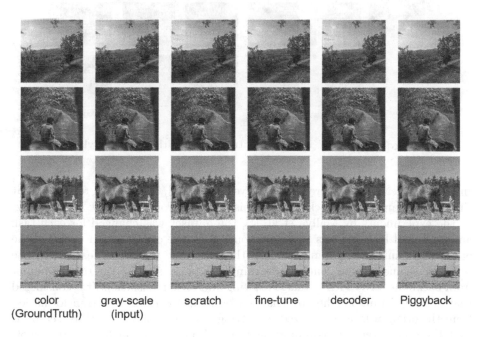

color   gray-scale   scratch   fine-tune   decoder   Piggyback
(GroundTruth)   (input)

Fig. 5. Results of Task 3 (gray image coloring)

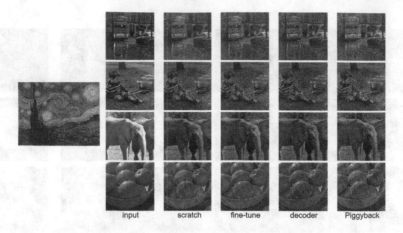

Fig. 6. Results of Task 4 (style translation with "Gogh" style)

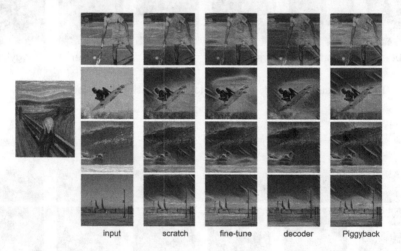

Fig. 7. Results of Task 5 (style translation with "Munk" style)

the types of features acquired by one layer decreased. The second reason is that the task of learning the backbone CNN is different from the newly learned task. In the experiments by Mallya et al. [1], the dataset was changed for each task and continual learning was performed only with the classification task. On the other hand, in the experiments, after the backbone CNN was learned by the semantic segmentation, continual learning was performed by the tasks different from backbone CNN such as image coloring and style translation. It is considered that the important weights are changed by tasks or a different transformation from the original layer is realized by setting 0 to the mask. The third reason is that the parameter that is important in CNN may be about 50% of the whole layer. Even if 50% of the pre-trained CNN parameters are pruned in advance by

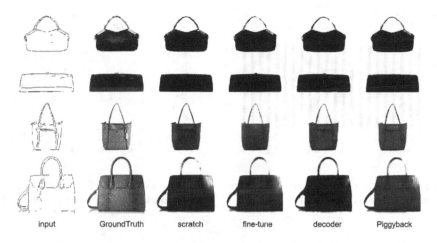

input   GroundTruth   scratch   fine-tune   decoder   Piggyback

Fig. 8. Results of Task 6 (edge image coloring)

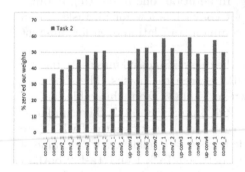

Fig. 9. The ratio of 0 in the trained binary masks for Task 2 (Pascal VOC segmentation)

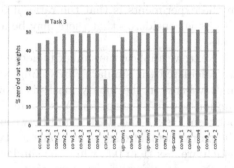

Fig. 10. The ratio of 0 in the trained binary masks for Task 3 (gray image coloring)

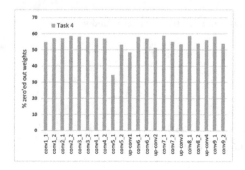

Fig. 11. The ratio of 0 in the trained binary masks for Task 4 (style translation with "Gogh")

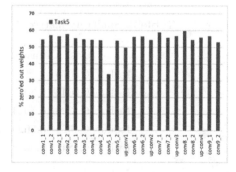

Fig. 12. The ratio of 0 in the trained binary masks for Task 5 (style translation with "Munck" style)

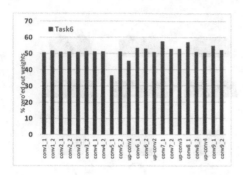

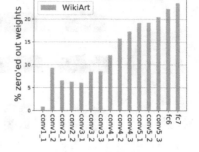

Fig. 13. The ratio of 0 in the trained binary masks for Task 6 (edge image coloring)

Fig. 14. The ratio of 0 in the trained binary masks for image classification [1].

Packnet [5] of Mallya et al., performance degradation of less than 1% is achieved from the accuracy of CNN before pruning. In addition, one interesting findings in the experiments is that the ratio of 0 for conv5_1 is low for all tasks. From this, it is considered that there may be a common translation even for different tasks.

Next, we analyze the similarity between the masks. Since all masks are binary, we calculated the similarity by taking XOR between the masks. The similarity matrix between the binary masks is shown in Table 3. From this table, all the similarity values except for Task 1 and Task 4, 5 are around 0.5. The similarity between Task 1 and Task 4 and between Task 1 and Task 5 both of which are the pair of semantic segmentation and style transfer are relative low the value of which are less than 0.5, while the similarity between Task 1 and Task 2 both of which are the semantic segmentation, and Task 4 and Task 5 both of which are the style translation are relatively high. From this, it is considered that important weights are different in each task, and common weights are used between similar tasks and different weights are used between different task types.

Table 3. Similarity matrix of binary masks between the tasks.

	Task 1	Task 2	Task 3	Task 4	Task 5
Task 2	0.5075	–	–	–	–
Task 3	0.5042	0.5054	–	–	–
Task 4	0.4326	0.5034	0.5020	–	–
Task 5	0.4529	0.5029	0.5025	0.5210	–
Task 6	0.4847	0.5063	0.5026	0.5093	0.5077

5   Conclusions

In this paper, we have explored continual learning of different image translation tasks. From the experimental results, it was found that applying Piggyback to the Encoder-Decoder CNN achieves the same performance as the baseline with minimum overhead in continual learning of semantic segmentation, gray image coloring and neural style transfer.

The experiments were conducted with only three kinds of the tasks, segmentation, coloring and style transfer. Therefore, we plan to conduct additional experiments with other kinds of image translation tasks used in this paper to confirm the versatility of "Piggyback". Hopefully we like to set up various kinds of image translation tasks like Visual Domain Decathlon in which the ten types of classification tasks are trained continuously[1] or the experiments conducted by Kokkinos at UberNet [10].

Acknowledgements. This work was supported by JSPS KAKENHI Grant Number 15H05915, 17H01745, 17H06100 and 19H04929.

References

1. Mallya, A., Davis, D., Lazebnik, S.: Piggyback: adapting a single network to multiple tasks by learning to mask weights. In: Ferrari, V., Hebert, M., Sminchisescu, C., Weiss, Y. (eds.) ECCV 2018. LNCS, vol. 11208, pp. 72 88. Springer, Cham (2018). https://doi.org/10.1007/978-3-030-01225-0_5
2. Robins, A.: Catastrophic forgetting, rehearsal and pseudorehearsal. J. Connect. Sci. **7**, 123–146 (1995)
3. Kirkpatrick, J., et al.: Overcoming catastrophic forgetting in neural networks. Proc. Natl. Acad. Sci. (PNAS). abs/1612.00796 (2016)
4. Rusu, A.A., et al.: Progressive neural networks. arXiv preprint arXiv:1606.04671 (2016)
5. Mallya, A., Lazebnik, S.: PackNet: adding multiple tasks to a single network by iterative pruning. In: Proceedings of the IEEE Computer Vision and Pattern Recognition (CVPR), pp. 7765–7773 (2018)
6. Ronneberger, O., Fischer, P., Brox, T.: U-Net: convolutional networks for biomedical image segmentation. In: Navab, N., Hornegger, J., Wells, W.M., Frangi, A.F. (eds.) MICCAI 2015. LNCS, vol. 9351, pp. 234–241. Springer, Cham (2015). https://doi.org/10.1007/978-3-319-24574-4_28
7. Johnson, J., Alahi, A., Fei-Fei, L.: Perceptual losses for real-time style transfer and super-resolution. In: Leibe, B., Matas, J., Sebe, N., Welling, M. (eds.) ECCV 2016. LNCS, vol. 9906, pp. 694–711. Springer, Cham (2016). https://doi.org/10.1007/978-3-319-46475-6_43
8. Isola, P., Zhu, J., Zhou, T., Efros, A.: Image-to-image translation with conditional adversarial networks. In: Proceedings of the IEEE Computer Vision and Pattern Recognition (CVPR) (2017)

---

[1] https://www.robots.ox.ac.uk/~vgg/decathlon/.

9. Gatys, L.A., Ecker, A.S., Bethge, M.: Image style transfer using convolutional neural networks. In: Proceedings of the IEEE Computer Vision and Pattern Recognition (CVPR), June 2016
10. Kokkinos, I.: UberNet: training a universal convolutional neural network for low-, mid-, and high-level vision using diverse datasets and limited memory. In: Proceedings of the IEEE Computer Vision and Pattern Recognition (CVPR), pp. 6129–6138 (2017)

A Real-Time Eye Tracking Method for Detecting Optokinetic Nystagmus

Mohammad Norouzifard[1,2]([✉]), Joanna Black[2], Benjamin Thompson[3], Reinhard Klette[1], and Jason Turuwhenua[2,4]

[1] School of Engineering, Computer and Mathematical Sciences,
Auckland University of Technology, Auckland, New Zealand
mohammad.norouzifard@aut.ac.nz
[2] Auckland Bioengineering Institute, University of Auckland, Auckland, New Zealand
[3] School of Optometry and Vision Science, University of Waterloo,
Waterloo, ON, Canada
[4] School of Optometry and Vision Science, University of Auckland,
Auckland, New Zealand

Abstract. Optokinetic nystagmus (OKN) is an involuntary repeated "beating" of the eye, comprised of sequences of slow tracking (slow phase) and subsequent quick re-fixation events (quick phase) that occur in response to (typically horizontally) drifting stimuli. OKN has a characteristic saw-tooth pattern that we detect here using a state-machine approach applied to the eye-tracking signal. Our algorithm transitions through the slow/quick phases of nystagmus (and a final state) in order to register the start, peak and end points of individual sawtooth events. The method generates duration, amplitude, velocity estimates for candidate events, as well as repetition estimates from the signal.

We test the method on a small group of participants. The results suggest that false positive detections occur as single isolated events in feature space. As a result of this observation we apply a simple criteria based on the repetitious "beating" of the eye. The number of true positives is high (94%) and false OKN detections are low (2%). Future work will aim to optimise and rigorously validate the proof-of-concept framework we propose.

Keywords: Vision testing · Optokinetic nystagmus · Eye tracking · Real-time

1 Introduction

Optokinetic nystagmus (OKN) is an involuntary motion of the eye that occurs as a person views a drifting stimulus. It is characterised by a slow tracking (the slow phase) and a subsequent resetting motion in the opposite direction (the quick phase) that allows the eye to fixate and track a different stimulus feature. The overall visual appearance of OKN, typically elicited by drifting vertical bars, is

S. Palaiahnakote et al. (Eds.): ACPR 2019, LNCS 12047, pp. 143–155, 2020.
https://doi.org/10.1007/978-3-030-41299-9_12

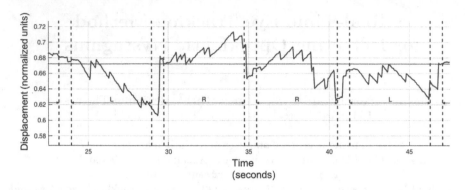

Fig. 1. An example of the horizontal displacement signal (coordinates normalised to the eye camera image) showing OKN. Also indicated on the figure is the direction of the stimulus (*i.e.*, in this case, L = leftward, R = rightward, synchronised to eye data.

a repeated "beating" of the eye, that appears as sawtooth eye movements in the horizontal displacement signal (see Fig. 1).

OKN is an established means to detecting deficits along the visual pathway [5,24,25]. The literature suggests [1,9,11,17,21] the presence/absence of OKN (in response to carefully designed stimulus) can be used to detect clinically significant deficits in *visual acuity*, the (self-reported) ability of the eye to see fine detail. Visual acuity can be difficult to obtain in non-verbal patients, such as young children [2] and the involuntary nature of OKN presents a potential method for rapidly and accurately assessing VA in these patients.

The automated detection of the slow/quick phases of the optokinetic and related vestibulo-ocular reflexes have been studied by a number of authors. Velocity threshold was used to determine the saccadic portion of the signal [13,18]. We found threshold of the velocity signal to be effective in an off-line situation in which a consumer grade camera was used to record video of the eye performing OKN in adult participants [28] as well as children [23]. Alternate approaches include a recursive digital filter that responds to the changes of phase of the signal obtained using *electronystagmography* (ENG) [13], and an system identification approach utilising ARX model for identifying the relationship between fast phase and slow phase velocity [22]. Recently, Ranjabaran et al. demonstrated a fully objective approach based on K-means clustering to provide an initial classification of data as belonging to fast/slow phase or non-slow phase followed by a system identification approach using ENG [20].

The major focus of this paper is to provide proof-of-concept for a simple approach to OKN detection; highly suited for real-time application. Our method takes as input the horizontal eye displacement obtained from a head-mounted eye tracking system. The method generates time points corresponding to the onset/peak and end of triangular "sawtooth" features characteristic of OKN from the incoming signal. The resulting feature vectors are filtered using heuristic rules to determine whether they are legitimate candidates for OKN. From this process, we find that the repetitive nature of OKN appears to be a discriminative factor for the presence/absence of OKN. Our overall finding is a high true positive rate and low false detection rate using this as a discriminating criteria.

The paper is organized as follows: Sect. 2 provides a general definition of OKN and an explanation of the background of our research. In Sect. 3, the experimental methods are described as used in the data collection stage of the research. Section 4 gives experimental results and evaluation. Section 5 provides a discussion of challenges for the detection of OKN along with some results. Section 6 concludes.

2 Background

We denote the input signal vector by $x(t) = (d(t), v(t), \gamma(t))$; the concatenation of the (horizontal) displacement of the eye $d(t)$, the velocity $v(t)$ and auxiliary information, $\gamma(t)$ (consisting of the direction of the stimulus and data quality). Our aim is to determine the start x_s, peak x_p and end points x_e of triangular features from $x(t)$, and to test the resulting feature vector (x_s, x_p, x_e) using reasonable decision criteria, to be described, to eliminate unlikely sawtooth candidates.

Consider a sample of $d(t)$ containing OKN as shown in Fig. 2. The onset of a sawtooth is given by the point $x_s = (t_s, d_s)$, the point where a rising displacement in the eye signal is first detected, the peak of the sawtooth by $x_p = (t_p, d_p)$ which occurs as the eye transitions to the quick resetting eye motion, and the end of the sawtooth by $x_e = (t_e, d_e)$ where the descending edge now transitions to rising or stationary. These points yield the (average) slow/quick phase velocities, v_{SP} and v_{QP}:

$$v_{SP} = \frac{\Delta d_{SP}}{\Delta t_{SP}} = \frac{d_p - d_s}{t_p - t_s} \tag{1}$$

$$v_{QP} = \frac{\Delta d_{QP}}{\Delta t_{QP}} = \frac{d_e - d_p}{t_e - t_p} \tag{2}$$

where Δt_{SP} and Δt_{QP} are the slow/quick phase durations, and Δd_{SP} and Δd_{QP} are slow/quick phase amplitudes. Figure 2 illustrates basic consistency constraints summarised by Table 1(a). The durations $(\Delta t_{SP/QP})$ should be positive and non-zero: the quick phase duration must be shorter than the slow phase duration. The slow/quick phase velocities $(v_{SP/QP}$ must be non-zero and of opposite sign. The magnitude of the quick phase velocity should not exceed that of the slow phase. Furthermore, additional empirically based thresholds were applied as summarised by Table 1(b). This thresholding was used to eliminate potential OKN candidates based on lower and upper estimates for quick/slow phase amplitude/duration and speed.

State-Machine Description of Algorithm
The optokinetic response is generated physiologically by independent slow phase and quick phase systems [30]. This behaviour naturally suggests a state-machine solution to detection; which is also highly suited for real-time implementation.

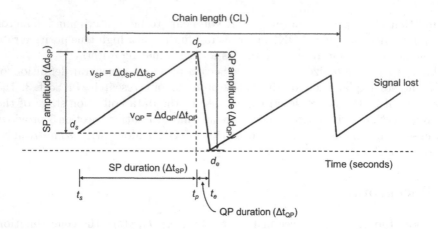

Fig. 2. Components of the OKN displacement signal

Table 1. Constraints applied to the detected sawtooth features

(a) Consistency constraints

Number	Constraint
1	$0 < \Delta t_{QP} \leq \Delta t_{SP})$
2	$v_{SP} \cdot v_{QP} < 0$
3	$|v_{SP}| < |v_{QP}|$

(b) Thresholded constraints

Number	Description	Variable	Lower limit	Upper limit
4	QP duration	Δt_{QP}	0.1	2.0
5	SP velocity[a]	$|v_{SP}|$	0.05	0.40
6	QP/SP amplitude	$|\Delta d_{QP/SP}|$	0.004	0.2

[a] Assumes the slow-phase velocity is the same as the stimulus direction

Our algorithm is now explained. In broad terms, the input data stream $x(t)$ drives our machine through three states (either slow phase detection, quick phase detection or finalise as shown in Fig. 3).

The purpose of the slow phase detection state is to find a rising edge of sufficient duration; if this is found then the start point of the rising edge is registered as the beginning of a sawtooth (t_s, d_s) and the machine transitions to the extend edge state. Whilst in the extension phase, the machine now looks for a falling edge to indicate the end of the slow phase, whereupon the end of SP/start of QP (t_p, d_p) will be registered. Any failure (for example, due to lost data caused by blinking) will discard the candidate and cause a reset to the find rising edge

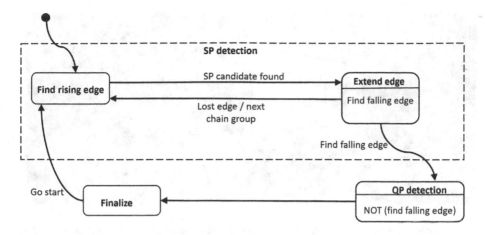

Fig. 3. State description of the algorithm.

state (and also increment a chain group counter to be explained presently). In any event, given a successful detection for the start of a QP, the machine will transition to the QP phase. The machine now looks for a non-decreasing edge indicating the end of the QP (t_e, d_e).

Once the end is found, and given the points of the sawtooth, the machine transitions to the finalize state: the purpose of which is to retain OKN like sawtooth features and discard candidates unlikely to be due to OKN. We do this by applying the criteria described in Table 1.

We measured also contiguous groupings of sawtooth features, using a chain group counter. This counter value was carried across to each new sawtooth candidate. In the case of a failure (*i.e.*, reset event) the group counter advanced thereby indicating a new group. This labelled each repeated sawtooth as belonging to a particular group, and the number of features belonging to a group gave the chain length as shown by variable CL in Fig. 2. In this example, the chain length is $CL = 2$ the number of complete sawtooth features detected.

Our aim in this work was to see whether there were readily discernible patterns in the measured parameters that would allow us to determine whether stimulus was present or absent. We were particularly interested in whether slow phase properties (velocity and duration) or chain-length could be of value in differentiating between OKN present or absent cases.

3 Experimental Methods

The study was approved by the Auckland University ethics committee and complied with all Helsinki declarations. All participants gave full, written informed consent.

Healthy participants (n = 6 adult volunteers) were recruited for this study. OKN was elicited for each participant using an array of drifting disks (see Fig. 5)

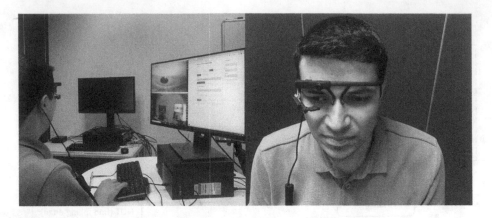

Fig. 4. *Left*: Experimental setup. *Right*: Used head-mounted eye tracking system [27].

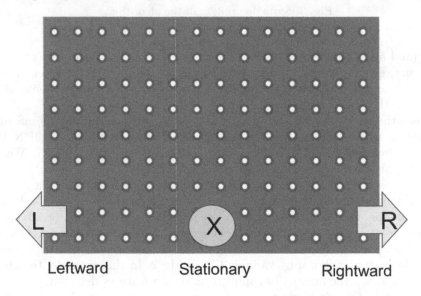

Fig. 5. The stimulus array shown to participants. The array was shown drifting left-ward/rightward or stationary.

presented on a 24" LCD display (AOC G2460PG with Ultra Low Motion Blur). The arrays comprised of disks with a central disk diameter chosen to ensure that the stimulus was easily seen (0.7 logMAR, 5 min of arc). The central intensity of the disks was $35\,\text{cd/m}^2$, the peripheral intensity of the disks was $9\,\text{cd/m}^2$ and the background intensity of the screen was $13\,\text{cd/m}^2$. The arrays were shown over intervals of 5 s, during which time it was shown either stationary (0 deg/sec) or drifting horizontally (±5 deg/sec) in a random left/right direction. Each state (left, right, stationary) was shown to the participant 5 times.

The experimental setup is shown in Fig. 4. Participants were asked to stare at the centre of the screen with one eye covered (the protocol was repeated for each eye). The viewing distance was 1.5 m. The eye displacement was recorded using a 120 fps, head mounted eye-tracker (Pupil-Labs, Berlin, Germany). The raw horizontal displacement was smoothed using a Savitsky-Golay (SG) filter. The SG filter preserves high frequency content, and is conveniently specified in the time-domain by polynomial order N and frame-length M [19]. Moreover, the velocity $v(t)$ is readily computed from these filter coefficients. In this work, the order was $M = 2$, the region frame-length was chosen as $f = 13$.

The algorithm was run with $(d(t), v(t), \gamma(t))$ data (across all participants and for each eye). The auxiliary function used in this work was $\gamma(t) = (\gamma_1(t), \gamma_2(t))$ comprising the known state of the stimulus ($\gamma_1(t) \in \{L, R, X, F\}$), and data quality measure ($0 \leq \gamma_2(t) \leq 1$) from the eye-tracker device. The latter measure was used to determine whether the data was of sufficient quality to be useful (maximum quality was 1 and minimum quality was 0). The former quantity passed the *a-priori* direction of the stimulus to the algorithm. Here, the potential values for $\gamma_1(t)$ were L = "leftward", R = "rightward", X = "stationary", or F = "fixation") (see Fig. 5).

The slow phase velocity v_{SP}, duration ($\triangle t_{SP}$) and the chain length (CL) were extracted and categorized as depending on whether they were obtained whilst the stimulus was moving (*i.e.*, trials labelled "L" or "R") or whilst the stimulus was stationary (*i.e.*, trials labelled "X"). The fixation trials labelled "F" were ignored. For trials in which the stimulus array was stationary, the algorithm was run twice. This avoided choosing a particular direction for these trials; and allowed an unbiased estimate of the false positive (FP) detection rate of the method.

The apparent direction of travel of the stimulus as perceived by the observer was recorded, for comparison purposes, by asking the participant to press keys to indicate the direction of travel as they watched the stimulus on-screen. This was used to confirm that the stimulus was perceived as expected by the observer. The eye tracking, experiment management and key-press collection where performed on a second computer running a custom web-server application (`node.js`) with an JavaScript/HTML5 interface. The display of the stimulus, ran on the primary display by accessing the server using the chrome web-browser. Figure 4 shows the experimental setup used.

4   Results

The stimulus was seen by all observers (n = 6) for all trials (n = 11 eyes) as indicated by correct key-presses measured during the trial. Data for one eye was excluded because stimulus direction information $\gamma_1(t)$ was lost. The detection algorithm completed successfully for all other test runs/participants. The output from the system is given by Fig. 6. The figure shows unshaded areas corresponding to intervals of time not identified as containing OKN, compared with shaded green areas where OKN was detected. The labels under the graph indicate the

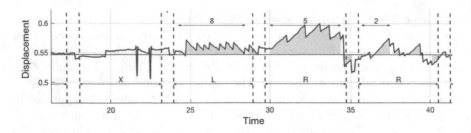

Fig. 6. The result of an example of horizontal displacement signal.

direction of travel as a function of time $\gamma_1(t)$, the numbered intervals above the graph show the chain length computed for the data (chain lengths of 1 are not indicated). Visual inspection (by the authors, who are experienced in identifying OKN) suggested the algorithm was effective in identifying regions containing OKN.

A total of $n = 661$ sawtooth features were detected for the moving stimulus category. A total of $n = 41$ sawtooth features were detected in the the stationary category. The FP rate at the level of features detected was 5.8% of all sawtooth detections. Figure 7a shows the distribution of results obtained for the slow phase speed $|v_{SP}|$. The mean slow phase velocity (mean $\pm$ 2SD) were 0.024 ± 0.014 and 0.017 ± 0.016 (s^{-1}) for the moving ($n = 661$) and stationary categories ($n = 41$) respectively. Figure 7b shows the distribution results obtained for the slow phase duration Δt_{SP}. The SP duration was 0.48 ± 0.52 s and 0.33 ± 0.50 s for the moving and stationary categories. The CL (*chain length*) is shown in Fig. 7c. In this instance the mean and standard deviations were 3.10 ± 4.44 and 1.03 ± 0.30 for moving ($n = 215$ chains) and stationary categories ($n = 40$ chains). A two-sample t-test rejected the null hypothesis (no difference between moving and still distributions) at 5% significance level for the three parameters.

Figure 8 summarizes all three features (SP velocity, SP duration and chain length CL) plotted on a single graph. The $n = 41$ false sawtooth detections appear as orange hued circles, compared to true detections ($n = 664$) shown in blue. Visual inspection of the data (*e.g.*, Fig. 8) indicated that these false detections, were shifted toward lower durations and speeds. Most visually significant was the observation that FP detections were clustered along the $CL = 1$ and $CL = 2$ planes; indicating that CL could be a discriminating factor for OKN present/absent. The performance of CL as an indicator of OKN present/absent is shown in Table 2. This table shows TP as a proportion of the total number of moving trials ($n = 110$) and FP as a proportion of total stationary trials ($n = 55$) shown to the observer. This table shows the reduction in FP rate for increasing CL as well as a drop in TP rate. The table suggests that a threshold of $CL = 2$ was the best balance between TP and FP for trial-by-trial detection of OKN. As a consequence of this result, the effects of SP duration and speed were not considered further, but would be the subject of future work.

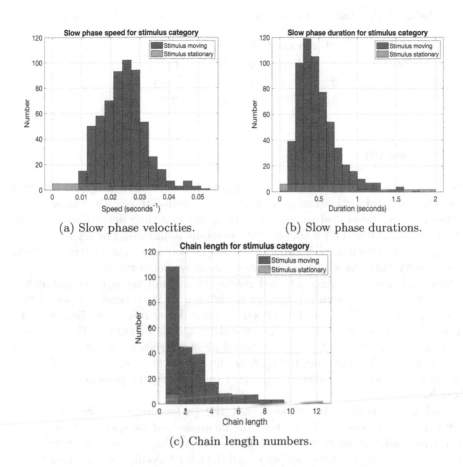

(a) Slow phase velocities.

(b) Slow phase durations.

(c) Chain length numbers.

Fig. 7. Eye displacement signal for the first sample.

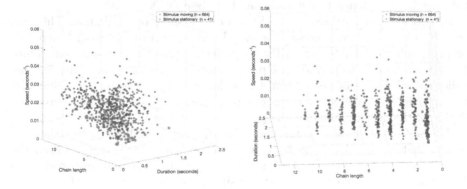

Fig. 8. Illustration of the distribution of three features from two views

Table 2. Per trial TP and FP rates as a function of the CL parameter.

CL threshold	TP rate	FP rate
1	109 (99%)	25 (46%)
2	102 (94%)	1 (2%)
3	89 (82%)	0 (0%)

5 Discussion

There is a clinical need for automated approaches able to identify the presence or absence of optokinetic nystagmus, particularly suited to real-time application. In this work we developed a method suitable for real-time detection of optokinetic nystagmus based on a simple state-machine approach. The algorithm was developed and run on a small cohort of adult participants, who watched drifting or stationary patterns whilst having their eye movements recorded.

Sawtooth patterns were detected readily in moving trials ($n = 664$ detections) but also during some stationary trials ($n = 41$ detections) (a per feature FP detection rate of 5.8%). The effects of false detections were eliminated by considering the *per trial* criteria that OKN should be repetitive. For example, we found that a criteria of $CL >= 2$ would identify trials with moving stimulus ($TPratewas94\%$) whilst eliminating false detections ($FPratewas2\%$). We suggest that chain-length may be a key factor in assessing the presence or absence of OKN.

Having said that, it is intended that the performance be evaluated more carefully in the future. We found that the mean slow-phase speed and durations for moving and stationary conditions were different, but we did not analyse this finding further. In this work, we presented data for the calibration set only, and we did not perform a robust validation analysis. We need to perform a full ROC analysis of the method which will be the subject of further work. It is emphasised that the aim here was to present the basic concept, which was the use of a state-machine approach to determine and analyse OKN.

In this work we utilised a web-browser to display our stimulus. This was facilitated by the jsPsych package (www.jsPsych.org), a web browser based API for psychophysical trials. The jsPsych package utilises a plug-in architecture that allows the user to perform pre-programmed tasks (*e.g.*, show a movie, play audio, record a reaction) or custom tasks that execute in a sequence defined by the experimenter. We wrote a custom plugin was written that facilitated the display of the disk stimulus for the purpose of web browser display.

The web-browser display was controlled from a second computer (the controller) that managed the experiment. Crucially the system synchronised the start and end of each trial of the experiment with the pupil labs eye tracker. In this work the server code was written using `node.js` and the interface to the server was written JavaScript/HTML5 thereby maintaining a non-platform specific implementation with the possibility of distributing more widely in future work. Furthermore, we developed a batch extraction of pupil location which can extract the location of the pupil for many participants simultaneously. This model can extract the (x, y) of pupil based on 2-D frame on the video stream. In the future, we are going to have an autonomous pupil detection based on cloud computing platform.

There are a number of limitations of the present study. It would be desirable to increase the number of participants tested which would allow a more detailed examination of the behaviour of participants during eye testing and to allow further generalization (if possible) of the threshold we have used already. As mentioned, we require to perform a more in-depth sensitivity-specificity analysis. We looked at a limited set of parameters in this work (essentially chain-length and slow phase duration/speeds), and a more in-depth analysis would be required to indicate optimal features for quantification of OKN. Our aim is to provide these methods for clinical use, and therefore future studies will quantify performance on target groups such as children. Future work will now look to determine whether the present protocols and processing approaches can be improved, and work is under-way to examine whether machine learning approaches will benefit the technique we have developed.

6 Conclusion

We have presented a method for detecting optokinetic nystagmus designed for real-time applications. We have obtained encouraging results for a cohort of adult participants (n = 11 eyes). Further research is warranted, and we will continue to improve upon and further validate the methods presented here. In a forthcoming publication, we will use machine learning model to detect OKN through signal processing and pattern recognition techniques.

References

1. Aleci, C., Scaparrotti, M., Fulgori, S., Canavese, L.: A novel and cheap method to correlate subjective and objective visual acuity by using the optokinetic response. Int. Ophthalmol., 1–15 (2017). https://doi.org/10.1007/s10792-017-0709-x
2. Anstice, N.S., Thompson, B.: The measurement of visual acuity in children: an evidence-based update. Clin. Exp. Optom. **97**(1), 3–11 (2014). https://doi.org/10.1111/cxo.12086
3. Barnes, G.R.: A procedure for the analysis of nystagmus and other eye movements. Aviat. Space Environ. Med. **53**(7), 676–682 (1982)
4. Connell, C.J.W., Thompson, B., Turuwhenua, J., Hess, R.F., Gant, N.: Caffeine increases the velocity of rapid eye movements in unfatigued humans. Psychopharmacology **234**(15), 2311–2323 (2017)
5. Dix, M.R.: The mechanism and clinical significance of optokinetic nystagmus. J. Laryngol. Otol. **94**(8), 845–864 (1980). https://doi.org/10.1017/S0022215100089611
6. Fife, T.D., et al.: Assessment: vestibular testing techniques in adults and children: report of the therapeutics and technology assessment subcommittee of the american academy of neurology. Neurology **55**(10), 1431–1441 (2000)

7. Garbutt, S., Harwood, M.R., Harris, C.M.: Comparison of the main sequence of reflexive saccades and the quick phases of optokinetic nystagmus. Br. J. Ophthalmol. **85**(12), 1477–1483 (2001)
8. Garrido-Jurado, S., Munoz-Salinas, R., Madrid-Cuevas, F.J., Marin-Jimenez, M.J.: Automatic generation and detection of highly reliable fiducial markers under occlusion. Pattern Recognit. **47**(6), 2280–2292 (2014). https://doi.org/10.1016/j.patcog.2014.01.005
9. Harris, P.A., Garner, T., Sangi, M., Guo, P., Turuwhenua, J., Thompson, B.: Visual acuity assessment in adults using optokinetic nystagmus. Investig. Ophthalmol. Vis. Sci. **60**(9), 5907 (2019)
10. Holmes, J.M., et al.: Effect of age on response to amblyopia treatment in children. Arch. Ophthalmol. **129**(11), 1451–1457 (2011)
11. Hyon, J.Y., Yeo, H.E., Seo, J.-M., Lee, I.B., Lee, J.H., Hwang, J.-M.: Objective measurement of distance visual acuity determined by computerized optokinetic nystagmus test. Investig. Ophthalmol. Vis. Sci. **51**(2), 752–757 (2010). https://doi.org/10.1167/iovs.09-4362
12. Jones, P.R., Kalwarowsky, S., Atkinson, J., Braddick, O.J., Nardini, M.: Automated measurement of resolution acuity in infants using remote eye-tracking. Investig. Ophthalmol. Vis. Sci. **55**(12), 8102–8110 (2014)
13. Juhola, M.: Detection of nystagmus eye movements using a recursive digital filter. IEEE Trans. Biomed. Eng. **35**(5), 389–395 (1988)
14. Kipp, M.: ANVIL - a generic annotation tool for multimodal dialogue (2001)
15. Kooiker, M.J.G., Pel, J.J.M., Verbunt, H.J.M., de Wit, G.C., van Genderen, M.M., van der Steen, J.: Quantification of visual function assessment using remote eye tracking in children: validity and applicability. Acta Ophthalmologica **94**(6), 599–608 (2016)
16. Levens, S.L.: Electronystagmography in normal children. Br. J. Audiol. **22**(1), 51–56 (1988)
17. Chang, L.Y.L., Guo, P., Thompson, B., Sangi, M., Turuwhenua, J.: Assessing visual acuity-test-retest repeatability and level of agreement between the electronic ETDRS chart (E-ETDRS), optokinetic nystagmus (OKN), and sweep VEP. Investig. Ophthalmol. Vis. Sci. **59**(9), 5789 (2018)
18. Pander, T., Czabanski, R., Przybyla, T., Pojda-Wilczek, D.: An automatic saccadic eye movement detection in an optokinetic nystagmus signal. Biomed. Eng./Biomedizinische Technik **59**(6), 529–543 (2014). https://doi.org/10.1515/bmt-2013-0137
19. Press, W.H., Teukolsky, S.A.: Savitzky-golay smoothing filters. Comput. Phys. **4**(6), 669 (1990)
20. Ranjbaran, M., Smith, H.L.H., Galiana, H.L.: Automatic classification of the vestibulo-ocular reflex nystagmus: integration of data clustering and system identification. IEEE Trans. Biomed. Eng. **63**(4), 850–858 (2016)
21. Reinecke, R.D., Cogan, D.G.: Standardization of objective visual acuity measurements: opticokinetic nystagmus us. Snellen acuity. AMA Arch. Ophthalmol. **60**(3), 418–421 (1958). https://doi.org/10.1001/archopht.1958.00940080436010
22. Rey, C.G., Galiana, H.L.: Parametric classification of segments in ocular nystagmus. IEEE Trans. Biomed. Eng. **38**(2), 142–148 (1991)
23. Sangi, M., Thompson, B., Turuwhenua, J.: An optokinetic nystagmus detection method for use with young children. IEEE J. Transl. Eng. Health Med. **3**, 1–10 (2015)

24. Valmaggia, C., Charlier, J., Gottlob, I.: Optokinetic nystagmus in patients with central scotomas in age related macular degeneration. Br. J. Ophthalmol. **85**(2), 169–172 (2001). https://doi.org/10.1136/bjo.85.2.169
25. Valmaggia, C., Gottlob, I.: Optokinetic nystagmus elicited by filling-in in adults with central scotoma. Investig. Ophthalmol. Vis. Sci. **43**(6), 1804–1808 (2002)
26. Valmaggia, C., et al.: Age related change of optokinetic nystagmus in healthy subjects: a study from infancy to senescence. Br. J. Ophthalmol. **88**(12), 1577–1581 (2004)
27. The future of eye tracking. pupil-labs.com/. Accessed 01 Aug 2019
28. Turuwhenua, J., Yu, T.Y., Mazharullah, Z., Thompson, B.: A method for detecting optokinetic nystagmus based on the optic flow of the limbus. Vis. Res. **103**, 75–82 (2014)
29. Wass, S.: The use of eye tracking with infants and children. In: Practical Research with Children. Routledge (2016)
30. Waddington, J., Harris, C.M.: Human optokinetic nystagmus: a stochastic analysis. J. Vis. **12**(12), 5 (2012). https://doi.org/10.1167/12.12.5
31. West, S., Williams, C.: Amblyopia in children (aged 7 years or less). BMJ Clin. Evid. **2016** (2016)
32. Yu, T.-Y., Jacobs, R.J., Anstice, N.S., Paudel, N., Harding, J.E., Thompson, B.: Global motion perception in 2-year-old children: a method for psychophysical assessment and relationships with clinical measures of visual function. Investig. Ophthalmol. Vis. Sci. **54**(13), 8408–8419 (2013)

Network Structure for Personalized Face-Pose Estimation Using Incrementally Updated Face-Shape Parameters

Makoto Sei[1,2(✉)], Akira Utsumi[1], Hirotake Yamazoe[2], and Joo-Ho Lee[2]

[1] ATR IRC Labs, Seika-cho, Sorakugun, Kyoto 619-0288, Japan
is0263xh@ed.ritsumei.ac.jp
[2] Ritsumeikan University, Kusatsu, Shiga 525-8577, Japan

Abstract. This paper proposes a deep learning method for face-pose estimation with an incremental personalization mechanism to update the face-shape parameters. Recent advances in machine learning technology have also led to outstanding performance in applications of computer vision. However, network-based algorithms generally rely on an off-line training process that uses a large dataset, and a trained network (e.g., one for face-pose estimation) usually works in a one-shot manner, i.e., each input image is processed one by one with a static network. On the other hand, we expect a great advantage from having sequential observations, rather than just single-image observations, in many practical applications. In such cases, the dynamic use of multiple observations will contribute to improving system performance. The face-pose estimation method proposed in this paper, therefore, focuses on an incremental personalization mechanism. The method consists of two parts: a pose-estimation network and an incremental estimation of the face-shape parameters (shape-estimation network). Face poses are estimated from input images and face-shape parameters through the pose-estimation network. The shape parameters are estimated as the output of the shape-estimation network and iteratively updated in a sequence of image observations. Experimental results suggest the effectiveness of using face-shape parameters in face-posture estimation. We also describe the incremental refinement of face-shape parameters using a shape-estimation network.

Keywords: Face-pose estimation · Network-based algorithm · Personalization mechanism · Face-shape parameter

1 Introduction

Much research has been conducted on face- or head-pose estimation, including work in such fields as user interfaces, human-robot interaction, behavior analysis, and driving assistance [13]. For example, in human-robot interaction, we can estimate gestures such as nodding or head shaking, user interests, and object of

© Springer Nature Switzerland AG 2020
S. Palaiahnakote et al. (Eds.): ACPR 2019, LNCS 12047, pp. 156–168, 2020.
https://doi.org/10.1007/978-3-030-41299-9_13

attention from face poses. Furthermore, in driving assistance, face direction is important for identifying drowsy or careless drivers. In our research, we assume the above situations in which one user employs a system for a certain period, and then a different user also uses it for a certain period. In such situations, we aim to achieve face-pose estimation that can adapt to the current user.

Many face-pose estimation methods have already been proposed, and these can be classified into appearance- and model-based methods. The former type focuses on the differences in face appearance with the changes in face pose and then estimates face poses based on such facial characteristics. Appearance-based methods can be further divided into several types. One is regression-based methods, which estimate face poses as a regression problem [8]. Another is manifold-embedding methods, which employ characteristics in which facial features exist on low-dimensional manifolds [6,7,11]. These appearance-based methods were not aimed at face-pose estimation that could be adapted to each user.

On the other hand, model-based methods focus on such facial feature points as the corners of the eye or mouth [1,2,12,15,19]. After detecting these facial features, the face poses are estimated by fitting facial models based on the facial feature positions. In many such methods, facial models are personalized during the face-pose estimation process for more accurate fitting of the face model. In recent years, deep learning methods have been proposed to estimate the face poses and deformation parameters of facial models, including facial expressions. However, deformation parameters for face models, which are estimated from each input image, increase the computational costs. In addition, if it is only possible to obtain images that pose difficulty for accurate estimation of face-model parameters, the accuracy of face-pose estimation may also decrease.

Therefore, in this paper, we solve these problems and thus achieve face-pose estimation that can adapt to particular users by combining two types of methods. In the proposed method, we separate face-pose estimation and the estimation of the deformation parameters of facial models rather than unifying the processes of face-pose estimation and face-shape parameter estimation. Our method consists of two modules: a Pose-estimation Network and a Shape-estimation Network. The Shape-estimation Network is a module that incrementally estimates the face-shape parameters (the deformation parameters of a facial model) with face-image sequences as input. The Pose-estimation Network estimates face poses from face images and face-shape parameters. We implemented appearance-based face-pose estimation using a simple CNN and compared the accuracy of face-pose estimation with and without face-shape parameters.

This paper provides two main contributions: (1) We proposed and designed a personalized face-pose estimation method that assumes situations occurring in typical interaction systems. (2) We confirmed the effectiveness of the incrementally estimated face-shape parameters in achieving accurate face-pose estimation.

In the next section, we describe related research. Section 3 explains our proposed approach, and Sect. 4 shows our method's implementation. Section 5 experimentally evaluates the effectiveness of the proposed method and discusses the results, and Sect. 6 concludes this paper.

2   Related Research

2.1   Appearance-Based Face-Pose Estimation Methods

Appearance-based methods focus on the differences in face appearance according to changes in face pose. In many cases, face poses can be estimated as a regression problem using face images as inputs. For example, Fanelli et al. proposed a head-pose estimation method with random regression forests for RGB-D images [8]. In addition, although facial images resemble high-dimensional features, these data reside on low-dimensional manifolds. Manifold-embedding methods estimate face poses by employing such characteristics [6,7,11,16]. For example, Drouard et al. proposed a method using HOG features and a mixture of linear regressions [7]. Although these methods could achieve basic face-pose estimation, they were not aimed at achieving user-independent face-pose estimation or, moreover, personalized face-pose estimation for adaptively handling multiple users.

2.2   Model-Based Face-Pose Estimation Methods

Model-based methods focus on facial features, such as the corners of the eye or mouth. Based on such facial features, facial models are fit to facial images to estimate face poses. An active appearance model (AAM) is one well-known scheme that models and learns the variations in 2D facial shapes and textures [5]. Many methods based on 3D facial models have been proposed [1,2,15]. For example, Zhu and Ramanan proposed a unified model for face detection, pose estimation, and facial landmark localization [15]. Recently, many deep learning methods have also been proposed [12,19]. Such methods can estimate face poses and localize facial landmarks as well as fit dense 3D facial models and estimate the deformation parameters of the facial models that correspond to a user's face and its expressions. However, such processes have been performed for each input image, which increases the computational costs. Furthermore, depending on the input image, we cannot always obtain accurate facial model parameters. In such cases, the accuracy of face poses is also decreased.

We solve the above problems by proposing a method that achieves personalized face-pose estimation.

2.3   Person-Specific Estimation/Recognition

In addition to face-pose estimation, various face-image processing methods, such as face-expression estimation or face recognition, have also been proposed. These methods achieved a person-specific method for improving estimation/recognition accuracy.

Many person-specific methods for estimating facial expressions have been proposed [3,4,18,21] For example, Baltrusaitis et al. proposed a person-specific feature-normalization technique that takes into account person-specific neutral expressions for action unit detection [3]. Chu et al. proposed a personalized facial expression analysis method that attempted to personalize a generic classifier for expression detection without needing additional labels from the test subject [4]. These modifications to the person-specific models can also be considered a kind of domain adaptation [14, 17] aimed at the ability to adapt to a specific user (i.e., domain).

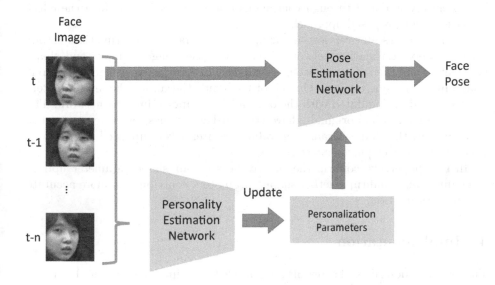

Fig. 1. Process flow of proposed method

Although there is no such example for face-pose estimation, many face-recognition methods, as well as facial expression estimation methods, implementing person-specific algorithms [9, 20] have been proposed.

3 Proposed Approach

Recent advances in machine learning technology have achieved outstanding performance exceeding human capabilities in many domains, including computer vision. As for face-image processing, many researchers have developed high-performance systems for face detection, identification, and pose estimation.

On the other hand, the current systems continue to suffer from limitations. In particular, the lack of a model-adaptation mechanism in network-based systems is a serious problem. Most network-based systems are constructed based on an off-line training process using a large dataset, and in a real-time process, the system utilizes constructed networks as a static model. However, since human faces have a wide variety of appearances based on different shapes, colors, and movements, covering all such variations in an off-line training process is difficult. In many practical situations, we have continuous observations of a target face, which is a reasonable and practical way to use multiple observations to adapt a network model to a particular target object. Accordingly, using multiple observations can enhance system performance.

To overcome the problem above, we propose a network system that has a built-in mechanism to detect and update target-specific parameters. We implement a face-pose estimation system utilizing face-shape parameters that are incrementally updated through continuous observations and confirm the effectiveness of our proposed approach.

Figure 1 shows an overview of our approach to face-pose estimation. In our system, a pose-estimation network accepts a single image as input and shape parameters and then outputs face-pose values. This process works for every input image in a one-shot manner. On the other hand, the above shape parameters are estimated and updated with the output of a shape-estimation network. The face-shape parameters originally have standardized values, but after sequential observations, the shape parameters' values are iteratively updated and personalized to the target person's face shape.

In this paper, we confirm the effectiveness of our shape-parameter update mechanism and confirm whether shape parameters contribute to more accurate face-pose estimation.

4   Implementation

This section describes the details of our system's implementation. First, we describe a face-shape parameter representation for use in the shape-estimation network. After that, network structures for both the shape-estimation network and pose-estimation network are explained.

4.1   Shape Parameter Representation

Since face data are typically large sets of 3-dimensional points, they are not suitable to use directly as an efficient representation of inter-personal variety. Therefore, we constructed a compact representation of face shapes through principal component analysis (PCA). Consequently, each face shape can be converted to a small vector by projecting the point data into a parameter space (Fig. 2). In this paper, we call that vector 'shape parameters'.

4.2   Shape-Estimation Network

The shape-estimation network receives a 150 × 150-pixel RGB color face image as input data. The network's output estimates the shape parameter for the input image data (Fig. 3). This network consists of eight layers (three sets of convolution and pooling layers and two fully connected layers near the output units). The averaged square errors are employed as an error function for training.

The output of the network is used for updating the shape parameters to be used for pose estimation. Currently, we use a simple average calculation for the update process. Refinement of this process should be a future work.

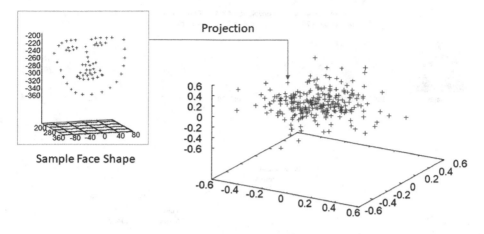

Fig. 2. Face-shape data and their representation in parameter space constructed by PCA

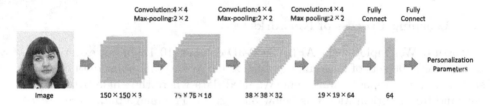

Fig. 3. Implementation of shape-estimation network

4.3   Pose-Estimation Network

Here, we explain the structure of the pose-estimation network (Fig. 4). Its input data are also a 150×150-pixel RGB color face image as well as the set of shape parameters described below. Three sets of convolution and pooling layers are applied to the input image, and their output is combined with the shape parameter. (In the current implementation, we employ a 7-dimensional vector to represent facial shapes). We attached a fully connected layer to the combined layer that eventually produced a 3D face-pose vector. The averaged square errors are employed as an error function for training.

The reason we employed pooling layers in the CNN part of this network is to reduce the size of the entire network's parameter space and thus avoid difficulty in the training process due to the limited variety of the training dataset. As for the performance of pose estimation, the pooling process may destroy the features that contain geometrical information of the target faces. We discuss a network structure without pooling layers in Sect. 5.4.

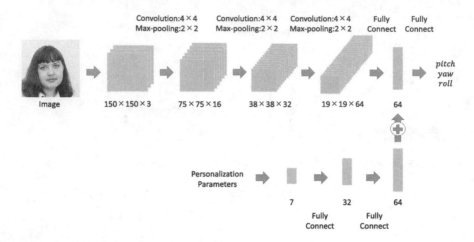

Fig. 4. Implementation of pose-estimation network

5 Experiment

5.1 Learning Process of Networks

Datasets. We employed the AFLW2000-3D dataset [10,19] in our experiments. It contains 2000 sets of facial images (450 × 450 pixels), 3D point data on facial surfaces, and face-pose parameters. We used data where the angle between the face and frontal orientation was below 30 degrees. The image data were resized to 150 × 150 pixels, and the 3D point data were converted to a smaller vector size through PCA as described in Subsect. 4.1. First, we selected 68 points corresponding to typical facial features from the data points and applied PCA to the selected point data for all 2000 images, thus obtaining an expression in a small number of dimensions. We found that a 7D expression can cover 85% of the data variability.

Learning of Pose-Estimation Network. In training the pose-estimation network, we used 1000 sets of randomly selected image data, 7D shape parameters, and 3D pose parameters (pitch, yaw, and roll). We used the following hyperparameters for the training process: learning rate: 0.0001; batch size: 64; iterations: 5000. The total training period was about 1 h.

Learning of Shape-Estimation Network. In training the shape-estimation network, we randomly selected 1000 sets of image data and 7D shape parameters. We selected the following hyperparameters for the training process: learning rate: 0.0001; batch size: 64; iterations: 10,000. The training required about two hours.

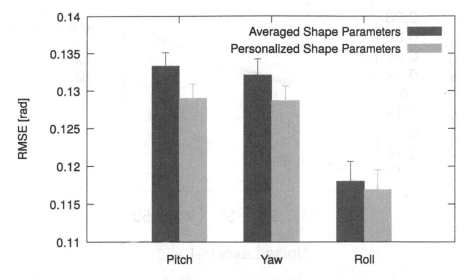

Fig. 5. Comparison between pose estimation with personalized shape parameters and pose estimation with common shape parameters (averaged) in terms of pose estimation performance. Vertical axis shows RMSE values of pose-estimation results. Here, we performed separate training and evaluation processes 10 times using different random data selections. Error bars denote standard errors among the processes.

5.2 Evaluation of Pose-Estimation Network

In this section, we confirm the performance of the pose-estimation network with the face-shape parameters as input. First, we compare the performance of our proposed network with/without appropriate face-shape parameters. Figure 5 shows the results of pose estimation: pitch, yaw, roll, and total. Here, in 'personalized shape parameters,' face image and shape parameters corresponding to the image are used for pose estimation. In 'averaged shape parameter,' common parameters (averaged parameter values of all evaluation data) are used for all face images. As can be seen, the RMSE of the personalized parameter case is lower than the averaged parameter case, suggesting solid contributions by the face-shape parameters to pose estimation.

Next, we confirm how the accuracy of face-shape parameters affects the pose-estimation performance. Figure 6 shows the performance of the pose estimation of the proposed method for different noise levels of face-shape parameter inputs. This result shows the importance of the shape parameters in the proposed method.

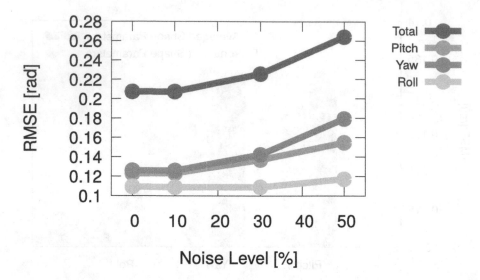

Fig. 6. Relation between accuracy of face-shape parameters and pose-estimation performance. Horizontal axis denotes level of noise added to shape parameter (average size of noise (normal distribution) as a percentage of maximum norm of shape vectors). Vertical axis shows RMSE values of pose-estimation results.

5.3 Parameter Personalization Using Shape-Estimation Network

In this section, we test the proposed process for incremental update of face-shape parameters. For this evaluation, we used the data of three people (one female and two males) whose multiple pose data were stored in the AFLW2000-3D dataset. These data were excluded from all of the above training sessions. Figure 7 shows the results of this incremental process.

As described in Sect. 3, the shape-estimation network outputs the estimation of a 7D face-shape vector for every frame of the image input. We calculated the RMSE values for every frame by comparing the estimated vector with the ground truth values derived from the dataset. We selected values for each target person's most frontal image in the dataset.

As shown in Fig. 7, for persons B and C, the RMSE values decrease with more images. These results suggest that a larger number of observations contribute to more accurate estimations of personalized shape parameters. In contrast, for person A, the RMSE range remains high, even with more than ten image observations. For this person, the shape parameters in the dataset also have large variability that might affect the results. Regarding this issue, we will investigate ways to achieve higher stability.

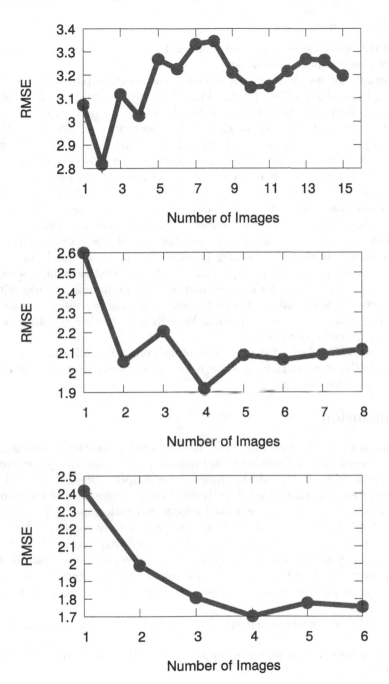

Fig. 7. Results of incremental shape-parameter estimations for three persons (top: person A, middle: person B, bottom: person C). Vertical axes show RMSE values for 7D face-shape vector, and horizontal axes show number of images used for estimation.

5.4 Discussion

Through the above experiments, we confirmed the effectiveness of the personalization of pose-estimation networks.

Unfortunately, several problems remain in the actual implementations. First, in the implementation of our pose-estimation network, although we added the personalized shape parameters to the network in the very final stage, we have to investigate a more effective approach from the perspective of personalization.

Although in this work we used PCA to extract the feature vectors from the face-shape data, it would be promising to investigate a more suitable approach that could select more significant parameters for pose estimation.

In the current implementation, we employed a standard CNN structure consisting of multiple convolution and pooling layers. Pooling layers are generally used for adapting a network to a variety of target sizes and positions in an image. Furthermore, it can help to reduce the size of the network parameters to be optimized through the learning process. On the other hand, introducing a pooling process causes geometrical features to be missed in input images, and thus it may be undesirable for pose estimation. Our preliminary results without pooling layers suggest that a network without a pooling process is promising in terms of performance in pose estimation. We will make further investigation into this matter in our future work.

In this study, we clarified the potential of network personalization. We plan to employ a similar framework for a wider range of domains such as the estimation of facial expressions and gaze.

6 Conclusion

We proposed a network-based algorithm to estimate face postures based on input images and personalized face-shape parameters. We estimated shape parameters as the output of an independently functioning shape-estimation network that can be updated and maintained iteratively through a sequence of image observations. In this paper, we implemented a pose estimation network as a combination of CNN-based image processing layers and a fully connected layer that integrates the image-processing results and the facial shape parameters. Experimental results show that using appropriate face-shape parameters improves performance in face-pose estimation. We also found that the iterative estimation of face-shape parameters using a shape-estimation network incrementally refines the estimation.

Future work will evaluate our method's performance using an online system.

Acknowledgement. This work was supported by JSPS KAKENHI Grant Number JP18H03269.

References

1. Abate, A.F., Barra, P., Bisogni, C., Nappi, M., Ricciardi, S.: Near real-time three axis head pose estimation without training. IEEE Access **7**, 64256–64265 (2019)

2. Asthana, A., Zafeiriou, S., Cheng, S., Pantic, M.: Incremental face alignment in the wild. In: Proceedings of the IEEE Conference on Computer Vision and Pattern Recognition, pp. 1859–1866 (2014)
3. Baltrušaitis, T., Mahmoud, M., Robinson, P.: Cross-dataset learning and person-specific normalisation for automatic action unit detection. In: 2015 11th IEEE International Conference and Workshops on Automatic Face and Gesture Recognition (FG), vol. 6, pp. 1–6. IEEE (2015)
4. Chu, W.S., De la Torre, F., Cohn, J.F.: Selective transfer machine for personalized facial expression analysis. IEEE Trans. Pattern Anal. Mach. Intell. **39**(3), 529–545 (2016)
5. Cootes, T.F., Edwards, G.J., Taylor, C.J.: Active appearance models. IEEE Trans. Pattern Anal. Mach. Intell. **6**, 681–685 (2001)
6. Diaz-Chito, K., Del Rincón, J.M., Hernández-Sabaté, A., Gil, D.: Continuous head pose estimation using manifold subspace embedding and multivariate regression. IEEE Access **6**, 18325–18334 (2018)
7. Drouard, V., Horaud, R., Deleforge, A., Ba, S., Evangelidis, G.: Robust head-pose estimation based on partially-latent mixture of linear regressions. IEEE Trans. Image Process. **26**(3), 1428–1440 (2017)
8. Fanelli, G., Gall, J., Van Gool, L.: Real time head pose estimation with random regression forests. In: CVPR 2011, pp. 617–624. IEEE (2011)
9. Hong, S., Im, W., Ryu, J., Yang, H.S.: SSPP-DAN: deep domain adaptation network for face recognition with single sample per person. In: 2017 IEEE International Conference on Image Processing (ICIP), pp. 825–829. IEEE (2017)
10. Köstinger, M., Wohlhart, P., Roth, P.M., Bischof, H.: Annotated facial landmarks in the wild: a large-scale, real-world database for facial landmark localization. In: 2011 IEEE International Conference on Computer Vision Workshops (ICCV Workshops), pp. 2144–2151, November 2011. https://doi.org/10.1109/ICCVW.2011.6130513
11. Lathuilière, S., Juge, R., Mesejo, P., Munoz-Salinas, R., Horaud, R.: Deep mixture of linear inverse regressions applied to head-pose estimation. In: Proceedings of the IEEE Conference on Computer Vision and Pattern Recognition, pp. 4817–4825 (2017)
12. Liu, Y., Jourabloo, A., Ren, W., Liu, X.: Dense face alignment. In: Proceedings of the IEEE International Conference on Computer Vision, pp. 1619–1628 (2017)
13. Murphy-Chutorian, E., Trivedi, M.M.: Head pose estimation in computer vision: a survey. IEEE Trans. Pattern Anal. Mach. Intell. **31**(4), 607–626 (2008)
14. Patel, V.M., Gopalan, R., Li, R., Chellappa, R.: Visual domain adaptation: a survey of recent advances. IEEE Signal Process. Mag. **32**(3), 53–69 (2015)
15. Ramanan, D., Zhu, X.: Face detection, pose estimation, and landmark localization in the wild. In: Proceedings of the 2012 IEEE Conference on Computer Vision and Pattern Recognition (CVPR), pp. 2879–2886. Citeseer (2012)
16. Ruiz, N., Chong, E., Rehg, J.M.: Fine-grained head pose estimation without keypoints. In: Proceedings of the IEEE Conference on Computer Vision and Pattern Recognition Workshops, pp. 2074–2083 (2018)
17. Wang, M., Deng, W.: Deep visual domain adaptation: a survey. Neurocomputing **312**, 135–153 (2018)
18. Zen, G., Porzi, L., Sangineto, E., Ricci, E., Sebe, N.: Learning personalized models for facial expression analysis and gesture recognition. IEEE Trans. Multimed. **18**(4), 775–788 (2016)

19. Zhu, X., Lei, Z., Liu, X., Shi, H., Li, S.Z.: Face alignment across large poses: a 3D solution. In: 2016 IEEE Conference on Computer Vision and Pattern Recognition (CVPR), pp. 146–155, June 2016. https://doi.org/10.1109/CVPR.2016.23
20. Zhu, Z., Luo, P., Wang, X., Tang, X.: Deep learning identity-preserving face space. In: Proceedings of the IEEE International Conference on Computer Vision, pp. 113–120 (2013)
21. Zong, Y., Huang, X., Zheng, W., Cui, Z., Zhao, G.: Learning a target sample regenerator for cross-database micro-expression recognition. In: Proceedings of the 25th ACM International Conference on Multimedia, pp. 872–880. ACM (2017)

Optimal Rejection Function Meets Character Recognition Tasks

Xiaotong Ji[1]([⊠]), Yuchen Zheng[1][iD], Daiki Suehiro[1,2][iD], and Seiichi Uchida[1][iD]

[1] Kyushu University, 744 Motooka Nishi-ku, Fukuoka 819-0395, Japan
{xiaotong.ji,yuchen,suehiro,uchida}@human.ait.kyushu-u.ac.jp
[2] RIKEN, 2-1 Hirosawa, Wako, Saitama 351-0198, Japan

Abstract. In this paper, we propose an optimal rejection method for rejecting ambiguous samples by a rejection function. This rejection function is trained together with a classification function under the framework of Learning-with-Rejection (LwR). The highlights of LwR are: (1) the rejection strategy is not heuristic but has a strong background from a machine learning theory, and (2) the rejection function can be trained on an arbitrary feature space which is different from the feature space for classification. The latter suggests we can choose a feature space which is more suitable for rejection. Although the past research on LwR focused only its theoretical aspect, we propose to utilize LwR for practical pattern classification tasks. Moreover, we propose to use features from different CNN layers for classification and rejection. Our extensive experiments of notMNIST classification and character/non-character classification demonstrate that the proposed method achieves better performance than traditional rejection strategies.

Keywords: Learning with Rejection · Optimal rejection function · Theoretical machine learning

1 Introduction

According to general performance improvement in various recognition tasks [1–3], many users may expect that most test samples should be classified correctly. In fact, recent classifiers use forced decision-making strategies, where any test sample is always classified into one of the target classes [4,5]. However, even with the recent classifiers, it is theoretically impossible to classify test samples correctly in several cases. For example, we cannot guarantee the correct recognition of a sample from the overlapping region of class distributions.

As a practical remedy to deal with those ambiguous samples, recognition with a rejection option has been used so far [6,7]. As shown in Fig. 1(a), there are three main targets for rejection: outliers, samples with incorrect labels, and samples in an overlapping area of class distributions. Among them, outliers and incorrect labels are detected by anomaly detection methods. The overlapping area is a more common reason to introduce the rejection option into the training

© Springer Nature Switzerland AG 2020
S. Palaiahnakote et al. (Eds.): ACPR 2019, LNCS 12047, pp. 169–183, 2020.
https://doi.org/10.1007/978-3-030-41299-9_14

stage as well as the testing stage, since the samples in the overlapping area are inherently distinguishable and thus better to be treated as "don't-care" samples.

In the long history of pattern recognition, researchers have tried to reject ambiguous samples in various ways. The most typical method is to reject samples around the classification boundary. Figure 1(b) shows a naive rejection method for a support vector machine (SVM) for rejecting samples in the overlapping area. If the distance between a sample and the classification boundary is smaller than a threshold, the sample is going to be rejected [8].

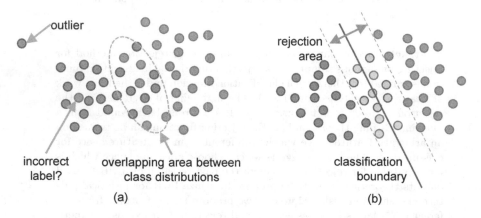

Fig. 1. (a) Three main targets for rejection. (b) Typical heuristics for rejecting the samples in the overlapping area of class distributions. This method assumes that we already have an accurate classification boundary—but, how can we reject ambiguous samples when we train the classification boundary?

Although this simple rejection method is often employed for practical applications, it has a clear problem that it needs an accurate classification boundary in advance to rejection. In other words, it can reject only test samples. Ambiguous samples in the overlapping area also exist in the training set and they will badly affect the classification boundary. This fact implies that we need to introduce a more sophisticated rejection method where the rejection criterion should be "co-optimized" with the classification boundary in the training stage.

In this paper, we propose an optimal rejection method for rejecting ambiguous training and test samples for tough character recognition tasks. In the proposed method, a rejection function is optimized (i.e., trained) together with a classification function during the training step. To realize this co-optimization, we follow the *learning-with-rejection* (LwR) framework proposed in theoretical machine learning research [9]. Figure 2 illustrates LwR framework. During the training step, several samples are determined to be rejected by a rejection function and they are not used to train the classifier function. It should be emphasized again this is a co-optimization framework where the rejection function and the classification function are optimized by solving a single optimization problem.

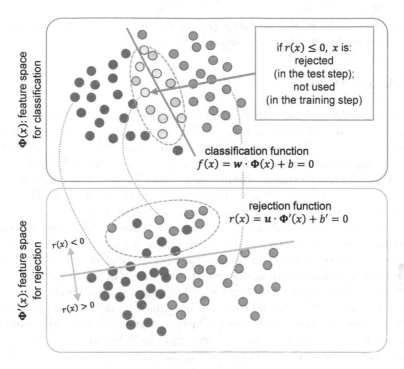

if $r(x) \leq 0$, x is:
rejected
(in the test step);
not used
(in the training step)

classification function
$f(x) = \boldsymbol{w} \cdot \boldsymbol{\Phi}(x) + b = 0$

$\boldsymbol{\Phi}(x)$: feature space for classification

rejection function
$r(x) = \boldsymbol{u} \cdot \boldsymbol{\Phi}'(x) + b' = 0$

$r(x) < 0$

$r(x) > 0$

$\boldsymbol{\Phi}'(x)$: feature space for rejection

Fig. 2. An overview of *learning-with-rejection* for a two-class problem. The rejection function $r(x)$ is co-trained with the classification function $f(x)$. Note that different feature spaces are used for optimizing $r(x)$ and $f(x)$.

LwR has many promising characteristics; for example, it has a strong theoretical support about its optimality (about the overfitting risk) and it can use a different feature space for its rejection function.

The main contributions of our work are summarized as follows:

- This is the first application of LwR to a practical pattern recognition task. Although the idea of LwR is very suitable to various pattern recognition tasks (including, of course, character recognition and document processing), no application research has been done so far, to the authors' best knowledge.
- For designing features for classification and rejection, we utilize a convolutional neural network (CNN) as a multiple-feature extractor; the classification function is trained by using the output from the final convolution layer and the rejection function is trained from a different layer.
- We conducted two classification experiments to show that LwR outperforms CNN with confidence-based rejection, and traditional SVM with the rejection strategy of Fig. 1(b).

The rest of the paper is organized as follows: Sect. 2 reviews related works and Sect. 3 is devoted to explain the problem formulation and the theoretical background of LwR. In Sect. 4, the experimental setting and the discussion of final results are shown, while conclusion is given in Sect. 5.

2   Related Work

Since data usually contains unrecognizable samples, data cleansing [10–12] has been an important topic in many fields such as medical image processing [13,14], document analysis [15–17] and commercial applications [18]. Among those applications, ambiguous samples in dataset might incur serious troubles at data processing stage. To handle this problem, many data cleansing methods have been studied [19,20] and the goal of all of them is to keep the data clean. To guarantee the data integrity, most of them apply repairing operation to the target data with inconsistent format. Instead of repairing those acceptable flaws, authors in [21] aim to remove erroneous samples from entire data, and in [22], Fecker et al. propose a method to reject query manuscripts without corresponding writer with a rejection option. These works show the necessity of rejection operation in data cleansing. For machine learning tasks, training data with undesirable samples (as shown in Fig. 1(a)) could lead to a ill-trained model, whose performance in the test phase is doomed to be poor. On the other hand, test data could also contains such unwelcome samples but are apparently better to be removed by the trained model than those data cleansing methods.

Other than data cleansing, rejection operation is also commonly used in many fields of machine learning such as handwriting recognition [23,24], medical image segmentation [25,26] and image classification tasks [27–29]. Here, the rejection operation are done during the test stage but not the training stage. In the handwriting recognition field, Kessentini et al. apply a rejection option to Latin and Arabic scripts recognition and gained significant improvements compared to other state-of-the-art techniques [30]. Mesquita et al. develop a classification method with rejection option [31] to postpone the final decision of non-classified modules for software defect prediction. In [32], He et al. construct a model with rejection on handwritten number recognition, where the rejection option nicely prevented misclassifications and showed high reliability. Mekhmoukh and Mokrani employ a segmentation algorithm that is very sensitive to outliers in [26] to effectively reject anomalous samples. Niu et al. in [33] present a hybrid model of integrating the synergy of a CNN and SVM to recognize different types of patterns in MNIST, achieved 100% reliability with a rejection rate of 0.56%. However, works mentioned above treat rejection operation as the extension of classification function. It is notable that dependence on classification function could endow potential limitation to rejection ability of the models.

Rejection operation also plays an important role in document processing field [34,35]. Bertolami et al. investigate rejection strategies based on various confidence measures in [36] for unconstrained offline handwritten text line recognition. In [37], Li et al. propose a new CNN based confidence estimation framework to reject the misrecognized character whose confidence value exceeds a preset threshold. However, to our knowledge, there still has no work on combining CNNs and independent rejection functions in document analysis and recognition field. Since it is well known that CNNs have powerful feature extraction capabilities, outputs of hidden layers in CNNs could be used instead of kernels as feature spaces for model training. In this paper, we propose a novel framework with a rejection function learned along with but not dependent on the classification function, using features extracted from a pre-trained CNN.

3   Learning with Rejection

In this section, we introduce the problem formulation of LwR. Let (x_1, y_1), ..., (x_m, y_m) denote training samples. Let Φ and Φ' denote functions that map x to different feature spaces. Let $f = w \cdot \Phi(x) + b$ be a linear classification function over Φ, while $r = u \cdot \Phi'(x) + b'$ be a linear rejection function over Φ'. Our LwR problem is formulated as an optimization problem of w, b, u and b' as follows. As shown in Fig. 2, a two-class classification problem with reject option uses the following decision rule :

$$\begin{cases} +1 \ (x \text{ is class1}) & \text{if } (f(x) > 0) \wedge (r(x) > 0), \\ -1 \ (x \text{ is class2}) & \text{if } (f(x) \le 0) \wedge (r(x) > 0), \\ \text{reject} & \text{if } r(x) \le 0. \end{cases}$$

To achieve a good balance between the classification accuracy and the rejection rate, we minimize the following risk:

$$R(f, r) = \sum_{i=1}^{m} \left(1_{(\text{sgn}(f(x_i)) \ne y_i) \wedge (r(x_i) > 0)} + c1_{(r(x_i) \le 0)} \right), \tag{1}$$

where c is a parameter which weights the rejection cost. The cost represents the penalty to the rejection operation.

As an example in document processing task, assuming digit recognition for bank bills. In this case, there is almost no tolerance to errors and thus the rejection cost c must be set at a lower value for rejecting more suspicious patterns to avoid possible mistakes. In contrast, in the case of character recognition of personal diary, c can be set at a larger value for accepting more ambiguous characters is acceptable.

Since the direct minimization of the risk R is difficult, Cortes et al. [9] proposed a convex surrogate loss L as follows:

$$L = \max \left(1 + \frac{\alpha}{2}(r(x) - yf(x)), c(1 - \beta r(x)), 0 \right),$$

where $\alpha = 1$ and $\beta = 1/(1 - 2c)$. In [9], it is proved that the minimization problem of L along with a typical regularization of $\boldsymbol{w}$ and $\boldsymbol{u}$ results in the following SVM-like optimization problem:

$$\min_{\boldsymbol{w},\boldsymbol{u},b,b'\boldsymbol{\xi}} \frac{\lambda}{2}\|\boldsymbol{w}\|^2 + \frac{\lambda'}{2}\|\boldsymbol{u}\|^2 + \sum_{i=1}^{m}\xi_i$$

$$\text{sub. to } \xi_i \geq c\left(1 - \beta\left(\boldsymbol{u}\cdot\boldsymbol{\Phi}'\left(x_i\right) + b'\right)\right),$$

$$\xi_i \geq 1 + \frac{\alpha}{2}\left(\boldsymbol{u}\cdot\boldsymbol{\Phi}'\left(x_i\right) + b' - y_i(\boldsymbol{w}\cdot\boldsymbol{\Phi}\left(x_i\right) + b)\right),$$

$$\xi_i \geq 0 \ (i = 1,\dots,m), \tag{2}$$

where λ and λ' are regularization parameters and ξ_i is a slack variable. This is a minimization problem of a quadratic objective function with linear constraints and thus its optimal solution can be easily obtained by using a quadratic programming (QP) solver.

It should be emphasized that LwR solution of the above-mentioned formulation has a strong theoretical background. Similarly to the fact that the standard SVM solution has theoretical guarantee of its generalization performance, the optimal solution of the optimization problem (2) in LwR mentioned above also has a theoretical support. See Appendix for its brief explanation. It should also be underlined that the original LwR research has mainly focused on this theoretical analysis and thus has never been used for any practical application. To the authors' best knowledge, this paper is the first application of LwR to a practical task.

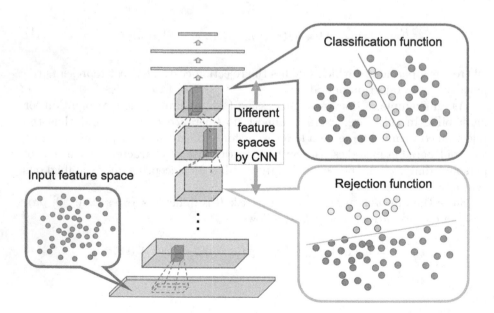

Fig. 3. Extracting different features from CNN for LwR.

4   Experimental Result

In this section, we demonstrate that our optimal rejection approach effectively works for character recognition tasks. More precisely, we consider two binary classification tasks: classification of similar characters and classification of character and non-character. Of course, in those tasks, it is better to achieve a high accuracy with a low reject rate.

4.1   Experimental Setup

Methods for Comparison. Through the experiments, we compare three methods: CNN classifier, SVM classifier[1], and LwR approach. The SVM classifier is set linear for deep CNN features we use already contain nonlinear compositions.

The performances of these three classification models are compared with shifting rejection cost parameters as $c = 0.1, \ldots, 0.4$. CNN and SVM output class probability $p(y|x)$, and thus we consider their final output $f(x)$ with reject option based on confidence threshold θ, that is,

$$
f(x) = \begin{cases} +1 \ (x \text{ is class1}) & \text{if } p(+1|x) > \theta, \\ -1 \ (x \text{ is class2}) & \text{if } p(-1|x) > \theta, \\ \text{reject} & \text{otherwise.} \end{cases}
$$

We determine the optimal confidence threshold θ of rejection according to the risk R using the *trained* $f(x)$ and a validation set, where the R of CNN and SVM can be defined as:

$$
R(f) = \sum_{i=1}^{m} \left(1_{((f(x_i)) \neq y_i) \wedge (f(x) \neq \text{reject})} + c 1_{(f(x) = \text{reject})} \right).
$$

For the LwR model, validation samples are used to tune hyperparameters λ and λ' for LwR could determine its rejection rule by using only training samples.

The learning scenario of these three methods in these experiments is introduced as follow. For each binary classification tasks, we divide the binary labeled training samples into three sets: (1) Sample set for training a CNN. This CNN is used as not only the feature extractor for LwR and SVM but also a CNN classifier. (2) Sample set for training SVM and LwR with the features extracted by the trained CNN. (3) Sample set for validating the parameters of each method.

Feature Extraction. As mentioned above, one of the key ideas of LwR is to use different feature spaces to construct rejection function. For this point, we use the outputs of two different layers of the trained CNN for the classifier $f(x)$ and rejection function $r(x)$ respectively, as shown in Fig. 3. Specifically, we use the final convolutional layer of the CNN for classification. Note that the SVM classifier also uses the same layer as its feature extractor. For the training stage

---

[1] We use an implementation of SVM that estimation of posterior class probabilities.

of rejection function, the second final layer of the CNN is used for we found it achieved good performance in our preliminary experiments. In addition, if the same layer is used for both classification and rejection functions, the LwR model will degrade to a SVM-like model with information from only one feature space.

Evaluation Criteria. Using sample set S (i.e. the test set) for each task, we evaluate the classification performance with reject option of these three models by following metrics: classification accuracy for non-rejected test samples (which we simply call *accuracy* for short) and *rejection rate* are defined as:

Fig. 4. Examples of "I" (upper row) and "J" (lower row) form notMNIST.

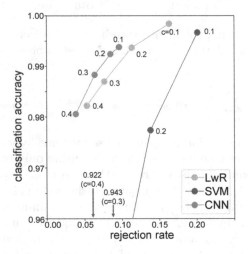

Fig. 5. The relationship between accuracy and reject rate for the classification task between "I" and "J" from notMNIST dataset.

$$\text{accuracy} = \frac{\sum_{x_i \in \{S - S_{rej}\}} 1_{f(x_i)=y_i}}{|S| - |S_{\text{rej}}|}, \quad \text{rejection rate} = \frac{|S_{\text{rej}}|}{|S|},$$

where S_{rej} is the set of rejected samples, and $f(x_i)$ is the classification function of each method.

Fig. 6. Examples from Chars74k (upper row) and CIFAR100 (lower row) datasets.

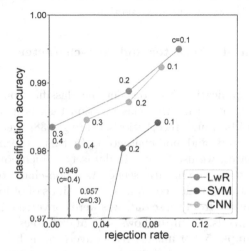

Fig. 7. The relationship between accuracy and reject rate for character and non-character classification using Chars74k and CIFAR100.

4.2 Classification of Similar Characters with Reject Option

In this section, we use a dataset named notMNIST[2] as both training and test data. The notMNIST dataset is an image dataset consists of A-to-J alphabets including various kinds of fonts, whose examples are shown in Fig. 4. We choose a pair of very similar letters "I" and "J" to organize a tough two-class recognition task, where reject operation of many ambiguous samples is necessary. We randomly selected 3,000 training samples of "I" and "J" respectively from notMNIST dataset. We also used 2,000 samples for CNN training, 2,000 samples for SVM and LwR training, and 2,000 samples for validations. We then obtain CNN classifier and SVM with reject option, and LwR. The CNN architecture in this experiment is as follows: two convolutional layers (3 kernels, kernel size 3×3 with stride (1), two pooling layers (kernel size 2×2 with stride (2) with Rectified Linear Units (ReLU) as the activation function, three fully-connected layers (3,136, 1,568, and 784 hidden units respectively), with minimization of cross-entropy loss.

---

[2] https://www.kaggle.com/lubaroli/notmnist/.

Figure 5 shows the relationship between accuracy and reject rate, evaluated by using the test dataset. LwR clearly shows significantly better performance than the standard SVM with rejection option. Surprisingly, CNN could show slightly better performance than LwR at the stage when the rejection rate is low (although LwR showed the highest accuracy in this experimental range). This might be because the classifier of CNN, i.e., fully-connected layers of CNN is trained (or tuned) together with the feature extraction part of CNN and thus CNN has an advantage by itself. We will see that LwR outperforms in a more realistic experiment shown in the next section.

4.3   Classification of Character and Non-character with Reject Option

As a more realistic application, we consider the classification of character and non-character. It is known that this classification task is a part of scene text detection task and still a tough recognition task (e.g., [38]) because of the ambiguity between characters and non-characters. In this experiment, as a scene character image dataset, we use Chars74k dataset[3], which contains large number of character images in the natural scene. As a non-character image dataset, we use CIFAR100 dataset[4] which contains various kinds of objects in the natural scene. The image samples from those datasets were converted into gray scale and resized into 32×32 pixels. Figure 6 shows several examples.

We randomly sample 5,000 images respectively from both CIFAR100 and Chars74k datasets. Among them, 6,000 are used for CNN training, 2,000 for SVM and LwR training, and 2,000 samples for validations. After training, we obtain CNN classifier and SVM with rejection thresholds, and LwR. CNN with the following architecture was used for this task: convolutional layers (32, 64, 128 kernels, kernel size 3×3, 5×5, 5×5, with stride 1, 2, 2 respectively, using ReLU as activation function), one max pooling layer (2×2 sized kernel with stride 2) and two fully-connected layers (512 and 2 units with ReLU and softmax), with minimization of cross entropy.

Figure 7 shows the test accuracy using 2,000 samples randomly chosen from Chars74k and CIFAR100[5]. We can see LwR could achieve the best compromise between accuracy and rejection rate among three methods. Surprisingly, when $c = 0.3$ and $c = 0.4$, LwR predicts the labels of all test samples (i.e., no rejected test samples) with keeping a higher accuracy than CNN. This result confirm the strength of theoretically formulated LwR that obtains the classification and rejection function for minimizing the risk R, while the rejection option for CNN and SVM is rather heuristic.

Figure 8 shows the image samples are to be misrecognized without the rejection function. Namely, if we only rely on $f(x)$, the sample x is misrecognized.

---

[3] http://www.ee.surrey.ac.uk/CVSSP/demos/chars74k/.

[4] https://www.cs.toronto.edu/~kriz/cifar.html.

[5] CIFAR100 has much more samples for testing, but unfortunately, Chars74k remains highly limited number of samples for testing.

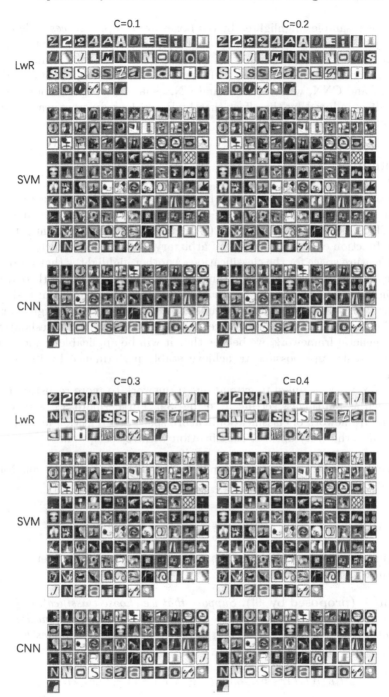

Fig. 8. Test samples misclassified by the LwR, the standard SVM and the CNN at different $c \in \{0.1, 0.2, 0.3, 0.4\}$. Rejected samples are highlighted in red. (Color figure online)

Among them, samples highlighted by red boxes are samples *successfully rejected* by the rejection function $r(x)$. Those highlighted samples represent that classification function $f(x)$ and rejection function $r(x)$ work in a complementary manner. In addition, we can find LwR has a lot of different rejected samples from SVM and CNN, whereas SVM and CNN share a lot of common samples. This result is induced by the effect that LwR employs different feature spaces for classification and rejection.

5   Conclusion

We propose an optimal rejection method for character recognition tasks. The rejection function is trained together with a classification function under the framework of Learning-with-Rejection (LwR). One technical highlight is that the rejection function can be trained on an arbitrary feature space which is different from the feature space for the classification. Another highlight is that LwR is not heuristic and its performance is guaranteed by a machine learning theory. From the application side, this is the first trial of using LwR for practical tasks. The experimental results show that the optimal rejection is useful for tough character recognition tasks than traditional strategies, such as SVM with a threshold. Since LwR is a general framework, we believe that it will be applicable to more tough recognition tasks and possible to achieve stable performance by its rejection function.

As future work, we plan to extend our framework to more realistic problem setting such as multi-class setting by using [39]. Furthermore, we will consider co-training framework of classification feature space and rejection feature space, which enables the truly optimal classification with reject option.

Acknowledgement. This work was supported by JSPS KAKENHI Grant Number JP17H06100 and JP18K18001.

Appendix

The optimal solution of (2) has a theoretical guarantee of the generalization performance as follows:

Theorem 1 (proposed by [9]**).** *Suppose that we choose classification function f from a function set F and choose rejection function r from a function set G. We denote R for test samples by R_{test}, and R for training samples by R_{train}. Then, the following holds:*

$$R_{test}(f, r) \leq R_{train}(f, r) + \Re(F) + (1 + c)\Re(G). \tag{3}$$

In above, $\Re$ is the Rademacher complexity [40], which is a measure evaluating the theoretical overfitting risk of a function set. Roughly speaking, this theorem says that, if we prepare a proper feature space (i.e., not-so-complex feature space) mapped by Φ and Φ', we will achieve a performance on test samples as high as on training samples. Thus, by solving the problem (2) of minimizing R_{train}, we can obtain theoretically optimal classification and rejection function which achieves the best balance between the classification error for not-rejected samples and the rejection cost [34].

References

1. Schroff, F., Kalenichenko, D. Philbin, J.: FaceNet: a unified embedding for face recognition and clustering. In: CVPR (2015)
2. Sun, Y., Liang, D., Wang, X., Tang, X.: DeepiD3: face recognition with very deep neural networks. arXiv preprint arXiv:1502.00873 (2015)
3. Rautaray, S.S., Agrawal, A.: Vision based hand gesture recognition for human computer interaction: a survey. Artif. Intell. Rev. **43**, 1–54 (2015)
4. Rouhi, R., Jafari, M., Kasaei, S., Keshavarzian, P.: Benign andmalignantbreast tumors classification based on region growing and cnn segmentation. Expert Syst. Appl. **42**(3), 990–1002 (2015)
5. Yang, Z., Yang, D., Dyer, C., He, X. Smola, A., Hovy, E.: Hierarchical attention networks for document classification. In: Proceedings of the 2016 Conference of the North American Chapter of the Association for Computational Linguistics: Human Language Technologies (2016)
6. Condessa, F., Bioucas-Dias, J., Kovačević, J.: Supervised hyperspectral image classification with rejection. IEEE J. Sel. Top. Appl. Earth Observ. Remote Sens. **9**, 2321–2332 (2016)
7. Marinho, L.B., Almeida, J.S., Souza, J.W.M., Albuquerque, V.H.C., Rebouças Filho, P.P.: A novel mobile robot localization approach based on topological maps using classification with reject option in omnidirectional images. Expert Syst. Appl. **72**, 1–17 (2017)
8. Bartlett, P.L., Wegkamp, M.H.: Classification with a rejectoption using ahinge loss. J. Mach. Learn. Res. **9**, 1823–1840 (2008)
9. Cortes, C., DeSalvo, G., Mohri, M.: Learning with rejection. In: Ortner, R., Simon, H.U., Zilles, S. (eds.) ALT 2016. LNCS (LNAI), vol. 9925, pp. 67–82. Springer, Cham (2016). https://doi.org/10.1007/978-3-319-46379-7_5
10. Hernández, M.A., Stolfo, S.J.: Real-world data is dirty: data cleansing and the merge/purge problem. DMKD **2**, 9–37 (1998)
11. Maletic, J.I., Marcus, A.: Data cleansing: beyond integrity analysis. In: IQ (2000)
12. Khayyat, Z., et al.: BigDansing: a system for big data cleansing. In: SIGMOD (2015)
13. Lehmann, T.M., Gonner, C., Spitzer, K.: Survey: interpolation methods in medical image processing. TMI **18**, 1049–1075 (1999)
14. McAuliffe, M.J., Lalonde, F.M., McGarry, D, Gandler, W., Csaky, K., Trus, B.L.: Medical image processing, analysis and visualization in clinical research. In: CBMS (2001)
15. Neumann, L., Matas, J.: Efficient scene text localization and recognition with local character refinement. In: ICDAR (2015)

16. Messina,R., Louradour, J.: Segmentation-free handwritten chinese text recognition with LSTM-RNN. In: ICDAR (2015)
17. Chen, K., Seuret, M., Liwicki, M., Hennebert, J., Ingold, R.: Page segmentation of historical document images with convolutional autoencoders. In: ICDAR (2015)
18. Rahm, E., Do, H.H.: Data cleaning: problems and current approaches. Data Eng. Bull. **23**(4), 3–13 (2000)
19. Mezzanzanica, M., Boselli, R., Cesarini, M., Mercorio, F.: A model-based approach for developing data cleansing solutions. JDIQ **5**(4), 1–28 (2015)
20. Lee, M.L., Lu, H., Ling, T.W., Ko, Y.T.: Cleansing data for mining and warehousing. In: Bench-Capon, T.J.M., Soda, G., Tjoa, A.M. (eds.) DEXA 1999. LNCS, vol. 1677, pp. 751–760. Springer, Heidelberg (1999). https://doi.org/10.1007/3-540-48309-8_70
21. Chen, H., Ku, W.-S., Wang, H., Sun, M.-T.: Leveraging spatio-temporal redundancy for RFID data cleansing. In: SIGMOD (2010)
22. Fecker, D., Asi, A., Pantke, W., Märgner, V., El-Sana, J., Fingscheidt, T.: Document Writer analysis with rejection for historical Arabic manuscripts. In: ICFHR (2014)
23. Pal, U., Sharma, N., Wakabayashi, T., Kimura, F.: Off-line handwritten character recognition of Devnagari Script. In: ICDAR (2007)
24. LeCun, Y., et al.: Handwritten digit recognition with a back-propagation network. In: NIPS (1990)
25. Olabarriaga, S.D., Smeulders, A.W.: Interaction in thesegmentation ofmedical images: a survey. Med. Image Anal. **5**, 127–142 (2001)
26. Mekhmoukh, A., Mokrani, K.: Improved fuzzy C-means based particle swarm optimization (PSO) initialization and outlier rejection with level setmethods for MR brain image segmentation. Comput. Methods Program. Biomed. **122**(2), 266–281 (2015)
27. Chan, T.-H., Jia, K., Gao, S., Lu, J., Zeng, Z., Ma, Y.: PCANet: a simple deep learning baseline for image classification? TIP **24**(12), 5017–5032 (2015)
28. Wang, J., Yang, J., Yu, K., Lv, F., Huang, T., Gong, Y.: Locality-constrained Linear Coding for Image Classification. In: CVPR (2010)
29. Cireşan, D., Meier, U., Schmidhuber, J.: Multi-column deep neural networks for image classification. arXiv preprint arXiv:1202.2745 (2012)
30. Kessentini, Y., Burger, T., Paquet, T.: A Dempster-Shafer theory based combination of handwriting recognition systems with multiple rejection strategies. Pattern Recogn. **48**, 534–544 (2015)
31. Mesquita, D.P., Rocha, L.S., Gomes, J.P.P., Neto, A.R.R.: Classification with reject option for software defect prediction. Appl. Soft Comput. **49**, 1085–1093 (2016)
32. He, C.L., Lam, L., Suen, C.Y.: A novel rejection measurement in handwritten numeral recognition based on linear discriminant analysis. In: ICDAR (2009)
33. Niu, X.-X., Suen, C.Y.: A novel hybrid CNN-SVM classifier forrecognizinghandwritten digits. Pattern Recogn. **45**(4), 1318–1325 (2012)
34. Maitra, D.S., Bhattacharya, U., Parui, S.K.: CNN based common approach to handwritten character recognition of multiple scripts. In: ICDAR (2015)
35. Serdouk, Y., Nemmour, H., Chibani, Y.: New off-line handwritten signature verification method based on artificial immune recognition system. Expert Syst. Appl. **51**, 186–194 (2016)

36. Bertolami, R., Zimmermann, M., Bunke, H.: Rejection strategiesfor offline hand-written text line recognition. Pattern Recogn. Lett. **27**(16), 2005–2012 (2006)
37. Li, P., Peng, L., Wen, J.: Rejecting character recognition errors using CNN based confidence estimation. Chin. J. Electron. **25**, 520–526 (2016)
38. Bai, X., Shi, B., Zhang, C., Cai, X., Qi, L.: Text/Non-text image classification in the wild with convolutional neural networks. Pattern Recognition **66**, 437–446 (2017)
39. Platt, J.C., Cristianini, N., Shawe-Taylor, J.: Large margin DAGs for multiclass classification. In: NIPS (2000)
40. Bartlett, P.L., Mendelson, S.: Rademacher and Gaussian complexities: risk bounds and structural results. JMLR **3**, 463–482 (2003)

Comparing the Recognition Accuracy of Humans and Deep Learning on a Simple Visual Inspection Task

Naoto Kato, Michiko Inoue, Masashi Nishiyama[(✉)], and Yoshio Iwai

Graduate School of Engineering, Tottori University, Tottori, Japan
nishiyama@tottori-u.ac.jp

Abstract. In this paper, we investigate the number of training samples required for deep learning techniques to achieve better accuracy of inspection than a human on a simple visual inspection task. We also examine whether there are differences in terms of finding anomalies when deep learning techniques outperform human subjects. To this end, we design a simple task that can be performed by non-experts. It required that participants distinguish between normal and anomalous symbols in images. We automatically generated a large number of training samples containing normal and anomalous symbols in the task. The results show that the deep learning techniques required several thousand training samples to detect the locations of the anomalous symbols and tens of thousands to divide these symbols into segments. We also confirmed that deep learning techniques have both advantages and disadvantages in the task of identifying anomalies compared with humans.

Keywords: Visual inspection · Deep learning · People

1 Introduction

In recent years, the automation of the manufacturing industry has been promoted to mitigate labor shortage [1,2,6,12]. We focus here on visual inspection, of the various tasks performed manually by workers in a factory. Visual inspection is the task of finding such anomalies in products as scratches, dents, and deformations on a manufacturing line. Deep learning techniques [5,9,13,14,18] are widely used to automate visual inspection, and have achieved better performance than humans on various applications, such as object recognition [4,8] and sketch search [19].

Network models of deep learning techniques to automatically predict anomalies are generated using training samples. To improve the accuracy of inspection using deep learning techniques, we need to prepare a large number of training samples containing stimulus images and labels indicating the presence or absence of anomalies. However, it is laborious to correctly assign anomaly-related labels to the stimulus images because the labels are manually assigned by experts through visual inspection. Furthermore, the number of these experts is small.

© Springer Nature Switzerland AG 2020
S. Palaiahnakote et al. (Eds.): ACPR 2019, LNCS 12047, pp. 184–197, 2020.
https://doi.org/10.1007/978-3-030-41299-9_15

In this paper, we consider a simple task where non-experts can assign anomaly-related labels. We design the simple task based on properties of the visual inspection of manufacturing lines. A minority of the stimulus images contained anomalies while the majority were normal. Some of these anomalies were easy to find whereas others were more challenging. In experiments using the simple task, we investigated the number of training samples needed for deep learning techniques to deliver higher accuracy than humans. To this end, we automatically generated labels indicating anomalies in stimulus images used for the simple task. We compared the accuracy of inspection of human subjects with that of deep learning techniques. This study is the first step in investigating knowledge we can obtain from a comparison of accuracy between humans and the deep learning techniques. While the simple task cannot comprehensively represent visual inspection by experts, we think that this study can provide new guidelines on data collection for deep learning techniques, especially manually assigning labels through the interaction between people and information systems.

The remainder of this paper is organized as follows: Sect. 2 describes the detail of the simple task considered here, and Sect. 3 presents the accuracy of inspection of human subjects. Section 4 presents the accuracy of inspection of deep learning techniques, and Sect. 5 contains our concluding remarks.

2   Design of the Simple Task

2.1   Overview

To design the simple task of visual inspection that non-experts can perform, we consider the property of a general task of visual inspection at a factory. In this task, the worker manually determines the presence or absence of anomalies, such as scratches, dents, or deformations, by observing products on a manufacturing line. We assume that a certain surface of a product has many symbols arranged in a grid. The simple task is to check the symbols to determine whether there are anomalies. A label indicating an anomaly means that part of the symbol is defected. We call the symbol pattern on the grid the stimulus image. We assume that the position where the symbol is placed on the grid is fixed but rotation is not. Figure 1 shows an example of the simple task. Details of the stimulus images of the simple task are described below.

2.2   Generation of Stimulus Images

We set $8 \times 4 = 32$ symbols on a grid in a stimulus image. To generate an anomalous symbol, we altered part of a normal symbol. A stimulus image is generated by the following steps:

S1: We randomly set the maximum number of anomalous symbols to zero, two, four, and six in a stimulus image. Note that zero means that all symbols are normal. Six anomalies are determined by considering the relationship among the numbers 4 ± 1 [3] and 7 ± 2 [10] of the short-term memories of people.

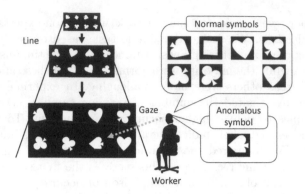

Fig. 1. Overview of the simple task of visual inspection by non-experts.

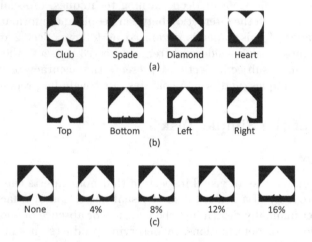

Fig. 2. Examples of the parameters used to generate anomalous symbols. We show the suits in (a), the defective positions in (b), and the rate of defect in (c).

S2: We determine the parameters of a given suit, rate of defect, defective position, and angle of rotation to generate an anomalous or a normal symbol.

- We randomly select a suit from among club, spade, diamond, and heart. Figure 2(a) shows examples of the suits.
- We randomly determine whether the given symbol is defected. Note that a symbol is not defective when the given number of anomalous symbols is the maximum determined in S1.
- We set the rate of defect and the defective position to generate the anomalous symbol.
  - We randomly set the position from among top, bottom, left, and right. Figure 2(b) shows examples of defective positions.
  - We randomly set the rate of defect to 4%, 8%, 12%, or 16%. Figure 2(c) shows examples of the rate of defect.

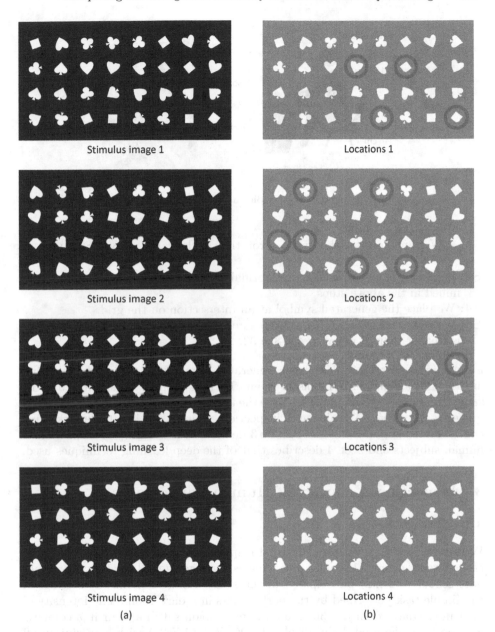

Stimulus image 1   Locations 1

Stimulus image 2   Locations 2

Stimulus image 3   Locations 3

Stimulus image 4   Locations 4

(a)   (b)

Fig. 3. Examples of stimulus images. The red circles indicate the locations of anomalous symbols. (Color figure online)

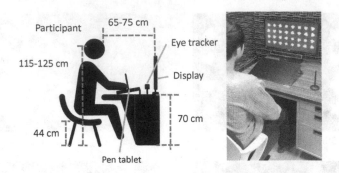

Fig. 4. Setting of the simple task for the participants.

- We randomly set the angle of rotation in the range -180–$180°$ in steps of $1°$.

S3: We generate an anomalous or a normal symbol using the parameters determined in the above steps.

S4: We place the generated symbol at an intersection on the grid.

S5: We repeat S2, S3, and S4 until the number of symbols in the stimulus image is smaller than or equal to $4 \times 8 = 32$.

Figure 3(a) shows examples of the generated stimulus images and (b) shows locations of anomalous symbols in them. There is a small possibility that the same stimulus images reappear because the total number of variations in stimulus images is 2.3×10^{21}. We calculated inspection accuracy on the simple task using the generated stimulus images. Section 3 describes the accuracy of inspection of human subjects and Sect. 4 describes that of the deep learning techniques used.

3   Inspection Accuracy of Human Participants

3.1   Setting

We investigated the accuracy of visual inspection of non-experts on the simple task. Twenty people (15 males, five females, average age, 22.2 ± 1.0 years, graduate school students) participated in the study. Figure 4 shows the settings of the simple task performed by the participants in a dark room. The intensity of light in the dark room was 360 ± 5 lx. A participant sat in a chair in a comfortable posture. We used 24-in. display (AOC G2460PF 24) with a resolution off 1920×1080 pixels to show the stimulus image. We measured the gaze locations of the participant using a standing eye tracker (Gazepoint GP3 Eye Tracker, sampling rate 60 Hz) because gaze has a potential capability to increase the accuracy of various recognition tasks [7,11,16,17]. To record anomalous symbols found by the participant, a pen tablet (Wacom Cintiq Pro 16) was used.

3.2   Experimental Procedures

We asked the participants to perform the simple visual inspection task using the following procedure:

P1: We randomly selected a participant.

P2: We explained the experiment to the participant.

P3: The participant performed visual inspection using a stimulus image on the display. We simultaneously measured the gaze locations of the participant.

P4: The participant recorded locations of anomalous symbols on the pen tablet by marking them.

P5: We repeated P3 and P4 until all 12 stimulus images had been examined by the participant.

P6: We repeated P1 to P5 until all 20 participants had finished the simple task.

The details of P2, P3, and P4 are described blow.

P2: Explanation of the Instruction. We explained to the participants the rule of the simple task, the procedure of gaze measurement, and how to use the pen tablet for marking the symbols. In this procedure, the participant was allowed to practice simple visual inspection task. Once the participant had completed the example task, we informed them of the correct answers for the locations of the anomalous symbols. Note that we did not provide the answers to the participants in the procedures below.

P3: Visual Inspection for Measuring Gaze Locations. The participants performed the simple visual inspection task by viewing a stimulus image on the display. Figure 5 illustrates the procedure of P3. We guided the initial gaze of the participant before he/she viewed the stimulus image by inserting blank image 1, containing a fixation point, on the center of the image. The blank image 1 was shown for two seconds. We then showed a stimulus image on the display for 30 s, which was considered sufficient time for the participant to check all symbols. We measured the gaze locations of the participants while they observed the stimulus image. We showed blank image 2 to the participant for five seconds after he/she had completed the task. While the participant was viewing the stimulus image on the display, we turned off the pen tablet.

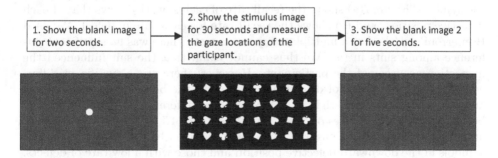

Fig. 5. Procedure of the participant in P3 viewing the stimulus image.

P4: Marking Anomalous Symbols. Each participant indicated the locations of the anomalous symbols by marking symbols on the pen tablet. We showed the same stimulus image displayed in P3 on the pen tablet. The participant circled anomalous symbols using a pen. Figure 6 shows examples of the circled symbols. While the participant was marking symbols in the stimulus image, we turned off the display. We embedded an eraser function in case the participant accidentally circled something and wanted to remove it. The maximum time allowed for marking anomalous symbols was 30 s. As soon as the marking had been finished, we moved to the next procedure.

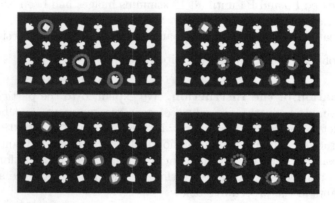

Fig. 6. Examples of symbols circled by the participants in the marking procedure.

3.3   Results

Table 1 shows the F-measure, precision, recall, and accuracy of the simple task performed by the participants. Precision was high at 0.97, and the participants rarely made a mistake in identifying a normal symbol as an anomalous symbol. On the contrary, the recall rate was 0.85, indicating that the participants had missed a large number of anomalous symbols when identifying them.

To check for anomalous symbols that had been incorrectly identified as normal symbols, we calculated the recall rate of each parameter used to generate the symbols. Figure 7(a) shows the recall rate of each suit, (b) shows that of each defective position, and (c) shows the recall rate for each rate of defect. We used Bonferroni's method as a multiple comparison test. There was no significant difference among suits in (a). We thus cannot claim that the suit influenced the inspection accuracy of the participants. However, there was a significant difference between the results of defective positions for the bottom and top, bottom and left, and bottom and right in (b). Significant differences were also observed between the results of rates of defect for 4% and 8%, 4% and 12%, 4% and 16%, and 8% and 16% in (c). Thus, the participants frequently missed the anomalous symbols in the downward defective position and those with a low rate of defects.

Table 1. The inspection accuracy of participants on the simple task.

F-measure	Precision	Recall	Accuracy
0.91	0.97	0.85	0.97

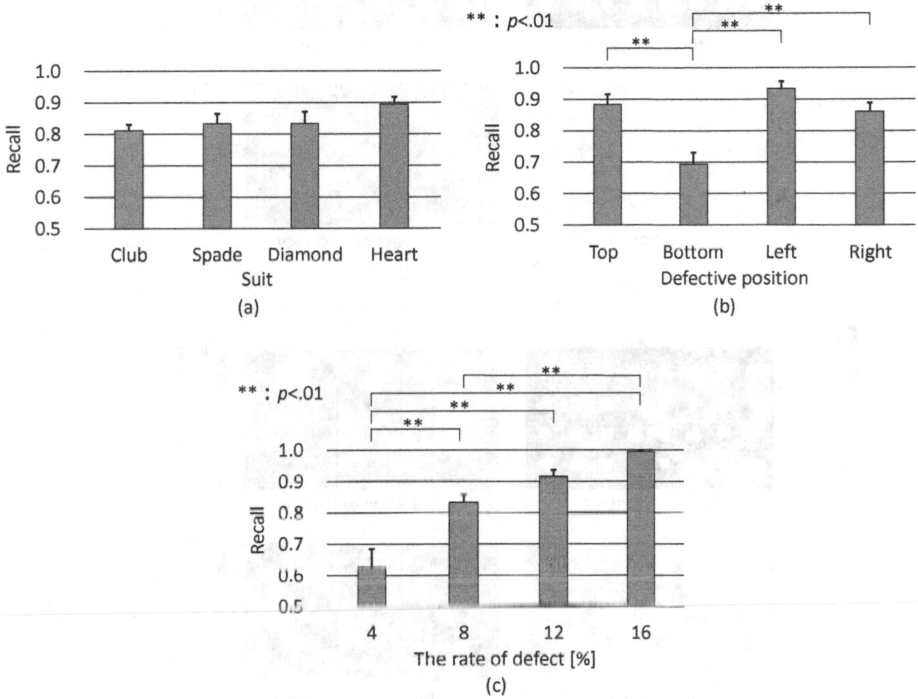

Fig. 7. Recall rates of the participants by suit, defective position, and the rate of defects.

3.4 Analysis of Gaze Locations of the Participants

We analyzed gaze locations while the participants performed the simple visual inspection task by recording the duration of gaze fixation of all participants. Figure 8 shows heatmaps of the duration of fixation and inspection accuracy on stimulus images of the simple task. In the heatmap, a deeper red means longer gaze fixation. In the heatmap of inspection accuracy, a deeper red indicates better performance. The results are summarized below.

- Gazes of the participants lingered around anomalous symbols for a longer time than on normal symbols.
- Certain anomalous symbols on which the participants recorded low accuracy featured long durations when their gazes were fixed. We think that the participants could simply not decide whether the given symbol contained anomalies.

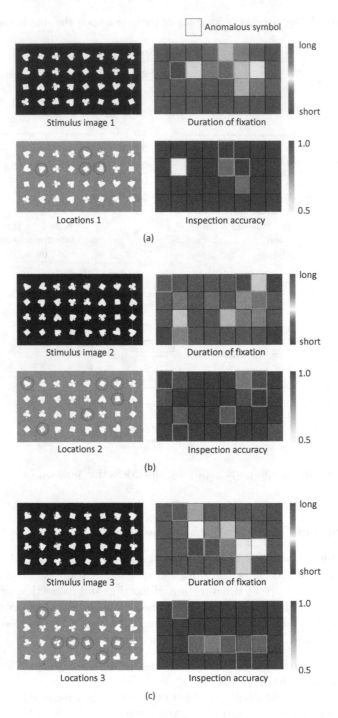

Fig. 8. Heatmaps of the duration of fixation and inspection accuracy on the stimulus images of the simple task.

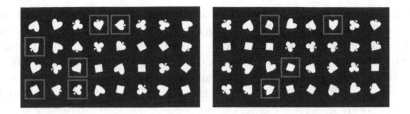

Fig. 9. Examples of outputs predicted using the SSD.

– Normal symbols with high accuracy of recognition featured a short duration
  of fixed gazes. We think that in these cases, the participants quickly judged
  that there were no anomalies.

4   Inspection Accuracy of Deep Learning Techniques

4.1   Overview

We investigated the inspection accuracy of representative deep learning
techniques—the single-shot multibox detector (SSD) [9] and U-Net [14]—on the
simple task. SSD was designed for localization tasks and U-Net for segmentation
task. To prepare a large number of training samples for the deep learning techniques, we used the stimulus images and labels generated by the steps described
in Sect. 2.2. The results of the SSD are described in Sect. 4.2 and those of U-Net
in Sect. 4.3.

4.2   Visual Inspection Using SSD

Experimental Conditions. We used bounding boxes of the anomalous symbols as labels to train the SSD model to predict their locations in the stimulus
image. Figure 9 shows examples of the outputs predicted by the SSD. We used
the VGG16 model [15] for the base network of the SSD. A total of 2,500 to
10,000 samples were prepared. For the test samples, we generated 1,000 stimulus images that were not used for training samples of the SSD. We repeated
these procedures three times to evaluate inspection accuracy.

Inspection Accuracy of SSD. Figure 10(a) shows the F-measure of the simple task using the SSD for different numbers of training samples. Bonferroni's
method was used as a multiple test. There was a significant difference in the
results between 2,500 and 5000, 2,500 and 7,500, and 2,500 and 10,000 samples.
We confirmed that the inspection accuracy using the SSD was satisfactory when
the number of training samples was more than or equal to 5,000. We also confirmed that the accuracy of the SSD of 0.95 was superior than that of the human
participants of 0.91 when using 5,000 training samples.

Figures 10(b), (c), and (d) show the recall rates for the suits, defective positions, and the rate of defects. There was a significant difference between diamond and club, diamond and spade, and diamond and heart in (b). The SSD did not perform well on diamond. On the contrary, there was no significant difference among the defective positions in (c). Furthermore, there was a significant difference between 4% and 8%, 4% and 12%, and 4% and 16% in (d). The SSD performed poorly at a rate of defect of 4%.

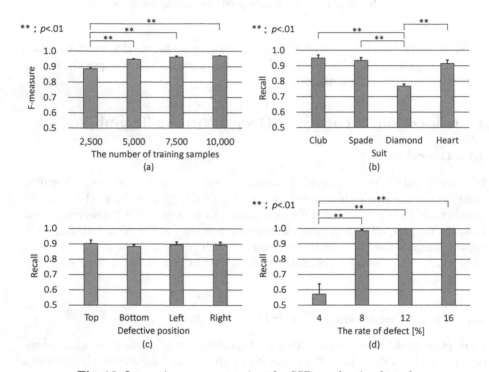

Fig. 10. Inspection accuracy using the SSD on the simple task.

4.3   Visual Inspection Using U-Net

Experimental Conditions. To train the U-Net model, we used label images in which only anomalous symbols appeared. The model was trained to predict segments of the only anomaly regions in the stimulus image. Figure 11 shows examples of the outputs predicted by U-Net. We used nine downsampling layers and nine upsampling layers, and used 5,000 to 20,000 training samples at intervals of 5,000. For the test samples, we generated 1,000 stimulus images not used as training samples of the U-Net model. We repeated these procedures three times to evaluate inspection accuracies.

Fig. 11. Examples of outputs predicted using U-Net.

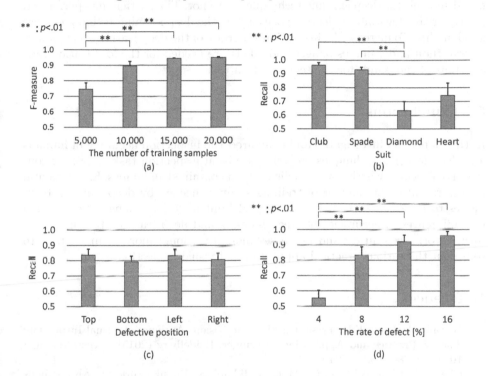

Fig. 12. Inspection accuracy using U-Net on the simple task.

Inspection Accuracy of U-Net. Figure 12(a) shows the F-measure of U-Net on the simple task for different numbers of training samples. Bonferroni's method was used as multiple test. There was a significant difference in the results between 5,000 and 10,000, 5,000 and 15,000, and 5,000 and 20,000 samples. We confirmed that inspection accuracy using U-Net was good when the number of training samples was more than or equal to 10,000. We also confirmed that the accuracy of U-Net of 0.94 was superior than that of the human participants of 0.91 when using 15,000 training samples.

Figures 12(b), (c), and (d) show the recall rates by suit, defective position, and the rate of defects. There was a significant difference between diamond and club, and diamond and spade in (b). U-Net performed poorly on diamond. On the contrary, there was no significant difference by defective position in (c), but there was a significant difference between results for 4% and 8%, 4% and 12%, and 4% and 16% in (d). U-Net performed poorly at a rate of defect of 4%.

4.4 Comparison with Human Participants

The participants as well as the deep learning techniques delivered poor results at a small rate of defects. The participants were good at identifying defects in diamond suit but the deep learning techniques were not. The participants performed poorly at the downward defective position but the deep learning techniques were good at this. To increase the inspection accuracy of the deep learning techniques, several thousand stimulus images and labels are needed for the localization task and tens of thousands for the segmentation task.

5 Conclusions

In this paper, we investigated and compared the inspection accuracy of humans and deep learning techniques on a simple visual inspection task. The task consisted of checking whether a symbolic pattern contained anomalies. Experimental results revealed the number of training samples needed by deep learning techniques to match or exceed the accuracy of human subjects. They also revealed the differences in accuracies between humans and deep learning techniques. In future work, we will expand our assessment to a long, more complex task to represent the various practical applications of visual inspection.

References

1. Beyerer, J., Leon, F.P., Frese, C.: Machine Vision: Automated Visual Inspection: Theory, Practice and Applications. Springer, Heidelberg (2015). https://doi.org/10.1007/978-3-662-47794-6
2. Cha, Y.J., Choi, W., Suh, G., Mahmoudkhani, S., Buyukozturk, O.: Autonomous structural visual inspection using region-based deep learning for detecting multiple damage types. Comput. Aided Civil Infrastruct. Eng. **33**(9), 731–747 (2018)
3. Cowan, N.: The magical number 4 in short-term memory: a reconsideration of mental storage capacity. Behav. Brain Sci. **24**(1), 87–114 (2001)
4. Dodge, S., Karam, L.: A study and comparison of human and deep learning recognition performance under visual distortions. In: Proceedings of the 26th International Conference on Computer Communication and Networks, ICCCN, pp. 1–7 (2017)
5. He, K., Zhang, X., Ren, S., Sun, J.: Deep residual learning for image recognition. In: Proceedings of the IEEE Conference on Computer Vision and Pattern Recognition, CVPR, pp. 770–778 (2016)
6. Huang, S.H., Pan, Y.C.: Automated visual inspection in the semiconductor industry: a survey. Comput. Ind. **66**, 1–10 (2015)
7. Wu, J., Zhong, S., Ma, Z., Heinen, S.J., Jiang, J.: Gaze aware deep learning model for video summarization. In: Proceedings of Pacific Rim Conference on Multimedia, pp. 285–295 (2018)

8. Kheradpisheh, S.R., Ghodrati, M., Ganjtabesh, M., Masquelier, T.: Deep networks can resemble human feed-forward vision in invariant object recognition. Sci. Rep. 6(32672), 1–24 (2016)
9. Liu, W., Anguelov, D., Erhan, D., Szegedy, C., Reed, S., Fu, C.-Y., Berg, A.C.: SSD: single shot multibox detector. In: Leibe, B., Matas, J., Sebe, N., Welling, M. (eds.) ECCV 2016. LNCS, vol. 9905, pp. 21–37. Springer, Cham (2016). https://doi.org/10.1007/978-3-319-46448-0_2
10. Miller, G.A.: The magical number seven, plus or minus two: some limits on our capacity for processing information. Psychol. Rev. 63(2), 81 (1956)
11. Murrugarra-Llerena, N., Kovashka, A.: Learning attributes from human gaze. In: Proceedings of Winter Conference on Applications of Computer Vision, pp. 510–519 (2017)
12. Newman, T.S., Jain, A.K.: A survey of automated visual inspection. Comput. Vis. Image Underst. 61(2), 231–262 (1995)
13. Rekabdar, B., Mousas, C.: Dilated convolutional neural network for predicting driver's activity. In: Proceedings of International Conference on Intelligent Transportation Systems, pp. 3245–3250. IEEE (2018)
14. Ronneberger, O., Fischer, P., Brox, T.: U-Net: convolutional networks for biomedical image segmentation. In: Navab, N., Hornegger, J., Wells, W.M., Frangi, A.F. (eds.) MICCAI 2015. LNCS, vol. 9351, pp. 234–241. Springer, Cham (2015). https://doi.org/10.1007/978-3-319-24574-4_28
15. Simonyan, K., Zisserman, A.: Very deep convolutional networks for large-scale image recognition. In: Proceedings of the International Conference on Learning Representations (2015)
16. Tavakoli, H.R., Rahtu, E., Kannala, J., Borji, A.: Digging deeper into egocentric gaze prediction. In: Proceedings of Winter Conference on Applications of Computer Vision, pp. 273–282. IEEE (2019)
17. Qiao, T., Dong, J., Xu, D.: Exploring human-like attention supervision in visual question answering (2018)
18. Wang, X., Gao, L., Song, J., Zhen, X., Sebe, N., Shen, H.T.: Deep appearance and motion learning for egocentric activity recognition. Neurocomputing 275, 438–447 (2018)
19. Yu, Q., Yang, Y., Liu, F., Song, Y.Z., Xiang, T., Hospedales, T.M.: Sketch-a-net: a deep neural network that beats humans. Int. J. Comput. Vis. 122(3), 411–425 (2017)

Improved Gamma Corrected Layered Adaptive Background Model

Kousuke Sakamoto[1], Hiroki Yoshimura[2], Masashi Nishiyama[2],
and Yoshio Iwai[2(✉)] (iD)

[1] Graduate School of Sustainability Science, Tottori University, Tottori, Japan
s142024@eecs.tottori-u.ac.jp
[2] Cross-Informatics Research Center, Tottori University, Tottori, Japan
{yosimura,nishiyama,iwai}@tottori-u.ac.jp

Abstract. This paper proposes a method for pixel-based background subtraction with improved gamma correction and a layered adaptive background model (IGLABM). The main problems of background subtraction are background oscillation and shadow. To solve these problems, we have proposed the gamma corrected layered adaptive background model (GLABM), however the performance of GLABM is not sufficient for real scenes. We hence improve the gamma estimation and prepossessing step of GLABM in this study using the covariance matrix of each pixel. We demonstrate the performance of the proposed improved method by comparing it with GLABM and other pixel-based background subtraction methods.

Keywords: Blind gamma correction · Background subtraction · Adaptive background model

1 Introduction

In recent years, given the drastic changes in our information-based society, it is necessary to pay close attention to the public safety and security of society. Against this social background, the security camera market has grown, and the need for moving object detection has also grown. Moving object detection is an important task in various practical security systems, such as person identification and traffic monitoring.

Many studies for moving object detection have been performed, and this has substantially improved detection accuracy. Object detection methods use various types of features, inter-frame differences, and background subtraction. In particular, background subtraction is frequently used for its simplicity. However, to deal with dynamic scene changes, background models should be adaptive to changes at pixel, region, or frame level. To deal with dynamic scene changes, we have previously proposed methods that employ adaptive background models (ABMs) [1,2,4]. An ABM is a two-color reflection model for dynamic illuminance changes and shadow removal [1]. The layered ABM (LABM) has multiple

© Springer Nature Switzerland AG 2020
S. Palaiahnakote et al. (Eds.): ACPR 2019, LNCS 12047, pp. 198–209, 2020.
https://doi.org/10.1007/978-3-030-41299-9_16

ABMs for background oscillations such as waving leaves or fluttering flags [2]. The gamma-corrected LABM (GLABM) adds gamma estimation to deal with various images captured by unknown cameras [4]. However, the performance of the gamma estimation implemented in the GLABM is not sufficient for use in real scenes.

In this paper, we propose a method for pixel-based background subtraction called the improved GLABM (IGLABM). The IGLABM method can deal with image sequences containing dynamic scene changes such as illuminance changes, waving leaves, or fluttering flags that were captured by an unknown camera. We demonstrate the performance of the proposed method by comparing it with other pixel-based background subtraction methods.

2   Related Work

As described in the previous section, many related methods have been proposed [5], and background subtraction is often used for detecting moving objects. The simplest model uses the first frame as the background and subtracts all subsequent frames, but this does not remove background oscillation. Lai and Yung uses a weighted moving average over several temporal frames for the background model [6]. For background subtraction, the background model is the key to success. There are many approaches for background modeling such as a statistical model [10], non-parametric model [7], fuzzy model [9], and low-rank sparse decomposition [8].

One of successful approaches to background modeling is statistical modeling. The distribution of pixels is frequently modeled as a univariate Gaussian, even if the input images are in RGB color space [10]. Pixels in the background are represented by six parameters: the sample mean and the standard deviation for each color components. A single Gaussian model, however, is not able to handle the background oscillation problem. Stauffer and Grimson proposed mixture of Gaussian (MoG) models for the background [11] to solve this problem. Kawe-TraKulPong and Bowden improved the update rules for the MoG model to solve dynamic illumination changes [14].

Our approach to background modeling is also based on statistical modeling, but our base model is the ABM [1], not a single Gaussian model. As described in the previous section, the ABM has two univariate Gaussians to model sunlight and ambient light (sky color), and LABM also has multiple layered ABMs, like a MoG [2]. These properties enable us to handle the background oscillation and remove shadows.

The implicit assumption of image processing is that the color space must be linear RGB space. This is true for images captured by industrial cameras, but not true for images captured by web or commercial cameras because these images are gamma corrected. If the input color space is not a linear RGB color space, the performance of background subtraction decreases. To solve this problem, many methods for estimating gamma have also been proposed [15,16]. We also tackled this problem by adding a prepossessing stage for blind gamma estimation into LABM [4].

3 IGLABM

Before we explain the detail of IGLABM, we briefly explain the concept of ABM. We assume a static camera capturing images of outdoor scenes.

3.1 ABM, Adaptive Background Model [1]

The adaptive background model is a two-color reflection model of ambient light and sunlight, which is similar to a dichromatic reflection model [3]. The pixel value $E(x, t)$ is given by the following equation:

$$E(x,t) = L_a(x,t) + S_d(x,t)L_d(x,t), \tag{1}$$

where L_a and L_d are the reflections of the ambient light and sunlight, respectively. Moreover, $S_d(x,t)$ represents the degree of brightness of the sunlight at point x at time t and ranges from 0 to 1. Additionally, L_a and L_d are assumed to be Gaussian processes.

3.2 GLABM, Gamma Corrected Layered Adaptive Background Model [4]

As described in the previous section, ABM can deal with dynamic illuminance changes and shadow removals because of reflection parameter S_d. However, the ABM has two disadvantages, one is the color space and the other is background oscillation such as waving of leaves or fluttering of flags. We have address these problems using gamma estimation and a layered background model, respectively. Hence, our method, GLABM, has a prepossessing step for gamma estimation and a (multiple) layered ABM (LABM), as shown in Fig. 1.

In this paper, we improve the prepossessing step and gamma estimation step of GLABM from the method in [4] and compare the performance of the proposed method (IGLABM) with those of other existing methods. We explain the details of IGLABM in the next section.

3.3 Proposed Method

Preprocessing for Gamma Estimation. We use ABM for the background model, so the color space of the input image must be linear RGB. We estimate the gamma value of the input images, and then apply inverse gamma correction to them. The true values of gamma are unknown because a camera setting changes in various ways when recording or capturing images.

We evaluate the linearity of background pixels after inverse gamma correction and determine the best gamma value. For better estimation, we eliminate foreground objects from the input images as much as possible. If these objects' pixels are treated as background pixels, the linearity of the background pixels decreases. We therefore first apply LABM to the input images $I(x)$ and generate synthesized background images $B(x)$ by replacing the foreground pixels with

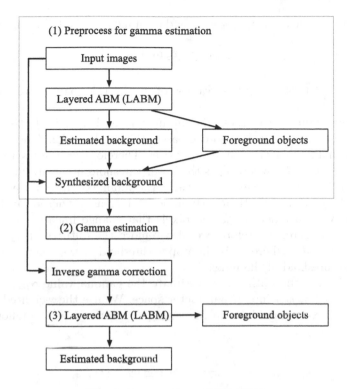

Fig. 1. Process flow of IGLABM

background pixels $S(x)$ estimated by LABM. A synthesized background image $B(x)$ is expressed by the following equation:

$$B(x) = \begin{cases} I(x) & \text{if } x \text{ is foreground,} \\ S(x) & \text{otherwise.} \end{cases}$$

Gamma Estimation. We first estimate the gamma values of each pixel. We calculate the covariance matrices of intensity at each pixel for a certain number of frames after inverse gamma correction with various values in Γ, and then estimate the linearity of each pixel from the eigenvalues of the corresponding covariance matrix.

If the illuminance of each pixel is changed linearly in the direction of sunlight, the most significant eigenvalue of the covariance matrix increases and other eigenvalues decrease or even become zero. From this observation, to determine the best gamma value, we use the index L, which is defined by the following equation:

$$L(x) = \frac{\lambda_2 + \lambda_3}{\lambda_1},$$

where $\lambda_{1,2,3}$ denotes the eigenvalues of the covariance matrix at x in descending order.

The best gamma value $\gamma_{x,best}$ at x is determined by the following equation:

$$\gamma_{x,best} = \arg\min_{\gamma \in \Gamma} L,$$

where Γ is a set of candidate values of γ. In this paper, we use $\Gamma = \{\gamma | 0.5 \leq \gamma \leq 2.5\}$.

When the gamma decreases, the input images becomes darker and the most significant eigenvalue λ_1 also decreases. Hence, the index L decreases and becomes unreliable when the gamma decreases. Therefore, we remove the outliers of $\gamma_{x,best}$ using $\lambda_1 \leq T_1$, where T_1 is the lower threshold of eigenvalue λ_1.

Moreover, if a pixel changes non-linearly, the volume of the covariance matrix is larger and $\lambda_{2,3}$ increases. Hence, the index L increases when a pixel changes non-linearly. We must determine the pixels that change linearly, so the index should be nearly zero. Therefore, we also remove the outliers of $\gamma_{x,best}$ using $\lambda_1 > T_2$, and $L < T_3$, where T_2 is the upper threshold of the eigenvalue λ_1 and the T_3 is the threshold of the index L.

After removing the outliers, we estimate the gamma value γ_{best} of the all images by voting $\gamma_{x,best}$ into Γ parameter space. We use the weighted mean as γ_{best} of the top N-th candidates in the voting Γ parameter as the follows:

$$\gamma_{best} = \frac{\sum\limits_{i=1}^{N} n_i \gamma_i}{\sum\limits_{i}^{\#\Gamma} n_i} \quad (n_1 \geq n_2 \geq \cdots),$$

where n_i is the number of votes, γ_i is the corresponding gamma value in Γ, and $\#\Gamma$ is the number of elements in Γ. In this paper, we use $N = 5, \#\Gamma = 21$.

LABM. Now that we have the estimated gamma value γ_{best} for all input images, we can apply an inverse gamma correction to the input images to obtain images in linear RGB color space. Then, we extract foreground regions by using LABM, which has multiple ABM layers. Each layer has a likelihood for the background pixels, and is maintained in descending order of likelihood in the layer list. The idea of a layer structure has been previously proposed [11]; however, our main contribution is that we apply the concept to our ABM instead of a MoG model. The system flow of LABM is depicted in Fig. 2.

4   Experimental Results

We evaluate the performance of the proposed method and compare the method with existing pixel-based methods. We prepared our datasets "Tree" and "Flag," which were captured in linear RGB, where $\gamma = 1$, to evaluate the performance of gamma estimation. The Tree dataset contains background oscillation consisting of waving of leaves and object shadows. The Flag dataset also contains

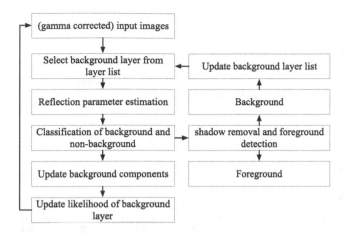

Fig. 2. System flow of LABM

tree   flag

Fig. 3. Examples of the Tree and Flag datasets

background oscillation consisting of a fluttering of a flag and object shadows. Examples of these datasets are shown in Fig. 3.

We also used several image sequences collected from the benchmark datasets: "CDW-2012" [12] and "PETS2001" [13]. We use "BusStation," "Overpass," "Fountain02," "Backdoor," and "Canoe" from CDW-2012 for evaluation, because these image sequences are outdoor scenes. In total, we use six datasets for experiments. We also used "dataset3 testing camera1" of PETS2001, which is referred as "PETS" in this paper.

4.1 Validity of Background Image Synthesis

To evaluate the validity of synthesizing background images, we compare the accuracy of gamma estimation on synthesized background images and input images. We choose 100 images randomly from each dataset, and then applied gamma correction to them from $\gamma = 0.5$ to $\gamma = 2.5$. The results of gamma estimation are shown in Fig. 4. The horizontal axis is threshold value T_3 of the index L, and the vertical axis is the mean absolute error of gamma estimation.

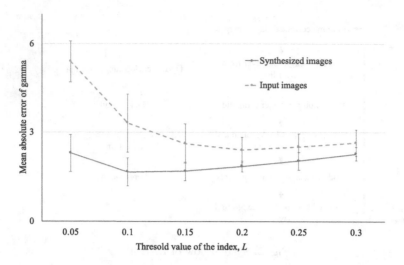

Fig. 4. Accuracy of gamma estimation

The estimation gamma values change when the other thresholds $T_{1,2}$ change. The values of T_1 in this experiment are 5, 10, 15, and 20. The values of T_2 in this experiment are 100, 200, 300, 400, 500, and 1,000. The standard deviations are also given in the figure.

The figure shows that the accuracy of gamma estimation using the synthesized background is better than that of gamma estimation using the input images. This is because the concept of our gamma estimation is based on the linearity of illuminance changes. If a pixel of interest is background, in other words, from a static object, the linearity is satisfied, but if the pixel is not from a static object, the linearity is violated. This result indicates that our proposed method requires the synthesized background images for better estimation.

Hereafter, we use the best parameter values, $T_1 = 10, T_2 = 200$, and $T_3 = 0.1$, of this experiment for other experiments in this study.

4.2 Comparison of Gamma Estimation Accuracy with Existing Methods

We compare the accuracy of gamma estimation with other gamma estimation methods: blind inverse gamma correction (BIGC) [15], and MV-Gamma [16], and GLABM [4]. As shown in Table 1, the proposed method (IGLABM) achieves the highest gamma estimation accuracy.

4.3 Performance Comparison of Background Estimation with Existing Methods

Next, we compare the performance of the proposed method with other pixel-based background subtraction methods: MOG2 [17] and MOG. MOG2 is an

Table 1. Absolute estimation error of gamma values

Dataset	IGLABM	GLABM	MV-Gamma	BIGC
Tree	**0.00**	0.18	0.30	2.14
Flag	**0.10**	0.41	0.50	0.36
PETS	**0.09**	0.20	0.30	1.69
BusStation	0.30	**0.03**	0.20	1.07
Overpass	**0.10**	0.25	0.20	0.84
Fountain02	**0.38**	0.72	0.50	0.53
Backdoor	**0.07**	0.57	0.40	0.55
Canoe	0.22	0.30	0.40	0.12
Average	**0.16**	0.33	0.35	0.91

Table 2. Comparative analysis of F_1-score

Dataset	IGLABM	GLABM	LABM	MOG	MOG2
Tree	**90.78**	90.64	**90.78**	87.69	84.15
Flag	79.31	**81.27**	79.94	69.75	63.54
PETS	51.14	49.65	49.65	**61.32**	29.34
BusStation	65.89	**66.35**	**66.35**	44.60	58.13
Overpass	**66.38**	64.06	60.24	48.80	51.89
Fountain02	**65.84**	63.17	63.17	57.37	26.84
Backdoor	76.21	71.47	76.48	**78.85**	67.10
Canoe	82.00	79.30	**82.17**	44.90	56.45
Average	**72.19**	70.74	71.10	61.66	54.68

improved version of MOG, which chooses the number of the mixture adapted for each pixel. We used the OpenCV library for MOG and MOG2.

Examples of the estimation results for foreground detection in this experiments are shown in Figs. 5 and 6. The results of IGLABM, GLABM, and LABM are almost the same because the gamma value of Tree is 1.0 and the gamma estimation errors are small. MOG and MOG2 cannot remove shadows from the foreground object, but LABM, GLABM, and IGLABM can remove shadows clearly. MOG2 cannot remove background oscillation such as waving leaves perfectly, but other methods can remove the background oscillation.

We show the F_1-score of each method in Table 2. Here, the proposed method also achieves the highest accuracy of all methods.

input image (frame No. 80)   Ground truth

IGLABM   GLABM

LABM

MOG   MOG2

Fig. 5. Examples of foreground detection results for Tree

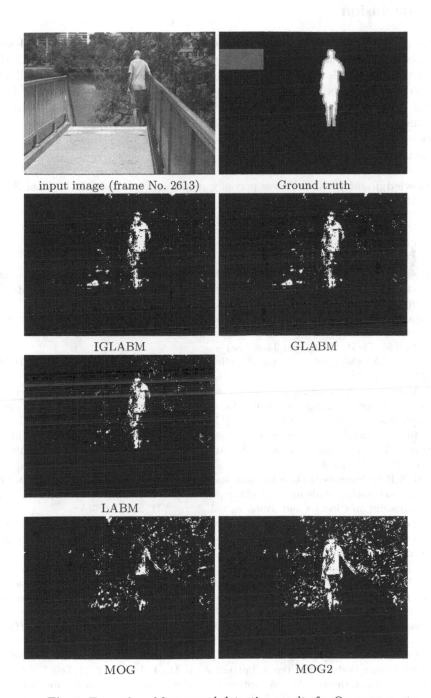

Fig. 6. Examples of foreground detection results for Overpass

5   Conclusion

We proposed the IGLABM method for foreground detection based on pixel-wise background subtraction with improved gamma correction and layered ABMs. Gamma estimation is effective for pixel-wise subtraction but assumes a linear RGB color space. The experimental results show that the performance of foreground detection of IGLABM is superior to those of existing methods from the experimental results. In future work, we will apply our method to various open datasets to improve its performance and extend it to a block-based classification method.

Acknowledgment. This work was partially supported by MIC SCOPE Grant Number 172308003, and JSPS KAKENHI Grant Number JP17K00238, JP18H04114.

References

1. Yoshimura, H., Iwai, Y., Yachida, M.: Object detection with adaptive background model and margined sign cross correlation. In: Proceedings of the International Conference on Pattern Recognition, vol. III, pp. 19–23 (2005)
2. Toyoda, M., Yoshimura, H., Nishiyama, M., Iwai, Y.: Background subtraction robust for shadow and dynamic background in outdoor scene. IEICE Tech. Rep. **116**(366), 71–76 (2016). (in Japanese)
3. Shafer, S.A.: Using color to separate reflection components. Color Res. Appl. **10**(4), 210–218 (1985)
4. Sakamoto, K., Yoshimura, H., Nishiyama, M., Iwai, Y.: GLABM: gamma corrected layered adaptive background model for outdoor scenes. In: 2018 IEEE 7th Global Conference on Consumer Electronics (GCCE), Nara, pp. 66–67 (2018)
5. Setitra, I., Larabi, S.: Background subtraction algorithms with postprocessing: a review. In: Proceedings of the International Conference on Pattern Recognition, pp. 2436–2441 (2014)
6. Lai, A.H.S., Yung, N.H.C.: A fast and accurate scoreboard algorithm for estimating stationary backgrounds in an image sequence. In: Proceedings of the International Symposium on Circuits and Systems, vol. 4, pp. 241–2444 (1998)
7. Elgammal, A., Harwood, D., Davis, L.: Non-parametric model for background subtraction. In: Vernon, D. (ed.) ECCV 2000. LNCS, vol. 1843, pp. 751–767. Springer, Heidelberg (2000). https://doi.org/10.1007/3-540-45053-X_48
8. Wright, J., Ganesh, A., Rao, S., Peng, Y., Ma, Y.: Robust principal component analysis: exact recovery of corrupted low-rank matrices via convex optimization. In: Proceedings of the Advances in Neural Information Processing Systems, pp. 2080–2088 (2009)
9. Kim, W., Kim, C.: Background subtraction for dynamic texture scenes using fuzzy color histograms. IEEE Signal Process. Lett. **19**(3), 127–130 (2012)
10. Wren, C.R., Azarbayejani, A., Darrell, T., Pentland, A.: Pfinder: realtime tracking of the human body. IEEE Trans. Pattern Anal. Mach. Intell. **19**(7), 780–785 (1997)
11. Stauffer, C., Grimson, W.: Adaptive background mixture models for real-time tracking. In: Proceedings of Conference on Computer Vision and Pattern Recognition, pp. 246–252 (1999)

12. Goyette, N., Jodoin, P.-M., Porikli, F., Konrad, J., Ishwar, P.: Changedetection.net: a new change detection benchmark dataset. In: Proceedings of IEEE Workshop on Change Detection at CVPR-2012, pp. 16–21 (2012)
13. Youngand, D., Ferryman, J.: PETS metrics: online performance evaluation service. In: Proceedings of the IEEE International Workshop on Visual Surveillance and Performance Evaluation of Tracking and Surveillance, pp. 317–324 (2005)
14. Kaewtrakulpong, P., Bowden, R.: An improved adaptive background mixture model for real-time tracking with shadow detection. In: Remagnino, P., Jones, G.A., Paragios, N., Regazzoni, C.S. (eds.) Video-Based Surveillance Systems. Springer, Boston (2002). https://doi.org/10.1007/978-1-4615-0913-4_11
15. Farid, H.: Blind inverse gamma correction. IEEE Trans. Image Process. **10**(10), 1428–1433 (2001)
16. Mahamdioua, M., Mohamed, B.: New mean-variance gamma method for automatic gamma correction. Int. J. Image Graph. Signal Process. **9**(3), 41–54 (2017)
17. Zivkovic, Z.: Improved adaptive Gaussian mixture model for background subtraction. In: International Conference on Pattern Recognition, vol. 2, pp. 28–31 (2004)

One-Shot Learning-Based Handwritten Word Recognition

Asish Chakrapani Gv[1]($\boxtimes$), Sukalpa Chanda[2], Umapada Pal[3],
and David Doermann[4]

[1] Electronics and Communication Department, Manipal University Jaipur,
Jaipur, India
asish.chakrapani@gmail.com
[2] Centre for Image Analysis, Department of Information Technology,
Uppsala University, Uppsala, Sweden
sukalpa@ieee.org
[3] Computer Vision and Pattern Recognition Unit, Indian Statistical Institute,
Kolkata, India
umapada@isical.ac.in
[4] University at Buffalo, SUNY, Buffalo, USA
doermann@buffalo.edu

Abstract. One-Shot and Few-shot Learning algorithms have emerged
as techniques that can imitate a humans ability to learn from very few
examples. This is an advantage over traditional deep networks which
require a lot of training samples and lack of robustness due to their excessive domain specific discriminators. In this paper, we explore a one-shot
learning approach to recognizing handwritten words using Siamese networks to classify the handwritten images at the word level. The Siamese
network's ability to compute similarities between two images is learned
using a supervised metric but the fully trained Siamese network can be
used to classify new data that has previously not been used to train the
network. The model learns to discriminate inputs from a small labelled
support set. By using a convolutional architecture we were able to achieve
robust results. We also expect that training the system over a larger
distributions of data will result in improved general handwritten word
classification. Accuracy as high as 92.4% was obtained while performing
5-way one-shot word recognition on a publicly available dataset which is
quite high in comparison to the state-of-the-art methods.

Keywords: One-shot learning · Handwriting recognition · Siamese
Networks · Image classification

1 Introduction

Humans possess a unique ability to learn, recognize and classify objects in real
time and at a rapid rate [11,22]. Studies have shown that children learn 10–30
thousand object categories of objects by the age of six [3]. Humans are known

© Springer Nature Switzerland AG 2020
S. Palaiahnakote et al. (Eds.): ACPR 2019, LNCS 12047, pp. 210–223, 2020.
https://doi.org/10.1007/978-3-030-41299-9_17

to adapt/comprehend new information/concepts and are able to learn key variations in them. This ability is due to the following factors:

(1) The capacity to synthesize, learn and classify new data from information derived while learning in the past.
(2) A learning mechanism which evolves over time. i.e. the capability of learning "How to learn".

Recent deep learning methods for image classification require training over hundreds of thousands of samples to achieve accuracy comparable to humans and on encountering new data may require extensive re-training to learn and accommodate new parameters. Humans however can learn from very few examples. Since properly annotated data is resource dependent (requires money and human intervention), a significant number of real-world learning problems demand an ability to draw valid inferences from small amounts of data and to rapidly adjust to new information [20]. These scenarios pose a challenge for deep learning, which typically relies on slow, incremental parameter changes and also requires significant amounts of annotated data. Using a simple gradient-based solution to completely re-learn the parameters from the little data available at the moment would be prone to poor learning.

One-Shot Learning refers to the class of learning algorithms which can learn from just one sample per unique class. The main objective of the paper is to investigate and devise methodologies using deep neural networks which are based on the idea of learning from a very small amount of data and use the acquired knowledge for classification of a test sample in the very first instance. In our case, we leverage a deep Siamese network to learn to recognize handwritten word embeddings and classify them in real time. A Siamese Network is a pair of deep convolutional networks with same weights tied together by a matching function in order to compute the similarity between a pair of input images. Deploying a traditional deep network for recognizing/classifying handwritten words would require a large dataset, which covers the entire set of possible words used in that script. This is not feasible in a real life scenario. Procuring such a huge corpus is time-consuming and also, demand human intervention which makes it a costly and overwhelming task. Moreover, the system would not be able to recognize a word class if it is not present in the corpus. This situation can be countered using One-Shot learning framework. Such a One-Shot learning-based handwritten word recognition system could be used for various tasks like document indexing and retrieval, postal automation system etc. This article investigates one-shot learning using deep Siamese networks in the context of the handwritten word recognition problem and explores the possibility of using it on two different kinds of datasets.

2   Related Work

In recent years, deep learning has attracted a lot of attention and found pivotal applications in the Computer Vision and Image recognition [19], speech-recognition [8], language translation [26] and game-playing [21]. But while it works well for some applications, there are still many real-world applications require learning or drawing inferences from small amounts of data and adjusting to a constant inflow of new information. This had led to the concept of One-shot or Few-shot learning. The remarkable results achieved by deep neural networks are usually contingent upon the availability of large amounts of input data. Although recently the concept of One-shot learning has attracted a great deal of attention, the research efforts are still at a primitive stage. Some of the key milestones in one shot learning are discussed in this section.

The work in this area was stimulated by Fei-Fei et al. [10,12] where the authors presented a generative object category model and variational Bayesian framework for representing and learning of visual object categories from a handful of training examples from each class. More recently, a number of publications have proposed methods which adopt the "learning to learn" concept. Santoro et al. [20] proposed a memory augmented neural network with a meta-learning capability which leverages a memory bank working in coordination with a neural network to rapidly learn and make inferences. Woodward et al. [25] introduced a recurrent neural network based action-value function and uses it in a classification task where a decision must be made to either predict a label or pay to receive the correct label. Another prominent work is one-shot imitation learning [9] proposing a framework that maps a single successful demonstration of a task to an effective policy that solves the task in a new situation. One shot learning has found applications in vision and image classification, as well as lowering the amount of data required to make meaningful predictions in drug discovery using the iterative refinement long short-term memory, and the learning of distance metrics on small-molecule space [1]. Maas and Kemp [18] have developed Bayesian networks which learns a distribution-wise hyperparameter to predict attributes for Ellis Island passenger data.

Not much research in handwriting recognition using one-shot learning exists. Some recent deep-learning based approach on handwritten word recognition and word spotting using limited data can be found in Chanda et al. [24] and Wilkinson et al. [6]. While Howe [15] proposed a non-deep learning based technique using part structured models for handwritten word spotting. Lake et al. [4] investigates a 20 way one shot classification of handwritten characters using a generative model of character composition from strokes, where the knowledge from previous characters helps to infer the latent strokes in novel characters. Another recent state-of-the-art method for performing one shot image classification is Matching networks Vinyals et al. [23] where the authors define a non parametric method defined at two stages. The first being, deploying a deep neural network with augmented memory enabling rapid learning and the second is defining a training strategy where the training and testing conditions match for each batch thus tailoring to the needs of one shot learning.

Similar one shot classification tasks on the Omniglot dataset have been published using Deep convolutional Siamese Networks by Koch et al. [17] and Memory augmented networks by Santoro et al. [20]. The concept of Siamese Networks was initially introduced by Bromley et al. [5] for the signature verification problem. But these works explore the individual character based classification task which might not be useful in applications like Handwritten word recognition. The concept of Siamese networks for word image spotting has been explored only once before by Barakat et al. [2]. However, the architecture and discriminating metric of our proposed network is different and we have outperformed their results especially in the case of GW dataset [13]. In our method, completely disjointed sets were used during the training and testing phase, whereas in Barakat et al. this was not always followed. Their objective was word spotting and the training and test data could have common classes & images. In this paper, we propose a model using a Deep Siamese Net for handwritten word image classification.

3 Methodology

Siamese Networks are useful in tasks that involve finding similarity between two inputs [7] For example, In our case the inputs would be a pair of images consisting of a reference image and a sample to compare it against. In our work, images are represented using non-linear features from a deep network. Later those features are learned using a supervised metric-based approach to be able to classify images in a one-shot learning framework. The architecture adopted for this purpose is a modified version of the deep Siamese architecture proposed by Koch et al. [17]. The Convolutional Siamese network is designed to learn features of the input images regardless of prior domain knowledge with very few samples from a given distribution. This model was also adopted because the twin networks share weights resulting in fewer parameters to train on and a lower tendency of over-fitting.

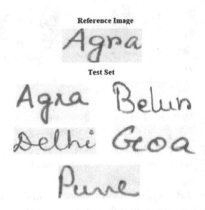

Fig. 1. Example of 5-way one shot task

For experiments, we used two small labeled support sets one containing consisting of 74 training and validation classes and 15 test classes of handwritten words depicting Indian city names and another consisting of a subset of 65(50 train and 15 test) classes of word images from the Washington dataset were used. During training, the network takes a pair of images as the input where it learns to discriminate between them based on their class labels and features. The task is achieved by generating probability scores which detect whether they belong to the same class or different classes. For evaluation of n way one-shot tasks, the network is provided with pairs of images consisting of a reference image and one sample image from each of the n unseen classes at each instance (Fig. 1). The label from the pair with the highest probability is then given to the reference image, i.e. the pairs of images are then considered to be of the same class. We also believe that ideally learning with a large distribution with a decent variation the model would predict with higher accuracy.

4   Siamese Network Model

Siamese networks are a class of neural network architectures that contain two identical sub-networks that use the same weights while taking two distinct inputs and are joined by a comparative function. These networks are usually used for verification purposes and in the context of one-Shot learning compare each of 'n' sample images to a reference image. The output vector of the network with input as the reference image is a baseline reference and can be compared against the output vector of the sample image. This method of measuring similarity can be leveraged for recognizing handwritten words.

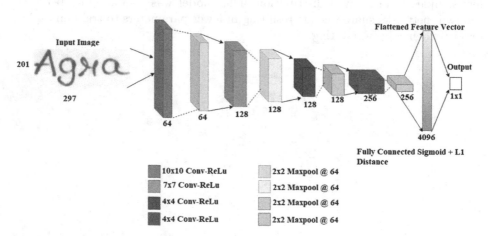

Fig. 2. Schematic diagram of our customized Convolutional Siamese Network architecture as used in our experiment (twin network not depicted).

The diagram of the adapted model shown in Fig. 2 is a deep Siamese convolutional neural network consisting of a sequence of alternating single channel convolutional and max-pooling layers followed by a fully connected layer to generate a high-level feature representation of both input images. The twin networks each produce a feature vector for the given input images and are both joined after the 4096 unit fully connected layer where they are given as input to the comparative function layer (L1 distance). Each of the convolutional layers uses rectified linear (ReLU) units and the fully connected layer uses sigmoidal units. After the network optimization each of the convolutional layers has convolutional filters in multiples of 16 and consecutively decreasing filter sizes and a fixed stride of 1, while the max-pooling layers have a stride of 2.

4.1 Model Adaptation

The output of the final convolutional and max pool layer pair is then flattened and given to a fully connected layer with 4096 units, generating the feature vector for an image. The network is then fed with the images from the pair to generate these vectors. When the two feature vectors are obtained, these are given to an L1 distance layer which computes the induced distance between the vectors and then the output is passed on to a single sigmoidal unit. The L1 distance layer computes an absolute difference between the feature vectors and the final layer uses this metric to provide a score or similarity index for the similarity of the input vectors. It provides value in a range between 0 and 1 where a score close to 0 indicates different class and a score close to 1 indicates the same class. The distance layer along with a single sigmoidal unit contributes to the comparative function that ties the twin networks.

In this paper, we made modifications to the Deep Siamese network architecture. We use the architecture proposed by Koch et al. [17] because of its good performance on the text based Omniglot dataset and its ability to generalize on features and perform classification on unseen classes. In our experiments, we found that an additional Max-pooling layer with stride 2 after the last convolving layer gave better results. The Max pool layer down samples the features extracted in the last convolving layer giving a more compressed representation of the feature vector. The reduction in the size of the feature map also helps with reduction in the number of parameters, computation in the network and helps control overfitting. We had to make the adaptations to accommodate the shift in image size from 105 * 105 to 201 * 297. Additionally, we follow an iteration based method where we use a smaller batch size to better generalize upon the dataset [16].

5 Model Parameters

5.1 Loss Function

Like the original model, we impose a regularized cross-entropy objective on the binary classifier. The governing equations are as follows

$$L(x_1^i, x_2^i) = y(x_1^i, x_2^i) log P(x_1^i, x_2^i)$$
$$+ (1 - y(x_1^i, x_2^i)) log(1 - P(x_1^i, x_2^i)) \qquad (1)$$
$$+ \lambda^N |W|^2$$

where i represents the i'th index of the current batch, $y(x_1^i, x_2^i)$ is a vector of length M consisting of labels. We assume that it equals 1 in case of same class and 0 in case of different class for iteration N.

5.2 Optimization

We take a batch size of 8 with a constant learning rate η_j for all the layers, along with a linearly evolving layer-wise momentum μ_j for the jth layer, and L2 regularization penalization, weights for each iteration N. So the weight update rule for iteration N is:

$$W_{kj}^N(x_1^i, x_2^i) = W_{kj}^N + \Delta W_{kj}^N(x_1^i, x_2^i) + 2\lambda_j |W_{kj}|$$
$$W_{kj}^N(x_1^i, x_2^i) = -\eta_j \nabla W_{kj}^N + \mu_j \Delta W_{kj}^{N-1} \tag{2}$$

where ΔW_{kj} is the partial derivative with respect to the weight between the jth neuron in a given layer and the kth neuron in the next layer.

5.3 Weights

All the network weights in the convolutional layers, as well as the fully connected layers, were initialized using the Glorot uniform initializer. The initializer draws samples from the uniform distribution of $[-a, a]$ where a is given by the equation

$$a = sqrt(\frac{6}{(fan_{in} + fan_{out})}) \tag{3}$$

where fan_{in} is the number of input units in the weight tensor and fan_{out} is the number of output units in the weight tensor [14]. The biases were initialized using the default setting of zeros in all the layers.

5.4 Learning

In general, we prefer a learning rate that results in a steep negative slope in the network's loss function. Since we know that too high of a learning rate can cause the parameter updates to skip the ideal minima and subsequent updates will either result in a continued noisy convergence in the general region of the minima, or in more extreme cases, may result in divergence from the minima. So by decaying the learning rate, the model will be able to converge closer to the minima.

We use a constant learning rate for all the layers while following a step based decay method decaying at a uniform rate of 1% at every 500 iterations. Validation accuracy is calculated after every 1000 iterations and the model with the best accuracy is saved. The model will be trained for a maximum for 100,000 iterations, or until the validation accuracy does not show improvement over a 10,000 iterations. The initial momentum is taken as 0.5 for each layer and evolves with a predefined linear momentum slope until reaching a final momentum of 0.9 for each layer.

6   Experimental Framework

6.1   Dataset

The experiments were conducted on two datasets and the results are reported for the same. The two datasets under consideration were the George Washington Dataset [13] and the Indian City Names Dataset.

George Washington (GW) Dataset. The Washington database contains word images alongside with their ground truth transcription. For our experiments we used a subset of the original GW database. We restricted our dataset to the word classes which had atleast 10 samples for the Siamese network to train and generalize upon. This resulted in total of 65 classes which were split into 50 train classes and 15 test classes accordingly where each class contained 10 images.

Indian City Names Dataset. The dataset contains a total of 89 classes where each class represents a city name. For the experiment of city name recognition, we collected a total of 1780 English handwritten city name samples. These city name samples are collected from handwritten address blocks of Indian postal documents as well as from some individuals using some specially designed forms. Each class in the dataset consists of 20 images of the city name, each written in unique handwriting. All the images in the dataset are cropped and resized to a dimensions of 201 * 297 to change the original aspect ratios as little as possible. We split the dataset into 74 train-validation classes and 15 evaluation classes. Furthermore, in the train-validation set we use the 80–20 split rule to partition the 74 classes. In this set 60 classes are used for learning the image representation and the remaining 14 are used for fine-tuning the hyper-parameters and calculating validation one-shot tasks, while the evaluation set is used only to measure the final one-shot classification accuracies on unseen classes.

The training task is performed by randomly pairing images from the same and different classes from the train set. Furthermore, to increase the robustness and to prevent overfitting we use additional copies of the image pairs by adding affine distortions. The affine distortions include transformations such as rotation, shear, zoom and shift with ranges of $[-15, 15]$, $[-0.3 * 180/\pi, 0.3 * 180/\pi]$, $[0.8, 2]$ and $[-5, 5]$ respectively. These affine transformations are performed with a probability which is set to 0.5. We performed an n way one shot task on the validation classes after every 1000 iterations where n represents the validation classes present in a set. The model was trained until there was no improvement in the validation accuracy or a maximum of 100k iterations. The models along with the weights are saved whenever there is an improvement in the validation accuracy.

6.2   Evaluation

After the training of the Siamese network, it is used to perform one-shot classification tasks on unseen classes. We use a fold-wise method where in each fold a

5-way one shot task performed on 5 support classes in three disjoint sets. Each time for evaluating the performance in the evaluation set two draws producing 5 samples each were taken, each one of the samples produced in the first draw are taken as test images and compared against all 5 samples of the second draw. This process was done twice for each evaluation set i.e. for each set of 5 classes, we perform 20 different one-shot tasks, thus performing a total of 60 tasks for the 3 sets. We also observe the individual set accuracy and a mean global accuracy for the model while evaluating its performance.

7   Results and Discussions

7.1   Results on George Washington Database

The performance of the model as a general word classification system was evaluated using the "GW Database" [13]. The results reported are obtained from observations conducted across 4 different train-evaluation runs. For our experiments, the subset of the Washington database consisting of 65-word classes with 10 samples in each class was used. Those 65 classes were selected since rest of the other classes have less than 10 samples. The new dataset was split into 50–15 train-validation & evaluation classes, where the train set was further split according to an 80–20% split resulting in 40 classes for training and 10 classes for validation. The evaluation was conducted using the same 5-way one shot tests on the 15 classes from the evaluation set. This was run 3 times, where the 15 test classes were different in each occasion. The Siamese Network was able to generalize well and performed better on this dataset as shown in Table 1.

Table 1. Accuracy of 5-way One shot Tasks on Washington Dataset using the proposed method.

Fold number	Best accuracy with respect to 3 sets	Mean accuracy
Fold 1	82.50%	57.50%
Fold 2	90.00%	78.33%
Fold 3	92.50%	83.33%
Fold 4	90.00%	84.16%

Further, we also performed a 15-way one shot test on this dataset. We obeserved the peroformance of the model across 4 train-evaluation runs where the evaluation was conducted on completely disjoint sets. The best accuracy of siamese network was observed to be 70% and an average accuracy of 63.12% was recorded across all the runs as shown in Table 2.

Table 2. Accuracy of 15-way One shot Tasks on Washington Dataset using the proposed method

Fold number	Accuracy
Fold 1	67.50%
Fold 2	70.00%
Fold 3	57.50%
Fold 4	57.50%

7.2   Results on Indian City Names Dataset

The results reported here are obtained from observations conducted across 6 different train-evaluation runs. To ensure the robustness of the model, we divided the model into 6 folds, where in each fold 3 sets of 5 classes each were used to perform the one-shot evaluation task with one class overlapping twice and the remaining 74 classes were used to train the model. We observed the best accuracy in a single 5-way one shot test in each fold along with the mean accuracy of the fold. The best 5-way one-shot results across various folds was 65% and the best mean accuracy was observed at 53.33%. The standard deviation across all the 6 accuracies is 3.2% which shows the performance consistency of the model.

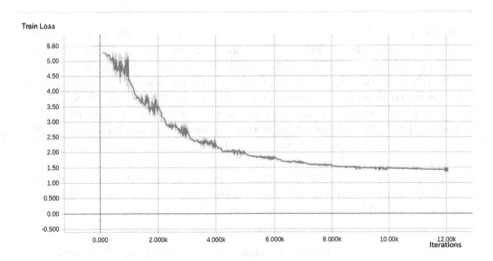

Fig. 3. Train loss with respect to iterations

Furthermore, standard metrics such as train loss, train accuracies, and validation accuracies are noted with respect to the iteration number. Ideally, the training loss is expected to reduce with the increase in the number of batches the model trains upon increases. An example of the negatively sloped train loss curve on one of our best performing models is shown in Fig. 3.

7.3   Comparison with Other One-Shot Recognition Frameworks

One-shot image classification tasks can also be achieved by recent one shot learning methods such as "Memory Augmented Neural Network" (MANN) [20] and Matching Networks [23]. Some experiments were also conducted using MANN and Matching Networks for one shot tasks to obtain baseline results for comparison with respect to our proposed approach. The data used for all three different methods were exactly the same and the same 5-way One-Shot tasks were performed on the model for evaluation.

Table 3. Comparison of One-shot classification accuracy on Indian City Name dataset

Method	Accuracy
Best accuracy using our proposed method on one shot tasks	**65.00%**
MANN Best accuracy on One-Shot [20]	17.88%
Matching Network Best accuracy on One-Shot	54.60% [23]

On the Indian City Names dataset, the best accuracy achieved by MANN and Matching networks on One-Shot task has been depicted in the Table 3 and in Table 4, performance of the methods on GW dataset has been stated. We noted that MANN requires 6-Shot learning strategy to achieve similar performance as of our proposed Siamese network on One-Shot tasks. "Matching Networks" by Vinyals et al. [23] defines a differentiable nearest neighbor loss contributed by the cosine similarities of embeddings, this embedding is produced by a CNN. The best accuracy for one shot tasks by Matching Networks observed during our experiments was 54.60% which is $\approx 10\%$ less than our achieved accuracy on the same dataset. It is evident from the comparison, that our method performs better than other two methods in terms of a One-Shot recognition task. However, on few-shot tasks this phenomenon might not always hold true.

Table 4. Comparison of One-shot classification accuracy on GW dataset

Method	Accuracy
Best accuracy using our proposed method on one shot tasks	**92.50%**
MANN Best accuracy on One-Shot tasks [20]	20.94%
Matching Network Best accuracy on One-Shot tasks [23]	85.41%

7.4 Comparison with Respect to a Pre-trained CNN

We were curious to see how a regular CNN reacts to the problem of very few training samples per class so we have also compared our results against a traditional deep network INCEPTION V3 which has shown state-of-the-art performance in image classification. INCEPTION V3 is a widely-used image recognition model introduced by Google that has been shown to attain greater than 78.1% accuracy on the ImageNet dataset. The model itself is made up of symmetric and asymmetric building blocks, including convolutions, average pooling, max pooling, concerts, dropouts, and fully connected layers. For our experiment, we used a transfer learning technique. Here, we build a new model to classify our dataset (in this case the Indian City Name) by freezing weights of the initial layers in the pre-trained network and update only a few top layers close to the classification part with the new data. We use transfer learning since we don't have much annotated data per class. In this experiment, the network was trained with 5 classes from our dataset each containing 20 samples and evaluated the performance by observing the probability scores assigned to the test image. For each test image, we obtained five scores each consisting of the probability of it belonging to the respective class. After multiple trials we observed that the probability score of the correct class varied between 3–7% and the highest probability observed was 20%. The low results prove that one-Shot learning framework is a good strategy to counter an image classification problems with very less annotated training data.

8 Conclusion and Future Work

This paper proposes a Siamese network-based one-shot learning framework for handwritten word recognition. The proposed method could counter the issue of limited annotated data often faced by a traditional Deep Neural Network and can recognize word images even with a very limited number of annotated data for training. Moreover, the network could also classify completely unseen class images with reasonable accuracy. Comparison with respect to another existing approach for one shot image classification (Using MANN and Matching Networks methods) and a state-of-the-art deep network Inception V3 has also been provided. Encouraging results were obtained while dealing with a 5-way one shot task on small labeled George Washington database and Indian City Name dataset. Future research directions include an ablation study on errors conceived by the proposed algorithm and accordingly amend the system.

References

1. Altae-Tran, H., Ramsundar, B., Pappu, A.S., Pande, V.: Low data drug discovery with one-shot learning. ACS Central Sci. **3**(4), 283–293 (2017)
2. Barakat, B., Alaasam, R., El-Sana, J.: Word spotting using convolutional siamese network, pp. 229–234 (2018). https://doi.org/10.1109/DAS.2018.67

3. Biederman, I.: Recognition-by-components: a theory of human understanding. Psychol. Rev. **31**(2), 115–147 (1987)
4. Lake, B., Salakhutdinov, R., Gross, J., Tenenbaum, J.: One shot learning of simple visual concepts. In: Proceedings of the 33rd Annual Conference of the Cognitive Science Society, Boston, USA, vol. 172 (2009)
5. Bromley, J., Guyon, I., LeCun, Y., et al.: Signature verification using a "siamese" time delay neural network. Int. J. Pattern Recogn. Artif. Intell. **7**(04), 669–688 (1993)
6. Chanda, S., Baas, J., Haitink, D., Hamel, S., Stutzmann, D., Schomaker, L.: Zero-shot learning based approach for medieval word recognition using deep-learned features. In: 16th International Conference on Frontiers in Handwriting Recognition, ICFHR, Niagara Falls, NY, USA, pp. 345–350 (2018)
7. Chopra, S., Hadsell, R., LeCun, Y.: Learning a similarity metric discriminatively, with application to face verification. In: 2005 IEEE Computer Society Conference on Computer Vision and Pattern Recognition (CVPR 2005), 20–26 June 2005, San Diego, CA, USA, pp. 539–546 (2005)
8. Deng, L., Hinton, G., Kingsbury, B.: New types of deep neural network learning for speech recognition and related applications: an overview. In: Proceedings of International Conference on Acoustic, Speech, and Signal processing, Vancouver, Canada, pp. 8599–8603, May 2013
9. Duan, Y., et al.: One-Shot Imitation Learning. In: NIPS, Long Beach, USA, December 2017
10. Fei Fei, L., Fergus, R., Perona, P.: One-shot learning of object categories. IEEE Trans. Pattern Anal. Mach. Intell. **28**(4), 594–611 (2006)
11. Fei-Fei, L.: Knowledge transfer in learning to recognize visual object classes. In: International Conference on Development and Learning (ICDL), Bloomington, Indiana, USA (2006)
12. Fei-Fei, L., Fergus, R., Perona, P.: A Bayesian approach to unsupervised one-shot learning of object categories. In: Proceedings of 9th IEEE International Conference on Computer Vision, Nice, France, pp. 1134–1141, October 2003
13. Fischer, A., Keller, A., Frinken, V., Bunke, H.: Lexicon-free handwritten word spotting using character HMMs. Pattern Recogn. Lett. - PRL **33**, 934–942 (2012)
14. Glorot, X., Bengio, Y.: Understanding the difficulty of training deep feedforward neural networks. In: Proceedings of the Thirteenth International Conference on Artificial Intelligence and Statistics, AISTATS 2010, 13–15 May 2010, Chia Laguna Resort, Sardinia, Italy, pp. 249–256 (2010)
15. Howe, N.R.: Part-structured inkball models for one-shot handwritten word spotting. In: 2013 12th International Conference on Document Analysis and Recognition, pp. 582–586 (2013)
16. Keskar, N.S., Mudigere, D., Nocedal, J., Smelyanskiy, M., Tang, P.T.P.: On Large-Batch Training for Deep Learning: Generalization Gap and Sharp Minima. arXiv preprint arXiv:1609.04836 (2016)
17. Koch, G., Zemel, R., Salakhudtdinov, R.: Siamese neural networks for one-shot image recognition. In: Proceedings of the 32nd International Conference on Machine Learning, Lille, France, vol. 37, July 2015
18. Maas, A., Kemp, C.: One-shot learning with Bayesian networks. In: Proceedings of Cognitive Science Society, Amsterdam, Netherlands, August 2009
19. Russakovsky, O., et al.: Imagenet large scale visual recognition challenge. Int. J. Comput. Vis. (IJCV) **115**, 211–252 (2015)

20. Santoro, A., Bartunov, S., Botvinick, M., Wierstra, D., Lillicrap, T.: Meta-learning with memory-augmented neural networks. In: Proceedings of the International Conference on Machine Learning (ICML), New York City, USA, pp. 1842–1850, June 2016
21. Silver, D.: Mastering the game of Go with deep neural networks and tree search. Nature **529**, 484–489 (2016)
22. Thorpe, S., Fize, D., Marlot, C.: Speed of processing in the human visual system. Nature **381**, 520–522 (1996)
23. Vinyals, O., Blundell, C., Lillicrap, T.P., Kavukcuoglu, K., Wierstra, D.: Matching networks for one shot learning. CoRR abs/1606.04080 (2016). http://arxiv.org/abs/1606.04080
24. Wilkinson, T., Lindström, J., Brun, A.: Neural Ctrl-F: segmentation-free query-by-string word spotting in handwritten manuscript collections. In: IEEE International Conference on Computer Vision, ICCV, Venice, Italy, pp. 4443–4452 (2017)
25. Woodward, M., Finn, C.: Active one-shot learning. In: NIPS, Deep Reinforcement Learning Workshop, Barcelona, Spain, December 2016
26. Wu, Y., et al.: Googles Neural Machine Translation System: Bridging the Gap between Human and Machine Translation. arXiv preprint arXiv:1609.08144 (2016)

First-Person View Hand Parameter Estimation Based on Fully Convolutional Neural Network

En-Te Chou[1], Yun-Chih Guo[1], Ya-Hui Tang[1], Pei-Yung Hsiao[2], and Li-Chen Fu[1,3(✉)]

[1] National Taiwan University, Taipei 10617, Taiwan
nderchou@gmail.com, {r07922094,lichen}@ntu.edu.tw
[2] National University of Kaohsiung, Kaohsiung 81148, Taiwan
pyhsiao@nuk.edu.tw
[3] NTU Research Center for AI and Advanced Robotics, Taipei 10617, Taiwan
http://robotlab.csie.ntu.edu.tw/
http://ai.robo.ntu.edu.tw

Abstract. In this paper, we propose a real-time framework that can not only estimate location of hands within a RGB image but also their corresponding 3D joint coordinates and their hand side determination of left or right handed simultaneously. Most of the recent methods on hand pose analysis from monocular images only focus on the 3D coordinates of hand joints, which cannot give a full story to users or applications. Moreover, to meet the demands of applications such as virtual reality or augmented reality, a first-person viewpoint hand pose dataset is needed to train our proposed CNN. Thus, we collect a synthetic RGB dataset captured in an egocentric view with the help of Unity, a 3D engine. The synthetic dataset is composed of hands with various posture, skin color and size. We provide 21 joint annotations including 3D coordinates, 2D locations, and corresponding hand side which is left hand or right hand for each hand within an image.

Keywords: Hand pose estimation · Convolutional neural network · Synthetic data

1 Introduction

Hand image understanding aims to estimate spatial configuration of hands from RGB images or depth images. Many applications such as human computer interaction, virtual reality and augmented reality are looking forward to a fast and

This research was supported by the Joint Research Center for AI Technology and All Vista Healthcare under Ministry of Science and Technology of Taiwan, and Center for Artificial Intelligence & Advanced Robotics, National Taiwan University, under the grant numbers of 108-2634-F-002-016 and 108-2634-F-002-017.

S. Palaiahnakote et al. (Eds.): ACPR 2019, LNCS 12047, pp. 224–237, 2020.
https://doi.org/10.1007/978-3-030-41299-9_18

robust hand pose estimation system to provide a natural way for users to interact with virtual world. Hand pose estimation is a very challenge task due to the
various appearances, self-occlusions and high degree of freedom of our human
hands. Although there have been many existing works studying in hand image
understanding, most of them require depth cameras or multi-view RGB images
as their input to handle this difficult task. However, most of the devices for VR
or AR nowadays are not equipped with those sensors, so those works are hard
to be implemented onto such devices. Thus, a RGB image based hand image
understanding system is urgently needed.

Previous works that are working on the hand image understanding are focusing on only the pose estimation itself, which aims to predict the coordinates of
hand joints in 3D space, but not other information. However, we claim that only
the 3D coordinates of hand joints cannot convey the full story of hands in the
images. To be applied to applications like virtual reality or augmented reality,
we often need more information to fully reconstruct a hand. For example, the
detection of hand is a crucial part since we have to define where the hand is in
the very first part of a hand image understanding system. Also, a hand image
understanding system should be able to distinguish left hands from right hands
since applications often give different rendering settings to left and right hands.

As a result, we present a novel real-time framework that can not only estimate
location of hands within a RGB image but also their corresponding 3D joint coordinates and their hand side determination of left or right handed simultaneously.
We employ a fully convolutional neural work with a two stream architecture for
estimating 3D joint coordinates of left hand and right hand individually. Our
proposed network is an one-stage network, which means that we do not need
other procedures for pre-cropping hands out of an image, and we can also train
it end-to-end. Experiments show that our approach has a significant boost of
FPS compared with previous methods and can run in real-time with a single
GPU.

On the other hand, there are still limitations to be applied on first-person
view applications. The most deeply affected issue is the lack of large-scale dataset
that is captured in a first person view. Existing datasets which are used for hand
pose estimation are almost captured in a third person view, making them hard
to be implemented onto virtual reality or augmented reality applications which
are usually used in a first person view. Therefore, we determine to collect a large
scale RGB synthetic dataset which images are captured in a first person view
with the help of Unity. Overall, the main advantages of our proposed framework
are listed below:

- We propose a hand parameter estimation system that not only estimates
  the 3D hand joints coordinates but also their bounding boxes and hand side
  determination of left or right handed to truly fit the demand of applications.
  Different from previous works, our proposed method can provide more abundant information that is needed for applications.

- We collect a RGB synthetic dataset which images are captured from egocentric viewpoint to better meet our demand for posting our system on a head mount device.
- Moreover, our proposed network can also be trained in an end-to-end procedure, and since it is an one-stage procedure, it will be fast enough for further usage of mobile devices in the future.

2 Related Work

Hand pose estimation has been a popular issue in recent years and lots of researchers have be dedicate to it, and many methods and algorithms have been proposed so far. For those methods, we can classify them into different categories. Regarding the input data, there are RGB images and depth images, whereas regarding the output representation, there are regression-based methods that directly estimate 3D coordinates and detection-based methods that predict heatmaps to speculate 3D coordinates with some post processing algorithms.

RGB Image Based Hand Pose Estimation. For 3D hand pose estimation from RGB images, previous methods such as [4,17,19] estimated hand pose from RGB videos. Gorce et al. [4] proposed a method that estimated 3D hand pose through minimizing the objective function. Sridhar et al. [15] adopted multi-view RGB images with depth data for estimating 3D hand pose with local optimization. Zimmermann et al. [22] proposed a two stage network for localizing hands and estimating 3D coordinates of hand joints with a two stream network. Mueller et al. [7] proposed a network that regresses both 2D heatmaps and 3D joint locations from a single cropped RGB hand image and apply additional skeleton fitting algorithm to give physical constraints to the hand structure. Mueller et al. [7] proposed a network that regresses both 2D heatmaps and 3D joint locations from a single cropped RGB hand image and apply additional skeleton fitting algorithm to give physical constraints to the hand structure.

Depth Image Based Hand Pose Estimation. For 3D hand pose estimation from depth image, Oberweger et al. [10] added a bottleneck layer, which contains fewer neurons than the output fully connected layer, to the second last layer with a purpose of forcing CNN to learn things more efficiently. Zhou et al. [21] divided their hand pose estimation system into three parts, which are thumb, index finger, and other fingers, to better estimate 3D coordinates of hand joints and better solve the relationship among every finger. Wu et al. [18] proposed a skeleton-difference layer which allows CNN to learn the shape and the physical constraints of the hand. Ting et al. [16] proposed a 3D residual network that estimates 3D coordinates of a cropped hand in a depth image with fully connected layers and apply a creative loss function called skeleton steadying loss to give physical constraints of the bone length ratio in each fingers and finger length ratio in a hand.

3 Hand Parameter Estimation via Two Stream Convolution

We propose a real-time hand parameter estimation system which can not only estimate location of hands but also their corresponding 3D joint coordinates and their hand side determination of left or right handed simultaneously. Different from other previous works, we do not need any predefined bounding box annotations provided from dataset or other software, our proposed system only takes a single RGB image as input. First, we send a RGB image into an encoder-decoder network to obtain heatmaps for initially locating hand joints in a frame. Then, we concatenate the high-level feature map with previous layer for reservation of more information, including color and textural information. Second, we pass the feature map into a deeper network, which is called a posture encoder, to estimate confidence map, bounding box coordinates and their corresponding hand side. Simultaneously, the high-level feature map output from posture encoder will be sent through two individual fully convolution blocks to obtain the 3D coordinates we need for supervision. Figure 1 shows the system of our proposed network.

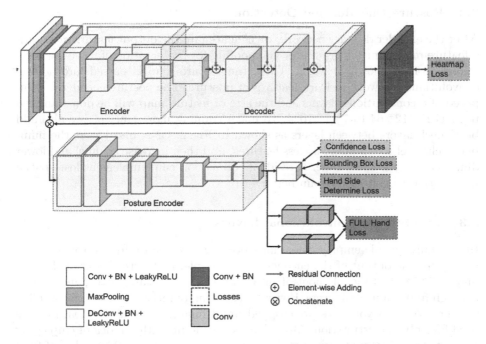

Fig. 1. The system overview of our proposed work. Our proposed network is mainly constructed by three part: encoder-decoder network for initially locating hand joints, posture encoder for understanding posture, and two stream convolution blocks for estimating 3D joint coordinates.

3.1 Encoder-Decoder

In recent years, researchers have been dedicated to pose estimation such as Moon *et al.* [6], Newell *et al.* [9], and Yuan *et al.* [20], and the experiment results show that an encoder-decoder architecture can outperform other encoder-only methods. However, the computational consumption must be highly concerned while applying an encoder-decoder network since the deconvolution procedure is often time consuming. Inspired by those works, we propose an encoder-decoder network at the first stage of our system, using it as an initial heatmap estimation to better localize where each joint in an image is. The encoder-decoder network is composed of only 4 convolutional layers and 4 deconvolutional layers due to the consideration of time consumption. Inside the network, we employ residual connections to better preserve low level features. Enlightened by recent methods as mentioned above, we employ this network to predict the 2D-Gaussian like heatmaps with K channels, representing the number of joints that constructs a hand. To generate heatmaps for supervision, we take 2D keypoints annotations from datasets, setting them as mean value for gaussian distribution with a variation of 3.

3.2 Posture Encoder and Detection

After the encoder decoder network, we concatenate the output feature map with a feature map from the layer after maxpooling for preservation of more low level information as shown in Fig. 1. Then, the feature map is passed into a deep convolutional network to learn features of posture. The posture encoder is composed of 8 convolutional layers and the size of feature map will be downsampled from 112* 112* 64 to 7* 7* 1024. To make training more efficient, we also add bottleneck layers between layers as shown in Fig. 2. CNN first lowers the number of channel to eliminate useless features, and then increase channel and lower dimension at the same time. A residual connection can supplement lost spatial information to help CNN train better.

3.3 Two Stream Convolutional Layers

In generally, people employ global max pooling layer and fully connected layers right after the output of the encoder. For example, if the size of output feature map is 7* 7* 1024, people pools it into a 1* 1* 1024 feature vector, and send it through fully connected layers. However, this kind of method can only be applied to scenario when you have pre-cropped hand images since all the features are used for only one estimation. After that, we come up with an idea of employing fully connected layers to every grid we have, so we will have 49 branches of fully connected layers. However, we discover that hands may not precisely appear within only one grid in most of the situation, and it may lead to a lack of information if using only features from one grid. That is, a convolutional layer will be a good choice for us, since the kernel will go through the feature map for obtaining more spatial information, it will be more possible for a convolutional

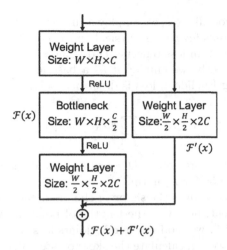

Fig. 2. Our designed bottleneck layer.

layer to collect all features that are needed. Also, since we don't crop the hand region out, we cannot modify the hand image into the same hand side. Thus, we set two individual convolutional blocks, one of which is responsible for estimating 3D joints coordinates of left hands and another is in charge of right hands. With such design, CNN can learn features from different hands more precisely to improve the performance.

3.4   Loss Functions

In this section, we will detail the designs of our loss functions to train the convolutional neural network. For the first part of our system, there is an encoder-decoder network which is supervised by heatmaps generated by 2D ground truth annotations. We generate ground truth heatmaps via Gaussian distribution as mentioned above. Thus, we employ binary cross entropy loss at encoder-decoder network for training as shown in below:

$$L_{HM} = \frac{-1}{N} \sum_{K=1}^{21} \sum_{i=1}^{h} \sum_{j=1}^{w} [H^*_{Kij} \log(H_{Kij}) + (1 - H^*_{Kij}) \log(1 - H_{Kij})] \quad (1)$$

where H and H^* are the predicted heatmap and ground truth heatmap, and w and h are the width and height of heatmap, 21 defines the number of joints, and N stands for the number of hands within an image. We divide the loss with N in order to balance gradients obtained from different batches.

For the posture encoder part, there are four different kinds of outputs which are confidence map (L_{conf}), bounding box coordinates (L_{bbox}), hand side determination (L_{hs}), and 3D joint coordinates (L_{coord}). For the confidence map and bounding box coordinates, we employ mean-squared-error for supervision. For the hand side determination part, we employ cross entropy loss for supervision.

For the 3D joints coordinates, we employ a full hand loss including Euclidean loss and some other losses for giving physical constraints to a hand, and angle of rotations. Since a Euclidean loss reflects 3D distances between predicted results and ground truths directly, we employ it as a base loss of our joints loss. The formula for computing Euclidean loss is described below:

$$L_e(H_j) = \sum_{j=1}^{J} \|J_K - J_K^*\|_2 \tag{2}$$

where J and J^* represent predicted 3D coordinates and the ground truth of the Kth joint and $\|.\|_2$ stands for L_2 norm.

We discover that sometimes the shape of the predicted hand may be roughly correct to certain extent; however, the lengths of bones between joints may be diverse. Inspired by [16], we modify their 16-joints loss to a 21-joints version in our work. As a result, we can calculate the skeleton steadying using the following equation:

$$L_{sk} = L_b + L_f \tag{3}$$

where L_b represents bone length ratio loss and L_f represents finger length ratio loss.

Also, when it comes to virtual reality applications nowadays, to prevent bone lengths from variation, people tend to apply vectors onto bones within a predefined hand model to reconstruct a hand. That is, we agree that the knowledge of vectors between joints should be taken into consideration for further usages. To achieve such goal, we employ loss as shown below to make sure the predicted vectors can be as close to ground truth as possible:

$$L_v = \sum_{B=1}^{20} (1 - \frac{V_B \cdot V_B^*}{\|V_B\| \|V_B^*\|}) \tag{4}$$

where V and V^* are the predicted vectors of between joints and the ground truth vectors, and there are 20 bones constructed by the 21 joints within a hand. Thus, losses for joint locations can be integrated as shown below:

$$L_{coord} = \alpha L_e + \beta L_{sk} + \gamma L_v \tag{5}$$

where we set $\alpha = 40$, $\beta = 1$, and $\gamma = 0.05$ in our experiment. In conclusion, we sum up all the losses mentioned above and give them different weightings to help CNN train better, where the total loss is shown in the following:

$$L_{total} = \omega_1 L_{HM} + \omega_2 L_{conf} + \omega_3 L_{bbox} + \omega_4 L_{hs} + \omega_5 L_{coord} \tag{6}$$

where we set $\omega_1 = 1$, $\omega_2 = 5$, $\omega_3 = 1$, $\omega_4 = 1$, and $\omega_5 = 1000$ in our experiment.

3.5   Implementation Detail

The input size of our proposed network is 224* 224 on our synthetic dataset and 320* 320 on RHD dataset. Inspired by YOLOv2 [13], we set 5 default anchor boxes in our network to help detection. We train our proposed network end-to-end using Adam optimizer [3] with an initial learning rate of 0.001 and batch size 24. We implement our system with PyTorch [12] and train it on a single GPU which is NVIDIA RTX 2080Ti. For data augmentation, we scale images with a range of [0.8, 1.4], rotation for [−45°, 45°].

3.6   NTU Synthetic Dataset

In this paper, we aim to propose a hand parameter estimation system which can be applied in an egocentric viewpoint. However, there are still very few amount of hand datasets captured in egocentric viewpoint. SynthHand dataset [8] is one of a rare dataset captured in an egocentric view, while others are almost in third person views. However, the postures in SynthHand dataset are kind of too monotonous to be applied in real-world applications. In order to efficiently learn the egocentric viewpoint feature, we decide to generate a synthetic dataset captured from egocentric view with the help of Unity.

First, we collect some realistic hand models which can be fully rigged, meaning that we can define rotation and translation of each joint to construct any postures we want. However, not all the hand models that we find on the internet can be formulated to meet what we want. Thus, we employ Maya [1], which is a computer animation and modeling software published by Adobe, for customizing structures we need. Also, in order to make hand postures continuously cover all the distribution in real-world applications, we define many animations of hand models to make postures of hand vary linearly along with time.

Second, we set images which are randomly captured from COCO 2017 dataset [5] as background for our synthetic data to increase diversity of RGB features. To increase the diversity of skin color, we manually make some modification of RGB values using Photoshop. Also, we randomly modify the saturation and lightness of our light source in Unity. As shown in Fig. 3 our proposed dataset contains hands in various poses, locations and rotation. Moreover, the background and lightness are much different from frame to frame to help CNN learn better hand information. Also, in consideration of the limitation of human limbs, we will not place hands too far away since it is not reasonable for scenario when users are under an egocentric view. To sum up, we collect 23712 image frames for training and 1318 image frames for testing.

4   Experimental Results

In this chapter, we will summarize experiments of our proposed hand parameter estimation system on datasets. We will first introduce the datasets we employ for our system. Next, we will show the experimental results on the two datasets and compare with other state-of-the-art.

Fig. 3. Some examples in our NTU synthetic dataset.

4.1 Datasets

The public dataset used here is Rendered Handpose Dataset [22]. And we also make some evaluation on our proposed NTU synthetic dataset. The introduction is listed below:

- Rendered Handpose Dataset (RHD): RHD is a public and challenging dataset. Images from this dataset are collected in third person view for hand pose estimation, and the dataset shows hands with various locations, postures, and sizes within images. RHD contains 41258 images for training and 2728 images for testing. Each hand in an image is annotated with 21 keypoints including four joints in each finger and a wrist joint. RHD provides their 2D coordinates, 3D coordinates, and a visibility indicator. To make sure that there are enough information for CNN to extract from an image, we set a threshold ($= 10$) for visible joints to filter out hands that are cut by margins of images.
- NTU Synthetic Dataset is a synthetic dataset collected by ourselves. This dataset contains 23712 images for training and 1318 images for testing. Each frame provides a 320* 240 pixels RGB image, and there is a single hand within it. Each hand is annotated with the same 21 joints as RHD, and providing their 3D coordinates and the 2D coordinates in an image frame. Also, the corresponding bounding boxes are annotated in formation of (x_{min}, y_{min}, x_{max}, y_{max}), and the hand side determination of left or right handed. Figure 3 shows some examples from our NTU synthetic dataset.

4.2 Evaluation Metric

There are two kinds of evaluation metrics we are using in our experiment for examining our network. The evaluation methods are introduced as below:

– Mean Average Precision (mAP): This evaluation metric is to compute the successful rate of the hand detection part. A successful detection here means that the Intersection-Over-Union (IOU) of the predicted bounding box and ground truth bounding box is higher than a threshold, and the predicted hand side is also correct. We will set the IOU threshold at 0.5 for calculating our mean average precision in our paper.
– Mean End-Point-Error (mean EPE): This evaluation metric simply calculates the Euclidean distance error between the predicted joints and their corresponding ground truth joints, and average them with joints number.

4.3 Results

To evaluate our system, we make several comparisons between different configuration of settings. For *Baseline*, we directly regress the 3D coordinates of joints together with the detection results via convolution layer, meaning that we increase the number of channel from $(1+4+2)$ to $(1+4+2+63)$, and the 63 channels are responsible for the prediction of 3D coordinates of 21 keypoints. For *Baseline+Heatmap*, we add the encoder-decoder part into *Baseline*. For *Ours (one stream)*, we set up only one stream of the convolutional block for estimating 3D coordinates of both hands. Table 1 shows the evaluation results using different configurations on RIID.

Table 1. The comparison under different configuration on RHD.

Method	mAP (%)	Mean EPE (mm)	FPS
Baseline	19.89	42.16	**172.22**
Baseline + Heatmap	34.53	37.48	147.82
Ours (one stream)	82.57	34.02	132.43
Ours	**83.91**	**28.30**	132.09

Table 2. Comparison between state-of-the-art and ours on Rendered Handpose Dataset.

Method	Mean EPE (mm)
Zimmermann *et al.* [22]	20.90
Spurr *et al.* [14]	**19.73**
Ours	28.30

Also, to evaluate our proposed method, we have some comparison between state-of-the-art works as shown in Table 2. Zimmermann *et al.* [22] proposed a two stage system for estimating 3D hand pose, and they achieve 20.90 mm mean EPE. Spurr *et al.* [21] propose a hand pose estimation network using the pre-cropped hand images provided by dataset, and they achieve 19.73 mm mean EPE. Although the mean EPE of our proposed method is a little bit higher than the other two works, our proposed method has a large promotion in speed as shown in Table 3. Paschalis *et al.* [11] employ openpose [2] as their hand detector and their overall system achieves 18 FPS. The system of Zimmermann *et al.* achieves 29.81 FPS in our implement. While our method can achieve about

Table 3. Comparison of FPS between state-of-the-art and ours on Rendered Handpose Dataset.

Method	FPS
Paschalis *et al.* [11]	18.00
Zimmermann *et al.* [22]	29.81
Ours	**132.09**

Table 4. Comparison between state-of-the-art and ours on NTU synthetic dataset.

Method	mAP (%)	Mean EPE (mm)	FPS
Zimmermann *et al.* [22]	–	7.92	42.73
Ours	97.62	8.47	146.04

Fig. 4. Some examples of the input and the predicted result of hand detection. We can see that many difficult tasks within this dataset have been conquered.

132.09 FPS, which is about 4.5 times faster than theirs. Table 4 shows the comparison on NTU synthetic dataset.

For the visualization results, Fig. 4 shows some results of the hand detection with our propose network. There are many difficult tasks, including hands in margin, low light source, small hands, and clustering background. However, our proposed model can still detect them well. Figure 5 shows the predicted results of 3D hand joints and their corresponding ground truth (Fig. 6).

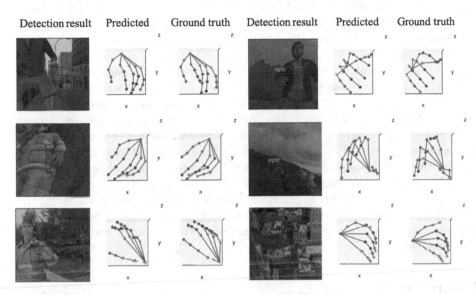

Fig. 5. Some examples of the input and the predicted result of 3D hand pose estimation. We will only show one hand once a time for better visualization. Left column: predicted bounding boxes and corresponding hand side determination. Middle column: predicted 3D joints coordinates. Right column: ground truth 3D joints coordinates.

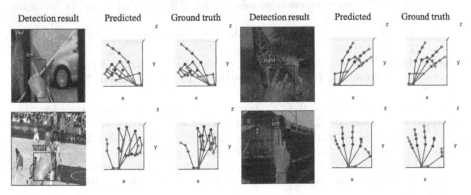

Fig. 6. Some examples of the predicted results of on our NTU Synthetic Dataset. Left column: predicted bounding boxes and corresponding hand side determination. Middle column: predicted 3D joints coordinates. Right column: ground truth 3D joints coordinates.

5   Conclusion

In this paper, we propose a real-time hand parameter estimation system which can not only estimate location of hands but also their corresponding 3D joint coordinates and their hand side determination of left or right handed simultaneously, using a single RGB image as input. For previous works, most of them employ two stage methods for estimation only hand pose, however, we claim that it is not a full story for hand image understanding. That is, we aim to build a system which can truly fit the demand of applications like virtual reality and augmented reality. Moreover, due to the lack of datasets which images are captured in an egocentric viewpoint, we also collect a synthetic dataset using the 3D engine, Unity. To collect our dataset, we collect some hand meshes from internet, and giving them skeletons using MAYA, so that we can rig their articulation to any postures that we want them to be. And we set up animations to make the hand meshes vary linearly with time going. Finally, we can capture our dataset with various light sources, postures, skin colors and meshes. Overall speaking, our proposed system is capable of detecting hands from RGB images and estimating their hand pose with a single stage convolutional neural net-work. Also, with the help of our first person view synthetic dataset, our system can overcome the scenario for VR or AR applications which are often used in an egocentric viewpoint. Therefore, we expect that our proposed system can help provide a real-time and accurate way for the interaction between human and computer in our future life.

References

1. Alias Systems Corporation: Maya—computer animation & modeling software—autodesk. https://www.autodesk.com/products/maya/overview
2. Cao, Z., Hidalgo, G., Simon, T., Wei, S.E., Sheikh, Y.: OpenPose: realtime multi-person 2D pose estimation using Part Affinity Fields. arXiv preprint arXiv:1812.08008 (2018)
3. Kingma, D.P., Ba, J.: Adam: a method for stochastic optimization. CoRR abs/1412.6980 (2014)
4. de La Gorce, M., Fleet, D.J., Paragios, N.: Model-based 3D hand pose estimation from monocular video. IEEE Trans. Pattern Anal. Mach. Intell. **33**, 1793–1805 (2011)
5. Lin, T.-Y., et al.: Microsoft COCO: common objects in context. In: Fleet, D., Pajdla, T., Schiele, B., Tuytelaars, T. (eds.) ECCV 2014. LNCS, vol. 8693, pp. 740–755. Springer, Cham (2014). https://doi.org/10.1007/978-3-319-10602-1_48
6. Moon, G., Chang, J.Y., Lee, K.M.: V2V-PoseNet: voxel-to-voxel prediction network for accurate 3D hand and human pose estimation from a single depth map. In: 2018 IEEE/CVF Conference on Computer Vision and Pattern Recognition, pp. 5079–5088 (2018)
7. Mueller, F., et al.: GANerated hands for real-time 3D hand tracking from monocular RGB. In: 2018 IEEE/CVF Conference on Computer Vision and Pattern Recognition, pp. 49–59 (2017)

8. Mueller, F., Mehta, D., Sotnychenko, O., Sridhar, S., Casas, D., Theobalt, C.: Real-time hand tracking under occlusion from an egocentric RGB-D sensor. In: Proceedings of International Conference on Computer Vision (ICCV) (2017). http://handtracker.mpi-inf.mpg.de/projects/OccludedHands/

9. Newell, A., Yang, K., Deng, J.: Stacked hourglass networks for human pose estimation. In: Leibe, B., Matas, J., Sebe, N., Welling, M. (eds.) ECCV 2016. LNCS, vol. 9912, pp. 483–499. Springer, Cham (2016). https://doi.org/10.1007/978-3-319-46484-8_29

10. Oberweger, M., Wohlhart, P., Lepetit, V.: Hands deep in deep learning for hand pose estimation. ArXiv abs/1502.06807 (2015)

11. Panteleris, P., Oikonomidis, I., Argyros, A.A.: Using a single RGB frame for real time 3D hand pose estimation in the wild. In: 2018 IEEE Winter Conference on Applications of Computer Vision (WACV), pp. 436–445 (2017)

12. Paszke, A., et al.: Automatic differentiation in pytorch (2017)

13. Redmon, J., Farhadi, A.: YOLO9000: better, faster, stronger. In: 2017 IEEE Conference on Computer Vision and Pattern Recognition (CVPR), pp. 6517–6525 (2017)

14. Spurr, A., Song, J., Park, S., Hilliges, O.: Cross-modal deep variational hand pose estimation. In: 2018 IEEE/CVF Conference on Computer Vision and Pattern Recognition, pp. 89–98 (2018)

15. Sridhar, S., Oulasvirta, A., Theobalt, C.: Interactive markerless articulated hand motion tracking using RGB and depth data. In: 2013 IEEE International Conference on Computer Vision, pp. 2456–2463 (2013)

16. Ting, P.-W., Chou, E.-T., Tang, Y.-H., Fu, L.-C.: Hand pose estimation based on 3D residual network with data padding and skeleton steadying. In: Jawahar, C.V., Li, H., Mori, G., Schindler, K. (eds.) ACCV 2018. LNCS, vol. 11365, pp. 293–307. Springer, Cham (2019). https://doi.org/10.1007/978-3-030-20873-8_19

17. Wöhlke, J., Li, S., Lee, D.: Model-based hand pose estimation for generalized hand shape with appearance normalization. ArXiv abs/1807.00898 (2018)

18. Wu, M.Y., Tang, Y.H., Ting, P.W., Fu, L.C.: Hand pose learning: combining deep learning and hierarchical refinement for 3D hand pose estimation. In: BMVC, vol. 1, p. 3 (2017)

19. Wu, Y., Lin, J., Huang, T.S.: Analyzing and capturing articulated hand motion in image sequences. IEEE Trans. Pattern Anal. Mach. Intell. **27**, 1910–1922 (2005)

20. Yuan, S., et al.: Depth-based 3D hand pose estimation: from current achievements to future goals. In: 2018 IEEE/CVF Conference on Computer Vision and Pattern Recognition, pp. 2636–2645 (2017)

21. Zhou, Y., Lu, J., Du, K., Lin, X., Sun, Y., Ma, X.: HBE: hand branch ensemble network for real-time 3D hand pose estimation. In: Ferrari, V., Hebert, M., Sminchisescu, C., Weiss, Y. (eds.) Computer Vision – ECCV 2018. LNCS, vol. 11218, pp. 521–536. Springer, Cham (2018). https://doi.org/10.1007/978-3-030-01264-9_31

22. Zimmermann, C., Brox, T.: Learning to estimate 3D hand pose from single RGB images. Technical report. arXiv:1705.01389 (2017). https://lmb.informatik.uni-freiburg.de/projects/hand3d/

Dual-Attention Graph Convolutional Network

Xueya Zhang, Tong Zhang$^{(\boxtimes)}$, Wenting Zhao, Zhen Cui, and Jian Yang

Key Lab of Intelligent Perception and Systems for High-Dimensional Information
of Ministry of Education, School of Computer Science and Engineering,
Nanjing University of Science and Technology, Nanjing, China
tong.zhang@njust.edu.cn

Abstract. Graph convolutional networks (GCNs) have shown the powerful ability in text structure representation and effectively facilitate the task of text classification. However, challenges still exist in adapting GCN on learning discriminative features from texts due to the main issue of graph variants incurred by the textual complexity and diversity. In this paper, we propose a dual-attention GCN to model the structural information of various texts as well as tackle the graph-invariant problem through embedding two types of attention mechanisms, i.e. the connection-attention and hop-attention, into the classic GCN. To encode various connection patterns between neighbour words, connection-attention adaptively imposes different weights specified to neighbourhoods of each word, which captures the short-term dependencies. On the other hand, the hop-attention applies scaled coefficients to different scopes during the graph diffusion process to make the model learn more about the distribution of context, which captures long-term semantics in an adaptive way. Extensive experiments are conducted on five widely used datasets to evaluate our dual-attention GCN, and the achieved state-of-the-art performance verifies the effectiveness of dual-attention mechanisms.

Keywords: Dual-attention · Graph convolutional networks · Text classification

1 Introduction

Text classification is an active research field of natural language processing and multimedia, and has attracted increasing attention in recent years. For those given text sequences, the purpose of text classification is to annotate them with appropriate labels which accurately reflect the textual content. As a fundamental problem of text analysis, text classification has become an essential component in many applications, such as document organization, opinion mining, and sentiment analysis.

© Springer Nature Switzerland AG 2020
S. Palaiahnakote et al. (Eds.): ACPR 2019, LNCS 12047, pp. 238–251, 2020.
https://doi.org/10.1007/978-3-030-41299-9_19

To achieve classification based on texts of irregular structure, numerous algorithms have been proposed for dealing with the text classification task. Traditional methods employ hand-crafted feature, i.e. TF-IDF, bag-of-words and n-grams [25] for text content representation, and then use widely used classifies such as support vector machine (SVM) and logistic regression (LR) for classification. However, these methods suffer from the limited feature learning ability. Deep neural network based algorithms have achieved great success in various tasks, and some studies apply them to text classification. Convolutional neural networks (CNNs) [12] and recurrent neural networks (RNNs) [16] are quite representative, which extract multi-scale features and compose them to obtain higher expressive representations. Especially, recursive RNN show better performance with the advantages of modeling sequences. However, these deep neural networks cannot well model the irregular structure of texts, which is crucial for text recognition task. Recently, graph convolutional networks (GCNs) [3,13] have been proposed with a lot of success in various tasks, and also applied in feature representation of texts. On the other hand, due to the difficulty in modeling data variance, the attention mechanism [2,18,23] is proposed and widely embedded in multiple models, achieving promising results on a variety of tasks.

Promising performance has been achieved on text classification by aforementioned methods, especially those GCN-based frameworks. However, challenges still exist in discriminative feature representation for describing the semantics when adapting GCN on a large number of texts. Basically, the main issue comes from the graph variants incurred by the complexity and diversity of texts, where the variants are mainly manifested in two aspects: (i) the local connection patterns of neighbour words vary with sentences, which can not be well modeled by the uniform connection weights defined by the adjacency matrix; (ii) the features of various connection scopes, which are extracted from each hop during the graph diffusion process, may contribute differently for capturing the long-term semantics in diverse texts, which make it difficult to learn the distribution of context by imposing fixed weights on them (as what is done by classic GCNs).

In this paper, we propose a dual-attention GCN framework to deal with text classification. The proposed method can learn discriminative features from texts through inference on graphs, as well as solve the graph-invariant problem by leveraging attention mechanisms. For mining the underlying structural information of text, we construct graph models based on text sequences and further conduct graph convolution for capturing contextual information through diffusion on graphs. Furthermore, considering the graph invariants incurred by the complexity and diversity of texts, we specifically propose two different types of attention mechanisms, i.e. the connection-attention and hop-attention, and integrate them with GCN as an whole deep framework. In view of various connection patterns between neighbour words in texts, we apply connection-attention to capture the short-term dependencies by adaptively imposing different weights specified to neighbourhoods of each word. Moreover, considering to model the long-term semantics in texts during the graph diffusion process, we propose the hop-attention which applies scaled coefficients to different scopes to make the model learn more about the distribution of context in an adaptive way. For evaluating the performance of our proposed dual-attention GCN, extensive experiments are conducted on five widely used datasets, and the experimental results show our competitive performance comparing with those state-of-the-art methods and verify the effectiveness of the dual-attention mechanism.

2   Related Work

Mainly two lines of research are related to our work: text classification methods from the view of application line, and graph convolution as well as attention-based methods from the view of technical line. Below we briefly overview them.

Text Classification. Traditional methods for text classification usually concentrate on two important steps, which are split into feature engineering and classification model. For feature extraction, some hand-craft features such as TF-IDF, bag-of-words and n-grams [25] are very common. To classify the texts, classical machine learning methods such as logistic regression (LR) and support vector machine (SVM) did play an important role. However, the representation of text is high-dimension and the neural network isn't good at processing such data, which limits the ability of feature learning. Surprisingly, deep learning methods have been proposed and successfully applied to text classification. Mikolov et al. [17] come to focus on the model based on word embeddings and recently Shen et al. [21] conduct a study between Simple Word-Embedding-based Models, which show the effectiveness of word embeddings. At the same time, the principle of some deep learning models such as CNN [15] and RNN [10]are employed to text classification. Kim et al. [12] led a breakthrough by directly apply CNN model to text classification. Lai et al. [16] successfully use a specified model LSTM to text classification, which means that CNNs and RNNs that can extract multi-scale localized spatial features and compose them to obtain higher expressive representations are suitable for the task of text classification. Effective as they are, some shortcomings are exposed immediately. They mainly capture local information so that ignore much global information such as word co-occurance.

Graph Convolutional Network. In recent years, graph convolutional networks (GCNs) gain more attention because of some unique advantages. Representatively, Bruna et al. [4] consider possible generalizations of CNNs, which extends convolution networks to graph domains. However, the expensive computation and non-localized filter are existing problems. To address this problem, Henaff et al. [9] develop an extension of Spectral Network, paying effort to spatially localizing through parameterizing spectral filters. They consider the question how to construct deep architectures with low requirements for the complexity of learning on non-Euclidean domains. On the basis of previous work, then Defferrard et al. [6] proposed a fast spectral filter, which use the Chebyshev polynomial approximation so that they are the same linear complexity of computation and classical CNNs, and especially are suitable for any other graph structure. Subsequently, Kipf et al. [13] change the filter to a linear function so

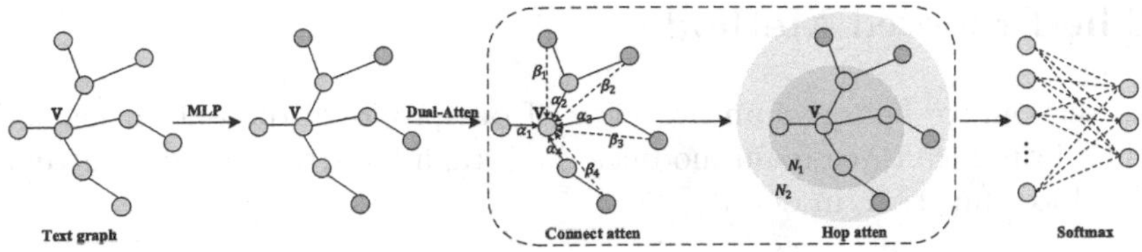

Fig. 1. The working process of our dual-attention GCN. In this text subgraph, we take the node v as the center node. After a multilayer perceptron, we show the dual-attention mechanism. The connection-attention and hop-attention respectively assign weights from width and depth. α and β represents the coefficients of connection-attention for different hop nodes. N represents the hop. We show the details in Sect. 3.

that the performance of model won't decrease. In addition, some non-spectral methods [7,8] like DCNN [1] and GraphSAGE [8] make operations spatially on close neighbors.

There are some research coming to explore the graph convolutional work that are more suitable for text classification. Firstly GCNs are used to capture the syntactic structure in [3], which produce representations of words and show the improvement. The method [13] mentioned in the last paragraph apply GCN to text classification, but it can't naturally support edge features. Some other methods like [27] regard documents or sentences as the graphs of words. Differently, Yao et al. [26] propose a new way to construct the graph by regarding both documents and words as nodes, which performs quite well with GCN.

Attention mechanism. The attention mechanism was first proposed in the field of visual images, and [18] led this mechanism to become popular in the true sense. Bahdanau et al. [2] use the mechanism similar to attention to simultaneously translate and align on machine translation task, which can be regarded as firstly proposing the application of the attention mechanism to NLP field. Then the Attention-based RNN model begin to be widely used in NLP, not just sequence-to-sequence models, but also for various classification problems. This mechanism can directly and flexibly capture global and local connections, and each step of the calculation does not depend on the calculation results of previous steps. Immediately, the self-attention attract people and this mechanism [23] also shows its effectiveness. Inspired by previous work, Veličković et.al [24] propose the graph attention network applied to graph nodes with different degrees, and assign arbitrary weights that are specified to neighbors so the learning model can capture related information more precisely.

In total, we also want to learn more hidden information across edges or more effective representation of nodes, so we should consider larger scale, which means considering more contextual information. The dual-attention graph convolution network we proposed, on the one hand, the connection-attention assign different weights to nodes automatically, on the other hand, the hop-attention take some hidden information of context into account by controlling the probability of sampling pairs of nodes within some distance.

3 The Proposed Method

In this section, we first give an overview of our proposed dual-attention GCN, and then introduce three main modules, i.e. graph construction, dual-attention layer and loss function, in detail.

3.1 Overview of Dual-Attention GCN

The whole architecture of the proposed dual-attention GCN framework is shown in Fig. 1, where the input is a graph based on a given text. For graph construction, we adopt the method proposed in [26] which removes useless words in texts first and then models both the text and its words as nodes. This process is described in detail in Sect. 3.2. Based on the constructed heterogeneous text graph, we conduct graph inference by passing it though our designed dual-attention GCN, where the central unit is a novel graph convolutional layer embedded with dual-attention mechanism. During the inference process, the connection-attention and hop-attention adaptively assign different weights to those neighbours of each node and the features of different scopes, respectively, to solve the graph-invariant problem (see Sects. 3.3 and 3.4 in detail). After graph inference, the obtained features are passed through a softmax classifier and finally cross entropy loss is calculated for network optimization (see Sect. 3.5).

3.2 Graph Construction

For a given text, the corresponding graph denoted as $\mathcal{G} = (\mathcal{V}, \mathcal{E}, \mathbf{A})$ is constructed for the content description, where $\mathcal{V}, \mathcal{E}$ denote the sets of vertex and edges separately, and $\mathbf{A} \in \mathbb{R}^{n \times n}$ is the adjacency matrix describing the connection relationship between each pair of nodes. Two types of nodes are involved in $\mathcal{V}$: one type is constructed by the words in texts and the other type is constructed by the whole text itself. To describe the relationship between these nodes, including both the connection between the nodes of words and the connection across the nodes of a word and the text, a corpus is first built based on the training texts and the PMI and TF-IDF values are calculated based on the statistics of the corpus, which are defined as follows:

$$\text{PMI}(i, j) = log \frac{p(i, j)}{p(i)p(j)},$$

and

$$\text{TF-IDF}_{ij} = \text{TF}_{ij} * \text{IDF}_i,$$

where

$$\text{TF}_{ij} = n_{ij}, \text{IDF}_i = log \frac{|\mathbf{D}|}{\{i : t_j \in \mathbf{d}_i\}}.$$

For PMI values, $p(i, j) = \frac{\text{W}(i,j)}{\text{W}}$ and $p(i) = \frac{\text{W}(i)}{\text{W}}$, where the statistical set W is the number of sliding windows and W(i) is the number of sliding windows that contains the word i. W(i, j) is the number of sliding windows in which word i and word j appear simultaneously. And for calculating TF-IDF values, n_{ij} is the number of times the word j appears in the document i, $\mid \mathbf{D} \mid$ is the number of documents, $\{i : t_j \in \mathbf{d}_i\}$ is the the number of documents that contains the word i.

Based on the defined PMI and TF-IDF values, the adjacency matrix can be obtained, which is represented as follows:

$$
A_{ij} = \begin{cases} \text{PMI}(i,j) & i \text{ and } j \text{ are words and PMI}(i,j) > 0 \\ \text{TF-IDF}_{i,j} & i \text{ is a document and } j \text{ is a word} \\ 1 & i = j \\ 0 & \text{otherwise} \end{cases}
$$

3.3 Connection-Attention

After obtaining the feature matrix $\mathbf{H} = [\mathbf{h}_1, ..., \mathbf{h}_i, ..., \mathbf{h}_j..., \mathbf{h}_N], \mathbf{h}_i \in \mathbb{R}^d$, we introduce the connection-attention mechanism to build latent representation for a specified hop . Firstly we apply a weight matrix $\mathbf{W} \in \mathbb{R}^{d' \times d}$ to each node, which plays a role in shared learnable linear transformation. So we obtain sufficient expressive power to transform the input features into higher-level features. We mark the new features as $\mathbf{H}' = [\mathbf{h}'_1, ..., \mathbf{h}'_i, ..., \mathbf{h}'_j, ..., \mathbf{h}'_N], \mathbf{h}'_i \in \mathbb{R}^{d'}$, where d' is the new dimension of feature vectors. The connection-attention is a shared mechanism $\mathbb{R}^{d'} \times \mathbb{R}^{d'} \rightarrow \mathbb{R}$ computing the attention coefficients. The matrix $\mathbf{C} \in \mathbb{R}^{n \times n}$ is applied to indicate the connection-attention coefficients performing on every nodes, where $e_{ij}^{(k)} \in \mathbf{C}$ is the element referring to the influence of node j's features to node i :

$$
e_{ij}^{(k)} = a_{connection}(\mathbf{h}'_i, \mathbf{h}'_j), j \in \mathcal{N}_i^k
$$

It means every node is allowed to attend the other nodes, and assigned more or less attention. Here we compute $e_{ij}^{(k)}$ for node i's neighborhood in the graph, and then add a softmax function to normalize the coefficients:

$$
\alpha_{ij}^{(k)} = softmax_j(e_{ij}^{(k)}) = \frac{exp(e_{ij}^{(k)})}{\sum_{j \in \mathcal{N}_i^k} exp(e_{ij}^{(k)})}
$$

In detail, attention mechanism $a_{connection}$ is expressed as:

$$
\alpha_{ij}^{(k)} = \frac{exp(LeakyReLU(\mathbf{a}^T[\mathbf{h}'_i \| \mathbf{h}'_j]))}{\sum_{j \in \mathcal{N}_i^k} exp(LeakyReLU(\mathbf{a}^T[\mathbf{h}'_i \| \mathbf{h}'_j])}
$$

where and $\|$ is the concatenation operation and $\mathbf{a}$ is a weight vector. Next we apply a nonlinearity δ:

$$
\mathbf{h}_i^{'(k)} = \delta(\sum_{j \in \mathcal{N}_i^k} \alpha_{ij}^{(k)} \mathbf{h}'_j)
$$

Now we obtain the updated feature vector $\mathbf{h}_i^{'(k)}$ with k-hop neighborhoods.

3.4   Hop-Attention

On the basis of feature vector $\boldsymbol{h}_i{}^{(k)}$, we consider the hop-attention by adding a constraint on the hop. In this case, we artificially fix a coefficient set whose coeffients are according to Chebyshev inequality. We define the coeffients of hop-attention $\mathbf{Q} = (q_1, q_2, ...q_c)$, a c-dimensional vector regarded as the context distribution with $q_k > 0$. Where c is the number of hops and there is another limitation $\sum_k q_k = 1$. For attention mechanism a_{hop}, the $\mathbf{Q}$ add weights to different range of neiborhoods, which will take more context distribution into account in a received field. For the hop-attention layer, we also add a sotfmax for regularizing:

$$\mathbf{Q} = softmax(q_1, q_2, ...q_c)$$

For $q_k \in Q$,

$$q_k = 1 - \frac{k-1}{c}$$

The connection-attention and hop attention work together on the nodes' features, then the feature vectors are updated:

$$\boldsymbol{h}_i^{''} = \sum_{k=1}^{c} q_k \boldsymbol{h}_i^{(k)}$$

For the stability of learning process, we can adopt multi-head. In summary, we define it as:

$$\boldsymbol{h}_i{}^{new} = \|_{m=1}^{M} \delta\left(\sum_{j \in \mathcal{N}_i^k} \alpha_{ij}^{(k,m)} \boldsymbol{h}_j^{'} \right)$$

where $\boldsymbol{h'}_j = \mathbf{W}^m \boldsymbol{h}_j$, $\alpha_{ij}^{(k,m)}$ represents m-th dual-attention mechanism for k-hop neighbors. M is the total number of heads and $\mathbf{W}^m$ represents the corresponding weight to the m-th attention mechanism. So we can obtain a new feature vectors with the dimension of Md'.

3.5   The Loss Function

In this task of text classification, the documents are annotated with a single label. If the final layer, we just map $\mathbf{H}$ to the dimension of the number of classifications, then fed it into a softmax classifier.

$$\mathbf{Z} = softmax(\mathbf{H}^{new})$$

We define the loss function by using cross-entropy as

$$\mathcal{L} = - \sum_{d \in \mathcal{Y}_D} \sum_{f=1}^{F} Y_{df} ln \mathbf{Z}_{df}$$

where $\mathcal{Y}_D$ is the set of document indices that have labels and F is the dimension of the output features. Y is the label indicator, and we add a L2 regularization.

4   Implementation Details

In experiments, we use pre-trained embedding features from TextGCN [26] with the size of 200. In the process of constructing the graph, we set the PMI window size as 20 to be more comparable. We set the learning rate as 0.05, dropout as 0.3 if not stated separately. If using the multi-head dual-attention, the dual-attention layer consists of 8 heads computing $d' = 64$ dimension features and in total are 512 features. For different datasets, we fine tune the parameters. We store the graph with the form of index instead of adjacency matrix. We select fixed number nodes in specified neighborhood every time. Especially we set the batchsize as 10 and subgraph size as 200 for 20ng and MR because of the large number of nodes. The number of neighborhoods we choose 200 and learning rate set as 0.01. For the rest we select 70 nodes in the one-hop neighborhood and gather connected nodes if we want to select two or more hop nodes. For example, the number of two-hop nodes will be the square of original data, and cube for the three-hop. Talking of the following activation we choose an exponential linear unit (ELU) [5] nonlinearity. We apply the Momentum optimizer [20] and models are trained to minimize cross-entropy with 300 epochs.

5   Experiments

5.1   Datasets

We also ran our experiment on the five used benchmark corpora, including 20-Newsgroups, Ohsumed, R52, R8 and Movie Review(MR). R52 and R8 are two subsets of the Reuters 21578 dataset. The datasets processed are same as [26], and we summarize the interesting characteristics of them in Table 1.

The 20NG consists of 18846 documents from 20 different newsgroups. In this dataset, training set includes 11314 documents and test set includes 7532 documents. The Ohsumed is a bibliographic dataset of medical literature. We just focus on the single-labeled documents from 23 disease categories. There are 3357 documents in the training set and 4043 documents in the test set. R52 and R8 are selected from the Reuters 21578 dataset. They have 52 and 8 categories respectively. R52 is divided documents to training set and documents to test set. R8 has the training set of 5485 documents and the test set of 2189 documents. For MR, it's a movie review dataset that only contains two classification. The MR is split to 7108 training documents and 3554 test documents. All the datasets were processed by cleaning the text, where stop words defined in NLTK were removed. Additionally, the words appear less than five times for 20NG, Ohsumed, R52 and R8 are also taken away except MR. Because of the short document, we keep the words appearing less than 5 times. As shown in Table 1, we summarize the division of each data set.

In experiments, the method is applied to the five datasets to complete the task of text classification. Additionally, we explore the effectiveness of our dual-attention GCN by comparing the results with ourselves, and experiment on the hop K to determine what value is appropriate.

Table 1. Details of datasets

Dataset	Train	Words	Test	Nodes	Classes
20ng	11314	42757	7532	61603	20
mr	7108	18764	3554	29426	2
Ohsumed	3357	14157	4043	21557	23
R52	6532	8892	2568	17992	52
R8	5485	7688	2189	15362	8

5.2 Results and Comparisons

Table 2. Performance (%) on five datasets: 20NG, MR, Ohsumed, R52 and R8. "–" donates the original paper didn't report the results.

METHODS	20NG	MR	Ohsumed	R52	R8
TF-IDF+LR	83.19	74.59	54.66	86.95	93.74
CNN-rand	76.93	74.98	43.87	85.37	94.02
CNN-non-static	82.15	**77.75**	58.44	87.59	95.71
LSTM	65.71	75.06	41.13	85.54	93.68
LSTM (pre-trained)	75.43	77.33	51.10	90.48	96.09
PV-DBOW	74.36	61.09	46.65	78.29	85.87
PV-DM	51.14	59.47	29.50	44.92	52.07
PTE	76.74	70.23	53.58	90.71	96.69
fastText	79.38	75.14	57.70	92.81	96.13
fastText (bigrams)	79.67	76.24	55.69	90.99	94.74
SWEM	85.16	76.65	63.12	92.94	95.32
LEAM	81.91	76.95	58.58	91.84	93.31
Graph-CNN-C	81.42	77.22	63.86	92.75	96.99
Graph-CNN-S	–	76.99	62.82	92.74	96.80
Graph-CNN-F	–	76.74	63.04	93.20	96.89
TextGCN	86.34	76.74	68.36	93.56	97.07
OURS (dual-attention)	**87.00**	77.14	**69.19**	**93.58**	**97.36**

We compare our proposed method dual-attention GCN with multiple state-of-the-art text classification and embedding methods by following, including TF-IDF+LR [26], CNN [12], LSTM [16], Bi-LSTM, PV-DBOW [14], PV-DM [14], PTE [22], fastText [11], SVEM [21], LEAM [19], Graph-CNN-C [6], Graph-CNN-S [4], Graph-CNN-F [9] and TextGCN. TF-IDF+LR is the bag-of-words model set term frequency-inverse document frequency as weights with Logistic Regression classifier. CNN is the Convolutional Neural Network and in experiment

and explored with CNN-rand and CNN-non-static. CNN-rand uses the word embeddings initialized randomly and CNN-non-static uses the word embeddings pre-trained. The word embeddings of LSTM is processed the same as CNN. Bi-LSTM is a bi-directional LSTM using pre-trained word embeddings. PV-DBOW and PV-DM are paragraph vector models and followed the Logisitic Regression classifier. The obvious difference is that the former considers the orders of the words but the latter does not. PTE is predictive text embedding, which using the graph included word, documents and labels and later regarding the average of word embeddings as document embeddings. The fastText also use the average of word or n-grams embeddings to generalize document embeddings, and in experiment we try the bigrams and non-bigrams. SWEM is a word embedding model and LEAM is a label-embedding attentive model. For Graph-CNN-C, Graph-CNN-S and Graph-CNN-F, they are all graph CNN models that operate on graphs with word embeddings. The difference from the three is that they use different filters, respectively, Chebyshev filter, Spline filter and Fourier filter. TextGCN aims to construct a heterogeneous text graph containing words and documents. For our dual-attention GCN, we use embedding features and run 10 times. We show the mean of 10 results, and especially compare the results of our model without dual-attention mechanism.

The details of comparison results are reported in Table 2. We show observations as follows:

- The TF-IDF+LR shows good performance and especially on the 20NG. It even performs better than some deep learning models. The simple method that increases words' importance with the number of times they appear in the file seems to be more suitable for the long texts. But not reflecting the position information of the word also limits the continued growth of accuracy.
- For the CNN and LSTM, it's obvious that two models were enhanced by using pre-trained word embedding features. CNN with randomly initialized embeddings and LSTM using the last hidden state as the representation of text perform not as well as using pre-trained word embeddings. One thing they have in common is that they perform better than TF-IDF+LR on short texts but worse than long texts.
- Conversely, the performance of PV-DBOW seems to be better on the long texts like 20NG and PV-DM seems terrible. It's likely to be the reason that PV-DBOW sampled words randomly from the output paragraph and ignoring the word orders. But PV-DM shows effect on MR with taking word orders into account and exactly the word orders are more necessary to focus on.
- The performance of PTE and fastText are more satisfied, which might because PTE is a semi-supervised representation learning method for text data and fastText is supervised. However the CNN with pre-trained embeddings still outperforms and might because CNN model can handle labeled information more effectively by utilizing word orders in the local context and solve the ambiguity of the word sense.
- There is a significant improvement in SWEM and LEAM, the simple word-embedding based model and the joint embedding of words and labels model,

which indicate the pooling operations and considering nonlinear interaction between phrase and labels do play a role.

- The graph CNN model with three kinds of filters show more competitive results on the five datasets. The results demonstrate that these supervised models are really suitable for the graph or node-focused applications. Except the long text 20NG, the overall perform is very well on the other four datasets.
- In contrast, our proposed method is superior to multiple state-of-the-art on the datasets 20NG, Ohsumed, R52 and R8. At the same time, it's also show the competitive performance on the dataset MR. Especially, the results are improved significantly on 20NG and Ohsumed. With the embedding features including the relations of word-word and document-word, our dual-attention mechanism can not only dynamically assign weights to related nodes and learn the edges, but also emphasize the context distribution. So it outperforms shown methods on both long texts and short texts. It's because that it equals to selecting the nodes that are more worthy of attention and update self by using their features.

Comparisons on the Hop K

For the hop-attention K, determining the hop K is equivalent to determining the size of the receptive field, that is, the length of the context. So the most appropriate size of K should be a problem that needs to be concerned. We experiment on the dataset Ohsumed and R52 to explore the impact of K on classification. To make the difference more obvious, we set one-hop neighbor nodes to 10.

As shown in Table 3, we can find that on Ohsumed the model performs best when K is 3, which means context distribution is ignorable. The performance can be improved about 3% between the hop of one and two. But for the hop three, the difference isn't obvious as before. For example, in the sentence "I am in my study, surrounded by books.", we need the directly adjacent word "my" to confirm "study" is a noun but we can't know it means learning or the room. "Surrounded by books" helps to understand the real meaning of "study" and these words are enough to help understand the meaning of word "study". For the shorter text R52, we can see that the model shows best performance when the hop K is 2. According to this result, we can note that two hops are more suitable for R52 and it might because the text is shorter than Ohsumed. When capturing information from other nodes, excess information will disturb

Table 3. Comparisions on the hop K on datasets Ohsumed and R52

K	Ohsumed (%)	R52 (%)
1	52.44	89.71
2	55.16	**90.34**
3	**55.47**	88.94

the classification. The results above demonstrate the effectiveness of our hop-attention, which concerns nodes differently at different distances.

5.3 Ablation Study

To better show the effectiveness of our models, we continue to do a comparison experiment, that is, keep the other settings unchanged and only remove the dual-attention mechanism on the five data sets to compare. We first conduct the experiment with a convolution layer by commenting out the dual-attention section. Then we recovery our dual-attention mechanism to compare. As shown in Table 4, the performance is crucial for the model without dual-attention mechanism. Although the dataset R8 has reached a high accuracy rate, our method still pulls it up a bit. In total, the performance can be improved by about 1% and 5% for the task of text classification.

In contrast, there is a significant difference between adding and not adding dual-attention mechanism to long texts or short texts, which demonstrates that our dual-attention GCN can capture both short-term dependences and long-term dependences well. It indicates that our dual-attention GCN model the structural information of various texts well. The connection-attention adaptively assign weights to related words and the hop-attention learn more about the distribution of context, so whether the long texts or short texts are both classified well.

Table 4. Self comparison on 20NG, MR, Ohsumed, R52 and R8

Dataset	Dual-attention	Convolution
20NG	**87.00**	84.46
MR	**77.14**	74.25
Ohsumed	**69.19**	64.95
R52	**93.58**	89.17
R8	**97.36**	96.71

6 Conclusion

In this paper, we propose a dual-attention graph convolutional network and apply it to text classification. We aim to encode various connection patterns between related nodes and learn more about the context distribution. To this end, we adopt the connection-attention and the hop-attention, one dynamically assigns weights, and one considers the importance of context. Our dual-attention graph convolutional network model the structural information of various texts and adaptively learn the representation of text. In experiment, we verified the effectiveness of our model on five widely used dataset, and further compare with the model removing the dual-attention. The comparison also shows that our model is very powerful in contrast with the non-dual-attention model and is very suitable for both long and short texts. In total, they are effective and can achieve the performance of state-of-the-art.

Acknowledgements. This work was supported by the National Natural Science Foundation of China (Grants Nos. 61772276, 61972204, 61906094), the Natural Science Foundation of Jiangsu Province (Grant No. BK20190452), the fundamental research funds for the central universities (No. 30919011232).

References

1. Atwood, J., Towsley, D.: Diffusion-convolutional neural networks. In: Advances in Neural Information Processing Systems, pp. 1993–2001 (2016)
2. Bahdanau, D., Cho, K., Bengio, Y.: Neural machine translation by jointly learning to align and translate. In: ICLR (2015)
3. Bastings, J., Titov, I., Aziz, W., Marcheggiani, D., Sima'an, K.: Graph convolutional encoders for syntax-aware neural machine translation. arXiv preprint arXiv:1704.04675 (2017)
4. Bruna, J., Zaremba, W., Szlam, A., Lecun, Y.: Spectral networks and locally connected networks on graphs. Comput. Sci. (2014)
5. Clevert, D.A., Unterthiner, T., Hochreiter, S.: Fast and accurate deep network learning by exponential linear units (ELUs). Comput. Sci. (2015)
6. Defferrard, M., Bresson, X., Vandergheynst, P.: Convolutional neural networks on graphs with fast localized spectral filtering. In: Advances in Neural Information Processing Systems, pp. 3844–3852 (2016)
7. Duvenaud, D.K., et al.: Convolutional networks on graphs for learning molecular fingerprints. In: Advances in Neural Information Processing Systems, pp. 2224–2232 (2015)
8. Hamilton, W., Ying, Z., Leskovec, J.: Inductive representation learning on large graphs. In: Advances in Neural Information Processing Systems, pp. 1024–1034 (2017)
9. Henaff, M., Bruna, J., Lecun, Y.: Deep convolutional networks on graph-structured data. Comput. Sci. (2015)
10. Hochreiter, S., Schmidhuber, J.: Long short-term memory. Neural Comput. **9**(8), 1735–1780 (1997)
11. Joulin, A., Grave, E., Bojanowski, P., Mikolov, T.: Bag of tricks for efficient text classification. arXiv preprint arXiv:1607.01759 (2016)
12. Kim, Y.: Convolutional neural networks for sentence classification. arXiv preprint arXiv:1408.5882 (2014)
13. Kipf, T.N., Welling, M.: Semi-supervised classification with graph convolutional networks. arXiv preprint arXiv:1609.02907 (2016)
14. Le, Q., Mikolov, T.: Distributed representations of sentences and documents. In: International Conference on Machine Learning, pp. 1188–1196 (2014)
15. LeCun, Y., Bottou, L., Bengio, Y., Haffner, P., et al.: Gradient-based learning applied to document recognition. Proc. IEEE **86**(11), 2278–2324 (1998)
16. Liu, P., Qiu, X., Huang, X.: Recurrent neural network for text classification with multi-task learning. arXiv preprint arXiv:1605.05101 (2016)
17. Mikolov, T., Sutskever, I., Chen, K., Corrado, G., Dean, J.: Distributed representations of words and phrases and their compositionality. In: Advances in Neural Information Processing Systems, vol. 26, pp. 3111–3119 (2013)

18. Mnih, V., Heess, N., Graves, A., et al.: Recurrent models of visual attention. In: Advances in Neural Information Processing Systems, pp. 2204–2212 (2014)
19. Peng, H., et al.: Large-scale hierarchical text classification with recursively regularized deep graph-CNN. In: Proceedings of the 2018 World Wide Web Conference on World Wide Web, pp. 1063–1072. International World Wide Web Conferences Steering Committee (2018)
20. Qian, N.: On the momentum term in gradient descent learning algorithms. Neural Netw. **12**(1), 145–151 (1999)
21. Shen, D., et al.: Baseline needs more love: on simple word-embedding-based models and associated pooling mechanisms. arXiv preprint arXiv:1805.09843 (2018)
22. Tang, J., Qu, M., Mei, Q.: PTE: predictive text embedding through large-scale heterogeneous text networks. In: Proceedings of the 21th ACM SIGKDD International Conference on Knowledge Discovery and Data Mining, pp. 1165–1174. ACM (2015)
23. Vaswani, A., et al.: Attention is all you need. In: Advances in Neural Information Processing Systems, pp. 5998–6008 (2017)
24. Veličković, P., Cucurull, G., Casanova, A., Romero, A., Lio, P., Bengio, Y.: Graph attention networks. arXiv preprint arXiv:1710.10903 (2017)
25. Wang, S., Manning, C.D.: Baselines and bigrams: simple, good sentiment and topic classification. In: Proceedings of the 50th Annual Meeting of the Association for Computational Linguistics: Short Papers-volume 2, pp. 90–94. Association for Computational Linguistics (2012)
26. Yao, L., Mao, C., Luo, Y.: Graph convolutional networks for text classification. arXiv preprint arXiv:1809.05679 (2018)
27. Zhang, Y., Liu, Q., Song, L.: Sentence-state LSTM for text representation. arXiv preprint arXiv:1805.02474 (2018)

Chart-Type Classification Using Convolutional Neural Network for Scholarly Figures

Takeo Ishihara[1], Kento Morita[1], Nobu C. Shirai[1], Tetsushi Wakabayashi[1(✉)], and Wataru Ohyama[2]

[1] Mie University, 1577 Kurima-machiya, Tsu 514, Japan
{ishihara,morita,waka}@hi.info.mie-u.ac.jp, shirai@cc.mie-u.ac.jp
[2] Saitama Institute of Technology, 1690 Fusaiji, Fukaya, Japan
ohyama@sit.ac.jp

Abstract. Text-to-speech conversion by smart speakers is expected to assist visually handicapped people who are near total blindness to read documents. This research supposes a situation where such a text-to-speech conversion is applied to scholarly documents. Usually, a page in scholarly documents consists of multiple regions, i.e. ordinary text, mathematical expressions, tables, and figures. In this paper, we propose a method which classifys chart-type of scholarly figures using a convolutional neural network. The method classifies an input figure image into line charts or others. We evaluated the accuracy of the method using scholarly figures dataset collected from actual academic papers. The classification accuracy of the proposed method achieved 97%. We also compared the performance of the proposed method with that of hand-crafted features and support vector machine. The results suggest that the proposed CNN classification outperforms the conventional approach.

Keywords: Chart · Image recognition · Image classification · Document recognition

1 Introduction

Commercial use of interactive voice operation devices like smart speakers for text-to-speech conversion becomes common. These text-to-speech conversion technologies are applicable to reading assistance for visually handicapped people in near-total blindness. The text-to-speech conversion technologies require input documents in digital text format to read it up automatically. To apply the text-to-speech conversion techniques to general and physical documents, automatic document analysis and recognition are needed.

This research supposes to a situation where we apply such text-to-speech conversion to scholarly documents, for instance, academic theses or textbooks. Applying these technologies to scholarly documents, handicapped people can obtain support to access academic or scientific information.

© Springer Nature Switzerland AG 2020
S. Palaiahnakote et al. (Eds.): ACPR 2019, LNCS 12047, pp. 252–261, 2020.
https://doi.org/10.1007/978-3-030-41299-9_20

Usually, a page in scholarly documents consists of multiple document components, i.e. ordinary text, mathematical expressions, tables, and figures. A standard document recognition process pipeline employs layout analysis [1,2] to extract each document component region. The extracted document components are handed over to specific processes that depend on the type of components. For instance, a Optical Character Recognition (OCR) software can recognize characters in ordinary text and mathematical expression areas for converting them to digital format. Diagrams can be divided into areas of natural images, charts, and tables. They can be assessed separately. Areas of natural images can be described by the image captioning techniques [3,4]. Table areas are recognized by OCR for table form recognition [5]. However, automatic recognition of chart areas is more challenging. To the best of our knowledge, there are no practical methods for describing chart areas, and only a few research reports are available [11–13].

In order to realize the text-to-speech conversion for scholarly documents including charts, the explanation of the charts should be given in sentences including the words related to the elements of the charts such as axis, axis label, and shape. Thus, we need to recognize the elements of chart to generate the explanation. While there are several challenges to realize the generation of explanation for scholarly charts, one of the principal problems involves that scholarly charts in academic documents contain multiple types of which appearance and containing information are drastically different. Figure 1 shows some examples of common chart in academic documents. As shown in the figure, there are several types of chart are used in these documents. This suggests that it is quite challenging to build a single universal method that can properly handle these multiple chart types.

In this paper, we focused on chart types in academic papers. We propose an approach to classify the chart types. We collected charts images from actual academic documents to create a dataset for training the proposed chart type classifier. We are using the term "chart" to refer to a visual representation of the relationship between two variables of data as shown in Fig. 1. The proposed method employs a dataset containing line charts, scatter plots, and bar charts. In the field of image classification, deep neural network architectures have been employed due to their promising performance and high adaptivity. Krizhevsky et al. [10] proposed a method called convolutional neural network (CNN) for a general image classification problem. In this research, we also employ a method of CNN for the chart type classification.

The main contributions of this research include; (1) we propose a method to classify chart images, (2) we create a dataset that consists of chart images collected from actual scholarly documents and chart type annotations for each images.

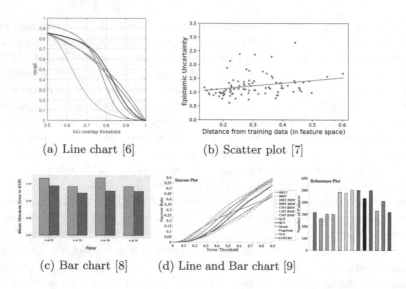

(a) Line chart [6]   (b) Scatter plot [7]

(c) Bar chart [8]   (d) Line and Bar chart [9]

Fig. 1. Examples of scholarly chart images. Each subfigure is actual example in the dataset we collected for performance evaluation. The detail of the dataset is described in Sect. 4.1.

2   Related Work

Image captioning, which is the task that generates text descriptions for an input image, has attracted attentions of many researchers. Some methods proposed for image captioning can be divided into bottom-up and top-down approaches.

The bottom-up approaches recognize objects in images, and process using language models. Kulkarni et al. [11] and Elliott et al. [12] proposed generation of caption using template sentences. Kiros et al. [13] present more powerful language models. However, their approaches are impossible to form novel sentences.

Top-down approaches introduced end-to-end models [3,4] using deep learning. As such it is possible to generate novel sentences for input images. These models extract features from images using CNN and generate image caption using recurrent neural network (RNN) with long-short term memory (LSTM).

For the chart images, the bottom-up approaches are conventionally used. The methods recognize chart images and create text sentences using templates [14]. Generally, chart images in scholarly documents are categorized into some classes, i.e. bar charts, flow charts, line charts, scatter charts, and pie charts [15]. Manolis Savva et al. [16] proposed a classification method which categorizes a chart in area graphs, bar graphs, curve plots, maps, pareto charts, pie charts, radar plots, tables, and Venn diagrams. A different type of text representation is needed to describe the given chart according to its chart type. For example, the explanation of line charts should include the transition of data plots and bar charts should include values of maximum and minimum for each bin. Thus, we need to classify chart types to generate the description of the chart suitable for each chart type.

3   Proposed Method

Figure 2 shows the outline of the proposed method. The proposed method inputs one still color image and performs preprocessing, classifying the image into line charts or the others by using CNN. The following sections describe the classification method in detail.

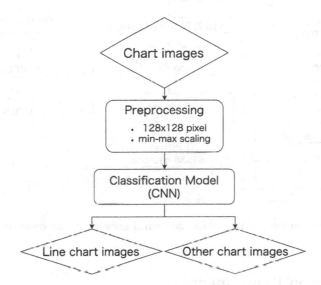

Fig. 2. Overview of the proposed method.

3.1   Chart Type Classification

In this paper, we focus on extracting line charts from various chart types. In some cases, the line charts are plotted multiple lines in one window. In other cases, the line charts are plotted together with bar charts or scatter plots. Taking into account those complicated cases, it is a difficult problem to classify line charts and the other charts. Thus, we need to focus on line charts. The proposed method classifies chart images into two classes; line charts and other charts. Figure 3 shows the structure of CNN we used. The first convolutional layer has 32 filter kernels of size 3 × 3, followed by the batch normalization operation. The second convolutional layer has 32 filter kernels of size 3 × 3, followed by the max-pooling and drop out operations. The third convolutional layer has 64 filter kernels of size 3 × 3 and the fourth convolutional layer has 64 filter kernels of size 3 × 3 followed by the batch normalization and max-pooling, drop out operations. Converts the output 32 × 32 × 64 size element to a 1 dimensional vector. Next, the fully connected layers have 512 and drop out operation. The output layers have 2 neurons and classify 2 class employing sigmoid function. We use ReLU function as the activation function.

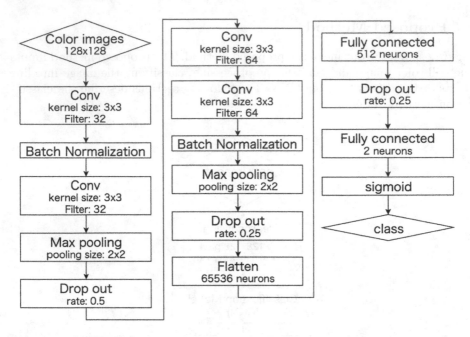

Fig. 3. The structure of the convolutional neural network in the proposed implementation.

4   Evaluation Experiments

We conducted experiments to evaluate the effectiveness of the proposed method. We also performed comparison experiments that employed the grayscale gradient feature [17] and support vector machine (SVM).

4.1   Dataset

We used the preprint repository called arXiv [18] as an academic paper database. We selected "Computer Vision and Pattern Recognition" in the fields in arXiv and employed chart images in academic papers of the field. The chart images were submitted the period from January 1, 2017, to March 31, 2017. And we annotated "line chart" or "other chart". Figure 1 shows examples of the dataset. Figure 1(a) is an example of "line chart". (b), (c), and (d) correspond "other chart". The "other chart" class includes chart images containing multiple sub-charts such as Fig. 1(d). The dataset consists of 2,978 line chart images and 1,740 other chart images.

We release the dataset employed in this paper as a CSV file. In the CSV file, we organized the article identifier of arXiv and file name of chart images, annotated line charts as 1 and other charts as 0.

4.2   Outline of Experiment

Training data contains 3,538 images and test data contains 1,180 images. The sizes of chart images are normalized to 128 × 128 pixels and the pixel value is normalized to 0 to 255 using min-max scaling. We employed four-fold cross-validation for performance evaluation. The dataset was randomly divided into four groups. Three were used for training and the remaining one was used as a test set. We conducted this evaluation four times and calculated the mean performance.

We evaluated the overall performance of the proposed method using accuracy A and recall R, precision P, and F-measure F. These are defined by the following:

$$A = \frac{TP + TN}{TP + FP + FN + TN} \tag{1}$$

$$P = \frac{TP}{TP + FP} \tag{2}$$

$$R = \frac{TP}{TP + FN} \tag{3}$$

$$F = \frac{2P * R}{P + R} \tag{4}$$

For the above calculation, TP, FP, TF, and FN denote the number of images that classified as line charts correctly, the number of images that misclassified as line charts, the number of images that are correctly classified as other charts. The number of images that misclassified as other charts, respectively.

4.3   Comparative Experiment

We verified comparative experiments to evaluate the effectiveness of the proposed method. In the field of image feature extraction, the grayscale gradient feature is often used due to its promising performance and high adaptivity. Also, we used Support Vector Machine (SVM) to perform the two-class classification using data applied min-max scaling.

Grayscale Gradient Feature. The preprocessing method converted color images to grayscale and normalized to 128 × 128 pixels. The gradient feature vector is extracted from the preprocessed chart image. The gradient feature vector is composed of the directional histogram of the gradient of the image.

Chart image is segmented into blocks and the 72-dimensional feature vector is composed of the local directional histograms. The gradient feature extraction is performed as in the following steps:

1. Apply the Gaussian filter that defined by the following equation to the local area of size $N \times N$ centered on the pixel (x, y) of the preprocessed image $I = I(x, y)$.

$$I'(x, y) = \frac{\displaystyle\sum_{i=-N/2}^{N/2} \sum_{j=-N/2}^{N/2} f(i, j) I(x+i, y+j)}{\displaystyle\sum_{i=-N/2}^{N/2} \sum_{j=-N/2}^{N/2} f(i, j)} \tag{5}$$

$$f(i, j) = \frac{1}{2\pi\sigma^2} \exp\left(-\frac{i^2 + j^2}{2\sigma^2}\right) \tag{6}$$

Where, N is the kernel size and σ is the weighting factor, 5 and 3, respectively.

2. The grayscale image obtained in Step 1 is normalized so that the mean grayscale becomes zero with variance value 1.
3. The normalized image is initially segmented into 5(width) 5(height) blocks.
4. The strength $G(x, y)$ and direction $\theta(x, y)$ is defined as follows:

$$G(x, y) = \sqrt{G_x(x, y)^2 + G_y(x, y)^2} \tag{7}$$

$$\theta(x, y) = \tan^{-1} \frac{G_y(x, y)}{G_x(x, y)} \tag{8}$$

the horizontal component $G_x(x, y)$ and the vertical component $G_y(x, y)$ of the grayscale gradient features at the pixel (x, y) is calculated by Sobel filter. $\theta(x, y)$ is initially quantized into 16 directions and the strength of the gradient is accumulated for each quantized direction.

5. Histograms for the values of the 16 quantized directions are computed in each block.
6. A directional histogram consisting of 16 directions is downsampled into 8 directions using the weighting filter [1 4 6 4 1].

Finally, the $3 \times 3 \times 8 = 72$ dimensional feature vector is obtained.

5   Results and Discussion

Table 1 shows the quantitative evaluation and comparison of the classification performance.

In the results, we can observe that the proposed method using CNN outperforms the comparative method using grayscale gradient features and SVM for all evaluation criteria. From these results, we can confirm the effectiveness of the proposed method. Tables 2 and 3 show confusion matrix performed four-fold cross-validation of the evaluation experiments. The effectiveness of the proposed method can be confirmed by the confusion matrix.

Figure 4 shows some examples of classification by the proposed method. Figure 4(a) and (b) correspond to examples of successful classification for the line and the other classes, respectively. Almost all samples of such typical scholarly charts are classified successfully in the experiment. In contrast, (c) and (d) are examples of false classification.

Table 1. Comparison of the classification performance.

	CNN	GFV and SVM
Accuracy	0.97	0.81
Precision	0.99	0.90
Recall	0.97	0.82
F-measure	0.98	0.86

Table 2. Confusion matrix of the classification results by the proposed CNN.

		True label	
		Line chart	Other chart
Predicted label	Line chart	2949	31
	Other chart	88	1236

Table 3. Confusion matrix of the classification results by SVM.

		True label	
		Line chart	Other chart
Predicted label	Line chart	2666	312
	Other chart	575	1165

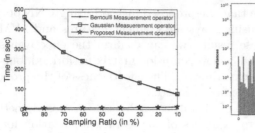
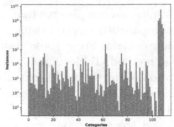

(a) Successful classification for line chart. [19]

(b) Successful classification for other chart. [20]

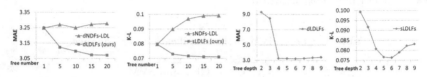

(c) False classification for line chart. [21]

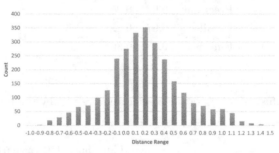

(d) False classification for other chart. [22]

Fig. 4. Examples of classification result.

6   Conclusion and Future Works

In this paper, we proposed an approach classifying chart images into line charts and other charts. Our approach achieves the classification accuracy of 97%. Also, we annotated chart images shown in academic papers and created dataset from them. Further study topic includes chart component recognition and generation of the description of charts.

References

1. Nagy, G., Seth, S., Viswanathan, M.: A prototype document image analysis system for technical journals. Computer **25**(7), 10–22 (1992)
2. Namboodiri, A.M., Jain, A.K.: Document structure and layout analysis. In: Chaudhuri, B.B. (ed.) Digital Document Processing. Advances in Pattern Recognition, pp. 29–48. Springer, London (2007). https://doi.org/10.1007/978-1-84628-726-8_2

3. Vinyals, O., et al.: Show and tell: a neural image caption generator. In: CVPR, pp. 3156–3164, June 2015

4. You, Q., Jin, H., Wang, Z., Fang, C., Luo, J.: Image captioning with semantic attention. In: The IEEE Conference on Computer Vision and Pattern Recognition (CVPR), pp. 4651–4659 (2016)

5. Kieninger, T.G.: Table structure recognition based on robust block segmentation. In: Document Recognition V, vol. 3305. International Society for Optics and Photonics (1998)

6. Jie, Z., et al.: Tree-structured reinforcement learning for sequential object localization. arXiv preprint arXiv: 1703.02710v1 (2017)

7. Kendall A., Gal, Y.: What uncertaintie do we need in bayesian deep learning for computer vision? arXiv preprint arXiv: 1703.04977v2 (2017)

8. Meier, R., et al.: Perturb-and-MPM: quantifying segmentation uncertainty in dense multi-label CRFs. arXiv preprint arXiv:1703.00312 (2017)

9. Siam, M., Singh, A., Perez, C., Jagersand, M.: 4-DoF tracking for robot fine manipulation tasks. arXiv preprint arXiv:1703.01698v2 (2017)

10. Krizhevsky, A., Sutskever, I., Hinton, G.E.: ImageNet classification with deep convolutional neural networks. In: Advances in Neural Information Processing Systems (2012)

11. Kulkarni, G., et al.: BabyTalk: understanding and generating simple image descriptions. IEEE Trans. Pattern Anal. Mach. Intell. **35**(12), 2891–2903 (2013)

12. Elliott, D., Keller, F.: Image description using visual dependency representations. In: EMNLP, pp. 1292–1302 (2013)

13. Kiros, R., Salakhutdinov, R., Zemel, R.S.: Unifying visual-semantic embeddings with multimodal neural language models. arXiv preprint arXiv 1411.2539 (2014)

14. Huang, W., Tan, C.L.: A system for understanding imaged infographics and its applications. In: Proceedings of the 2007 ACM symposium on Document engineering. ACM (2007)

15. Tang, B., et al.: DeepChart: combining deep convolutional networks and deep belief networks in chart classification. Signal Process. **124**, 156–161 (2016)

16. Savva, M., et al.: Revision: automated classification, analysis and redesign of chart images. In: Proceedings of the 24th Annual ACM Symposium on User Interface Software and Technology. ACM (2011)

17. Shi, M., Fujisawa, Y., Wakabayashi, T., Kimura, F.: Handwritten numeral recognition using gradient and curvature of gray scale image. Pattern Recogn. **35**(10), 2051–2059 (2002)

18. arXiv https://arxiv.org/. Accessed 16 Jan 2019

19. Elouard, C., et al.: Extracting work from quantum measurement in Maxwell demon engines. arXiv preprint arXiv:1702.01970v1 (2017)

20. Kemker, R., Salvaggio, C., Kanan, C.: Algorithms for semantic segmentation of multispectral remote sensing imagery using deep learning. arXiv preprint arXiv: 1703.06452v3 (2017)

21. Yang, S.: Propensity score weighting for causal inference with clustered data. arXiv preprint arXiv: 1703.06086v4 (2017)

22. Chen, L., Tang, W., John, N.W., Wan, T.R., Zhang, J.J.: Augmented reality for depth cues in monocular minimally invasive surgery. arXiv preprint arXiv: 1703.01243v1 (2017)

Handwritten Digit String Recognition
for Indian Scripts

Hongjian Zhan[1]($\boxtimes$), Pinaki Nath Chowdhury[2], Umapada Pal[2], and Yue Lu[1]

[1] Shanghai Key Laboratory of Multidimensional Information Processing, Department of Computer Science and Technology, East China Normal University, Shanghai 200241, China
ecnuhjzhan@gmail.com
[2] CVPR Unit, Indian Statistical Institute, Kolkata, India

Abstract. In many documents digits/numerals may touch each other and hence digit string recognition is necessary as segmentation of individual numeral from the touching string is difficult. In this paper, we propose a digit string recognition system for four Indian popular scripts. Here we consider strings of Kannada, Oriya, Tamil and Telugu scripts for our experiment. This paper has two contributions: (i) we have developed 4 datasets of digit string for each of these four scripts. Each dataset has 20000 numeral string samples for training and 30000 samples for testing. As there is no such dataset available, it will be helpful to the community (ii) we apply a RNN free CNN (Convolutional Neural Network) and CTC (Connectionist Temporal Classifica-tion) based architecture for numeral string recognition. Unlike normal text string, in string of digits has no contextual information among the digits and hence a digit may be followed by an arbitrary digit in a digit string. Because of such behaviors we apply a CNN and CTC based architecture without RNN for numeral string recognition. We tested our scheme on our different test datasets and results are provided.

Keywords: String recognition · Convolutional Neural Network · Connectionist Temporal Classification · Postal Automation

1 Introduction

Because of various applications, recognition of handwritten numeral string has been a popular area for many years to the researchers. Some of its potential application areas are Postal Automation, Bank cheque processing, etc. Although India is a multi-lingual and multi-script country (in India there are about 25 official languages and 11 different scripts are used to write these languages) but not much work is done towards the string recognition of Indian handwritten numerals [2,18,19].

There are two approaches for handwriting numeral string recognition. One is segmentation based and other is segmentation-free [1]. In many Indian documents, digits may touch in different manners like top touching, middle-touching

© Springer Nature Switzerland AG 2020
S. Palaiahnakote et al. (Eds.): ACPR 2019, LNCS 12047, pp. 262–273, 2020.
https://doi.org/10.1007/978-3-030-41299-9_21

and bottom touching. Such touching may be categorized as single point touching, multiple point touching, ligature point touching etc. Moreover, two, three, four and five digit touching strings are also available in Indian documents and hence it is very difficult for accurate segmentation of individual digits from such touching string.

Although there are many pieces of work on Indian isolated digit recognition, there is not much work on Indian digit string recognition. Related pieces of Indian isolated digit recognition work can be seen in [2–9]. As this work is on string recognition, we briefly described here the existing work on digit string recognition.

To recognize Indian pin code written in Pal et al. [18] proposed a segmentation free segmentation approach. Here, at first, binarization of the input document is done. Next, water reservoir concept is applied to pre-segment a pin code string into possible primitive components (individual digits or its parts). Pre-segmented components [10] of the pin code are then merged into possible digits to get the best pin code using dynamic programming (DP) and modified quadrat-ic discriminant function (MQDF) [11] classifier. In 2009, Pal et al. [19] also pro-posed a technique for Bangla, Hindi and English digit-string recognition system.

Although there are several pieces of work on isolated numeral recognition for Kannada, Oriya, Tamil and Telugu scripts [2], to the best of our knowledge there is no work on numeral string recognition for any of these four scripts. Hence in this paper we have considered digit string recognition of these four scripts for our experiments. Also no datasets are available for digit string for these four scripts and hence we have proposed several datasets for this work.

Unlike normal text string, in string of digits there is no contextual information among the digits as a digit may be followed by an arbitrary digit in a string of digits. Because of this, in this paper, we propose a new architecture which is based on CNN (Convolutional Neural Network) and CTC (Connectionist Temporal Classification) [13], without using RNN for numeral string recognition. Also to connect CNN with CTC, we transform the outputs of CNN to a two-dimension vector to meet the feeding requirement of CTC. Furthermore, we utilize dense blocks to build CNN part to extract efficient image features.

Rest of the paper is organized as follows. In Sect. 2 we discuss about the properties of Indian scripts considered here. Section 3 deals with dataset details. Proposed methodology is presented in Sect. 4. The experimental results are discussed in Sect. 5. Finally, conclusion is given in Sect. 6.

2 Properties Indian Scripts

Most of the Indian scripts are originated from Brahmi script through various transformations. Writing style of the Indian scripts considered in this paper is from left to right, and concept of upper/lower case is absent in these scripts.

Oriya is a popular language and script of India. This language is used mainly in the Odisha (formerly Orissa) state of India and also in West Bengal, Jharkhand, and Gujarat. Oriya is the official language of Odisha state.

Kannada is another popular script and it is the official language of the southern Indian state, Karnataka. Kannada is a Dravidian language mainly used by the people of Karnataka, Andhra Pradesh, Tamil Nadu and Maharashtra.

Telugu is the 3rd most popular scripts in India. It is the official language of the southern Indian state, Andhra Pradesh. Telugu is also spoken in Bahrain, Fiji, Malaysia, Mauritius, Singapore and the UAE. The Telugu script is closely related to the Kannada script.

Tamil is also a popular south Indian Language and one of the oldest languages in the world. It is the official language of the southern Indian state, Tamil Nadu. Apart from India, it is also one of the official languages in the countries like Singapore, Malaysia and Sri Lanka.

To get an idea about digit shapes of the four scripts considered in this paper, a set of handwritten samples of these scripts is shown in Fig. 1.

Script Numerals	Telugu	Oriya	Kannada	Tamil
1	౧	೮	∩	⑮
2	౨	୨	೨	௰
3	3	୩	೩	௫
4	౪	୪	౪	௪
5	౫	୫	౫	௫
6	౬	୬	౬	௬
7	౭	୭	౭	௭
8	౮	୮	౮	௮
9	౯	୯	౯	௯
0	౦	୦	౦	౦

Fig. 1. Handwritten numeral samples of four scripts.

The challenging part of Indian script handwritten recognition is the distinction between the similar shaped components. Sometimes a very small part is the distinguishing mark between two numerals. These small distinguishing parts increase the recognition complexity and decrease recognition accuracy. Because of the writing styles of different individuals, same numerals may take different shapes and conversely two or more different numerals of a script may take similar shape. These factors also increase the complexity of recognition method. To get the idea of similar shape numerals, we provide some examples in Fig. 2. Here in the first row both the numerals of the first pair of Telugu script look like zero. But the first numeral of this pair is Telugu 'one' and 2nd numeral is Telugu 'zero'. Similarly, in the second pair, it seems both the digits are similar. But first digit is '6' and second digit is '9'.

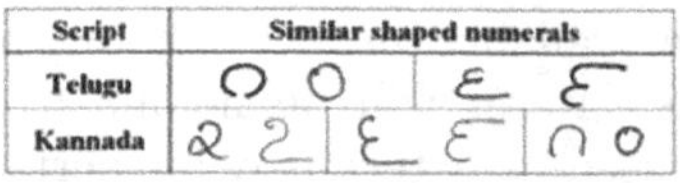

Script	Similar shaped numerals		
Telugu	○ ○	Ɛ Ɛ́	
Kannada	𝒬 2	Ɛ Ɛ̄	∩ ○

Fig. 2. Examples of some similar shaped numerals.

3 Data Collection

Deep Learning models requires large amount of data for training. Moreover, in many of the Indian scripts no large numeral database is available for deep learning purpose. As there are 11 different scripts are available for India, it is difficult to develop large datasets for each of these 11 scripts. Although some large handwritten numeral string datasets are available for Bangla and Hindi, no large handwritten numeral/digit string dataset is available for Kannada, Oriya, Tamil and Telugu. Hence, here we have developed 4 datasets of handwritten digit string for each of these four scripts. Each dataset has 20000 handwritten digit string samples for training and 30000 samples for testing. The dataset contains numeral string samples of length 6 digits to 10 digits uniformly distributed throughout the training and testing set.

To make the dataset of various complexities, we generate 4 datasets for each script and they are named as Dataset-1, Dataset-2, Dataset-3, and Dataset-4. The first dataset, i.e. "Dataset-1" has only non-touching or non-overlapping numeral whereas Dataset-2 consists of numeral string which may or may not be touching/overlapping. Both real digit string and synthetic digit string are there in these two datasets. By synthetic digit string we mean the digit strings that are generated through computer program from real isolated handwritten digits.

Other two datasets (i.e. "Dataset-3" and "Dataset-4") are completely synthetic and all the digits in a string are touching/overlapping. These datasets are generated through computer program from real isolated handwritten digits and the touching/overlapping area is different in these two datasets. "Dataset-3" has overlapping digit string where there is a maximum overlap of 1 stroke width and "Dataset-4" consists of digit strings having a maximum overlap of 2 stroke width. Here stroke width of the digit which is going to be included in the string during digit string formation is considered for touching. Examples of each of the datasets of each script are shown in Fig. 3(a–d).

During synthetic digit string generation, each digit strings were randomly generated ensuring that the individual digits in a particular sample have similar size and stroke width.

As mentioned earlier, we have four datasets in each scripts and hence there are total 16 datasets. As mentioned above, each dataset has 20000 handwritten digit string samples for training and 30000 samples for testing. That means we have a total of 50,000 samples, including training and testing set, for each dataset of a script. Thus, for 16 datasets we have a total of 800,000-digit string samples.

Now these datasets are available freely to the researchers. As there so such big datasets, we hope this dataset will be helpful to the researchers and the generated dataset can be used as a test bed for performance evaluation of on digit string recognition.

4   Methodology

We propose a new network for handwritten digit string recognition, which is shown in Fig. 4. In order to enhance the performance of DenseNet [12], we add residual connections between dense blocks, For the output layer, we apply a CTC [13] to calculate loss at the training phase and give the predictions at the testing phase.

4.1   Dense Block

A dense block is a stack of densely connected convolutional layers [12]. It can extract more efficient features than plain and residual convolutional networks. It consists of a group of layers, the batch normalization layer [15], ReLU layer [16] and a convolutional layer. The kernel size of the convolutional layer is 3*3, which can maintain the size of feature map in the whole dense block. After the convolutional layer, we apply a dropout layer [17] with rate 0.2.

4.2   Transition Block

The feature maps pass through a dense block will keep the size the same. But it is important to reduce the feature map size in a convolutional network, so we apply a transition block between two dense blocks to decrease the size. A typical transition block consists of a batch normalization layer, a 1*1 convolutional layer and an average pooling layer with kernel 2*2. The dropout layer has dropout rate to 0.2.

4.3   Residual Connections

There are two main differences between DenseNet [12] and ResNet [14]. In DenseNet, the connections between convolutional layers are densely and the way to combine feature maps is concatenation. We enhance these advantages in this paper. In transition block there is an average pooling layer to reduce the feature map size, in order to retain more information, we add max pooling residual connections, as shown in Fig. 4. And we follow the way in DenseNet, we concatenate the two branches features to feed into next layer.

Kannada	Oriya	Tamil	Telegu
100079	100122	100777	106996
10651	204081	111539	107121
107447	224313	112958	110326

(a) Dataset-1

Kannada	Oriya	Tamil	Telegu
100133	102342	104727	104777
108107	103622	111446	106175
110080	109375	113169	110744

(b) Dataset-2

Kannada	Oriya	Tamil	Telegu
101505	104033	104771	101184
105166	105146	105560	101208
108717	108273	105916	102086

(c) Dataset-3

Kannada	Oriya	Tamil	Telegu
103098	103133	104717	100188
103106	103719	112398	101226
103501	213233	116823	211209

(d) Dataset-4

Fig. 3. Samples of different datasets.

4.4 Dimension Adjustment

The output of convolutional layer always has three dimensions. However, the CTC we apply in our network requires the input data with two dimensions. The output of dense block is always a 3-D tensor, i.e., the *number*, *height* and *width* of feature maps (4-D if we consider the *batch − size* dimension). First we flatten the 3-D feature tensor to 2-D by expanding on the *number* dimension, then we apply a column-wise fully connected layer to reduce the *height* dimension to the assigned value. With these actions, we can generate the suitable input for the following CTC output layer.

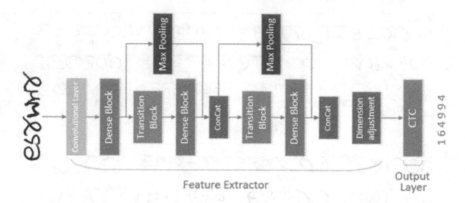

Fig. 4. The structure of the proposed network.

4.5 CTC Output Layer

Connectionist temporal classification [13] is a kind of output layer with two main functions, calculating the loss at training phase and generating the prediction results at testing phase.

For a string recognition task, the labels are drawn from a set A (in this paper, A is the ten digits). With an extra label named blank, we get a new set $A' = A \bigcup$ blank, which is used in reality. The input of CTC is a sequence $y = y_1, ..., y_T$, where T is the sequence length. The corresponding label denotes as I over A. Each y_i is a probability distribution on the set A'. We define a many-to-one function $F : A_T => A_{<=T}$ to resume the repeated labels and blanks. For example $F(11 - 6 - 49 - 999 - 44 - - - - - -) = 164994$. ($-$ indicates the 'blank' label).

Let $S = (X, I)$ is the training set, where X is the training image and I is the ground truth. So the CTC loss is calculated as:

$$\mathscr{O}(S) = - \sum_{(x,I) \in S} \log p(I|y) \tag{1}$$

where $p(I/y)$ is the conditional probability defined as the sum of probabilities of all predictions that are mapped by F onto I, I is the prediction result. Therefore, the network can be end-to-end trained on pairs of images and sequences, without the procedure of manually labeling individual components in training images.

5 Result and Discussions

Before going to present the results, we provide here parameters and system information. We apply ADADELTA to update the parameters. In all experiments, we train the network with 50 epochs. The hyper-parameters of three dense blocks are the same with the growth rate to 8 and the number of convolution layer to 8. Our experiments are performed on a Super-micro server with the GPU NVIDIA TITAN X. The software is the Caffe [20] framework with cuDNN V5 accelerated.

5.1 Global Recognition Results

Overall accuracies on 16 datasets (four datasets for each of the four scripts) obtained from the experiments for Kannada, Oriya, Tamil and Telugu scripts are shown in Table 1. Here both digit level as well as string level accuracies are presented here. From the table it can be seen that maximum accuracy for digit level is obtained from Telugu (98.90%) and it is for dataset 1. Minimum accuracy for digit level is obtained from Kannada (96.42%) and it is for dataset 4. Similarly, that maximum accuracy for string level is obtained from Tamil (91.78%) and it is for dataset 1. Minimum accuracy for string level is obtained from Kannada (75.94%) and it is for dataset 4. It can be seen that dataset 4 has relatively lower accuracy than other datasets and this is because dataset 4 is the most complex dataset having many types of touching.

Table 1. Digit level and string level accuracies of four scripts on different datasets.

Datasets & mode		Scripts			
		Kannada	Oriya	Tamil	Telugu
Dataset-1	Digit level Accuracy	97.43	98.00	98.87	98.90
	String level Accuracy	82.70	85.08	91.78	91.39
Dataset-2	Digit level Accuracy	97.57	97.20	98.69	98.39
	String level Accuracy	83.52	80.60	90.52	88.17
Dataset-3	Digit level Accuracy	97.26	96.97	98.33	98.55
	String level Accuracy	81.51	78.50	88.30	89.01
Dataset-4	Digit level Accuracy	96.42	96.64	98.34	98.21
	String level Accuracy	75.94	76.84	88.59	86.52

5.2 Confusing Numeral Pair Computation

We also noted the main confusing numeral pair of different scripts considered here and we observed that main reason of such confusion is shape similarity. Four confusing matrices for the four scripts on dataset-1 are presented in the Tables 2, 3, 4 and 5 for Kannada, Oriya, Tamil and Telugu scripts, respectively.

In Kannada main confusing numeral pair is numeral six and numeral seven. They confuse about 2.37% cases. Next pair of confusion in Kannada is numeral three and numeral seven with confusion rate 2.35%. For Oriya, maximum confusion is between is numerals two and seven they confuse about 3.04% cases. It can be seen that from Table 4, for Tamil, maximum confusion is between numerals four and six and they confuse about 1.19% cases. Similarly, it can be seen that from Table 5 for Telugu that numeral nine and six have maximum confusion it is 1.74%.

Table 2. Confusion matrix for Kannada.

Digit	Recognized as									
	0	1	2	3	4	5	6	7	8	9
0	98.31	1.01	0.22	0.15	0.01	0.17	0.0	0.0	0.12	0.01
1	0.01	99.89	0.02	0.05	0.0	0.02	0.0	0.0	0.0	0.0
2	0.01	0.0	99.46	0.19	0.01	0.14	0.0	0.01	0.03	0.14
3	0.0	0.05	0.01	99.35	0.01	0.44	0.0	0.03	0.0	0.1
4	0.0	0.02	0.0	0.2	98.42	1.09	0.07	0.03	0.07	0.09
5	0.0	0.04	0.49	1.19	0.21	97.96	0.0	0.01	0.06	0.05
6	0.01	0.0	0.04	0.26	0.43	0.07	94.59	2.37	2.02	0.19
7	0.03	0.03	0.41	2.35	0.07	0.06	0.27	95.97	0.76	0.06
8	0.23	0.04	0.11	0.12	0.01	0.15	0.02	0.0	98.91	0.42
9	0.0	0.1	0.0	0.02	0.02	0.0	0.42	0.01	0.17	99.26

Table 3. Confusion matrix for Oriya.

Digit	Recognized as									
	0	1	2	3	4	5	6	7	8	9
0	99.15	0.02	0.02	0.08	0.02	0.0	0.29	0.03	0.05	0.35
1	0.06	99.2	0.0	0.03	0.0	0.28	0.0	0.11	0.31	0.0
2	0.07	0.04	96.15	0.2	0.06	0.0	0.37	3.04	0.0	0.06
3	0.01	0.0	0.0	98.73	0.0	0.0	0.27	0.93	0.03	0.01
4	0.17	0.0	0.0	0.12	99.45	0.09	0.0	0.1	0.04	0.02
5	0.0	0.3	0.0	0.57	0.4	98.51	0.0	0.02	0.0	0.2
6	0.29	0.29	0.35	0.2	0.02	0.0	97.15	1.66	0.01	0.03
7	0.0	0.0	1.22	0.5	0.02	0.0	0.09	98.07	0.0	0.08
8	0.0	0.01	0.0	0.13	0.0	0.0	0.0	0.0	99.85	0.0
9	0.14	0.04	0.01	0.55	0.05	0.14	0.08	0.02	0.16	98.81

5.3 Erroneous Results

To get the idea about the digit-string samples where our system provides erroneous results we provide some examples in Fig. 5. Here four samples are given, In the first samples the actual string 1000018082 is recognized as 100018082. The first two samples show the mistake about losing one digit. In the second samples the actual string 10014801 is recognized as 1001480. In the third and fourth samples, there are some wrong predictions and this is mainly because of their shape similarity.

Table 4. Confusion matrix for Tamil.

Digit	Recognized as									
	0	1	2	3	4	5	6	7	8	9
0	99.55	0.03	0.0	0.0	0.01	0.0	0.0	0.37	0.0	0.03
1	0.0	99.73	0.01	0.0	0.01	0.0	0.14	0.0	0.0	0.1
2	0.04	0.02	98.66	0.31	0.0	0.08	0.02	0.01	0.86	0.0
3	0.0	0.06	0.03	99.77	0.0	0.11	0.0	0.04	0.0	0.0
4	0.0	0.53	0.0	0.0	98.02	0.0	1.19	0.03	0.23	0.0
5	0.0	0.01	0.0	0.0	0.0	99.94	0.03	0.0	0.0	0.0
6	0.0	0.0	0.0	0.0	0.02	0.01	99.74	0.05	0.0	0.18
7	0.0	0.0	0.0	0.01	0.0	0.01	0.18	99.73	0.05	0.0
8	0.01	0.0	0.0	0.0	0.01	0.0	0.11	0.0	99.88	0.0
9	0.1	0.02	0.01	0.21	0.02	0.34	0.7	0.02	0.0	98.59

Table 5. Confusion matrix for Telugu.

Digit	Recognized as									
	0	1	2	3	4	5	6	7	8	9
0	99.19	0.77	0.01	0.0	0.02	0.01	0.0	0.01	0.0	0.0
1	0.08	99.84	0.0	0.02	0.0	0.0	0.0	0.05	0.0	0.0
2	0.07	0.0	99.68	0.12	0.11	0.0	0.01	0.0	0.0	0.0
3	0.02	0.03	0.01	99.83	0.07	0.0	0.0	0.03	0.0	0.0
4	0.0	0.0	0.0	0.0	99.77	0.22	0.0	0.0	0.0	0.0
5	0.01	0.03	0.01	0.02	0.54	99.39	0.01	0.01	0.0	0.0
6	0.06	0.0	0.0	0.01	0.03	0.01	98.65	0.34	0.0	0.89
7	0.0	0.0	0.01	0.01	0.02	0.0	0.0	99.94	0.0	0.0
8	0.09	0.0	0.0	0.02	0.2	0.0	0.0	0.0	99.67	0.0
9	0.01	0.01	0.0	0.01	0.08	0.0	1.74	0.01	0.22	97.91

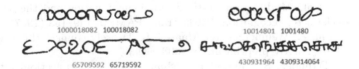

Fig. 5. Examples of some similar shaped numerals.

6   Conclusion

In this paper we apply a CNN-CTC architecture for four handwritten Indian script numeral string recognition. We also develop datasets of digit string for these four scripts. Each dataset has 20000 numeral string samples for training and 30000 samples for testing. As there is no such dataset available, it will be helpful to the community. Moreover, to the best of our knowledge there is no work on digit-string recognition of these four scripts, and hence this is the first work on work on these scripts.

References

1. Plamondon, R., Srihari, S.N.: On-line and off-line handwritten recognition: a comprehensive survey. IEEE Trans. PAMI **22**, 62–84 (2000)
2. Pal, U., Chaudhuri, B.: Indian script character recognition: a survey. Pattern Recogn. **37**, 1887–1899 (2004)
3. Bhowmick, T., et al.: An HMM based recognition scheme for handwritten Oriya numerals. In: Proceedings of the 9th International conference on Information Technology, pp. 105–110 (2006)
4. Sharma, N., Pal, U., Kimura, F.: Recognition of handwritten Kannada numerals. In: Proceedings of the 9th International Conference on Information Technology, pp. 133–136 (2006)
5. Hanmandlu, M., Ramana Murthy, O.: Fuzzy model based recognition of handwritten numerals. Pattern Recogn. **40**, 1840–1854 (2007)
6. Wen, Y., Lu, Y., Shi, P.: Handwritten Bangla numeral recognition system and its appli-cation to postal automation. Pattern Recogn. **40**, 99–107 (2007)
7. Bajaj, R., Dey, L., Chaudhury, S.: Devnagari numeral recognition by combining deci-sion of multiple connectionist classifiers. Sadhana **27**, 59–72 (2002)
8. Kumar, S., Singh, C.: A study of Zernike moments and its use in Devnagari handwrit-ten character recognition. In: Proceedings of the International conference on Cognition and Recognition, pp. 514–520 (2005)
9. Bhattacharya, U., et al.: Neural combination of ANN and HMM for handwritten Devnagari numeral recognition. In: Proceedings of the 10th International Workshop on Frontiers of Handwriting Recognition, pp. 613–618 (2006)
10. Otsu, N.: A Threshold selection method from grey level histogram. IEEE Trans. SMC **9**, 62–66 (1979)
11. Kimura, F., et al.: Modified quadratic discriminant function and the application to Chinese character recognition. IEEE Trans. PAMI **9**, 149–153 (1987)
12. Huang, G., Liu, Z., Weinberger, K., Maaten, L.: Densely connected convolutional networks (2016). arXiv preprint arXiv:1608.06993
13. Graves, A., Fernndez, S., Gomez, F., Schmidhuber, J.: Connectionist temporal classification: labelling unsegmented sequence data with recurrent neural networks. In: Proceedings of the 23rd International Conference on Machine learning, pp. 369–376 (2006)
14. He, K., Zhang, X., Ren, S., Sun, J.: Deep residual learning for image recognition. In: Proceedings of the 29th IEEE Conference on Computer Vision and Pattern Recognition, pp. 770–778 (2016)

15. Ioffe, S., Szegedy, C.: Batch normalization: accelerating deep network training by reducing internal covariate shift. In: Proceedings of International Conference on Machine Learning, pp. 448–456 (2015)
16. Glorot, X., Bordes, A., Bengio, Y.: Deep sparse rectifier neural networks, In: Proceedings of the 14th International Conference on Artificial Intelligence and Statistics, pp. 315–323 (2011)
17. Hinton, G., et al.: Improving neural networks by preventing co-adaptation of feature detectors (2012). arXiv preprint arXiv:1207.0580
18. Pal, U., Roy, K., Kimura, F.: Bangla handwritten pin code string recognition for indian postal automation. In: Proceedings of International Conference on Frontiers in Handwriting Recognition, pp. 290–295 (2008)
19. Pal, U., Roy, K., Kimura, F., Indian multi-script full pincode string recognition for postal automation, In: Proceedings of the 10th International Conference on Document Analysis and Recognition (ICDAR), pp. 456–460 (2009)
20. Jia, Y., et al.: Caffe: convolutional architecture for fast fea-ture embedding (2014). arXiv preprint arXiv:1408.5093

Spatial-Temporal Graph Attention Network for Video-Based Gait Recognition

Xinhui Wu[1]([⊠]), Weizhi An[1]([⊠]), Shiqi Yu[2,3]([⊠]), Weiyu Guo[4]([⊠]),
and Edel B. García[5]([⊠])

[1] College of Computer Science and Software Engineering, Shenzhen University,
Shenzhen, China
{wu,anweizhi2016}@email.szu.edu.cn,wu_xinhui09@163.com
[2] Department of Computer Science and Engineering,
Southern University of Science and Technology, Shenzhen, China
shiqi.yu@gmail.com
[3] Shenzhen Institute of Artificial Intelligence and Robotics for Society,
Shenzhen, China
[4] Watrix Technology Limited Co., Ltd., Beijing, China
guoweiyu168@hotmail.com
[5] Advanced Technologies Application Center (CENATAV), Havana, Cuba
egarcia@cenatav.co.cu

Abstract. Gait is a kind of attractive feature for human identification at a distance. It can be regarded as a kind of temporal signal. At the same time the human body shape can be regarded as the signal in the spatial domain. In the proposed method, we try to extract discriminative feature from video sequences in the spatial and temporal domains by only one network, Spatial-Temporal Graph Attention Network (STGAN). In spatial domain, we designed one branch to select some distinguished regions and enhance their contribution. It can make the network focus on these distinguished regions. We also constructed another branch, a Spatial-Temporal Graph (STG), to discover the relationship between frames and the variation of a region in temporal domain. The proposed method can extract gait feature in the two domains, and the two branches in the model can be trained end to end. The experimental results on two popular datasets, CASIA-B and OU-ISIR Treadmill-B, show the proposed method can improve gait recognition obviously.

Keywords: Graph Attention Network · Spatial-temporal graph · Temporal domain

1 Introduction

Gait feature is a kind of behavioral biometric features. Compared with other biometric features such as face, finger print, iris, etc., it has the unique advan-

The work is partly supported by the Science Foundation of Shenzhen (Grant No. 20170504160426188).

tage of being captured at a distance. For this reason, gait recognition has great potential in video surveillance. Many pioneer researchers have worked on it for two decades and proposed many gait recognition methods. Among these methods, Gait Energy Image (GEI) should be the most popular feature because of its effectiveness and robustness. But the temporal information is lost during the computation of GEI. It is reasonable to use both spatial information and temporal information in gait recognition since gait is a kind of behavioral features and contains some unique moving patterns of the subjects in the temporal domain.

In spatial domain, different positions of the silhouette images should contribute to recognition differently. The region contains identity information generally may be some parts of the silhouette image. To address this issue, some methods [24] extract these regions in advance or divided an image into several parts manually. We think the distinguished region selection can be implemented by a neuronal network automatically. We design an attention network to select the most distinguished regions and take them into the forward process again to increase the contribution of these regions.

There are some pioneer works for temporal feature extraction in the literature. One of them is C3D [6]. C3D is a 3D CNN method, so there are many parameters in the network. That makes the training process difficult. Another category of them is LSTM [8]. It is challenging to train a LSTM model with a high recognition rate. To improve the performance, some methods [30] combine multiple frames to multi-channel data as the input of a CNN model. ST-GCN [32] introduce a graph-based structure for temporal feature. The temporal information can be represented by the relationship between different nodes in the graph network. Inspired by those works, the proposed method was designed to extract gait feature effectively in spatial and temporal domains.

Our main contributions are listed as: (1) A Spatial-Temporal Graph Attention Network (STGAN) is proposed for gait feature extraction in spatial domain and temporal domain. (2) An attention based method is involved to select distinguished regions. (3) A Spatial-Temporal Graph (STG) is constructed to describe the relation between different frames and the variation of a region in temporal domain.

The following part of the paper is organized as follows. In the next section we will introduce some related work on gait feature extraction. The details of the proposed method are in Sect. 3, and the experimental results are presented in Sect. 4. Section 5 concludes the paper and gives some possible future directions.

2   Related Work

Most methods in gait recognition can be categorized into model-based methods and appearance-based ones. Model-based methods [2,25] normally build one 2D or 3D model for human body and joints. It is a challenging task especially when the resolution and image quality are not high. Compared with model-based methods, appearance-based [1,16,19,21,34] ones are normally easy to implement since they extract features directly from 2D human silhouettes. Such as in [21], Makihara et al. extract frequency-domain features first, construct view transformation model. In [16], Kusakunniran *et al.* consider correlations between gaits across views. Among the appearance-based methods, the gait energy image (GEI) [13] may be the most popular feature. GEI is the average of silhouettes in a gait cycle, and it is effective in gait recognition and robust to variations. But the temporal information is lost during the average operation of the feature.

To address the temporal information in the video sequences, one of the possible solutions is 3D convolution. C3D [6] proposed by Du *et al.* employs 3D convolution to extract temporal information. On the other hand, LSTM can intrinsically handle temporal information in the input data. In [8], a Recurrent Neural Network (RNN) is employed to extract feature from silhouette sequences. There are more related works [12,15,28,30] which use CNN to extract features in the temporal domain. Multiple frames of silhouettes construct a multiple channel data blob, that is sent to a CNN network to train a classifier. Wu *et al.* [30] achieved really good classification results with this method.

Yan *et al.* [32] regarded that temporal information can not be represented in a grid-like structure like pictures and CNN cannot utilize the temporal information effectively. The Graph Neural Network (GNN) [4,7,11,22] is another solution for temporal feature extraction. In [7], Duvenaud *et al.* apply a convolutional neural network on graphs directly. In [22], Niepert *et al.* propose a learning convolutional neural networks for arbitrary graphs. Another category of GNN based methods uses RNN to each node in the graph [9,17]. In [9], the authors try to update recurrently the hidden state of each node. In [17], Li *et al.* propose to capture joint dependencies between roles by GNN. ST-GCN [32] performed well on their task. Therefore, we introduced GNN and graph structure here to handle this kind of information.

Attention mechanism is popular in natural language processing. In [26], a self-attention method is proposed which can connect the information of different positions in a sequence and obtain a representation of this sequence. In [29], Wang *et al.* added attention mechanism to calculate the similarity between feature maps. Inspired by the idea of attention mechanisum, Velickovic *et al.* [27] add attention mechanism to Graph Convolutional Networks (GCN) to construct Graph Attention Networks (GAN). GAN can update the hidden representations for each node by using the information of its neighborhoods in the graph convolution process. It combines GCN and attention together for feature extraction. In the proposed method We follow the idea but in a different manner for gait recognition.

3   Proposed Method

3.1   The Network Architecture

A Spatial-Temporal Graph Attention Network (STGAN) is designed and shown in Fig. 1. The network can extract spatial and temporal features respectively. Our backbone network is the same with that in [5], and shown in Fig. 2. In the spatial branch, as shown in the upper part of Fig. 1, the most distinguished 3 regions can be selected. These regions correspond to the top 3 strongest activated pixels in the feature map. The feature map is the output of last convolution layers. The lower part of Fig. 1 is Spatial-Temporal Graph (STG). STG is also shown detailly in Fig. 4. It is designed for extracting the temporal information in sequences. Inspired by the idea of ST-GCN [32], we construct STG with the feature vector of regions as graph nodes and naturally connect in spatial and temporal as graph edges.

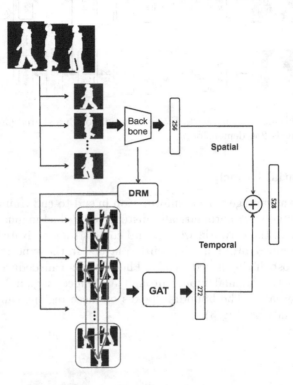

Fig. 1. The framework of the proposed method. The upper part is the spatial branch, the lower part is the temporal branch. Backbone network is shown in Fig. 2, DRM stands for Deduce Region Module, GAT is for Graph Attention Module. The input is T frames of silhouettes.

We use the Graph Attention module (GAT) [27] to process graph structure data. For each node in the graph, we only focus on its neighborhood nodes, and constantly update the node with the weight of its neighborhoods in the graph convolution process. The spatial branch and the temporal branch are not completely independent. Parameters are shared in the backbone network between these two branches. Input data is a set of consecutive silhouettes, which has a size of $T \times C \times H \times W$, where T is the number of frames in this set, C, H, W are the number of channels, height, width of the image respectively. C is set to 1 here because silhouettes are binary images and contain only one channel.

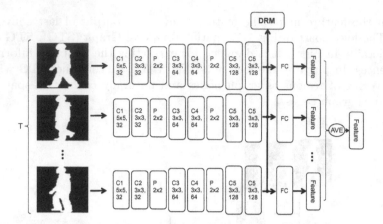

Fig. 2. The backbone CNN network. Every frame through a shared CNN respectively, the feature vector is 256 demension.

3.2 The Spatial Branch

In order to take advantage of local information in end-to-end training, we extract some regions by network automatically instead of labelled manually. In the first forward process of the network, we can find some strong activated pixels on the feature map obtained after four convolution layers. In the experiments we select top 3 as the most distinguished pixels. These pixels can deduce the receptive field regions on the original input through a Deduce Region Module (DRM) applied layer by layer. The bounding boxes of these regions can be calculated by following equations (Fig. 3).

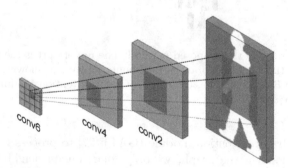

Fig. 3. Deduce Region Module. The blue region is feature map, the orange region represents receptive field. (Color figure online)

$$POS = [x_1^{l_n}, y_1^{l_n}, x_2^{l_n}, y_2^{l_n}]$$
$$[x_1^l, y_1^l] = -P^l + [x_1^{l+1}, y_1^{l+1}]S^l$$

$$[x_2^l, y_2^l] = (K^l - 1) - P^l + [x_2^{l+1}, y_2^{l+1}]S^l$$

where POS is the position of the strong activated pixels on the feature map after l_n layers, $[x_1^l, y_1^l]$ is the upper left corner point of the bounding box at l-th layer, $[x_2^l, y_2^l]$ is the lower right corner point of the bounding box at l-th layer. K^l, P^l, S^l represents the kernel size, padding and strike of the l-th convolution layer. The bounding box is $[x_1^0, x_2^0, y_1^0, y_2^0]$. The size of output data is $T \times N \times 5$, where T is the number of frames, N is the number of selected regions, and 5 corresponds to the four coordinates of bounding box plus the picture index. The bounding boxes may be out of the boundary because of the padding operation. We crop and resize these bounding boxes to same size.

The extracted strongest activation regions can be used in two ways. Firstly, It can be used to emphasis the contribution of these regions when feeding them into the network. Secondly, the feature vector obtained from spatial backbone network can be used as node features in a spatial-temporal graph.

3.3   The Temporal Branch

To extract the temporal information from sequences, a Spatial-Temporal Graph (STG) is designed. The idea of using STG is inspired by ST-GCN [32] which achieves good results in action recognition by automatically learning both the spatial and temporal patterns from data. In our STG, the graph is defined as $G = (V, E)$. The set of nodes $V = \{v_{ti} | t = 1, ..., T, i = 1, ...N\}$ where T is the number of frames, N is the number of selected regions, and v_{ti} represents the i-th selected region in the t-th frame. As shown in Fig. 4. $rank_i$ is the area corresponding to the i-th strongest activated point in the feature map. The feature vector $F(v_{ti})$ actually is the feature vector of v_{ti} region after linear transformation layer.

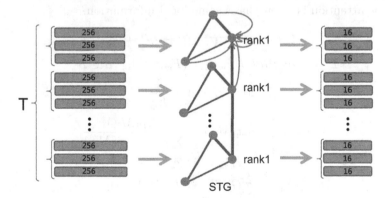

Fig. 4. Graph Attention Module. It shows the process of updating nodes in STG. Update a node by the correlation of its neighborhood nodes.

The set of edges E consists of two subsets. The first subset is spatial edges, which represents the spatial relationship, as shown in blue line of Fig. 4. At the

t_1 frame, the nodes $v_{t_1 i}$ are connected to each other. $E_s = \{v_{ti}v_{tj}\}$. The second subset is temporal edges, which represents the temporal relationship, as shown in green line of Fig. 4. At the $i_1(rank_1)$ region, the nodes v_{ti_1} are connected linearly. $E_t = \{v_{ti_1}v_{(t+1)i_1}\}$. We calculate the adjacency matrix A to build the relationship between nodes.

The Graph Attention module GAT [27] is originally from the Graph Convolution Network (GCN) [14] with the attention mechanism. When calculating graph structure data, GCN updates every node in the graph convolution operation by following equations.

$$\hat{A} = A + I$$
$$H^l = \sigma(D^{-\frac{1}{2}}\hat{A}D^{-\frac{1}{2}}H^{l-1}W^l)$$

where A is the adjacency matrix of undirected graph G, and I is the unit matrix and D is the degree matrix. $\hat{A}$ contains the information of adjacent nodes and itself. $D^{-\frac{1}{2}}\hat{A}D^{-\frac{1}{2}}$ normalizes the weight of adjacent matrix. W^l is the trainable weight matrix of the l-th layer, σ is activation function, and H^l is the activation matrix of the l-th layer. In the GCN structure, in the higher layer, each node contains the information of the multi-level adjacent nodes.

In the proposed method, GAT is used to execute convolution operation and update the state of each node in the graph, namely update the temporal feature, as shown in Fig. 4. GAT combines attention mechanism with GCN. When updating hidden representations for each node in the graph, attention matrix is involved for the relationships of neighborhood nodes. Different weights are assigned to these nodes. That is, attention mechanism is used to focus on those important nodes, and neglect some useless nodes in the neighborhood of this node. In [26,29], all nodes for the prior attention mechanism are weighted and integrated with global information. But in the proposed method the main purpose of the attention is to emphasis some local information.

$$F_w = F \times W, \quad W \in R^{D \times D_1} \tag{1}$$

$$E = LeakyReLU[F_w a_1 + (F_w a_2)^T], \quad a \in R^{D_1} \tag{2}$$

$$M = E \odot \hat{A} \odot \frac{1}{E_{ij}} \tag{3}$$

$$M = softmax(M_{ij}) = \frac{exp(M_{ij})}{\sum_{k \in N_n} exp(M_{ik})} \tag{4}$$

$$F_m = M F_w \tag{5}$$

$$F_{out} = \sigma[GAT_2(GAT_1(F, \hat{A}), \hat{A})] \tag{6}$$

The calculation process for node updating as following equations. First, the input data is the set of feature vectors of all nodes, $F = \{f_1, f_2, ..., f_{N_n}\}$, which is $N_n \times D$, where N_n is the number of nodes, $N_n = T \times N$, T is the number of frames, N is the number of selected regions, D is the length of feature vectors. W is trainable weight matrix. $\hat{A}$ is adjacency matrix of STG, which contains the information of adjacent nodes and itself.

4   Experimental Design and Result

4.1   Datasets

To evaluate the robustness of proposed method to variations, the datasets should contains some challenging variations at view, clothing, etc. The datasets CASIA-B [35] contains view, carrying and clothing variations. OU-ISIR Treadmill B [20] focus on clothing variations. They are both challenging datasets. Some sample images in the two datasets are shown in Fig. 5. CASIA-B is a large multi-view gait database, which is created in January 2005. The dataset was collected indoors. There are 13640 gait videos, 124 subjects, 10 sequences (NM01-06, BG01-02, CL01-02) for each person, and the gait data was captured from 11 views (000, 018, ..., 180). Three variations, namely view angle, clothing and carrying condition changes, are separately considered.

Fig. 5. Sample images in CASIA Dataset B (upper) and OU-ISIR Treadmill Dataset B (lower)

OU-ISIR Treadmill Dataset B is a gait dataset with challenging clothing variation. The data has been collected since March 2007 at the Osaka University. The dataset is basically in a form of silhouette sequences, and size-normalized to 88 by 128 pixels size. The dataset contains gait silhouette sequences of 68 subjects. There are 32 types of clothes combinations for each subject. Treadmill B dataset is a challenging dataset for clothing variation.

4.2   Experimental Design

In CASIA-B dataset, The first 74 subjects are put into the training set, and the remaining 50 subjects into the test set, as shown in Table 1. In the test set, we put the NM01-04 sequences into gallery set and the rest of the sequences are put into the probe set. For experiments using OU-ISIR Treadmill-B dataset, the first 20 subjects are put into the training set, and the remaining 48 subjects into the test set. In the test set, we take the 9 type sequences as gallery set, and the rest types as probe set. Input data is a set of consecutive silhouettes, we takes

30 frames for CASIA-B dataset, and takes 16 frames for Treadmill-B dataset. In the process of testing, we divide a video into multiple sequences, and then average the features of each sequence to obtain a feature of this video.

Table 1. The training set and test set of CASIA-B dataset

Training	Test	
	Gallery Set	Probe Set
ID: 001-074	ID: 075-124	ID: 075-124
NM01-NM06	NM01-NM04	NM05-NM06
& BG01-BG02		& BG01-BG02
& CL01-CL02		& CL01-CL02

4.3   Results and Analysis

To evaluate the effectiveness of different parts of the proposed method. The experiments can be carried out in four different conditions. They are (1) Base: backbone network, (2) SR: only three selected regions are used in the spatial branch, (3) GAT: the graph contains only relations between frames and no relations among regions, and (4) STGAN: the proposed method as illustrated in Fig. 1. The correct classification rates under these four conditions on CASIA-B daset and Treadmill-B dataset are given in Table 2. We observe that when we only extract activated regions or only construct graph between frames, the improvement is not very obvious on result. But in our proposed, we combine the feature of regions and graph constructing, there is a obvious improvement including clothing condition on CASIA-B daset and Treadmill-B dataset. It turns out the information of activated regions is very useful in the graph network. Similar conclusion can be draw from the results:

Table 2. Correct classification rates on CASIA-B and Treadmill-B dataset under four conditions, Base, SG, GAT and STGAN

Network	Accuracy(%)			
	CASIA-B			Treadmill-B
	NM	BG	CL	
Base	85.8	72.6	57	62.8
SR	85.9	73.3	54	62.15
GAT	86.3	74	55.2	65.5
STGAN (Proposed)	**87.2**	**76.8**	**58.2**	**67.0**

On Treadmill-B dataset, We compared the proposed method with some state-of-the-art methods, such as GEI [13], GEnI [3], GEINet [23], GERF [18]. The results are shown in Table 4. Compared with those methods, it can be found that the proposed method achieves best results. We also compared our proposed method with other methods like ViDP [12], C3A [31], AE [33], MGAN [10], CNN [30] on CASIA-B dataset, and the results are shown in Table 3. Except for CNN [30] in normal condition, the correct classification rate of the proposed method is better than other methods. We also noticed that [5] achieved a really high classification rate by using Horizontal Pyramid Pooling (HPM). But it is the first time we try to combine graph neural network in our task, we belive we can obtain further improvement later.

Table 3. Correct classification rates on CASIA-B dataset

Gallery NM01-04		$0°-180°$					Mean
Probe	Method	$0°$	$54°$	$90°$	$126°$	$180°$	
NM05-06	ViDP [12]	–	59.1	50.2	57.5	–	–
	C3A [31]	–	64.5	58.1	65.7	–	–
	AE [33]	49.3	63.6	58.1	66.5	44.0	59.3
	MGAN [10]	54.9	74.8	65.7	75.6	53.8	68.1
	CNN-3D [30]	**87.1**	**94.6**	**88.3**	**93.8**	**85.7**	**92.1**
	STGAN (Proposed)	76.7	90.6	82.3	91.9	73.0	87.2
BG01-02	AE [33]	29.8	40.5	37.5	42.7	28.5	37.2
	MGAN [10]	48.5	58.0	49.8	61.3	43.1	54.7
	CNN-LB [30]	64.2	76.9	63.1	76.9	61.3	72.4
	STGAN (Proposed)	**68.1**	**80.1**	**66.6**	**81.7**	**66.6**	**76.8**
CL01-02	AE [33]	18.7	25.1	26.3	30.0	19.0	24.2
	MGAN [10]	23.1	33.3	32.7	37.6	21.0	31.5
	CNN-LB [30]	37.7	61.1	54.6	59.1	39.4	54.0
	STGAN (Proposed)	**51.3**	**63.8**	**51.5**	**58.6**	**41.4**	**58.2**

Table 4. Correct classification rates on Treadmill-B dataset

Method	Accuracy (%)
GEI [13]	52.8
GEnI [3]	57.4
GEINet [23]	63.5
GERF [18]	65.9
STGAN (Proposed)	**67.0**

5   Conclusions and Future Work

The gait video sequences are signals in spatial and temporal domains. It is reasonable to design a model to extract gait features in the two domains. The spatial-temporal graph attention network that we designed is a graph model. It can model spatial and temporal relations at the same time. It can also be trained by an end to end manner. Comprehensive experiments on two popular datasets show it is feasible to extract robust gait feature by a graph model. It can also achieve better results when the temporal information is involved.

We think there is great potential to improve with the idea in the proposed method. Besides of a deeper CNN and a bigger dataset, we can build a more detailed model in spatial domain with the help of human pose estimation. The relationship between different frames can also be described in a more detailed way with the help of human joints matching.

References

1. An, W., Liao, R., Yu, S., Huang, Y., Yuen, P.C.: Improving gait recognition with 3D pose estimation. In: Zhou, J., et al. (eds.) CCBR 2018. LNCS, vol. 10996, pp. 137–147. Springer, Cham (2018). https://doi.org/10.1007/978-3-319-97909-0_15
2. Ariyanto, G., Nixon, M.S.: Model-based 3D gait biometrics. In: International Joint Conference on Biometrics, pp. 1–7 (2011)
3. Bashir, K., Xiang, T., Gong, S.: Gait recognition using gait entropy image. In: International Conference on Crime Detection and Prevention, pp. 1–6 (2010)
4. Bruna, J., Zaremba, W., Szlam, A., Lecun, Y.: Spectral networks and locally connected networks on graphs. CoRR abs/1312.6203 (2013)
5. Chao, H., He, Y., Zhang, J., Feng, J.: GaitSet: regarding gait as a set for cross-view gait recognition. In: AAAI (2019)
6. Du, T., Bourdev, L., Fergus, R., Torresani, L., Paluri, M.: Learning spatiotemporal features with 3D convolutional networks. In: IEEE International Conference on Computer Vision, pp. 4489–4497 (2015)
7. Duvenaud, D., et al.: Convolutional networks on graphs for learning molecular fingerprints. In: International Conference on Neural Information Processing Systems, pp. 2224–2232 (2015)
8. Feng, Y., Li, Y., Luo, J.: Learning effective gait features using LSTM. In: International Conference on Pattern Recognition, pp. 325–330 (2017)
9. Scarselli, F., Gori, M., Tsoi, A.C., Hagenbuchner, M., Monfardini, G.: The graph neural network model. IEEE Trans. Neural Netw. **20**(1), 61 (2009)
10. He, Y., Zhang, J., Shan, H., Wang, L.: Multi-task gans for view-specific feature learning in gait recognition. IEEE Trans. Inform. Forensics Secur. **14**(1), 102–113 (2018)
11. Henaff, M., Bruna, J., Lecun, Y.: Deep convolutional networks on graph-structured data. arXiv:abs/1506.05163 (2015)
12. Hu, M., Wang, Y., Zhang, Z., Little, J.J., Huang, D.: View-invariant discriminative projection for multi-view gait-based human identification. IEEE Trans. Inform. Forensics Secur. **8**(12), 2034–2045 (2013)
13. Han, J., Bir, B.: Individual recognition using gait energy image. IEEE Trans. Pattern Anal. Mach. Intell. **28**(2), 316 (2006)

14. Kipf, T.N., Welling, M.: Semi-supervised classification with graph convolutional networks. In: International Conference on Learning Representations (2016)
15. Kusakunniran, W., Wu, Q., Zhang, J., Li, H.: Support vector regression for multi-view gait recognition based on local motion feature selection. In: Computer Vision and Pattern Recognition (2010)
16. Kusakunniran, W., Qiang, W., Zhang, J., Li, H., Wang, L.: Recognizing gaits across views through correlated motion co-clustering. IEEE Trans. Image Process. **23**(2), 696–709 (2014)
17. Li, R., Tapaswi, M., Liao, R., Jia, J., Urtasun, R., Fidler, S.: Situation recognition with graph neural networks. In: IEEE International Conference on Computer Vision (ICCV), pp. 4183–4192 (2017)
18. Li, X., Makihara, Y., Xu, C., Muramatsu, D., Yagi, Y., Ren, M.: Gait energy response function for clothing-invariant gait recognition. In: Lai, S.-H., Lepetit, V., Nishino, K., Sato, Y. (eds.) ACCV 2016. LNCS, vol. 10112, pp. 257–272. Springer, Cham (2017). https://doi.org/10.1007/978-3-319-54184-6_16
19. Liao, R., Cao, C., Garcia, E.B., Yu, S., Huang, Y.: Pose-based temporal-spatial network (PTSN) for gait recognition with carrying and clothing variations. In: Zhou, J., et al. (eds.) CCBR 2017. LNCS, vol. 10568, pp. 474–483. Springer, Cham (2017). https://doi.org/10.1007/978-3-319-69923-3_51
20. Makihara, Y., et al.: The OU-ISIR gait database comprising the treadmill dataset. IPSJ Trans. Comput. Vis. Appl. **4**, 53–62 (2012)
21. Makihara, Y., Sagawa, R., Mukaigawa, Y., Echigo, T., Yagi, Y.: Gait recognition using a view transformation model in the frequency domain. In: Leonardis, A., Bischof, H., Pinz, A. (eds.) ECCV 2006. LNCS, vol. 3953, pp. 151–163. Springer, Heidelberg (2006). https://doi.org/10.1007/11744078_12
22. Niepert, M., Ahmed, M., Kutzkov, K.: Learning convolutional neural networks for graphs. In: the 33rd International Conference on Machine Learning, pp. 2014–2023 (2016)
23. Shiraga, K., Makihara, Y., Muramatsu, D., Echigo, T., Yagi, Y.: GEINet: view-invariant gait recognition using a convolutional neural network. In: International Conference on Biometrics, pp. 1–8 (2016)
24. Sun, Y., Liang, Z., Yi, Y., Qi, T., Wang, S.: Beyond part models: person retrieval with refined part pooling (and a strong convolutional baseline). In: 15th European Conference on Computer Vision (ECCV) (2018)
25. Urtasun, R., Fua, P.: 3D tracking for gait characterization and recognition. In: IEEE International Conference on Automatic Face and Gesture Recognition, pp. 17–22 (2004)
26. Vaswani, A., et al.: Attention is all you need. In: The 31st Conference on Neural Information Processing Systems (2017)
27. Velickovic, P., Cucurull, G., Casanova, A., Romero, A., Lio, P., Bengio, Y.: Graph attention networks. In: International Conference on Learning Representations (2018)
28. Wang, C., Zhang, J., Pu, J., Yuan, X., Wang, L.: Chrono-gait image: a novel temporal template for gait recognition. In: Daniilidis, K., Maragos, P., Paragios, N. (eds.) ECCV 2010. LNCS, vol. 6311, pp. 257–270. Springer, Heidelberg (2010). https://doi.org/10.1007/978-3-642-15549-9_19
29. Wang, X. Girshick, R., Gupta, A., He, K.: Non-local neural networks. In: CVPR (2018)
30. Wu, Z., Huang, Y., Wang, L., Wang, X., Tan, T.: A comprehensive study on cross-view gait based human identification with deep cnns. IEEE Trans. Pattern Anal. Mach. Intell. **39**(2), 209–226 (2017)

31. Xing, X., Wang, K., Yan, T., Lv, Z.: Complete canonical correlation analysis with application to multi-view gait recognition. Pattern Recogn. **50**(C), 107–117 (2016)
32. Yan, S., Xiong, Y., Lin, D.: Spatial temporal graph convolutional networks for skeleton-based action recognition. In: AAAI (2018)
33. Yu, S., Chen, H., Wang, Q., Shen, L., Huang, Y.: Invariant feature extraction for gait recognition using only one uniform model. Neurocomputing **239**(C), 81–93 (2017)
34. Yu, S., et al.: GaitGANv2: Invariant gait feature extraction using generative adversarial networks. Pattern Recogn. **87**, 179–189 (2019)
35. Yu, S., Tan, D., Tan, T.: A framework for evaluating the effect of view angle, clothing and carrying condition on gait recognition. In: International Conference on Pattern Recognition, pp. 441–444 (2006)

Supervised Interactive Co-segmentation Using Histogram Matching and Bipartite Graph Construction

Harsh Bhandari, Sarbani Palit[(✉)], and Bhabatosh Chanda

Indian Statistical Institute, 203, B.T. Road, Kolkata 700108, India
bhandariharsh1990@gmail.com, {sarbanip,chanda}@isical.ac.in

Abstract. The identification and retrieval of images of same or similar objects finds application in various tasks that are of prime importance in Image Processing and Computer Vision. Accurate and fast extraction of the object of interest from several images is essential for the construction of 3D models, image retrieval applications. The joint partitioning of multiple images having same or similar objects of interest into background and foreground parts is referred to as co-segmentation.

This article proposes a novel and efficient interactive co-segmentation method based on the computation of a global energy function and a local smooth energy function. Computation of global energy function from the scribbled regions of the images is based on histogram matching. This is used to estimate the probability of each region belonging either to foreground or background region. The local smooth energy function is used to estimate the probability of regions having similar colour appearance. To further improve the quality of the segmentation, a bipartite graph is constructed using the segments. The algorithm has been implemented on iCoseg and MSRC benchmark data sets. The extensive experimental results show significant improvement in performance compared to many state-of-the-art unsupervised co-segmentation and supervised interactive co-segmentation methods, both in computational time and accuracy.

Keywords: Co-segmentation · Histogram matching · Bhattacharya Distance · Bipartite graph

1 Introduction

With the steady growth of computing power and rapid decline in cost of memory, along with the development of high-end digital cameras and ever increasing access to internet, digital acquisition of visual information has become increasingly popular in recent years. It is common and easy to capture large numbers of images and share them on the internet using social networks like Facebook and Twitter or using WhatsApp. The availability of several related images of same objects, events or places makes it possible for researchers to exploit them

© Springer Nature Switzerland AG 2020
S. Palaiahnakote et al. (Eds.): ACPR 2019, LNCS 12047, pp. 287–306, 2020.
https://doi.org/10.1007/978-3-030-41299-9_23

for many tasks such as constructing a 3D model of a particular object or developing image retrieval applications [8]. In such tasks it is imperative to extract the foreground objects from all related images forming a group.

The idea of co-segmentation, first introduced in [1], refers to simultaneously segmenting two or more images, where the same (or similar) objects appear with different (or unrelated) backgrounds. Performing segmentation of similar regions (objects) in all the images using suitable attributes of the foreground is a very challenging task. The co-segmentation problem has attracted much focus in the last decade, most of the co-segmentation approaches [2,15,17,22] are motivated by Markov Random Field (MRF) based energy functions, generally solved by optimization techniques such as linear programming [17]. Results were improved by the introduction of interactive approaches [1,2,9]. However, they suffered from problems such as over sensitivity to user defined scribbles and inadequate performance with respect to either computational time or quality of the performance.

We present a new framework to perform image co-segmentation by first computing the probability of each pixel to be in the foreground or background regions by applying the Bhattacharya Distance on segments of the image. Next, to further increase the quality of the segmentation, we construct a bipartite graph by using the probability computed using the Bhattacharya Distance. Using the size of the segments we compute the matching coefficient which further reduces the misclassification of segments. Extensive experiments have been carried out in order to examine the strength of the proposed approach. Comparison of performance with existing algorithms with respect to accuracy of classification of each pixel into foreground or background as well as computational time, has been presented.

The proposed interactive algorithm adopts a two-phase approach with the phases being

- Estimation of the probability of each pixel to be in foreground/background region based on a novel segment classification method. Computation of the histograms of the segments of the image is based on a global scribbled unary energy function and a local pairwise smooth energy function subject to prior information.
- Bipartite graph generation using the scribbled foreground/background regions as well as the regions of the original image which need to be properly classified.

The salient contributions of this work include the achievement of good performance in terms of standard metrics in reasonably low computation time. Use of the Bhattacharya Distance and a total energy optimization approach for obtaining preliminary estimates of the pixel probabilities is followed by the construction of a bipartite graph to further improve the pixel classification. The deployment of low computational resources for obtaining this quality of performance is especially noteworthy.

The workflow of the proposed interactive co-segmentation framework using global and local energy function along with bipartite graph construction is shown in Fig. 1.

Fig. 1. The three topmost images on the left are scribbled images indicating red foreground region and green background region. The lowermost image is not scribbled since the proposed algorithm can segment some of the images without scribble. The right side images are result of co-segmenting four images in the Flower class in the MSRC dataset [23]. (Color figure online)

Section 2 presents the existing state-of-the-art techniques in this area. Section 3 gives the details of the proposed algorithm. Results of the experiments performed and the parameter values used in the simulations are provided in Sect. 4 with performance evaluation in Sect. 5. The article is concluded in Sect. 6.

2 Related Work

The identification of similar objects in more than one image is an important problem finding use in varied applications and has largely relied on the construction of models [7,24]. Most co-segmentation methods are derived from single-image segmentation methods by adding similar foreground constraints in the MRF based optimization framework. Early co-segmentation approaches [11,16–18] only used a pair of images as input making an assumption of sharing a common foreground object. Similar to image-segmentation, co-segmentation can be classified into two groups: unsupervised and interactive co-segmentation.

2.1 Unsupervised Co-segmentation

Many unsupervised approaches [5,9,10,13,14,19,21] have recently been developed to co-segment multiple images and have achieved more accurate results than any classic single-image segmentation algorithm. In [18] image co-segmentation method is introduced by combining MRF framework and global constraints with foreground histogram matching. Houchbaum and Singh [11] proposed a max-flow algorithm by modifying the histogram matching and using clustering for co-segmentation. Joulin [13] proposed a combination of normalized cuts and kernel methods to design a discriminative clustering co-segmentation framework. Recently, they extended their framework to multi-class co-segmentation [14]. Inspired by single-image interactive segmentation methods [3,4] several co-segmentation algorithms have been proposed in recent years. But these methods fail to perform well when the foreground and background are similar in images as it is difficult to find common objects automatically. The interactive co-segmentation methods alleviate these problems by using interactive scribble.

2.2 Interactive Co-segmentation

Interactive co-segmentation algorithm [1,2] added user scribbles in some input images to build two Gaussian Mixture Models (GMM) one for each of foreground and background classes. A graph cut algorithm is then used to co-segment these images. In contrast to co-segmentation approaches within MRF, some researchers [5] proposed an image co-segmentation method using random walker algorithm based on normalized Euclidean distance of pixel intensity. However the random walk optimization will make the co-segmentation results sensitive to the quantities and positions of the user scribbles [20].

In [8] another approach was proposed where global scribbled energy function using Gaussian Mixture Model and a local smoothness function using spline regression have been designed to improve the image co-segmentation performance. This study formulates the interactive co-segmentation problem in terms of Gibbs energy optimization followed by generating bipartite graphs of regions of images complementing the existing MRF segmentation framework. This improves the accuracy of co-segmentation of complex images having foreground objects with variations in colour and texture. Higher order energy optimization [12,25] has been widely used in many fields: computer vision and image processing like image de-noising and single image segmentation.

3 Proposed Algorithm

The proposed algorithm proceeds by classifying each pixel into foreground or background regions. A variety of symbols have been used throughout this article - which have been tabulated in Table 4, for the convenience of the reader.

Let $I = \{I^1, \cdots, I^n\}$ be a set of n images and let S be a subset of k images, where k $\ll$ n, selected from I with foreground and background scribbles marked.

Fig. 2. Scribbled images (Alaskan Brown Bear) from iCoseg dataset

Fig. 3. Scribbled images (Flower) from MSRC dataset

Let p_j^i denotes a pixel of image I^i with index j and $a_{j,l}^i$ be its probability for foreground/background region with l = 0 indicating the background region and l = 1 as the foreground region. To make the algorithm computationally efficient, each image I^i is divided into small segments $r_m^i \in R^i$ by using an over-segmentation method such as mean-shift [6]. Let $N(R^i)$ be the total number of segments in image I^i i.e. $1 \leq m \leq N(R^i)$ and $b_{m,l}^i$ be the probability of segment r_m^i belonging to region $l \in \{0,1\}$. In order to compute, $b_{m,l}^i$, two energy functions are designed: a unary second order global scribbled energy function E_{global} and a pairwise second order local smooth function E_{local}. Let S^0 be the set of scribbled background segments and S^1 be the set of scribbled foreground segments of set P.

3.1 Global Scribbled Function

The effective utilization of the user scribbles is the key for interactive co-segmentation. Examples of scribbled images from two well known datasets *viz.* iCoseg and MSRC are given in Figs. 2 and 3, respectively.

3.1.1   Description of the Approach

The global energy function is computed in order to obtain the probability $b^i_{m,l}$ of segment r^i_m belonging to foreground/background region, as computed below:

Let $c^i_{m,l}$ be the initial estimate of the probability of segment r^i_m belonging to region l where

$$c^i_{m,l} = 1 - e^i_{m,l} \tag{1}$$

Denoting a segment of S^0 as s^0 and a segment of S^1 as s^1, the initial estimates for the probability of segment r^i_m may be formulated as

$$e^i_{m,0} = \frac{\min_{s^0 \in S^0}(A(h^i_m, h^{s^0}))}{\min_{s^0 \in S^0}(A(h^i_m, h^{s^0})) + \min_{s^1 \in S^1}(A(h^i_m, h^{s^1}))} \tag{2}$$

$$e^i_{m,1} = \frac{\min_{s^1 \in S^1}(A(h^i_m, h^{s^1}))}{\min_{s^0 \in S^0}(A(h^i_m, h^{s^0})) + \min_{s^1 \in S^1}(A(h^i_m, h^{s^1}))}$$

Note that $A(x, y)$ stands for Bhattacharya Distance between segment x and y defined as

$$A(x, y) = -\log(D(x, y)) \quad \text{where,} \quad D(x, y) = \sum_{s=1}^{s=L} \sqrt{h^i_x(s).h^i_y(s)}$$

where h^i_q = normalized histogram of region r^i_q in image I^i and L = number of bins in a histogram.

Thus the unary global energy function is defined as

$$E_{global} = \sum_{m=1}^{N(R^i)} (b^i_{m,l} - c^i_{m,l})^2 = (\boldsymbol{b}^i_l - \boldsymbol{c}^i_l)^T * I_{N(R^i) \times N(R^i)} * (\boldsymbol{b}^i_l - \boldsymbol{c}^i_l) \tag{3}$$

where $\boldsymbol{b}^i_l$ and $\boldsymbol{c}^i_l$ are vectors of size $N(R^i)$ and I is an identity matrix of size $N(R^i) \times N(R^i)$.

Hence, E_{global} is defined by satisfying the constraint that the probability $b^i_{m,l}$ of each segment r^i_m belonging to the foreground/background region, tends to be close to $c^i_{m,l}$, which is estimated through the Bhattacharya Distance.

3.1.2   Justification

The unary global scribble function is responsible for computation of the initial probability of each segment of an image belonging to the foreground/background regions, obtained using the Bhattacharya Distance. Implementation of Mean Shift [6] with small values of spatial and range bandwidths provides reasonable over-segmentation. All the pixels in each segment of the image have similar colour to each other. Thus the expectation of the distribution of the pixel values is close to the average value of the segment.

The histogram of each segment is normalized so that it emulates the distribution of each segment. $D(p, q)$ checks the overlap of two histograms p and q. It outputs a value 1 to indicate that the histograms are identical while 0 indicates absence of any relation among them. The quantity $-(\log(D(p,q)))$ thus represent the distance between histogram p and q where $0 \leq -(\log(D(p,q))) < \infty$ with value 0 indicating identical segments. For segment r^i_m we look for the segment with red/green scribbles representing foreground/background segments. The label is segmented as foreground or background region depending on the minimum distance between r^i_m and that of the scribbled segment.

It follows that each segment r^i_m belongs to foreground region if its distance from a red scribble segment is lower than that of the green scribbled one. Thus, the segment r^i_m belongs to foreground region if its distance with that of a *red* scribbled segment is minimum compared to that of *green* scribbled segment as the probability which is $1 - e^i_{m,l}$ is higher for foreground then background. Hence we can write

$$c^i_{m,0} + c^i_{m,1} = 1 \qquad (4)$$

Thus the global scribbled function designed using the Bhattacharya Distance provides a good prediction for each segment to belong to either the foreground or background region.

3.2   Local Smooth Function

The local smooth energy function considers the smoothness of segments i.e. it provides interactive constraint that all segments in the image have same probability of belonging to foreground/background if they possess a similar appearance with respect to colour.

3.2.1   Description of the Approach

To compute the local energy function, a graph $G^i(V^{(i)}, E^{(i)})$ is constructed for image I^i where $V^{(i)} = b^i_{1,l}, \ldots, b^i_{N(R^i),l}$ and

$E^{(i)} = (b^i_{1,l}, b^i_{1,l}), (b^i_{1,l}, b^i_{2,l}), \ldots, (b^i_{N(R^i),l}, b^i_{N(R^i),l})$ *i.e.* each vertex is connected to all other vertices including itself with weight $w^i_{p,q}$ computed as,

$$w^i_{p,q} = e^{D(h^i_p(s), h^i_q(s)) - 1} \qquad (5)$$

where $0 \leq w^i_{p,q} \leq 1$, and p and q are the segments of the image. For the ith image, E_{local} defines a constraint that all segments having similar colour appearance will have same probability of getting classified into either the foreground ($l = 1$) or the background ($l = 0$). Thus the pairwise local function can be defined as

$$E_{local} = \sum_{r,s=1}^{N(R^i)} w^i_{r,s} * (b^i_{r,l} - b^i_{s,l})^2 \qquad (6)$$

Equation 6 can be solved by constructing Laplacian Le of graph $G^i(V^{(i)}, E^{(i)})$, which is a positive semi-definite matrix of size $N(R^i) \times N(R^i)$ represented as:

$$Le = \begin{bmatrix} \sum_{r=1}^{N(R^i)} w_{1,r}^i & 0 & \cdots & 0 \\ 0 & \sum_{r=1}^{N(R^i)} w_{2,r}^i & \cdots & 0 \\ \vdots & \vdots & \ddots & \vdots \\ 0 & 0 & \cdots & \sum_{r=1}^{N(R^i)} w_{N(R^i),r}^i \end{bmatrix} - \begin{bmatrix} w_{1,1}^i & w_{1,2}^i & \cdots & w_{1,N(R^i)}^i \\ w_{2,1}^i & w_{2,2}^i & \cdots & w_{2,N(R^i)}^i \\ \vdots & \vdots & \ddots & \vdots \\ w_{N(R^i),1}^i & w_{N(R^i),2}^i & \cdots & w_{N(R^i),N(R^i)}^i \end{bmatrix}$$

$$E_{local} = (\boldsymbol{b}_l^i)^T * Le_{N(R^i) \times N(R^i)} * (\boldsymbol{b}_l^i) \tag{7}$$

3.2.2   Justification

The purpose of constructing a pairwise local smoothness function is to have segments of the image with similar colours possess similar probability of belonging to foreground or background region. This maintains the consistency of the segmentation labels resulting in a more effective co-segmentation result. Implementation of the function is carried out by the construction of Laplacian of the graph constructed using the segments of the image with an edge of weight $w_{r,s}^i$ between vertices r_r^i and r_s^i where $w_{r,s}^i = e^{(D(h_{r,l}^i, h_{s,l}^i)) - 1}$ indicates similarity measure between the segments.

3.3   Total Energy Optimization

The total energy can be formulated as the sum of the global scribbled energy, E_{global} and local smooth energy, E_{local}.

$$E_{Total}^i = E_{global} + E_{local} \tag{8}$$

$$E_{Total}^i = (\boldsymbol{b}_l^i - \boldsymbol{c}_l^i)^T * I_{N(R^i) \times N(R^i)} * (\boldsymbol{b}_l^i - \boldsymbol{c}_l^i) + (\boldsymbol{b}_l^i)^T * Le_{N(R^i) \times N(R^i)} * (\boldsymbol{b}_l^i) \tag{9}$$

Here, $\boldsymbol{b}_l^i$ is the parameter and can be found by differentiating E_{Total}^i with respect to $\boldsymbol{b}_l^i$. Thus

$$I_{N(R^i) \times N(R^i)} * (\boldsymbol{b}_l^i - \boldsymbol{c}_l^i) + Le_{N(R^i) \times N(R^i)} * \boldsymbol{b}_l^i = 0 \tag{10}$$

Hence,

$$\boldsymbol{b}_l^i = \boldsymbol{c}_l^i [I_{N(R^i) \times N(R^i)} + Le_{N(R^i) \times N(R^i)}]^{-1} \tag{11}$$

Now, $I_{N(R^i) \times N(R^i)} + Le_{N(R^i) \times N(R^i)}$ can be expressed as $A_{N(R^i) \times N(R^i)}$ where

$$A_{N(R^i) \times N(R^i)} = \begin{bmatrix} \sum_{r=1}^{N(R^i)} w_{1,r}^i & w_{1,2}^i & \cdots & w_{1,N(R^i)}^i \\ w_{2,1}^i & \sum_{r=1}^{N(R^i)} w_{2,r}^i & \cdots & w_{2,N(R^i)}^i \\ \vdots & \vdots & \ddots & \vdots \\ w_{N(R^i),1}^i & w_{N(R^i),2}^i & \cdots & \sum_{r=1}^{N(R^i)} w_{N(R^i),r}^i \end{bmatrix}$$

$$\boldsymbol{b}_l^i = \boldsymbol{c}_l^i [A_{N(R^i) \times N(R^i)}]^{-1} \tag{12}$$

Note that $a_{j,l}^i = b_{m^j,l}^i$, where m^j indicates the segment $r_{m^j}^i$ that pixel p_j^i belongs to.

3.4   Bipartite Graph Construction

The proposed energy function usually provides a good estimate for the foreground/background region in each image. However, for images having complex textures and with foreground and background regions having very close colour appearances, such regions may get misclassified since the initial estimate may come close to 0.5.

To further increase the accuracy of the co-segmentation results, for each image I^i, a set X^i is constructed containing segments r_r^i having $b_{r,1}^i \in \{0.5 - \epsilon_1, 0.5 + \epsilon_2\}$. All segments having $b_{r,1}^i > 0.5 + \epsilon_2$, are classified into foreground region and those having $b_{r,1}^i < 0.5 - \epsilon_1$ is classified into background region. Set $Y_{r,l}^i$ is constructed which contains all adjacent segments of r_r^i classified into region l. To obtain a better estimate of the segments in X^i, matching coefficient $d_{r,l}^i$ is computed as stated below:

$$d_{r,l}^i = \frac{N(r_{r,l}^i) * b_{r,l}^i + u_{r,l}^i + v_{r,l}^i}{N(r_{r,l}^i) + N(r_{tv,l}^i) + N(r_{tu,l}^i)} \tag{13}$$

where $u_{r,l}^i$ indicates the highest similarity value between segment r_r^i and its adjacent segments in $Y_{r,l}^i$ and $v_{r,l}^i$ indicates the highest similarity value between segment r_r^i and S^l. $N(r_r^i)$ indicates number of pixels in segment r_r^i.

3.5   Computation of Similarity Value $v_{r,l}^i$

Let $G_{b,v}^i(X^i \bigcup S^l, E_v)$ be a complete bipartite graph with X^i and S^l being two disjoint set of vertices. E_v is the set of edges between X^i and S^l having weight $\varepsilon_{r,t,l}^i$ between segment $r_r^i \in X^i$ and segment $r_{tv,l} \in S^l$. Hence

$$\varepsilon_{r,tv,l}^i = \frac{\sum_{j=1}^{L}((h_r^i(j)) - \frac{1}{L})((h_{tv,l}^i(j)) - \frac{1}{L})}{\sqrt{(\sum_{j=1}^{L}(h_r^i(j))^2 - \frac{1}{L^2})(\sum_{j=1}^{L}(h_{tv,l}^i(j))^2 - \frac{1}{L^2})}}, \quad v_{r,l}^i = N(r_{tv,l}^i) * \max(\max_{r_{tv,l} \in S^l}(\varepsilon_{r,tv,l}^i, 0)) \tag{14}$$

$N(r_{tv,l}^i)$ indicates number of pixels in segment $r_{tv,l}$, which means a large segment will have large weight and $\max(\max_{r_{tv,l} \in S^l}(\varepsilon_{r,tv,l}^i, 0))$ considers the highest correlation that the segment can have with the scribbled foreground/background segments and thus improves the segmentation quality.

3.6   Computation of Similarity Value $u_{r,l}^i$

Let $G_{b,u}^i(X^i \bigcup Y_{r,l}^i, E_u)$ be a complete bipartite graph with X^i and $Y_{r,l}^i$ be two disjoint set of vertices. E_u is the set of edges between segment r_r^i and $Y_{r,l}^i$ having weight $\varepsilon_{r,tu,l}^i$ between segment r_r^i and segment $r_{tu,l} \in Y_{r,l}^i$. Hence, we can write

$$\varepsilon^i_{r,tu,l} = \frac{\sum_{j=1}^{L}((h^i_r(j)) - \frac{1}{L})((h^i_{tu,l}(j)) - \frac{1}{L})}{\sqrt{(\sum_{j=1}^{L}(h^i_r(j))^2 - \frac{1}{L^2})(\sum_{j=1}^{L}(h^i_{tv,l}(j))^2 - \frac{1}{L^2})}}, \quad u^i_{r,l} = N(r^i_{tu,l}) \max(\max_{r_{tu,l} \in S^l}(\varepsilon^i_{r,tu,l}, 0)) \tag{15}$$

These values can now be substituted in (13) to calculate the matching coefficient $d^i_{r,l}$ and

$$a^i_{j,l} = d^i_{r_j,l} \tag{16}$$

All pixels p^i_j with $a^i_{j,1} > 0.5$ are classified into foreground region and rest to background region.

Algorithm 1. Interactive Co-segmentation

Input: $I = \{I^1, \ldots, I^n\}$, set of n images, S = set of scribbled images,
Output: $a^i_{j,1}$
1: Perform Mean Shift over-segmentation on I.
2: Select segments from S.
3: Classify scribbled segments into foreground/background based on red/green scribble respectively.
4: Compute $b^r_{m,l}$.
5: Set ϵ_1 and ϵ_2 parameters.
6: Construct Bipartite Graph.
7: Compute $u^i_{r,l}$ and $v^i_{r,l}$.
8: Compute $d^i_{r,l}$ using $u^i_{r,l}$ and $v^i_{r,l}$.
9: Set $a^i_{j,l}$.
10: Select $a^i_{j,1} > 0.5$ as foreground pixel.
11: **return**

---

4   Results

The performance of the algorithm is evaluated by making a qualitative and quantitative study of two benchmark datasets: MSRC dataset [23] and iCoseg dataset [1] which have been widely used for the evaluation of the performance of image co-segmentation methods. The iCoseg dataset consists of 38 groups with a total of 643 images such that each group of images has a common object. The experiment has been evaluated by randomly selecting 35 such group of images for perform evaluation. Similarly the MSRC dataset contains 20 group of images from which we randomly selected 9 such group for performance evaluation. Quantitative evaluation includes the performance metric Accuracy (A), which indicates the ratio of correctly classified pixels into foreground and background region. Performance has also been reported on a subset of the Internet database as in [26] and compared with their state-of-the-art CNN-based approach [26].

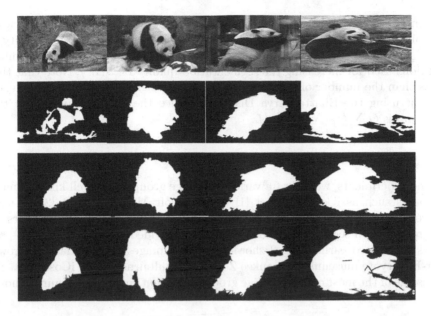

Fig. 4. Co-segmentation results on the Panda set of images from iCoseg. The first row: input images. The second row: result of the proposed algorithm without bipartite graph. The third row: co-segmentation results by full algorithm. The last row: ground truth.

4.1 Parameter Settings

Some important observations regarding how to provide user scribbles are:

- Some images with complex backgrounds should be provided scribbles first as it provides required information for making good estimation of the background regions and to provide sparse scribbles to images with simple background.
- The scribble should cover as many colour variations as possible, implying that the regions with variable colour inside foreground/background are a good choice to put the scribbles on.
- The regions with similar colours in foreground and background, must be scribbled. The user should add scribbles until these scribbles provide the maximum possible colour information.

Once the scribbles are provided, the values of the mean shift over-segmentation parameters *viz.* spatial bandwidth $hs = 8$, range bandwidth $hr = 7$ and minimum size of the segment $M = 20$ are set experimentally based on training images. Unless mentioned otherwise, the following parameters are used in the proposed algorithm, $\epsilon_1 = 0.1$ and $\epsilon_2 = 0.1$.

4.2 Computational Environment

MATLAB 2013 has been used for the experiments using an Intel Core $i5 - 6200U$ CPU @ 2.3 GHz with 4 GB RAM. The complexity of the proposed algorithm can be computed separately over the two phases. For determining the probability of each pixel belonging to either foreground/background regions, the complexity is $O(\sum_{i=1}^{n}(N(R^i))^2 * L$. In case of bipartite graph construction, let the number of regions used for increasing the segmentation quality be $S(R^i)$. Note that this is less than the number of segments used for computing the probability of each segment using the Bhattacharya Distance. Hence the time complexity of the algorithm is $O(\sum_{i=1}^{n}(N(R^i))^2 * L$.

4.3   Co-segmentation Results

For the experiments, we collect a variety of image groups from well known image databases such as iCoseg dataset [1] or Microsoft MSRC dataset [23]. These datasets are the most popular ones for image co-segmentation experiments as the ground truth segmentation masks are also provided.

As mentioned earlier, Fig. 2 shows scribbled images of 002 Alaskan Brown Bear-Eukaryote museums Milwaukee Zoo 2006-Cmlburnett from iCoseg dataset as it is one of the most complex image groups in the entire dataset. Figure 3 shows

Fig. 5. Co-segmentation results (Alaskan Brown Bear): using the proposed algorithm (Color figure online)

the scribbled images of Flower class from MSRC dataset with a wide variation of flowers types and colours. The corresponding co-segmentation results using the proposed algorithm are shown in Figs. 5 and 6 respectively. The improvement obtained using Bipartite graph construction is clearly perceived from Fig. 4.

5 Evaluation of Performance

We present here the evaluation of performance of the proposed algorithm with respect to accuracy and also its computational complexity. The comparison is done by comparing pixel by pixel of the mask of ground truth with that obtained from using the algorithm. In Fig. 7 a comparison of the proposed algorithm is made with Co-segmentation by Composition algorithm [9] using iCoseg dataset and in Fig. 8 using the MSRC dataset.

Fig. 6. Co-segmentation results (Flower): using the proposed algorithm

5.1 Comparison of Accuracy

In this section, we list the precision statistics i.e Accuracy (A) for each of iCoseg dataset group of images in Table 1 and MSRC dataset group of images in Table 2 selected for evaluation. We have also compared the corresponding statistics with state-of-the-art co-segmentation by composition algorithm [9]. In Tables 1 and 2, the first column indicates the name of the group image followed by number of images in each group. The third column indicates the accuracy of our proposed algorithm. To evaluate the performance of our algorithm, the results of Co-segmentation using the Composition algorithm [9] has been presented in the last

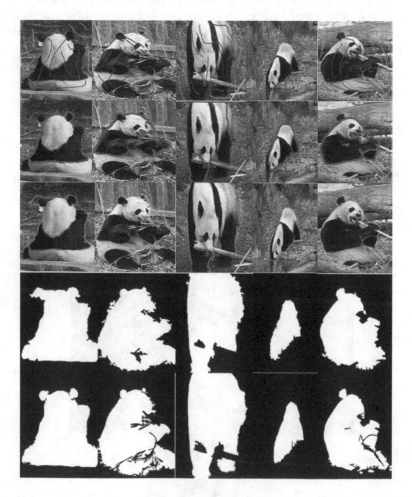

Fig. 7. Comparison results: the first row contains the scribbles from ICOSEG [1] dataset. The second row is the co-segmentation results obtained by Co-segmentation by Composition [9]. Third row contains co-segmentation results by the proposed algorithm. Fourth row contains the mask obtained from the proposed algorithm and fifth row contains the ground truth.

column. From the accuracy values, it can be seen that the proposed algorithm greatly outperforms state-of-the-art algorithm [9]. The average accuracy of the proposed algorithm on iCoseg dataset is 96.33% and that of MSRC dataset is 91.19%. It may be noted that the accuracy values achieved on the MSRC dataset is less compared to iCoseg as the images in iCoseg are less complex than that of MSRC dataset.

Accuracy was also computed for the Internet images of Car, Horse and Airplane and found to be 91.25, 80.32 and 90.22 while the corresponding values in [26] were reported to be 93.9, 92.4 and 94.1. Since [26] employs a CNN-based approach requiring far greater computational resources, the fact that all the values except for 'Horse' are comparable, bears testimony to the efficiency of the proposed approach which is available at much smaller computational expense.

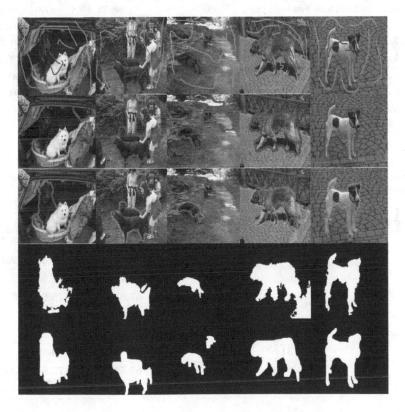

Fig. 8. Comparison results: the first row contains the scribbles of Dog class images from MSRC dataset [23]. The second row is the co-segmentation results obtained by Co-segmentation by Composition [9]. Third row contains co-segmentation results by the proposed algorithm. Fourth row contains the mask obtained from the proposed algorithm and fifth row contains the ground truth.

Table 1. Performance evaluation of the co-segmentation result obtained after the implementation of the proposed algorithm with respect to the ground truth based on accuracy (A) (iCoseg dataset)

Name of group-image	#images	A (our algorithm)	A [9]	A [19]	A [13]	A [21]
002 Alaskan Brown Bear	19	97.50	72.56	86.4	74.8	90.0
006 Red Sox Players	25	96.54	81.93	90.5	73.0	90.9
009 Stonehenge1	5	96.51	72.70	87.3	56.6	63.3
012 Stonehenge2	18	96.26	54.09	88.4	86.0	88.8
015 Ferrari	11	95.86	67.51	84.3	85.0	89.9
017 Agra Taj Mahal	5	97.32	80.45	88.7	73.7	91.1
021 Elephants-safari	15	96.37	73.65	75.0	70.1	43.1
023 Pandas-Tai-Land	25	92.96	55.74	60.0	84.0	92.7
032 Kite-Brighton kite Festival	18	96.82	88.53	89.8	87.0	90.3
033 Kite-kitekid	10	96.51	61.24	78.3	73.2	90.2
036 Gymnastics-1	6	98.24	88.73	87.1	90.9	91.7
038 Skating-ISU	12	98.22	88.55	76.8	82.1	77.5
041 Hot Balloons-Skybird	23	98.80	84.68	89.0	85.2	90.1
042 Statue of Liberty	40	98.27	73.73	91.6	90.6	93.8
Brown bear	5	95.37	64.16	80.4	74.0	95.3

Table 2. Performance evaluation of the co-segmentation result obtained after the implementation of the proposed algorithm with respect to the ground truth based on accuracy (A) (MSRC dataset)

Name of group-image	#images	A (our algorithm)	A [9]	A [19]	A [13]	A [21]
Bird	34	95.26	79.29	–	62.2	95.3
Cat	24	86.29	67.56	77.1	74.40	92.3
Cows	30	92.91	71.75	80.10	81.6	94.2
Dog	30	95.42	67.68	–	75.6	93
Face	30	85.19	70.09	76.30	84.30	–
Flower	32	88.80	56.61	–	–	–
Horse	$\sqrt{31}$	93.56	55.50	74.90	80.10	–
Plane	30	92.16	67.59	77.00	73.80	83.0
Sheep	30	94.25	68.89	–	92.5	94.0

5.2   Time Comparison

Table 3 provides the average run times on each group of images of the MSRC and iCoseg datasets.

All the run time values are measured in seconds on Intel Core $i5 - 6200U$ CPU @ 2.3 GHz and 4 GB RAM. It can be observed that time taken for co-segmentation in MSRC dataset is more than that of iCoseg dataset, as the images in MSRC are highly complex compared to that of iCoseg. Thus it produces a greater number of segments resulting in more time consumption. Comparison with the average run times in [8] shows that the proposed approach fares better for some of the images *e.g.* 7.5 s for 'Cows', 13.7 s for 'Flower' in MSRC and 8.1 s for 'Ferrari', 7.6 s for 'State of Liberty'- the times indicated being for [8] even though the computational facilities used there were higher, *viz.* a system with 64 GB RAM, 2.7 GHz. processor.

Table 3. Average run time of each image group in MSRC and iCoseg datasets

MSRC dataset			iCoseg dataset		
Name of image-group	#images	Time in seconds	Name of image-group	#images	Time in seconds
Bird	34	10.25	002 Alaskan Brown Bear	19	14.38
Cat	24	15.57	006 Red Sox Players	25	4.92
Cows	30	7.17	009 Stonehenge	5	4.69
Dog	30	10.43	012 Stonehenge	18	5.14
Face	30	9.40	015 Ferrari	11	7.81
Flower	32	11.40	017 Agra Taj Mahal	5	5.58
Horse	31	13.80	021 Elephants-safari	15	4.52
Plane	30	10.96	023 Pandas-Tai-Land	25	5.90
Sheep	30	6.86	032 Kite-Brighton kite Festival	18	4.17
			033 Kite-kitekid	10	15.79
			036 Gymnastics-1	6	5.77
			038 Skating-ISU	12	2.31
			041 Hot Balloons-Skybird	23	2.79
			042 Statue of Liberty	40	5.42
			Brownbear	5	5.98

Table 4. Symbols used

Symbol used	Stands for	Symbol used	Stands for
I	Set of images	S	Set of scribbled images
p_j^i	Pixel of image I^i and index j	$a_{j,l}^i$	Prob. for p_j^i belonging to background $(l = 0)$/foreground $(l = 1)$
R^i	I^i after mean-shift segmentation	r_m^i	mth segment of I^i
$N(r^i)$	Total # of segments in I^i	$b_{m,l}^i$	Prob. of r_m^i belonging to $l \in \{0, 1\}$
S^0	Set of scribbled background segments	S^1	Set of scribbled foreground segments
h_q^i	Normalized histogram of segment r_q^i	L	#bins in histogram h_q^i
$c_{m,l}^i$	Initial estimate of prob. of $r_m^i \in l$	$A(x, y)$	Bhattacharya Distance
$G^i(V^{(i)}, E^{(i)})$	Graph for I^i	Le	Laplacian of $G^i(V^{(i)}, E^{(i)})$
X^i	Set containing r_r^i with $b_{r,1}^i \in \{0.5 - \epsilon_1, 0.5 + \epsilon_2\}$	$Y_{r,l}^i$	Contains all adjacent segments of r_r^i classified into region l
$u_{r,l}^i$	Highest similarity between segment r_r^i and its adjacent segments in $Y_{r,l}^i$	$v_{r,l}^i$	Highest similarity between segment r_r^i and segments in S^l
$r_{tu,l}^i$	Segment in $Y_{r,l}^i$	$r_{tv,l}^i$	Segment in scribbled images
$N(r_r^i)$	# pixels in segment r_r		
$N(r_{tu,l}^i)$	# pixels in segment $r_{tu,l}$	$N(r_{tv,l}^i)$	# pixels in segment $r_{tv,l}$

6   Conclusions and Scope for Future Work

In this work, a new framework for solving the interactive co-segmentation problem has been presented. The superiority of the proposed algorithm has been amply established by the experimental results. The accuracy achieved is significantly higher than reached by existing methods. The proposed algorithm consists of one global unary function responsible for providing prior probability to each segment belonging to foreground/background region by matching each segment histogram to all the scribbled segments histograms using the Bhattacharya distance. The local function is responsible for maintaining smoothness of the outcome by providing similar probability to those segments which are similar in colour appearance. Bipartite graph further improves the segmentation quality by considering the segments which are prone to be misclassified due to similarity in both foreground/background colour regions.

Modifications of the proposed approach being explored are the construction of an alternative to histogram matching in order to incorporate structural information and auto-correction of user scribbles, as well, if necessary. Extension of the approach to provide co-segmentation on videos or multi class images is also being considered.

References

1. Batra, D., Kowdle, A., Parikh, D., Luo, J., Chen, T.: iCoseg: interactive co-segmentation with intelligent scribble guidance. In: Proceedings of 2010 IEEE Computer Society Conference on Computer Vision and Pattern Recognition, pp 3169–3176, June 2010
2. Batra, D., Kowdle, A., Parikh, D., Luo, J., Chen, T.: Interactively co-segmentating topically related images with intelligent scribble guidance. Int. J. Comput. Vis. **93**(3), 273–292 (2011)
3. Boykov, Y., Funka-Lea, G.: Graph cuts and efficient N-D image segmentation. Int. J. Comput. Vis. **70**(2), 109–131 (2006)
4. Cheng, B., Liu, G., Wang, J., Huang, Z., Yan, S.: Multitask low-rank affinity pursuit for image segmentation. In: Proceedings of 2011 International Conference on Computer Vision, pp. 2439–2446, November 2011
5. Collins, M.D., Xu, J., Grady, L., Singh, V.: Random walks based multi-image segmentation: quasiconvexity results and GPU based solutions. In: Proceedings of 2012 IEEE Conference on Computer Vision and Pattern Recognition, pp. 1656–1663, June 2012
6. Comaniciu, D., Meer, P.: Mean shift: a robust approach toward feature space analysis. IEEE Trans. Pattern Anal. Mach. Intell. **24**(5), 603–619 (2002)
7. Cootes, T.F., Taylor, C.J., Cooper, D.H., Graham, J.: Active shape models-their training and application. Comput. Vis. Image Underst. **61**(1), 38–59 (1995)
8. Dong, X., Shen, J., Shao, L., Yang, M.H.: Interactive cosegmentation using global and local energy optimization. IEEE Trans. Image Process. **24**(11), 3966–3977 (2015)
9. Faktor, A., Irani, M.: Co-segmentation by composition. In: Proceedings of 2013 IEEE International Conference on Computer Vision, pp. 1297–1304, December 2013
10. Fu, H., Xu, D., Lin, S., Liu, J.: Object-based RGBD image co-segmentation with mutex constraint. In: IEEE Conference on Computer Vision and Pattern Recognition (CVPR), pp. 4428–4436, June 2015
11. Hochbaum, D.S., Singh, V.: An efficient algorithm for co-segmentation. In: Proceedings of 2009 IEEE 12th International Conference on Computer Vision, pp. 269–276, September 2009
12. Kim, T.H., Lee, K.M., Lee, S.U.: Non parametric higher-order learning for interactive segmentation. In: Proceedings of 2010 IEEE Computer Society Conference on Computer Vision and Pattern Recognition (2010). https://doi.org/10.1109/CVPR.2010.5540078

13. Joulin, A., Bach, F., Ponce, J.: Discriminative clustering for image cosegmentation. In: Proceedings of 2010 IEEE Computer Society Conference on Computer Vision and Pattern Recognition, pp. 1943–1950, June 2010
14. Joulin, A., Bach, F., Ponce, J.: Multi-class cosegmentation. In 2012 IEEE Conference on Computer Vision and Pattern Recognition, pp. 542–549, June 2012
15. Lou, Z., Gevers, T.: Extracting primary objects by video co-segmentation. IEEE Trans. Multimedia **16**(8), 2110–2117 (2014)
16. Mu, Y., Zhou, B.: Co-segmentation of image pairs with quadratic global constraint in MRFs. In: Yagi, Y., Kang, S.B., Kweon, I.S., Zha, H. (eds.) ACCV 2007. LNCS, vol. 4844, pp. 837–846. Springer, Heidelberg (2007). https://doi.org/10.1007/978-3-540-76390-1_82
17. Mukherjee, L., Singh, V., Dyer, C.R.: Half-integrality based algorithms for cosegmentation of images. In: Proceedings of 2009 IEEE Conference on Computer Vision and Pattern Recognition, pp. 2028–2035, June 2009
18. Rother, C., Minka, T., Blake, A., Kolmogorov, V.: Cosegmentation of image pairs by histogram matching incorporating a global constraint into MRFs. In: Proceedings of 2006 IEEE Computer Society Conference on Computer Vision and Pattern Recognition, CVPR 2006, Washington, DC, USA, vol. 1, pp. 993–1000 (2006)
19. Rubio, J.C., Serrat, J., Lpez, A., Paragios, N.: Unsupervised co-segmentation through region matching. In: Proceedings of 2012 IEEE Conference on Computer Vision and Pattern Recognition, pp. 749–756, June 2012
20. Shen, J., Du, Y., Wang, W., Li, X.: Lazy random walks for superpixel segmentation. IEEE Trans. Image Process. **23**(4), 1451–1462 (2014)
21. Vicente, S., Rother, C., Kolmogorov, V.: Object cosegmentation. In: Proceedings of 2011 IEEE Conference on Computer Vision and Pattern Recognition, CVPR 2011, Washington DC, USA, pp. 2217–2224 (2011)
22. Wang, J.Z., Li, J., Wiederhold, G.: SIMPLIcity: semantics-sensitive integrated matching for picture libraries. IEEE Trans. Pattern Anal. Mach. Intell. **23**(9), 947–963 (2001)
23. Winn, J., Criminisi, A., Minka, T.: Object categorization by learned universal visual dictionary. In: Proceedings of 10th IEEE International Conference on Computer Vision, ICCV 2005, Washington DC, USA, vol. 2, pp. 1800–1807 (2005)
24. Yuille, A.L., Cohen, D.S., Hallinan, P.W.: Feature extraction from faces using deformable templates. In: Proceedings of Computer Vision and Pattern Recognition, CVPR 1989, pp. 104–109, June 1989
25. Zhu, H., Lu, J., Cai, J., Zheng, J., Thalmann, N.M.: Multiple foreground recognition and cosegmentation: an object-oriented CRF model with robust higher-order potentials. In: Proceedings of IEEE Winter Conference on Applications of Computer Vision, pp. 485–492, March 2014
26. Li, W., Hosseini Jafari, O., Rother, C.: Deep object co-segmentation. In: Jawahar, C.V., Li, H., Mori, G., Schindler, K. (eds.) ACCV 2018. LNCS, vol. 11363, pp. 638–653. Springer, Cham (2019). https://doi.org/10.1007/978-3-030-20893-6_40

Using Deep Convolutional LSTM Networks for Learning Spatiotemporal Features

Logan Courtney[✉] and Ramavarapu Sreenivas

University of Illinois, Urbana-Champaign, IL 61820, USA
{courtne2,rsree}@illinois.edu

Abstract. This paper explores the use of convolutional LSTMs to simultaneously learn spatial- and temporal-information in videos. A deep network of convolutional LSTMs allows the model to access the entire range of temporal information at all spatial scales. We describe our experiments involving convolutional LSTMs for lipreading that demonstrate the model is capable of selectively choosing which spatiotemporal scales are most relevant for a particular dataset. The proposed deep architecture holds promise in other applications where spatiotemporal features play a vital role without having to specifically cater the design of the network for the particular spatiotemporal features existent within the problem. Our model has comparable performance with the current state of the art achieving 83.4% on the Lip Reading in the Wild (LRW) dataset. Additional experiments indicate convolutional LSTMs may be particularly data hungry considering the large performance increases when fine-tuning on LRW after pretraining on larger datasets like LRS2 (85.2%) and LRS3-TED (87.1%). However, a sensitivity analysis providing insight on the relevant spatiotemporal temporal features allows certain convolutional LSTM layers to be replaced with 2D convolutions decreasing computational cost without performance degradation indicating their usefulness in accelerating the architecture design process when approaching new problems.

Keywords: Convolutional LSTMs · Deep learning · Video analytics · Lip Reading · Action recognition

1 Introduction

Learning from video sequences requires models capable of handling both spatial and temporal information. Due to the advent of large image datasets such as ImageNet [25], there has been significant progress in the development of convolutional-based architectures for learning spatial features [15,20,29,34]. It is not surprising that almost all methodologies for video sequences revolve around convolutional networks combined with additional temporal elements.

The origin of video based learning begins primarily with the task of human action recognition. [17] saw success by stacking frames together before passing them through a pretrained convolutional network as well as processing frames individually with some variant of temporal pooling applied to the network output. [28] saw larger improvements by passing RGB and optical flow images through pretrained networks. [22] leveraged

© Springer Nature Switzerland AG 2020
S. Palaiahnakote et al. (Eds.): ACPR 2019, LNCS 12047, pp. 307–320, 2020.
https://doi.org/10.1007/978-3-030-41299-9_24

the success of recurrent neural networks by using a stack of LSTMs [16] to process the outputs of a convolutional network. All of these methods for handling the temporal information used pretrained convolutional networks. The spatial and temporal features were, in some sense, handled separately.

[35] first explored the use of 3D convolutions for processing spatial and temporal information together. That is, the network was capable of learning spatiotemporal features. Research has continued predictably along this path with large action recognition datasets such as the Kinetics dataset [18]. [14] replaced 2D convolutions in common image-based architectures such as ResNet with 3D convolutions. [6] was able to expand the filters from a pretrained 2D network to a 3D convolution network capturing the benefits of the extra training data from ImageNet.

Lipreading is a technique for understanding speech using only the visual information of the speaker. It is a well-structured problem for looking into how deep networks learn spatiotemporal features. For a problem like action recognition, the current datasets have classes that can be identified from a single image alone (e.g. playing baseball versus swimming). In such instances, the temporal information is less important than the spatial information reducing the necessity for architectures capable of directly handling spatiotemporal features. High performing models still separate the spatial and temporal learning. [5] temporally post-processes learned features from a pretrained Inception-ResNet-v2 [33] achieving higher performance than any of the 3D convolution architectures from [14] on Kinetics.

On the other hand, lipreading from a single frame within a sequence provides little information about what is being said. It is necessary to utilize the temporal context. Lipreading certainly aligns with the concept of a spatiotemporal problem. This research explores the use of convolution LSTMs for lipreading. That is, 2D convolutions are used within the LSTM structure providing the network with the capacity to learn features at many combinations of spatial and temporal scales. The primary contributions of the paper are as follows.

- Successfully trains the first very deep network built primarily with convolutional LSTMs and achieves competitive performance on the Lip Reading in the Wild dataset [8]
- Convolutional LSTMs see the same improvements as 2D convolutions see in image classification tasks when upgrading from architectures like VGG to ResNet. This is similar to what [14] demonstrated for 3D convolutions.
- Presents an analytical technique providing insight on what spatiotemporal features are relevant to a particular problem and demonstrates how this information can be used to facilitate the architecture design process (Sect. 5.1)
- The model utilizes 2D convolutions along with convolutional LSTMs demonstrating the ability to intermix temporally capable modules with spatial-only processing modules to reduce the number of parameters and total computation without sacrificing performance

Section 2 reviews related work along with discussing the basics of spatiotemporal features. Section 3 describes our proposed methodology and model architectures. Section 4 describes the experiments and implementation. Section 5 discusses our results and provides empirical evidence explaining the similar performance between convolutional LSTM models and competing methods.

2 Background and Related Work

2.1 Spatiotemporal Features and Receptive Field

Each layer within a convolutional network has an output which can be interpreted as an image with channels made up of a prescribed number of features and a corresponding height/width related to the spatial dimensions. Spatial pooling and/or strided-convolutions are used throughout the network to reduce the height/width of these feature maps. This reduces computational costs due to fewer convolution operations per layer and increases the visual receptive field. Each pixel of the output is calculated based on a larger portion of the original input image. Additionally, it allows for deeper (i.e. more layers) as well as wider (i.e. more channels) networks which have been shown to increase the network's capability to learn complex visual tasks.

However, with each spatial pooling layer and/or strided convolution layer, a certain degree of spatial information is inevitably lost due to the reduction in resolution of the output. This can impede a network's ability to learn functions capable of discriminating high resolution visual information. The choice of when to apply spatial dimension reduction techniques within the network remains an "art-form" in the hands of the network designer. An application requiring the classification of large objects can utilize more spatial reduction layers to increase the visual receptive field without discarding relevant information. Using spatial reduction layers too early in the network may prevent the network from being able to detect the precise location of objects taking up a small portion of the input image. The receptive field has been well studied in the past [21].

For 1D sequence problems, such as those seen in natural language processing (NLP), there are typically two techniques used to capture sequence information: convolutions and recurrent neural networks. A series of 1D convolutions applied to sequence data gradually increases the temporal receptive field. This gradual exposure of temporal information with deep convolutional networks works well for tasks like document classification [10]. On the other hand, each layer of a network made up of LSTMs has access to the full sequence. That is, the temporal receptive field when calculating the output at a particular timestep includes all previous inputs. This technique works well when training language models for predicting the next word in the sequence [11].

Dealing with sequences of images creates an additional challenge due to the temporal information appearing at multiple spatial scales. Our work is meant to motivate a principled approach for applications that depend on detecting these spatiotemporal features. It blends methods that (a) involve just images and (b) involve just time series data. So far, methods for applications like action recognition or lipreading have utilized a straightforward recipe of using 3D convolutions together with time-tested techniques from the two individual fields. As shown in [35] for action recognition, utilizing a 3D convolution ResNet50 over a 2D convolution ResNet50 model with an LSTM shows an increase from 68.0% to 72.2%. Utilizing the temporal information after the final layer of a 2D convolution network may be too late to capture the relevant spatiotemporal features. In [31] for lipreading, including a 3D convolution at the front of the network sees an increase of 5.0% compared to the model without early temporal processing. At first glance, it may seem that any architecture capable of processing spatial and temporal features together is all that is needed.

However, related events may be separated by temporal gaps. 2D convolutions work with images due to the assumed symmetry between spatial dimensions and local connectivity of pixels. There is no inherent reason for treating the temporal dimension as an extra spatial dimension as is done with 3D convolutions.

Additionally, there are performance discrepancies with the use of 3D convolutions. In action recognition, [35] saw success by using a model with the temporal receptive field increasing at the same rate as the spatial receptive field. Lipreading [8] saw this same method achieve a 10% lower classification rate when compared to processing all of the temporal information at a single spatial scale. Between the two high performing models, there was an additional 4.0% improvement when the temporal processing was delayed until a later spatial scale. These discrepancies suggest different applications contain different spatiotemporal features and performance is dictated by which spatiotemporal features the model is designed to handle.

If the spatial resolution of the input doubled, would the model need to adapt the location of the temporal processing? If the temporal resolution of the input doubled, would a larger kernel size or more 3D convolutions at a particular spatial scale be necessary? There are many unanswered questions with both 3D convolutions and spatiotemporal features. As much as deep learning is about creating networks capable of learning generalized features relevant to a particular problem, it would be advantageous to create architectures capable of generalizing well across problems. This is the main focus of our work. The architecture presented in this paper and the empirical results suggest convolutional LSTMs hold promise in learning the spatiotemporal features relevant to the dataset without having to cater the design of the network in a particular way.

2.2   Lipreading in the Wild (LRW) Dataset

The Lip Reading in the Wild (LRW) [8] dataset consists of 500,000 videos with 29 frames each taken from BBC TV broadcasts. There are 500 target words (1,000 videos for each word) with 50 videos of each word for both the validation and test sets. There are often context words surrounding the target word. Although there are ambiguous classes (e.g. "weather" and "whether"), it is sometimes possible to distinguish between these based on the surrounding context. The videos are centered on the speaker with the speaker facing the camera.

2.3   Lipreading Sentences in the Wild (LRS2/LRS3-TED) Datasets

The LRW dataset has a constrained vocabulary with fixed input size making it a well structured sequence classification problem. The Lip Reading Sentences (LRS) [7] dataset contains variable length sequences with unique words appearing in the test set unseen in the training set. Additionally, there are moments of off-angle facial views up to 90° (side profile). The original LRS dataset is unreleased to the public and the

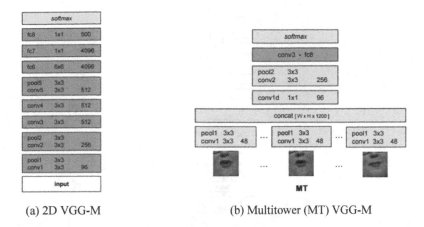

(a) 2D VGG-M   (b) Multitower (MT) VGG-M

Fig. 1. Modifications of the VGG-M model (left) were used in [8] with the release of the LRW dataset. The multitower (MT) model (right) set the initial benchmark at 61.1% by processing each frame individually with a 2D convolution before concatenating the features for temporal processing.

LRS2 dataset [1] is used in its place. The LRS3-TED [3] dataset, similar in structure to LRS2, is made from TED talks as opposed to BBC TV Broadcasts.

2.4 Related Work

The original work on LRW [8] tested multiple variations of the VGG-M model (seen in Fig. 1a). Their highest performing model (seen in Fig. 1b) processed each frame of the video with a 2D convolution before concatenating the outputs allowing the remainder of the spatial processing to have access to the full temporal receptive field. This outperformed the networks utilizing 3D convolutions which gradually increased the spatial and temporal receptive fields as the input passed deeper into the network.

The original work on LRS [7] used a similar VGG-M based model processing five frames at a time. This network acted as a sliding window across the sequence providing a separate output at each timestep. Long-term temporal information was managed by an LSTM encoder-decoder network with attention [37] to spell words one character at a time. The model was then fine-tuned on LRW achieving 76.2%. The 15.1% improvement over the model in Fig. 1b is due to some combination of the temporal receptive field and extra training data.

[31] replaced the VGG-M portion of the network with the deeper 34-layer ResNet [15]. This network (shown in Fig. 2) is separated into three parts: a spatiotemporal front-end with a 3D convolution, ResNet34 for processing the remaining spatial information, and a bidirectional LSTM. It was state of the art with 83.0% before being surpassed by a model replacing ResNet34 with ResNet18 [30] achieving 84.3%.

[2] uses the same 3D+ResNet spatiotemporal front-end while testing three back-end models for sequence transcription on LRS2. The self-attention based Transformer [36] outperformed the fully convolutional back-end and the bidirectional LSTM back-end. A

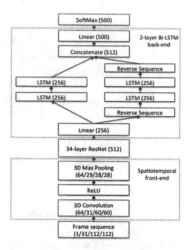

Fig. 2. 3D+ResNet+BiLSTM model from [31] sets the benchmark on LRW at 84.3% when ResNet34 is replaced with ResNet18 [30].

continuation of the work [1] compares performance of the Transformer back-end when trained as a seq2seq [32] model versus training with the CTC loss [12]. The seq2seq Transformer set the benchmark at 48.3% word error rate (WER) for LRS2 and 58.9% WER for LRS3-TED.

The above lipreading sentences work all use the same 3D+ResNet34 spatiotemporal front-end model and only compare back-end performance. The front-end model is pretrained on short sequences from the LRW and LRS datasets using the technique in [9]. The back-end is trained on extracted frozen features from the front-end. The results illustrate the importance of long-term context and language modeling when lipreading sentences. [1] shows over a 12% reduction in WER when testing on phrases containing more than three words and the Transformer model contains over three times as many parameters (65 million) as the spatiotemporal front-end (21 million). No improved results were reported for the LRW dataset.

This is mentioned to emphasize the convolutional LSTM models explored here can be used in conjunction with these techniques by swapping out the spatiotemporal front-end. The work here focuses on the spatiotemporal learning.

3   Proposed Technique

3.1   Convolution LSTM

The convolutional LSTM calculates an internal cell state c_t (cf. Eq. 1 below) and a hidden state h_t utilized as the output to subsequent layers and for state-to-state transitions. While processing a sequence of frames, c_t and h_t can be viewed as images of appropriate size maintained by the network with relevant information based on what it has seen in the past. Learnable filters $W_\bullet$ with bias terms $b_\bullet$ are used to handle a new frame x_t along with the past information h_{t-1} being used by learnable filters $U_\bullet$.

$$f_t = \sigma(W_f * x_t + U_f * h_{t-1} + b_f)$$
$$i_t = \sigma(W_i * x_t + U_i * h_{t-1} + b_i)$$
$$o_t = \sigma(W_o * x_t + U_o * h_{t-1} + b_o) \qquad (1)$$
$$c_t = f_t \circ c_{t-1} + i_t \circ tanh(W_c * x_t + U_c * h_{t-1} + b_c)$$
$$h_t = o_t \circ tanh(c_t)$$

Convolutional LSTMs have been used in the past in rather limited ways. [26] demonstrated success with predicting future precipitation maps. A shallow two-layer model is used at a single spatial scale. [39] explored learning spatiotemporal learning for gesture recognition using 3D convolutions followed by two layers of convolutional LSTMs for longer context. [13] used a bidirectional convolutional LSTM on the output of a 2D VGG13 [29] network to detect violence in videos. The majority of the spatial processing occurs before reaching the convolutional LSTM layers. The datasets for these applications are relatively small which may explain convolutional LSTMs limited use due to their tendency to overfit [38]. However, recent large-scale datasets are providing an opportunity to explore their use in new ways.

There is a fundamental difference between the temporal receptive of convolutional LSTMs and 3D convolutions. Successive 3D convolutions inherently limit long-term temporal information from being processed until deep in the network after the spatial dimension has been reduced. All layers in a network of convolutional LSTMs will have access to the full sequence at all spatial scales. The models here incorporate convolutional LSTMs throughout the entirety of the network and they demonstrate the existence of spatiotemporal features at multiple scales.

3.2   VGG-M ConvLSTM

The first convolutional LSTM model (Fig. 3a) is based on the VGG-M architecture seen in Fig. 1a. The number of output channels has been reduced to compensate for the larger number of weights in convolutional LSTM layers. The first layer is a 2D convolution which matches the highest performing multitower (MT) model from [8] (Fig. 1b).

3.3   ResNet ConvLSTM

Certain convolution layers can be replaced with convolutional LSTM layers to extend the ResNet architecture to learn spatiotemporal features. The model is shown in Fig. 3b. The four parameters of a **Res_ConvLSTM()** module represent the number of sub-blocks, the number of input channels, the number of intermediate channels, and the number of output channels. For any **Res_ConvLSTM()** module, the first sub-block is always of type A (shown in Fig. 4a) with the remaining sub-blocks of type B (shown in Fig. 4b). This is similar to the technique used in [14] to replace the 2D convolutions in ResNet with 3D convolutions.

There are two variants. The first uses an LSTM in place of the attention mechanism for training on the LRW dataset. The attention network [37] is used when pretraining on the LRS datasets with unconstrained vocabularies. The network is modified to output a character (or blank token) after every frame in order to spell words and uses the

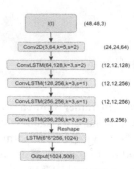

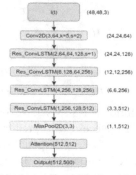

(a) Baseline convolutional LSTM architecture modeled after the VGG-M architecture used in [8].

(b) A residual based convolutional LSTM network. Each **Res_ConvLSTM()** is made up of multiple sub-blocks shown in Figure 4.

Fig. 3. Two different convolutional LSTM architectures. The output dimensions are shown in the margins.

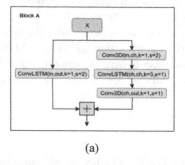

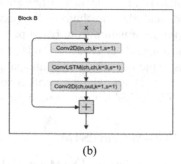

(a)

(b)

Fig. 4. Block type A contains two convolutional LSTMs with one of them in the path of the skip connection. This block has a stride of 2 for reducing the spatial dimensions. Block type B contains one convolutional LSTM with an open skip connection.

CTC loss function [12]. The convolutional LSTM based ResNet model has 14.5 million parameters. The bidirectional version has 29 million parameters which is roughly equivalent to the 23.5 million parameters of the competing model shown in Fig. 2. There are 48 total layers: 14 3 × 3 convolutional LSTMs, 4 1 × 1 convolutional LSTMs, 28 1 × 1 2D convolutions, and 1 LSTM before the final output layer.

3.4 Reduced ResNet ConvLSTM

The importance of temporal information in the context of lipreading varies depending on the spatial scale within the network. The model in Fig. 3b is modified by replacing select convolutional LSTMs determined to be irrelevant by the sensitivity analysis (see Sect. 5.1) with 2D convolutions. This reduces both the number of parameters and total computation without losing the necessary representation power for high performance.

The convolutional LSTM layers at scales 12×12 and 6×6 are replaced due to the hidden-to-hidden connections providing relatively small amounts of useful information here. The 3×3 scale is removed due to the temporal back-end's ability to capture the long-term context after the spatial resolution has been fully collapsed.

4 Experiments

4.1 Training Convolutional LSTMs on LRW

A random crop around the mouth between 48 and 64 pixels is attained for each frame and resized to a 48×48 input. During training, the frames are randomly flipped, randomly rotated $\pm 10°$, and have the brightness adjusted randomly by $\pm 10\%$. The models are trained on random subsequences of length 24 as opposed to the full 29 frames. Dropout for recurrent neural networks [38] (uses same dropout mask for entire sequence) is used with $p = 0.5$ before the final fully connected LSTM layer.

The internal cell states c_t and the hidden states h_t are reset to 0 before processing a new sequence. The model applies the Cross Entropy Loss to the output at each timestep allowing the use of truncated backpropagation. Gradients are calculated every eight frames (three times per sequence) with the parameter update performed once at the end of sequence. Temporal features longer than eight frames can still be learned due the internal cell states and hidden states carrying old information even after the gradient propagation has been truncated. The GPU memory use scales linearly with the length of truncated backpropagation implying the batch size can be increased for faster training.

The models take approximately three weeks to train with PyTorch [24] on a NVIDIA Titan X GPU with the Adam [19] optimizer. The learning rate is initialized to $1e^{-4}$ and reduces whenever validation performance stops progressing. Near the end of training, the sequence length is gradually increased to the full 29 frames in order to allow the model to take more advantage of the context.

A bidirectional version of the ResNet based model is created by separately training a reverse direction network initialized with the already trained forward direction network. This network converges more rapidly and increased the accuracy by 1.9% when tested together with the forward direction network.

4.2 Pretraining on LRS

As mentioned in Sect. 3.3, the output of the ResNet based convolutional LSTM model is modified to predict characters in order to leverage additional training data from the LRS datasets. These datasets contain labeled word boundaries allowing the use of Curriculum Learning [4] to speed up training and improve results [7]. Training begins on subsequences of length eight and gradually increases until a sequence length of 24. This pretraining stage takes approximately one month. The model is then fine-tuned on LRW for an additional week. Gradient clipping [23] (set to 5.0) is necessary due to occasionally large gradients.

Table 1. Results comparing the performance on the LRW dataset. The bidirectional ResNet with convolutional LSTMs achieves comparable results with the current state of the art. Pretraining on more data improves results significantly.

	Pretraining?	Top-1	Top-5	Top-10
VGG-M Multitower [8]	No	61.1%	–	90.4%
VGG-M+LSTM [7]	LRS	76.2%	–	–
2D+ResNet34+Conv [31]	No	69.6%	90.4%	94.8%
3D+ResNet34+Conv [31]	No	74.6%	93.4%	96.5%
3D+ResNet34+BiLSTM [31]	No	83.0%	96.3%	98.3%
3D+ResNet18+BiLSTM [30]	No	84.3%	–	–
VGG-M ConvLSTM	No	73.1%	92.5%	96.7%
ResNet ConvLSTM	No	81.5%	96.1%	98.2%
ResNet BiConvLSTM	No	83.4%	96.8%	98.5%
ResNet BiConvLSTM	LRS2	85.2%	97.4%	98.9%
Reduced ResNet BiConvLSTM	LRS2+LRS3	87.1%	97.7%	98.9%

5   Results

During inference, three crops of size 48×48, 56×56, and 64×64 with their flipped counterparts (six transformations total) are passed through the network with the outputs averaged for the final prediction. The results are summarized in Table 1. The bottom five entries are convolutional LSTM models trained here.

Convolutional LSTMs see the same improvements as 2D convolutions see with image classification (and 3D convolutions see with action recognition [14]) when upgrading from architectures like VGG to ResNet.

The 3D+ResNet34+BiLSTM model from [31] demonstrates the importance of capturing both the short-term spatiotemporal dynamics by using a 3D convolution early in the network (+5.0%) as well as the long-term context by using a bidirectional LSTM late in the network (+8.4%). The ResNet BiConvLSTM was able to slightly outperform the above (83.4% vs. 83.0%) without requiring special consideration for the placement of the temporally capable layers although this does fall short of the current state of the art [30] set at 84.3% which utilizes a ResNet18 architecture in place of Resnet34.

5.1   Spatiotemporal Features Sensitivity Analysis

The internal cell/hidden states are reset to 0 before processing a new sequence indicating the outputs from convolutional LSTM layers for the first frame have no prior information. The relative importance of temporal information at various spatial resolutions can be explored by artificially resetting the internal states during processing to cause these layers to respond as if they have no context.

The performance on a random subset of 5000 video sequences from the validation set is used as the metric for determining the importance of a particular spatiotemporal

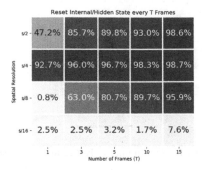

Fig. 5. Relative performance when convolutional LSTM internal states are reset every T frames at particular spatial scales.

scale. Every T frames for $T = \{1, 3, 5, 10, 15\}$, the convolutional LSTMs at a single spatial scale are reset to 0 with all other layers operating normally. The relative performance compared to normal operation is shown in Fig. 5. The layers at $s/4$ perform relatively well even when the internal state is reset every frame. The temporal connections are essentially unused. There is a gap between the high resolution, short-term dynamics learned at $s/2$ and the long-term context deeper in the network. A network of stacked 3D convolutions is not naturally geared for these spatiotemporal gaps considering they gradually extend the temporal receptive field. This may help explain the 3D VGG-M model's poor performance relative to its multitower 2D counterpart in [8] as well as why the 3D+ResNet+BiLSTM [30,31] performs so well due to the 2D ResNet being placed within this gap. This also sheds light on the similar performance of ResNet BiConvLSTM with 3D+ResNet+BiLSTM considering it converged to nearly the same model.

5.2 Importance of Data

The convolutional LSTM models trained here increase the state of the art to 85.2% when pretrained on LRS2 and 87.1% when pretrained on LRS2 and LRS3-TED. Although there is value in comparing performance of various spatiotemporal techniques on a dataset like LRW, there remain concerns about the strength of conclusions that can be drawn. The relative difference in performance when utilizing additional data is significantly larger than the differences in performance between techniques. The Reduced ResNet BiConvLSTM model pretrained on LRS2 (195 h) and LRS3 (444 h) is used to extract features from LRW (165 h) to fine-tune the temporal back-end. The back-end model is trained on reduced portions of the LRW dataset with the test accuracies shown in Fig. 6. Even with high-quality spatiotemporal features from a model pretrained on nearly 4× the amount of data, performance is heavily affected by the proportion of training data used. All of the models heavily overfit and easily achieve above 99% accuracy on the training set.

With the back-end prone to overfitting, it is difficult to imagine training a highly complex, spatiotemporal front-end to its full potential. Although Sect. 5.1 provides a

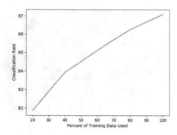

Fig. 6. Test accuracy on LRW after fine-tuning only the temporal back-end on a subset of the LRW training set.

possible explanation for the similarity in performance between competing models, there is potentially not enough data to appropriately train such models for a fair comparison. This is further supported by the results from [27]. Their model takes one month to train with 64 GPUs on a massive internal dataset of 3,886 h and outperforms the benchmark for LRS3-TED even *without* any fine-tuning. With such drastic dataset size increases, the cost for hyperparameter/architecture tuning is prohibitive.

However, convolutional LSTMs can be beneficial from a design perspective allowing a quicker approach for new applications when knowledge of the relevant spatiotemporal features is unknown ahead of time. It is easier to initially train a full convolutional LSTM network to figure out what is relevant from the data and then alter the design for the particular application based on the key spatiotemporal features. This is the motivation for the Reduced ResNet ConvLSTM model discussed in Sect. 3.4 which maintains high performance yet uses roughly half as many parameters. The modified model need not even use convolutional LSTMs as the same sensitivity analysis would support the design of the successful 3D+ResNet+BiLSTM model [30,31].

6   Conclusion

A deep convolutional LSTM model, the first of its kind, is successfully trained and achieves competitive performance for the Lip Reading in the Wild (LRW) dataset. The network is shown to have successfully learned relevant spatiotemporal features from the data without having to specifically cater the design of the network for the specific problem. The sensitivity analysis provides a technique for discovering relevant spatiotemporal features and it was demonstrated how this can facilitate the architecture design process when approaching new spatiotemporal problems. Additionally, the benefits of using improved convolutional architectures like ResNet are apparent for Convolutional LSTMs just as they have been shown in the past for 2D and 3D convolutions.

It remains an open question whether their ability to handle a larger array of spatiotemporal features is necessary for many real applications. However, as datasets become larger and more complex, we may need to rely on the capabilities discussed here or discover new techniques with similar capabilities. Future work will explore utilizing convolutional LSTMs on more general spatiotemporal problems for further verification.

References

1. Afouras, T., Chung, J.S., Senior, A., Vinyals, O., Zisserman, A.: Deep audio-visual speech recognition. arXiv e-prints, September 2018
2. Afouras, T., Chung, J.S., Zisserman, A.: Deep lip reading: a comparison of models and an online application. arXiv e-prints, June 2018
3. Afouras, T., Chung, J.S., Zisserman, A.: LRS3-TED: a large-scale dataset for visual speech recognition. arXiv e-prints, September 2018
4. Bengio, Y., Louradour, J., Collobert, R., Weston, J.: Curriculum learning (2009)
5. Bian, Y., et al.: Revisiting the effectiveness of off-the-shelf temporal modeling approaches for large-scale video classification. CoRR abs/1708.03805 (2017)
6. Carreira, J., Zisserman, A.: Quo vadis, action recognition? A new model and the kinetics dataset. In: 2017 IEEE Conference on Computer Vision and Pattern Recognition (CVPR), pp. 4724–4733 (2017)
7. Chung, J.S., Senior, A., Vinyals, O., Zisserman, A.: Lip reading sentences in the wild. In: IEEE Conference on Computer Vision and Pattern Recognition (2017)
8. Chung, J.S., Zisserman, A.: Lip reading in the wild. In: Lai, S.-H., Lepetit, V., Nishino, K., Sato, Y. (eds.) ACCV 2016. LNCS, vol. 10112, pp. 87–103. Springer, Cham (2017). https://doi.org/10.1007/978-3-319-54184-6_6
9. Chung, J.S., Zisserman, A.: Lip reading in profile. In: British Machine Vision Conference (2017)
10. Conneau, A., Schwenk, H., Barrault, L., LeCun, Y.: Very deep convolutional networks for natural language processing. CoRR abs/1606.01781 (2016)
11. Cotterell, R., Mielke, S.J., Eisner, J., Roark, B.: Are all languages equally hard to language-model? In: NAACL HLT (2018)
12. Graves, A., Fernández, S., Gomez, F., Schmidhuber, J.: Connectionist temporal classification: labelling unsegmented sequence data with recurrent neural networks. In: Proceedings of the 23rd International Conference on Machine Learning, ICML 2006, pp. 369–376 (2006)
13. Hanson, A., Pnvr, K., Krishnagopal, S., Davis, L.: Bidirectional convolutional LSTM for the detection of violence in videos. In: Leal-Taixé, L., Roth, S. (eds.) ECCV 2018. LNCS, vol. 11130, pp. 280–295. Springer, Cham (2019). https://doi.org/10.1007/978-3-030-11012-3_24
14. Hara, K., Kataoka, H., Satoh, Y.: Can spatiotemporal 3D CNNs retrace the history of 2D CNNs and imagenet? CoRR abs/1711.09577 (2017)
15. He, K., Zhang, X., Ren, S., Sun, J.: Deep residual learning for image recognition. CoRR abs/1512.03385 (2015)
16. Hochreiter, S., Schmidhuber, J.: Long short-term memory. Neural Comput. 9(8), 1735–1780 (1997)
17. Karpathy, A., Toderici, G., Shetty, S., Leung, T., Sukthankar, R., Fei-Fei, L.: Large-scale video classification with convolutional neural networks. In: Proceedings of International Computer Vision and Pattern Recognition (CVPR 2014) (2014)
18. Kay, W., et al.: The kinetics human action video dataset. CoRR abs/1705.06950 (2017)
19. Kingma, D.P., Ba, J.: Adam: a method for stochastic optimization. CoRR abs/1412.6980 (2014)
20. Krizhevsky, A., Sutskever, I., Hinton, G.E.: Imagenet classification with deep convolutional neural networks. In: Pereira, F., Burges, C.J.C., Bottou, L., Weinberger, K.Q. (eds.) Advances in Neural Information Processing Systems 25, pp. 1097–1105. Curran Associates, Inc. (2012)
21. Luo, W., Li, Y., Urtasun, R., Zemel, R.S.: Understanding the effective receptive field in deep convolutional neural networks. CoRR abs/1701.04128 (2017)

22. Ng, J.Y.H., Hausknecht, M., Vijayanarasimhan, S., Vinyals, O., Monga, R., Toderici, G.: Beyond short snippets: deep networks for video classification. In: Computer Vision and Pattern Recognition (2015)
23. Pascanu, R., Mikolov, T., Bengio, Y.: Understanding the exploding gradient problem. CoRR abs/1211.5063 (2012)
24. Paszke, A., et al.: Automatic differentiation in pytorch (2017)
25. Russakovsky, O., et al.: ImageNet large scale visual recognition challenge. Int. J. Comput. Vis. (IJCV) **115**(3), 211–252 (2015). https://doi.org/10.1007/s11263-015-0816-y
26. Shi, X., Chen, Z., Wang, H., Yeung, D., Wong, W., Woo, W.: Convolutional LSTM network: a machine learning approach for precipitation nowcasting. CoRR abs/1506.04214 (2015)
27. Shillingford, B., et al.: Large-scale visual speech recognition. CoRR abs/1807.05162 (2018)
28. Simonyan, K., Zisserman, A.: Two-stream convolutional networks for action recognition in videos. In: Ghahramani, Z., Welling, M., Cortes, C., Lawrence, N.D., Weinberger, K.Q. (eds.) Advances in Neural Information Processing Systems 27, pp. 568–576. Curran Associates, Inc. (2014)
29. Simonyan, K., Zisserman, A.: Very deep convolutional networks for large-scale image recognition. CoRR abs/1409.1556 (2014)
30. Stafylakis, T., Khan, M.H., Tzimiropoulos, G.: Pushing the boundaries of audiovisual word recognition using residual networks and LSTMs. CoRR abs/1811.01194 (2018)
31. Stafylakis, T., Tzimiropoulos, G.: Combining residual networks with LSTMs for lipreading. CoRR abs/1703.04105 (2017)
32. Sutskever, I., Vinyals, O., Le, Q.V.: Sequence to sequence learning with neural networks. In: Ghahramani, Z., Welling, M., Cortes, C., Lawrence, N.D., Weinberger, K.Q. (eds.) Advances in Neural Information Processing Systems 27, pp. 3104–3112. Curran Associates, Inc. (2014)
33. Szegedy, C., Ioffe, S., Vanhoucke, V.: Inception-v4, inception-resnet and the impact of residual connections on learning. CoRR abs/1602.07261 (2016)
34. Szegedy, C., et al.: Going deeper with convolutions (2015)
35. Tran, D., Bourdev, L.D., Fergus, R., Torresani, L., Paluri, M.: C3D: generic features for video analysis. CoRR abs/1412.0767 (2014)
36. Vaswani, A., et al.: Attention is all you need. CoRR abs/1706.03762 (2017)
37. Yang, Z., Yang, D., Dyer, C., He, X., Smola, A.J., Hovy, E.H.: Hierarchical attention networks for document classification. In: HLT-NAACL (2016)
38. Zaremba, W., Sutskever, I., Vinyals, O.: Recurrent neural network regularization. CoRR abs/1409.2329 (2014)
39. Zhang, L., Zhu, G., Shen, P., Song, J.: Learning spatiotemporal features using 3DCNN and convolutional LSTM for gesture recognition, pp. 3120–3128, October 2017. https://doi.org/10.1109/ICCVW.2017.369

Two-Stage Fully Convolutional Networks for Stroke Recovery of Handwritten Chinese Character

Yujung Wang[1](✉), Motoharu Sonogashira[2], Atsushi Hashimoto[3], and Masaaki Iiyama[2]

[1] Graduate School of Informatics, Kyoto University, Kyoto, Japan
wang.jung.64z@st.kyoto-u.ac.jp
[2] Academic Center for Computing and Media Studies, Kyoto University, Kyoto, Japan
[3] Graduate School of Education, Kyoto University, Kyoto, Japan

Abstract. In this paper, we propose a method to recover strokes from offline handwritten Chinese characters. The proposed method employs a fully convolutional network (FCN) to estimate the writing order of connected components in offline Chinese character images and a multi-task FCN to estimate the writing order and directions of strokes in each connected component. Online dataset CASIA-OLHWDB1.0 from the CASIA database is hired as the training set. Because the network produces discontinuous strokes, we refine the estimated writing orders using a graph cut (GC), in which the estimated directions are used for calculation of smoothness term. Experimental results with test dataset of CASIA-OLHWDB1.0tst demonstrate the effectiveness of our method.

Keywords: Handwriting trajectory recovery · Semantic segmentation · Fully convolutional networks · Graphcut

1 Introduction

Handwritten Chinese character recognition has been studied for the last two decades. This has been considered as a difficult problem, owing to target's complicated structures, leading to a large number of character classes, and the variability in writing style [1]. The approaches to handwriting recognition can be classified into two categories: offline recognition and online recognition. Offline recognition involves digital character images which are previously written on a paper and then captured by a scanner or a camera, while online recognition involves a sequence of coordinate points of pen trajectories which are captured on-line through special devices such as digitizer tablets.

In general, offline recognition is more difficult than online recognition. With character images only, offline recognition usually concentrates on complicated image processing for feature extraction [2]. On the other hand, using information of the pen trajectories, online recognition can dynamically analyze the structure

© Springer Nature Switzerland AG 2020
S. Palaiahnakote et al. (Eds.): ACPR 2019, LNCS 12047, pp. 321–334, 2020.
https://doi.org/10.1007/978-3-030-41299-9_25

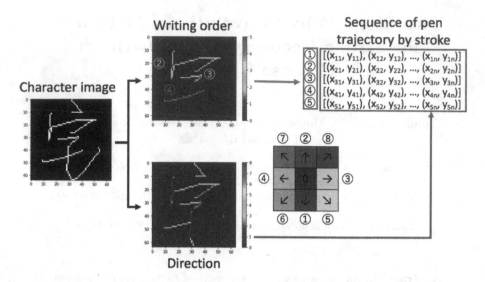

Fig. 1. Chinese characters can be represented as a sequence of strokes consisting of points with similar direction. The writing order is the order of strokes, and the direction is the direction a stroke written.

of characters, which results in a capability to recognize the cursive script and an increase in the recognition rate [3]. Therefore, by extracting the information of strokes from character images, an offline recognition problem can be virtually viewed as online recognition.

In this paper, we propose a method to recover stroke of handwritten Chinese character images by estimating the information of stroke such as writing order and direction. A Chinese character is composed of one or more subcharacters consisting one or more strokes, which are mostly straight lines with a few polylines. In other words, a Chinese character can be represented as a sequence of strokes, consisting of one or more sub-strokes with similar directions. This is shown in Fig. 1. We model the stroke recovery as a semantic segmentation problem, and employ two-stage fully convolutional networks (FCN) [4]. The first FCN estimates the writing order of connected components approximating the subcharacters from character images, and the second FCN estimates the writing order and directions of strokes of connected component images. Both FCN are trained using online data. Because the FCN could produce discontinuous strokes, a graph cut (GC) [5] is utilized to refine the estimated writing order by the estimated directions. The overview of our method is shown in Fig. 2.

The contributions of our method are as follows: first, we estimate the writing order and directions of strokes using FCNs. Previous solutions [7–9, 11, 12, 14, 15] typically involve pixelwise sequence-to-sequence models, which assume a two-stage solution of stroke segmentation and direction estimation. In other words, such an approach assumes that strokes can be reliably segmented in the first stage. In contrast, our method recovers the strokes from a raw character image.

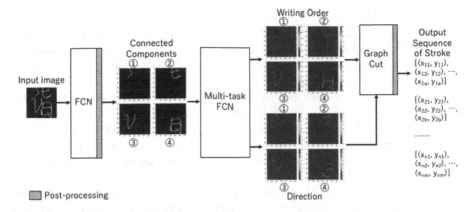

Fig. 2. Overview of the proposed method.

Stroke segmentation is done simultaneously with estimation of stroke direction by trained FCNs. The only required preprocessing is a connected component extraction, which is feasible in many cases. Second, we successfully recover the writing order and directions of strokes simultaneously for multi-stroke images. Because the writing order and direction relate to each other, the direction refines the connectivity problem of the writing order, and the writing order limits possible choices of the direction. As a result, both the writing order and the direction are improved.

2 Related Work

The recovery of online information from offline images has been studied for the last two decades [7–9,11–16]. Features such as endpoints, trajectories of strokes, and relations between strokes have been proven to be useful in online recognition [6]. Research on stroke recovery aims to extract these features from offline images.

Most previous research works on tracing the trajectory, or extracting directions of pixels in other words. Methods involving the continuity between a pixel and its neighbors are first considered. Lee and Pan [7] traced the offline signatures using hierarchical decision making with a set of heuristic rules. Boccignone et al. [8] reconstructed trajectories based on some continuity criteria. Lallican and Viard-Gaudin [11] recovered the trajectories using the Kalman filter.

After a number of studies, researchers have become to focus more on the continuity between strokes, which is proved to improve not only the performance of trajectory recovery on single-stroked character images but also multi-stroked character images by the following methods. Qiao et al. [12] restored the trajectories of single-stroked handwriting images within the framework of edge continuity relations. Kato and Yasuhara [14] traced the drawing order of offline multi-stroke handwriting images. Qiao and Yasuhara [15] recovered writing trajectories from multi-stroked images using a bidirectional dynamic search, to find the best matching with the writing paths of template strokes.

On the other hand, some previous research has worked on either of estimating the writing order of the strokes or segmenting strokes. Liu et al. [13] utilized a model-based structural method to estimate the writing order of offline Chinese characters. Kim et al. [16] segmented the strokes from hand drawings using neural networks without ordering.

Only a few previous research has worked on estimating both directions and the writing order. Abuhaiba et al. [9] proposed a method to extract temporal information from offline Arabic characters through stroke segmentation, straight-line approximation, and the learning of token strings extracted from strokes by a fuzzy sequential machine, transforming the character images into sequences of coordinate points [10].

Our method is designed to estimate the writing order and directions of strokes from offline multi-stroke character images using semantic segmentation, and then recover the pen trajectory using the writing order and directions of strokes. Moreover, with the multi-task FCN, our method estimate the writing order and directions simultaneously.

3   Proposed Method

In this section, we elaborate on our proposed stroke recovery method for offline handwritten Chinese characters.

A Chinese character is composed of one or more subcharacters, which are often common among many Chinese characters. Given subcharacters instead of the original Chinese character as the network input inherently decreases the number of character classes to learn. Therefore, we employ two-stage writing order recovery.

First, given bitmaps of offline character images, an FCN is utilized to estimate the writing order of the connected components approximating subcharacters in the character images. Next, a multi-task FCN is utilized to estimate the writing order and directions of strokes in each component. The FCNs are trained separately. Then, a graph cut (GC) is utilized to refine the estimation of the writing order by using the estimation of the direction for weighting. Finally, combining the results for the writing order and direction, the character images can be described in a sequence of strokes composed of coordinate points. The flow chart of the proposed method is presented in Fig. 2.

3.1   Estimation of Writing Order of Connected Components

As the first step, we intend to extract subcharacters from the offline image with its writing order. However, we have no ground truth to train such FCN. Since the strokes of a subcharacter are typically written continuously, we alternate subcharacter segmentation for a connected components extraction in a binarized character image. Then, we employ an FCN that predicts their writing order.

Given a bitmap of an offline character image, the first FCN outputs an esti-mation of the writing order at the level of connected components. The connected components are represented as $P = \{p_1, p_2, ..., p_k\}$, and collection of their writing order is represented as $C = \{c_1, c_2, ... c_k\}$. The first FCN estimates the writing order at the level of connected components by predicting the pixel-wise likelihood of label assignments $P(C|v)$ for a pixel v. After predicting the order $l_{cv} = \text{argmax}_{c_i} P(c_i|v)$ as the label with the maximum likelihood, we aggregate them to obtain a relative order of each connected component l_{p_i} by the following criteria:

$$l_{p_i} = \frac{1}{N(p_i)} \sum_{v \in p_i} l_{cv}, \tag{1}$$

where $N(p_i)$ is the number of pixels in the connected component p_i. We determine the writing order of P by matching l_{p_i} in ascending order to $c_1, c_2, \ldots, c_k$. Pixels in the connected component p_i are all reassigned to the label of writing order of p_i.

The architecture of the first FCN is illustrated in Fig. 3, which is adapted from [4]. The architecture of the first FCN consists of five blocks of convolutional layers; three split layers placed after the third, forth, and fifth blocks of layers; a merge layer merging the split layers; and an output layer. The Rectified Linear Unit (ReLU) activation function is adopted in each Conv2D layer, and the softmax activation function is adopted in the output layer. Then, the connected components are sorted based on Eq. (1) and its post-process. The ordered components are input to the second FCN independently.

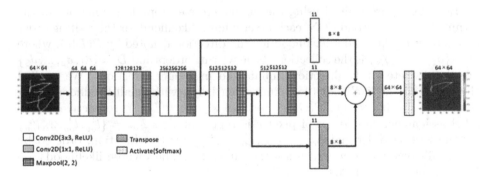

Fig. 3. Structure of the first FCN.

3.2 Estimation of Writing Order and Stroke Direction of Each Connected Components

The second FCN is a multi-task FCN, and predicts the writing order and directions of strokes in the connected components. The architecture of the second FCN is illustrated in Fig. 4. The layers are the same as in the first FCN, except that the layers are split for the writing order and the direction, respectively.

The sigmoid activation function is used in the output layer. Because the overlaps between strokes should have more than one labels, the sigmoid activation function, which gives probabilities ranging from 0 to 1 and not necessarily summed up to 1, can give these pixels more than one labels with similarly large probabilities.

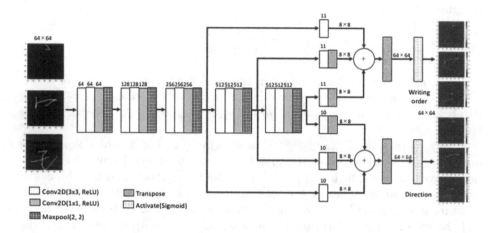

Fig. 4. Structure of the second FCN.

Both FCNs are trained using the binary cross-entropy loss function. In the estimation of the second FCN, each pixel v has a likelihood for the writing order denoted by $P(O|v)$ and a likelihood for the direction denoted by $P(D|v)$, where $O = \{o_1, o_2, ..., o_N\}$ is the collection of the writing orders and $D = \{d_1, d_2, ..., d_8\}$ are the discrete stroke directions, quantized into eight directions. We applied thresholds to the likelihoods for eliminating labels with small likelihoods. The thresholds are $0.8 \times \max(P(O|v))$ and $0.8 \times \max(P(D|v))$, respectively. The label assignment candidates of pixel v is represented as $L_{ov} = \{l_{ov}^1, l_{ov}^2, ..., l_{ov}^n\}$ for the writing order and $L_{dv} = \{l_{dv}^1, l_{dv}^2, ..., l_{dv}^m\}$ for the direction, ordered by the likelihoods, where n and m are the number of labels whose likelihoods are larger than the thresholds.

3.3   Label Smoothing with Graph Cut

Then, given the label assignment candidates L_{ov} and L_{dv} for all pixels v from the multi-task FCN as input, the graph cut (GC) [5] is utilized to smooth the estimation of the writing order. The energy function of GC for a graph $G = (V, E)$ in our method is defined as follow:

$$E(l) = \sum_{v \in V} E_v(l_v) + \sum_{(u,v) \in E} E_{uv}(L_{du}, L_{dv})\kappa \tag{2}$$

where vertices V and edges E correspond to pixels and the relations between pixels and their eight neighbors, respectively, κ is the parameter, $E_v(l_v)$ is the data term that measures the label assignments l_v, and $E_{uv}(L_{du}, L_{dv})$ is the smoothness term that measures the penalties of difference between the input label set of directions L_{du} and L_{dv}.

The GC is executed separately for each label of the writing order. In the ith execution, pixels with label i are considered as the foreground, and others as the background. Pixels with label i but resulted in background in the GC calculation will move to the next execution with label $i + 1$. The data term indicates the difference between the input label i and the assigned label l_v. Therefore, the data term is set as zero if the input label set L_{ov} of a pixel v contains the assigned label l_v, and a constant λ if it does not. However, in order to maintain the connectivity between pixels with the same assigned label, even though l_v is not in L_{ov}, data term of a pixel v is set as zero if there exists a neighbor pixel n with input label set L_{on} containing the assigned label l_v. The smoothness term is weighted as the difference between the directions of a pixel and those of its neighbors. The GC outputs the smoothed result of the estimation of the writing order. The equation of the data term and the smoothness term can be written as follows:

$$E_v(l_v) = \begin{cases} \lambda & \text{if } l_v \notin L_{ov} \text{ and } l_v \notin L_{on} \\ 0 & \text{otherwise} \end{cases} \tag{3}$$

$$E_{uv}(L_{du}, L_{dv}) = \begin{cases} w_{dd} & \text{if } (u, v) \in F \\ 9.0 & \text{otherwise} \end{cases}, \tag{4}$$

where w_{dd} is the weight of the difference between the directions of pixel u and pixel v, defined as

$$w_{dd} = \begin{cases} 1.0 & \text{if } u \to v \text{ and } v \to u \\ \min(w_a) & \text{if } u \to v \text{ or } v \to u \\ 9.0 & \text{otherwise} \end{cases} \tag{5}$$

$$w_a = \begin{cases} 1.0 & \text{if } \Delta\theta = 0° \\ 2.0 & \text{if } \Delta\theta = 45° \\ 3.0 & \text{otherwise} \end{cases} \tag{6}$$

where $u \to v$ indicates that the direction of pixel u pointing to pixel v is in the input label set L_{du}, and $\Delta\theta$ is the minimum angular difference between L_{du} and L_{dv}.

Finally, with the result of the writing order and directions of strokes, pixels with the same writing order are put into a sequence with corresponding writing order to form a stroke, and the directions of each pixels are presented as a list of label L_{dv} for each pixel v.

4 Experiments

4.1 Implementation

In this section, we demonstrate the effectiveness of our method in extracting the online information from offline Chinese character images.

The online character dataset OLHWDB1.0 in the CASIA Chinese handwriting database was utilized in the experiment. The online data of characters were composed of sequences of coordinate points as strokes. We constructed character images with online data, and resized them to a size of 64×64. The FCN networks were trained by 87,600 character images, with the corresponding online information containing 300 classes. For the multi-task FCN, the character images are cut into connected components for training. Because the numbers of connected components with the same number of strokes differ a lot, we limited the numbers of connected components to be the same. For the parameters of the GC, we set $\lambda = 1500.0$, and $\kappa = 5.0$ where the values were obtained by optimizing the results. Then, 21,700 character images were used for testing.

4.2 Evaluation

We evaluated the estimation as follow:

1. Direction accuracy: If the label of the direction of a point corresponds to the ground truth, the label assignment is considered to be correct. A pixel may have multiple labels owing to the overlaps, so the labels are counted separately.
2. Relative writing order accuracy: Because our method may cut a stroke into several strokes due to large changes in direction, which means that the number of strokes may be different from ground truth but the relative ordering of strokes is the same, relative writing order is used instead of absolute writing order. We employ two criteria for evaluating the writing order accuracy. One is Pearson correlation coefficient between the predicted writing order and its ground truth. The other is the relative ordering of each pixel v.
   (a) Pearson correlation coefficient: Pearson correlation coefficient globally represent the relation between the prediction and the ground truth. Value of coefficient near to 1 means that the number and the relative relation between strokes of prediction are similar to the ground truth, while value of coefficient near to 0 means that number or the relative relation between strokes of prediction are different from the ground truth. For value of coefficient smaller than 0, it usually means that the relative order between connected components of prediction are different from the ground truth.
   (b) Relative writing order accuracy: Relative writing order accuracy locally represent the pixel-based relation between the prediction and the ground truth. We calculate the mean of the predicted writing order of each stroke s_i with pixels within labeled with i in ground truth and compare it with the predicted writing order of each pixel. The mean of the predicted

writing order of pixels that belongs to the ith stroke s_i, which is denoted by l_{s_i}, is calculated as

$$l_{s_i} = \frac{1}{N(s_i)} \sum_{v \in s_i} l_{ov}^{S'} \tag{7}$$

where $N(s_i)$ is the number of pixels in the stroke s_i and $l_{ov}^{s'}$ is the corresponding label of pixel v in the estimated sequence of strokes S'. For a pixel v assigned with label $l_{ov}^{S'}$ and belongs to stroke s_i, the assignment is considered to be correct if $l_{ov}^{S'} < l_{s_{i+1}}$ and $l_{ov}^{S'} > l_{s_{i-1}}$, which is presented in Fig. 5(a). We compare the ordering in the level of stroke instead of the level of pixels because when multiple pixels are assigned with the same label, the relative ordering of pixels in the same stroke will all be counted as positive, which is indicated in Fig. 5(b). We evaluated the writing order in a relative manner for two reasons. First, as described above, some of the strokes are cut into several sub-strokes by our method, owing to large changes in direction, which means that the points of these strokes have the same order as in the online sequences of characters, but belong to different strokes in the estimations. These points can be correctly evaluated using the relative writing order. Second, for points belonging to the same strokes, the order between them is determined by the direction, which means that the correctness of the absolute ordering between the points depends on the direction accuracy. Therefore, for the writing order, we do not consider the ordering between points in the same strokes.

4.3   Result

First, our method achieved an accuracy of 96.97% in direction estimation, which means most of the directions of strokes are recovered. The trajectories of strokes are successfully extracted from the offline character images by semantic segmentation.

Second, with the Pearson correlation coefficient of 0.52, which means that the corresponding estimation of strokes is related to the ground truth, our method achieved an accuracy of 75.87% in relative writing order estimation. Several results are shown in Fig. 6. Result (a) to (c) are successful results, where strokes are successfully extracted and the ordering between the estimation and the ground truth are closely related. Result (d) and (e) are results with high accuracy but poor correlation coefficient, where strokes are well extracted but the ordering of continuous part does not match with the ground truth, which leads to the poor correlation. Result (c) and result (d) are differ from the ground truth but are actually more precise than the ground truth. Since our method recover strokes by the estimation of writing order and directions, it is more easier to find where to start and end with precise estimation.

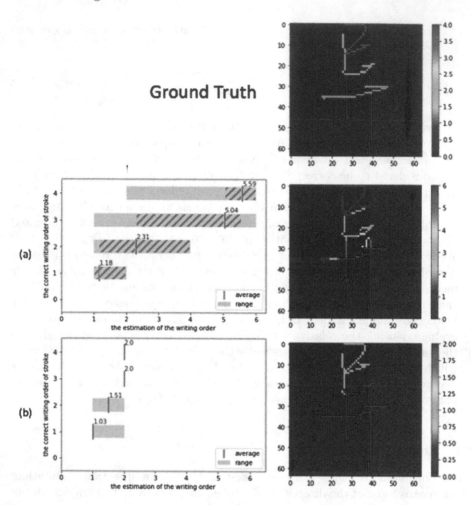

Fig. 5. Description of the relative writing order accuracy. In the left graphs, the vertical axis is the correct writing order of strokes s_1, s_2, s_3, s_4, the horizontal axis is the estimation of the writing order $l_{ov}^1, l_{ov}^2, ..., l_{ov}^6$, the blue bar is the range of the estimation in the same stroke, and the red line is the mean. The slash line area in (a) is the area counted as positive. We use the range instead of checking the label directly to prevent the situation of (b) which will make all points in s_3, s_4 to be correct. (Color figure online)

In addition, the peak of the relative writing order accuracy remains at 98% to 100% for characters with one to five strokes and 70% to 80% for characters with six to nine strokes. Our method performs promising results for characters with one to five strokes. For characters with six to nine strokes, though the

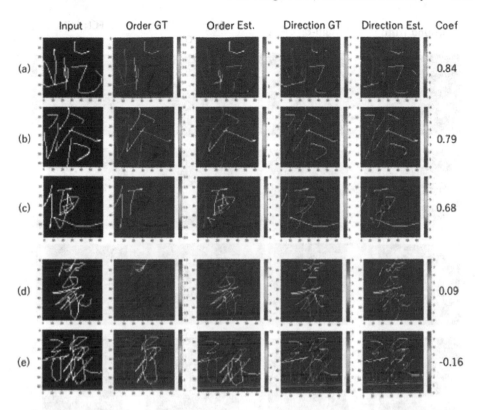

Fig. 6. Results of the proposed method. From left to right are the input, the ground truth of the writing order, the estimation of the writing order, the ground truth of directions, the estimation of directions, and the Pearson correlation coefficient. (a), (b), and (c) are successful results, and (d) and (e) are results with high relative writing order accuracy but poor correlation coefficient.

accuracy is not higher than characters with less strokes, the average correlation coefficient remains 0.58 to 0.61, indicating the close relation between the ground truth and the estimation. Therefore, our method is proved to work on characters with different numbers of strokes.

Third, Fig. 7 shows the results using only multi-task FCN, multi-task FCN with GC, and the proposed FCN-FCN with GC. The results of the proposed FCN-FCN with GC are obviously better than the others. Extracting a character into continuous part and then into strokes is proved to achieve better results than extracting strokes directly.

In total, our method successfully estimates online information from offline handwritten character images.

Ground Truth   Multi-task FCN   Multi-task FCN
+ GC   FCN-FCN
+ GC

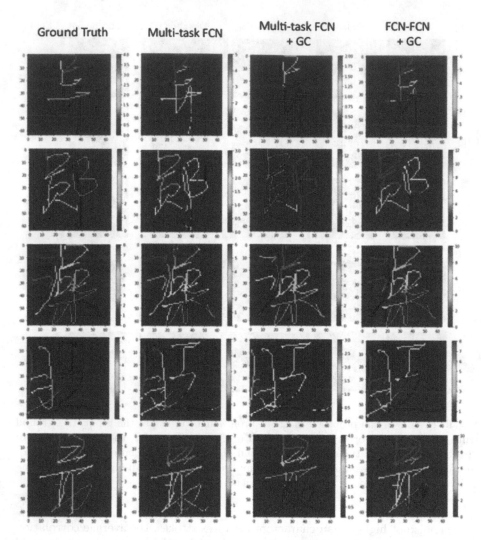

Fig. 7. Images of the ground truth, estimation using multi-task FCN only, estimation using multi-task FCN and GC, and the estimation using the proposed FCN-FCN model with GC. FCN-FCN model with GC obviously performs better than the others.

5   Conclusion

In this study, we introduced a deep learning method using FCN for stroke recovery of handwritten Chinese characters. Our method is based on semantic segmentation using two FCNs, respectively extracting connected components and strokes, and the estimations are refined by a GC. We demonstrated through the experiments that our method successfully extracts online information from offline character images.

However, for characters separated into left part and right part such as Fig. 6(e), in which the correct ordering is left to right, FCN sometimes gives the reversed results. In the future work, we plan to improve the performance on ordering, and compare the result of online recognition using output of our method as input with the result of offline recognition using the original character images as input to evaluate our method on improving the accuracy of character recognition.

Acknowledgment. This work was supported by JSPS KAKENHI Grant Number 17H06288.

References

1. Liu, C., Dai, R., Xiao, B.: Chinese character recognition: history, status and prospects. Front. Comput. Sci. **1**(2), 126–136 (2007)
2. Liu, C.-L., Jaeger, S., Nakagawa, M.: Online recognition of Chinese characters: the state-of-the-art. IEEE Trans. Pattern Anal. Mach. Intell. **26**(2), 198–213 (2004)
3. Zhang, X.-Y., Bengio, Y., Liu, C.-L.: Online and offline handwritten Chinese character recognition: a comprehensive study and new benchmark. Pattern Recogn. **61**, 348–360 (2017)
4. Long, J., Shelhamer, E., Darrell, T.: Fully convolutional networks for semantic segmentation. In: 2015 IEEE Conference on Computer Vision and Pattern Recognition (CVPR), Boston, MA, pp. 3431–3440 (2015)
5. Boykov, Y., Veksler, O., Zabih, R.: Fast approximate energy minimization via graph cuts. IEEE Trans. Pattern Anal. Mach. Intell. **23**(1), 1222–1239 (2001)
6. Doermann, D.S., Rosenfeld, A.: Recovery of temporal information from static images of handwriting. In: Proceedings 1992 IEEE Computer Society Conference on Computer Vision and Pattern Recognition, Champaign, IL, USA, pp. 162–168 (1992)
7. Lee, S., Pan, J.C.: Offline tracing and representation of signatures. IEEE Trans. Syst. Man Cybern. **22**(4), 755–771 (1992)
8. Boccignone, G., Chianese, A., Cordella, L.P., Marcelli, A.: Recovering dynamic information from static handwriting. Pattern Recogn. **26**(3), 409–418 (1993)
9. Abuhaiba, I.S.I., Holt, M.J.J., Datta, S.: Recognition of off-line cursive handwriting. Comput. Vis. Image Underst. **71**(1), 19–38 (1998)
10. Zhao, B., Yang, M., Tao, J.: Drawing order recovery for handwriting Chinese characters. In: ICASSP 2019 - 2019 IEEE International Conference on Acoustics, Speech and Signal Processing (ICASSP), Brighton, United Kingdom, pp. 3227–3231 (2019)
11. Lallican, P.M., Viard-Gaudin, C.: A Kalman approach for stroke order recovering from off-line handwriting. In: Proceedings of the Fourth International Conference on Document Analysis and Recognition, Ulm, Germany, vol. 2, pp. 519–522 (1997)
12. Qiao, Y., Nishiara, M., Yasuhara, M.: A framework toward restoration of writing order from single-stroked handwriting image. IEEE Trans. Pattern Anal. Mach. Intell. **28**(11), 1724–1737 (2000)
13. Liu, C.-L., Kim, I.-J., Kim, J.H.: Model-based stroke extraction and matching for handwritten Chinese character recognition. Pattern Recogn. **34**(12), 2339–2352 (2001)

14. Kato, Y., Yasuhara, M.: Recovery of drawing order from scanned images of multi-stroke handwriting. In: Fifth International Conference on Document Analysis and Recognition, pp. 261–264 (1999)
15. Qiao, Y., Yasuhara, M.: Recover writing trajectory from multiple stroked image using bidirectional dynamic search. In: 18th International Conference on Pattern Recognition (ICPR 2006), Hong Kong, pp. 970–973 (2006)
16. Kim, B., Wang, O., Öztireli, A.C., Gross, M.: Semantic segmentation for line drawing vectorization using neural networks. Comput. Graphics Forum **37**(2), 329–339 (2018)

Text Like Classification of Skeletal Sequences for Human Action Recognition

Akansha Tyagi, Ashish Patel$^{(\boxtimes)}$, and Pratik Shah

Indian Institute of Information Technology Vadodara, Vadodara, India
{201761001,201771002,pratik}@iiitvadodara.ac.in

Abstract. Human Action Recognition (HAR) has many applications in surveillance, gaming, animation and Active and Assisted Living (AAL). Several actions performed in daily life are composed of various poses arranged sequentially in time. Recognition of such actions is a difficult and challenging task. The classification approach proposed in this paper considers an analogy between actions and text, where an action is considered as a sentence and a single pose as a word. In the first stage, the poses are grouped based on their similarity and are then assigned labels. These labels are used for constructing label sequences representing motion. We propose Hierarchical Agglomerative Clustering (HAC) for clustering poses. Once the actions are modelled as the spatio-temporal evolution of key poses, we classify the actions using the Hidden Markov Model (HMM) and Hyper-dimensional Computing (HDC) classifiers. The experiments are performed on different datasets using both classifiers and the results are indicative of the effectiveness of the proposed approach in comparison with state-of-the-art methods.

Keywords: Hierarchical Agglomerative Clustering · Hidden Markov Model · Hyperdimensional Computing · Human Action Recognition · Text like classification

1 Introduction

Since the last two decades, HAR has attracted a lot of attention from researchers in computer vision and machine learning fields. A human action can be defined as a temporal sequence of body postures/poses, and a pose can be represented by coordinates of important markers/joints which form a human skeleton. HAR has applications in various domains like surveillance, human-computer interaction, animation, education, film making, sports etc. Actions are represented by high-dimensional time-series data and the motion databases are very large, thus there is a need for efficient classification methods for the same. The key point to efficiently classify the actions is to have an appropriate similarity metric and an appropriate proper motion representation scheme. An image is the basic unit of a video similarly, a pose can be considered as the basic unit of human action. The actions performed, certainly, have several common poses, for instance watching clock and grabbing phone would have several poses in common.

© Springer Nature Switzerland AG 2020
S. Palaiahnakote et al. (Eds.): ACPR 2019, LNCS 12047, pp. 335–350, 2020.
https://doi.org/10.1007/978-3-030-41299-9_26

The proposed method primarily involves three tasks: pose normalization, pose analysis/clustering and action classification. Poses are normalized using the idea proposed by Gaglio et al. [4] from the skeletal data followed by the clustering of similar poses using HAC and K-means clustering (used for baseline comparison). Next, the actions are transformed into text like sequences where the key poses form the word of the sentences. The text like classification is then performed on the action sequences using HMM and HDC. This work also provides a comparative study of different classification approaches (HMM, HDC, Long Short Term Memory (LSTM)) for human action recognition using skeletal data.

The process of Motion Classification is summarized in Fig. 1.

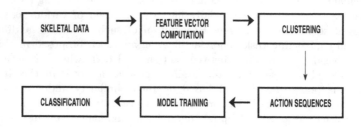

Fig. 1. Motion Classification process.

2   Literature Survey

Gaglio et al. [4] proposed a human activity recognition method based on three machine learning techniques: K-means, SVM (support vector machine) and HMM. The poses involved while performing an activity are detected, normalized and then classified. The authors used SVM to validate the poses which make the system quite complex. The approach proposed by authors uses K-means for clustering, which doesn't choose cluster heads from amongst the poses to be clustered thus there is a high probability that a cluster head might not be a pose in itself which may lead to misclassification. Cippitelli et al. [3] proposed an activity recognition algorithm mainly for AAL (Active and Assisted Living) activities. The algorithm uses a feature vector representing the whole activity computed by using key poses obtained after running the K-means clustering algorithm.

Liu et al. [12] proposed a two-phase human motion retrieval process by incorporating TA-BoW (temporal adjacent bag of words) and DNP-DL (discriminative neighbourhood preserving dictionary learning) for human motion retrieval. Theodorakopoulos et al. [19] proposed DS-SRC (a sparse representation based classification in dissimilarity space) scheme for action recognition in which vectors of dissimilarities to a set of prototype actions are computed. A dynamic time warping based rapid motion retrieval technique using a simple and discrete representation of feature vectors is suggested by Pantuwong et al. [15]. In this method distance between any two frames is found as an element of similarity matrix by keeping the bit length of the frame representation fixed.

Lan et al. [11] proposed a key-pose extraction and hierarchical clustering based generic text-like motion representation. It uses LDA (Latent Dirichlet Allocation) to find motion topics for motion retrieval. Mokari et al. [14] suggested a model which uses Fisher Linear Discriminant Analysis (LDA), Mahalanobis distance and HMM for human action recognition. This method defines the body states and represents each action as a sequence of these body states. Zi Hau Chin et al. in [2] suggested a method that uses statistical parameters, energy spectral density and correlation for extracting features followed by a few feature selection techniques to classify the human motion data using Support Vector Machine, K-Nearest Neighbours and Random Forest.

In [7], the authors proposed temporal convolutional neural networks for human action recognition. Yan et al. in [20] proposed a novel approach to learn temporal and spatial patterns from the data for action recognition using Spatial-temporal graph convolutional networks. In [9] the authors suggested a deep learning-based action recognition method that uses skeletal joint coordinates, joint line distances and Grassmannian Pyramids. In [8] used HDC based approach for human activity recognition from smartphone sensors data was proposed. Barmpoutis et al. [1] proposed a method that uses third-order tensor representation and spatiotemporal inter-correlations among skeletal joints of different actions to recognize human actions. The model uses the neural network for identification of patterns in action. Zhang et al. [22] also proposed a method for skeleton based human action recognition using view adaptive neural networks. In the next section, we describe the proposed approach to action recognition.

3 Methodology

The human pose is typically represented using a skeletal model. For example Figs. 2 and 3 represent 20 and 15 joint skeletal models respectively. Different datasets use different skeletal models to represent a pose. UPCV dataset consists of a 20-joint skeletal model and KARD, CAD-60 consist of a 15-joint skeletal model. The activities performed by an individual can be represented as a sequence of poses recorded over a series of time intervals as shown in Fig. 4.

Fig. 2. 20 joints.

Fig. 3. 15 joints.

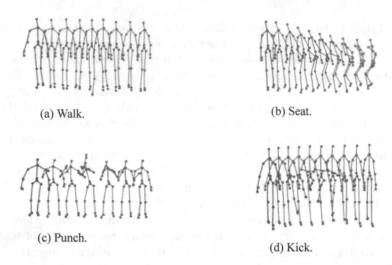

(a) Walk.

(b) Seat.

(c) Punch.

(d) Kick.

Fig. 4. Different action sequences from UPCV action dataset.

3.1 Pose Representation

Actions performed by different subjects vary in terms of physical appearance (height) and thus normalization of the poses for actions is required as suggested in [4]. Each skeletal pose is transformed into a normalized feature vector. The joints of a pose are translated to a new coordinate system fixed at torso/hip centre joint (considering left-right hip axis in the x-direction). All the features are then scaled according to a reference distance h (distance between the neck and the torso joints of the reference skeleton). The details of the feature vector generation are summarized next for the sake of completeness.

Pose Normalisation: Let us consider a skeleton for the pose p where J_i be the real coordinates vector of joint i and j_i is its respective vector containing the 3-D normalized coordinates of the same joint. The feature vector V is defined as

$$V = [j_1, j_2, j_3, \ldots\ldots, j_t] \tag{1}$$

where t is the number of joints. The vector j_i is calculated as:

$$j_i = \frac{J_i}{s} + T. \tag{2}$$

where s is the scaling factor given by:

$$s = \frac{\| J_1 - J_2 \|}{h}, \tag{3}$$

and T is the translation matrix, required to set the torso as the origin of the new coordinate system, and J_1 and J_2 are torso and neck joints respectively.

3.2   Pose Analysis

Several actions performed like "scratch head", "watch clock", "chopping", "stirring" etc have several overlapping poses. The mo-cap data isn't continuous, thus there are poses with minor variations which play a vital role in action recognition. Clustering results in the replacement of similar poses with a key pose. Two clustering approaches HAC (proposed) and K-means (used for baseline comparison) are used:

Hierarchical Agglomerative Clustering (HAC): In this clustering approach, each pose is considered as the singleton cluster in the beginning and then pairs of clusters are successively merged till a single cluster is obtained. We have used two linkage variants namely: "Ward" and "Complete" which uses "euclidean" and "cosine" affinity respectively. The input is given in the form of feature vectors. The dendrogram obtained after clustering is cut at various levels for different values of N and labels are obtained corresponding to the same.

Figure 5 represents the partial dendrogram obtained after using HAC on poses of UPCV dataset. Each node represents the set of similar poses. The first level consists of two nodes where the left node consists of "standing poses" and the right node of a mixture of "standing" and "bend" poses. The right child of the node at the fourth level further bifurcates into two nodes with "arms up" and "cross arms" poses. The occurrence of "anonymous" poses at third level is due to the presence of some bad poses in the dataset. The aim of constructing this diagram is to find out the main poses (amongst all the actions) in which the dataset can be divided.

K-means: It uses Euclidean distance to find similarity between the poses and groups all the poses into K clusters. It is used for clustering of poses with a varying number of centres N. The input is given in the form of feature vectors corresponding to the poses and labels corresponding to the same are obtained as output.

The action sequences are formed from the labels, thus actions are transformed into sequences of labels similar to sentences comprising of words. Further, these sequences are used for the classification of actions.

3.3   Action Classification

The task of action recognition involves the identification of each action. Since the action sequences are temporal and differ in the number of poses thus a classical approach HMM and a recently proposed HDC approach are used for classification:

Hidden Markov Model: An HMM is a Markov model with hidden states, which is used to compute the probability of the sequence of events that are not visible [17]. In this paper, different actions performed by humans are classified using the discrete HMM method.

An HMM has k states $Q = \{q_1, q_2, q_3, \ldots\ldots, q_k\}$, and N observation symbols $Y = \{y_1, y_2, y_3, \ldots\ldots, y_N\}$. It is characterized by three parameters $\lambda = \{A, B, \pi\}$. Let the state at time t be q_t, then $k \times k$ state transition probability matrix is given by $A = \{a_{ij}\}$ where,

$$a_{ij} = P(q_j(t+1)|q_i(t)), \quad i, j \in [1, k] \tag{4}$$

and $N \times k$ observation probability matrix is given by $B = \{b_{ij}\}$ where

$$b_{ij} = P(y_t = i|q_j(t)), \quad i \in [1, N] \text{ and } j \in [1, k] \tag{5}$$

In this paper, we encode each action as a sequence of poses and build the corresponding HMM. The hidden states are denoted by k ranging from 2 to 6,

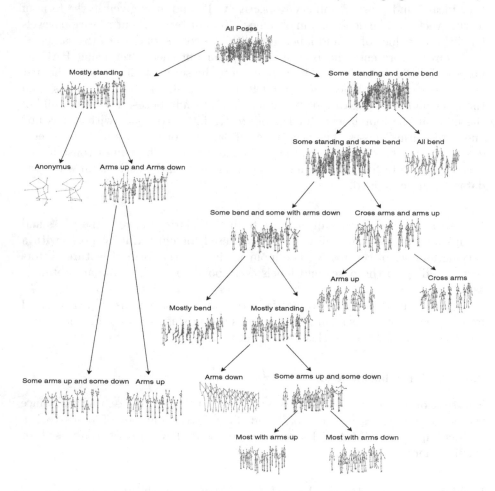

Fig. 5. Partial dendrogram (tree like structure) by applying HAC on poses (feature vectors) of UPCV dataset.

Q is the set of hidden states, N is the number of observation symbols or cluster heads ranging from 10 to 100 and Y is the set of cluster heads.

The action for a test sequence is predicted as follows:

$$output = \arg\max_{i=1,2,....,n\_actions}\{L_i\} \tag{6}$$

where n_actions denotes the number of actions in the dataset.

$$L_i = P(O|\lambda_i) \tag{7}$$

where L_i is the posterior probability of ith HMM, O is the observation sequence to be classified and output is the predicted action.

Hyperdimensional Computing: Hyperdimensional Computing (HDC) is a brain-inspired technique which represents an entity of interest in space as a high dimensional vector (around 10,000 dimensions). Authors in [6] discussed the idea of creating high dimensional vectors (hypervectors). By construction, these hypervectors are orthogonal to each other. In the fields of linguistics, an n-gram is considered as a feature with a contiguous sequence of n items from a given sample of text. For action/observation sequence O, pose tri-gram is considered as a feature. For generating tri-gram using HDC, every observation symbol $\{y_1, y_2, y_3, ..., y_N\}$ is assigned to a unique hypervector. For an observation sequence $O = \{y_2, y_2, y_3, y_1, y_3\}$ tri-grams are $(y_2, y_2, y_3), (y_2, y_3, y_1), (y_3, y_1, y_3)$. An action sequence is represented by a collection of tri-grams present in the action. The HDC representation of an action sequence is generated by operating the hypervectors. The mathematical operations like binding (∗), superposition (+) and permutation (ρ) are well defined in HDC. When these three operations are performed over the hypervectors, it generates a vector with the same dimension. The resultant of the addition of the two vectors shows some similarity with the vectors added together. Binding of the two vectors results into a new hypervector. Rotation of the hypervector also gives a new hypervector. Rotation is generally used for preservation of the order in a sequence.

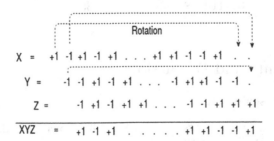

Fig. 6. Tri-gram vector generation using HDC

In HDC tri-gram hypervector $H(y_2, y_2, y_3)$ is represented using $\rho^2(H(y_2)) * \rho(H(y_2)) * H(y_3)$. Here $H(x)$ represents the hypervector of an entity x and

$\rho^i(H(x))$ represents the rotation of a hypervector $H(x)$ with i number of bits. Figure 6 illustrates the generation of tri-gram of a sequence XYZ. In tri-gram ρ^i preserves order i of the entity. All tri-grams of the same action are superimposed together to generate the action hypervector. Similarly, the test action sequence generates the test hypervector. The smallest distance between test hypervector and action hypervectors determines the class associated with the test sequence. In [16] authors have used HDC for classification. We imitate the same for the action sequences.

LSTM: It is an artificial recurrent neural network (RNN) architecture used in the field of deep learning. LSTM is used for classifying, processing and making a prediction of temporal data [13]. For experiments, basic LSTM is implemented and used for comparison with HMM and HDC classification approaches. The training data comprises of the unclustered skeletal poses (feature vectors) corresponding to each action. The main parameters used in the training of the LSTM model are the number of hidden states, epochs and input size. The values of the parameters used in UPCV and KARD datasets corresponding to the highest classification accuracy achieved are mentioned in Table 1. The results obtained after using LSTM on CAD-60 dataset were not promising due to the small size of the dataset, thus the results for the same aren't provided.

Table 1. Parameter values for LSTM Training

Datasets		Parameters		
		Hidden states	Epochs	Input size
UPCV		100	300	60
KARD	AS1	50	200	45
	AS2	50	300	45
	AS3	50	200	45
	ACTIONS	50	250	45
	GESTURES	50	350	45

4 Experiments and Discussion

4.1 Dataset Description

We have performed experiments on three publicly available datasets namely CAD-60 [18], KARD [4] and UPCV [19].

CAD-60: It contains 10 activities. Each activity is performed by four subjects: two males, two females and among them one is left-handed. The activities are listed as follows: "rinsing mouth", "brushing teeth", "wearing contact lens", "talking on the phone", "drinking water", "opening pill container", "cooking (chopping)", "cooking (stirring)", "talking on the couch", "relaxing on the couch", "writing on the whiteboard", "working on a computer". The activities are performed in 5 different environments namely: "bedroom", "bathroom", "livingroom", "kitchen" and "office". It uses a 15 joint skeletal model.

UPCV: This dataset consists of skeletal data featuring different human activities, performed by several subjects [19]. The depth sequences of the actions performed were captured by using Microsoft Kinect. It contains data corresponding to 10 actions performed by 20 different individuals (10 males and 10 females) in two separate sessions. The actions performed are "walk", "grab", "watch clock", "scratch head", "phone", "cross arms", "seat", "punch", "kick", "wave". This dataset uses a 20 joint skeletal model. It comprises of 40,692 poses and 400 action sequences.

KARD: This dataset contains 18 activities. Each activity is performed 3 times by 10 different subjects. The activities are listed as follows "Horizontal arm wave", "High arm wave", "Two hand wave", "Catch Cap", "High throw", "Draw X", "Draw Tick", "Toss Paper", "Forward Kick", "Side Kick", "Take Umbrella", "Bend", "Hand Clap", "Walk", "Phone Call", "Drink", "Sit down", "Stand up". From this dataset, 63,630 poses are obtained and 540 action sequences are formed. It uses a 15 joint skeletal model which is further reduced to 11 and 7 joints for performing experiments.

4.2   Experiments and Results

All the experiments are performed on an Intel Core^{TM} i7 2.8 GHz processor with 16-GB RAM. Clustering and HDC are implemented in python, HMM in R and LSTM in Matlab. Grid search is used to find the best N and k by performing experiments for different values of the number of centres (N) and hidden states (k). The best N and k is considered for the HMM model and the results obtained are compared with the HDC model. We have used comparison metrics like accuracy, precision, recall and F-score for the comparison of different classification methods.

UPCV: Two-third of the samples of each subject are used for training and rest for testing. The experiments are repeated ten times, randomly splitting the training and testing sequences. Table 2 provides the classification accuracies obtained using different clustering and classification methods. The results indicate that HMM provides better classification accuracy than HDC and LSTM. The results obtained using HAC (Ward) clustering are better than those obtained using K-means. Table 3 shows the comparison of results obtained with state-of-the-art methods for UPCV dataset.

An additional experiment is performed for evaluating the robustness of the proposed model which involves the addition of Gaussian noise to the testing skeletal poses. The noise is added in proportion to mean 0 and standard deviation (σ) where σ is the Euclidean distance between neck and hip centre joint of the reference skeleton. Figure 7 shows that the accuracy drops down as the value of the noise added is increased.

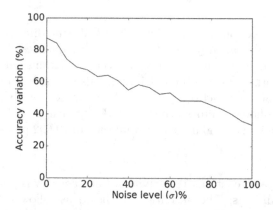

Fig. 7. Variation in classification accuracy with different noise levels in skeleton joints.

KARD: This dataset is divided into several sub-datasets and three types of experiments are performed on each sub-dataset. The experiments vary in the number of joints and training sequences. $P = (7, 11, 15)$ represents the number of joints. Experiments A, B, C consists of 70%, 50% and 30% training data respectively. Each of the experiments are repeated ten times, randomly choosing the action sequences for training and testing sets.

Table 2. Comparison of Accuracy(%) of proposed method using HMM and HDC for UPCV dataset.

N	K-means		HAC (Complete)		HAC (Ward)	
	HMM	HDC	HMM	HDC	HMM	HDC
10	64.75	63.75	67.25	55.83	66.08	56.83
20	74.41	74.5	74.17	74.5	74.5	65.58
30	79.66	78.83	74.42	76.0	76.91	76.58
40	84.41	81.75	75.58	78.08	84.33	80.16
50	83.25	81.08	78.42	74.0	85.58	81.41
60	85.25	80.08	82.25	76.16	85.5	80.16
70	86.08	80.58	82.75	76.08	88	83.83
80	86.50	84.08	82.92	77.0	90	84.5
90	87.33	85.5	83.42	78.41	88.33	86.08
100	86.91	86.66	83.25	80.91	87.25	85.75

Table 3. Comparison of Accuracy (%) with state-of-the-art methods for UPCV dataset.

Algorithm	Accuracy
Ilias et al. [19] using DTW	89.25
Ilias et al. [19] using MNPD	88.50
Proposed method using K-means and HMM	**87.33**
Proposed method using K-means and HDC	**85.50**
Proposed method using HAC (Ward) and HMM	**90.0**
Proposed method using HAC (Ward) and HDC	**86.5**
Proposed method using HAC (Complete) and HMM	**83.42**
Proposed method using HAC (Complete) and HDC	**80.91**
Classification using LSTM	**78.91**

The statistics obtained for different experiments varying in the number of training-testing sequences and number of joints are shown in Tables 6, 8 and 7. Table 5 shows the accuracy statistics obtained for different values of N using K-means and HMM for different sub-datasets of KARD dataset. The highest classification accuracy for AS1, AS2, AS3 are obtained corresponding to $N - 30$, for ACTIONS ($N = 40$) and for GESTURES ($N = 70$). The difference in the values of N is due to the different number and complexity of actions present in different sub-datasets.

Tables 6, 8, 7 shows the different accuracy statistics obtained for HMM and HDC classification approaches using K-means, HAC (Ward) and HAC (Complete) clustering methods. Overall the results indicate that the highest classification accuracies are obtained corresponding to 15 joints and 70% training data. HMM performs better than HDC classification approach. HAC (Ward) clustering method performs better than K-means and HAC (Complete). The statistics obtained corresponding to HMM, HDC for classification and K-means, HAC (Complete) and HAC (Ward) for clustering are compared with state of the art methods, shown in Table 4.

CAD-60: "New-Person" test is performed in which the model is trained on three out of the four people and tested on the left out. The training and testing action sequences have been randomly chosen. The statistics obtained corresponding to different performance metrics (Precision, Recall and F-score) obtained for CAD-60 are mentioned in Table 10. From the results, it is evident that HMM performs better than HDC and HAC (Ward) provides better results than K-means. K-means fails to distinguish between "relaxing on the couch", "talking on the couch" and "drinking water" and "talking on the phone" whereas HAC (Ward) is capable of doing the same. Table 9 shows a comparison with state-of-the-art methods. The different values of N are subject to the complexity of the actions present in different locations (sub-datasets).

Table 4. Comparison of Accuracy (%) of proposed method with state-of-the-art methods for KARD dataset, E refers to the experiments.

Method	E	AS1	AS2	AS3	ACTIONS	GESTURES
Gaglio et al. [4]	A	97.77	98.75	86.66	90.97	86.11
	B	96.08	98.25	81.58	90.75	85.53
	C	92.5	96.19	79.76	88.63	83.80
Proposed method using K-means and HMM	A	98.33	99.17	90.56	93.33	93.56
	B	98.17	98.75	87.50	93.58	90.87
	C	97.74	98.63	83.27	90.89	89.33
Proposed method using K-means and HDC	A	97.91	95.83	86.25	90.0	91.66
	B	98.17	94.75	86.0	89.83	88.66
	C	94.10	92.50	79.82	86.96	83.61
Proposed method using HAC (Complete) and HMM	A	98.19	99.64	93.06	92.64	95.78
	B	97.50	99.25	92.42	92.08	94.73
	C	96.37	99.29	90.36	92.44	93.33
Proposed method using HAC (Complete) and HDC	A	95.83	97.36	85.27	91.52	89.33
	B	95.58	97.50	85.33	89.52	88.13
	C	95.25	97.44	86.38	87.14	83.47
Proposed method using HAC (Ward) and HMM	A	98.61	99.44	88.89	94.72	90.10
	B	97.19	99.08	87.08	92.50	86.53
	C	98.10	98.21	85.71	90.71	83.95
Proposed method using HAC (Ward) and HDC	A	99.02	96.66	88.19	89.02	90.1
	B	98.16	95.91	85.41	86.33	86.53
	C	96.60	95.11	82.32	83.75	83.95
Classification using LSTM	A	97.22	93.33	77.78	87.50	82.22
	B	86.67	90.28	51.67	77.50	69.33
	C	61.90	70.83	50.60	64.29	65.52

Table 5. Accuracy (%) using K-means and HMM with varying number of centres for experiment A for KARD dataset.

N	AS1	AS23	AS3	ACTIONS	GESTURES	ALL
10	89.02	99.3	70.83	90.13	84.78	67.7
20	98.33	99.16	86.38	92.63	87.55	86.54
30	98.88	99.44	89.16	93.05	87.77	89.01
40	98.47	99.02	87.36	95.97	88.22	85.55
50	97.63	97.63	85.13	94.58	91.22	88.45
60	96.94	97.77	86.25	93.33	91.88	85.98
70	96.52	97.63	87.36	94.44	93.33	85.06
80	96.52	97.5	87.77	94.44	92.44	86.11
90	96.25	97.5	86.94	94.58	89.66	88.39
100	95.83	97.36	87.22	92.91	90.44	86.79

Table 6. Accuracy (%) of proposed method using HAC (Complete) for clustering and HMM, HDC for classification for KARD dataset.

J	E	AS1			AS2			AS3			ACTIONS			GESTURES		
		C	HMM	HDC	C	HMM	HDC	C	HMM	HDC	C	HMM	HDC	C	HMM	HDC
7	A	90	98.06	95.83	30	99.31	95.69	90	93.06	85.27	60	91.39	88.88	80	95.44	90.22
	B	90	97.50	95.58	30	98.50	94.41	90	92.42	84.75	60	91.17	87.41	80	94.73	84.46
	C	90	96.01	91.90	20	98.39	97.44	90	90.36	78.86	60	88.88	85.41	80	93.33	80.0
11	A	50	98.19	92.77	20	99.17	97.36	100	90.56	81.48	60	92.64	91.52	60	93.44	90.33
	B	90	97.25	91.25	20	99.25	97.50	90	89.58	78.0	60	92.08	89.52	60	93.20	88.13
	C	90	96.37	95.25	20	99.29	97.20	90	86.19	74.52	60	92.44	87.14	50	91.19	83.47
15	A	90	97.50	94.16	50	99.64	96.8	100	91.67	80.83	40	91.39	86.8	80	95.78	88.11
	B	90	96.75	93.83	20	98.83	97.41	100	92.0	85.33	40	89.58	87.02	80	94.73	86.66
	C	90	96.07	91.78	20	98.81	96.54	90	87.38	86.38	40	87.50	85.92	80	93.29	81.23

Table 7. Accuracy (%) of proposed method using K-means for clustering and HMM, HDC for classification for KARD dataset.

J	E	AS1			AS2			AS3			ACTIONS			GESTURES		
		N	HMM	HDC	N	HMM	HDC	N	HMM	HDC	N	HMM	HDC	N	HMM	HDC
7	A	30	98.61	97.91	30	99.72	95.69	30	90.14	85.0	30	92.92	89.86	40	92.89	89.45
	B	30	96.75	96.5	30	98.83	93.08	40	88.33	83.5	30	90.25	89.83	40	93.33	88.66
	C	30	96.31	93.39	30	98.21	91.42	40	85.06	78.0	30	89.52	86.96	40	89.38	80.66
11	A	30	97.5	97.08	30	98.89	95.83	30	90.69	86.25	30	95.14	88.19	40	92.97	91.66
	B	30	97.33	95.33	30	98.75	93.41	40	87.08	86.0	30	92.58	87.58	40	90.67	88.4
	C	30	96.96	94.10	30	97.62	91.30	30	83.93	79.82	30	90.95	85.41	40	88.62	83.61
15	A	30	98.33	97.36	30	99.17	94.58	30	90.56	81.25	30	93.33	90.0	40	93.56	87.88
	B	30	98.17	95.66	30	98.75	94.75	40	87.50	78.66	30	93.58	88.58	40	90.87	84.73
	C	30	97.74	93.45	30	98.63	92.50	30	83.27	75.23	30	90.89	86.13	40	89.33	81.57

Table 8. Accuracy (%) of proposed method using HAC (Ward) for clustering and HMM, HDC for classification for KARD dataset, J refers to the number of joints.

J	E	AS1			AS2			AS3			ACTIONS			GESTURES		
		N	HMM	HDC	N	HMM	HDC	N	HMM	HDC	N	HMM	HDC	N	HMM	HDC
7	A	20	96.67	95.83	30	99.44	95.83	30	88.89	88.19	80	93.47	88.19	50	94.11	90.0
	B	20	97.17	96.33	30	99.08	95.91	30	87.08	85.41	80	92.50	83.75	50	93.27	86.53
	C	20	95.89	94.52	30	98.15	94.64	30	85.71	82.32	80	88.15	79.04	40	90.62	83.95
11	A	20	98.61	99.02	30	99.03	96.66	50	86.39	81.66	80	93.61	89.02	50	91.0	90.1
	B	20	97.67	98.16	30	98.58	95.41	30	83.92	80.16	80	91.33	86.33	50	89.27	85.4
	C	20	97.26	96.07	20	98.21	95.11	30	83.63	79.58	50	90.71	81.9	30	87.90	83.23
15	A	20	98.47	97.22	20	98.75	94.86	40	86.39	80.13	90	94.72	84.58	80	89.0	86.33
	B	20	97.92	97.16	30	98.75	94.41	30	85.25	85.0	80	92.17	85.91	30	87.47	86.2
	C	20	98.10	96.60	30	97.32	90.41	30	84.52	78.27	30	90.60	83.75	50	85.19	76.9

Table 9. Precision (%) and Recall (%) of state-of-the-art methods on CAD-60.

Method	Precision	Recall
Sung et. al. [18]	67.9	55.5
Koppuls et. al. [10]	80.8	71.4
Yang and Tian et. al. [21]	71.9	66.6
Gupta et. al. [5]	78.1	75.4
Gaglio et. al. [4]	77.3	76.7
Cippitelli et al. [3]	93.9	93.5
Proposed using K-means and HMM	**70.67**	**75.41**
Proposed using K-means and HDC	58.4	58.4
Proposed using HAC (Complete) and HMM	**84.66**	**83.75**
Proposed using HAC (Complete) and HDC	66.73	64.13
Proposed using HAC (Ward) and HMM	**92.25**	**90.83**
Proposed using HAC (Ward) and HDC	64.6	64.0

Table 10. Precision (%), Recall (%) and F-score (%) of proposed methods using K-means, HAC for clustering and HMM, HDC for classification in the five environments of CAD-60.

Location	Activity	K-means N	K-HMM Precision	K-HMM Recall	K-HMM F-score	K-HDC Precision	K-HDC Recall	K-HDC F-score	HAC(C) N	C-HMM Precision	C-HMM Recall	C-HMM F-score	C-HDC Precision	C-HDC Recall	C-HDC F-score	HAC(W) N	W-HMM Precision	W-HMM Recall	W-HMM F-score	W-HDC Precision	W-HDC Recall	W-HDC F-score
Bathroom	brushing teeth	20	80.0	100.0	88.0	100.0	70.0	82.0	20	100.0	100.0	100.0	83.0	100.0	91.0	10	100.0	100.0	100.0	100.0	100.0	100.0
	wearing contact lens		100.0	100.0	100.0	100.0	100.0	100.0		100.0	100.0	100.0	70.0	100.0	82.0		100.0	100.0	100.0	100.0	80.0	89.0
	rinsing mouth		100.0	75.0	85.71	77.0	100.0	87.0		100.0	100.0	100.0	91.0	100.0	95.0		100.0	100.0	100.0	83.0	100.0	91.0
	Average		93.33	91.66	92.49	92.0	90.0	90.0		100.0	100.0	100.0	91.0	90.0	90.0		100.0	100.0	100.0	94.0	93.0	93.0
Bedroom	drinking water	20	50.0	100.0	66.66	25.0	20.0	22.0	10	100.0	100.0	100.0	29.0	20.0	24.0	10	80.0	100.0	88.88	44.0	40.0	42.0
	opening container		100.0	100.0	100.0	40.0	40.0	40.0		100.0	100.0	100.0	38.0	50.0	43.0		100.0	100.0	100.0	43.0	30.0	35.0
	talking on phone		0.0	0.0	0.0	75.0	90.0	82.0		100.0	100.0	100.0	100.0	100.0	100.0		100.0	75.0	85.71	71.0	100.0	83.0
	Average		50.0	66.66	55.55	47.0	50.0	48.0		100.0	100.0	100.0	55.66	56.66	55.66		93.33	91.66	91.53	53.0	57.0	54.0
Kitchen	cooking(chopping)	30	100.0	100.0	100.0	100.0	78.0	88.0	30	66.66	100.0	80.0	42.0	80.0	55.0	20	66.66	100.0	80.0	70.0	70.0	70.0
	drinking water		100.0	50.0	66.66	100.0	67.0	80.0		100.0	50.0	66.66	100.0	50.0	67.0		100.0	100.0	100.0	60.0	60.0	60.0
	opening container		66.66	100.0	80.0	100.0	86.0	92.0		66.66	100.0	80.0	91.0	100.0	95.0		100.0	100.0	100.0	82.0	90.0	86.0
	cooking(stirring)		100.0	100.0	100.0	78.0	70.0	74.0		100.0	50.0	66.66	40.0	20.0	27.0		100.0	50.0	66.66	78.0	70.0	74.0
	Average		91.66	87.5	89.53	84.0	82.0	82.0		83.33	75.0	73.33	68.0	62.0	61.0		91.66	87.5	86.665	72.0	72.0	72.0
Living room	drinking water	20	36.36	100.0	53.33	31.0	40.0	35.0	10	33.33	50.0	40.0	30.0	30.0	30.0	10	75.0	75.0	75.0	43.0	60.0	50.0
	talking on phone		100.0	50.0	66.66	30.0	30.0	30.0		0.0	0.0	0.0	33.0	30.0	32.0		80.0	100.0	88.88	38.0	30.0	33.0
	talking on couch		0.0	0.0	0.0	33.0	30.0	32.0		100.0	100.0	100.0	82.0	90.0	86.0		100.0	75.0	85.71	73.0	80.0	76.0
	relaxing on couch		100.0	75.0	85.71	50.0	40.0	44.0		100.0	100.0	100.0	100.0	100.0	100.0		100.0	100.0	100.0	71.0	50.0	59.0
	Average		59.09	56.25	51.425	36.0	35.0	35.0		58.33	62.5	60.0	61.0	62.0	62.0		88.75	87.50	87.39	56.0	55.0	55.0
Office	working on computer	20	80.0	100.0	88.88	33.0	20.0	25.0	20	100.0	100.0	100.0	40.0	40.0	57.0	10	100.0	100.0	100.0	57.0	40.0	47.0
	drinking water		57.14	100	72.72	23.0	30.0	26.0		60.0	75.0	66.66	20.0	20.0	20.0		75.0	75.0	75.0	21.0	30.0	25.0
	talking on phone		0.0	0.0	0.0	12.0	10.0	11.0		66.66	50.0	57.14	29.0	40.0	33.0		75.0	75.0	75.0	22.0	21.0	21.0
	writing on whiteboard		100.0	100.0	100.0	62.0	80.0	70.0		100.0	100.0	100.0	83.0	100.0	91.0		100.0	100.0	100.0	90.0	90.0	90.0
	Average		59.28	75.0	65.40	33.0	35.0	33.0		81.66	81.25	80.95	58.0	50.0	50.0		87.5	87.5	87.5	48.0	45.0	46.0
Overall Average			70.67	75.41	70.87	58.4	58.4	57.6		84.66	83.75	82.85	66.73	64.13	63.73		92.25	90.83	90.62	64.6	64.0	64.0

5   Conclusion

In this work, we have proposed a text like classification approach for human action recognition (skeletal data) which uses HAC (Ward) algorithm for the extraction of key poses and construction of action sequences followed by the classification using HMM. This work also provides a comparative study of different clustering algorithms (HAC, K-means) and classification methods (HMM, HDC and LSTM) for human action recognition using skeletal data. The experiments are performed on three publicly available datasets namely UPCV, KARD and CAD-60. For all the datasets HMM provides better classification results than HDC because in HDC only a pose tri-gram is considered for feature construction. Moreover, HAC (Ward) performs better than K-means because in

K-means the cluster heads are not from amongst the poses to be clustered, thus may lead to misclassification and HAC works better than K-means for clusters with non-spherical shapes. LSTM doesn't provide comparable results for UPCV and KARD datasets. The number of cluster-heads (N) corresponding to the maximum classification accuracy varies with the dataset due to the difference in the number of actions. We intend to build a motion retrieval system for the same as a part of future work where the skeletal sequences are given as an input query and the system returns a ranked list of similar skeletal sequences to the user.

References

1. Barmpoutis, P., Stathaki, T., Camarinopoulos, S.: Skeleton-based human action recognition through third-order tensor representation and spatio-temporal analysis. Inventions **4**, 9 (2019)
2. Chin, Z.H., Ng, H., Yap, T.T.V., Tong, H.L., Ho, C.C., Goh, V.T.: Daily activities classification on human motion primitives detection dataset. Computational Science and Technology. LNEE, vol. 481, pp. 117–125. Springer, Singapore (2019). https://doi.org/10.1007/978-981-13-2622-6_12
3. Cippitelli, E., Gasparrini, S., Gambi, E., Spinsante, S.: A human activity recognition system using skeleton data from RGBD sensors. Comput. Intell. Neurosci. (2016)
4. Gaglio, S., Re, G.L., Morana, M.: Human activity recognition process using 3-D posture data. IEEE Trans. Hum.-Mach. Syst. **45**, 586–597 (2015)
5. Gupta, R., Chia, A.Y.S., Rajan, D.: Human activities recognition using depth images. In: Proceedings of the 21st ACM International Conference on Multimedia, MM 2013 (2013)
6. Kanerva, P.: Hyperdimensional computing: an introduction to computing in distributed representation with high-dimensional random vectors. Cogn. Comput. **1**, 139–159 (2009)
7. Kim, T.S., Reiter, A.: Interpretable 3D human action analysis with temporal convolutional networks. In: 2017 IEEE Conference on Computer Vision and Pattern Recognition Workshops (CVPRW) (2017)
8. Kim, Y., Imani, M., Rosing, T.S.: Efficient human activity recognition using hyperdimensional computing. In: Proceedings of the 8th International Conference on the Internet of Things (2018)
9. Konstantinidis, D., Dimitropoulos, K., Daras, P.: Skeleton-based action recognition based on deep learning and Grassmannian pyramids. In: 2018 26th European Signal Processing Conference (EUSIPCO) (2018)
10. Koppula, H.S., Gupta, R., Saxena, A.: Learning human activities and object affordances from RGB-D videos. CoRR (2012)
11. Lan, R., Sun, H., Zhu, M.: Text-like motion representation for human motion retrieval. In: Yang, J., Fang, F., Sun, C. (eds.) IScIDE 2012. LNCS, vol. 7751, pp. 72–81. Springer, Heidelberg (2013). https://doi.org/10.1007/978-3-642-36669-7_10
12. Liu, X., He, G., Peng, S., Cheung, Y., Tang, Y.Y.: Efficient human motion retrieval via temporal adjacent bag of words and discriminative neighborhood preserving dictionary learning. IEEE Trans. Hum.-Mach. Syst. **47**, 763–776 (2017)

13. Mahasseni, B., Todorovic, S.: Regularizing long short term memory with 3D human-skeleton sequences for action recognition. In: Proceedings of the IEEE Conference on Computer Vision and Pattern Recognition, pp. 3054–3062 (2016)
14. Mokari, M., Mohammadzade, H., Ghojogh, B.: Recognizing involuntary actions from 3D skeleton data using body states. CoRR (2017)
15. Pantuwong, N., Takahara, K., Sugimoto, M.: A rapid motion retrieval technique using simple and discrete representation of motion data. In: 2015 7th International Conference on Information Technology and Electrical Engineering (ICITEE) (2015)
16. Patel, A., Shah, P.: IIITV@INLI-2018: Hyperdimensional Computing for Indian Native Language Identification. INLI track at Forum for Information Retrieval Evaluation DAIICT, Gandhinagar (2018)
17. Stamp, M.: A revealing introduction to hidden Markov models (2004)
18. Sung, J., Ponce, C., Selman, B., Saxena, A.: Unstructured human activity detection from RGBD images. In: Proceedings - IEEE International Conference on Robotics and Automation, July 2011
19. Theodorakopoulos, I., Kastaniotis, D., Economou, G., Fotopoulos, S.: Pose-based human action recognition via sparse representation in dissimilarity space. J. Vis. Commun. Image Represent. **25**(1), 12–23 (2014)
20. Yan, S., Xiong, Y., Lin, D.: Spatial temporal graph convolutional networks for skeleton-based action recognition. In: Thirty-Second AAAI Conference on Artificial Intelligence (2018)
21. Yang, X., Tian, Y.: Effective 3D action recognition using EigenJoints. J. Vis. Commun. Image Represent. **25**, 2–11 (2014)
22. Zhang, P., Lan, C., Xing, J., Zeng, W., Xue, J., Zheng, N.: View adaptive neural networks for high performance skeleton-based human action recognition. IEEE Trans. Pattern Anal. Mach. Intell. **41**, 1963–1978 (2019)

Background Subtraction Based
on Encoder-Decoder Structured CNN

Jingming Wang and Kwok Leung Chan[(⊠)]

Department of Electrical Engineering, City University of Hong Kong, Kowloon, Hong Kong
jingmwang9-c@ad.cityu.edu.hk, itklchan@cityu.edu.hk

Abstract. Background subtraction is commonly adopted for detecting moving objects in image sequence. It is an important and fundamental computer vision task and has a wide range of applications. We propose a background subtraction framework with deep learning model. Pixels are labeled as background or foreground by an Encoder-Decoder Structured Convolutional Neural Network (CNN). The encoder part produces a high-level feature vector. Then, the decoder part uses the feature vector to generate a binary segmentation map, which can be used to identify moving objects. The background model is generated from the image sequence. Each frame of the image sequence and the background model are input to the CNN for pixel classification. Background subtraction result can be erroneous as videos may be captured in various complex scenes. The background model must be updated. Therefore, we propose a feedback scheme to perform the pixelwise background model updating. For the training of the CNN, the input images and the corresponding ground truths are drawn from the benchmark dataset Change Detection 2014. The results show that our proposed architecture outperforms many well-known traditional and deep learning background subtraction algorithms.

Keywords: Foreground detection · Background subtraction · Deep learning · Background modeling · Convolutional neural network

1 Introduction

Background subtraction is one commonly adopted method for moving objects detection in video. It is one of the most important and fundamental computer vision tasks. Its result can be further used in other areas such as object classification, part segmentation, motion tracking, action recognition, etc. A vision-based system capable of inferring meaning from video can have many applications such as video surveillance [1], multimedia applications [2], human motion recognition [3], anomaly event detection [4], video indexing and retrieval [5], etc. Background subtraction is generally formulated as a pixelwise binary classification problem. Pixels in each image frame are identified as background if they are similar to the background model. The pixels that are not similar to the background model are labeled as foreground. A background subtraction framework often contains background modeling, joint background/foreground classification, and background model updating. Literature survey on background subtraction techniques can be found in [6, 7].

© Springer Nature Switzerland AG 2020
S. Palaiahnakote et al. (Eds.): ACPR 2019, LNCS 12047, pp. 351–361, 2020.
https://doi.org/10.1007/978-3-030-41299-9_27

In reality, most background subtraction methods proposed use rule-based model with hand-crafted features. The background scene can be represented by statistical model [8, 9]. Elgammal et al. [10] proposed an algorithm for estimating the pdf of the background directly from previous pixels using kernel estimator. Recently, more researches employ background model generated using real observed pixel values. Barnich et al. [11] proposed a fast sample-based background subtraction algorithm. Some algorithms model background characteristics on larger domain. In [12], the background is modeled by dynamic texture features obtained in the center and surrounding windows. Heikkilä and Pietikäinen [13] proposed to model the background of a pixel by local binary pattern histograms estimated around that pixel. Liao et al. [14] proposed the scale invariant local ternary pattern which can tackle illumination variations. State-of-the-art algorithms may employ foreground modeling [15].

Recently computer vision has advanced rapidly through the use of deep learning. In contrast to rule-based algorithm, deep learning belongs to machine learning based on learning data representations. It is inspired by the way nervous system processes and communicates information. With multiple layers of processing, data are transformed and abstract features are extracted. For image and video analysis, one popular deep learning model is convolutional neural network (CNN). Wang et al. [16] proposed a multi-resolution CNN with a cascaded architecture for object segmentation. In [17], object segmentation is formulated as background subtraction and is performed by an encoder-decoder network. In [18], a single CNN is used for background subtraction that can handle various scenes without the need for re-training.

Moving object detection can be very difficult under various complex circumstances – camouflage, illumination changes, background motions, intermittent object motion, shadows, camera jitter, etc. Camouflage can produce false negative error. Background motions (e.g. waving trees and rippling water) and shadows can produce false positive error. Object detection can be improved via background model updating. Many background subtraction methods update parameters of matched background model with a fixed learning factor. In [11], a random policy is employed for updating the background model at the pixel location and its neighbor. Van Droogenbroeck and Paquot [19] introduced some modifications, such as inhibiting the update of neighboring background model across the background-foreground border, for improving ViBe. The detected foreground often suffers from distorted shape and holes. Kim et al. [20] proposed a PID tracking control system for foreground segmentation refinement.

In this paper, we propose a background subtraction framework with deep learning model. Pixels are labeled as background or foreground by an Encoder-Decoder Structured CNN. Our first contribution is to adopt a background model generation method in the framework. While some background subtraction algorithms extract one image frame as the background model, our adopted a method generates background model by analyzing a short image sequence of the original video. It is the first time that such background model generation algorithm is incorporated into a deep learning background subtraction framework. Each frame of the image sequence and the background model are input to the CNN for pixel classification. It is important to maintain the validity of the background model in order to perform background subtraction for long duration.

Therefore, our second contribution is to propose a feedback scheme to perform the pixelwise background model updating. Our results show that the deep learning framework together with the feedback scheme is effective in producing very accurate background subtraction result that outperforms many background subtraction algorithms.

2 Background Subtraction Framework

The background subtraction framework is shown in Fig. 1. The background model is generated from the original image sequence. We adopted the background frame generation method LaBGen-P [21]. Each frame of the image sequence and the background model are input to the CNN for pixel classification. We adopted ResNet [22] as our CNN architecture. For the training of the neural network, the input video images and ground truths are drawn from the benchmark dataset. We propose a feedback scheme to perform the pixelwise background model updating via continuous analyzing the background subtraction results.

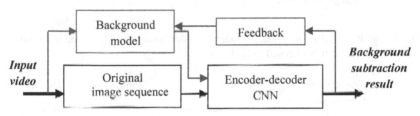

Fig. 1. Overview of background subtraction framework.

2.1 Background Modeling

It is essential to generate an accurate background model for background subtraction. However, the quality of background model will be bad when the background scene is not static. Noise can be prominent in the background/foreground segregation when simple background modeling method such as median filtering is used. In order to solve this problem, we adopted the LaBGen-P algorithm, which generates background frame from the original image sequence based on color information and temporal features. LaBGen-P can generate accurate background frame in various types of dynamic background scenes.

2.2 Encoder-Decoder Structured CNN

Deep residual network (ResNet) was the winner of ImageNet Classification Challenge in 2015. Instead of directly learning some "mapping functions", the author reformulated the layers as explicitly learning residual functions with reference to the inputs. Including these "residual blocks" with skip connections makes it easier to optimize much deeper CNN than before (as deep as hundreds of layers). Recently, GPU acceleration with the support of CUDA software makes training deep learning models more and more efficient. Furthermore, it has been a trend for large organizations or companies to turn their deep learning frameworks into open source projects. Implementing and validating a new model has been a lot more straightforward.

In this paper, we use Python-based Tensorflow, which is by far the most popular deep learning framework. As for the hardware platform, since the Tensorflow support GPU acceleration, we use Tesla K80 12G on Google Server for training. In order to train our CNN, we choose part of the data from the CDnet 2014 dataset [23], which includes video frames and its corresponding ground truth in different challenging scenes. Since we use Tensorflow to implement our model, we need to transform the data into a binary file called tfrecords, which can improve data processing efficiency on a large scale. As for the choice of training set, we use the video frames which do not challenge the background image so that the CNN can better learn the representative features, so the "Pan-tilt-zoom Camera" subset is excluded. Therefore, we select the training set and test set from the rest 10 categories.

As shown in Table 1, the whole convolutional neural network uses the encoder-decoder structure. This architecture is inspired by the work of Noh et al. [24], which was used for semantic segmentation.

Table 1. The architecture detail of our Encoder-Decoder Structured CNN, and the detail of the pre-trained Resnet_50 can be checked in [22].

	Filter size/Pooling window size	Stride	Input size	Output size
Pre-conv	1 * 1 * 6 * 3	1	321 * 321 * 6	321 * 321 * 3
Resnet_50	–	–	321 * 321 * 3	21 * 21 * 2048
3D avg_pool	1 * 1 * 48	40	21 * 21 * 2048	21 * 21 * 51
Deconv_1	3 * 3 * 32 * 51	2	21 * 21 * 51	43 * 43 * 32
3D max_pool	3 * 3 * 2	2 for depth; 1 for height and width	43 * 43 * 32	41 * 41 * 16
Deconv_2	3 * 3 * 8 * 16	2	41 * 41 * 16	83 * 83 * 8
2D max_pool	3 * 3	1	83 * 83 * 8	81 * 81 * 8
Deconv_3	3 * 3 * 4 * 8	2	81 * 81 * 8	163 * 163 * 4
2D max_pool	3 * 3	1	163 * 163 * 4	161 * 161 * 4
Deconv_4	3 * 3 * 1 * 4	2	161 * 161 * 4	323 * 323 * 1
2D max_pool	3 * 3	1	323 * 323 * 1	321 * 321 * 1
conv	1 * 1 * 1 * 1	1	321 * 321 * 1	321 * 321 * 1

For the encoder part, it consists of 1 pre-convolutional layer and the residual network which has 50 layers. Firstly, since the residual network only takes input that has 3 channels, convolutional operation with 1 * 1 filters are used to map the input data cube into a 3-channel feature map. Secondly, the 3-channel feature map runs through pre-trained fully-convolutional ResNet for feature extraction. Then we add three deconvolutional layers to up-sample the feature map and then use one convolution layer to resize the final output feature map to 321 * 321 * 1. We also use the pooling layer to reduce the spatial size of the representation based on our needs step by step, and it also controls the overfitting according to the past experience. For the output layer, we use sigmoid function (Eq. (1)) as the activation function to do the normalization, and then apply the median filter with 3 * 3 kernel. The median filter not only eliminates lots of noise, but also produces a smooth foreground.

$$S(x) = \frac{1}{1 + e^{-x}} \tag{1}$$

Next, in order to map the pixel to either 0 or 1, we use the threshold function as the following (Eq. (2)), where R is the threshold value. Afterwards, we use the output of the threshold function to multiply 255, and this become the final binary foreground mask.

$$g(z; R) = \begin{cases} 1 & if \ z > R \\ 0 & otherwise \end{cases} \tag{2}$$

Finally, the output the whole CNN can be used to compute the loss, and then the coefficient inside can be updated iteratively, which is the so-called CNN training.

As for the CNN training, we first run through the whole dataset to generate all the background images. These background images together with the origin frame images and corresponding ground truths will be used to train the model. The whole CNN model we illustrated above is based on Tensorflow, and the pre-trained ResNet model can be implemented through using TF-slim, which is the high level Application Program Interface (API) of the Tensorflow.

We train the network with mini batches of size 16 via Adam optimizer. The training runs for 10,000 iterations, for the first 500 steps, the learning rate is 1e−3. For step from 500 to 9,000, the learning rate is reduced to 0.5e−3. Then the learning rate is reduced to 1e−4 and kept unchanged until the end of the training. For the lost function, we choose the classical binary cross-entropy loss function (Eq. 3) so as to update the coefficient

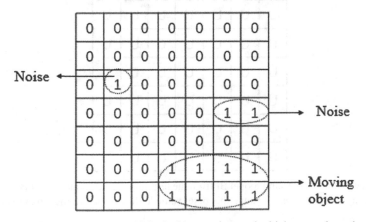

Fig. 2. Test method to determine which pixels are noises and which are real moving objects.

inside the neural network more efficiently. In the process of training and testing, pixels outside the Region of Interest (ROI) and boundaries of foreground objects are marked by gray value in the ground truth segmentation result. These gray pixels are ignored while training and testing.

$$C = -\frac{1}{n} \sum_x [y \ln a + (1 - y)\ln(1 - a)] \tag{3}$$

2.3 Background Model Updating

The feedback mechanism is inspired by SuBSENSE [15]. When we get a binary result, the segmentation map informs which parts of the frame are changed. We will get the binary result at time t and $t - 1$, which means the results of current frame and previous frame. Then we do the XOR calculation between these two binary result images and get the compute result Br_{xt} by

$$Br_{xt} = Br_t \otimes Br_{t-1} \tag{4}$$

The positions on the Br_{xt} whose pixel values are 1 mean that these positions get changed compared with previous time. The positions whose pixel values are 0 mean that these positions get unchanged. Meanwhile the pixels get changed can be the real moving object, or it can be the noise point. Therefore, we need to classify which positions changed are noises and which positions are moving objects. Here if the corresponding pixel at the segmentation map is 1, we will calculate the sum of surrounding pixels in a window with size 3 * 3 pixel. As shown in Fig. 2, if the sum of the pixels in the window is smaller or equal to 3, we consider that position is a noise point. Then we will change the pixel value at that position on the background model to the pixel value at that position of

U	U	U	U	U	U	U
U	U	U	U	U	U	U
U	C	U	U	U	U	U
U	U	U	U	U	C	C
U	U	U	U	U	U	U
U	U	U	U	U	U	U
U	U	U	U	U	U	U

Fig. 3. Generate the background model B_t from the last frame background model B_{t-1}, which U means unchanged and C means change this pixel value to the original frame's pixel value at that position.

this original image frame. Other pixels of the background model will stay unchanged. Then the new background model image for the next frame is generated.

Figure 3 shows the background model B_t which is updated based on the analysis of the background subtraction results in Fig. 2. Figure 4 shows some examples of background model updating. The noises in the first background model are marked with a red ellipse. It can be seen that the noises are mostly corrected in the updated background model.

Fig. 4. Background models – original image frames (first column), the first background model (second column), the updated background model (last column). (Color figure online)

3   Result

As we mentioned before, the background subtraction benchmark dataset called Change Detection 2014 (CDnet 2014) is adopted. It is used not only for the training of our proposed CNN model, but also for performance evaluation. The dataset, as shown in Table 2, contains 11 video categories with 4 to 6 video sequences in each category, which is a collection of the video data from the real scenarios. Those videos include almost all of the challenges like illumination change, shadow and dynamic background. These complications always happen in the real world. Furthermore, each video also comes with large amount of ground truth images, which are annotated by human.

Table 2. All of the categories and its video scenes in the Change Detection 2014 dataset.

CDnet 2014 dataset categories	Video scenes
Bad Weather	"blizzard" "skating" "snowFall" "wetSnow"
Baseline	"highway" "office" "pedestrians" "PETS2006"
Camera Jitter	"badminton" "boulevard" "sidewalk" "traffic"
Dynamic Background	"boats" "canoe" "fountain01" "fountain02" "overpass" "fall"
Intermittent Object Motion	"abandonedBox" "parking" "streetLight" "sofa" "tramstop" "winterDriveway"
Low Framerate	"port_0_17fps" "tramCrossroad_lfps" "tunnelExit_0_35fps" "turnpike_0_5fps"
Night Videos	"bridgeEntry" "busyBoulvard" "fluidHighway" " streetCornerAtNight" "tramStation"
PTZ	"continuousPan" "intermittentPan" "twoPositionPTZCam" "zoomlnZoomOut"
Shadow	"backdoor" "bungalows" "busStation" "cubicle" "peoplelnShade" "copyMachine"
Thermal	"corridor" "library" "diningRoom" "lakeSide" "park"
Turbulence	"turbulence0" "turbulence 1" "turbulence2" "turbulence3"

Table 3. The comparison result of F-measure for different background subtraction algorithms in various video scenes of CDnet 2014 dataset (the number in bold means it is the highest in each category).

Category	GMM	PAWS	FTSG	SuBSENSE	Shared Model	Our method
Baseline	0.852	0.938	0.934	0.963	0.969	**0.982**
Dynamic background	0.7	0.942	0.951	0.944	0.936	**0.966**
Camera jitter	0.571	0.844	0.719	0.815	0.764	**0.906**
shadow	0.749	0.873	0.871	**0.899**	0.87	0.860
Intermittent object motion	0.592	0.819	0.617	0.657	0.773	**0.825**
Thermal	0.677	0.615	0.369	**0.817**	0.753	0.781
Bad weather	0.745	0.774	0.741	0.862	0.723	**0.899**
Low frame rate	0.591	**0.906**	0.658	0.645	0.901	0.885
Night videos	0.451	0.626	0.739	0.559	0.763	**0.824**

In order to do the performance evaluation and compare the result with the conventional background subtraction algorithm, we need to choose the test set from the 10 categories of the CDnet 2014 dataset. In this case, we focus on the scenario that camera is not moving significantly, so we will not use category "PTZ Camera" for performance

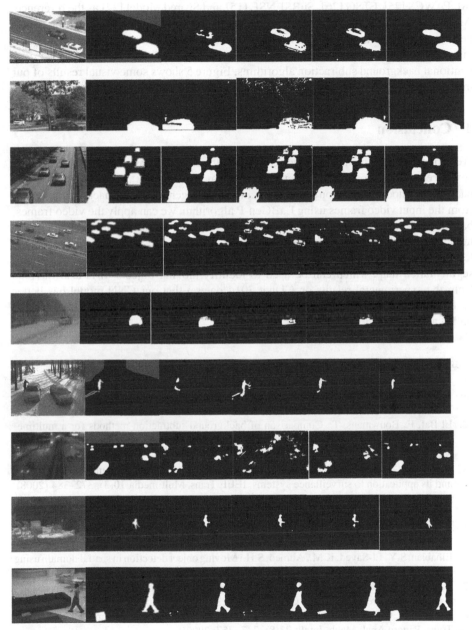

Fig. 5. Background subtraction results – original image frames (first column), ground truths (second column), results obtained by SuBSENSE (third column), results obtained by GMM (fourth column), results obtained by FTSG (fifth column), results obtained by our method (last column).

evaluation. Besides, the video scenes which have been used for training set of our CNN model cannot be applied for performance test.

We choose five typical conventional background subtraction algorithms called GMM [8], PAWCS [25], FTSG [26], SuBSENSE [15] and Shared Model [27] as the reference. As we can see from Table 3, it is apparent that our proposed CNN model has better performance than other conventional approaches in most of the video scenes. Besides, our proposed method also has the best overall performance comparing with other conventional background subtraction algorithms. Figure 5 shows some visual results of our method and other background subtraction algorithms.

4   Conclusion

We proposed a new method for background subtraction task based on encoder-decoder structured convolutional neural network. This method mainly consists of two parts, background model generation and CNN training. After generating the background model from the input video frames using LaBGen-P algorithm, we can apply the video frames, background model and its corresponding ground truth for the training of our CNN model. As for the encoder-decoder structured CNN, the encoder part produces a high-level feature vector, and then the decoder uses the feature vector to generate binary segmentation map, which can be used to identify moving objects. It is also obvious from the experimental results on CDnet 2014 dataset that our CNN model has better performance than other conventional background subtraction algorithms such as GMM, FSTG and Shared Model.

References

1. Bouwmans, T.: Background subtraction for visual surveillance: a fuzzy approach. In: Handbook on Soft Computing for Video Surveillance. Taylor and Francis Group (2012)
2. El Baf, F., Bouwmans, T.: Comparison of background subtraction methods for a multimedia learning space. In: Proceedings of International Conference on Signal Processing and Multimedia (2007)
3. Hsieh, J.-W., Hsu, Y.-T., Liao, H.-Y.M., Chen, C.-C.: Video-based human movement analysis and its application to surveillance systems. IEEE Trans. Multimedia 10(3), 372–384 (2008)
4. Lu, W., Tan, Y.-P.: A vision-based approach to early detection of drowning incidents in swimming pools. IEEE Trans. Circuits Syst. Video Technol. 14(2), 159–178 (2004)
5. Visser, R., Sebe, N., Bakker, E.: Object recognition for video retrieval. In: Proceedings of International Conference on Image and Video Retrieval, pp. 250–259 (2002)
6. Elhabian, S.Y., El-Sayed, K.M., Ahmed, S.H.: Moving object detection in spatial domain using background removal techniques – state-of-art. Recent Pat. Comput. Sci. 1, 32–54 (2008)
7. Bouwmans, T.: Recent advanced statistical background modeling for foreground detection - a systematic survey. Recent Pat. Comput. Sci. 4(3), 147–176 (2011)
8. Stauffer, C., Grimson, W.E.L.: Learning patterns of activity using real-time tracking. IEEE Trans. Pattern Anal. Mach. Intell. 22(8), 747–757 (2000)
9. Zivkovic, Z.: Improved adaptive Gaussian mixture model for background subtraction. In: Proceedings of International Conference on Pattern Recognition, pp. 28–31 (2004)

10. Elgammal, A., Duraiswami, R., Harwood, D., Davis, L.S.: Background and foreground modeling using nonparametric kernel density estimation for visual surveillance. Proc. IEEE **90**(7), 1151–1163 (2002)
11. Barnich, O., Van Droogenbroeck, M.: ViBE: a powerful random technique to estimate the background in video sequences. In: Proceedings of International Conference Acoustics, Speech and Signal Processing, pp. 945–948 (2009)
12. Mahadevan, V., Vasconcelos, N.: Spatiotemporal saliency in dynamic scenes. IEEE Trans. Pattern Anal. Mach. Intell. **32**(1), 171–177 (2010)
13. Heikkilä, M., Pietikäinen, M.: A texture-based method for modeling the background and detecting moving objects. IEEE Trans. Pattern Anal. Mach. Intell. **28**(4), 657–662 (2006)
14. Liao, S., Zhao, G., Kellokumpu, V., Pietikäinen, M., Li, S.Z.: Modeling pixel process with scale invariant local patterns for background subtraction in complex scenes. In: Proceedings of IEEE Conference on Computer Vision and Pattern Recognition, pp. 1301–1306 (2010)
15. St-Charles, P.-L., Bilodeau, G.-A., Bergevin, R.: SuBSENSE: a universal change detection method with local adaptive sensitivity. IEEE Trans. Image Process. **24**(1), 359–373 (2015)
16. Wang, Y., Luo, Z., Jodoin, P.-M.: Interactive deep learning method for segmenting moving objects. Pattern Recogn. Lett. **96**, 66–75 (2017)
17. Lim, L.A., Keles, H.Y.: Foreground segmentation using a triplet convolutional neural network for multiscale feature encoding. arXiv:1801.02225 [cs.CV] (2018)
18. Babaee, M., Dinh, D.T., Rigoll, G.: A deep convolutional neural network for video sequence background subtraction. Pattern Recogn. **76**, 635–649 (2018)
19. Van Droogenbroeck, M., Paquot, O.: Background subtraction: experiments and improvements for ViBE. In: Proceedings of IEEE Workshop on Change Detection at IEEE Conference on Computer Vision and Pattern Recognition, pp. 32–37 (2012)
20. Kim, S.W., Yun, K., Yi, K.M., Kim, S.J., Choi, J.Y.: Detection of moving objects with a moving camera using non-panoramic background model. Mach. Vis. Appl. **24**, 1015–1028 (2013)
21. Lauguard, B., Piérard, S., Van Droogenbroeck, M.: LaBGen-P: a pixel-level stationary background generation method based on LaBGen. In: Proceedings of IEEE International Conference on Pattern Recognition, pp. 107–113 (2016)
22. He, K., Zhang, X., Ren, S., Sun, J.: Deep residual learning for image recognition. In: Proceedings of the IEEE Conference on Computer Vision and Pattern Recognition, pp. 770–778 (2016)
23. Wang, Y., Jodoin, P.-M., Porikli, F., Konrad, J., Benezeth, Y., Ishwar, P.: CDnet 2014: an expanded change detection benchmark dataset. In: Proceedings of IEEE Workshop on Change Detection at CVPR-2014, pp. 387–394 (2014)
24. Noh, H., Hong, S., Han, B.: Learning deconvolution network for semantic segmentation. In: Proceedings of International Conference on Computer Vision (2015)
25. St-Charles, P.-L., Bilodeau, G.-A., Bergevin, R.: A self-adjusting approach to change detection based on background word consensus. In: Proceedings of IEEE Winter Conference on Applications of Computer Vision, pp. 990–997 (2015)
26. Wang, R., Bunyak, F., Seetharaman, G., Palaniappan, K.: Static and moving object detection using flux tensor with split gaussian models. In: Proceedings of IEEE Conference on Computer Vision and Pattern Recognition Workshops (2014)
27. Chen, Y., Wang, J., Lu, H.: Learning sharable models for robust background subtraction. In: Proceedings of IEEE International Conference on Multimedia and Expo (2015)

Multi Facet Face Construction

Hamed Alqahtani[1(✉)] ⓘ and Manolya Kavakli-Thorne[2]

[1] King Khalid University, Abha, Saudi Arabia
hsqahtani@kku.edu.au
[2] Macquarie University, Sydney, Australia

Abstract. To generate a multi-faceted view, from a single image has always been a challenging problem for decades. Recent developments in technology enable us to tackle this problem effectively. Previously, Several Generative Adversarial Network (GAN) based models have been used to deal with this problem as linear GAN, linear framework, a generator (generally encoder-decoder), followed by the discriminator. Such structures helped to some extent, but are not powerful enough to tackle this problem effectively.

In this paper, we propose a GAN based dual-architecture model called DUO-GAN. In the proposed model, we add a second pathway in addition to the linear framework of GAN with the aim of better learning of the embedding space. In this model, we propose two learning paths, which compete with each other in a parameter-sharing manner. Furthermore, the proposed two-pathway framework primarily trains multiple sub-models, which combine to give realistic results. The experimental results of DUO-GAN outperform state of the art models in the field.

Keywords: GAN · Multi-faceted face construction · Neural network · Machine learning

1 Introduction

Constructing a multi-faceted image from a single image is a well-investigated problem and has several real-life applications. Essential applications of creating a multi-posed image from a single image are its use for identification purposes, detecting malicious, criminals in public, capturing the identity of people in general etc. Constructing multi-posed image is a challenging task comprising of imagining the objects might looking like, constructed from another pose [3]. It requires the construction of unknown possibilities and hence requires a very rich embedding space so that the constructed view of the object should have the same identity and should be relevant in context.

Several research efforts have been made to address this problem using different models like synthesis based models, and data-based models [16,19]. These GAN based models consist of linear framework and encoder-decoder followed by Discriminator to address this issue. Here, the main purpose of the encoder(E) is to map the input images to the latent space(Z), which are fed into the decoder(G) after some manipulation for generating multi-faceted images [1,2].

© Springer Nature Switzerland AG 2020
S. Palaiahnakote et al. (Eds.): ACPR 2019, LNCS 12047, pp. 362–370, 2020.
https://doi.org/10.1007/978-3-030-41299-9_28

But, it is found empirically that the linear framework isn't powerful enough to learn appropriate embedding space. The linear framework generates an output for creating a multi-faceted image isn't clear enough and doesn't preserve identity across various posed images. Learning incomplete embedding space leads to incomplete generalization on test images or unseen images. The primary reason of incapability of linear frameworks in learning complete presentation is that during training the encoder part of G only sees a fraction of Z and while testing, very likely model come across samples corresponding to unseen embedding space. This results in poor generalization.

In order to tackle this problem, Tian et al. [14] proposed a dual-pathway architecture, termed as Complete-Representation (CR-GAN). Unlike linear framework, the authors of CR-GAN have used dual pathway architecture. Besides the typical re-construction path, they introduced another generation path for constructing multi-faceted images from embeddings, randomly sampled from Z. In the proposed architecture, they used the same G, which aids the learning of E and discriminator (D). In their proposed model, E is forced to be an inverse of G, which theoretically should yield complete representations that should span the entire Z space.

However, the experiments conducted in this work demonstrate that one encoder is not convincing to span the entire Z space. Therefore, in order to address this challenge, we propose DUO-GAN with *dual encoder* to learn complete representation for a multi-facet generation. The primary purpose is to distribute the task of spanning the entire Z space, across two encoders instead of one as proposed in the previous work. We empirically demonstrate that dual encoder architecture produces many realistic results in comparison to prior work in this field.

2 Related Work

Several researchers contributed to constructing a multi-faceted image from a single image. The significant work in this field is presented as follows.

Goodfellow et al. [5] first introduced GAN to learn models with generative ability via an adversarial process. In the proposed model, a two-player min-max game is played between generator (G) and discriminator (D). Competing with each other in the game, both G and D tend to improve themselves. GAN has been used in various fields like image synthesis, super-resolution image generation etc. Every model proposed with the help of GAN manipulates constraints on Z and attempt to cover more and more embedding space for a better synthesis of images.

Hassner et al. [8] proposed a 3D face model in order to generate a frontal face for any subject. Sagonas et al. [13] used a statistical model for creating joint frontal face reconstruction, which is quite useful. The reported results were not very useful, as frontal face generation from a side view is a very challenging task. Because of occlusion and variation in spatial feature from side view face pictures.

Yan et al. [16] solved the problem of multi-pose generation to a certain level by using projection information by their Perspective Transformer Nets. Whereas, Yang et al. [17] proposed a model which incrementally rotated faces in fixed yaw angles. For generating multi-poses, Hinto et al. [9] tried generating images with view variance by using auto-encoder. Tian et al. [14] proposed dual pathway architecture CR-GAN for constructing multiple poses. However, all the above-mentioned system fail to construct realistic images in an unseen wild condition. In comparison, DUO-GAN spans embedding space in a much more exhaustive manner using it's multi-path architecture and produces higher-quality images than previously proposed models.

Preserving identity synchronously across images with numerous positions is a very active research area. Previously DR-GAN [15] attempted to solve this problem, by providing pose code along with image data, while training. Li et al. [12] attempted this challenge by using *Canonical Correlation Analysis* for comparing the difference between the sub-spaces of various poses. Tian et al. [14] tried solving this problem with dual pathway architecture. We propose dual encoder dual-pathway architecture, which results in a much better generation of multi-faceted images.

3   The Proposed Method

Most of the previous research on this field involves a linear network, *i.e.* an encoder-decoder generator network, followed by Discriminator network. As empirically found, such linear network is incapable of spanning entire embedding space, which leads to incomplete learning as a single encoder can only span limited space, irrespective of the variance and quantity of data. So while testing, when an unseen image is passed through the G, it is very likely that the unseen input will be mapped to un-covered embedding space, which consequently leads to the poor generation of images.

Yu et al. [14] proposed CR-GAN, which uses dual-pathway architecture to cover embedding space more extensively than a linear framework. It's primary uses a second-generation path, with the aim to map the entire Z space to corresponding targets. However, we empirically found that single encoder used in dual pathway architecture is not powerful enough to span the entire embedding dimension. This fact motivates us to use dual encoder architecture for spanning embedding space more extensively. Figure 1 illustrates the comparison between our proposed model, CR-GAN and other linear networks. The proposed model consists of two paths, namely Generator path, and Reconstruction path, described in following subsections.

3.1   Generator Path

This path is similar to the Generator path proposed in CR-GAN [14]. Here both the encoder are not involved, and G is trained to generate with random noise. Here we give a view-label v and random noise z. Aim is to produce very realistic

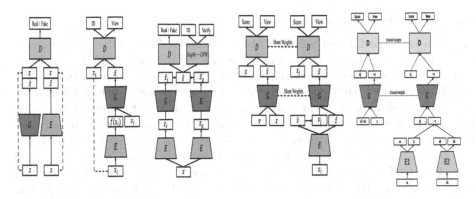

Fig. 1. Comparison of models: BiGAN, DR-GAN, TP-GAN, CR-GAN, and the proposed model

image $G(v, z)$ with view-label v. And like in GANs aim of D is to distinguish the output of G's from real. G tries to minimize Eqs. 1 and 2.

$$\mathbb{E}_{z \sim \mathbf{P_z}}[D_s(G(v, z))] - \mathbb{E}_{x \sim \mathbf{P_x}}[D_s(\mathbf{x})] + \mathbf{C_1}\mathbb{E}_{\hat{x} \sim \mathbf{P_{\hat{x}}}}[(\| \nabla_{\hat{x}}D(\hat{x}) \|)_2 - 1)^2] - \mathbf{C_2}\mathbb{E}_{x \sim \mathbf{P_x}}[\mathbf{P}(\mathbf{D_v}(x) = \mathbf{v})] \quad (1)$$

Here, $\mathbf{P_x}$ represents the distribution of data, and $\mathbf{P_z}$ represents the uniform noise distribution. Further, $\mathbf{P_{\hat{x}}}$ represents the interpolation between the data constructed form different images. In the proposed model, we randomly pass either v_i or v_k, as we want to learn G to generate high quality images either from $\hat{x}$ which is interpolation of x_i and x_k as further discussed in Sect. 3.2. We also experimentally found that feeding in $\hat{x}$ in first phase of training did not give good results, possibly because of noise, formed due to interpolation.

$$\mathbb{E}_{z \sim \mathbf{P_z}}[D_s(G(v, z))] + \mathbf{C_3}\mathbb{E}_{z \sim \mathbf{P_z}}[P(D_v(G(v, z)) = v)] \quad (2)$$

The proposed algorithm for training our model in phase 1 and phase 2, with batch-size b and time-steps t is described as below.

Algorithm

Input: Sets of images X.
Result: Trained architecture, G, D, E1, E2.

1. Sample $\mathbf{z_1} \sim \mathbf{P}_z$, $\mathbf{x}_i \sim \mathbf{P}_x$ with $\mathbf{v}_i$ and $\mathbf{x}_k \sim \mathbf{P}_x$ with $\mathbf{v}_k$;
2. $\hat{x} \leftarrow \mathbf{G}(\mathbf{v_i}, \mathbf{z}) n \mathbf{G}(\mathbf{v_k}, \mathbf{z})$;
3. Update $\mathbf{D} by Eq. 1, and \mathbf{G}$ with Eq. 2;
4. Sample $\mathbf{x}_j$ with $\mathbf{v}_j$ (where $\mathbf{x}_j$, $\mathbf{x}_i$ and $\mathbf{x}_k$ share the same identity);
5. $(\hat{\mathbf{v}}_i, \hat{\mathbf{z}}_i) \leftarrow \mathbf{E_1}$;
6. $(\hat{\mathbf{v}}_k, \hat{\mathbf{z}}_k) \leftarrow \mathbf{E_2}$;
7. $\hat{\mathbf{x}}_j \leftarrow G(\mathbf{v_j}, \mathbf{z})$;
8. Updated D by Eq. 3, and E by Eq. 4;

3.2   Reconstruction Path

We train both the **E1** and **E2** and **D** but not the **G**. In reconstruction path we make **G** generate image from the features extracted from **E1** and **E2** re-generate images, which makes them both inverse of **G**. Passing different poses in both **E1** and **E2** makes sure they cover different embedding space, which in turns leads to complete learning of latent embedding space. Further, the output generated from the **E1** and **E2** is combined using the interpolation between the data points from each of encoders, which are in spirit the same as $\hat{x}$ in generation part.

For making sure the re-constructed images by **G** from the features extracted from **E1** and **E2** share the same identity we use the cross reconstruction task, in order to make **E1** and **E1** preserve identity. To be more precise, we pass in image of same identity in both **E1** and **E2** having different poses. As primary goal is to re-construct an image x_j with interpolation of images x_i and x_k. So in order to do this, **E1** takes x_i and **E2** takes x_k, both of these encoders output an identity preserved $\bar{z}_i$ and $\bar{z}_k$ with respective view estimation $\bar{v}_i$ and $\bar{v}_k$.

G takes $\bar{z}$ and view v_i as input, and is trained to reconstruct the image of the same person with view v_i with the help of interpolated $\bar{z}$. Here $\bar{z}$ should help $\mathbb{G}$ to preserve identity and carry out essential latent features of the person. **D** here is trained to differentiate between the fake image $\hat{x}_j$ from the real one $\hat{x}_i$ or $\hat{x}_k$. Thus, **D** minimizes the Eq. 3.

$$\mathop{\mathbb{E}}_{x_i,x_j,x_k \sim P_x} [2 \times D_s(\hat{x}_j) - D_s(x_i) - D_s(x_k)] + C_1 \mathop{\mathbb{E}}_{\hat{x} \sim P_{\hat{x}}} [(\| \nabla_{\hat{x}} D(\hat{x}) \|)_2 - 1)^2]$$
$$-C_2 \mathop{\mathbb{E}}_{x_i \sim P_x} [P(D_v(x_i) = v_i)] \quad (3)$$

Here, $\tilde{x} = G(v_j, E_z(x_i))$. **E** helps **G** to generate realistic image, with v_j. Basically, **E1** and **E2** maximizes Eq. 4.

$$\mathop{\mathbb{E}}_{x_i,x_j,x_k \sim P_x} [D_s(\hat{x}_j) + C_3 P(D_v(\tilde{x}_j = v_j) - C_4 L j_1(\tilde{x}_j, x_j) - C_5 L_v(E_v(x_i), v_i)) \quad (4)$$

Here, L_1 is the loss to ensure $\tilde{x}_j$ is reconstructed property from x_j. L_v is the loss estimated from cross-entropy of the ground and estimated views, for **E1** and **E2**.

This dual-dual-pathway network efficiently spans complete embedding space. In the first path of the algorithm, **G** learns how to better produce image, from the random noise, which in time, when produced through the **E1** leads to better output.

In comparison to previously proposed linear-networks, the proposed double-dual pathway network helps better solve the problem of multi-facet construction in following ways:

– It leads to better covering of latent embedding space, which in turns leads to better generation of multi-faceted pictures.
– Once trained on good quality images, model seems to work pretty well even for low quality images, probably because of expansive embedding space covered.

4   Experiments and Results

This section describes the experimental setup, benchmark dataset, experimental results and compares the results with existing state of the art in the field. Also, considering the fact that we can not separate just the encoder part of the model, we can not just compare the feature encoding capability of respective models, so we decided it would be better if we can just compare the output of the model, and the ability to reconstruct images. So we've compared the output of images by two models, and calculated root mean square value (RMSE) value for the constructed images.

4.1   Experimental Settings

- **Benchmark Dataset:** In this experimental work, we used primary dataset as, Multi-PIE [6] and 300wLP [18]. These datasets are labelled datasets collected in an artificially constructed environment. The dataset consists of 250 subjects, with 9 poses within ±60°, two expressions and twenty illuminations. For the training purpose, we choose the first 200 for training and the remaining 50 for test. 300wLP contains view labels that are used to extract images with yaw angles from –60° to +60°, dividing them into 9 intervals. So, they can synchronize with Multi-PIE dataset after feeding into the model.
- **Implementation Details:** The network-implementation is modified from CR GAN, where each of E shares the dual-pathway architecture with the G. The main structure for our model is adopted from the res-net (residual networks) as proposed in WGAN-GP [7], where E shares a similar network structure with D. During training v is set to 9 dimensional one-hot vectors where $z \in [-1,1]^1 19$ in the latent embedding space. The batch-size we chose for our model is 20. We used Adam optimizer [11] with the learning rate of 0.0001 and momentum of [0.01, 0.89]. Choosing rest of the parameters of CR-GAN as default, we have $C_1 = 10$, $C_2 - C_4 = 1$, and $C_5 = 0.01$. Finally, we train the model for 50 epochs.

4.2   Results and Discussion

The primary aim of the proposed model - DUO-GAN is to learn complete representation by using its dual-encoder architecture and dual-pathway architecture to span entire embedding space. We conduct experiments to evaluate these contributions with respect to CR-GAN. The comparative results are shown in Table 1. We can see how the model performs in the wild settings in Fig. 2.

Table 1. Average RMSE(in mm) using dual encoder architecture, validated against CR-GAN.

Values	DUO-GAN	CR-GAN
Female subject	2.342 (±0.501)	2.64 (±0.491)
Male subject	2.4757 (±0.143)	2.795 (±0.52)

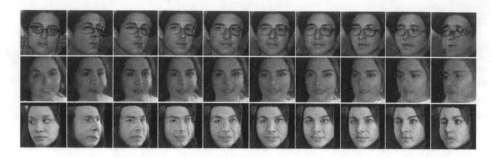

Fig. 2. Sample output on test images

In order to demonstrate the applicability of the proposed model, we compare it with four GANs, namely, BiGAN, DR-GAN, TP-GAN, and CR-GAN as depicted in Fig. 1. **CR-GAN** [14] used a dual-architecture for spanning embedding space, and learning better representation. Authors used a second reconstruction-pathway in order to make the encoder inverse of the generator. However, in practice, the Encoder doesn't seem to be powerful enough to span the entire embedding space. Comparatively, DUO-GAN uses two encoders in order to span the entire embedding space, which learns the representation comparatively more efficiently. The output produced by the proposed model is presented in Fig. 3.

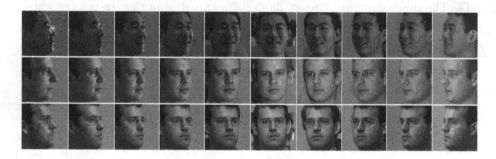

Fig. 3. Sample output on similar, but unseen images

DR-GAN [15] also tackled the problem of generating multi-view images from a single image, through a linear network. Like in a linear network, input of decoder is the output of encoder, the model is not very robust to images outside the dataset. Comparatively, we use a second generation path, which leads to better learning and generalization.

TP-GAN [10] also used a dual-architecture for solving this problem. However, unlike our model, it uses two separate structure, i.e. these two structures don't share parameter, unlike our architecture. Further, these two independent architectures in TP-GAN aims to learn different set of features. Where as our architecture aims to learn collectively.

Bi-GAN [4] aims to learn collectively a **G** and an **E**. Theoretically, **E** should be an inverse of **G**. Because of their linear network, Bi-GAN leads to poor learning and doesn't lead to good generation especially for unseen data.

5 Conclusion

In this paper, we investigated different models and compared them for constructing multi-facet images from a single image. We propose a dual architecture model called DUO-GAN, which uses double duo-pathway framework for better learning the representation. The proposed model leverages the architecture to span latent embedding space in a better way and produces higher quality images in comparison to existing models.

Acknowledgement. I would like to express my special thanks of gratitude to my friend, Mr. Shivam Prasad who helped me in doing a lot in finalizing this paper within the limited time frame.

References

1. Alqahtani, H., Kavakli-Thorne, M.: Adversarial disentanglement using latent classifier for pose-independent representation. In: International Conference on Image Analysis and Processing (ICIAP) (2019)
2. Alqahtani, H., Kavakli-Thorne, M., Kumar, G · An analysis of evaluation metrics of gans. In: International Conference on Information Technology and Applications (ICITA) (2019)
3. Alqahtani, H., Kavakli-Thorne, M., Liu, C.Z.: An introduction to person re-identification with generative adversarial networks. arXiv preprint arXiv:1904.05992 (2019)
4. Dumoulin, V., et al.: Adversarially learned inference. arXiv preprint arXiv:1606.00704 (2016)
5. Goodfellow, I., et al.: Generative adversarial nets. In: Advances in Neural Information Processing Systems, pp. 2672–2680 (2014)
6. Gross, R., Matthews, I., Cohn, J., Kanade, T., Baker, S.: Multi-pie. Image Vis. Comput. **28**(5), 807–813 (2010)
7. Gulrajani, I., Ahmed, F., Arjovsky, M., Dumoulin, V., Courville, A.C.: Improved training of wasserstein gans. In: Advances in Neural Information Processing Systems, pp. 5767–5777 (2017)
8. Hassner, T., Harel, S., Paz, E., Enbar, R.: Effective face frontalization in unconstrained images. In: Proceedings of the IEEE Conference on Computer Vision and Pattern Recognition, pp. 4295–4304 (2015)
9. Hinton, G.E., Krizhevsky, A., Wang, S.D.: Transforming auto-encoders. In: Honkela, T., Duch, W., Girolami, M., Kaski, S. (eds.) ICANN 2011. LNCS, vol. 6791, pp. 44–51. Springer, Heidelberg (2011). https://doi.org/10.1007/978-3-642-21735-7_6
10. Huang, R., Zhang, S., Li, T., He, R.: Beyond face rotation: global and local perception gan for photorealistic and identity preserving frontal view synthesis. In: Proceedings of the IEEE International Conference on Computer Vision, pp. 2439–2448 (2017)

11. Kingma, D.P., Ba, J.: Adam: A method for stochastic optimization. arXiv:1412.6980 (2014)
12. Li, Y., Yang, M., Zhang, Z.: Multi-view representation learning: a survey from shallow methods to deep methods. arXiv preprint arXiv:1610.01206 (2016)
13. Sagonas, C., Tzimiropoulos, G., Zafeiriou, S., Pantic, M.: 300 faces in-the-wild challenge: the first facial landmark localization challenge. In: Proceedings of the IEEE International Conference on Computer Vision Workshops, pp. 397–403 (2013)
14. Tian, Y., Peng, X., Zhao, L., Zhang, S., Metaxas, D.N.: Cr-gan: learning complete representations for multi-view generation. arXiv preprint arXiv:1806.11191 (2018)
15. Tran, L., Yin, X., Liu, X.: Disentangled representation learning gan for pose-invariant face recognition. In: Proceedings of the IEEE Conference on Computer Vision and Pattern Recognition, pp. 1415–1424 (2017)
16. Yan, X., Yang, J., Yumer, E., Guo, Y., Lee, H.: Perspective transformer nets: learning single-view 3d object reconstruction without 3d supervision. In: Advances in Neural Information Processing Systems, pp. 1696–1704 (2016)
17. Yang, H., Mou, W., Zhang, Y., Patras, I., Gunes, H., Robinson, P.: Face alignment assisted by head pose estimation. arXiv preprint arXiv:1507.03148 (2015)
18. Zhu, X., Lei, Z., Liu, X., Shi, H., Li, S.Z.: Face alignment across large poses: a 3d solution. In: Proceedings of the IEEE Conference on Computer Vision and Pattern Recognition, pp. 146–155 (2016)
19. Zhu, Z., Luo, P., Wang, X., Tang, X.: Multi-view perceptron: a deep model for learning face identity and view representations. In: Advances in Neural Information Processing Systems, pp. 217–225 (2014)

Multi-media and Signal Processing and Interaction

Automated 2D Fetal Brain Segmentation of MR Images Using a Deep U-Net

Andrik Rampun[(✉)][iD], Deborah Jarvis, Paul Griffiths, and Paul Armitage

Academic Unit of Radiology, Department of Infection, Immunity and Cardiovascular Disease, University of Sheffield, Sheffield S10 2RX, UK
y.rampun@sheffield.ac.uk

Abstract. Fetal brain segmentation is a difficult task yet an important step to study brain development *in utero*. In contrast to adult studies automatic fetal brain extraction remains challenging and has limited research mainly due to arbitrary orientation of the fetus, possible movement and lack of annotated data. This paper presents a deep learning method for 2D fetal brain extraction from Magnetic Resonance Imaging (MRI) data using a convolution neural network inspired from the U-Net architecture [1]. We modified the network to suit our segmentation problem by adding deeper convolutional layers allowing the network to capture finer textural information and using more robust functions to avoid overfitting and to deal with imbalanced foreground (brain) and background (non-brain) samples. Experimental results using 200 normal fetal brains consisting of over 11,000 2D images showed that the proposed method produces Dice and Jaccard coefficients of $92.8 \pm 6.3\%$ and $86.7 \pm 7.8\%$, respectively providing a significant improvement over the original U-Net and its variants.

Keywords: Fetal brain segmentation · U-Net · Deep learning · MRI · CNN

1 Introduction

Segmentation tasks in medical imaging particularly in MRI, are difficult due to obscure/absent/overlapping anatomical structure, artefacts and noise. Automated fetal brain segmentation is an important first step towards providing a detail neuroimaging analysis workflow for studying fetal brain development. A robust segmentation method would enable accurate measurement of morphological brain structures which can be used for monitoring and characterising fetal brain development, earlier diagnosis and intervention. According to the recent review conducted in 2018 by Makropoulos *et al.* [2], despite many segmentation methods being developed for brain MRI, only limited studies have attempted to extract fetal brain. The majority of methods developed so far are either optimised for neonatal brain segmentation or for tissue segmentation of the fetus' brain. Fetal brain segmentation is significantly more challenging as the boundary

© Springer Nature Switzerland AG 2020
S. Palaiahnakote et al. (Eds.): ACPR 2019, LNCS 12047, pp. 373–386, 2020.
https://doi.org/10.1007/978-3-030-41299-9_29

of the fetus is less visible, particularly towards the bottom and top slices, which can lead to under-segmentation. On the other hand, in more central slices, the fetal and skull boundaries might be overlapping, resulting in over-segmentation. Furthermore, in fetal MRI, the brain often occupies a relatively small portion of the imaging field of view (FOV) and its location and orientation within the FOV can be highly variable.

From a method development point of view, the majority of existing studies in the literature are based on conventional image processing (e.g. superpixel and region growing techniques) and machine learning (e.g. Support Vector Machine or Random Forests) techniques. However, in the last few years several studies have used deep learning (e.g. convolutional neural network (CNN)) to address fetal brain segmentation and it is becoming a methodology of choice within the research community. Indeed, the main challenge of developing a deep learning based method is the requirement for a large number of annotated data, which is not always available especially in the field of medicine. Nevertheless, this issue can be countered via data augmentation and transfer learning techniques. As a result, many studies reported that deep learning based methods could potentially produce results close to human performance.

In this paper, we propose a deep learning method using a CNN to tackle the challenges in fetal brain segmentation by modifying the original U-Net to make the network much deeper, allowing it to capture finer textural information at different resolution without the loss of spatial information. Our motivation for using a U-Net based architecture are:

1. It was designed to provide accurate segmentation with a small amount of training data, whereas most of the architectures in the literature require large amounts of training data (e.g. hundreds of thousands, or even over a million). Biomedical image annotation requires expert in the relevant anatomy we always need experts and this is either very expensive or unavailable.
2. The original U-Net was designed to tackle the medical image segmentation problem, combining rich global and local features extracted from different resolutions without the loss of spatial information. In medical image segmentation, spatial information is very important for localisation of the region of interest. The U-Net based architecture solves this problem by concatenating features extracted at the downsampling part with the corresponding layer at the upsampling part. Furthermore, the brain region of a fetus can vary significantly between 2D slices within a single 3D MI volume. To learn this variability of brain sizes, the U-Net architecture provides the contraction path, which extracts contextual information at different scales.

This study has the following contributions

1. To the best of our knowledge, our study is the largest study in the literature, covering 200 fetal brains consisting of over 11,000 2D images.
2. We propose a deep learning approach based on the original U-Net architecture. We modified the original U-Net by making it deeper (e.g. deeper convolutional layers) and employed more robust activation and optimiser functions to achieve better learning generalisation across patients.

3. We make quantitative comparisons using a large dataset to the original U-Net and its variants namely wide U-Net [21] and U-Net++ (with and without deep supervision) [21].

1.1   Summary of Existing Related Studies

As mentioned in the previous section, all of the early fetal brain segmentation methods were based on conventional image processing and machine learning techniques. Anquez *et al.* [3] proposed a method that relies on detecting the eyes. However this method is not robust as the fetus' eyes are sometimes not visualised completely. Taleb *et al.* [4] proposed a method based on age-dependent template/atlas matching. Firstly, it defines a region of interest (ROI) using the intersection of all scans and then registers this ROI to an age dependent fetal brain template, before refining the segmentation with a fusion step. Alansary *et al.* [7] proposed a segmentation method using superpixel graphs and reported brain detection accuracy of 94.6% with dice score of 73.6% based on 55 cases. Firstly, the authors extract all superpixels using the simple linear iterative clustering (SLIC) [8] technique and calculate texture descriptors for each superpixel. Subsequently, superpixels with similar features were merged to create a more robust representation of each class (e.g. brain versus non-brain). Finally, they employed the random forest classifier to differentiate brain and non-brain regions. In a larger study of 199 normal cases, Link *et al.* [5] proposed a semi-automatic method where a user draws a circle around the brain and a region growing technique is used to estimate the brain region followed by post processing to remove over segmented areas. Their experimental results reported a correlation value of the brain volume $r^2 = 0.91$ when compared with manual segmentation.

In a more complex method, Ison *et al.* [6] developed a method consisting of four phases; (i) Binary classification using the Random Forest (RF) classifier with 3D Haar-like descriptors to roughly estimate the location of the fetus' head, (ii) Candidate selection within the estimated location of the fetus' head by taking centroids with high probabilities, (iii) Candidate optimisation using the Markov Random Field (MRK) appearance model and (iv) Brain segmentation using Otsu's thresholding technique based on the probability image optimised in the previous step. The method was tested on 43 fetal brains and produced 82% and 89% sensitivity and specificity, respectively. Keraudren *et al.* [9] proposed a similar pipeline with the following steps: (i) Employ the Maximally Stable Extremal Regions (MSER) technique to find all candidates of the fetus' head, (ii) Filter all candidates by size based on prior knowledge (e.g. gestational age), (iii) Extract Scale-Invariant Feature Transform (SIFT) features for each of finalised candidates and classify them into brain or non-brain using the Support Vector Machine classifier, (iv) Use the RANdom SAmple Consensus (RANSAC) procedure to find the best fitting cube around the fetal head, resulting in a bounding box (only candidates within the bounding box considered as final brain candidates), (v) Use patch-based segmentation to find the most similar patches within an atlas template to get a rough segmentation of the brain and (vi) For each pixel in the bounding box extract SIFT features and classify them using the RF classifier. The resulting probability map is then converted to a binary segmentation using a Conditional Random Field (CRF). The authors evaluate the performance of their method on 66 fetal brains and reported 93% dice score. In contrast, Kainz et al. [10] developed a method which bypasses the head detection step by directly estimating the fetal brain via the ensemble forests classifier using the spherical 3D Gabor features.The resulting probability map was further processed via thresholding to select all possible brain candidates and the largest connected region is chosen as the initial brain segmentation. Finally, a 2D level-set is initialised from the thresholded probability map and evolved with a few hundred iterations. The authors tested the proposed method on 50 cases and reported a dice score of 90%.

Although these methods reported promising results, we identify the following disadvantages:

1. Most of the developed methods are indirect (or user dependent) which means they try to initially detect other areas of the fetus to aid localisation of the brain, such as the eye or the head. In many cases the eyes are invisible and the skull might be obscured particularly in the bottom/top slices. Moreover, since subsequent steps rely on this initial step, it is essential for the first step to achieve 100% localisation accuracy which is often difficult, or time consuming in the case of manual region placement.
2. The phase to generate initial candidates typically results in many false positives. Therefore, in subsequent steps it is necessary to have additional functionality for false positive reduction.
3. The 2D level-set and region growing methods often perform poorly when the brain boundary is unclear or obscured. Unfortunately, this is a frequent occurrence in fetal MRI especially in the bottom/top slices. While the methods might work successfully on the central slices, in some cases the brain and the skull boundaries can be overlapping, resulting in over-segmentation. Moreover, these methods do not generally work well with low contrast images.

In the last two years, many computer scientists have used deep learning to address classification and segmentation problems in the medical domain [11–14]. However, in fetal brain segmentation the number of studies in the literature is small mainly due to lack of data availability. Salehi et al. [15] proposed a network consisting of 3-pathway inspired from the U-Net [1] architecture. Each pathway has four convolutional layers and receives different patch sizes as well as different kernal sizes to collect local and global context information. The output was processed using morphological operations. The authors reported a dice score of 95.9% based on 75 cases. In another study, Rajchl et al. [16] extended the GrabCut [17] method to include machine learning by training a neural network classifier from bounding box annotations (which is called Deepcut).They also proposed variants of the DeepCut method and compare them with a naïve approach to CNN training under weak supervision. Experimental evaluation produced dice scores of 90.3% and 94.1% for the extended GrabCut and DeepCut methods, respectively.

2   Methodology

This section presents a summary of the segmentation pipeline, data preprocessing to limit the amount of memory required for training and descriptions of the proposed network used in this study.

2.1   Segmentation Pipeline

The following pipeline is applied: (1) data pre-processing, which includes cropping the brain region and noise reduction, (2) data augmentation to teach the network the desired invariance and robustness properties of the brain and (3) segmentation of the fetal brain using the proposed method.

2.2   Data Preprocessing

Following the studies of [10] and [9], all images were automatically cropped to 256×256 from the center of the brain region of the mask (512×512). This ensures that every new image contains the brain region. Subsequently, every cropped image is filtered using a small window of 3×3 median filter. Our motivation of cropping the original image to half of its original size is to limit the amount of memory required during training as well as to speed up the training process.

2.3   Data Augmentation

Although the number of images used in our study is considerably more than most existing studies in the literature, the data augmentation phase is always useful to enrich the variation of brain appearance, ensuring that the network is trained with as much contextual information of the brain as possible, hence improving the network's generalisation across different patients. In this study we applied the following data augmentation: (i) random rotation range of up to 180°, (ii) zooming in and out with a range of 0.1 to 2.0, and (iii) horizontal and vertical flips.

2.4   Network Architecture

The original U-Net [1] was developed for biomedical image data and based on the fully convolutional network [18]. Its architecture contains two main parts namely contracting/downsampling and expanding/upsampling. The contracting path captures the textural information of the input image by performing the convolution operation and downsampling the input image via the *Max pooling* operation. The expanding part enables precise localisation combined with contextual information from the contracting path.

Figure 1 shows the architecture of our proposed network. In contrast to the original U-Net network [1], our architecture has the following modifications:

1. We have added more convolution blocks, making our network deeper and able to capture more coarse and fine contextual information. The U-Net consists of four block in the downsampling and upsampling parts, whereas our network has seven blocks in each part.
2. The original U-Net uses a combination of the pixel-wise softmax with cross-entropy loss function, whereas our network uses a combination of balanced cross-entropy in conjunction with dice coefficient loss function which is more robust to deal with imbalance between brain and non-brain pixels.
3. Our network employs the exponential linear unit ('ELU') [19] and 'RMSprop' to achieve better learning generalisation of the brain across patients, whereas the original U-Net uses the 'Relu' and 'Adam' for activation and optimiser functions, respectively.
4. To avoid overfitting, we have added a batch normalisation function at the end of every block, both at the downsampling and upsampling parts. Furthermore, we have added a drop-out layer at the end of the last three blocks in the downsampling part, whereas the original U-Net only uses a drop-out layer at the end of the downsampling part.
5. The original U-Net starts with 64 up to 1024 number of filters (n_f) during the downsampling process, whereas our network starts with lower $n_f = 32$ up to $n_f = 512$.

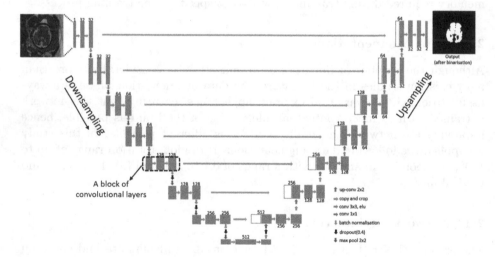

Fig. 1. The architecture of the proposed deep U-Net. The arrows denote the different operations. For each convolutional layer in both parts we use a small window size of 3×3.

2.5   Training

Let the input training data set be denoted by $T_R = (I_n, \hat{I}_n), n = 1, ..., N$, where I_n and $\hat{I}_n$ are the training image and its corresponding ground truth binary image, respectively. The training images and their corresponding masks (ground truth) are used to train the network with the Root Mean Square Propagation (RMSprop) [20] implementation in Keras with Tensorflow backend. The initial learning rate (lr) and gradient moving average decay factor(ρ) are 0.0003 and 0.8, respectively. To maximize the speed of the network and to have a better estimation of the gradient of the full dataset (hence it converges faster), we favor a maximum batch size $(bs = 32)$ that our machine can accommodate, rather than large input tiles as implemented in the original paper [1]. The number of iterations used per epoch (E) is based on the number of samples divided by batch size $(\frac{\#n}{bs})$. We monitor the Dice and Jaccard coefficients, set $E = 1000$ and employ the '*EarlyStopping*' mechanism on the validation set to automatically stop the training when the loss function value did not change after 200 epochs. To minimise covariance shifts, for each convolutional block we employed the batch normalisation procedure, which further speeds up the learning process.

The loss function [21] is computed as a combination of binary cross-entropy (our study can be seen as a binary classification: brain versus non-brain pixels) and dice coefficient, which is described as:

$$L(I, \hat{I}) = -\frac{1}{N} \sum_{n=1}^{N} \left(\frac{1}{2} \cdot I_n \cdot \log \hat{I}_n + \frac{2 \cdot I_n \cdot \hat{I}_n}{I_n + \hat{I}_n} \right) \tag{1}$$

In our proposed network, with many convolutional layers and different paths within the network, a good initialisation of the weights is essential. Otherwise, parts of the network might give excessive activations, while other parts never contribute. For this purpose, we follow the weight initialisation implemented in the original paper [1] using the Gaussian distribution with a standard deviation of $\sqrt{2/\eta}$, where η is the number of incoming nodes of one neuron.

3 Experimental Results

3.1 Data Descriptions

More information on the dataset used in this study can be found in [22,24]. For all experiments, our dataset consisted of MR images of 200 healthy/normal fetuses with gestational age is ranging between 18 and 37 weeks. The pregnant women provided written, informed consent with the approval of the relevant ethics board, from the following sources; either funded extension to the MERIDIAN [23] study or through a second research study sponsored independently by the University of Sheffield.

All studies were performed on a 1.5 T whole body scanner (Signa HDx, GE Healthcare) with an 8-channel cardiac coil positioned over the maternal abdomen, with the mother in the supine or supine/oblique position. All images are from the 3D FIESTA sequence, which is a 3D gradient echo steady state free precision sequence (TR $= 5.1$ ms, TE $= 2.5$ ms, $\alpha = 60°$, slice thickness used varies according to gestational age/size of fetal head so can range between 2.0 mm and 2.6 mm, acquisition time $= 40$ s. Some of the images contain motion artefacts that are typical for such acquisitions. For each image, the brain was manually segmented by a radiographer (trained by an expert radiologist) using the '3D Slicer' software.

3.2   Evaluation Metrics

The following metrics are used to quantitatively evaluate the performance of the method:

$$Jaccard(J) = \frac{TP}{TP + FN + FP} \tag{2}$$

$$Dice(D) = \frac{2 \times TP}{FN + (2 \times TP) + FP} \tag{3}$$

$$Accuracy(A) = \frac{TP + TN}{FN + TP + FP + TN} \tag{4}$$

$$Sensitivity(Sn) = \frac{TP}{FN + TP} \tag{5}$$

$$Specificity(Sp) = \frac{TN}{FP + TN} \tag{6}$$

$$Precision(P) = \frac{TP}{FP + TP} \tag{7}$$

where TP, TN, FP and FN are true positive, true negative, false positive and false negative, respectively. The J measures the ratio of the number of overlapping elements to the number of union elements from the segmented region of the proposed method and ground truth (manual segmentation). The D measures the ratio of twice the common number of overlapping elements to the total number of elements from both automatic segmentation and manual segmentation.

3.3   Quantitative Results

Table 1 shows the average results of the proposed method based on a stratified five runs 3-fold cross validation. We include in the same group (e.g. training set) all slices/images of the same fetus. Experimental results show that the proposed method outperformed the other U-Nets in the literature across different evaluation metrics except Sn and Sp where our method produced 1% and 0.3% lower than wU-Net and U-Net++ ws, respectively. Note that all the other U-Nets in Table 1 produced significantly larger standard deviation. This is due to these networks mostly failing to capture brain regions located in bottom and top slices. The proposed method produced Jaccard and Dice metrics that were at least 7.5% and 6% higher compared to the other U-Net architectures, respectively. In terms of accuracy and specificity, all architectures produce similar results with a maximum difference of 0.9%. For sensitivity, our method produced close ($Sn = 95.8\%$) to the result of wU-Net which is 96.8%. The proposed method outperformed the other U-Nets in terms of precision with a score value of 91.2%. The original U-Net and U-Net++ ws architectures produced competitive precision scores of 84.1% and 85.6%, respectively.

Table 1. Quantitative results of the proposed method, U-Net, wU-Net, U-Net++ ws (with deep supervision) and U-Net++ wos (without deep supervision). Each numerical value represents the average result (%± standard deviation) of 3-fold cross validation.

	J	D	A	Sn	Sp	P
Proposed U-Net	**86.7 ± 7.8**	**92.8 ± 6.3**	**99.3 ± 0.4**	95.8 ± 6.2	99.4 ± 0.4	**91.2 ± 8.1**
U-Net	79.2 ± 17.1	86.8 ± 16.4	99.2 ± 0.7	92.4 ± 14.9	99.3 ± 0.5	84.1 ± 16.2
wU-Net	74.9 ± 21.6	83.5 ± 19.5	98.6 ± 0.8	**96.8 ± 7.6**	98.6 ± 0.7	77.6 ± 21.8
U-Net++ ws	66.3 ± 26.5	75.6 ± 66.8	98.8 ± 0.9	71.7 ± 28.4	**99.7 ± 0.3**	85.6 ± 24.8
U-Net++ wos	74.9 ± 23.2	83.7 ± 22.6	99.1 ± 0.5	88.7 ± 24.3	99.2 ± 0.5	80.2 ± 24.3

Figure 2 shows examples of segmentation results produced by the proposed network. From our experience, we found that the proposed method tends to produce very good results when segmenting brains from the middle slices. The main challenge for brains in the middle slices is when the brain boundary is overlapping with the skull (e.g. in the first column in Fig. 2, the upper part of the brain boundary is overlaping with the skull (within the red line)). In this case, the proposed method is robust producing $J = 93.4\%$ and $D = 96.7\%$. In cases where the brain's region and boundary are clearly defined such as in the third and fourth rows, our method produced dice scores of 95.6% and $D = 97.1\%$, respectively. In the fifth row, the brain structure is more complex compared to the other examples and our network produced $J = 87.4\%$ and $D = 93.3\%$.

Figure 3 shows examples of segmentation results produced for bottom slices which were particularly challenging due to the brain boundary is diffuse (all cases in Fig. 3), low contrast (e.g. second row), the area of the brain is very small (e.g. first row) and the area of the brain is obscure (e.g. third row). Despite these challenges, from our analysis we found that the majority of the cases, the proposed method produced $80\% \leq D \leq 89.9\%$.

4   Discussion

Figure 4 shows four examples of segmentation results from four different cases. The first row represents a case where the image has good contrast balance and the brain boundary is clearly visible/defined. In this case, all methods produced very similar results with $J > 85\%$ and $D > 90\%$. The second row shows a case where the brain region is very small and obscure. Our method produced $J = 80.1\%$ and $D > 88.3\%$ which were better than the other networks investigated and summarised in Table 1. The third row shows an example where the image suffers from low contrast resulting in some parts of the brain being obscure or having

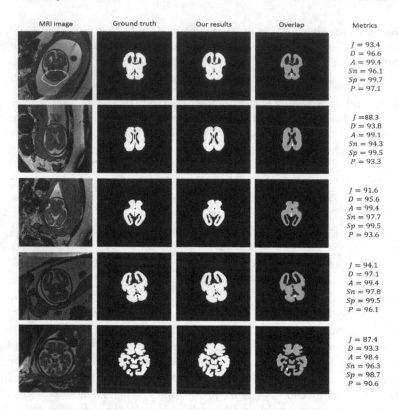

Fig. 2. Segmentation results for middle slices. In the fourth row the ground truth and segmentation result are overlapping. Cyan, red and yellow represents TP, FP and FN, respectively. (Color figure online)

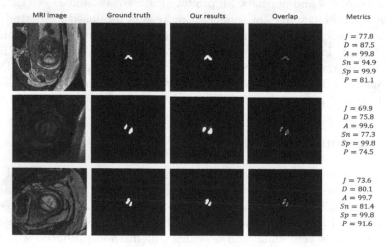

Fig. 3. Examples of segmentation results for bottom slices. Note that each image is the overlapping between the ground truth image the automatic segmentation image. Cyan, red and yellow represents TP, FP and FN, respectively. (Color figure online)

similar appearance to the with background. In this case, our method performed significantly better with $J = 82.9\%$ and $D = 90.6\%$ in comparison to U-Nets ++, which produced a high number of false negatives. The original U-net and wU-net produced similar results to our method with $D = 88.7\%$ and $D = 90.4\%$, respectively. Finally, the fourth row represents a case where there is a small region (red arrow) within the head with very similar appearance to the brain. The original U-Net and U-Net++ wos segmented this region resulting in significant false positives. Our method produced $D = 81.2\%$ and $D = 89.6\%$ closely followed by the U-Net++ ws with $D = 80.2\%$ and $D = 89.1\%$.

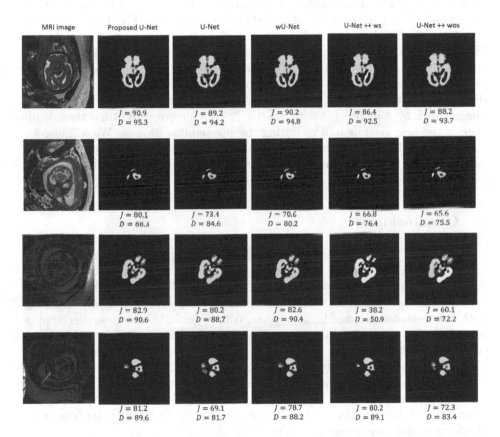

Fig. 4. Quantitative and qualitative comparison with our method, the original U-Net, wU-Net, U-Net ++ ws and U-Net ++ wos. Cyan, red and yellow represents TP, FP and FN, respectively. (Color figure online)

In this study, the proposed method produced significantly better results compared to the original U-Net and its variants (wU-Net, U-Net++ ws and U-Net++ wos) due to our U-Net architecture being deeper, hence able to capture finer contextual information which is particularly important when segmenting brain regions which are complex in appearance and obscure. We found cases like these frequently appear in the bottom and top slices of the MRI volume. In comparison, the original U-Net and wU-Net are shallow with more coarse features being extracted than fine features. As a result, they perform reasonably well when the brain is well separated from the skull, but perform poorly when the brain boundary is overlapping with the skull (e.g. first row in Fig. 2), the image suffers from low contrast (e.g third row in Fig. 4) and when the brain region is very small and obscure (e.g. second row in Fig. 4). Furthermore, the U-Net++ (both with and without deep supervision) architectures extracted too many unnecessary features from its series of nested and dense skip pathways. As a result, we found the U-Net++ based architectures suffer from overfitting for our segmentation task. This can be clearly observed in the third row of Fig. 4. Nevertheless, simplifying the U-Net++ architecture may produce competitive results in the future but this is not within the scope of our study. Our network also employed the 'ELU' [19] and 'RMSprop' functions to achieve better learning generalisation of the brain across different fetuses. In addition, we use batch normalisation and drop-out functions to avoid overfitting. As for future work, we plan to extend our network by combining it with SegNet [25] architecture to utilise the advantage of the max pooling indices during the upsampling process. We will also be focusing on 3D segmentation to enable us to estimate brain volume more accurately, which may also overcome some of the issues associated with the small brain volume found in the top and bottom slices.

5   Conclusions

In conclusion, we have presented a CNN network inspired by the U-Net [1] architecture. Experimental results show that the proposed method produced significantly better results compared to the original U-Net and its variants. We tested all methods with different scenarios, or challenges, such as images suffering with low contrast, obscure brain regions, diffuse brain boundaries, overlapping boundaries between the brain and skull, complex brain structures and very small regions located in the bottom and top slices. Quantitative results show that our network produced on average $J = 86.7\%$ and $D = 92.8\%$ based on a stratified five runs 3-fold cross validation of 200 cases, which suggest that the proposed method produced competitive results across different scenarios. Nevertheless, segmenting brain regions located in the bottom and top slices remains very challenging and more research is needed to address this issue.

References

1. Ronneberger, O., Fischer, P., Brox, T.: U-Net: convolutional networks for biomedical image segmentation. In: Navab, N., Hornegger, J., Wells, W.M., Frangi, A.F. (eds.) MICCAI 2015. LNCS, vol. 9351, pp. 234–241. Springer, Cham (2015). https://doi.org/10.1007/978-3-319-24574-4_28
2. Makropoulos, A., Counsell, S.J., Rueckert, D.: A review on automatic fetal and neonatal brain MRI segmentation. NeuroImage **170**, 231–248 (2018)

3. Anquez, J., Angelini, E.D., Bloch, I.: Automatic segmentation of head structures on fetal MRI. In: 2009 IEEE International Symposium on Biomedical Imaging: From Nano to Macro, Boston, MA, pp. 109–112 (2009). https://doi.org/10.1109/ISBI.2009.5192995

4. Taleb, Y., Schweitzer, M., Studholme, C., Koob, M., Dietemann, J.L., Rousseau, F.: Automatic template-based brain extraction in fetal MR images (2013)

5. Link, D., et al.: Automatic measurement of fetal brain development from magnetic resonance imaging: new reference data. Fetal Diagn. Ther. **43**(2), 113–122 (2018)

6. Ison, M., Dittrich, E., Donner, R., Kasprian, G., Prayer, D., Langs, G.: Fully automated brain extraction and orientation in raw fetal MRI. In: MICCAI Workshop on Paediatric and Perinatal Imaging 2012 (PaPI 2012) (2012)

7. Alansary, A., et al.: Automatic brain localization in fetal MRI using superpixel graphs. In: Proceedings Machine Learning Meets Medical Imaging, pp. 13–22 (2015)

8. Achanta, R., Shaji, A., Smith, K., Lucchi, A., Fua, P., Susstrunk, S.: Slic superpixels compared to state-of-the-art superpixel methods. IEEE Trans. Pattern Anal. Mach. Intell. **34**(11), 2274–2281 (2012)

9. Keraudren, K., et al.: Automated fetal brain segmentation from 2D MRI slices for motion correction. NeuroImage **101**, 633–643 (2014)

10. Kainz, B., Keraudren, K., Kyriakopoulou, V., Rutherford, M., Hajnal, J.V., Rueckert, D.: Fast fully automatic brain detection in fetal MRI using dense rotation invariant image descriptors. In: 2014 IEEE 11th International Symposium on Biomedical Imaging (ISBI), Beijing, pp. 1230–1233 (2014). https://doi.org/10.1109/ISBI.2014.6868098

11. Rampun, A., et al.: Breast pectoral muscle segmentation in mammograms using a modified holistically-nested edge detection network. Med. Image Anal. **57**, 1–17 (2019)

12. Hamidinekoo, A., Denton, E., Rampun, A., Honnor, K., Zwiggelaar, R.: Deep learning in mammography and breast histology, an overview and future trends. Med. Image Anal. **76**, 45–67 (2018)

13. Litjens, G., et al.: A survey on deep learning in medical image analysis. Med. Image Anal. **42**, 60–88 (2017)

14. Rampun, A., Scotney, B.W., Morrow, P.J., Wang, H.: Breast mass classification in mammograms using ensemble convolutional neural networks. In: 2018 IEEE 20th International Conference on e-Health Networking, Applications and Services (Healthcom), Ostrava, pp. 1–6 (2018)

15. Mohseni Salehi, S.S., Erdogmus, D., Gholipour, A.: Auto-context convolutional neural network (auto-net) for brain extraction in magnetic resonance imaging. IEEE Trans. Med. Imag. **36**(11), 2319–2330 (2017)

16. Rajchl, M., et al.: DeepCut: object segmentation from bounding box annotations using convolutional neural networks. IEEE Trans. Med. Imag. **36**(2), 674–683 (2017). https://doi.org/10.1109/TMI.2016.2621185

17. Rother, C., Kolmogorov, V., Blake, A.: Grabcut: Interactive foreground extraction using iterated graph cuts. ACM Trans. Graph. (TOG) **23**(3), 309–314 (2004)

18. Long, J., Shelhamer, E., Darrell, T.: Fully convolutional networks for semantic segmentation. In: Proceedings of the IEEE Conference on Computer Vision and Pattern Recognition, pp. 3431–3440 (2015)

19. Clevert, D.A., Unterthiner, T., Hochreiter, S.: Fast and accurate deep network learning by exponential linear units (elus) (2015). arXiv preprint arXiv:1511.07289

20. Hinton, G.: Neural Networks for Machine Learning - Lecture 6a - Overview of mini-batch gradient descent. https://www.cs.toronto.edu/~tijmen/csc321/slides/lecture_slides_lec6.pdf. Accessed 4 July 2019
21. Zhou, Z., Rahman Siddiquee, M.M., Tajbakhsh, N., Liang, J.: UNet++: a nested U-net architecture for medical image segmentation. In: Stoyanov, D., et al. (eds.) DLMIA/ML-CDS -2018. LNCS, vol. 11045, pp. 3–11. Springer, Cham (2018). https://doi.org/10.1007/978-3-030-00889-5_1
22. Jarvis, D.A., Finney, C.R., Griffiths, P.D.: Normative volume measurements of the fetal intra-cranial compartments using 3D volume in utero MR imaging. Eur Radiol. **29**(7), 3488–3495 (2019)
23. Griffiths, P.D., et al.: A on behalf of the MERIDIAN collaborative group. (2017) Use of MRI in the diagnosis of fetal brain abnormalities in utero (MERIDIAN): a multicentre, prospective cohort study. Lancet **389**, 538–546 (2017)
24. Griffiths, P.D., Jarvis, D., McQuillan, H., Williams, F., Paley, M., Armitage, P.: MRI of the foetal brain using a rapid 3D steady-state sequence. Br. J. Radiol. **86**(1030), 20130168 (2013)
25. Badrinarayanan, V., Kendall, A., Cipolla, R.: SegNet: a deep convolutional encoder-decoder architecture for image segmentation. IEEE Trans. Pattern Anal. Mach. Intell. **39**(12), 2481–2495 (2017)

EEG Representations of Spatial and Temporal Features in Imagined Speech and Overt Speech

Seo-Hyun Lee, Minji Lee, and Seong-Whan Lee[✉]

Korea University, Seoul 02841, Republic of Korea
{seohyunlee,minjilee,sw.lee}@korea.ac.kr

Abstract. Imagined speech is an emerging paradigm for intuitive control of the brain-computer interface based communication system. Although the decoding performance of the imagined speech is improving with actively proposed architectures, the fundamental question about 'what component are they decoding?' is still remaining as a question mark. Considering that the imagined speech refers to an internal mechanism of producing speech, it may naturally resemble the distinct features of the overt speech. In this paper, we investigate the close relation of the spatial and temporal features between imagined speech and overt speech using electroencephalography signals. Based on the common spatial pattern feature, we acquired 16.2% and 59.9% of averaged thirteen-class classification accuracy (chance rate = 7.7%) for imagined speech and overt speech, respectively. Although the overt speech showed significantly higher classification performance compared to the imagined speech, we found potentially similar common spatial pattern of the identical classes of imagined speech and overt speech. Furthermore, in the temporal feature, we examined the analogous grand averaged potentials of the highly distinguished classes in the two speech paradigms. Specifically, the correlation of the amplitude between the imagined speech and the overt speech was 0.71 in the class with the highest true positive rate. The similar spatial and temporal features of the two paradigms may provide a key to the bottom-up decoding of imagined speech, implying the possibility of robust classification of multiclass imagined speech. It could be a milestone to comprehensive decoding of the speech-related paradigms, considering their underlying patterns.

Keywords: Imagined speech · Overt speech · Brain-computer interface · Common spatial pattern · Electroencephalography

1 Introduction

Imagined speech is one of the most recent paradigms indicating a mental process of imagining the utterance of a word without emitting sounds or articulating facial movements [1]. It is first-person movement imagery consisting of the internal pronunciation of a word [2]. Imagined speech may play a role as an intuitive paradigm for brain-computer interface (BCI) communication system because it contains the subject intention of producing language. BCI had been developed as a medium of communication for locked-in or paralyzed patients in various forms; P300 speller, steady state visual evoked potential

© Springer Nature Switzerland AG 2020
S. Palaiahnakote et al. (Eds.): ACPR 2019, LNCS 12047, pp. 387–400, 2020.
https://doi.org/10.1007/978-3-030-41299-9_30

(SSVEP) speller, and motor imagery (MI) [3–6]. However, they require external stimuli or an additional display equipment to convey the user intention [7]. As a result, there had been active attempts in decoding the imagined speech to be utilized in the conversation system. It holds potential of a powerful paradigm that directly conveys the user intention when decoded robustly [8]. Qureshi et al. [9] acquired 32.9% classification performance in classifying five-class imagined speech. Nguyen et al. [10] showed 50.1% in three-class, and Sereshkeh et al. [11] reported 54.1% in three-class classification. Also, Zhao et al. [12] showed over 90% classification performance in the binary condition.

Although various methods and architectures were suggested showing robust classification performance, the question 'why are they classified?' is still remaining unknown [13]. In case of MI, the spatial regional origins are well known as contralateral hemisphere being activated when performing the left or right hand movement imagery [14]. Therefore, the spatial feature extraction methods such as common spatial pattern (CSP) are commonly applied in the MI data analysis [15, 16]. Also, the attenuation or enhancement of rhythmical synchrony over the sensorimotor cortex in the specific frequency bands of μ-rhythm (8–13 Hz) and β-rhythm (14–30 Hz) is well-known as a fundamental neuro-physiological phenomena of MI [15, 17].

Compared to the conventional paradigms such as MI, there are limited evidence supporting the underlying components of the imagined speech. DaSalla et al. [18] reported the speech related potentials (SRP) peaking 350 ms after the speech onset. However, the differences in the potentials among different word classes haven't been discerned. Previous studies have shown that speech processing involves the activation of posterior-inferior frontal gyrus (Broca's area), and posterior-superior temporal lobe (Wernicke's area) [19, 20]. However, the background evidence supporting the different spatial origination of the different word classes of imagined speech is yet remaining unknown. Nevertheless, various works applied CSP for the feature extraction method when classifying different classes of imagined speech [10].

Although the underlying components that discriminate the imagined speech are yet undiscovered, we can naturally come to our sense that the imagined speech could have similar features as the overt speech. In fact, it is known that the subject should feel as though they are producing speech rather than simply talking to themselves when performing the imagined speech [1]. Therefore, the features of the imagined speech would naturally resemble the overt speech. Iotzov et al. [21] measured the correlation of the sound amplitude with the evoked brain responses, contending the speech evoked brain potential correlates with the sound amplitude. However, they do not compare the brain signals of imagined speech and the real speech, but only compare the sound amplitude and the evoked brain potential.

There were several attempts of evaluating the brain signals of imagined speech together with overt speech. Zhao et al. [12] classified phonemes of both imagined speech and overt speech. They observed the correlation between the EEG signals of imagined speech and audio sound, but did not investigate the correlation between the EEG signals of imagined speech and overt speech. Martin et al. [22] showed the relationship between overt and imagined speech by reconstructing the imagined speech using overt speech trained model. Although it supports the partially shared neural substrates of imagined speech and overt speech, the direct components that made the classification or the reconstruction possible hasn't been discovered.

Therefore, directly comparing the features of the imagined speech with the overt speech features holds necessity to be further investigated. In this work, we investigated the spatial and temporal features of the different word classes of the imagined speech and the overt speech. Since EEG signal is a time × channel data, we focused on the spatial and temporal features. Based on the fact that the speakers should feel as though they are producing speech while performing the imagined speech [1], we analyzed the imagined speech in the perspective of the overt speech. We compared the spatial features of the imagined speech and overt speech by plotting their CSP features, and observed the changes in the grand averaged amplitude in the same class of the imagined speech and overt speech. Overall, we focused on comparing the spatial and temporal features of imagined speech and overt speech. Therefore, we aimed at investigating the relevant features that make the decoding of different classes of imagined speech be discriminated from each other. Here are the main contributions:

- We classified thirteen classes of the imagined speech and overt speech, showing the distinctively discriminated classes among the whole classes.
- We investigated the spatial features of the imagined speech and overt speech of the identical class at the same time. Therefore, we could verify the spatial features of the identical classes of the imagined speech and the overt speech showing the similar patterns.
- We analyzed the grand averaged epochs in each class of imagined speech and overt speech in the aspect of the temporal features. We observed the correlation between the potentials of the same class of imagined speech and overt speech.

2   Materials and Methods

2.1   Data Acquisition

Seven subjects (six males; mean age: 24.43 ± 1.13, range: 23–26) participated in the study. Every subject had no history of neurological disease or language disorder. All of them received more than 15 years of English education, furthermore, English is their commonly and currently used language. The study was approved by the Korea University Institutional Review Board [KUIRB-2019-0143-01] and was conducted in accordance with the Declaration of Helsinki. Informed consent was obtained from all subjects. EEG signals of the imagined speech and the overt speech were measured in two separate sessions.

Subjects performed imagined speech and overt speech of twelve words/phrases ('ambulance', 'clock', 'hello', 'help me', 'light', 'pain', 'stop', 'thank you', 'toilet', 'TV', 'water', and 'yes'). The vocabulary used in the experiment was chosen in order to provide key words for the patients' communication [23, 24]. They were selected from a communication board used in the hospital for paralyzed or aphasia patients. Rest state was also measured as a separate class for the comparison with the words/phrases.

In the imagined speech session, subjects were instructed to imagine as if they were pronouncing the given words/phrases without moving their articulators nor making sounds [1]. On the other hand, the subjects were to pronounce out loud the given words/phrases in the overt speech session. The experimental paradigm is presented in Fig. 1. In both sessions, one of the twelve word/phrase or rest class was randomly provided as an auditory cue for 2 s. Then, a cross mark was presented in a random time between 0.8–1.2 s. The subjects were instructed to perform the imagined speech or overt speech one time as soon as the cross mark disappears. They were given 2 s for performing the speech and another cross mark appears in the screen. For one cue, four times of cross mark and speech performing trial was repeated. After four times of repeated speech phase, 3 s of relaxation phase was given, followed by another random cue of twelve words/phrases or rest. This block of four speech phase was repeated twenty-two times per every twelve words/phrases and a rest class in each session. Four times of repeated speech phase was designed based on the previous imagined speech works [10].

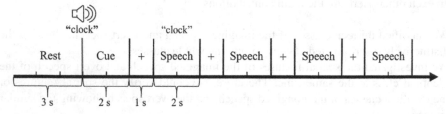

Fig. 1. Experimental process of the imagined speech and the overt speech. All the process was consistent in the two experiments, other than the subjects performing imagined speech in the imagined speech session and pronouncing out loud in the overt speech session. Each block containing four times of speech was randomly repeated 22 times for per every twelve words/phrases and rest class. Cross mark (+) indicates the fixation point in the intervals.

We used 64-channel EEG cap with active Ag/AgCl electrodes placement following the international 10–20 system. We used FCz channel as a reference electrode and FPz channel as a ground electrode. All impedances of the electrodes were maintained below 10 kΩ during the experiment. EEG signals were recorded with ActiCap system, BrainAmp amplifiers (Brain Product GmbH, Germany), and Matlab 2018a software. The data were collected at 1000 Hz sampling rate and down-sampled into 256 Hz. We applied the fifth order Butterworth filter in the frequency range of 0.5–40 Hz. The EEG signals were segmented into 2 s epochs from the beginning of each trial. Baseline correction was performed by subtracting the average value of 200 ms before the trial onset.

2.2 Feature Extraction and Classification

We used CSP for the feature extraction method. CSP is a feature extraction method that finds the optimal spatial filters of the given samples [25]. The logarithmic variances of the first and last three were used as an input feature for the classifier. After the feature extraction, we performed thirteen-class classification in order to discriminate each word/phrase. We performed the thirteen-class classification by the shrinkage regularized linear discriminant analysis (RLDA). The shrinkage RLDA is a classification method adding a regularization term to a covariance matrix using an optimal shrinkage parameter [26]. Since CSP is basically a binary feature extraction method, we applied one-versus-rest (OVR) strategy for the multiclass classification [25]. OVR strategy assumes the multiclass setting into binary setting, assigning one class and the remaining class as a binary class. We tested the classification performance by 10-fold cross validation. A k-fold cross-validation is a method randomly dividing the data into k-number of groups and testing k-times by setting each of k group as a test set and remaining k-1 groups as a training set [9]. In our case, we divided the data into ten groups and tested ten times using each group as a test set. Based on the classification result, we plotted the confusion matrix in order to determine the distinctive classes according to the true positive rate. True positive rate is the percentage of correctly identified trials [27].

2.3 Spatial and Temporal Feature Analysis

We examined the CSP features of the imagined speech and the overt speech for the spatial feature analysis. In order to investigate the distinctive region of the different classes of the imagined speech and the overt speech, we conducted the CSP feature extraction method aiming to differentiate between the three classes with the highest true positive rate according to the confusion matrix in each subject. We observed the first and the last CSP patterns of the identical classes of the imagined speech and the overt speech. The first and the last CSP patterns correspond one-by-one to the CSP filters that increase the variance of the one-versus-rest condition [10].

We further examined the distinct potentials among dissimilar classes for the time domain feature analysis. Like as the spatial feature analysis, we analyzed the top two classes with the highest true positive rate. We calculated the average of all 88 epochs for each class in each subject. We plotted the potential changes of each epochs starting from 500 ms before the speech onset to 300 ms after the end of the trial. We examined the potential peaks of the identical classes in qualitative analysis, and computed the Pearson correlation between the two potentials for a quantitative analysis. The correlation analysis between the imagined speech and the overt speech features was performed based on the previous works [22].

2.4 Statistical Analysis

Statistical analysis was performed on the thirteen-class classification accuracy of seven subjects in order to estimate the significance of the result. We performed Friedman's test between the classification accuracy of the imagined speech and the overt speech. Friedman's test is a nonparametric version of balanced two-way analysis of variance (ANOVA) [28]. It is a statistical method usually used to determine the significant differences between the variances of two datasets [29].

We additionally calculated the confidence boundary in order to confirm the significance of the classification result on the basis of Müller-Putz et al. [30]. Confidence boundary demonstrates the reliability of the classifier performance compared to the random chance rate. It is calculated based on the number of trials, significance level, and the chance rate. Classification performance over the confidence boundary indicates that the classifier can significantly differ the data from a random one.

3 Results

3.1 Multiclass Classification

The classification performance of the thirteen-class imagined speech and overt speech is presented in Table 1. The averaged 10-fold classification accuracy of all subjects was 16.2% for imagined speech and 59.9% for overt speech. Overall, the overt speech showed significantly higher classification accuracy than the imagined speech (p-value = 0.008). The highest 10-fold cross validation performance among subjects was 22.8% for the imagined speech in subject 3. The lowest performance was 10.9% in subject 5 but, nonetheless, exceeded the chance rate 7.7% as well as confidence bound 9.2% (88 trials). In the overt speech session, the highest and the lowest 10-fold cross validation performance was 82.6% and 30.6% for subject 5 and subject 1, respectively. Every subject exceeded the chance rate as well as the confidence boundary. The classification performance of the overt speech showed high variance among subjects (19.8%). Subject 5 showed the lowest performance in the imagined speech, however, performed the overt speech in the highest accuracy.

Table 1. Classification performance of 10-fold cross validation on thirteen-class imagined speech and overt speech using RLDA shrink classifier. (RLDA = Regularized linear discriminant analysis)

	Imagined speech	Overt speech
Subject 1	16.27 ± 0.98	30.61 ± 0.81
Subject 2	15.54 ± 0.57	34.12 ± 0.85
Subject 3	**22.76 ± 1.06**	78.89 ± 0.45
Subject 4	18.57 ± 1.03	51.93 ± 0.70
Subject 5	10.91 ± 0.52	**82.64 ± 0.73**
Subject 6	13.43 ± 0.70	65.38 ± 1.06
Subject 7	15.36 ± 0.77	75.37 ± 0.55
Average	**16.16 ± 3.46**	**59.85 ± 19.76**

Mean accuracy ± standard deviation (%)

Subject 3 showed the highest classification performance in the imagined speech session and performed second highest accuracy in the overt speech session. Therefore, we plotted the confusion matrix of imagined speech and overt speech for subject 3. Confusion matrix of imagined speech and overt speech is presented in Fig. 2. The diagonal value of the confusion matrix indicates the true positive rate of each class. The highest true positive rate among the words/phrases of the imagined speech was 33.2% in the word 'ambulance'. The highest true positive rate of the overt speech was 91.3% in

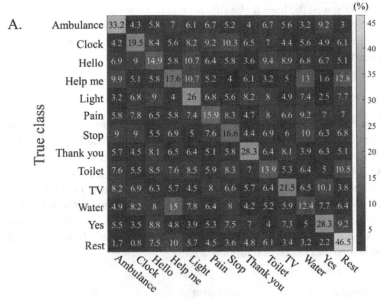

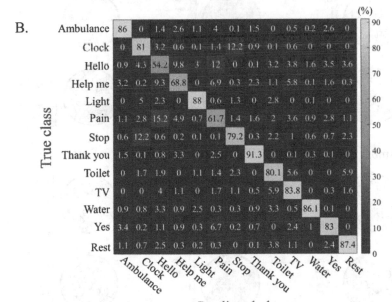

Fig. 2. Confusion matrix of imagined speech (A) and overt speech (B) for subject 3. The feature extraction was performed by CSP and classification was done by RLDA shrink. (CSP = Common spatial pattern, LDA = Linear discriminant analysis)

'thank you'. Subject 5 showed the lowest classification performance in the thirteen class classification. The word with the maximum true positive rate was 'hello' with 14.7%. The word 'clock' showed the lowest true positive rate, 5.6%. The highly distinguished class differed depending on the subjects.

3.2 CSP Feature Analysis

Figure 3 shows the first and last CSP patterns distinguishing among the three-class ('ambulance', 'thank you', and rest) imagined speech and overt speech for subject 3 who had the highest classification performance in the imagined speech. Similar spatial patterns are found between the identical classes of the imagined speech and the overt speech. The distinct brain areas shown in the first and the last CSP pattern contained the left inferior frontal lobe, left superior temporal lobe and motor cortex in the imagined speech classes. The word 'ambulance' showed distinctive area in the left temporal lobe in both imagined speech and the overt speech, however, 'thank you' showed distinct pattern in the frontal lobe and the right temporal area.

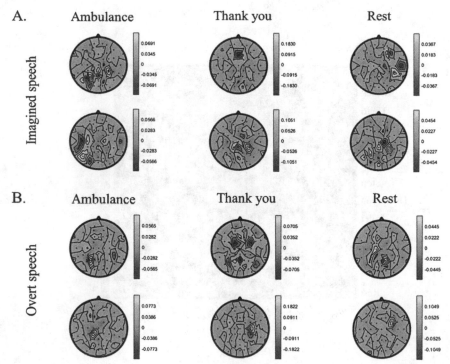

Fig. 3. The first and last CSP patterns for subject 3 in discriminating three-class ('ambulance', 'thank you', and rest) imagined speech (A) and overt speech (B). The distinct regions of the identical classes are shown to be similarly activated. Up = First CSP pattern, down = last CSP pattern (CSP = Common spatial pattern)

However, subject 5 who showed the lowest classification performance in the imagined speech classification showed weakly dissimilar spatial patterns between the imagined and the overt speech. We analyzed the two classes with the highest true positive rate ('hello' and 'help me') along with the rest class. Subject 5 showed distinctive patterns in the frontal and occipital lobe in the imagined speech of 'hello', which had the highest true positive rate. However, there was activation on the left temporal and occipital region in the overt speech of the identical class. Also, there was distinct spatial pattern in the right inferior temporal lobe for the class 'help me' in the imagined speech, however, showed activation in the right inferior temporal lobe for the overt speech. The distinctive area of identical classes in the imagined speech and the overt speech showed relatively consistent results across the subjects.

3.3 Speech Related Potentials

We observed the distinct changes in the time domain by discriminating the prominent SRPs among different classes of the imagined speech and the overt speech. The grand averaged amplitude changes between the same classes of the imagined speech and the overt speech resembled each other.

For a quantitative analysis, we performed Pearson correlation analysis between the equivalent classes of the imagined speech and the overt speech in all subjects. The correlation values of the top two classes with the highest true positive rate in each subject

Table 2. Pearson correlation of the top two classes with the highest true positive rate in each subject. Class 1 indicates the class with the highest true positive rate, and class 2 indicates the class with the second highest true positive rate. The significance level was set at $\alpha = 0.05$.

	Class 1 Class 2	True positive (%)	*rho*	*p*-value
Subject 1	Ambulance	**19.1**	**−0.25**	<0.001
	Clock	16.1	0.12	0.10
Subject 2	Hello	21.0	**0.15**	0.03
	TV	17.5	0.13	0.07
Subject 3	Ambulance	33.2	**0.42**	<0.001
	Thank you	28.3	**0.42**	<0.001
Subject 4	Help me	46.6	**0.71**	<0.001
	Ambulance	24.9	**0.44**	<0.001
Subject 5	Hello	14.7	0.05	0.51
	Help me	14.3	**0.14**	0.04
Subject 6	TV	19.7	**0.27**	<0.001
	Pain	19.3	0.07	0.31
Subject 7	Light	20.1	0.01	0.89
	Thank you	19.1	0.03	0.67

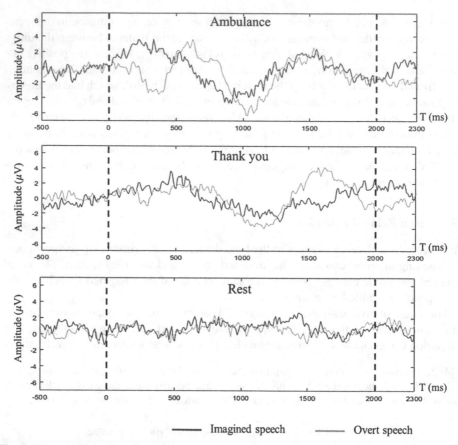

Fig. 4. The grand averaged potentials in each word/phrase trials for subject 3. The blue line indicates the imagined speech and the red line indicates the overt speech. Vertical dotted line indicates the onset of the trial as well as the end of the epoch. (CSP = Common spatial pattern) (Color figure online)

are presented in Table 2. Subject 3 and subject 4 showed positive correlation between imagined speech and overt speech in both word/phrase classes, and subject 2, 5, 6 showed positive correlation in only one class. In subject 1, word class 'ambulance' showed significantly negative correlation, while the other class showed no significant correlation between the imagined speech and the over speech. Subject 7 had no correlation in both classes.

Overall, subjects with the higher classification performance in the imagined speech classification showed relatively higher positive correlation between the same classes of the imagined speech and the overt speech. Figure 4 shows the grand averaged epochs of the three-class ('ambulance', 'thank you', and 'rest') imagined speech and overt speech performed by subject 3. According to the Fig. 2, 'ambulance' and 'thank you' were the most well distinguished word/phrase classes in subject 3. The 'ambulance' showed positive peak near 300 ms and 1500 ms and a negative peak near 1000 ms from the onset in the imagined speech. Also in the overt speech, the 'ambulance' showed positive peak near 600 ms and 1500 ms near the onset associating a negative peak near 1000 ms from the onset. The potential changes in the identical classes of the imagined speech and overt speech resembled each other.

4 Discussion and Conclusion

In this study, we investigated the similarity of the spatial and temporal features in the imagined speech and the overt speech. The averaged thirteen-class classification performance of the overt speech significantly outperformed the imagined speech in all subjects. This shows the robustness of decoding real speech only by brain signal. The classification result implies that if the features of the imagined speech and the overt speech have some relationship between each other, the robustness of the overt speech can be utilized to improve the robustness of decoding the imagined speech. This may function as a base for Martin et al. [22], reconstructing the imagined speech based on the overt speech.

According to the confusion matrix, we could find out the more distinguishable word/phrase classes. Although the distinguishable classes differed by subjects, the result showed consistently relative high true positive rate in the classes 'ambulance', 'thank you', and 'help me' across subjects. This can be due to the length of the words/phrases. Nguyen et al. [10] reported that the long words ('independent' and 'cooperative') are more highly distinguished than the short words ('in', 'out', and 'up'). Therefore, we could interpret that our result showed consistent manner with the previous works. Further investigation on the length or phonetic characteristics affecting the classification performance will act as a valuable attempt.

The CSP pattern shows the most distinguishable spatial regions of the given classes. CSP features were used as an input vector for the classifier, therefore, is regarded as a factor of discriminating among different classes. Although there is insufficient evidence of the regional features of imagined speech compared to the MI [14], we observed the significant thirteen-class classification accuracy exceeding the confidence bound. Therefore, it implies that although it is known that the speech related paradigms are processed in the narrow region of the whole brain, Broca's area and Wernicke's area associated with speech processing, it can still be classified by spatial features. Further study could improve the results by using a more detailed EEG with more electrodes in analyzing the relevant narrow region of the brain.

Previous studies also applied CSP method in classifying different words of imagined speech [9, 10]. However, analyzing the spatial patterns of imagined speech and overt speech together hasn't been reported. Our result showed the potential of discriminating the different classes of imagined speech based on the overt speech features. The result that several subjects did not show similar spatial patterns between the imagined speech and the overt speech may be due to the imbalanced classification performance of imagined and overt speech. Also, small movement artifacts in the overt speech session might also have been the reason of interpretation. The result might improve in the quality if we remove the movement artifacts using an additional electromyography channel while performing the overt speech.

However, the similar spatial features of identical classes of imagined speech and overt speech found in the high performance imply the potential of the existing common spatial features of the different classes of the imagined and overt speech. This implies that the factor that leads to the significant classification performance may have been an underlying spatial pattern of the imagined speech that is similar to the spatial pattern when speaking out in real. It suggests the potential of the CSP feature vector truly representing the intrinsic features of each word.

There have been previous works that the brain signals of motor imagery and motor execution have similarity in their spatial features between each other [31]. There are also previous works in visual imagery paradigm [32] that investigates the common features of imagined and real perception. This may support our finding that the imagined speech feature and overt speech feature resembling each other. For more quantitative analysis, the method of projecting the feature vectors and calculating the distances between the clusters can be necessary to strengthen our findings.

In the temporal feature analysis, the imagined speech showed correlation with the same class word/phrase of the overt speech. DaSalla et al. [18] reported the SRP that is significant in all classes of the imagined speech, however, our findings have discrepancy in the contribution of revealing the different temporal features by different classes, comparing with the overt speech. Itozov et al. [21] showed the evoked responses of speech and the sound amplitude having relevant correlation. Our result can be interpreted in line with the previous works.

Subject 1 showed significantly negative correlation in the grand averaged potentials between the highest distinguished class of the imagined speech and the overt speech. This may be due to the time shifting occurred by the different onset point of the two paradigms. If we apply more precise method in comparing the two amplitude, such as dynamic time warping, would help to improve the correlation analysis result. Dynamic time warping is an algorithm for measuring similarity between two temporal features that vary in speed [33]. Also, further correlation analysis on all classes in accordance with the confusion matrix would more strongly support our results.

As a result, we could conclude that equivalent classes of the two speech paradigms share certain part of the common characteristics. The decoding performance of the imagined speech holds a potential to be improved when adjusting the recent machine learning methods considering its underlying features. However, adjusting it to our brain signal domain with small data may require the precise comprehension of the distinct features of the decoding paradigms [34]. Further analysis on the brain connectivity while performing mental imagery may also contribute to revealing the underlying components and mechanism of human speech performing [35]. Our result may contribute to revealing the similar features embedded in the imagined speech and overt speech, leading a step forward to the development of a robust BCI communication system based on the imagined speech.

Overall, our finding could be an important background in decoding imagined speech based on the brain signal. Further analysis on the imagined speech and the overt speech will help improve the robustness of the decoding performance, therefore, lead to a practical communication system using BCI.

Acknowledgements. This work was supported by the Institute for Information & Communications Technology Planning & Evaluation (IITP) grant funded by the Korea government (No. 2017-0-00451; Development of BCI based Brain and Cognitive Computing Technology for Recognizing User's Intentions using Deep Learning). The authors thank D.-K. Han for the useful discussion of the data analysis.

References

1. García-Salinas, J.S., Villaseñor-Pineda, L., Reyes-García, C.A., Torres-García, A.A.: Transfer learning in imagined speech EEG-based BCIs. Biomed. Signal Process. Control **50**, 151–157 (2019)
2. Schultz, T., Wand, M., Hueber, T., Krusienski, D.J., Herff, C., Brumberg, J.S.: Biosignal-based spoken communication: a survey. IEEE Trans. Audio Speech Lang. Process. **25**(12), 2257–2271 (2017)
3. Rezeika, A., Benda, M., Stawicki, P., Gembler, F., Saboor, A., Volosyak, I.: Brain-computer interface spellers: a review. Brain Sci. **8**(4), 1–38 (2018)
4. Kwak, N.-S., Müller, K.-R., Lee, S.-W.: A convolutional neural network for steady state visual evoked potential classification under ambulatory environment. PLoS ONE **12**(2), e0172578 (2017)
5. Yeom, S.-K., Fazli, S., Müller, K.-R., Lee, S.-W.: An efficient ERP-based brain-computer interface using random set presentation and face familiarity. PLoS ONE **9**(11), e111157 (2014)
6. Won, D.-O., Hwang, H.-J., Dähne, S., Müller, K.-R., Lee, S.-W.: Effect of higher frequency on the classification of steady-state visual evoked potentials. J. Neural Eng. **13**(1), 016014–016024 (2015)
7. Nicolas-Alonso, L.F., Gomez-Gil, J.: Brain computer interfaces, a review. Sensors **12**(2), 1211–1279 (2012)
8. Gerven, M.V., et al.: The brain-computer interface cycle. J. Neural Eng. **6**(4), 041001–041011 (2009)
9. Qureshi, M.N.I., Min, B., Park, H.J., Cho, D., Choi, W., Lee, B.: Multiclass classification of word imagination speech with hybrid connectivity features. IEEE Trans. Biomed. Eng. **65**(10), 2168–2177 (2018)
10. Nguyen, C.H., Karavas, G.K., Artemiadis, P.: Inferring imagined speech using EEG signals: a new approach using Riemannian manifold features. J. Neural Eng. **15**(1), 016002–016018 (2018)
11. Sereshkeh, A.R., Trott, R., Bricout, A., Chau, T.: EEG classification of covert speech using regularized neural networks. IEEE/ACM Trans. Audio Speech Lang. Process. **25**(12), 2292–2300 (2017)
12. Zhao, S., Rudzicz, F.: Classifying phonological categories in imagined and articulated speech. In: 40th International Proceedings on Acoustics. Speech and Signal Processing, pp. 992–996. IEEE, Brisbane (2015)
13. Lee, S.-H., Lee, M., Jeong, J.-H., Lee, S.-W.: Towards an EEG-based intuitive BCI communication system using imagined speech and visual imagery. In: Proceedings of the IEEE International Conference on Systems, Man and Cybernetics, pp. 4409–4414. IEEE, Bari (2019)
14. Sitaram, R., et al.: Temporal classification of multichannel near-infrared spectroscopy signals of motor imagery for developing a brain-computer interface. Neuroimage **34**(4), 1416–1427 (2007)

15. Suk, H.-I., Lee, S.-W.: Subject and class specific frequency bands selection for multiclass motor imagery classification. Int. J. Imag. Syst. Tech. **21**(2), 123–130 (2011)
16. Jeong, J.-H., Shim, K.-H., Cho, J.-H., Lee, S.-W.: Trajectory decoding of arm reaching movement imageries for brain-controlled robot arm system. In: Proceedings of the IEEE Engineering in Medicine and Biology Society, pp. 1–4. IEEE, Berlin (2019)
17. Lee, M., et al.: Motor imagery learning across a sequence of trials in stroke patients. Restor. Neurol. Neurosci. **34**(4), 635–645 (2016)
18. DaSalla, C.S., Kambara, H., Sato, M., Koike, Y.: Single-trial classification of vowel speech imagery using common spatial patterns. Neural Netw. **22**(12), 1334–1339 (2009)
19. Leuthardt, E.C., et al.: Using the electrocorticographic speech network to control a brain-computer interface in humans. J. Neural Eng. **8**(3), 036004–036014 (2011)
20. Towle, V.L., et al.: ECoG gamma activity during a language task: differentiating expressive and receptive speech areas. Brain **131**(8), 2013–2027 (2008)
21. Iotzov, I., Parra, L.C.: EEG can predict speech intelligibility. J. Neural Eng. **16**(3), 036008–036018 (2019)
22. Martin, S., et al.: Decoding spectrotemporal features of overt and covert speech from the human cortex. Front. Neuroeng. **7**(14), 1–15 (2014)
23. Patak, L., Gawlinski, A., Fung, N.I., Doering, L., Berg, J., Henneman, E.A.: Communication boards in critical care: patients' views. Appl. Nurs. Res. **19**(4), 182–190 (2006)
24. Kitzing, P., Ahlsen, E., Jonsson, B.: Communication aids for people with aphasia. Logoped Phoniatr Vocol. **30**(1), 41–46 (2005)
25. Wu, W., Gao, X., Gao, S.: One-Versus-the-Rest (OVR) algorithm: an extension of Common Spatial Patterns (CSP) algorithm to multi-class case. In: Proceedings of the IEEE Engineering in Medicine and Biology Society, pp. 2387–2390. IEEE, Shanghai (2005)
26. Blankertz, B., Lemm, S., Treder, M., Haufe, S., Muller, K.R.: Single-trial analysis and classification of ERP components–a tutorial. Neuroimage **56**(2), 814–825 (2011)
27. Lepeschkin, E., Surawicz, M.: Characteristics of true-positive and false-positive results of electrocardiographs master two-step exercise tests. N. Engl. J. Med. **258**(11), 511–520 (1958)
28. Theodorsson-Norheim, E.: Friedman and Quade tests: basic computer program to perform nonparametric two-way analysis of variance and multiple comparisons on ranks of several related samples. Comput. Viol. Med. **17**(2), 85–99 (1987)
29. Ruxton, G.D.: The unequal variance t-test is an underused alternative to Student's t-test and the Mann-Whitney U test. Behav. Ecol. **17**(4), 688–690 (2006)
30. Müller-Putz, G.R., Scherer, R., Brunner, C., Leeb, R., Pfurtscheller, G.: Better than random? A closer look on BCI results. Int. J. Bioelectromagnetism **10**(1), 52–55 (2008)
31. Pfurtscheller, G., Neuper, C.: Motor imagery and direct brain-computer communication. Proc. IEEE **89**(7), 1123–1134 (2001)
32. Kosslyn, S.M., Tompson, W.L.: Shared mechanisms in visual imagery and visual perception: insights from cognitive neuroscience. New Cogn. Neurosci., 975–986 (2000)
33. Berndt, D.J., Clifford, J.: Using dynamic time warping to find patterns in time series. In: Proceedings of KDD Workshop, pp. 359–370. ACM (1994)
34. Roh, M.-C., Shin, H.-K., Lee, S.-W.: View-independent human action recognition with volume motion template on single stereo camera. Pattern Recogn. Lett. **31**(7), 639–647 (2010)
35. Lee, M., et al.: Connectivity differences between consciousness and unconsciousness in non-rapid eye movement sleep: a TMS–EEG study. Sci. Rep. **9**(1), 5175 (2019)

GAN-based Abnormal Detection
by Recognizing Ungeneratable Patterns

Soto Anno[1](✉) and Yuichi Sasaki[2](✉)

[1] Neural Pocket Inc. Intern; Tokyo Institute of Technology, Meguro City, Japan
anno.s.aa@m.titech.ac.jp
[2] Neural Pocket Inc., 1-1-2, Yurakucho, Chiyoda-ku, Tokyo, Japan
y_sasaki@neuralpocket.com

Abstract. One of the approaches of image anomaly detection is to build a model to generate normal images and distinguish ungeneratable images as abnormal. Recently multiple studies reported the effectiveness of applying generative adversarial networks (GANs) in the task. A GAN trained on normal images is unable to generate abnormal image which is not included in the manifold of normal images, hence the generated image shows a substantial difference from the original images. Most of the previous studies measure the difference by pixel-wise residual loss, where the essential difference between normal and abnormal is often smeared out by inevitable pixel-level reconstruction error of Generator. In this report, a new GAN-based semi-supervised anomaly detection model, *AnoGAN*, is proposed, in which the pixel-wise residual loss is replaced by a classifier network that is trained to recognize the essential difference between normal and abnormal. Proposed model achieves state-of-the-art performance in image anomaly detection using public datasets.

Keywords: Generative adversarial networks · Anomaly detection · Image recognition

1 Introduction

Anomaly image detection has a wide range of applications in many industries, such as medical and manufacturing. Approaches of anomaly detection are studied in two major streams. One is *supervised learning* approach, in which a neural network model is trained on both normal and abnormal images with label specified [17]. This approach is effective when sufficient number of abnormal images ($\sim 1k$) are available, however, the condition is rarely met in the real world. Another is *unsupervised learning* approach, where model is trained on normal images only [18,19]. Many studies using Auto Encoder (AE) and Variational Auto-Encoders (VAE) [3–8], as well as Generative Adversarial Networks (GAN) [1,1,2,10,16,20–22] have been intensively proposed since they fit well in the real world situations where sufficient number of abnormal images are not obtainable.

© Springer Nature Switzerland AG 2020
S. Palaiahnakote et al. (Eds.): ACPR 2019, LNCS 12047, pp. 401–411, 2020.
https://doi.org/10.1007/978-3-030-41299-9_31

Among them, *AnoGAN* proposed in *Schlegl et al.* [1] invoked a bunch of studies in GAN-based anomaly detection. Once trained on normal images, Generator G works as a good mapping function from a prior noise z to the manifold of normal images, therefore an optimized noise z_γ can be found which satisfies $G(z_\gamma) \sim x$ for any normal query image x. On the other hand, since G is never trained on abnormal images, Generator should not be able to generate abnormal images from any noise z. Therefore, abnormal images can be identified by comparing the reconstructed image $G(z_\gamma)$ and original query image x. The comparison is primarily conducted on pixel-wise residual loss, however, due to the limitation from representation power of Generator, there usually happens a lot of pixel-level mis-representation. The pixel-level mis-representation adds unexpected noise into pixel-wise residual loss which smears out the distribution of normal images thus degrading the discrimination power.

In place of pixel-wise residual loss, we propose to add a shallow classifier network which takes a stacked input of query image x and reconstructed image $G(z_\gamma)$ to discriminate if a substantial difference exists between x and $G(z_\gamma)$. Thanks to the simpler problem setting than supervised learning, the classifier is modeled as a very shallow network thus required quantity of abnormal images for training can be minimum. Given we need abnormal images for classifier training, this approach falls in one of the semi-supervised learning. Most of the anomaly detection task, however, can provide some numbers of abnormal images as examples, thus the requirement does not overly limit the applicable situations.

Contributions of our approach are three folds:

- We introduce a classifier network to GAN-based anomaly detection model which mitigates the noise in pixel-wise residual loss to enhance the discrimination power.
- We propose the classifier network to take a stacked input of query and reconstructed images, which simplifies the problem setting and reduces the required number of abnormal images for training. (Table 1 describes the simpleness of patterns which the classifier recognizes and the robustness to imbalanced patterns compared to "classifier only.")
- We conducted a comprehensive study of the proposed anomaly detector *Anno-GAN* on various open datasets to show the state-of-the-art performance in anomaly detection tasks.

2   Related Works

GAN [9] opened many deep learning applications, especially in image generation tasks. It consists of two sets of adversarial networks, Generator G and Discriminator D. G learns to generate an image from noise z sampled from a prior distribution. D learns to discriminate the generated images ("Fake") from original training images ("Real"). Entire objective is formulated as follow:

$$\min_G \max_D V(D, G) = \mathbb{E}_{x \sim p_d} \left[\log D(x) \right] + \mathbb{E}_{x \sim p_z} \left[\log \left(1 - D(G(z)) \right) \right], \quad (1)$$

where p_d and p_z are the distributions of original training images and the noise, respectively. When objective $V(D, G)$ is minimized, G learns to generate images that look similar to real images.

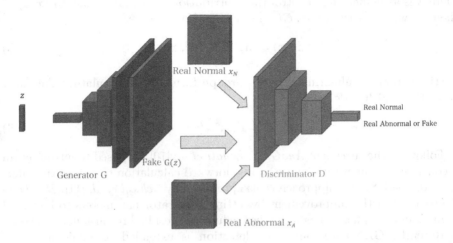

Fig. 1. Training procedure of proposed model. Similar to vanilla GAN, Discriminator is trained on Real Normal images x_N and generated images $G(z)$ labeled as "Real" and "Fake", respectively. In addition, we include available Real Abnormal images x_A as "Fake".

Recently, GAN started to be used in unsupervised anomaly detection. *AnoGAN* [1] proposed that once trained on normal images, Generator G works as a good mapping function from a prior noise z to generated image $G(z)$ for normal images. If a query image x is within the high-dimensional manifold of normal images, a noise z_γ can be found within the prior distribution which satisfies $x' = G(z_\gamma) \sim x$. On the other hand, if an abnormal image is queried, the calculated noise z_γ is not included in the prior distribution hence generated image $x' = G(z_\gamma)$ may be different from query image x. *AnoGAN* introduced two loss functions to measure the difference:

Residual Loss is sum of the pixel-wise difference between query image x and generated image $G(z_\gamma)$. In this paper, for the sake of clarity, we call this loss as *pixel-wise residual loss*:

$$\mathcal{L}_R(x, z) = \sum |x - G(z)|. \tag{2}$$

Discrimination loss based on feature matching is sum of the difference between query image x and generated image $G(z_\gamma)$ in feature space $f(x)$ of Discriminator D:

$$\mathcal{L}_D(x, z) = \sum |f(x) - f(G(z))|. \tag{3}$$

Total loss $\mathcal{L}$ is then calculated by summing the two losses above:

$$\mathcal{L}(x, z) = (1 - \lambda) \cdot \mathcal{L}_R(x, z) + \lambda \cdot \mathcal{L}_D(x, z), \tag{4}$$

where λ is a coefficient to control the contributions of $\mathcal{L}_R(x, z)$ and $\mathcal{L}_D(x, z)$. Then, z_γ which satisfies $x' = G(z_\gamma) \sim x$ is find out by

$$z_\gamma(x) = \arg \min_z \mathcal{L}(x, z). \tag{5}$$

Discrimination of abnormal images are performed by calculating Anomaly score $A(x)$ which is defined as:

$$A(x) = \mathcal{L}(x, z_\gamma(x)). \tag{6}$$

Following the success of *AnoGAN*, *Zenati et al.* [10] proposed to introduce an Encoder that estimates the noise z_γ in a forward calculation to reduce the inference overhead. Similar approach is also proposed by *Schlegl et al.* [11]. *Deecke et al.* proposed further improvement by letting Generator parameters to be tuned during inference phase so as to further reduce the residual reconstruction errors.

Recently, GAN-based anomaly detection is extended to semi-supervised learning, in which a few abnormal samples are accessible in training. [2] included abnormal images as "Fake" in GAN training to ensure that Generator does not map a noise z to around the abnormal images. New training objective function $V'(G, D)$ is then defined as:

$$V'(G, D) = \beta \cdot V(G, D) + (1 - \beta) \cdot \mathbb{E}_{x \sim p_{Ano}} \left[\log(1 - D(x)) \right], \tag{7}$$

where β is a coefficient to control the contributions of the terms. A noise $z_\gamma(x)$ which corresponds to query image x is found by:

$$z_\gamma(x) = \arg \min_z \mathcal{L}'(x, z), \tag{8}$$

where, $\mathcal{L}'(x, z)$ is defined as:

$$\mathcal{L}'(x, z) = \rho \cdot \mathcal{L}(x, z) + (1 - \rho) \cdot (1 - D(G(x)). \tag{9}$$

Here, ρ is a coefficient.

3   Proposed Approach

Pixel-wise residual loss is commonly used in GAN-based abnormal detection. Though it is an intuitive metric to measure the image difference between query image x and reconstructed image $G(z_\gamma(x))$, it is susceptible to inevitable pixel-level noise that is resulted from the limitation of Generator's representation power even for the normal images. Figure 2 gives some examples of this problem.

To mitigate the problem, we introduce classifier $C(x, y)$ which takes stacked inputs of query image x and reconstructed image $y = G(z_\gamma(x))$. Reconstructed images y_N, y_A are respectively estimated for both normal image x_N and abnormal image x_A, then stacked, and fed into training of the classifier. It learns a substantial difference between the sets of (x_N, y_N) and (x_A, y_A), i.e. a reconstruction failure in abnormal images. Overall model structure is summarized in Fig. 3 with *AnoGAN* as a reference.

Since a shallow model is used for the classifier, no precaution for data imbalance problem is required in our study. However, well-known measures such as upsampling and downsampling [13] should be considered in the other general cases to further improve the performance.

As mentioned, GAN is trained following the procedure proposed in [2]. Real images are split into "Real Normal" and "Real Abnormal" classes and also "Fake" images are generated from noise z. "Real Normal" is fed into the classifier C with "Real" image label. "Real Abnormal" and "Fake" are fed with "Fake" label. By adding "Real Abnormal" as "Fake", we make sure abnormal-like images are not included in $G(z_\gamma)$. This training procedure is summarized in Fig. 1.

4   Experiments

Our model is evaluated on MNIST and CIFAR-10. In both evaluations, the same network structure and data split policy are applied. For hyper-parameters, we use $\lambda = 0.1, \beta = 0.5, \rho = 0.1$.

4.1   Networks

Network consists of two parts, GAN and the classifier. For GAN part, we adopt DCGAN [12] model for its stability in training. The classifier consists of a simple network of one layer of 3×3 convolution network with 32-dim., followed by Global Average Pooling [14] and full-connection network. Global Average Pooling layer is adopted to prevent overfitting by reducing the number of parameters. Note that the classifier network is much shallower than the one used in supervised anomaly detection due to simpleness of the problem setting.

4.2   Datasets and Problem Definitions

We evaluated our model in two public open datasets.

MNIST includes 28×28 pixels monochrome images of handwritten digits (0–9). Total 70 k images are divided into 60 k training samples and 10 k test samples. Following [2], we select one class y_c as normal and use the other classes as abnormal. While taking all normal images from class y_c, abnormal images are sampled randomly from the other classes. In training, multiple sampling rates of 10%, 1%, 0.5% and 0.1% are separately evaluated so as to investigate the model performance on the scarcity of abnormal images. For test, we balanced the number of both images.

Original   Reconstruct   diff

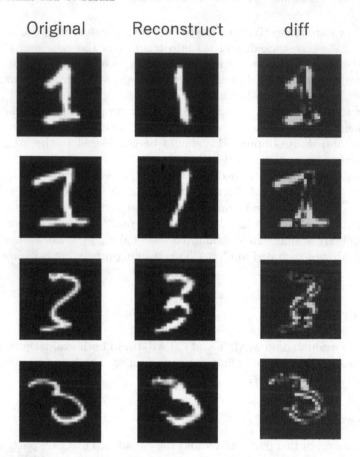

Fig. 2. Visualization of query images x (left), reconstructed images $G(z_\gamma(x))$ (center), and pixel-wise difference (right). *AnoGAN* is trained on digit "3" of MNIST. Abnormal images ("1") are not reconstructed well. Normal images ("3") are better reconstructed, however, there still remains pixel-level reconstruction error.

CIFAR10 includes 32×32 pixels color images of 10 object categories (e.g. horse, dog, ship, plane). Total 60 k images are divided into 50 k training images and 10 k test images. The same settings are applied as above. Since intra-class variety is much larger, this dataset provides a harder anomaly detection problem.

4.3   Training

GAN and Classifer are trained separately. For GAN training, a noise $z \in \mathbb{R}^{100}$ is sampled from uniform distribution of $[-1, +1]$. *Adam* optimizer [15] is used with learning rate of 0.0002. Trained model at epoch $1,000$ for MNIST and epoch $4,000$ for CIFAR10 is used in the following steps. To train the classifier, reconstructed image $G(z_\gamma(x))$ is calculated for each of query image x so as to build stacked input. Following Eq. 8, appropriate noise z_γ is searched using *Adam*

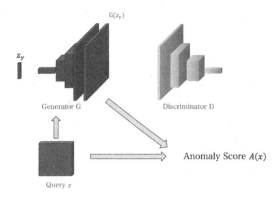

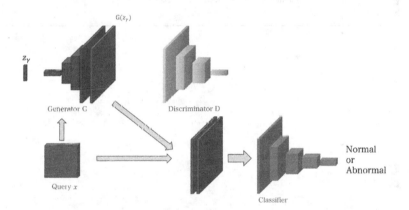

Fig. 3. Difference of *AnoGAN* (Top) and proposed model (Bottom). *AnoGAN* calculates the pixel-wise residual loss between query image x and reconstructed image $G(z_\gamma(x))$. Instead, proposed model adds a shallow classifier which takes stacked input of x and $G(z_\gamma(x))$ and outputs classification result.

optimizer with 100 iterations and learning rate of 0.0002. The classifier is then trained on the stacked input using *Adam* optimizer with learning rate of 0.0002. Training converged at approximately $1,500$ epoch, then we take the model in the following evaluations. No augmentation is adopted.

5   Results

Figure 4 shows output score distributions of normal (blue) and abnormal (green) images for *AnoGAN* and proposed model. As seen from smaller overlap between output distributions, the classifier output of proposed model shows a clear

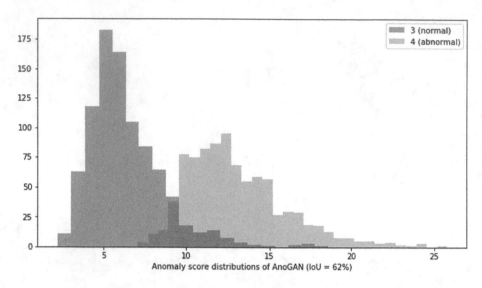

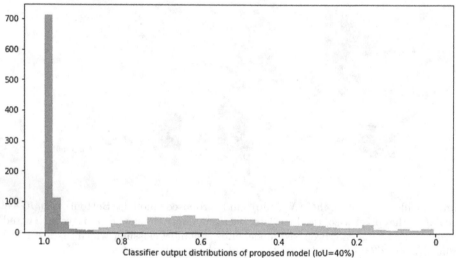

Fig. 4. Comparison of pixel-wise residual loss of *AnoGAN* (Top) and the classifier output of proposed model (Bottom). A case of MNIST is shown here with "3" used as normal and "4" used as abnormal. IoU (Intersecxtion over Union) of the normal and abnormal histograms is calculated for quantitative comparison. Proposed model exhibits smaller IoU.

Table 1. Performance comparison of the proposed classifier with stacked inputs ("Proposed model") and a normal classifier with single inputs ("classifier only"). ROC-AUC is summarized for different abnormal sampling rates. Out model keeps better performance even for small sampling rates than "classifier only". Class "3" is chosen as normal (y_c).

DATASET	y_c	Sampling rate	Proposed model ROC-AUC	Classifier only ROC-AUC
MNIST	3	10%	**0.983**	0.870
		1%	**0.980**	0.857
		0.5%	**0.984**	0.828
		0.1%	**0.968**	0.822
CIFAR-10	3	10%	**0.849**	0.784
		1%	**0.766**	0.720
		0.5%	**0.724**	0.710
		0.1%	**0.637**	0.627

improvement in discrimination power. Proposed model shows a narrower peak for normal images than *AnoGAN* thanks to the classifier's insensitivity to reconstruction noise of Generator.

Table 1 shows ROC-AUC scores with different abnormal sampling rates. Along side of our proposed model, performance of a supervised anomaly detection model is shown as "classifier only" for comparison in which the same model structure as the classifier is used with single image input. As the sampling rates become smaller, models suffer from overfitting. For example, 0.1% corresponds to only 54 images of abnormal images for MNIST dataset, which easily causes performance drop as shown in "classifier only". On the other hand, our proposed model keeps a good performance.

Table 2 shows performance comparison for various normal classes y_c. F1-score and AUC-ROC scores are evaluated for proposed model. AUC-ROC of *AnoGAN* and *ADGAN* [1] are cited aside. Since these models are pure unsupervised models, these models just provide reference figures. However, we can see a good performance gain by adopting proposed approach.

Table 2. Summary of proposed model performance for different normal class selection (y_c) with abnormal sampling rate of 10%. F1-score and ROC-AUC are evaluated for our model. For reference, ROC-AUC of other unsupervised anomaly detection models are cited as AnoGAN and ADGAN.

DATASET	y_c	Proposed model		AnoGAN	ADGAN
		F1-score	ROC-AUC	ROC-AUC	ROC-AUC
MNIST	0	0.982	**0.996**	0.990	0.995
	1	0.986	0.996	0.998	**0.999**
	2	0.934	**0.984**	0.888	0.936
	3	0.931	**0.983**	0.913	0.921
	4	0.923	**0.975**	0.944	0.936
	5	0.949	**0.984**	0.912	0.944
	6	0.948	**0.979**	0.925	0.967
	7	0.940	**0.981**	0.963	0.968
	8	0.935	**0.978**	0.883	0.854
	9	0.931	**0.971**	0.958	0.957
	avg.	0.945	**0.982**	0.937	0.947
CIFAR-10	0	0.815	**0.881**	0.610	0.632
	1	0.667	**0.617**	0.565	0.529
	2	0.768	**0.819**	0.648	0.580
	3	0.761	**0.849**	0.528	0.606
	4	0.806	**0.878**	0.670	0.607
	5	0.835	**0.912**	0.592	0.659
	6	0.887	**0.952**	0.625	0.611
	7	0.826	**0.894**	0.576	0.630
	8	0.859	**0.927**	0.723	0.744
	9	0.854	**0.924**	0.582	0.644
	avg.	0.807	**0.865**	0.612	0.624

6   Conclusion

In this study, we introduce a classifier network to GAN-based anomaly detection model which mitigates the noise in pixel-wise residual loss to enhance the discrimination power. Proposed model shows a state-of-the-art performance on MNIST and CIFAR-10.

References

1. Schlegl, T., Seeböck, P., Waldstein, S.M., Schmidt-Erfurth, U., Langs, G.: Unsupervised anomaly detection with generative adversarial networks to guide marker discovery. In: Niethammer, M., et al. (eds.) IPMI 2017. LNCS, vol. 10265, pp. 146–157. Springer, Cham (2017). https://doi.org/10.1007/978-3-319-59050-9_12

2. Kimura, M., et al.: Semi-supervised anomaly detection using GANs for visual inspection in noisy training data. In: Asian Conference on Computer Vision (2018)
3. An, J., Cho, S.: Variational autoencoder based anomaly detection using reconstruction probability. Technical Report, SNU Data Mining Center (2015)
4. Edmunds, R., Feinstein, E.: Deep semi-supervised embeddings for dynamic targeted anomaly detection (2017)
5. Estiri, H., Murphy, S.: Semi-supervised encoding for outlier detection in clinical observation data. bioRxiv, 334771 (2018)
6. An, J., Cho, S.: Variational autoencoder based anomaly detection using reconstruction probability. Spec. Lect. IE **2**, 1–18 (2015)
7. Suh, S., Chae, D.H., Kang, H.G., Choi, S.: Echo-state conditional variational autoencoder for anomaly detection. In: 2016 International Joint Conference on Neural Networks (IJCNN), pp. 1015–1022. IEEE (2016)
8. Xu, H., et al.: Unsupervised anomaly detection via variational auto-encoder for seasonal kpis in web applications. In: Proceedings of the 2018 World Wide Web Conference on World Wide Web, pp. 187–196. International World Wide Web Conferences Steering Committee (2018)
9. Goodfellow, I., et al.: Generative adversarial nets. In: Advances in Neural Information Processing Systems, pp. 2672–2680 (2014)
10. Zenati, H., Foo, C.S., Lecouat, B., Manek, G., Chandrasekhar, V.R.: Efficient GAN-based anomaly detection. arXiv:1802.06222 (2018)
11. Schlegl, T., et al.: Fast unsupervised anomaly detection with generative adversarial networks. Med. Image Anal. **54**, 30–44 (2019)
12. Radford, A., Metz, L., Chintala, S.: Unsupervised representation learning with deep convolutional generative adversarial networks. In: ICLR (2015)
13. Buda, M., Maki, A., Mazurowski, M.A.: A systematic study of the class imbalance problem in convolutional neural networks. Neural Netw. **106**, 249–259 (2018)
14. Lin, M., Chen, Q., Yan, S.: Network in network. In: ICLR (2014)
15. Kingma, P., et al.: Adam: a method for stochastic optimization. In: ICLR (2014)
16. Wang, H.-G., Li, X., Zhang, T.: Generative adversarial network based novelty detection usingminimized reconstruction error. Front. Inf. Technol. Electron. Eng. **19**(1), 116–125 (2018)
17. Gornitz, N., Kloft, M., Rieck, K., Brefeld, U.: Toward supervised anomaly detection. J. Artif. Intell. Res. **46**, 235–262 (2013)
18. Kiran, B.R., Thomas, D.W., Parakkal, R.: An overview of deep learning based methods for unsupervised and semi-supervised anomaly detection in videos. arXiv preprint arXiv:1801.03149 (2018)
19. Min, E., Long, J., Liu, Q., Cui, J., Cai, Z., Ma, J.: SU-IDS: a semi-supervised and unsupervised framework for network intrusion detection. In: Sun, X., Pan, Z., Bertino, E. (eds.) ICCCS 2018. LNCS, vol. 11065, pp. 322–334. Springer, Cham (2018). https://doi.org/10.1007/978-3-030-00012-7_30
20. Kliger, M., Fleishman, S.: Novelty detection with gan. arXiv preprint arXiv:1802.10560 (2018)
21. Sabokrou, M., Khalooei, M., Fathy, M., Adeli, E.: Adversarially learned one-class classifier for novelty detection. In: Proceedings of the IEEE Conference on Computer Vision and Pattern Recognition, pp. 3379–3388 (2018)
22. Lawson, W., Bekele, E., Sullivan, K.: Finding anomalies with generative adversarial networks for a patrolbot. In: CVPR Workshops, pp. 484–485 (2017)

Modality-Specific Learning Rate Control for Multimodal Classification

Naotsuna Fujimori[✉], Rei Endo, Yoshihiko Kawai, and Takahiro Mochizuki

Science & Technology Research Laboratories, Japan Broadcasting Corporation (NHK),
Setagaya-ku, Tokyo 157-8510, Japan
{fujimori.n-fe,endou.r-mm,kawai.y-lk,mochizuki.t-fm}@nhk.or.jp

Abstract. Multimodal machine learning is an approach to performing tasks with inputs containing multiple expressions for a single subject. There are many recent reports on multimodal machine learning using the framework of deep neural networks. Conventionally, a common learning rate has been used for the network of all modalities. This has led, however, to a decrease in the overall accuracy due to overfitting in some modality models in cases when the convergence rate and generalization performance differ among modalities. In this paper, we propose a method that solves this problem by constructing a model within the framework of multitask learning, which simultaneously learns modality-specific classifiers as well as a multimodal classifier, to detect overfitting in each modality and carry out early stopping separately. We evaluated the accuracy of the proposed method using several datasets and demonstrated that it improves classification accuracy.

Keywords: Deep neural network · Multimodal machine learning · Training algorithm · Learning rate control

1 Introduction

The main approach in the field of media analysis using machine learning has been to use a single modality, such as image or speech recognition. However, methods that take a multimodal approach have been recently proposed to improve accuracy. Multimodal machine learning involves building models to perform tasks by inputting multiple expressions, such as images, speech, text, etc., for a single subject. For instance, the accuracy of speech recognition can be improved using the movements of the mouth as well as someone's voice as input. Multimodal approaches have been used in various fields. In particular, recent developments in deep learning have made it possible to project various modality inputs into a feature space based on learning, which has played a major role in making it easier to construct models that integrate features from multiple modalities. There are a wide variety of multimodal machine learning approaches using deep learning, such as classification using text and image [1, 2], speech recognition [3, 4], visual question answering (VQA) [5, 6], and gesture recognition [7, 8]. The goal of multimodal machine learning is to achieve higher accuracy than when using a single modality by integrating mutually complementary information from multiple modalities. Most of the

© Springer Nature Switzerland AG 2020
S. Palaiahnakote et al. (Eds.): ACPR 2019, LNCS 12047, pp. 412–422, 2020.
https://doi.org/10.1007/978-3-030-41299-9_32

previous research on multimodal machine learning targets the improvement of feature integration or model construction.

Multimodal machine learning entails simultaneously learning characteristically different types of information using a single model. Since multiple modalities generally have different input distributions and model capacities for each modality, the convergence rate and generalization performance can differ among modalities. In particular, this problem can be serious when modalities have remarkable differences in the volume of noise in their data or difficulty of performing the task. Therefore, learning all modalities at the same learning rate results in decreased overall accuracy due to underfitting in some modalities and overfitting in others. This problem prevents the benefits of integrating multiple modalities from being fully maximized.

To solve the above problem, in this paper, we propose a method to restrain the decrease in overall accuracy of the model by detecting the occurrence of overfitting in each modality and individually controlling the learning process. The proposed method supports various modality combinations and enables the construction of a learning model that fully maximizes the benefits of integrating multiple modalities.

The rest of the paper is organized as follows: Sect. 2 presents related research. Section 3 presents the proposed method. Section 4 presents the results of the experiments and discusses the effectiveness of the proposed method. Section 5 presents the conclusions.

2   Related Work

2.1   Multimodal Machine Learning

Various methods of feature integration have been proposed through research on multimodal machine learning using neural networks and deep learning. They include methods using simple elementwise summation or multiplication, as well as those aimed at using relationships between multiple modalities in a more complex manner. In the Gated Multimodal Unit (GMU) [9], multiple modalities are integrated by using learnable weights that determine the influence of each modality on the activation of the unit. The Central-Net [10], on the other hand, integrates features in multiple stages in order to incorporate the characteristics of both the traditional early fusion and late fusion strategies. These methods, however, do not consider differences in how learning is carried out in each modality. There is therefore a need to look into algorithms that determine the most suitable learning schedule for each modality to improve the accuracy of multimodal learning.

2.2   Learning Rate Control

Various methods have been proposed to control the learning rate of deep neural networks. Among them are those for statically adjusting the learning rate according to a preset schedule [11, 12], and those for adaptively adjusting the learning rate for each parameter, such as AdaDelta [13] and Adam [14]. Since these methods, however, control the learning rate without reference to results of an estimation using validation data, they do not necessarily prevent overfitting. Early stopping, on the other hand, is often used as a regularization method to avoid overfitting [15]. In this method, part of the training data is used as validation data, and an evaluation is performed on the validation data at intervals, e.g. after every epoch; the learning process is terminated once the validation loss increases, which is interpreted as the start of overfitting. Applying early stopping to the general multimodal learning model shown in Fig. 1 means stopping the learning once the loss function L_M calculated with the validation data increases. If the loss function increases as a result of overfitting of the feature extractor of some of the modalities, the entire learning process is stopped, even if the progress of learning in the feature extractors of other modalities is still insufficient. Therefore, effectively using early stopping in multimodal learning entails a mechanism for detecting overfitting of the feature extractor of each modality. This paper proposes a way to incorporate such a mechanism.

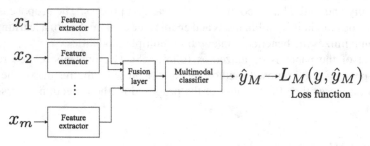

Fig. 1. Structure of the model traditionally used for multimodal machine learning

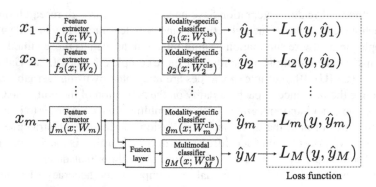

Fig. 2. Structure of the model used in the proposed method

3   Proposed Method

3.1   Network Structure

The structure of the model used in this paper is shown in Fig. 2. x_i and $\hat{y}_i$ respectively express the input and output of the ith modality, while $\hat{y}_M$ expresses the output using the fused features. f_i and g_i respectively denote the feature extractor and modality-specific classifier for the ith modality, with W_i and W_i^{cls} as the learnable parameters. g_M is the multimodal classifier whose input is a feature integrated through the fusion layer, with W_M^{cls} as a learnable parameter. L_i and L_M respectively represent the loss function for the modality-specific classification by the ith modality and for the multimodal classification. The overall loss function L is defined as

$$L = L_M + \sum_{i=1}^{m} L_i \tag{1}$$

3.2 Learning Algorithm

In the above model, multitask learning is carried out using the classification results for each modality $\hat{y}_i$, and it has a mechanism for improving the classification performance of the integrated model by using L_i for detecting overfitting in each modality, as well as maintaining the classification performance in each modality. An increase in L_i is interpreted as the start of overfitting of f_i, upon which learning of f_i is stopped to prevent overfitting of the integrated model, while the learning of other feature extractors and the multimodal classifier continues. In the following, we will always assume that the validation loss is calculated after every epoch. We define the loss L_i at epoch t as $L_i[t]$. Because the validation loss generally does not evolve smoothly and has many local minima, we use $\hat{L}_i$, the validation loss smoothed with the weight w_s, defined as

$$\begin{cases} \hat{L}_i[0] = L_i[0] \\ \hat{L}_i[t] = w_s L_i[t-1] + (1-w_s)L_i[t] \text{ for } t \geq 1 \end{cases} \tag{2}$$

Note that using too large a value of w_s delays the attainment of a minimum of $\hat{L}_i$ and thus delays the detection of overfitting. We empirically set $w_s = 0.6$ in this paper since that value appears to offer a reasonable balance between smoothness of $\hat{L}_i$ and delay in overfitting detection. We also define patience p as the number of epochs without updates to the minimum value of the validation loss after which training will be stopped. The learning algorithm is listed below.

1. Train the model and record $\hat{L}_i$ and $\hat{L}_M$ after every training epoch.
2. Save the current model parameters M_i^{best} and M^{best} every time $\hat{L}_i$ and $\hat{L}_M$ updates the minimum value, respectively.
3. If the minimum value of $\hat{L}_i$ is not updated during p epochs, stop the training and load the parameters M_i^{best}. We assume that M_i^{best} is saved at epoch t_i.
4. Resume learning with W_i fixed and discard model M_j^{best} and $\hat{L}_j$ for j other than i that are recorded after epoch t_i.
5. Repeat steps (1) to (4) until W_i is fixed for all modalities.
6. Terminate learning if the minimum value of $\hat{L}_M$ is not updated during p epochs.
7. M^{best} is the parameter for the final model.

We call this algorithm Modality-Specific Early Stopping (MSES).

3.3 Influence of Learning Rate on Validation Loss

Accurately measuring the extent of overfitting for each modality would make it possible to adjust the learning rate appropriately. We therefore carried out the following formularizations to examine the effects of the learning rate on the validation loss.

We define $L_i^{\text{train}} = L_i(W_i, x_{\text{train}}, y_{\text{train}})$ and $L_i^{\text{val}} = L_i(W_i, x_{\text{val}}, y_{\text{val}})$, which are the training loss with the training data $(x_{\text{train}}, y_{\text{train}})$ and the validation loss with the validation data $(x_{\text{val}}, y_{\text{val}})$. After updating W_i by using the gradient descent algorithm with the learning rate μ, the validation loss is written as

$$L_i\left(W_i - \mu \nabla L_i^{\text{train}}, x_{\text{val}}, y_{\text{val}}\right) \tag{3}$$

where $\nabla L_i^{\text{train}} = \partial L_i^{\text{train}} / \partial W_i$, which is the gradient of the training loss on W_i. Applying a Taylor expansion to Eq. (3) for $\mu = 0$, we get:

$$L_i\left(W_i - \mu \nabla L_i^{\text{train}}, x_{\text{val}}, y_{\text{val}}\right) = L_i^{\text{val}} - (\Delta\mu_i)\mu + \mathcal{O}\left(\mu^2\right) \tag{4}$$

where

$$\Delta\mu_i = -\frac{\partial L_i(W_i, x_{\text{val}}, y_{\text{val}})}{\partial W_i} \cdot \frac{\partial\left(W_i - \mu \nabla L_i^{\text{train}}\right)}{\partial \mu} = \nabla L_i^{\text{val}} \cdot \nabla L_i^{\text{train}} \tag{5}$$

∇L_i^{val} is defined as $\nabla L_i^{\text{val}} = \partial L_i^{\text{val}} / \partial W_i$, which is the gradient of the validation loss on W_i. Equation (5) means that the validation loss decreases when the gradient of the training loss and the gradient of the validation loss are in the same direction and it increases when they are in the opposite direction—which is consistent with intuition. Therefore, $\Delta\mu_i$ can be used for measuring the extent of overfitting of the ith modality. We will discuss the possibility of using $\Delta\mu_i$ for controlling the learning rate in Sect. 4.3.

4 Experiments

We demonstrated the effectiveness of the method by applying it to a classification problem using multimodal datasets. We use three different integration methods as fusion layers in our experiments and compare results with and without our proposed method. (a) "Elementwise Sum" is the simple integration layer whose output is the elementwise summation of the outputs of feature extractors. (b) "Gated Multimodal Unit (GMU)" is implemented following Arevalo et al. [9]. (c) "CentralNet" is implemented following Vielzeuf et al. [10]. When applying Modality-Specific Early Stopping (MSES) to CentralNet, we also freeze the weight of modality-specific classifiers as well as that of feature extractors since CentralNet has a structure in which the outputs of the middle layers of modality-specific classifiers are input to the middle layers of the multimodal classifier.

4.1 HM-MNIST

The heterogeneous multimodal MNIST (HM-MNIST) is a multimodal dataset artificially prepared based on MNIST. It adds 10% salt and pepper noise to each 28×28 image of MNIST, and divides the dataset into 28×12 and 28×16 images, assigning them as modality A and B. Further, modality A is input to the model as a one-dimensional vector of size 336. In this dataset, the information content is intentionally made to be different according to modality; i.e., the information content of modality A is smaller than that of modality B and is more prone to overfitting. Out of the 60,000 training data images, 10% were used as validation data to control the learning rate, and 10,000 test images were used to evaluate accuracy.

The structure of the model for learning HM-MNIST is shown in Table 1. "Dense" layers are fully-connected layers followed by a batch normalization layer and a ReLU activation. "Pred" layers are fully-connected layers followed by a softmax activation. "Conv" layers are convolution layers followed by a batch normalization layer, a ReLU activation, and a max pooling layer with stride 2 and size 2. We used a dropout layer with a drop rate of 50% on the fully-connected layers in the classifier. The model was optimized using SGD with a batch size of 128. Momentum was set to 0.9, the learning rate to 0.01, and patience p to 10. Cross entropy was used as the loss function.

Table 1. Architecture of the model for the HM-MNIST dataset

	Modality A		Multimodal		Modality B	
	Type	Size	Type	Size	Type	Size
Feature Extractor	Dense	1024	–	–	Conv	$5 \times 5 \times 32$
	Dense	1792			Conv	$5 \times 5 \times 64$
Classifier	Dense	1024	Dense	1024	Dense	1024
	Pred	10	Pred	10	Pred	10

We first trained the unimodal models A and B, which are the models that respectively use only modality A and B, and the conventional multimodal model shown in Fig. 1. We used the Elementwise Sum fusion layer. The transition of the validation loss is shown in Fig. 3. The validation loss of the unimodal model A starts to increase at an earlier epoch than that of the unimodal model B. The validation loss of the multimodal model starts to increase at the same timing as the unimodal model A. This implies that the feature extractor of modality A in the multimodal model started overfitting when that of modality B had still not been trained enough. Figure 4 shows the transition of the validation loss of the multimodal model shown in Fig. 2, in which the fusion layer is Elementwise Sum, with and without MSES. W_A and W_B, which represents the weights of feature extractor of modality A and B, were respectively fixed at epoch 5 and 25, where L_A and L_B started to increase. We can see that L_M continues to decrease after fixing each weight. This means each feature extractor and the multimodal classifier has been trained enough with MSES before L_M starts to increase.

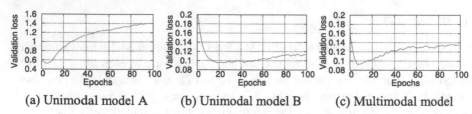

(a) Unimodal model A   (b) Unimodal model B   (c) Multimodal model

Fig. 3. Validation loss of the models

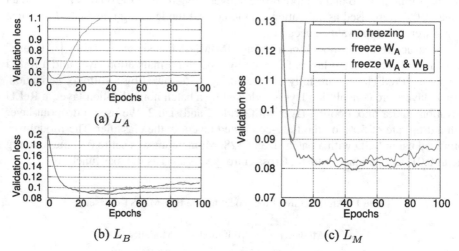

(a) L_A

(b) L_B (c) L_M

Fig. 4. Validation loss of the multimodal model with and without MSES

Table 2 shows the number of errors out of HM-MNIST 10000 test set images for different models. The models with every fusion layer improve the accuracy compared to

Table 2. Number of errors on the HM-MNIST test set for different methods

Method	Errors
Unimodal model A	1522
Unimodal model B	262
Elementwise Sum	228
with MSES	**159**
GMU	202
with MSES	**152**
CentralNet	192
with MSES	**132**

each unimodal model. Using MSES further decreases the number of errors by about 25–31%, which means MSES could suppress the effect of overfitting in feature extractors of each modality.

4.2 MM-IMDb

The multimodal IMDb (MM-IMDb) dataset [9] comprises 25,959 movies along with their plot, posters, genre, and more than 50 other additional metadata, such as year released, language, etc. The experiment involved learning of a model that predicts a movie genre based on the plot and poster. Since one movie can have more than one genre, this is a multilabel classification task.

We used the features provided by the authors [9] as inputs of the model. The image modality input is a feature vector with a size of 4096 extracted by VGG-16 pretrained on ImageNet [16]. The text modality input is a feature vector with a size of 300 obtained using a fine-tuned word2vec encoder.

The structure of the model for learning MM-IMDb is shown in Table 3. "Dense" layers and "Pred" layers have the same settings as in the model for HM-MNIST except that a sigmoid activation was used in the "Pred" layers. The model was optimized using SGD with a batch size of 128. Momentum was set to 0.9, the learning rate to 0.01, and patience p to 10. Binary cross entropy was used as the loss function; and the positive weight was set to 2 for all categories to balance precision and recall.

Table 3. The architecture of the model for the MM-IMDb dataset

	Image		Multimodal		Text	
	Type	Size	Type	Size	Type	Size
Feature extractor	Dense	2048	–	–	Dense	2048
Classifier	Dense	512	Dense	512	Dense	512
	Pred	23	Pred	23	Pred	23

Table 4. F1 scores on the MM-IMDb test set for different methods

Method	Micro	Macro	Weighted	Samples
Unimodal image model	0.486	0.276	0.440	0.477
Unimodal text model	0.634	0.541	0.625	0.630
Elementwise Sum	0.630	0.539	0.624	0.626
with MSES	**0.650**	**0.573**	**0.644**	**0.645**
GMU	0.626	0.510	0.616	0.621
with MSES	**0.650**	**0.573**	**0.645**	**0.647**
CentralNet	0.635	0.554	0.629	0.630
with MSES	**0.644**	**0.571**	**0.640**	**0.639**

Table 4 shows the results of the experiments on MM-IMDb for different models. The unimodal text model clearly has higher score than the unimodal image model. Multimodal models with Elementwise Sum and GMU have worse accuracy than the unimodal text model. This is because the feature extractor of the image modality starts to overfit at an earlier epoch than that of the text modality, and the text feature extractor couldn't be trained enough. On the other hand, when using MSES, we can train both feature extractors adequately, improving the F1 scores of the models with any fusion layer to values higher than that of the unimodal text model.

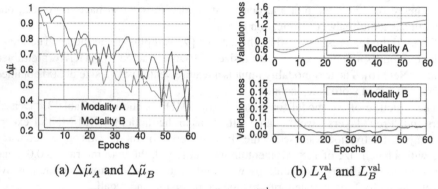

(a) $\Delta\tilde{\mu}_A$ and $\Delta\tilde{\mu}_B$ (b) L_A^{val} and L_B^{val}

Fig. 5. Comparison of $\Delta\tilde{\mu}_i$ and L_i^{val} on HM-MNIST

4.3 Relationship Between $\Delta\mu_i$ and Overfitting

We conducted an experiment to verify the hypothesis stated in Sect. 3.3. In this experiment, ∇L_i^{train} and ∇L_i^{val} are calculated as the average of all samples before carrying out learning for each epoch. To deduct the effect of the norm of the gradient due to difference in number of parameters, we normalized each gradient and get their inner product, which is written as

$$\Delta\tilde{\mu}_i = \frac{\nabla L_i^{val}}{\left|\nabla L_i^{val}\right|} \cdot \frac{\nabla L_i^{train}}{\left|\nabla L_i^{train}\right|} \tag{6}$$

Figure 5 shows the transition of $\Delta\tilde{\mu}_A$, $\Delta\tilde{\mu}_B$, L_A^{val} and L_B^{val} of the multimodal model, in which the fusion layer is Elementwise Sum, when trained on HM-MNIST without MSES. These are smoothed by $w_s = 0.6$. L_A^{val} and L_B^{val} attained their minima at epochs 4 and 20. Both $\Delta\tilde{\mu}_A$ and $\Delta\tilde{\mu}_B$ tend to decrease and in particular before epoch 10, $\Delta\tilde{\mu}_A$ goes down more steeply than $\Delta\tilde{\mu}_B$. This is considered to correspond to the result that the feature extractor of modality A starts overfitting earlier than that of modality B.

Figure 6 shows the transition of $\Delta\tilde{\mu}_{Text}$, $\Delta\tilde{\mu}_{Image}$, L_{Text}^{val} and L_{Image}^{val} of the multimodal model that have the same setting as the previous experiment. L_{Text}^{val} and L_{Image}^{val} attained their minima at epochs 76 and 16. The difference between the transition of $\Delta\tilde{\mu}_{Text}$ and $\Delta\tilde{\mu}_{Image}$ is clearer than the previous result. $\Delta\tilde{\mu}_{Image}$ starts to accelerate its decrease at about epoch 15, and L_{Text}^{val} at about epoch 50. This difference again corresponds to the difference in the onset of overfitting in the two feature extractors.

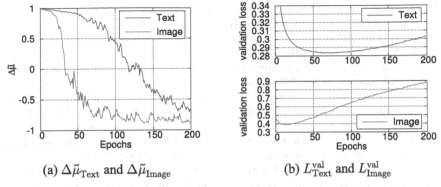

(a) $\Delta \tilde{\mu}_{\text{Text}}$ and $\Delta \tilde{\mu}_{\text{Image}}$ (b) $L_{\text{Text}}^{\text{val}}$ and $L_{\text{Image}}^{\text{val}}$

Fig. 6. Comparison of $\Delta \tilde{\mu}_i$ and L_i^{val} on MM-IMDb

Despite the difference between the timing when $\Delta \tilde{\mu}_i$ became negative and the timing when L_i^{val} increased, there was a high correlation between the two events. As such, this value can serve as a useful index for more flexibly adjusting the learning rate than early stopping, and will be the subject of future research.

5 Conclusions

We proposed a learning method for preventing overfitting by separately controlling the learning rate for each modality and maximizing the contribution of each modality to classification performance. Our method carries out learning of both classifiers that use combined features and those that use individual features of each modality, wherein the learning of each modality's feature extractor is terminated once the minimum validation loss of each modality is achieved. The method is effective when there is a difference in the progress of learning between modalities. We demonstrated through experiments on several datasets that controlling the learning rate using the proposed method improves the classification accuracy. We formularized the effect of the learning rate on the validation loss and discussed the relationship between the inner product of the gradients obtained from the training and validation data and the extent of overfitting of the model based on the results of the experiments. We also demonstrated the possibility of using the inner product of the gradients for controlling the learning rate.

References

1. Gallo, I., Calefati, A., Nawaz, S.: Multimodal classification fusion in real-world scenarios. In: Proceedings of the 14th IAPR International Conference on Document Analysis and Recognition, pp. 36–41 (2017)
2. Kim, E., McCoy, K.F.: Multimodal deep learning using images and text for information graphic classification. In: Proceedings of the 20th International ACM SIGACCESS Conference on Computers and Accessibility, pp. 143–148 (2018)
3. Ngiam, J., Khosla, A., Kim, M., Nam, J., Lee, H., Ng, A.Y.: Multimodal deep learning. In: Getoor, L., Scheffer, T. (eds.) Proceedings of the 28th International Conference on Machine Learning, pp. 689–696. Omnipress, USA (2011)

4. Palaskar, S., Sanabria, R., Metze, F.: End-to-end multimodal speech recognition. In: Proceedings of the IEEE International Conference on Acoustics, Speech and Signal Processing, pp. 5774–5778 (2018)

5. Yu, Z., Yu, J., Fan, J., Tao, D.: Multi-modal factorized bilinear pooling with co-attention learning for visual question answering. In: Proceedings of the International Conference on Computer Vision, pp. 1839–1848 (2017)

6. Peng, G., et al.: Dynamic fusion with intra- and inter- modality attention flow for visual question answering. In: Proceedings of the IEEE Conference on Computer Vision and Pattern Recognition, pp. 6639–6648 (2019)

7. Wu, D., et al.: Deep dynamic neural networks for multimodal gesture segmentation and recognition. IEEE Trans. Pattern Anal. Mach. Intell. **38**(8), 1583–1597 (2016)

8. Abavisani, M., Joze, H.R.V., Patel, V.M.: Improving the performance of unimodal dynamic hand-gesture recognition with multimodal training. In: Proceedings of the IEEE Conference on Computer Vision and Pattern Recognition, pp. 1165–1174 (2019)

9. Arevalo, J., Solorio, T., Montes-y-Gómez, M., González, F.A.: Gated multimodal units for information fusion. In: 5th International Conference on Learning Representations Workshop (2017)

10. Vielzeuf, V., Lechervy, A., Pateux, S., Jurie, F.: CentralNet: a multilayer approach for multimodal fusion. In: Leal-Taixé, L., Roth, S. (eds.) ECCV 2018. LNCS, vol. 11134, pp. 575–589. Springer, Cham (2019). https://doi.org/10.1007/978-3-030-11024-6_44

11. Loshchilov, I., Hutter, F.: SGDR: stochastic gradient descent with warm restarts. In: Proceedings of the International Conference on Learning Representations (2017)

12. Smith, L.N.: Cyclical learning rates for training neural networks. In: Proceedings of the IEEE Winter Conference on Applications of Computer Vision, pp. 464–472 (2017)

13. Zeiler, M.D.: ADADELTA: an adaptive learning rate method. arXiv preprint arXiv:1212.5701 (2012)

14. Kingma, D.P., Ba, J.: Adam: a method for stochastic optimization. In: The 3rd International Conference on Learning Representations (2015)

15. Prechelt, L.: Early stopping—but when? In: Montavon, G., Orr, G.B., Müller, K.-R. (eds.) Neural Networks: Tricks of the Trade. LNCS, vol. 7700, pp. 53–67. Springer, Heidelberg (2012). https://doi.org/10.1007/978-3-642-35289-8_5

16. Russakovsky, O., et al.: ImageNet large scale visual recognition challenge. Int. J. Comput. Vis. **115**(3), 211–252 (2015)

3D Multi-frequency Fully Correlated Causal Random Field Texture Model

Michal Haindl[1,2(✉)] [iD] and Vojtěch Havlíček[1]

[1] Institute of Information Theory and Automation,
The Czech Academy of Sciences, Prague, Czechia
{haindl,havlicek}@utia.cz
[2] Faculty of Management, University of Economics, Jindřichův Hradec, Czechia
http://www.utia.cz/

Abstract. We propose a fast novel multispectral texture model with an analytical solution for both parameter estimation as well as unlimited synthesis. This Gaussian random field type of model combines a principal random field containing measured multispectral pixels with an auxiliary random field resulting from a given function whose argument is the principal field data. The model can serve as a stand-alone texture model or a local model for more complex compound random field or bidirectional texture function models. The model can be beneficial not only for texture synthesis, enlargement, editing, or compression but also for high accuracy texture recognition.

1 Introduction

The visual appearance of surface materials and object shapes are crucial for visual scene understanding or interpretation. Visual aspects of surface materials which manifest themselves as visual textures even if there is still missing a rigorous definition of the texture [12]. Thus solid visual scene modeling or interpretation cannot avoid a sound and physically correct texture model quality. The correct material modeling is hindered by the considerable variability of a physical appearance and thus its corresponding textural representation based on changing observation conditions. Several texture modeling methods were published, but most of them do not account for simultaneously variable illumination and viewing conditions.

Real surface material visual appearance is a very complex physical phenomenon which intricately depends on the incident and reflected spherical angles, time, and light spectrum among other at least 16 physical quantities [12]. Although, the general and physically correct material reflectance function should be at least sixteen dimensional [12] which is recently unmeasurable, and even if some simplifying assumptions have to be inevitably accepted, the essential dependencies have to be respected. Among them, these are spectral, illumination, and viewing parameters. Its state-of-the-art approximation, which allows expressing spectral, spatial, illumination angle, and observation angle

© Springer Nature Switzerland AG 2020
S. Palaiahnakote et al. (Eds.): ACPR 2019, LNCS 12047, pp. 423–434, 2020.
https://doi.org/10.1007/978-3-030-41299-9_33

visual dependencies of a measured material is the Bidirectional Texture Function (BTF). BTF significantly improves the visual realism of a modeled object at the expense of non-trivial measurements and mathematical modeling of these vast BTF data spaces. Static random field-based BTF texture modeling demands complex seven-dimensional models. It is far from being a straightforward generalization of any 3D model (required for customary static color textures) with supplementing just four additional dimensions.

Compound Markov random field models (CMRF) consist of several submodels each having different characteristics along with an underlying structure model which controls transitions between these submodels [18]. The exceptional CMRF [7,14] models allow analytical synthesis at the cost of a slightly compromised compression rate due to the non-parametric control field data. Methods based on different Markov random fields [3–5,11,13,15] combine an estimated range-map with synthetic multiscale smooth texture using Markov models. Any of the above CMRF or BTF-CMRF model are build from a set of simpler two or three dimensional random field models. Such a efficient novel three dimensional model is presented in this contribution.

The ideal synthetic texture should be visually indiscernible for any observation or illumination directions from the given measured natural texture sample. A qualitative analysis of modeling results requires a still non-existing reliable mathematical criterion or very impractical and expensive visual psycho-physics. Our results [8] illustrate that neither the standard image quality criteria (MSE [23], VSNR [1], VIF [21], SSIM [22], CW-SSIM [24]) nor the STSIM [25] or ζ [19] texture criteria can be reliably used for texture quality validation. Our main contributions are the following:

- Introduction of a new efficient multispectral descriptive texture model which can be applied for high-quality material appearance modeling or recognition.
- Analytical Bayesian solution for the model parameters.
- Fast recursive model synthesis.

2   3D Multi-spectral Multi-frequency CAR Texture Model

The seven-dimensional bidirectional texture function (BTF) reflectance model is currently the state-of-the-art general reflectance function model, approximation which can be reliably measured [2,12]. However, to model such a function, we need its factorization to a set of lower two or three-dimensional models because they can be represented with significantly fewer amount of parameters to be estimated. The proposed multispectral model represents such a three-dimensional building factor for a high-quality BTF texture model.

The texture region (not necessarily continuous) is represented by a multi-frequency generalization (3DmfCAR) of the adaptive 3D causal auto-regressive random (3DCAR) field model [6,10] which combines the principal random field

Y with the auxiliary field $\tilde{Y}$. This model can be analytically estimated as well as easily synthesized. The 3DmfCAR model is defined:

$$\breve{Y}_r = \sum_{s \in I_r^c} A_s \breve{Y}_{r-s} + \breve{e}_r \qquad \forall r \in I \qquad (1)$$

where $\breve{Y}_r = \begin{bmatrix} Y_r \\ \tilde{Y}_r \end{bmatrix}$ is a $\breve{d} \times 1$ vector, $\breve{d} = d + \tilde{d}$, $A_s = \begin{bmatrix} A_{s,1} \vdots A_{s,2} \\ \cdots \cdots \\ A_{s,3} \vdots A_{s,4} \end{bmatrix}$ are $\breve{d} \times \breve{d}$

parametric matrices, Y_r $(d \times 1)$ is a pixel from image Y, I is a finite rectangular lattice, $r = [r_1, r_2, r_3]$ is a multiindex $r \in I$ where r_1, r_2, r_3 are row, column and spectral indices, and d is a number of spectral bands of the image Y. $\tilde{Y}_r$ $(\tilde{d} \times 1)$ is a pixel created by some processing of the measured image pixels

$$\tilde{Y}_r = f(Y_s \forall s \in I_r^c), \qquad (2)$$

where the only restriction we assume is that the argument of the function $f(\ )$ is limited to pixels from the causal contextual neighborhood I_r^c.

Such an example can be a convolution with some convolution filter kernel h, i.e., $\tilde{Y}_r = h * y_r$ with zero padding for $s \notin I_r^c$ such as a low-pass filtered or upsampled Gaussian pyramid rough level. In this paper, we will further assume the function $f(\ )$ to be a median (3DmdCAR). The model (1) can be expressed in the matrix form:

$$\breve{Y}_r = \gamma \breve{X}_r + \breve{e}_r, \qquad (3)$$

where γ is the $\breve{d} \times \breve{d}\eta_A$ parameter matrix $\gamma = [A_1, \ldots, A_{\eta_A}]$, $\eta_A = card(I_r^c)$, I_r^c is a causal neighborhood, $\breve{e}_r = \begin{bmatrix} e_r \\ \tilde{e}_r \end{bmatrix}$ is a Gaussian white noise vector with zero mean and a constant but unknown $\breve{\Sigma}$ covariance matrix.

$$\breve{X}_r = [\breve{Y}_{r-s}^T : \ \forall s \in I_r^c], \qquad (4)$$

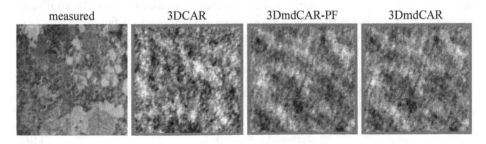

| measured | 3DCAR | 3DmdCAR-PF | 3DmdCAR |

Fig. 1. Measured bark texture and its 3DCAR and 3DmdCAR synthesis, respectively.

2.1 Parameter Estimation

The model parameters $\gamma, \check{\Sigma}$ are estimated from measured texture Y_{sample} and its modified version $\tilde{Y}_{sample}$ using Bayesian estimations.

These Bayesian parameter estimates (conditional mean values)

$$\hat{\gamma}_{r-1}^T = V_{xx(r-1)}^{-1} V_{xy(r-1)} \tag{5}$$

and

$$\hat{\Sigma}_{r-1} = \frac{\lambda_{(r-1)}}{\beta(r)}. \tag{6}$$

can be accomplished using fast, numerically robust and recursive statistics [10], given the known 3DmfCAR process history

$$\check{Y}^{(t-1)} = \left\{ \check{Y}_{t-1}, \check{Y}_{t-2}, \ldots, \check{Y}_1, \check{X}_t, \check{X}_{t-1}, \ldots, \check{X}_1 \right\} :$$

$$\hat{\gamma}_{t-1}^T = V_{xx(t-1)}^{-1} V_{xy(t-1)}, \tag{7}$$

$$V_{t-1} = \tilde{V}_{t-1} + V_0, \tag{8}$$

$$\tilde{V}_{t-1} = \begin{pmatrix} \sum_{u=1}^{t-1} \check{Y}_u \check{Y}_u^T & \sum_{u=1}^{t-1} \check{Y}_u \check{X}_u^T \\ \sum_{u=1}^{t-1} \check{X}_u \check{Y}_u^T & \sum_{u=1}^{t-1} \check{X}_u \check{X}_u^T \end{pmatrix} = \begin{pmatrix} \tilde{V}_{yy(t-1)} & \tilde{V}_{xy(t-1)}^T \\ \tilde{V}_{xy(t-1)} & \tilde{V}_{xx(t-1)} \end{pmatrix}, \tag{9}$$

where t is the traversing order index of the sequence of multi-indices r and is based on the selected model movement in the lattice I, V_0 is a positive definite initialization matrix (see [10]) which is in our experiment the identity matrix.

The optimal causal functional contextual neighborhood I_r^c can be solved analytically by a straightforward generalization of the Bayesian estimate in [10].

2.2 Model Synthesis

The principal multispectral texture field synthesis can be computed using three possible simple and fast alternatives. The auxiliary random field can be similarly separately synthesized if there is a knowledge of the principal field. To simplify our notation, we will not differentiate measured Y_r (in Sect. 2.1) and synthesized Y_r from the model (this section) because their usage is clear from the context.

Complete Model Synthesis. The complete model $\check{Y}$ synthesis uses direct application of the 3DmfCAR model Eq. (1)

$$\check{Y}_r = \hat{\gamma} \check{X}_r + \mathcal{N}(0, \hat{\Sigma}), \tag{10}$$

for simultaneous synthesis of the principal Y and auxiliary $\tilde{Y}$ random fields. Both random fields are causal thus the required contextual neighbors $Y_{r-s}, \tilde{Y}_{r-s} \; \forall s \in I_r^c$ are known from previous model synthesis steps.

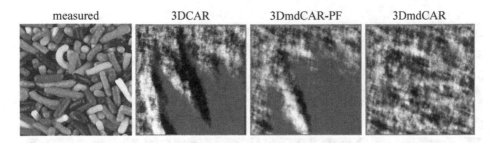

measured   3DCAR   3DmdCAR-PF   3DmdCAR

Fig. 2. Measured sweets texture and its unstable 3DCAR and 3DmdCAR-PF synthesis, respectively.

Complete Model Combined Synthesis. The alternative complete model $\check{Y}$ synthesis (3DmfCAR-PF) combines the principal field Y synthesis using the model equation:

$$Y_r = \sum_{s \in I_r^c} (A_{s,1} Y_{r-s} + A_{s,2} \tilde{Y}_{r-s}) + e_r \qquad \forall r \in I \qquad (11)$$

while the auxiliary field $\tilde{Y}$ is separately computed from (2) using just estimated neighbors $Y_s \in I_r^c$ as arguments for the corresponding function $f(\ )$ for computing $\tilde{Y}_r$ pixels.

Principal Model Synthesis. In most applications the auxiliary field $\tilde{Y}$ is not needed, and the model (11) is sufficient for the principal field Y synthesis.

Auxiliary Model Synthesis. If there is a need to synthesize the auxiliary field to a known principal field, it can be easily done using the corresponding part of the 3DmfCAR model equation:

$$\tilde{Y}_r = \sum_{s \in I_r^c} (A_{s,3} Y_{r-s} + A_{s,4} \tilde{Y}_{r-s}) + \tilde{e}_r \qquad \forall r \in I. \qquad (12)$$

The auxiliary random field can be alternatively synthesized directly from the measured data ($Y_{measured}$) if there are used in (12) instead of estimated principal pixels Y_r, but then this field cannot be enlarged.

3 Texture Measurement Database

We verified the model on color textures cutouts from our large (more than 2000 high-resolution 7.9 MB (4288×2848) color textures categorized into 14 thematic classes and 20 subclasses) Prague color texture database [9,16]. All these textures are natural textures or man-made material textures. Some tested color textures are also from the VisTex database [20] where all textures have 512×512 size.

measured	3DCAR	3DmdCAR-PF	3DmdCAR

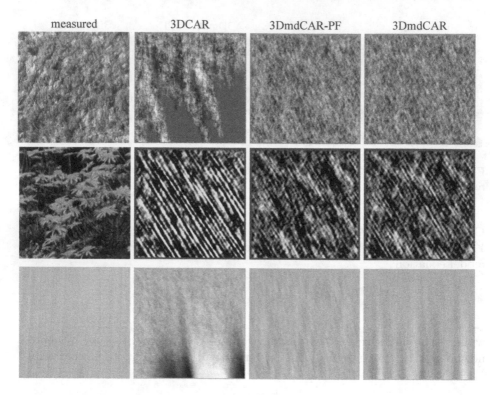

Fig. 3. Measured textile, flowers, and wood textures (first column), their unstable 3DCAR synthesis (second column) and the stabilizing effect of the additional auxiliary field in the 3DmdCAR model application.

4   Results

All our experiments were provided on color texture sets. Some textures were modeled using a single 3DmdCAR model, while others more complicated textures used a combination of several such models using the concept of the random compound field. Figure 1 illustrates the difference between simple 3DCAR model synthesis of a bark texture and both versions of the 3DmdCAR synthesis with all model using the same contextual neighborhood I_r^c. The 3DmdCAR model tends to stabilize some unstable 3DCAR model as is illustrated in three examples in Fig. 3 where all models in each row share the same contextual neighborhood I_r^c with the corresponding unstable 3DCAR models. Figure 2 suggests the stronger stabilizing effect of the complete model synthesis (3DmdCAR) over its combined synthesis (3DmdCAR-PF) alternative. Although the possible instability problem can be easily solved by just increasing the model order, this stabilizing tendency is the advantage of this otherwise more complex proposed model. Figure 4 shows visual quality improvement of the proposed 3DmdCAR model over our previously published 3DCAR model on the plant's texture example. The influence of the median window size on the visual synthetic texture appearance in a range of

Table 1. Spectral modeling per pixel error ζ (13) compared to the measured original.

	3DCAR	3DmdCAR-PF	3DmdCAR
Figure 1 bark	6.49	7.36	7.97
Figure 2 sweets	41.88	64.29	29.07
Figure 3 textile	23.54	8.30	8.89
Figure 3 flowers	16.62	10.91	10.80
Figure 3 wood	51.30	5.81	7.60
Figure 4 plants	16.96	8.93	8.61

median windows 2×2–11×11 is illustrated on the lichen texture in Fig. 5. The larger is the median estimation window; the more enhanced is the low frequencies in the model.

Fig. 4. Measured plants texture (left), its synthesis using 3DCAR (upper right) and 3DmdCAR (lower right) models.

The application of the presented 3DmdCAR model in the more complex compound model is illustrated in Fig. 6. The compound Markov random field models (CMRF) consist of several sub-models each was having different characteristics along with an underlying structure model which controls transitions between these sub-models. The non-parametric control field is estimated using the iterative method [15], and six local Markovian models are the presented 3DmdCAR model.

Table 2. Spectral modeling per pixel error ζ (13) compared to the measured original.

Median	3DCAR	3DmdCAR						
		2×2	3×3	4×4	5×5	7×7	9×9	11×11
Figure 5 lichen	16.46	10.87	10.87	11.19	17.20	10.90	10.75	11.03

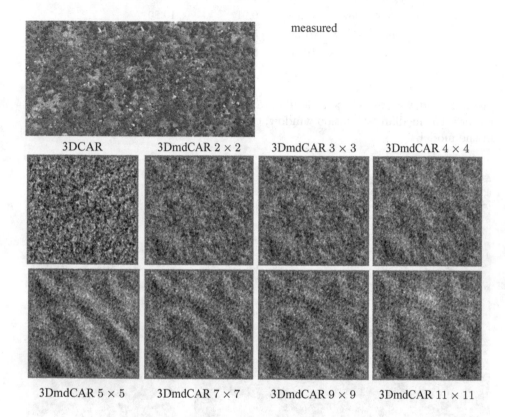

Fig. 5. Measured lichen texture, its 3DCAR synthesis, and several 3DmdCAR syntheses with gradually growing median filter windows.

The 3DmdCAR model can also be beneficially used in texture recognition (supervised or unsupervised) applications. Some preliminary results on the unsupervised bidirectional texture function segmentation can be checked on the Prague texture segmentation data-generator and benchmark [9]. The 3DmdCAR model with 7×7 median there achieves the best average rank over 21 benchmark criteria while the comparable 3DCAR model won only three individual segmentation criteria. Detailed results in these applications will be published elsewhere.

measured $CMRF^{3DmdCAR}$

Fig. 6. Measured lichen and stone textures (left), and their enlarged synthesis (lichen upper, stone bottom) using a compound model with several 3DmdCAR submodels.

4.1   Texture Spectral Similarity

Although there is not any reliable texture similarity criterion, we can quantitatively measure spectral texture similarity. We have recently proposed [17] new reliable criterion for image spectral composition comparison. This ζ criterion simultaneously considers texture spectral similarity as well as the mutual ratios of similar pixels based on the mean exhaustive minimum distance:

$$\downarrow \zeta(A,B) = \frac{1}{M} \sum_{(r_1,r_2) \in \langle A \rangle} \min_{(s_1,s_2) \in U} \left\{ \rho \left(Y^A_{r_1,r_2,\bullet}, Y^B_{s_1,s_2,\bullet} \right) \right\} \qquad \geq 0, \qquad (13)$$

where $Y^A_{r_1,r_2,\bullet}$ represents the pixel at location (r_1, r_2) in the image A, $\bullet$ denotes all the corresponding spectral indices, and similarly for $Y^B_{s_1,s_2,\bullet}$. Further, ρ is the maximum vector metric. U is the set of unprocessed pixel indices of B, $M = \min\{\sharp\{A\}, \sharp\{B\}\}$, $\sharp\{A\}$ is the number of pixels in A, and similarly for $\sharp\{B\}$. We define $\min\{\emptyset\} = 0$.

Table 1 illustrates color modeling quality presented in 3DmdCAR model examples. While the bark synthesis from the simpler 3DCAR model has comparable quality, their 3DmdCAR counterparts, textile, flowers, wood, and plants achieved much better appearance from the novel model. This can be expected for the unstable 3DCAR models (Fig. 3 - textile, wood) or unstable 3DmdCAR-PF (Fig. 2 sweets) model, but in the case of stable plants texture (Fig. 4) it is a less obvious conclusion. The ζ criterion value shows that the best spectral modeling quality was achieved for the 3DmdCAR-PF wood texture synthesis (Table 1) moreover, bark, textile, and plants results are slightly worse. The sweets texture, on the contrary, is the most complicated for this model. Similarly, Table 2 indicates the 9×9 median size to be optimal for the lichen texture although faster and smaller $2 \times 2, 3 \times 3$ medians can be used with slight compromise as well.

5   Conclusions

The presented 3DmfCAR model in its median version 3DmdCAR exhibits outstanding modeling capability on a wide range of natural or artificial color textures representing visual properties of surface materials. The model is inherently multispectral, and thus it can be used to model any number of spectral or hyperspectral bands. The major advantage of the model is that it can be analytically estimated as well as easily synthesized and used for seamless texture enlargement to fill any required size. The model tends to stabilize the simpler 3DCAR model of the same order. Visual properties of the model can be easily changed using a modification of the auxiliary field properties. The drawback of the proposed model is more parameters that have to be estimated, and thus also a large learning set required. However, even this more extensive learning set is negligible with the learning set required for convolutional neural networks.

The proposed model can be used either in its stand-alone version or as a primary factor for a more complex BTF or compound BTF models.

Acknowledgments. The Czech Science Foundation project GAČR 19-12340S supported this research.

References

1. Chandler, D.M., Hemami, S.S.: VSNR: a wavelet-based visual signal-to-noise ratio for natural images. IEEE Trans. Image Process. **16**(9), 2284–2298 (2007)
2. Dana, K.J., Nayar, S.K., van Ginneken, B., Koenderink, J.J.: Reflectance and texture of real-world surfaces. In: CVPR, pp. 151–157. IEEE Computer Society (1997)
3. Haindl, M., Filip, J.: Fast BTF texture modelling. In: Chantler, M. (ed.) Texture 2003. Proceedings, pp. 47–52. IEEE Press, Edinburgh, October 2003
4. Haindl, M., Filip, J.: A fast probabilistic bidirectional texture function model. In: Campilho, A., Kamel, M. (eds.) ICIAR 2004. LNCS, vol. 3212, pp. 298–305. Springer, Heidelberg (2004). https://doi.org/10.1007/978-3-540-30126-4_37
5. Haindl, M., Filip, J., Arnold, M.: BTF image space utmost compression and modelling method. In: Kittler, J., Petrou, M., Nixon, M. (eds.) Proceedings of the 17th IAPR International Conference on Pattern Recognition, vol. III, pp. 194–197. IEEE Press, Los Alamitos, August 2004. http://dx.doi.org/10.1109/ICPR.2004.1334501
6. Haindl, M., Havlíček, V.: A multiscale colour texture model. In: Kasturi, R., Laurendeau, D., Suen, C. (eds.) Proceedings of the 16th International Conference on Pattern Recognition, pp. 255–258. IEEE Computer Society, Los Alamitos, August 2002. http://dx.doi.org/10.1109/ICPR.2002.1044676
7. Haindl, M., Havlíček, V.: A compound MRF texture model. In: Proceedings of the 20th International Conference on Pattern Recognition, ICPR 2010, pp. 1702–1705. IEEE Computer Society CPS, Los Alamitos, August 2010. https://doi.org/10.1109/ICPR.2010.442. http://doi.ieeecomputersociety.org/10.1109/ICPR.2010.442
8. Haindl, M., Kudělka, M.: Texture fidelity benchmark. In: 2014 International Workshop on Computational Intelligence for Multimedia Understanding (IWCIM), pp. 1–5. IEEE Computer Society CPS, Los Alamitos, November 2014. https://doi.org/10.1109/IWCIM.2014.7008812. http://ieeexplore.ieee.org/stamp/stamp.jsp?tp=&arnumber=7008812&isnumber=7008791
9. Haindl, M., Mikeš, S.: Texture segmentation benchmark. In: Lovell, B., Laurendeau, D., Duin, R. (eds.) Proceedings of the 19th International Conference on Pattern Recognition, ICPR 2008, pp. 1–4. IEEE Computer Society, Los Alamitos, December 2008. https://doi.org/10.1109/ICPR.2008.4761118. http://doi.ieeecomputersociety.org/10.1109/ICPR.2008.4761118
10. Haindl, M.: Visual data recognition and modeling based on local markovian models. In: Florack, L., Duits, R., Jongbloed, G., Lieshout, M.C., Davies, L. (eds.) Mathematical Methods for Signal and Image Analysis and Representation. Computational Imaging and Vision, vol. 41, pp. 241–259. Springer, London (2012). https://doi.org/10.1007/978-1-4471-2353-8_14
11. Haindl, M., Filip, J.: Extreme compression and modeling of bidirectional texture function. IEEE Trans. Pattern Anal. Mach. Intell. **29**(10), 1859–1865 (2007). https://doi.org/10.1109/TPAMI.2007.1139. http://doi.ieeecomputersociety.org/10.1109/TPAMI.2007.1139
12. Haindl, M., Filip, J.: Visual Texture. Advances in Computer Vision and Pattern Recognition. Springer, London (2013). https://doi.org/10.1007/978-1-4471-4902-6

13. Haindl, M., Havlíček, M.: Bidirectional texture function simultaneous autoregressive model. In: Salerno, E., Çetin, A.E., Salvetti, O. (eds.) MUSCLE 2011. LNCS, vol. 7252, pp. 149–159. Springer, Heidelberg (2012). https://doi.org/10.1007/978-3-642-32436-9_13. http://www.springerlink.com/content/hj32551334g61647/

14. Haindl, M., Havlíček, V.: A plausible texture enlargement and editing compound markovian model. In: Salerno, E., Çetin, A.E., Salvetti, O. (eds.) MUSCLE 2011. LNCS, vol. 7252, pp. 138–148. Springer, Heidelberg (2012). https://doi.org/10.1007/978-3-642-32436-9_12. http://www.springerlink.com/content/047124j43073m202/

15. Haindl, M., Havlíček, V.: BTF compound texture model with non-parametric control field. In: The 24th International Conference on Pattern Recognition (ICPR 2018), pp. 1151–1156. IEEE, August 2018. http://www.icpr2018.org/

16. Haindl, M., Mikeš, S.: A competition in unsupervised color image segmentation. Pattern Recogn. **57**(9), 136–151 (2016). https://doi.org/10.1016/j.patcog.2016.03.003. http://www.sciencedirect.com/science/article/pii/S0031320316000984

17. Havlíček, M., Haindl, M.: Texture spectral similarity criteria. IET Image Process. **13**(6), 1998–2007 (2019). https://doi.org/10.1049/iet-ipr.2019.0250

18. Jeng, F.C., Woods, J.W.: Compound Gauss-Markov random fields for image estimation. IEEE Trans. Signal Process. **39**(3), 683–697 (1991)

19. Kudělka, M., Haindl, M.: Texture fidelity criterion. In: 2016 IEEE International Conference on Image Processing (ICIP), pp. 2062–2066. IEEE, September 2016. https://doi.org/10.1109/ICIP.2016.7532721. http://2016.ieeeicip.org/

20. Pickard, R., Graszyk, C., Mann, S., Wachman, J., Pickard, L., Campbell, L.: Vistex database. Technical report, MIT Media Laboratory, Cambridge (1995)

21. Sheikh, H., Bovik, A.: Image information and visual quality. IEEE Trans. Image Process. **15**(2), 430–444 (2006)

22. Wang, Z., Bovik, A.C., Sheikh, H.R., Simoncelli, E.P.: Image quality assessment: from error visibility to structural similarity. IEEE Trans. Image Process. **13**(4), 600–612 (2004). https://doi.org/10.1109/TIP.2003.819861

23. Wang, Z., Bovik, A.: Mean squared error: love it or leave it? A new look at signal fidelity measures. IEEE Signal Process. Mag. **26**(1), 98–117 (2009)

24. Wang, Z., Simoncelli, E.P.: Translation insensitive image similarity in complex wavelet domain. In: Proceedings of the IEEE International Conference on Acoustics, Speech, and Signal Processing, ICASSP 2005, pp. 573–576 (2005)

25. Zujovic, J., Pappas, T., Neuhoff, D.: Structural texture similarity metrics for image analysis and retrieval. IEEE Trans. Image Process. **22**(7), 2545–2558 (2013). https://doi.org/10.1109/TIP.2013.2251645

A Segment Level Approach to Speech Emotion Recognition Using Transfer Learning

Sourav Sahoo[1]([✉]) [iD], Puneet Kumar[2] [iD], Balasubramanian Raman[2] [iD], and Partha Pratim Roy[2] [iD]

[1] Department of Electrical Engineering,
Indian Institute of Technology Madras, Chennai 600036, India
sourav.sahoo@smail.iitm.ac.in
[2] Department of Computer Science and Engineering,
Indian Institute of Technology Roorkee, Roorkee 247667, India
pkumar99@cs.iitr.ac.in, {balarfcs,proy.fcs}@iitr.ac.in

Abstract. Speech emotion recognition (SER) is a non-trivial task considering that the very definition of emotion is ambiguous. In this paper, we propose a speech emotion recognition system that predicts emotions for multiple segments of a single audio clip unlike the conventional emotion recognition models that predict the emotion of an entire audio clip directly. The proposed system consists of a pre-trained deep convolutional neural network (CNN) followed by a single layered neural network which predicts the emotion classes of the audio segments. The predictions for the individual segments are finally combined to predict the emotion of a particular clip. We define several new types of accuracies while evaluating the performance of the proposed model. The proposed model attains an accuracy of 68.7% surpassing the current state-of-the-art models in classifying the data into one of the four emotional classes (*angry, happy, sad* and *neutral*) when trained and evaluated on IEMOCAP audio-only dataset.

Keywords: Emotion recognition · Affective computing · Deep learning · Mel spectrograms · Computational paralinguistics

1 Introduction

Speech is one of the most natural means of communication. The semantics as well as the emotional prosody of speech are both essential for conveying any information through it. Despite the remarkable advances made in speech related tasks such as speech recognition [1] and text-to-speech synthesis [22], *natural* emotion

---

Electronic supplementary material The online version of this chapter (https://doi.org/10.1007/978-3-030-41299-9_34) contains supplementary material, which is available to authorized users.

© Springer Nature Switzerland AG 2020
S. Palaiahnakote et al. (Eds.): ACPR 2019, LNCS 12047, pp. 435–448, 2020.
https://doi.org/10.1007/978-3-030-41299-9_34

understanding is still an unaccomplished capability for the computational systems. Speech emotion recognition is essential in the domains that require a significant amount of man-machine interaction. In the recent years, conversational interfaces or voice assistants have become ubiquitous through smartphones and home automation [3]. These systems will perform better in certain situations if they can capture and process both the semantics as well as the emotional content of speech. Speech emotion recognition is challenging due to a number of reasons. It is difficult to strictly categorize different emotions because the very definition of emotion is obscure [14]. Large scale annotated emotional datasets are required for the training of complex emotion recognition systems. However, creating such large datasets is cost prohibitive due to the extensive human efforts involved, which is another significant challenge.

In this work, we attempt to address the data insufficiency challenge with a simple yet an elegant solution. Following two ideas are presented to overcome the same: (1) train and test the classification model on *multiple segments*, both overlapping and non-overlapping segments, of the audio clip rather than the entire audio clip as a whole and (2) use *transfer learning* to improve the model performance. In the recent years, transfer learning has successfully tackled data insufficiency challenge up to a great extent. In this study, we specifically use *inductive transfer learning*, a transfer learning method in which the horizon of possible models is reduced by implementing a model trained on a different but related task [21]. We propose a new emotion recognition model which uses Google VGGish [11], a deep convolutional neural network followed by a single layered neural network for classification. We conducted multiple experiments to investigate various architectures and hyperparameters. The model when trained and evaluated on overlapping segments, achieves an accuracy of 68.7% and outperforms the current state-of-the-art model [33] by 6.3% relative (4.1% absolute) accuracy in speech emotion recognition on IEMOCAP audio-only dataset.

The rest of the paper is organized as follows. Existing techniques in the context of speech emotion recognition have been reviewed in Sect. 2. Section 3.1 depicts two methods to partition the audio clips in the dataset into multiple segments. The model predicts the emotion class for multiple segments of a single audio clip rather than predicting the emotion class of the entire audio clip at once. The individual predictions are finally incorporated for predicting the emotion class of the entire clip. Several new types of accuracies are defined in Sect. 3.3 to evaluate the performance of the model. In Sect. 4.4, we discuss the outcome of our experiments and compare the performance of the proposed model with the existing models in speech emotion recognition and finally conclude in Sect. 5.

2 Related Work

Speech emotion recognition is a well studied research area in which several architectures, techniques and approaches have been deployed. In this section we briefly review the existing work in this domain.

2.1 Traditional Machine Learning Approaches

Traditional machine learning methods such as hidden markov models (HMM), support vector machines (SVM) and decision-trees etc. have been utilized for speech emotion recognition problems [16,26,27]. A recent work by Sahu [24] has shown that an ensemble of multiple traditional machine learning methods can achieve performance as good as the latest models in emotion recognition. All these methods extensively explored various features that determine the emotion contained in speech. However, a major drawback of traditional machine learning techniques is that a prior knowledge of all the necessary features that influence emotion recognition like fundamental frequency (F0), energy etc. is required.

2.2 Deep Learning Approaches

Deep neural networks were deployed for automatic extraction of high-level features from audio and were shown to be successful for speech emotion recognition [10]. Since then, several neural network architectures have been deployed for this task. Zheng et al. [35] did an experimental study on the use of convolutional neural network (CNN) for speaker independent emotion recognition system. Their system determined that deep learning methods outperform traditional machine learning techniques for SER. Variants of recurrent neural networks (RNN) like bidirectional long-short term memory (BLSTM) have proven to be successful in emotion recognition [17]. In an another work, Trigeorgis et al. [32] deployed a combination of CNN and RNN to efficiently recognize emotions in speech samples.

2.3 Audio Segmentation Based Approaches

It was demonstrated that a speech segment longer than 0.25 s carries sufficient information for detecting the emotion present in it [23]. Since then, various research attempts have been made to detect emotion from multiple segments of audio clip instead of processing the clip at once. A natural advantage of using segments instead of clips is that the model learns the salient features that determine the presence of a particular emotion in speech in a more elaborated manner. On the other hand, a potential downside of using segments is that slicing the clip into non-overlapping segments causes loss of correlation and flow of the speech. In this context, a study by Shami and Kamel [28] combined the use of segment level and utterance level features for emotion recognition. Satt et al. [25] presented a system which detected emotion at segment level whose performance was comparable to the state-of-the-art model in SER.

2.4 Transfer Learning in SER

Transfer learning has been applied in SER in multiple ways. One of the approaches is learning features from one emotion dataset and applying it on another emotion dataset. Song et al. [30] had presented a similar approach for use of transfer learning for cross-corpus speech emotion recognition. Since many paralinguistic tasks are closely related, a different approach is learning the features from other paralinguistic tasks such as - speaker or gender recognition and applying it on emotion recognition [9]. Both the approaches were successful but the idea relies on paralinguistic datasets which are currently very limited in terms of number as well as size. Badshah et al. [2] made an attempt to use pre-trained model in emotion recognition where they compared the performance of fine-tuned AlexNet [15] and a CNN trained from scratch. The freshly trained CNN outperformed the fine-tuned AlexNet which was not very surprising considering that AlexNet is trained on ImageNet [7], a large-scale image database.

In this paper, segments are used rather than entire clips for emotion detection. To investigate the loss of correlation between segments, both overlapping and non-overlapping segmentations are experimented. We use mel spectrograms as opposed to numerous low-level hand-crafted features which were practiced in traditional machine learning approaches. We use a model pre-trained on a generic audio database instead of paralinguistic or emotional database that the previous papers utilized for transfer learning.

3   The Proposed Method

The audio clips are segmented using two different methods i.e. overlapping and non-overlapping audio segmentation. These segments are given as inputs to the proposed system. The proposed system comprises of a generator which generates mel spectrogram from raw audio input which is passed into a pre-trained deep CNN. The CNN produces a 128-dimensional embedding from the mel spectrogram which passes through a single layered neural network which finally predicts

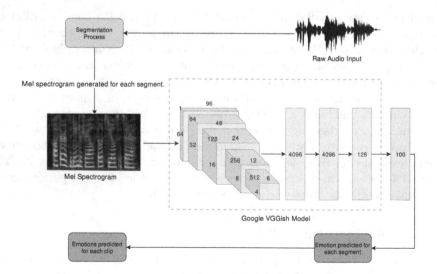

Fig. 1. Visual representation of the proposed method

the emotion class. The entire methodology has been shown in Fig. 1 and elaborated in the following subsections.

3.1 Audio Segmentation

Non-overlapping Segmentation. We extract non-overlapping segments of one-second duration and pad the last segment with silence to also make its length one second. By applying this process for the entire dataset, we get a total of 49795 segments out of which we use 27935 segments that belong to the classes that are relevant in this context. Table 1 may be referred for detailed sample distribution of the dataset after non-overlapping segmentation.

Overlapping Segmentation. We extract overlapping segments of one-second duration and pad the last segment with silence to also make its length one second. The overlapping duration is 0.5 s for all the segments. Overlapping segmentation serves two main purposes: (1) it captures better correlation among the segments of the clip (2) it increases the number of data points. By applying the process for the entire dataset, we get a total of 91017 segments out of which we use 51180 segments that belong to the classes that are relevant in this context. The detailed sample distribution of the dataset after overlapping segmentation has been presented in Table 1.

For both types of segmentations, we hypothesize that if an utterance belongs to class X, then each segment of the utterance also belongs to the same class X. A visual representation of the segmentation of a toy example has been shown in Fig. 2.

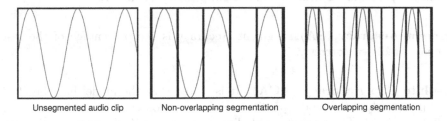

Unsegmented audio clip   Non-overlapping segmentation   Overlapping segmentation

Fig. 2. Segmentation process. The first image shows an unsegmented audio clip. The second image shows the five segments obtained after non-overlapping segmentation. The third image shows the segments obtained after overlapping segmentation. The duration of each segment is *same* in both the cases. The zero padding which is done during segmentation process is visible in the last segment in case of overlapping segmentation.

3.2 Spectrogram Generator

Mel spectrogram generation method, as adopted from Google VGGish paper [11] has been described as follows. For each 1000 ms segment, Short-Time Fourier

Transform (STFT) magnitude is computed using a 25 ms length window, 10 ms hop and Hann window function. Mathematically, the expression for Hann window is:

$$w[k] = 0.5 \left[1 - cos\left(\frac{2\pi k}{N}\right) \right] \tag{1}$$

where $N =$ window length and $k = 0, 1, 2 \dots N - 1$. The log mel-spectrogram of 96×64 patches has been obtained from the resulting spectrogram by integrating it into the 64 mel-spaced frequency bins. A small offset of 0.01 has been added to avoid numerical issues and then the magnitude of each bin is log transformed.

3.3 Evaluation Metrics

The unsegmented dataset is represented as $\mathcal{D}_{\mathcal{US}} = \{(c_0, y_0), \dots, (c_{N-1}, y_{N-1})\}$ where c_i is the ith clip, y_i is the corresponding emotion class and N is the number of relevant utterances in total i.e. 5531 in this case. Let $\mathcal{N} = \{n_0, \dots n_{N-1}\}$ where n_i is the number of segments of clip c_i for $i = 0, 1, \dots, N-1$. After segmentation, the dataset is $\mathcal{D}_{\mathcal{S}} = \{(s_{0,0}, y_0), \dots (s_{0,n_0}, y_0), (s_{1,0}, y_1), \dots, (s_{N-1,n_{N-1}}, y_{N-1})\}$ where $s_{i,j}$ represents the jth segment of the ith clip and y_i represents the corresponding emotion class. Let the segmented test dataset be $\mathcal{T}_{\mathcal{S}} \subset \mathcal{D}_{\mathcal{S}}$ and $\mathcal{Y}_{\mathcal{S}} = \{y_0, \dots, y_{T-1}\}$ be the correct emotion classes of the segments in $\mathcal{T}_{\mathcal{S}}$, where $T = |\mathcal{T}_{\mathcal{S}}|$. Suppose a clip $c \in \mathcal{T}_{\mathcal{S}}$ consists of k segments $s_0, \dots s_{k-1}$ and $\mathcal{P} = \{p_0, \dots p_{k-1}\}$ be the corresponding prediction, where p_i is the predicted emotion class for s_i for $i = 0, 1, \dots, k - 1$. Let $\mathcal{M}$ be the set of emotions that are predicted for maximum number of segments of c i.e. the set of elements that appear maximum number of times in $\mathcal{P}$. Explanation of various types of accuracies is provided in the subsequent sections and more details about their calculation procedure is available in the supplementary material.

Segment Accuracy (SA). Segment Accuracy is the percentage of the test segments predicted correctly.

Absolute Clip Accuracy (ACA). It is stated that the model has classified the clip correctly if the model predicts correct emotion class for *each* s_i in c for $i = 0, 1, \dots, k - 1$. The percentage of clips that are classified correctly using the aforementioned criterion is defined as Absolute Clip Accuracy.

Standard Clip Accuracy (SCA). If $|\mathcal{M}| = 1$ and the emotion in $\mathcal{M}$ is the correct emotion class of the clip, then it is stated that the model has classified correctly. Standard Clip Accuracy is the percentage of clips that are classified correctly using the aforementioned criterion.

Average Logits Clip Accuracy (ALCA). The final layer of classification model, also called the logits layer, gives an n-length array of floating point values called logits, where n is the number of classes (which is 4 in this context).

We compute the average value of logits over all the segments of a particular clip and state the argument of the maximum value in the average n-length array as the predicted class. The percentage of clips that are classified correctly using the aforementioned criterion is called Average Logits Clip Accuracy.

Best Clip Accuracy (BCA). If the correct emotion of the clip $e \in \mathcal{M}$, then we state that the model has classified the clip correctly. The percentage of clips that are classified correctly using this criterion is termed as Best Clip Accuracy.

4   Experiments and Results

4.1   Experimental Setup

The experiments have been performed on NVIDIA Quadro P5000 graphics processing unit (GPU) and Intel Xeon central processing unit (CPU) using Tensor-Flow Deep Learning library[1]. The model training utilized mini-batch size of 32 and Adam optimizer [13] with learning rate of 10^{-6}. The learning rate is kept very low as compared to the default value of 10^{-3} because we are fine-tuning a pre-trained model instead of training it from scratch. The model is prone to overfitting due to lack of sufficient data as well as very high number of parameters. So, early stopping [6] is used to counter overfitting by fixing the value of patience to be 50.

4.2   Dataset

The proposed model is trained and evaluated on the Interactive Emotional Dyadic Motion Capture (IEMOCAP) dataset [5]. We choose IEMOCAP dataset to conduct our experiments because (1) the dataset is more diverse and larger than other popular emotional database like Emo-DB [4] or RAVDESS [18] and (2) most of the state-of-the-art models like [33,34] have been trained and tested on this database. The dataset consists of five recorded sessions of conversations, each containing utterances from two speakers (one female and one male). The dataset contains audio, audio + video and corresponding transcriptions. In this paper, the audio-only dataset is used. The audio clips are sampled at 16 kHz. Each of 10039 utterances has been labeled as one of the following classes - angry, happy, sad, neutral, frustrated, excited, fear, surprise, disgust and others by expert annotators. We use only four emotion classes i.e. *angry, happy, sad* and *neutral* for consistent comparison with the previous works that used IEMOCAP dataset [33,34] and utterances labelled as *excited* are merged with those labelled as *happy*. So, the final dataset contains 5531 utterances. The dataset is divided into train, validation and test sets in the ratio 8:1:1. Table 1 may be referred for detailed sample distribution of the dataset.

---

[1] https://www.tensorflow.org/.

Table 1. Sample distribution of IEMOCAP dataset before and after segmentation

	Before seg.	Non-overlapping seg.	Overlapping seg.
Total	10039	49795	91017
Relevant	5531	27935	51180
Rel. Frac.	55.10%	56.10%	56.23%
Angry	19.94%	19.80%	19.75%
Happy	29.58%	30.04%	30.06%
Sad	19.60%	23.25%	23.76%
Neutral	30.88%	26.91%	26.44%

where Seg. = Segmentation, Rel. Frac. = Relevance Fraction i.e. percentage of the entire dataset that we are considering for this paper. It is to be noted that *happy* includes both *happy* and *excited* data.

4.3 Model Architecture

A baseline model is defined against which we compare our results. The proposed architecture has two components: (a) VGGish, which is a deep CNN and (b) a single layered neural network, also called the classification model. The detailed architecture of the models is discussed in the following sections.

Baseline Model. The input to the baseline model is 96×64 dimensional mel spectrogram of non-overlapping segments of audio clips in the training set. A fully connected neural network of N layers with M units in each layer is considered. We try $N = [2, 3, 4]$ and $M = [100, 200]$ and choose the best performing model. A dropout [31] layer with $p = 0.5$ and batch normalization [12] layer is used between every fully connected layer. All the fully connected layers use ReLU [19] activation function. We get the best results with $N = 2$ layers, $M = 200$ units which has been considered as the baseline model.

Proposed Architecture

– **VGGish Model:** The Google VGGish [11] model is a deep convolutional neural network which has an architecture very similar to that of VGG [29] model that was designed for large scale image classification. The VGGish model has been pre-trained on AudioSet [8], a collection of ~2M human labelled ten second length audio clips from YouTube videos spread over ~600 sound classes. The VGGish model takes a 96×64 dimensional mel spectrogram as input. The VGGish architecture comprises of four blocks of two dimensional convolution and max-pooling layers. The final max-pooling layer is followed by 2 fully connected layers each comprising of 4096 units and finally a fully connected layer of 128 units which generates the embedding vector. All the convolution and fully connected layers use ReLU activation function. The model has ~72 million parameters. The architecture of VGGish model has been described in Table 2.

- **Classification Model:** The 128-dimensional embedding vector passes into to a single layered neural network comprising of a fully connected layer with N units followed by the final fully connected layer, the logits layer, which predicts the emotion class of each segment of the audio clip. We have tried different values of $N = [100, 200, 400]$ to find the best performing model.

Table 2. VGGish network architecture

Layer	Activation size
Input	$1 \times 96 \times 64$
$64 \times 3 \times 3$ conv, stride 1	$64 \times 96 \times 64$
2×2 maxpool, stride 2	$64 \times 48 \times 32$
$128 \times 3 \times 3$ conv, stride 1	$128 \times 48 \times 32$
2×2 maxpool, stride 2	$128 \times 24 \times 16$
$256 \times 3 \times 3$ conv, stride 1	$256 \times 24 \times 16$
$256 \times 3 \times 3$ conv, stride 1	$256 \times 24 \times 16$
2×2 maxpool, stride 2	$256 \times 12 \times 8$
$512 \times 3 \times 3$ conv, stride 1	$512 \times 12 \times 8$
$512 \times 3 \times 3$ conv, stride 1	$512 \times 12 \times 8$
2×2 maxpool, stride 2	$512 \times 6 \times 4$
Flatten	1×12288
Fully connected I	1×4096
Fully connected II	1×4096
Output	1×128

where $C \times H \times W$ conv denotes a 2D convolutional layer with C filters of size $H \times W$. $H \times W$ maxpool denotes a max-pooling layer of pooling size $H \times W$.

4.4 Results and Discussion

The model is trained and evaluated six times on IEMOCAP audio-only dataset and the mean accuracy and standard deviation is reported. The single layered neural network that follows the VGGish model consists of $N = 200$ units because the value of N corresponding to the model with best performance is observed to be 200. The performance of the proposed system has been compared for overlapping and non-overlapping segments using the evaluation metrics mentioned in Sect. 3.3 in Table 3.

In Table 4, we compare the performance of the model when overlapping segments are given as input to three different models with $N = 100, N = 200$ and $N = 400$ units in the penultimate layer of the model. The performance of the proposed model is compared with various existing state-of-the-art models in SER in Table 5. We have calculated multiple evaluation metrics as defined in Sect. 3.3. However, *Average Logits Clip Accuracy* has been primarily used for the comparison because existing studies on segment level approach in SER have used similar metrics to compare their models with the conventional models [25].

Table 3. Performance of model with overlapping and non-overlapping segmentation

	Non-overlapping seg.	Overlapping seg.
SA	**0.564 ± 0.006**	0.561 ± 0.011
ACA	**0.196 ± 0.010**	0.125 ± 0.010
SCA	0.559 ± 0.009	**0.621 ± 0.019**
ALCA	0.668 ± 0.014	**0.687 ± 0.019**
BCA	**0.707 ± 0.011**	0.703 ± 0.017

where SA: Segment Accuracy, ACA: Absolute Clip Accuracy, SCA: Standard Clip Accuracy, ALCA: Average Logits Clip Accuracy, BCA: Best Clip Accuracy.

Table 4. Comparison of models with different values of N on overlapping segments

	$N = 100$	$N = 200$	$N = 400$
SA	0.560 ± 0.012	**0.561 ± 0.011**	0.559 ± 0.017
ACA	**0.132 ± 0.012**	0.125 ± 0.010	0.131 ± 0.010
SCA	0.616 ± 0.017	**0.621 ± 0.019**	0.617 ± 0.022
ALCA	0.684 ± 0.020	**0.687 ± 0.019**	0.674 ± 0.018
BCA	**0.703 ± 0.013**	0.703 ± 0.017	0.699 ± 0.020

where SA: Segment Accuracy, ACA: Absolute Clip Accuracy, SCA: Standard Clip Accuracy, ALCA: Average Logits Clip Accuracy, BCA: Best Clip Accuracy.

Table 5. Comparison of different state-of-the-art models on IEMOCAP dataset

Model name	Modality	Accuracy
Baseline	A	0.511
ARE [34]	A	0.546
ACNN [20]	A	0.561
Ensemble [24]	A	0.562
audio-BRE [33]	A	0.646
Ensemble [24]	T	0.631
TRE [34]	T	0.635
Proposed (Non-overlap)	A	0.668 ± 0.014
Proposed (Overlap)	A	**0.687 ± 0.019**

where ARE: Audio Recurrent Encoder, ACNN: Attentive CNN, Ensemble: Ensemble of six traditional machine learning methods, BRE: Bidirectional Recurrent Encoder, TRE: Text Recurrent Encoder. A = Audio-only, T = Text-only

Discussion. Even though SA of overlapping segments is 0.3% lower than that of non-overlapping segments, the SCA of the model trained on overlapping segments is 6.2% better than the model trained on non-overlapping segments. Similarly, the ALCA in case of overlapping segments is 1.9% better than non-overlapping segments. This is expected because overlapping segments capture the correlation between the different segments of the clip which is not present in non-overlapping segments. The sharp difference in SCA of overlapping and non-overlapping segments proves that the model learns to recognize the emotions precisely in case of overlapping segments because by definition, SCA is the percentage of clips that are said to be classified correctly when the model predicts *exactly one* emotion for majority of the segments of a clip. The difference between SCA and BCA in case of overlapping segments (8.2%) is much lower than that of non-overlapping segments (14.8%). This observation shows that instead of predicting multiple classes with equal frequency for the segments of a particular clip, the model can pick out a *single* class with maximum frequency when trained on overlapping segments. There is a 7.1% decrease in ACA in case of overlapping segments which is anticipated due to the increased number of segments (almost 2x) of the clip because of which it is difficult to predict the correct class for *every* segment of a clip.

In Table 4, we observe the performance of model when trained on overlapping segments as N varies from 100 to 400. As we increase N from 100 to 400, we do not notice any strict trend in the performance of the model. However, all the experiments resulted in a slightly better ALCA of the model with $N = 200$ units which we use in Table 5 for comparison with the existing models in SER. Apart from ALCA, SA and SCA are also better for $N = 200$ than $N = 100$ or $N = 400$.

As per the surveyed literature, the current state-of-the-art model for SER on IEMOCAP audio-only dataset is audio-BRE, a bidirectional recurrent encoder. The proposed model shows higher performance than the audio-BRE by 6.3% relative (0.646 to 0.687 absolute) accuracy. Most of the existing works on IEMO-CAP dataset have shown that emotion recognition systems show better performance on transcriptions than the audio-only dataset [24,33,34]. Although our model is trained on audio-only dataset, it shows better performance than some of the models trained on text-only dataset, as demonstrated in Table 5.

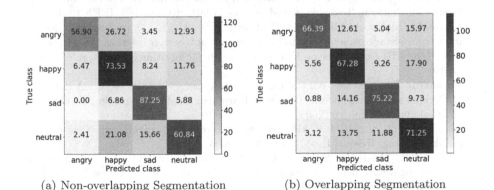

(a) Non-overlapping Segmentation   (b) Overlapping Segmentation

Fig. 3. Confusion matrix

We compute the confusion matrix for both overlapping and non-overlapping cases which is presented in Fig. 3. In Fig. 3a, we observe that model incorrectly classifies most examples of *angry* as *happy* (26.72%). The true positives in case of non-overlapping is low for *angry* (56.9%) and *neutral* (60.84%) as compared to the other emotions. These numbers are improved when we observe the case of overlapping segments i.e. 66.39% for *angry* and 71.25% for *neutral*. Although Neumann and Vu [20] and Yoon et al. [34] have shown that most of the emotions are gotten confused with *neutral* class during SER because it lies in the centre of activation-valence space, the proposed model shows much less confusion when trained on overlapping segments.

5   Conclusion

We present a segment level approach for speech emotion recognition using transfer learning in this paper. The proposed approach consisting of a single layered neural network on top of a pre-trained CNN outperformed the current state-of-the-art emotion classification model on IEMOCAP audio-only dataset by a relative accuracy of 6.3%. The improved performance proves the applicability of transfer learning for SER. In future, we will focus on incorporating transcriptions and audio-visual data to design a model with better performance in emotion recognition. A different research could focus on developing an intelligent segmentation process instead of using fixed segment length or fixed overlapping duration.

References

1. Amodei, D., et al.: Deep speech 2: end-to-end speech recognition in English and mandarin. In: International Conference on Machine Learning (ICML), pp. 173–182 (2016)
2. Badshah, A.M., Ahmad, J., Rahim, N., Baik, S.W.: Speech emotion recognition from spectrograms with deep convolutional neural network. In: International Conference on Platform Technology and Service (PlatCon), pp. 1–5. IEEE (2017)
3. Braun, M., Mainz, A., Chadowitz, R., Pfleging, B., Alt, F.: At your service: designing voice assistant personalities to improve automotive user interfaces. In: Proceedings of the CHI Conference on Human Factors in Computing Systems, p. 40. ACM (2019)
4. Burkhardt, F., Paeschke, A., Rolfes, M., Sendlmeier, W.F., Weiss, B.: A database of German emotional speech. In: Ninth European Conference on Speech Communication and Technology (2005)
5. Busso, C., et al.: IEMOCAP: interactive emotional dyadic motion capture database. Lang. Resour. Eval. **42**(4), 335 (2008)
6. Caruana, R., Lawrence, S., Giles, C.L.: Overfitting in neural nets: backpropagation, conjugate gradient, and early stopping. In: Advances in Neural Information Processing Systems, pp. 402–408 (2001)
7. Deng, J., Dong, W., Socher, R., Li, L.J., Li, K., Fei-Fei, L.: ImageNet: a large-scale hierarchical image database. In: Computer Vision and Pattern Recognition (CVPR). IEEE (2009)

8. Gemmeke, J.F., et al.: Audio set: an ontology and human-labeled dataset for audio events. In: Proceedings of the International Conference on Acoustics, Speech and Signal Processing (ICASSP), pp. 776–780. IEEE (2017)

9. Gideon, J., Khorram, S., Aldeneh, Z., Dimitriadis, D., Provost, E.M.: Progressive neural networks for transfer learning in emotion recognition. arXiv preprint arXiv:1706.03256 (2017)

10. Han, K., Yu, D., Tashev, I.: Speech emotion recognition using deep neural network and extreme learning machine. In: Fifteenth Annual Conference of the International Speech Communication Association (INTERSPEECH), pp. 223–227. ISCA (2014)

11. Hershey, S., et al.: CNN architectures for large-scale audio classification. In: International Conference on Acoustics, Speech and Signal Processing (ICASSP), pp. 131–135. IEEE (2017)

12. Ioffe, S., Szegedy, C.: Batch normalization: accelerating deep network training by reducing internal covariate shift. arXiv preprint arXiv:1502.03167 (2015)

13. Kingma, D.P., Ba, J.: Adam: a method for stochastic optimization. arXiv preprint arXiv:1412.6980 (2014)

14. Kleinginna, P.R., Kleinginna, A.M.: A categorized list of emotion definitions, with suggestions for a consensual definition. Motiv. Emot. 5, 345–379 (1981)

15. Krizhevsky, A., Sutskever, I., Hinton, G.E.: Imagenet classification with deep convolutional neural networks. In: Advances in Neural Information Processing Systems, pp. 1097–1105 (2012)

16. Lee, C.C., Mower, E., Busso, C., Lee, S., Narayanan, S.: Emotion recognition using a hierarchical binary decision tree approach. Speech Commun. 53(9–10), 1162–1171 (2011)

17. Lee, J., Tashev, I.: High-level feature representation using recurrent neural network for speech emotion recognition. In: Sixteenth Annual Conference of the International Speech Communication Association (INTERSPEECH), pp. 1537–1540. ISCA (2015)

18. Livingstone, S.R., Russo, F.A.: The ryerson audio-visual database of emotional speech and song (RAVDESS): a dynamic, multimodal set of facial and vocal expressions in North American English. PLoS ONE 13(5), e0196391 (2018)

19. Nair, V., Hinton, G.E.: Rectified linear units improve restricted Boltzmann machines. In: Proceedings of the 27th International Conference on Machine Learning (ICML), pp. 807–814 (2010)

20. Neumann, M., Vu, N.T.: Attentive convolutional neural network based speech emotion recognition: a study on the impact of input features, signal length, and acted speech. arXiv preprint arXiv:1706.00612 (2017)

21. Pan, S.J., Yang, Q.: A survey on transfer learning. Trans. Knowl. Data Eng. 22(10), 1345–1359 (2009)

22. Ping, W., et al.: Deep voice 3: scaling text-to-speech with convolutional sequence learning. arXiv preprint arXiv:1710.07654 (2017)

23. Provost, E.M.: Identifying salient sub-utterance emotion dynamics using flexible units and estimates of affective flow. In: International Conference on Acoustics, Speech and Signal Processing (ICASSP), pp. 3682–3686. IEEE (2013)

24. Sahu, G.: Multimodal speech emotion recognition and ambiguity resolution. arXiv preprint arXiv:1904.06022 (2019)

25. Satt, A., Rozenberg, S., Hoory, R.: Efficient emotion recognition from speech using deep learning on spectrograms. In: Eighteenth Annual Conference of the International Speech Communication Association (INTERSPEECH), pp. 1089–1093. ISCA (2017)

26. Schuller, B., Rigoll, G., Lang, M.: Hidden Markov model-based speech emotion recognition. In: Proceedings of the International Conference on Acoustics, Speech, and Signal Processing (ICASSP), pp. II-1. IEEE (2003)
27. Seehapoch, T., Wongthanavasu, S.: Speech emotion recognition using support vector machines. In: 5th International Conference on Knowledge and Smart Technology (KST), pp. 86–91. IEEE (2013)
28. Shami, M.T., Kamel, M.S.: Segment-based approach to the recognition of emotions in speech. In: International Conference on Multimedia and Expo, pp. 4-pp. IEEE (2005)
29. Simonyan, K., Zisserman, A.: Very deep convolutional networks for large-scale image recognition. arXiv preprint arXiv:1409.1556 (2014)
30. Song, P., Jin, Y., Zhao, L., Xin, M.: Speech emotion recognition using transfer learning. IEICE Trans. Inf. Syst. **97**(9), 2530–2532 (2014)
31. Srivastava, N., Hinton, G., Krizhevsky, A., Sutskever, I., Salakhutdinov, R.: Dropout: a simple way to prevent neural networks from overfitting. J. Mach. Learn. Res. **15**(1), 1929–1958 (2014)
32. Trigeorgis, G., et al.: Adieu features? End-to-end speech emotion recognition using a deep convolutional recurrent network. In: International Conference on Acoustics, Speech and Signal Processing (ICASSP), pp. 5200–5204. IEEE (2016)
33. Yoon, S., Byun, S., Dey, S., Jung, K.: Speech emotion recognition using multi-hop attention mechanism. In: International Conference on Acoustics, Speech and Signal Processing (ICASSP), pp. 2822–2826. IEEE (2019)
34. Yoon, S., Byun, S., Jung, K.: Multimodal speech emotion recognition using audio and text. In: Spoken Language Technology Workshop (SLT), pp. 112–118. IEEE (2018)
35. Zheng, W., Yu, J., Zou, Y.: An experimental study of speech emotion recognition based on deep convolutional neural networks. In: International Conference on Affective Computing and Intelligent Interaction (ACII), pp. 827–831. IEEE (2015)

A Decomposition Based Multi-objective Genetic Programming Algorithm for Classification of Highly Imbalanced Tandem Mass Spectrometry

Samaneh Azari[1]($\boxtimes$), Bing Xue[1], Mengjie Zhang[1], and Lifeng Peng[2]

[1] School of Engineering and Computer Science, Victoria University of Wellington,
PO Box 600, Wellington 6140, New Zealand
{samaneh.azari,bing.xue,mengjie.zhang}@ecs.vuw.ac.nz
[2] Centre for Biodiscovery and School of Biological Sciences,
Victoria University of Wellington, PO Box 600, Wellington 6140, New Zealand
lifeng.peng@vuw.ac.nz

Abstract. Preprocessing tandem mass spectra to classify the signal and noise peaks plays a crucial role for improving the accuracy of most peptide identification algorithms. As a CID tandem mass spectra dataset is highly imbalanced with high noise ratio and a small number of signal peaks (low signal to noise ratio), a classification strategy which is able to maintain the performance trade-off between the minority (signal) and the majority (noise) class accuracies prior to peptide identification is required. Therefore, this paper proposes a Multi-Objective Genetic Programming (MOGP) approach based on the idea of MOEA/D, named MOGP/D, to evolve a Pareto front of classifiers along the optimal trade-off surface that offers the best compromises between objectives. In comparison with an NSGA-II base MOGP method, called NSGP, with decreasing the signal to noise ratio, MOGP/D produces better solutions in the region of interest (centre of the Pareto front) according to the hypervolume indicator on the training sets. Moreover, the best compromise solution achieved by the proposed method is compared with the best single objective GP and the best of NSGP, and the results show that MOGP/D retains a reasonable number of signal peaks and filters more noise peaks compared to the other two methods. To further evaluate the effectiveness of MOGP/D, the preprocessed MS/MS data is submitted to the mostly used *de novo* sequencing software, PEAKS, to identify the peptides. The results show that the proposed multi-objective GP method improves the reliability of peptide identification compared to the single objective GP.

Keywords: Genetic programming · Multi-objective optimisation · Imbanalced binary classification · Tandem mass spectrometry

© Springer Nature Switzerland AG 2020
S. Palaiahnakote et al. (Eds.): ACPR 2019, LNCS 12047, pp. 449–463, 2020.
https://doi.org/10.1007/978-3-030-41299-9_35

1   Introduction

Effective identification of peptides and their modifications enables scientists to study the genetic diseases. Tandem mass spectrometry (MS/MS) is the most widely used technique to peptide identification. However, MS/MS spectra include a massive number of noise peaks compared to the signal peaks and this results in poor peptide identification as noise peaks could be the cause of false discovery in identification results. Low confidence peptide identification results in less reliable protein inference since the proteins are inferred by assembling a set of identified peptide sequences. To overcome the problems imposed by the noise, a preprocessing step to denoise the MS/MS spectra in order to increase the overall confidence of peptide identification is required [1].

In our previous works [2,3], genetic programming (GP) was successfully used in solving imbalanced classification problems aiming at improving the reliability of peptide identification. Being a population-based problem solving technique, GP has been proved to be more stable compared to the other classification algorithms including decision tree (DT), k-nearest neighbour (K-NN), multilayer perceptron (MLP), naive Bayes (NB), random forest (RF), and support vector machines (SVMs) when the signal to noise (S/N) ratio in the MS/MS data decreases. GP starts with a random population of individuals/programs, which is then iteratively updated via genetic operators (selection, crossover, mutation and elitism) to search for the optimal solution until a stopping criterion is met. However, working with imbalanced data is difficult as uneven distribution of class examples in the train dataset could leave the learning algorithm with a performance bias, resulting a high majority class accuracy and a poor performance on the minority class [4]. Moreover, as the objective preference information from the decision maker is usually *a priori* built into the learning algorithm, in practice it is very difficult to obtain sufficient preference information and accurately represent the decision maker's preferences. Subsequently after finding the best satisfying solution, any change to the decision maker's preference requires to start the search process with the new preference information again.

Evolutionary multi-objective optimisation (EMO) as *a posteriori* method provides a set of Pareto optimal solutions prior to the decision maker's preference. So during a single run, a set of non-dominated solutions along the trade-off surface is found and then the decision maker can choose one of them based on his/her preference. There have been successful attempts to use GP and Pareto dominance-based algorithms to solve the class imbalance problem by maximising two conflicting objectives, the classification accuracy of the minority and majority classes [5]. While Pareto dominance-based algorithms usually produce non-dominated solutions around the centre of the Pareto front, decomposition-based EMO algorithms benefit from having the ability of differently allocating resources to better approximate the Pareto front [6].

Decomposing a multi-objective optimisation problem (MOP) into a set of scalar sub-problems and simultaneous optimising them, MOEA/D (multi-objective evolutionary algorithm based on decomposition) is an efficient framework for EMO. Neighbourhood is an essential property of MOEA/D. It uses evolutionary operators to combine good solutions of neighbouring problems for better convergence. MOEA/D has been previously applied on several benchmark problems such as feature selection for classification problems [6]. To our best knowledge, this is the first try of EMO and particularly MOEA/D to the solution of classification of highly imbalanced MS/MS data.

As GP proved to be a promising tool in MS/MS analysis, its potential for further improvement in handling two conflicting objectives of majority and minority classes using EMO has not been investigated. This paper aims to extend our previous work by developing a multi-objective GP (MOGP) approach using the MOEA/D framework to evolve a set of solutions which maintain the best trade-off between these two conflicting objectives.

Research Goals

The main goal of this study is to develop an MOGP approach based on MOEA/D, named MOGP/D, to solve the class imbalance problem by maximising two conflicting objectives, the accuracy of minority and majority class in imbalanced MS/MS spectra. It is expected that the proposed MOGP/D algorithm can evolve a Pareto front of classifiers along the optimal trade-off surface that offers the best compromises between the conflicting objectives. The classifier with the best trade-off can be used to preprocess the MS/MS spectra prior to peptide identification by any existing tool. The following objectives are investigated:

1. Exploring an appropriate modification to the MOEA/D framework to allocate the computation/search resources differently in order to produce the desired solutions in the region of interest.
2. Investigating the stability of the proposed MOGP/D method with the decrease in the S/N ratio in the MS/MS data in terms of convergence to the Pareto front and comparing with MOGP based on NSGA-II (named NSGP).
3. Analysing the classification performance of the best compromise solutions evolved by MOGP/D and NSGP and comparing them with the best solutions evolved by the single objective GP (SGP) approach.
4. Analysing the effectiveness of the proposed GP method in terms of the improvement in the reliability of peptide identification with existing tools.

2   Background

2.1   Multi-objective Optimisation Problem (MOP)

Generally speaking, in an MOP, N_{obj} conflicting objectives are required to be optimised simultaneously, while a set of inequality and equality constraint functions are satisfying. Therefore, a multi-objective minimisation problem can be formulated based on Eq. (1).

$$\text{minimise } F(x) = (f_1(x), f_2(x), .., f_{N_{obj}}(x)) \tag{1}$$

$$\text{subject to } g_i(x) \leq 0, \quad i = 1, 2, ..., k$$
$$h_i(x) = 0, \quad i = 1, 2, ..., l$$

where $F(x)$ is an objective vector representing N_{obj} objectives. $f_i(x)$ is the i-th objective of $F(x)$ and x is the decision vector. $g_i(x)$ and $h_i(x)$ represent the inequality and equality constraint functions of the optimisation problem.

In single objective optimisations, the optimal solution is usually unique, while the optimal solution in MOP is often a set of non-dominated solutions due to the conflict between different objective functions. Since the quality of each solution is based on the compromise between objectives, knowing the concepts of dominance is necessary. A solution y dominates z (denoted by y $\prec$ z) if and only if: $f_i(y) \leq f_i(z)$ for all f_i functions in F, and there is **at least one** j such that $f_j(y) < f_j(z)$. The solution y is called a Pareto optimal if it is not dominated by any other feasible solutions. The set of all Pareto optimals is called the Pareto front, representing the trade-off surface in the objective space. An EMO algorithm is expected to evolve a set of non-dominated solutions to approximate the Pareto front.

2.2   MOEA/D

In order to find the set of non-dominated solutions for Pareto front approximation, MOEA/D decomposes an MOP into a set of N (equal to the population size) scalar objective optimisation sub-problems, each with the objective of the aggregation of all objective functions. MOEA/D attempts to optimise these N scalar optimisation sub-problems simultaneously instead of solving MOP directly in a single run. Tchebycheff is one of the most widely used decomposition approaches [7]. In this approach, the fitness function of each single objective sub-problem is defined by a weight vector λ. This approach represents the j-th scalar optimisation sub-problem in the following form:

$$\text{minimise } g^{te}(x|\lambda^j, z^*) = \max_{1 \leq i \leq N_{obj}} \{\lambda_i^j |f_i(x) - z_i^*|\} \tag{2}$$

where $\lambda^j = \left(\lambda_1^j, ..., \lambda_{N_{obj}}^j\right)^T$, z^* is the reference point, and z_i^* in a minimisation MOP is the minimum value of each objective function. The major motivation behind MOEA/D is the concept of neighbourhood.

In this approach the neighbourhood of λ^j is defined as a set of the T closest weight vectors in $\{\lambda^1, \lambda^2, ..., \lambda^N\}$ and the Euclidean distance between these weight vectors defines the neighbourhood relation. It is expected that any information from the neighbouring sub-problems should be helpful for optimising the current sub-problem. In summary, each Pareto optimal point, x^*, with a weight vector of λ, which is the optimal solution of Eq. (2), is a Pareto optimal solution of Eq. (1) and belongs to Pareto front.

3 The Proposed Approach

In this section, the proposed MOGP method based on the MOEA/D framework is explained and the new algorithm is named MOGP/D. The two conflicting objectives and the modification applied on MOEA/D to effectively initialising and allocating the weight vectors are also described. The pseudo-code of the evolutionary search algorithm is presented in Algorithm 1.

3.1 Objective Functions

The two conflicting objectives in the classification of highly imbalanced MS/MS spectra include accuracy of minority class (sensitivity) and accuracy of majority class (specificity) which both are presented in Eq. (3).

$$\text{sensitivity} = \frac{TP}{TP + FN} \quad ; \quad \text{specificity} = \frac{TN}{TN + FP} \tag{3}$$

where TP and FN count the number of correctly and incorrectly classifying signal peaks. TN and FP represent the same concept for the noise class.

In order to convert the multi-objective classification to a minimisation problem, the two objective functions in Eq. (3) are normalised into the following form: $f_1(x) = 1 - \text{sensitivity}$; $f_2(x) = 1 - \text{specificity}$.

3.2 Problem-Specific Weight Vector Initialisation

Reference point, which represents an idealised solution, is one of the most effective ways to give preference information to the EMO algorithm. The preference information can be interpreted as the preferred goal that the decision-maker is wanting to get. As in standard MOEA/D, preference information is usually provided by using uniformly distributed weight vectors with the same reference point, here we use a problem-specific method to initialise the weight vectors in order to allocate more resources to the region of interest.

Based on the results of our previous works [2,3], we noticed that in order to improve the peptide identification, the classifier should not have specificity and sensitivity less than 0.5. Therefore, here we introduce a problem-specific initialisation method in Eq. (4) in order to put more weight vectors in the centre of Pareto front where hopefully a good set of solutions with the desired trade-off could be found. It is more effective to specifically allocate resources with respect to the difficulty of the problem on both conflicting and non-conflicting regions to obtain desired Pareto optimal solutions [6]. Starting with an effective set of weight vectors can ensure generating a good approximation of the Pareto front. Therefore, the initial weight vectors are designed to satisfy the following conditions:

$$\sum_{i=1}^{N_{obj}} w_i = 0.5, \text{ and } w_i \in \{0, \frac{0.5 \times 1}{N}, \frac{0.5 \times 2}{N}, .., 0.5\} \tag{4}$$

where N is the number of sub-problems which equals to the population size.

As previously mentioned, the pseudo-code in Algorithm 1 presents the overall framework of the MOGP/D algorithm for classification of highly imbalanced MS/MS spectra. The input to the algorithm is the information of MOP (based on Eq. (1)) and a set of parameters, and the output is an external population (EP) used to store the solutions of the Pareto front.

Algorithm 1. Pseudo-code of the proposed MOGP/D approach

Input: MOP; NGen: number of generation as stopping criterion; N: number of sub-problems; a set of uniformly distributed N weight vectors; T: number of neighbours; σ: probability of selecting the parents from the neighbourhood.

Output: an external population (EP) as the final optimal Pareto front.

1: **Initialisation:**
2: Set EP $= \emptyset$; Generate initial weight vectors based on Equation (4) and calculate the Euclidean distance between any two vectors;
3: For i $= 1, .., N$, find T closest weight vectors to the weight vector of the i-th sub-problem, and denote B(i) as its neighbouring set;
4: Randomly initialise each GP individual to create the population P where each individual in P is the candidate solution of the i-th sub-problem;
5: Initialise the reference point z^*;$gen \leftarrow 0$
6: **while** $gen \leq maxGen$ **do**
7:     **for** $i = 1$ to Popsize **do**
8:         **Reproduction:** $Ne = \begin{cases} B(i), & \text{if } rand < \sigma \\ P, & \text{otherwise} \end{cases}$

        Randomly select two solutions from Ne (either from the neighbouring set, $B(i)$, or from the whole population, P) to generate a new solution y by using the genetic operators;
9:         **Update of z:** for each j=1,..., N_{obj} if $f_j(y) < z_j$ then $z_j = f_j(y)$;
10:         **Update of Neighbouring Solutions:** update solutions of neighbouring sub-problems if the fitness value of y, (F(y), based on Equation (2)) is better than the solutions of sub-problem.
11:         **Update of EP:** remove all the weight vectors dominated by F(y) from EP, and add F(y) to EP if it is not dominated by any vector in EP.
12:     **end**
13:     $gen \leftarrow gen + 1$
14: **end**
15: **return** EP

3.3 MOGP/D Setup and Evolutionary Parameters

To represent the MOGP/D individuals, the tree-based GP structure is used. The spectral features along with randomly generated floating point numbers are used as the GP terminal set. Four arithmetic operators ($+, -, \times$, protected/(dividing by zero gives the result of 1)) are used as the function set of GP. For the purpose of binary classification strategy, if the output of the GP program, which is a floating point number, is positive, the instance under investigation is classified into the minority class (signal class), otherwise as majority class.

The MOGP/D parameters used in this study are as the following. For initialising the population, the ramped-half-and-half method is used. The population size is 600 and the evolutionary process runs for a maximum of 100 generations. The crossover and mutation rates are 80% and 20%, similar to the parameter setting in our SGP work [2]. The maximum program depth is restricted to 8 in order to prevent bloating. For MOGP/D, the number of neighbours, T, for each sub-problem is set to $\frac{N}{10}$ where N is the population size. The maximum number of individuals replaced by each child is 1, and the probability that parent solutions are selected from neighbourhoods, σ, is 0.85. The proposed MOGP/D method is implemented in Python 3.6 and uses DEAP (Distributed Evolutionary Algorithms in Python) package [8].

4 Experiment Design

4.1 MS/MS Datasets

The spectra used in this study are selected from the comprehensive full factorial LC-MS/MS benchmark dataset [9] particularly designed for benchmarking the MS/MS analysis tools. The dataset contains collision-induced dissociation (CID) spectra from 50 protein samples extracted individually from *Escherichia coli* K12 (*E. coli*). The MS/MS spectra are in the Mascot generic peak list with known identifications using Mascot database search v2.2 searched against a curated Refseq *E. coli* database. The synthetic dataset is used to investigate the stability of MOGP/D across different S/N ratios. Totally six pairs of training and test sets having different S/N ratios are created. This dataset contains four commonly used intensity-based spectral features. The golden standard dataset is used to compare the classification performance of the best compromise solutions evolved by the multi-objective GP approach with the best GP evolved program of the single objective approach. This dataset contains a set of 40 spectral features. Moreover, the test set of this dataset is used to evaluate the effectiveness of the best compromise solution evolved by MOGP/D and the results are compared with the SGP method.

4.2 Benchmark Algorithms

NSGA-II [10], the popular elitist non-dominated sorting genetic algorithm, is an extension of the genetic algorithm for multiple objective optimisation, which previously has been successfully used for binary class imbalance problems [5]. The NSGA-II framework as a dominance-based algorithm is used to compare with the MOEA/D framework as a decomposition based multi-objective evolutionary algorithm to investigate which is more appropriate with GP for binary classification problems with imbalanced data.

Table 1. Datasets details

Datasets		No. of spectra	No. of signal peaks	No. of noise peaks
Synthetic dataset	Train	10	270	$n \times 270$, $n = \{1, 2, 4, 6, 13, 20\}$
	Test	5	130	$n \times 130$, $n = \{1, 2, 4, 6, 13, 20\}$
Golden standard dataset	Train	2,630	42,960	1,687,230
	Test	1,674	38,707	1,189,822

To evaluate the performance of the multi-objective algorithms, the hypervolume indicator (HV) [11] is used to obtain a single figure indicating the convergence of the evolved Pareto front. Moreover, in order to offer a single solution from the Pareto front to the decision maker, the concept of the best compromise solution [12] is used to select the best compromise solution which later is to compared with SGP in terms of average accuracy presented in Eq. (5).

$$average\ accuracy = 0.5 \times sensitivity + 0.5 \times specificity \qquad (5)$$

4.3   Experiments

A set of experiments on the synthetic dataset are conducted in order to investigate the performance of MOGP/D and NSGP across various ratios of S/N peaks on MS/MS data. Since the ratio of imbalance between the MS/MS datasets is different, these experiments are used to figure out which EMO algorithm produces better solutions along the Pareto front across different imbalance ratios. A set of six pairs of training and test sets each pair having different S/N ratios including 1:1, 1:2, 1:4, 1:6, 1:12, 1:20 from the spectra in the dataset of Table 1 are created. 1:1 indicates a balanced dataset, while 1:20 implies 20 times of noise peaks over signal peaks in the dataset. The performance of each EMO algorithm is evaluated based on HV and average accuracy.

After investigating the stability of the EMO algorithms on the synthetic dataset, another experiment to investigate the classification performance of MOGP/D and NSGP on the golden standard dataset is conducted and the results of the best compromise solutions of the two MOGP algorithms are compared with the SGP approach. Finally, the effectiveness of the proposed MOGP/D method in terms of improvement in the reliability of peptide identification with PEAKS [13], which is a benchmark *de novo* sequencing tool for *de novo* sequencing MS/MS spectra, is investigated and the results are discussed.

Table 2. Average ($\pm$ standard deviation) HV of the *all* evolved Pareto fronts obtained by MOGP/D and NSGP on the *training sets* of the synthetic dataset across different S/N ratios over the 30 MOGP runs.

	1:1	1:2	1:4	1:6	1:12	1:20
MOGP/D	0.163 ± 0.001	0.160 ± 0.001	$\mathbf{0.156 \pm 0.001}$	$\mathbf{0.154 \pm 0.001}$	$\mathbf{0.144 \pm 0.001}$	$\mathbf{0.140 \pm 0.001}$
	($\downarrow$)	(o)	($\uparrow$)	($\uparrow$)	($\uparrow$)	($\uparrow$)
NSGP	$0.0.166 \pm 0.007$	0.160 ± 0.002	0.154 ± 0.003	0.153 ± 0.001	0.142 ± 0.001	0.135 ± 0.001

5 Results and Discussions

5.1 Analysis of the Overall Pareto Front Behaviour in Terms of HV

Table 2 presents the results of the average HV values of the *all* evolved Pareto fronts obtained by MOGP/D and NSGP on the six training sets of the synthetic dataset with each set having a different S/N ratio. To calculate the HV values, the reference point (0.5, 0.5) is used. To compare the performance of the two EMO algorithms over the 30 GP runs, the statistical t-test with 95% confidence interval and two-tailed P value less than 0.0001 is considered. The signs below the HV values show the significance test results, where ($\uparrow$)/($\downarrow$) indicates that MOGP/D is significantly better/worse than NSGP. Also (o) sign is used to represent that the result of MOGP/D is not significantly different from NSGP. The overall trend in the results of Table 2 shows that with the decrease in the S/N ratio in the MS/MS data on the training sets, the HV values of both EMO algorithms decrease. This indicates that higher imbalanced data makes the classification problem more difficult. Moreover, in terms of statistically comparing the performance of the two EMO algorithms, it can be seen that with the decrease in the S/N ratio in the training sets, MOGP/D achieves significantly better HV values than NSGP. This result could indicate that MOGP/D is more stable that NSGP in approximating the Pareto front when the S/N ratio is getting higher. As the problem is becoming more difficult, MOGP/D shows better convergence than NSGP. However, the performance of NSGP when the dataset is balanced is statistically significantly better than MOGP/D. This is interesting and needs further analysis, which is done by visualising the performance of the evolved solutions by the two EMO methods over the 30 independent GP runs on the test sets. To have an overview, three scenarios are selected when: the dataset is balanced (1:1), the S/N ratio is reasonable (1:6), and the S/N ratio is high (1:20). Based on the results of Table 2, although the performance of NSGP on the balanced training set (1:1) is significantly better than MOGP/D, the plots in Fig. 1 show that the results of NSGP on both balanced (1:1) and imbalanced (1:6 and 1:20) test sets are worse than MOGP/D. This means that most solutions of NSGP are dominated by those of MOGP/D on test sets. This indicates that the evolved solutions by NSGP have lower generalisation performance than those of MOGP/D. Further analysis on the size of the GP program solutions evolved by NSGP are conducted, but due to the page limit detailed results are not presented here. To summarise the analysis of the GP size, an unnecessary growth of the GP

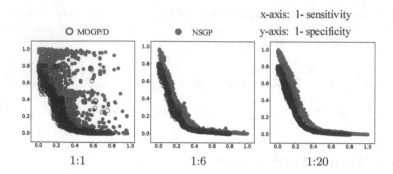

Fig. 1. Classification performance of the *all* evolved solutions using MOGP/D and NSGP on the *test sets* of the synthetic dataset across different S/N ratios.

program known as bloat or code growth is seen. This problem could be resolved by considering the GP program size as a third objective besides the other two objectives related to the program functionality, since a number of researches have been successfully used Pareto-based approaches for bloat controlling. However, without focusing on bloat control as the third objective, the plots in Fig. 1 shows that MOGP/D is able to indirectly handle the bloat problem, resulting in a significant improvement in the objective values when the test sets are used.

5.2   Analysis of the Overall Best Compromise Solutions

In the multi-objective classification of MS/MS data, after obtaining a set of non-dominated solutions, the decision maker needs to select a single classifier for preprocessing the MS/MS data prior to peptide identification. Therefore, here we use a fuzzy membership technique to find the best compromise solution among all the evolved solutions. So, with respect to each Pareto front across the 30 runs, one single best compromise solution is obtained for each EMO algorithm. For each single best compromise solution, the average accuracy according to Eq. (5) is calculated. Table 3 presents the results of the average accuracy of all best compromise solutions with respect to the 30 independent MOGP runs for two EMO algorithms on the training sets with different S/N ratios.

The overall trends in the results of Table 3 show that the decrease in the S/N ratio results in the decrease in the classification performance of evolved solutions

Table 3. Average (± standard deviation) accuracy of the best compromise solutions selected from the *all* evolved Pareto fronts obtained by MOGP/D and NSGP on the *training sets* of the synthetic dataset over the 30 MOGP runs.

	1:1	1:2	1:4	1:6	1:12	1:20
MOGP/D	0.829 ± 0.002	0.828 ± 0.004	0.825 ± 0.001	**0.824 ± 0.002**	**0.822 ± 0.001**	**0.819 ± 0.001**
	(↓)	(◦)	(◦)	(↑)	(↑)	(↑)
NSGP	**0.831 ± 0.013**	0.827 ± 0.007	0.824 ± 0.004	0.822 ± 0.003	0.811 ± 0.003	0.802 ± 0.002

Table 4. The results on the *test sets* of the synthetic dataset.

	1:1	1:2	1:4	1:6	1:12	1:20
MOGP/D	0.704 ± 0.072	0.757 ± 0.027	0.796 ± 0.009	0.804 ± 0.005	0.805 ± 0.004	0.804 ± 0.004
	(↑)	(↑)	(↑)	(↑)	(↑)	(↑)
NSGP	0.556 ± 0.101	0.700 ± 0.032	0.751 ± 0.015	0.750 ± 0.009	0.761 ± 0.007	0.754 ± 0.005

by both MOGP/D and NSGP. From the results in Table 3, it can be seen that similar to the results of Table 2, with the decrease in the S/N ratio, the best compromise solutions evolved by MOGP/D outperform those of NSGP in terms of the average accuracy on the training sets. Looking more closely at the results of both algorithms when the dataset is balanced, (1:1), NSGP is statistically significantly better than MOGP/D, however based on the test results in Table 4 the performance of NSGP at (1:1) dramatically drops. The results on the test sets show that MOGP/D has better performance than NSGP across all different S/N ratios.

Finally, to compare the results of MOGP with SGP, the best compromise solution of each EMO algorithm from its non-dominated front after combining all 30 Pareto fronts over the 30 GP runs are obtained. Figure 2 presents the classification performance of the best compromise solutions evolved by MOGP/D and NSGP along with the best SGP solution from the previous study [2]. The results of classification in terms of sensitivity, specificity and average accuracy are plotted. Starting from the sensitivity results on the training set, it can be seen that all the three methods have a steady decline from 1:1 to 1:6 followed by a sharp drop at 1:20. This shows that decreasing the S/N ratio results in misclassification of more signal peaks. However, the results show that MOGP/D and NSGP outperform SGP in terms of sensitivity on the training sets. Also on the test set, the same decline in the sensitivity results of all the three methods can be seen. So in summary, MOGP/D outperforms both methods in terms of sensitivity on both the training and test sets. This is very important in classification of peaks in MS/MS spectra, as retaining the signal peaks as much as possible is more important than filtering out the noise peaks.

The specificity results on the training sets in Fig. 2 shows that as the number of noise peaks increases, the classification performance of all three method on the majority class increases as well. Comparing the plots of sensitivity and specificity on both the training and test sets obviously shows the conflict between these two objectives. Similar to the results of sensitivity plots, with the decrease in the S/N ratio, MOGP/D outperforms NSGP and SGP in terms of specificity on the training and test sets.

Finally, the results of the average accuracy show that MOGP/D and NSGP outperform SGP on the training sets. On the test sets, SGP and NSGP almost have the same performance, while MOGP/D outperforms both of them. Overall, with the decrease in the S/N ratio of the MS/MS data in the training set MOGP/D shows constant accuracy at >83% whereas these values for NSGP and SGP are at >81% and >80%, respectively.

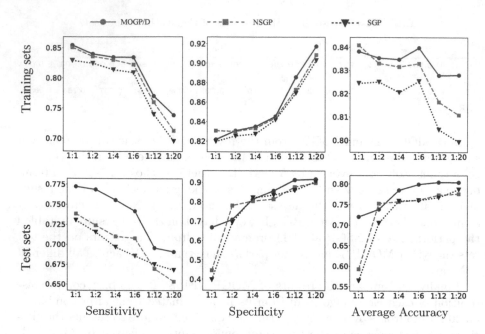

Fig. 2. Classification performance of the best SGP and the best compromise solution of MOGP/D and NSGP on the synthetic dataset.

5.3 Classification Performance on the Golden Standard Dataset

Figure 3 presents the classification performance of the best compromise solutions evolved by MOGP/D and NSGP along with the best SGP on the golden standard dataset. As the S/N ratio in the golden standard dataset is almost 1:30, the classification results on the training set in Fig. 3 are consistent with those in Fig. 2 at 1:20 S/N ratio, where the both MOGP methods outperform SGP in terms

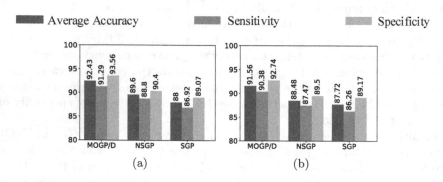

Fig. 3. Classification performance of the best SGP and the best compromise solution of MOGP/D and NSGP on the golden standard dataset. (a) Results on the training set (b) Results on the test set

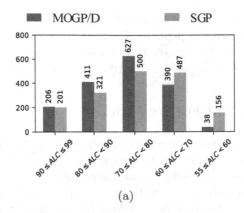

Fig. 4. The results of peptide identification by PEAKS on the test set of the golden standard dataset preprocessed by MOGP/D and SGP.

of average accuracy, sensitivity, and specificity. Comparing the both MOGP algorithms, it can be seen that MOGP/D outperforms NSGP by 2.49% and 2.91% of retaining more signal peaks on the training and test set of the golden standard dataset, receptively.

In summary, the MOGP/D algorithm has shown to be promising in evolving better solutions with compromises between the two conflicting objectives of sensitivity and specificity in the classification of highly imbalanced MS/MS data. In the next section, the preprocessed MS/MS data by the best compromise solution evolved by MOGP/D is submitted to PEAKS for peptide identification and the results are analysed to evaluate the effectiveness of the prepossessing method.

5.4   Performance Evaluation

The test set of the golden standard dataset which is already preprocessed by MOGP/D and SGP in previous section is submitted to PEAKS for peptide identification. The result of PEAKS is a set of identified peptides with each having an ALC score which indicates the confidence of the match. The ALC score in PEAKS, which ranges between 0 and 99, reflects the average correct ratio of the predicted amino acids in a peptide sequence. In this software, any predicted peptide with score at 55% or above is considered as a confident match.

The results of impact of MOGP/D in terms of improvement in the reliability of peptide identification on the test set of the golden standard dataset is presented in Fig. 4. The results are classified into 5 groups of different ALC ranges. More number of identified peptides in high score ranges is desired.

Overall, MOGP/D has significantly improved the peptide identification compared to SGP, since MOGP/D retains more signal peaks and removes more noise peaks compared to SGP. Although there is not a significant difference between the results of the two methods with ALC higher than 90%, for $70 \leq ALC < 90$, the results shows that MOGP/D obviously improves the reliability of peptide identification by helping PEAKS to identify more high confident peptides rather than SGP. Adding up all the peptide identified with ALC higher than 70, the results show that MOGP/D improves the reliability of peptide identification by 21.72% ($= \frac{(206+411+627)-(201+321+500)}{(201+321+500)} \times 100$) compared to SGP.

6   Conclusions and Future Work

The goal of this paper was to develop an effective MOGP method based on the idea of MOEA/D (MOGP/D) to evolve a set of non-dominated solutions, along the optimal trade-off surface that offers the best compromises between the two conflicting objectives of sensitivity and specificity. The non-dominated solutions are used for classification of peaks in the MS/MS spectra. The goal has been successfully achieved by applying a suitable modification to MOEA/D weight vector initialisation to allocate the resources more efficiently in the region of interest which is the centre of the Pareto front. Compared with NSGP, an NSGA-II based MOGP method, MOGP/D evolves more solutions in the middle of Pareto front, pushing this front outwards better minority (signal) and the majority (noise) class accuracies.

As the MS/MS spectra is highly imbalanced, the stability of the proposed MOGP method with the decrease in the S/N ratio in the MS/MS data was investigated and the results were compared with NSGP in terms of convergence to the Pareto front. The HV value was used as a single figure for the purpose of comparison. The results showed that with decreasing the signal to noise (S/N) ratio, MOGP/D outperformed NSGP in terms of the HV values of the evolved Pareto fronts on both the training and test sets.

For selecting a single solution from the evolved Pareto front, a fuzzy membership approach was used to obtain the best compromise solution. The single classifiers from MOGP/D and NSGP were compared with SGP. The results showed that MOGP/D outperformed NSGP and SGP in terms of sensitivity, specificity and average accuracy on both the training and test sets. In conclusion, MOGP/D has shown to be more suitable for evolving a classifier that has the best trade-off between the two conflicting objectives of the majority and minority class accuracies in the problem of classification of imbalanced MS/MS spectra. The evolved classifier can be used to preprocess the MS/MS spectra prior to peptide identification with any existing peptide identification tools. This reduces the search space of all possible peptide sequences, decreasing the time complexity of the peptide identification and improves the reliability of the results. Further experiments also showed that the proposed multi-objective GP-based preprocessing method improved the reliability of peptide identification, and increased the number of high confident peptides by 21.72% compared to the single objective GP method.

In our future work, we will use MS/MS dataset from different mass spectrometers with more number of features to investigate how MOGP/D is able to maintain a trade-off between the majority and minority class accuracies and the number of selected features.

References

1. Sheng, Q., et al.: Preprocessing significantly improves the peptide/protein identification sensitivity of high-resolution isobarically labeled tandem mass spectrometry data. Mol. Cell. Proteomics **14**(2), 405–417 (2015)
2. Azari, S., Zhang, M., Xue, B., Peng, L.: Genetic programming for preprocessing tandem mass spectra to improve the reliability of peptide identification. In: Vellasco, M. (ed.) 2018 IEEE Congress on Evolutionary Computation (CEC), Rio de Janeiro, Brazil, 8–13 July 2018. IEEE (2018)
3. Azari, S., Xue, B., Zhang, M., Peng, L.: Preprocessing tandem mass spectra using genetic programming for peptide identification. J. Am. Soc. Mass Spectrom. **30**, 1–14 (2019)
4. Bhowan, U., Johnston, M., Zhang, M., Yao, X.: Reusing genetic programming for ensemble selection in classification of unbalanced data. IEEE Trans. Evol. Comput. **18**(6), 893–908 (2013)
5. Bhowan, U., Johnston, M., Zhang, M., Yao, X.: Evolving diverse ensembles using genetic programming for classification with unbalanced data. IEEE Trans. Evol. Comput. **17**(3), 368–386 (2012)
6. Nguyen, B.H., Xue, B., Andreae, P., Ishibuchi, H., Zhang, M.: Multiple reference points-based decomposition for multiobjective feature selection in classification: static and dynamic mechanisms. IEEE Trans. Evol. Comput. **1**(1), 170–184 (2020). https://doi.org/10.1109/TEVC.2019.2913831
7. Ma, X., Zhang, Q., Tian, G., Yang, J., Zhu, Z.: On tchebycheff decomposition approaches for multiobjective evolutionary optimization. IEEE Trans. Evol. Comput. **22**(2), 226–244 (2017)
8. Fortin, F.-A., De Rainville, F.-M., Gardner, M.-A., Parizeau, M., Gagné, C.: DEAP: evolutionary algorithms made easy. J. Mach. Learn. Res. **13**, 2171–2175 (2012)
9. Wessels, H.J.C.T., et al.: A comprehensive full factorial LC-MS/MS proteomics benchmark data set. Proteomics **12**(14), 2276–2281 (2012)
10. Deb, K., Pratap, A., Agarwal, S., Meyarivan, T.A.M.T.: A fast and elitist multiobjective genetic algorithm: NSGA-II. IEEE Trans. Evol. Comput. **6**(2), 182–197 (2002)
11. Riquelme, N., Von Lücken, C., Baran, B.: Performance metrics in multi-objective optimization. In: 2015 Latin American Computing Conference (CLEI), pp. 1–11. IEEE (2015)
12. Paul, S., Das, S.: Simultaneous feature selection and weighting-an evolutionary multi-objective optimization approach. Pattern Recogn. Lett. **65**, 51–59 (2015)
13. Ma, B., et al.: PEAKS: powerful software for peptide de novo sequencing by tandem mass spectrometry. Rapid Commun. Mass Spectrom. **17**(20), 2337–2342 (2003)

Automatic Annotation Method for Document Image Binarization in Real Systems

Ryosuke Odate[✉]

Hitachi Ltd. Research & Development Group,
1-280, Higashi-koigakubo, Kokubunji-shi, Tokyo 185-8601, Japan
`ryosuke.odate.qs@hitachi.com`

Abstract. The accuracy of optical character recognition (OCR) has significantly improved recently through the use of deep learning. However, when OCR is used in real applications, the shortage of annotated images often makes training difficult. To solve this problem, there are automatic annotation methods. However, many of these methods are based on active learning, and operators need to confirm generated annotation candidates. I propose a practical automatic annotation method for binarization, which is one of the components of OCR. The purpose with the proposed method is to automatically confirm the quality of annotation candidates. This method consists of three simple processes to achieve this. First, cropping a text from a whole image. Second, applying binarization to the cropped image at all thresholds. Third, recognizing all binarized cropped images and matching the recognition results and correct character database. If the characters match, the cropped binary image is correctly binarized. The method selects that cropped binarized image as an annotation for binarization. Cropping coordinates and the correct character database (DB) can be obtained from a practical OCR system. Because users of such a system usually input corrections for misrecognition of OCR to the system, the system can obtain the correct characters and coordinates. The experimental results indicate that the annotations generated with the proposed method can improve the performance of deep-learning-based binarization. As a result, the normalized edit distance between the recognized text and grand truth text can be reduced by 38.56% on the Find it! receipt image dataset.

Keywords: Character recognition · OCR · Binarization · Annotation · Practical systems

1 Introduction

Optical character recognition (OCR) is used in many applications and systems such as machine translation, receipts recognition, and automated teller machines. As shown in Fig. 1, OCR for documents is generally composed of multiple steps

© Springer Nature Switzerland AG 2020
S. Palaiahnakote et al. (Eds.): ACPR 2019, LNCS 12047, pp. 464–477, 2020.
https://doi.org/10.1007/978-3-030-41299-9_36

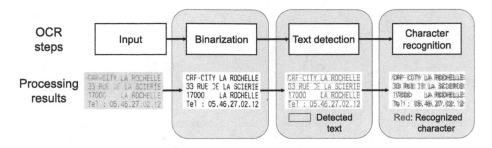

Fig. 1. OCR processing steps and output of each step. Document image is from [2].

including image binarization [8] to accurately remove noises accurately and maintain strings, text detection [5], and character recognition [20]. In recent years, the OCR performance has been greatly improved recently by using deep learning for each component. As a result, hard-to-read documents, as shown in Fig. 2, can be read using deep learning-based OCR trained on large-scale datasets [12].

If there is sufficient training data, deep learning improves the performance of each OCR step and improves the overall recognition accuracy. In fact, OCR has succeeded in processing hard-to-read documents such as historical documents (top of Fig. 2) with the enrichment of datasets and increase in research [12].

However, it is difficult to train deep-learning-based OCR when the number of training samples is not enough. To solve the shortage of training samples for OCR, two approaches are mainstream: image generation and annotation.

In image generation-based approaches, training images and their annotations are artificially created using generative adversarial networks (GANs) [3] or variable autoencoders (VAEs) [14]. SyntheText [6] can generate images specialized for text. This is not a GAN-based approach and creates text-overlay scene images by taking into account local 3D scene geometry. However, Ravuri et al. [13] reported that training images generated using GANs or VAEs contribute very little to the performance improvement of classifiers.

Automatic annotation-based approaches use real images instead of generated images. I define real images as images not synthesized and acquired using a scanner or camera. For example, the image-segmentation annotation method [4] and keyword-assignment method [19] have been proposed. Many automatic annotation methods for real images are based on active learning [4,10]. In other words, many automatically generate annotation candidates, then users judge the quality of them and make corrections if necessary. These methods work well when real images are abundant. Therefore, they are useful in constructing applications and systems in which real images of recognition targets can be obtained. The problems with them is that the accuracy of annotation candidates is uncertain, and the cost for human judgment and correction is high since the processing is not completely automatic.

I propose a practical automatic annotation method to improve the performance of binarization, which is one of the steps of OCR. Specifically, I automate

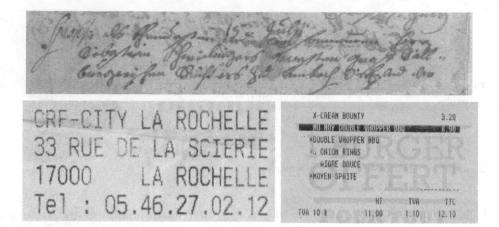

Fig. 2. Various hard-to-read documents. Top: stained historical document [12]; Bottom left: receipt with crease [2]; Bottom right: receipt with watermarks [2].

the judgment with good and bad annotations, which is the cost-based weakness of active learning-based annotation methods. I also examine changes in binarization performance by using the annotations generated with the proposed method.

The following three reasons are why I specifically aimed at improving binarization. First, as shown in Fig. 1, binarization is the first pre-processing to remove noise and correct foreground; thus, binarization is important, and the performance of subsequent steps should be affected by it. The second is that binarization requires specialized training specifically for the target document. Figure 2 shows examples of various hard-to-read documents. Noise such as creases, watermarks, and hatchings, on various types of documents should be removed by binarization. To remove them, specialized binarization training is needed. Therefore, deep-learning-based binarization models should be trained with real images similar to a recognition target. The third is that the annotation cost of binarization is especially high. Binarization is a type of segmentation, so it requires pixel-wise annotation of the foreground and background, while text detection and character recognition only require bounding boxes and texts, respectively.

The rest of this paper is organized as follows. In Sect. 2, I give a brief review of related work on automatic annotation and document image binarization. In Sect. 3, I discuss the proposed automatic training annotation method based on multiple OCR results and the correctness estimation for those results. In Sect. 4, I describe the experimental results and provide discussion. Finally, I conclude the paper in Sect. 5.

2   Related Works

I introduce two research fields related to this paper. I first introduce document-image binarization in Sect. 2.1 then automatic annotation in Sect. 2.2.

2.1 Document Image Binarization

Document-image binarization is the first step in OCR, as shown in Fig. 1. This technique makes the foreground in an image black and makes the background white to remove noise and correct the foreground. Therefore, if the binarization fails, the accuracy of the subsequent processing steps, i.e., text detection and character recognition, will deteriorate. At that time, processing time may also increase because of detecting the non-text region incorrectly.

Binarization has been studied for several decades, and various methods have been proposed. Otsu's method [11] and Sauvola's method [16] are well-known conventional rule-based methods. Both methods set thresholds for binarization without training. Otsu's method sets the global threshold to minimize intra-class intensity variance and maximize inter-class variance. Sauvola's method sets the adaptive thresholds by computing the average and variance of luminance in a local area in an image. These methods do not require large amount of training data and the processing speed is fast. In addition, users can easily control the performance of these methods by parameter adjustment. Thus, users still often use these methods.

Binarization methods using deep learning have been actively studied recently. Many such methods use fully convolutional networks (FCNs) [9] as a basic structure [17]. FCN-based methods use binarization as one of the image segmentation tasks that segment images into foreground and background. These methods can accurately binarize hard-to-read documents such as historical documents if there are sufficient training data [12]. However, it is difficult for users to paint foreground and background accurately with black and white for each pixel to obtain annotated training data. Therefore, the training for the binarization of real documents with poor annotated images is often difficult.

2.2 Automatic Annotation

Automatic annotation methods reduce the annotation cost to real images. Since deep learning requires many annotated images, the demand for automatic annotation is increasing. Since the annotation to be attached to images are diverse, there are methods for each type. For example, a review paper [19] shows keyword-assignment methods for image retrieval. To obtain the annotation of image segmentation, Clemens [4] used Otsu's method as a simple segmentation method.

Many of these methods are based on active learning [4,10]. In other words, many automatic annotation methods automatically generate annotation candidates, then users judge the quality of the candidates and correct them if necessary. Perhaps many researchers and engineers unconsciously use this type of method. To use such a method continuously in a real system, it is desirable to automate tasks such as judgments and correction of annotation candidates.

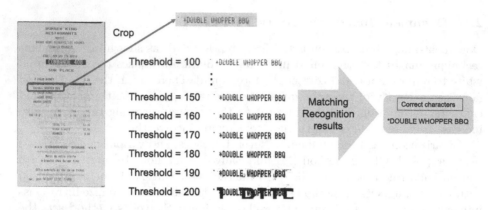

Fig. 3. Schematic diagram of proposed method: Region segmentation and multiple hypothesis binarization. Document image is from [2].

3   Proposed Method

3.1   Overview

I first give an overview of the proposed method in this section, give details of the method's algorithm in Sect. 3.2, then describe a practical OCR system in Sect. 3.3.

The purpose with the proposed method is to automatically select annotation candidates. In this case, it is necessary to make the selection more reliable, similar to human confirmation with other annotation methods. Figure 3 shows an outline of the proposed method. The proposed method performs three processes for reliable annotation generation. The first is cropping. The second is a simple threshold binarization to the cropped area. This is done by all the threshold values. Then, thanks to cropping, correct binarization can often be carried out at a specific threshold, as shown in Fig. 3. Finally, each binary image is recognized, and the recognition result is collated with the correct characters. If the recognition result matches the correct characters, it is regarded as being correctly binarized and selected for annotation. The bottleneck of these processing is to obtain crop coordinates and correct characters, but this is often not a problem in practical OCR systems. I give an example of obtaining coordinates and correct characters in a real OCR system in Sect. 3.3.

3.2   Algorithm

The algorithm of the proposed method, Algorithm 1 shows the details of annotation-candidate selection. In this algorithm, inputs are document image X, correct character strings G_N, and coordinates of correct character strings C_N, and N_G is the number of correct strings. The output is an annotation T, which is the generated binarized image for X. Intermediate output is annotation candidates $A = \{a_i\}_{i=1}^{N_G}$, where a_i is a binarized image for each string area defined by C_N. Parameter α is the initial parameter of D_{th}, which is the threshold of edit distance, X_{iB} is a binarized image of cropped image X_i, R_{iB} is a recognized character of X_{iB}, and D_{iB} is the normalized edit distance between R_{iB} and G_i. This algorithm repeats cropping for an image and binarizes the cropped image with all threshold values. Then, when the recognized character matches the correct character string, the binary image X_{iB} is stored in A.

Algorithm 1. Annotation-candidate selection

Input: X, $G_N = \{g_i\}_{i=1}^{N_G}$, $C_N = \{c_i\}_{i=1}^{N_G}$
Output: T

1 $T = \mathrm{Otsu}(X)$
2 $A = \{a_i\}_{i=1}^{N_G}$
3 **for** i **in** $range(N_G)$ **do**
4 $\quad$ $D_{th} = \alpha$
5 $\quad$ $X_i = \mathrm{crop}(X, C_i)$
6 $\quad$ $a_i = \mathrm{crop}(T, C_i)$
7 $\quad$ **for** j **in** $range(255)$ **do**
8 $\quad\quad$ $X_{iB} = \mathrm{binarize}(X_i, threshold = j)$
9 $\quad\quad$ $R_{iB} = \mathrm{recognize}(X_{iB})$
10 $\quad\quad$ $D_{iB} = \mathrm{editdistance}(R_{iB}, G_i)/\mathrm{len}(G_i)$
11 $\quad\quad$ **if** $D_{iB} =< D_{th}$ **then**
12 $\quad\quad\quad$ $D_{th} = D_{iB}$
13 $\quad\quad\quad$ $a_i = X_{iB}$
14 $\quad\quad$ end
15 $\quad$ end
16 end
17 paste A on T
18 return T

3.3   Practical System Using OCR

Figure 4 shows the configuration of a data-entry system using OCR. The numbers in the figure denote the order of processing. The advantage of such a system is that the user receives the recognition results of OCR and corrects them through the interface if there is a recognition error. In this case, the system can store the corrected characters and coordinates of those characters in the correct character DB. Conversely, if no correction is input, it is determined that the recognition result of OCR is correct, and this result can be stored directly in the correct character DB. Many data-entry systems using OCR reduce the risk of recognition errors by using the configuration illustrated in Fig. 4. Therefore, building a correct character DB with a practical OCR system incurs very low cost.

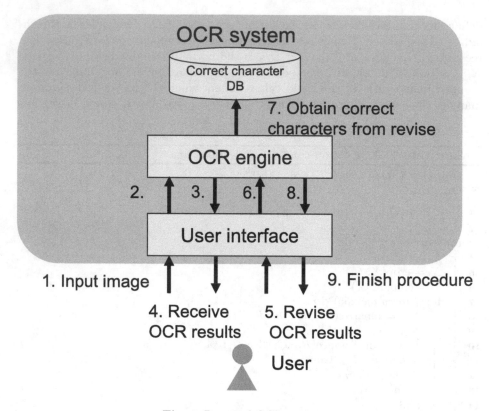

Fig. 4. Practical OCR system

4   Experiments

4.1   Setup

Dataset

I evaluated the proposed method on the Find it! [2] receipt image dataset to evaluate its performance. Find it! was a contest held at ICPR 2018 aimed at detecting forged receipts. The dataset for task 1 contains 1000 scanned real receipt images and 1000 script files including printed strings in each image. The dataset is divided into two sets of 500 images in advance. One is for training and the other is for testing. I maintained the data partition and used this dataset for the binarization experiments. To evaluate the proposed method, I added the coordinates at which the text in the image exists to the script file. In subsequent experiments, I applied the proposed method to training images and trained a deep-learning-based binarization model with the generated annotations. I then binarized 500 images test images with the trained model and finally recognizes the binary image using an existing classifier.

The following three reasons are why I chose the Find it! dataset. First, the demand for receipt recognition is increasing as a practical application. For example, ICDAR 2019 will hold the Robust Reading Challenge on Scanned Receipts OCR and Information Extraction (SROIE) [1]. Second, this dataset is appropriate for the purpose of our study: annotating a large number of real images. These receipt images are useful for binarization training if I annotate for binarization, but labor is a problem. Third, since there are script files including printed strings according to each image, I can easily obtain a correct character DB. Follow-up experiments by others will also be easy to conduct for the same reason.

Evaluation Protocol

To measure the effectiveness of the annotations generated with the proposed method, I evaluated the accuracy of binarization trained with these annotations. I also evaluated the accuracy of the final character recognition through this binarization.

First, I evaluated binarization in terms of recall, precision, and F-measure. The formulas of them are shown below. These are often used in the evaluation of binarization and segmentation [18]. I randomly selected 10 images (images 13, 72, 119, 156, 337, 552, 716, 830, 882, and 1243) from the test set and manually created the ground truth of binarization. There were about 50 million pixels in total for the 10 images. Since the test set of the binarization competition DIBCO2016 had about 13 million pixels, our test set had a sufficient amount.

$$Recall = \frac{T_P}{T_P + F_N} \tag{1}$$

$$Precision = \frac{T_P}{T_P + F_P} \tag{2}$$

$$F_{measure} = \frac{2 \times Recall \times Precision}{Recall + Precision} \tag{3}$$

where T_P is the number of correctly blackened pixels, F_P is the number of incorrectly blackened pixels, and F_N is the number of incorrectly whitened pixels.

I then evaluated the character-recognition accuracy in terms of correctly recognized words ratio (C.R.W.) and total edit distance (T.E.D.), both are case-insensitive. The formulas of them are shown below. To calculate them, I used the ICDAR2013 [7] evaluation tool.

$$C.R.W = \frac{N_C}{N_W} \tag{4}$$

$$T.R.D. = \sum_{(X,Z) \in S} \frac{ED(h(X), Z)}{|Z|} \tag{5}$$

where N_C is the number of correctly recognized words, N_W is the total number of words, S is the test set, h is the classifier, $ED(p, q)$ is the edit distance between two sequences p and q, X is a sample, and Z is a grand truth. Since I wanted to evaluate the pure recognition results of the classifier, I did not use post-processing such as natural language processing. I consistently used the same text-detection method and character-recognition method to evaluate final character-recognition accuracy in the all experiments. The text-detection method is a conventional rule-based method based on connected components, and the character-recognition method is also a conventional method using nearest neighbor search.

I used a calculator which had 8 GB RAM, Core i7-6700K @ 4.00 GHz, and GTX1080.

4.2   Performance Evaluation of Parameter α

In this experiment, The proposed method automatically generated annotations on various α. I then used the model called deeply-supervised nets (DSN) binarization proposed by Vo et al. [18] as a deep-learning-based binarization model trained by the generated annotations.

Table 1 lists the performance of DSN binarization and the final character recognition for each α of the proposed method. I compared $\alpha = 0.0$ to 0.4. I also compared DSN trained using the results of Otsu's method as a baseline. This is a similar training situation to that [1] without humans. When $\alpha = 0.3$, T.E.D. was the smallest. This is because unmatching with the proposed method due to misrecognition by the classifier decreased and correct annotation candidates could be selected easily. When trained by Otsu's binarization, T.E.D was especially low, although F-measure was high. Even if the F-measure of binarization is high, the final character-recognition accuracy is not necessarily high. It can be said that if it is more than a certain degree of binarization accuracy, the binarization result suitable for the classifier to be used is desirable. In this experiment, even when the recall was low, there was no problem in the final character recognition accuracy. Therefore, there was less noise, and the binarization in which characters become thinner improved.

It took about 120 h to create annotations of 500 images. Since it took about 40 h to create annotations of 10 images by a person, it would take approximately 2000 h to make them for 500 images. The proposed method was about 16 times faster than humans.

The proposed method can automatically generate annotations more suitable for training than a conventional method [1]. Because I found that $\alpha = 0.3$ was good in this experiment, I used $\alpha = 0.3$ in the subsequent experiments.

4.3   Performance Evaluation on Various Networks

In this experiment, I used three deep-learning-based binarization models, an FCN-based one (hereafter just FCN) [9], U-Net [15], and DSN. The purpose of this experiment was to make sure that the annotations generated with the proposed method are useful for training any deep-learning models. Table 2 lists the performance of binarization and the final character recognition. I used the annotations generated using proposed method with $\alpha = 0.3$ for all the trained models. The results indicate that DSN was the best. Moreover, the annotations generated using the proposed method are effective for training regardless of the deep-learning model.

Table 1. Accuracy comparison of binarization and character recognition for each α.

α	Recall	Precision	F-measure	C.R.W	T.E.D.
Otsu (baseline)	0.98	0.86	0.92	0.27	7099.30
0.0	0.96	0.92	0.93	0.36	5343.79
0.1	0.94	0.94	0.94	0.39	4747.33
0.2	0.93	0.95	0.94	0.40	4511.07
0.3	0.92	0.95	0.94	0.41	4362.69
0.4	0.91	0.96	0.93	0.40	4430.91

Table 2. Accuracy comparison of binarization and character recognition among deep-learning-based binarization models

Method	Recall	Precision	F-measure	C.R.W	T.E.D.
FCN	0.93	0.91	0.92	0.37	5242.65
U-Net	0.93	0.94	0.93	0.42	4766.51
DSN	0.92	0.95	0.94	0.41	4362.69

4.4   Comparison with Other Methods

Finally, I compared DSN trained by the annotations generated using the proposed method with a DSN trained by historical documents obtained from the DIBCO competition dataset [12]. I also fine-tuned this DSN with the annotations generated using the proposed method. The experimental results of the above two models, rule-based binarization methods, and DSN with $\alpha = 0.3$ are listed in Table 3 and illustrated in Fig. 5.

Table 3. Accuracy comparison of binarization and character recognition for proposed and other binarization methods

Method	Recall	Precision	F-measure	C.R.W	T.E.D.
Otsu (baseline)	0.98	0.86	0.92	0.27	7099.30
Sauvola	0.85	0.66	0.74	0.14	9165.06
Adaptive mean	0.94	0.56	0.70	0.21	7342.31
Adaptive gaussian	0.95	0.62	0.75	0.30	6072.39
Trained by DIBCO	0.94	0.74	0.83	0.36	5433.91
Proposed fine tune	0.92	0.95	0.94	0.41	4384.08
Proposed	0.92	0.95	0.94	0.41	4362.69

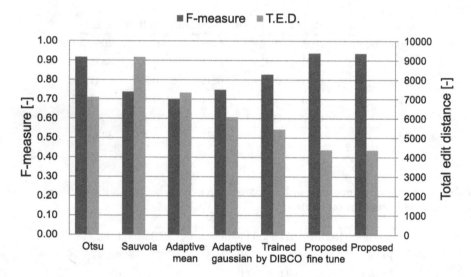

Fig. 5. OCR processing steps and output of each step

The DSN trained by the annotations generated using the proposed method was the best for both the F-measure and T.E.D.. The proposed method reduced T.E.D. by 38.56% compared to the baseline. The T.E.D of the DSN trained by historical documents was good. However, the precision was low because the tendency of the characters and the noise differed between historical documents and receipts. The fine-tuned DSN achieved comparable performance to that trained using the proposed method. This result more strongly indicates the usefulness of the annotations created using the proposed method. The actual binarized images are shown in Fig. 6.

Fig. 6. Receipts from previous study [2]. Upper left: Original image, Upper right: Binarized image using Otsu's method, Lower left: Binarized using DSN trained with DIBCO dataset, Lower right: Binarized image using DSN trained with annotations generated using proposed method ($\alpha = 0.3$).

5   Conclusion

I proposed a practical automatic annotation method to improve the performance of binarization, which is one of the steps of OCR. The advantage of this method is that it automates the selection of annotation candidates, which is a cost-based weakness of active-learning-based automatic annotation methods because due to labor costs. To automate this selection, the proposed method applies various binarizations to the local regions of a document image and treats the binarization result of correctly reading the character as the correct annotation. The experimental results indicate that the annotations generated using the proposed method contribute to the training of binarization models using deep learning and improves the final character-recognition accuracy. The proposed method created annotations about 16 times faster than humans thus far.

In the future, I plan to apply the proposed method to annotation for other functions of OCR, such as text detection.

References

1. ICDAR 2019 Robust Reading Challenge on Scanned Receipts OCR and Information Extraction - ICDAR 2019 RobustReading Competition. https://rrc.cvc.uab.es/
2. Artaud, C., Sidère, N., Doucet, A., Ogier, J., Yooz, V.P.D.: Find it! Fraud detection contest report. In: 2018 24th International Conference on Pattern Recognition (ICPR), pp. 13–18, August 2018
3. Brock, A., Donahue, J., Simonyan, K.: Large Scale GAN Training for High Fidelity Natural Image Synthesis. arXiv:1809.11096, September 2018
4. Clemens, W.: Using Otsu's method to generate data for training of deep learning image segmentation models. https://www.microsoft.com/developerblog/2018/05/17/using-otsus-method-generate-data-training-deep-learning-image-segmentation-models/
5. Gruning, T., Leifert, G., Straub, T., Michael, J., Labahn, R.: A two-stage method for text line detection in historical documents. Int. J. Doc. Anal. Recogn. (IJDAR) **22**(3), 285–302 (2019)
6. Gupta, A., Vedaldi, A., Zisserman, A.: Synthetic Data for Text Localisation in Natural Images, pp. 2315–2324. IEEE, June 2016
7. Karatzas, D., et al.: ICDAR 2013 robust reading competition. In: 2013 12th International Conference on Document Analysis and Recognition, pp. 1484–1493, August 2013
8. Karthika, M., James, A.: A novel approach for document image binarization using bit-plane slicing. Procedia Technol. **19**, 758–765 (2015)
9. Long, J., Shelhamer, E., Darrell, T.: Fully convolutional networks for semantic segmentation. In: 2015 IEEE Conference on Computer Vision and Pattern Recognition (CVPR), pp. 3431–3440, June 2015
10. Marvasti, N., Yoruk, E., Acar, B.: Computer-aided medical image annotation: preliminary results with liver lesions in CT. IEEE J. Biomed. Health Inf. **22**(5), 1561–1570 (2017)
11. Otsu, N.: An automatic threshold selection method based on discriminant and least squares criteria. Trans. Inst. Electron. Commun. Eng. Jpn. **63**, 349–356 (1980)

12. Pratikakis, I., Zagoris, K., Barlas, G., Gatos, B.: ICDAR2017 competition on document image binarization (DIBCO 2017). In: 2017 14th IAPR International Conference on Document Analysis and Recognition (ICDAR), vol. 01, pp. 1395–1403, November 2017
13. Ravuri, S., Vinyals, O.: Classification Accuracy Score for Conditional Generative Models. arXiv:1905.10887, May 2019
14. Razavi, A., Oord, A.V.D., Vinyals, O.: Generating Diverse High-Fidelity Images with VQ-VAE-2. arXiv:1906.00446, June 2019
15. Ronneberger, O., Fischer, P., Brox, T.: U-Net: Convolutional Networks for Biomedical Image Segmentation. arXiv:1505.04597, May 2015
16. Sauvola, J., Pietikainen, M.: Adaptive document image binarization. Pattern Recogn. **33**(2), 225–236 (2000)
17. Tensmeyer, C., Martinez, T.: Document image binarization with fully convolutional neural networks. In: 2017 14th IAPR International Conference on Document Analysis and Recognition (ICDAR), vol. 1, pp. 99–104, November 2017
18. Vo, Q.N., Kim, S.H., Yang, H.J., Lee, G.: Binarization of degraded document images based on hierarchical deep supervised network. Pattern Recogn. **74**, 568–586 (2018)
19. Zhang, D., Islam, M.M., Lu, G.: A review on automatic image annotation techniques. Pattern Recogn. **45**(1), 346–362 (2012)
20. Zhang, X.Y., Bengio, Y., Liu, C.L.: Online and Offline Handwritten Chinese Character Recognition: A Comprehensive Study and New Benchmark. arXiv:1606.05763, June 2016

Meaning Guided Video Captioning

Rushi J. Babariya[1] and Toru Tamaki[2(✉)]

[1] BITS Pilani, Pilani, India
[2] Hiroshima University, Higashihiroshima, Japan
tamaki@hiroshima-u.ac.jp

Abstract. Current video captioning approaches often suffer from problems of missing objects in the video to be described, while generating captions semantically similar with ground truth sentences. In this paper, we propose a new approach to video captioning that can describe objects detected by object detection, and generate captions having similar meaning with correct captions. Our model relies on S2VT, a sequence-to-sequence model for video captioning. Given a sequence of video frames, the encoding RNN takes a frame as well as detected objects in the frame in order to incorporate the information of the objects in the scene. The following decoding RNN outputs are then fed into an attention layer and then to a decoder for generating captions. The caption is compared with the ground truth by learning metric so that vector representations of generated captions are semantically similar to those of ground truth. Experimental results with the MSDV dataset demonstrate that the performance of the proposed approach is much better than the model without the proposed meaning-guided framework, showing the effectiveness of the proposed model. Code are publicly available at https://github. com/captanlevi/Meaning-guided-video-captioning-.

Keywords: Video captioning · Sequence-to-sequence · Object detection · Sentence embedding

1 Introduction

The task of describing a video with a text has been receiving a great attention in recent years. The mapping from a sequence of frames to a sequence of words was first introduced with a sequence-to-sequence model [1], then a variety of models [2] have been proposed. However, these approaches often suffer from some common problems. First, captions should reflect objects in the scene while generated captions may not include terms indicating such objects. This issue is caused by captioning models that take frames for capturing features, not for detecting objects in the scene. Second, generated captions are evaluated with ground truth captions by using loss functions, which typically compare two sentences in a word-by-word manner. This may not reflect a semantic similarity between sentences because the change of a single word in a sentence could lead

© Springer Nature Switzerland AG 2020
S. Palaiahnakote et al. (Eds.): ACPR 2019, LNCS 12047, pp. 478–488, 2020.
https://doi.org/10.1007/978-3-030-41299-9_37

to a completely opposite meaning, but the loss might be small due to the small difference of the single word.

In this paper, we propose a meaning-guided video captioning model in cooperating with an object detection module. Our model uses encoder LSTMs to learn the mapping from a video to a description, of which back born network is the sequence-to-sequence video-to-text model, called S2VT [1]. Upon this base network, we feed object detection results [3] of each frame into the encoder LSTMs to extract the most dominant object in each frame. Our model further incorporates attention in decoding LSTMs for enhancing information of frames that characterize the given video. In addition, we proposes a new approach to train the proposed video-to-text model. Instead of a classical training using a word-by-word loss, we train the model to learn the meaning of captions, or semantic similarity of captions. To this end, we propose a metric leaning model to embed captions so that distances between a semantically similar pair of captions becomes smaller than a dissimilar pair.

2 Related Work

There are many works on video captioning. The early model was a sequence-to-sequence model (S2VT) [1]. This was inspired by a sequence-to-sequence translation model that takes a text in one language and output a text in another language. Instead, the S2VT model takes a sequence of video frames as input to encoder LSTMs, and outputs a sequence of words through decoder LSTMS. Later a 3DCNN was used to extract video features [2] to generate texts describing videos, and also attention has been used [4] to find which part of the video are more informative.

Image captioning [5–7] is a closely related task describing images, instead of videos. Some works for image captioning have been aware of the issue—generated captions may miss objects in the scene [8]—however not well studied for the video captioning task. This could be alleviated by the help of object detection [3]. We therefore use object detectors to find objects in the scene and then reflect the object information in generated captions.

Designing the loss function is a key for many captioning models to success, and for video captioning we need a loss to compare generated and ground truth captions. This is common for many text-related tasks such as image and video captioning and visual question generation (VGQ) [9]. A problem is that a loss usually compares texts word-by-word, which is fragile to a little difference of words in sentences. Furthermore, a typical dataset for captioning has several different captions as ground truth of a single video, which is another cause for the word-by-word loss to be confused. In the proposed model, we propose a loss using sentence embedding and metric leaning so that semantically similar captions have small distances while different captions are far apart from each other.

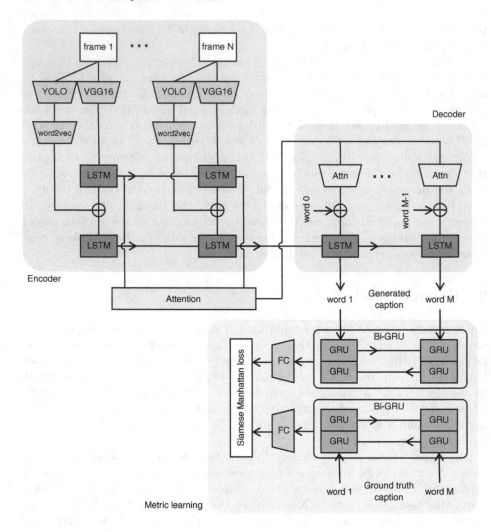

Fig. 1. Our model

3 Encoder-Decoder Model

Our proposed model is built on top of a baseline sequence-to-sequence video-to-text generator, S2VT [1]. The baseline model uses a stacked 2-layer LSTM encoder-decoder model that takes a sequence of RGB frames $f_1, f_2, \ldots, f_N$ as input and produces a caption or a sequence of words $w_1, w_2, \ldots, w_M$. Frame features are extracted by using the VGG16 pre-trained model. The lower LSTM in the encoder takes the output of the upper LSTM, encoding the visual information, concatenated with padding due to the absence of text information. In contrast, in the decoder the upper LSTM is fed padding due to the lack of video frames, and the lower LSTM takes the concatenation of padding and word information.

The proposed model is shown in Fig. 1. It consists of encoder, decoder, and metric learning components.

3.1 Encoder

The encoder LSMT now takes the concatenation of holistic visual information and scene object information (instead of padding). VGG16 features of 2048 dimension is extracted from a current RGB frame as holistic visual information. However it would not reflect objects in frames, and therefore we use the YOLOv3 object detector [3]. It may find many objects in a frame, however we focus on the dominant object in each frame. Specifically, we pick up the object having the highest objectness score in the YOLO detector, and find the string describing the category of the object (e.g., 'person' or 'cat'). The string is embedded with word2vec [10,11], pre-trained on a part of Google News Dataset [1], to convert it to an embedding vector of 300 dimension.

The upper LSTM in the encoder takes the 2048-d visual vector of the frame, then the hidden state of 1000 dimension is passed to the lower LSTM after concatenating with the object embedding vector, resulting in a 1300-d vector to be fed to the lower LSTM.

The lower LSTM outputs a 1000-d hidden state vector that is passed through the encoder and to the decoder LSTM. In contrast, 1000-d hidden states of the upper LSTM are not passed to the decoder, but to the attention layer.

3.2 Attention

Let $h_1, \ldots, h_N$ be the hidden states of the upper encoder LSTM. These are stacked in column-wise to make a matrix

$$H = (h_1, \ldots, h_N) \quad \in R^{1000 \times N}, \tag{1}$$

where N is the number of video frames encoded. This is used as attention [4] for $s_t \in R^{1000}$, a given output of the decoder LSTM at time step t for $t = 1, \ldots, M$. To do so, we construct an activation energy vector of the following form [12]

$$\lambda_t = \text{softmax}(s_t^T W H), \tag{2}$$

where $W \in R^{1000 \times 1000}$ is a trainable linear layer. Using λ_t, we have the attention vector of 1000 dimension as $a_t = H\lambda_t$.

This attention vector is used as input at time step t to the decoder LSTM after a linear layer keeping dimension and concatenation with the word embedding w_t of 300 dimension at time t.

---

[1] https://code.google.com/archive/p/word2vec/.

3.3   Decoder

The decoder LSTM takes the 1000-d hidden state from the lower decoder LSTM, and the input (the concat of attention a_t and word embedding w_t). The output is 1000 dimension and fed into a linear layer to convert a vector of vocabulary size of 25231 words (in the case of the experiments below). This is then passed to a softmax layer to obtain the word probability p_t at time t.

3.4   Word-by-Word Loss

This probability p_t is used to compute the cross entropy with the word w_t in the ground truth caption. The sum of these word-wise cross entropy values for $t = 1, \ldots, M$ is used as a loss to train the network.

This loss has been typically used for train networks to compare generated and ground truth captions in the literature [9,13]. However, it compares texts word-by-word, which is fragile against a little difference of words in sentences. Furthermore, a typical dataset for captioning has several different captions as ground truth of a single video, which is another cause for this word-by-word loss to be confused.

Therefore, in the training procedure, we first use this loss to train the network until convergence, then switch to another loss that captures the semantic similarity between captions, which is described next.

4   Metric Learning Component for Captions

4.1   Soft-Embedding Sequence Generation

In order to construct a loss comparing generated and ground truth captions, our model generates captions during training. To this end, a possible way might be sampling the next word by using the word probability p_t. This is however not useful for training because the sampling procedure cuts the computation graph and back propagation doesn't go back through the decoder LSTM.

Instead, we propose to use the probability p_t as weights for the next word. If it was a one-hot vector, then finding the next word is simply picking up the corresponding column of the 300×25231 word embedding matrix E, or equivalently multiplying the one-hot vector to E. As the similar way, we construct a single word embedding by $s_t = E p_t$. This is actually not any of words in the vocabulary, but should reflect a "soft" word choice of the decoder LSTM.

This is passed to the attention in the next time step, then the next weighted embedding word s_{t+1} is computed. Eventually, the decoder LSTM outputs the sequence $s_1, \ldots, s_M$ as a generated caption for the given video.

4.2   Meaning-Guided Loss

Now we have two sequences; generated and ground truth captions. In our metric learning component, these captions are first embedded with a sentence-to-vector model. This is a bi-directional GRU [14] with 1000-d hidden states each, resulting in 2000-d output. Then a linear layer is used to reduce the dimension to 1000.

To compare two 1000-d vectors v_1 and v_2, corresponding to generated and ground truth captions, we use the Siamese Manhattan loss [15]

$$L_{\text{sim}}(v_1, v_2) = 1 - \exp(\|v_1 - v_2\|_1). \tag{3}$$

This loss should be small for v_1 and v_2 because these two captions should be similar and the vectors should also be close to each other. This assumes that the model generates a reasonable caption v_1 that should be semantically similar to v_2. However in the early stage of the training, the model might be giving a very different caption from the ground truth, and if this is the case then the network might learn an identity mapping where it says that any caption pairs be semantically similar. To prevent this, we also use two different captions v_3 and v_4, and minimize the following loss as well;

$$L_{\text{dis}}(v_3, v_4) = \exp(\|v_3 - v_4\|_1). \tag{4}$$

The overall loss for this is given by

$$L = E_{v_1, v_2 \sim \text{training sample pair}}[L_{\text{sim}}(v_1, v_2)] + \\ E_{v_3, v_4 \sim \text{dissimilar sentence pair}}[L_{\text{dis}}(v_3, v_4)]. \tag{5}$$

We need to pre-train the model for this to work, as we will describe later.

4.3 Intra-batch Training

A possible drawback of the metric leaning component described above is that datasets for captioning do not provide any dissimilar sentence pairs. In the followings, we describe tricks to train the proposed model efficiently.

The first trick is to use a mini-batch for dissimilar pair sampling (Fig. 2 (top)). Suppose we are given a batch consists of 50 ground truth captions for 50 different videos in a dataset, and then we have corresponding 50 generated captions. Among these 100 captions, the 50 training pairs are used for L_{sim}. In addition, there are many more dissimilar caption pairs because different ground truth captions can be considered as different sentences. Therefore we can sample different caption pairs inside the batch for L_{dis}.

However, a naive sampling is inefficient because a ground truth caption is encoded in a vector several times; once as a training (similar) pair, and more as a dissimilar pair. This leads to encoding the same caption multiple times with the network having the same weights. This is a waste of resource because the encoding results are the same before computing the loss and backprop.

Our second trick is to do this efficiently (Fig. 2 (bottom)). Again suppose we are given a 50-sample batch, and we have generated and ground truth caption embeddings as two matrices of size 50×1000 (each row is a 1000-d embedding vector). Let V_{gen} is the matrix of generated captions, and V_{gt} is the matrix of ground truth captions. We split them to two to obtain four 25×1000 matrices;

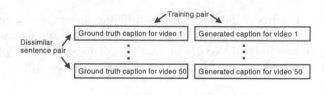

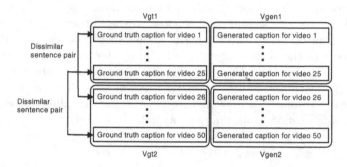

Fig. 2. Intra-batch sampling. (top) Sampling two ground truth sentences in a batch as dissimilar sentence pairs. (bottom) Use corresponding ground truth pairs as dissimilar sentence pairs.

$V_{\text{gen1}}, V_{\text{gt1}}$ and $V_{\text{gen2}}, V_{\text{gt2}}$. Now each row in V_{gen1} has nothing related to any row in V_{gen2} due to random sampling from the training dataset, and the same for $V_{\text{gt1,2}}$. Therefore, we can use i-th rows of $V_{\text{gen2}}, V_{\text{gt2}}$ as the dissimilar pair for i-th rows of $V_{\text{gen1}}, V_{\text{gt1}}$, and vice versa. This can be done by keeping these embedding vectors just before computing the Siamese Manhattan loss.

Our third trick is to do it more efficiently. As a concept, i-th training pair and i-th dissimilar pair are used for computing the loss, for $i = 1, \ldots$ over training samples in the batch. However, usually the loss is aggregated for training samples in the batch, to compute the loss value of the batch. Here the order doesn't matter; we can compute and aggregate the loss L_{sim} for training samples first. Then we can add the loss L_{dis} to compute the final loss value of the batch.

The final trick is to use two different optimizers. For computing the loss L_{sim} for training samples, video frames are input the network to generate a caption. However, the loss L_{dis} is only used for learning the metric learning component, not for the encoder-decoder LSTMs. In other words, only v_1 connects the loss and the encoder and decoder components; v_2 is a given ground truth, and v_3, v_4 are only used for the metric learning component. Therefore we use different optimizers (updaters) for two losses. For training with the loss L_{sim}, one optimizer updates all networks weights. For training with the loss L_{dis}, another optimizer updates weights in the metric learning component only.

5   Experiments

5.1   Dataset

The dataset used is the Microsoft Video Description corpus (MSVD) [16]. This is a collection of Youtube clips (1970 in total), average length of videos is about 6 s. Each video has descriptions given by several annotators who describe the video in one sentence (40 captions per video on average). The data is split in the following way [1]; 1200 videos are used for training, and 100 for validation. The remaining 670 are used for testing. For a single video, we used up to $N = 80$ frames as input to the encoder LSTM.

5.2   Training Procedure

The metric learning component uses the Siamese Manhattan loss, however we use a triplet loss for pre-training the metric learning component. Here we can use the similar trick explained in the section of intra-batch training. A triplet loss takes three arguments; reference, positive, and negative samples. Given a batch of 50 samples, we have 50 pairs of generated and ground truth captions. For one of these pairs, the other 49 samples can be considered as negatives for the triplet loss. This is much more efficient than a naive training.

For the encoder and decoder components, we use the word-by-word loss without the metric learning component. Once the pre-training phase has been done, then we use the both losses for training in a stochastic manner. Specifically, given a batch, we randomly select if the metric leaning component is used (and then the Siamese Manhattan loss) in chance of 70%, or not (the word-by-word loss is used) in 30%.

5.3   Results

Table 1 shows results of the baseline and proposed models. There are three different settings for the proposed models; O stands for the case using object information only (hence the attention and metric leaning component are not used), OA stands for the case when object information and attention are used but not the metric learning component, and OAM stands for the full model including the object, attention, and metric learning components.

The use of the object information clearly improve the performance against the baseline (original sequence-to-sequence) model. This is because main objects in each frame are explicitly used in the encoder. Attention further improves the results (except CIDEr), which is expected as many results reported with better performance with attention. Our full model, shown as OAM in Table 1, can further boost the performance by 1.1% in BLUE4. This is not as large as improvements with object information (by 10.7%) and attention (2.9%), but this results suggest that the proposed metric learning component can be used to improve results by adding to any models other than the sequence-to-sequence architecture.

Table 1. Results of baseline and proposed models on the MSVD dataset. Symbols stand for; O – Objects, A – Attention, M – Meaning model.

Model	BLEU4	METEOR	CIDEr
Baseline (S2VT)	0.288	0.246	–
O	0.395	0.295	0.641
OA	0.424	0.312	0.641
OAM	0.435	0.316	0.649

GT	A man and woman ride a motorcycle.
O	a man is dancing.
OA	two people are dancing.
OAM	a man and a woman are riding a motorcycle.

GT	A man is lifting the car.
O	a man is lifting a car.
OA	a man is lifting the back of a truck.
OAM	a man is lifting a truck.

GT	An animal is eating.
O	a cat is licking a lollipop.
OA	a dog is eating.
OAM	a dog is eating.

Fig. 3. Examples of generated captions. Left images are frames of videos, and right texts are ground truth and generated captions. Symbols stand for; O – Objects, A – Attention, M – Meaning model.

Figure 3 shows some examples generated by the proposed model. In the first video (top row in the figure), the scene changes frequently because of the movie editing; the video is composed of several cuts from different angles. Therefore the model without the metric learning component is confused and generated "dancing" instead of "riding".

In the last video (bottom row of the figure), the caption generated by the full model is considered as wrong because the ground truth caption mentions the animal as "animal" and it is obviously not a dog, while the generated caption says that it is a "dog". This is a limitation of the proposed model because the mistakes of the object detector ("dog" for "animal") directly affect the encoder.

Table 2 shows a comparison with other recent methods. While the proposed method doesn't perform as like recent methods, the proposed meaning-guided loss and tricks for intra-batch training are expected to work for boosting other methods.

Table 2. Results of the proposed model and other methods.

Model	BLEU4	METEOR	CIDEr
OAM (ours)	0.435	0.316	0.649
[17]	0.523	0.341	0.698
[18]	0.479	0.350	0.781

6   Conclusions

We have proposed a model for video captioning guided by the similarity between captions. The proposed model has three components; the 2-layer LSTM encoder involving scene object information, the LSTM decoder with attention, and the metric learning for comparing two captions in a latent space. Experimental results show that the proposed model outperforms the baseline, a sequence-to-sequence model, by introducing the metric learning component in addition to attention mechanism and object information.

Our future work includes using the proposed metric learning component in other state of the-art caption models where word-by-word losses are used [17,18], and incorporating other corpuses for sentence-level similarity computation, such as [19].

Acknowledgement. This work was supported by International Linkage Degree Program (ILDP) in Hiroshima University (HU). This work was also supported by JSPS KAKENHI grant number JP16H06540.

References

1. Venugopalan, S., Rohrbach, M., Donahue, J., Mooney, R., Darrell, T., Saenko, K.: Sequence to sequence - video to text. In: 2015 IEEE International Conference on Computer Vision (ICCV), pp. 4534–4542, December 2015
2. Aafaq, N., Gilani, S.Z., Liu, W., Mian, A.: Video description: a survey of methods, datasets and evaluation metrics. CoRR abs/1806.00186 (2018)
3. Redmon, J., Farhadi, A.: YOLOv3: an incremental improvement. CoRR abs/1804.02767 (2018)
4. Vaswani, A., et al.: Attention is all you need. In: Guyon, I., et al. (eds.) Advances in Neural Information Processing Systems 30, pp. 5998–6008. Curran Associates, Inc. (2017)
5. Bai, S., An, S.: A survey on automatic image caption generation. Neurocomputing **311**, 291–304 (2018)
6. Hossain, M.Z., Sohel, F., Shiratuddin, M.F., Laga, H.: A comprehensive survey of deep learning for image captioning. ACM Comput. Surv. **51**(6), 118:1–118:36 (2019)

7. Liu, X., Xu, Q., Wang, N.: A survey on deep neural network-based image captioning. Vis. Comput. **35**(3), 445–470 (2019)
8. Cornia, M., Baraldi, L., Cucchiara, R.: Show, control and tell: a framework for generating controllable and grounded captions. In: The IEEE Conference on Computer Vision and Pattern Recognition (CVPR), pp. 8307–8316, June 2019
9. Li, Y., et al.: Visual question generation as dual task of visual question answering. In: The IEEE Conference on Computer Vision and Pattern Recognition (CVPR), pp. 6116–6124, April 2018
10. Mikolov, T., Sutskever, I., Chen, K., Corrado, G., Dean, J.: Distributed representations of words and phrases and their compositionality. In: Proceedings of the 26th International Conference on Neural Information Processing Systems, NIPS 2013, vol. 2, pp. 3111–3119. Curran Associates Inc., New York (2013)
11. Mikolov, T., Chen, K., Corrado, G., Dean, J.: Efficient estimation of word representations in vector space. In: 1st International Conference on Learning Representations, ICLR 2013, Scottsdale, Arizona, USA, 2–4 May 2013, Workshop Track Proceedings (2013)
12. Luong, T., Pham, H., Manning, C.D.: Effective approaches to attention-based neural machine translation. In: Proceedings of the 2015 Conference on Empirical Methods in Natural Language Processing, Lisbon, Portugal, pp. 1412–1421. Association for Computational Linguistics, September 2015
13. Vinyals, O., Toshev, A., Bengio, S., Erhan, D.: Show and tell: a neural image caption generator. In: The IEEE Conference on Computer Vision and Pattern Recognition (CVPR), June 2015
14. Schuster, M., Paliwal, K.K.: Bidirectional recurrent neural networks. IEEE Trans. Signal Process. **45**(11), 2673–2681 (1997)
15. Mueller, J., Thyagarajan, A.: Siamese recurrent architectures for learning sentence similarity. In: Proceedings of the Thirtieth AAAI Conference on Artificial Intelligence, AAAI 2016, pp. 2786–2792. AAAI Press (2016)
16. Chen, D.L., Dolan, W.B.: Collecting highly parallel data for paraphrase evaluation. In: Proceedings of the 49th Annual Meeting of the Association for Computational Linguistics: Human Language Technologies, HLT 2011, vol. 1, pp. 190–200. Association for Computational Linguistics, Stroudsburg (2011)
17. Wang, B., Ma, L., Zhang, W., Liu, W.: Reconstruction network for video captioning. In: The IEEE Conference on Computer Vision and Pattern Recognition (CVPR), June 2018
18. Aafaq, N., Akhtar, N., Liu, W., Gilani, S.Z., Mian, A.: Spatio-temporal dynamics and semantic attribute enriched visual encoding for video captioning. In: The IEEE Conference on Computer Vision and Pattern Recognition (CVPR), June 2019
19. Bowman, S.R., Angeli, G., Potts, C., Manning, C.D.: A large annotated corpus for learning natural language inference. In: Proceedings of the 2015 Conference on Empirical Methods in Natural Language Processing (EMNLP). Association for Computational Linguistics (2015)

Infant Attachment Prediction Using Vision and Audio Features in Mother-Infant Interaction

Honggai Li[1]([✉]), Jinshi Cui[1]([✉]), Li Wang[2], and Hongbin Zha[1]

[1] Key Laboratory of Machine Perception, Peking University, Beijing 100871, China
lihonggai@pku.edu.cn , cjs@cis.pku.edu.cn
[2] School of Psychological and Cognitive Sciences, Peking University,
Beijing 100871, China

Abstract. Attachment is a deep and enduring emotional bond that connects one person to another across time and space. Our early attachment styles are established in childhood through the interaction between infants and caregivers. There are two attachment types, secure and insecure. The attachment experience affects personality development, particularly a sense of security, and research shows that it influences the ability to form stable relationships throughout life. It is also an important aspect of assessing the quality of parenting. Therefore, attachment has been widely studied in psychology research. It's usually acquired by Ainsworth's Strange Situation Assessment (SSA) through tedious observation. As far as we know, there is no computational method to predict infant attachment type. We try to use the Still-Face Paradigm (SFP) video and audio as input to predict attachment types through machine learning methods. In the present work, we recruited 64 infant-mother participants, collected videos of SFP when babies are 5–8 months of age and identified their attachment types including secure and insecure by SSA when those infants are almost 2 years old. For the visual part, we extract motion features and apply a RNN network with LSTM units model for classification. For the audio part, speech enhancement is conducted as data pre-processing, pitch frequency, short-time energy and Mel Frequency Cepstral Coefficient feature sequences are extracted. Then SVM is deployed to explore the patterns in it. The experiments show that our method is able to discriminate between the 2 classes of subjects with a good accuracy.

Keywords: Attachment · Still-Face paradigm · Affective computing · Motion analysis · Audio analysis

1 Introduction

Attachment describes an emotional bond that connects one person to another, especially the long-term bonds, such as parent-child relationship and intimate partners. Bowlby defined attachment as a lasting psychological connectedness

© Springer Nature Switzerland AG 2020
S. Palaiahnakote et al. (Eds.): ACPR 2019, LNCS 12047, pp. 489–502, 2020.
https://doi.org/10.1007/978-3-030-41299-9_38

between human beings [1]. It is a special emotional relationship that involves an exchange of comfort, care and pleasure between people. Initial research typically focused on infant attachment, however, attachment theory is related to relationships that we engage in as adults. These intimate and/or romantic relationships are also directly related to our attachment styles formed in the childhood and the care we received from our primary caregivers [2]. Our early attachment styles are established in childhood through the infant/caregiver relationships. A baby forms indiscriminate attachments at 6 weeks to 7 months and specific attachment at 7 to 9 months of age. The way how to play and interact with infants plays an important role in forming attachment. Hence, it is important for parents to know the attachment type of their baby as early as possible so that they can adjust the interaction with infants.

There are four types of attachment, including secure attachment, ambivalent attachment, avoidant attachment and disorganized attachment. The latter three types known collectively as insecure attachment. In psychology studies, one of the ways to assess infant attachment is through observation of Ainsworth's Strange Situation Assessment (SSA) for children over 11 months. However, the procedure and observation is tedious and infant attachment type has been relatively stable at that time. As far as we know, there is no computational method to identify infant attachment type.

Thus, we aim at proposing a novel approach of recognizing infant attachment between secure and insecure using vision-based motion features and audio features through machine learning methods. The underlying idea and rationale come from the psychological studies claiming that the relevance between infants and caregivers behavior pattern during the Still-Face Paradigm (SFP) and the attachment type assessed by SSA [3,4]. Still-Face paradigm (SFP) is used to examine the ways in which infants modulate their affect and attention under a stress situation [5]. In many ways, the SFP contains similar elements as those involved in the Strange Situation. Both procedures involve separation and reunion episodes [6]. Thus, the SFP may be a particularly salient procedure to recognize infant attachment type through analysis of reaction and action of infants when parents disengage their attention from their infants and then reestablish interaction [4]. In addition, maternal sensitivity and responsiveness is also important for the prediction of infant attachment. Maternal sensitivity is the mother's ability to perceive and infer the meaning behind her infant's behavioural signals, and to respond to them promptly and appropriately. Through analyzing mothers' touch, spatial orientation and vocal affect can assess the quality of infant-mother interaction and synchrony between them [3].

In present work, we build the first infant attachment dataset consisting of videos and audios of mother and infant during the SFP and labels of infant attachment types (secure or insecure) acquired by the Strange Situation. For the video part, the attachment prediction problem can be regarded as action analysis problem. We design a deep learning architecture that only takes SFP videos as input for early attachment type classification. Specifically, we devise and test a LSTM network to sequentially deal with features extracted by CNN.

Our method has proven to be feasible since the experiments indicate that we achieve high accuracy on our dataset. On the other hand, in consideration of the importance of voice in expressing emotion needs, we extract pitch frequency, short-time energy and Mel Frequency Cepstral Coefficient audio features for classification to modify our architecture. Before that, we conduct speech enhancement to remove noise using a fully convolutional neural network. The experiments demonstrate that the audio channel is effective for the classification of attachment type. The entire study proposes some novel ideas of computation of attachment and further verifies the feasibility through experiments. The rest of the paper is organized as follows: Sect. 2 reviews the related work in the attachment theory and mother-infant interaction analysis, action analysis and affect computing. Section 3 presents the infant attachment dataset and its acquisition protocol. In the Sect. 4 the details of the proposed method are described followed by the experiments results and discussion provided in Sect. 5. Finally, Sect. 6 draws conclusions and sketches future work.

2   Related Work

The visual-audio attachment prediction system consists of three important parts: vision-based action analysis and audio feature extraction and multimodality fusion. In following parts, we review related works focusing on mother-infant interaction analysis and the first two parts of system, respectively.

2.1   Mother-Infant Interaction Analysis

Studying the quality and dynamics of early parent-child interactions is crucial for the prediction of attachment. It requires the perception and integration of multimodal social signals and the understanding of synchrony between two interactive partners. Automatic computational methods focused on some cues like head movements [11], motion [12], facial expression [13], motherese [14], vocalization, speech turns [15], gaze and motion synchrony [16].

2.2   Vision-Based Action Analysis

From recent studies, we can identify three main categories of deep neural network architectures used in action recognition, namely multiple stream networks, 3D convolutional networks and deep generative models using LSTM [17]. The original model of multiple stream network is Two-Stream Convolutional Networks [18]. The original model of 3D convolutional networks is C3D network [19]. After that, there are many variants of these two models which show good performance on standard action recognition benchmarks. LSTM networks have been proven to effectively codify temporal information of video sequences, especially when fed with powerful deep features [20]. In recent studies, the model based on LSTM has been used to detect Autism Spectrum Disorder (ASD) through gesture videos with a good accuracy [21].

2.3   Audio Feature Extraction

The widely-used audio affective features can be categorized into prosody features, voice quality features, and spectral features [22]. Pitch, energy, and duration time are popular prosody features, since they can reflect the rhythm of spoken language. Mel Frequency Cepstral Coefficient (MFCC) is the most well-known spectral features since it is used to model the human auditory perception system [23]. These audio features are widely used in infant cry analysis [24] and emotion recognition [25,26].

3   Dataset

3.1   Participants

We recruited 95 infant-mother participants from Hebei Province of China. There are 63 secure attachment and 32 insecure attachment children in our dataset. In consideration of the balance, we selected 32 secure subjects randomly. The total number of dataset is 64, including 32 baby boys and 32 baby girls (Mean and standard deviation of age is 6.4 and 0.1 months respectively).

3.2   Paradigm and Procedure

We collected videos of SFP when babies were 5 to 8 months old, and then we identified their attachment type (secure or insecure) by Ainsworth's Strange Situation Assessment as ground truth when they are almost 2 years old. The procedure of collecting SFP videos is as follows. Firstly, we collected videos in a quiet room. An infant was placed in a safety seat and his/her mother sat face to face with the baby. Secondly, as shown in Fig. 1(a) and (b), the mother was instructed to conduct three episodes of SFP, which are free-play, still face and reunion. During the free-play episode, the mother interacts beautifully and normally with the infant. Then the mother is told to not move and hold a "still face" without any expressions. The infant has different tactics to re-engage the mother, like pointing, trying to touch the mother, and screaming. Besides, the infant will try to self comfort himself, which in the still face experiment is classified as sucking, self-clasping, rocking or escape from the safety seats. He will also try to shift his focus, on objects and on his self. After that, the mother is told to reestablish the connections with infant through touch, hug, smile and verbal positive expression. The infant will show different responses to mother's effort which is related to infant personality and the quality of relationship between them.

We recorded the whole experiment from two viewpoints using AXIS M1054 video camera (resolution: 1280 × 800 pixels, 25/30 frames/sec), as shown in Fig. 1(c) and (d). In the end, we removed the free play part at the beginning of the video. The video clips containing still face and reunion episodes were saved as our data. The duration of each data was 4 min, including 2 min still face episode and 2 min reunion episode. After that, we extracted the audio files in wav format from video files of infant viewpoint through ffmpeg. Sampling frequency is 44.1 KHz which means a recording with a duration of 60 s will contain 2646000 samples.

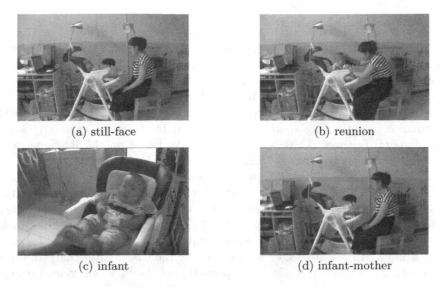

(a) still-face

(b) reunion

(c) infant

(d) infant-mother

Fig. 1. Sample frames of two episodes and two viewpoints from SFP videos.

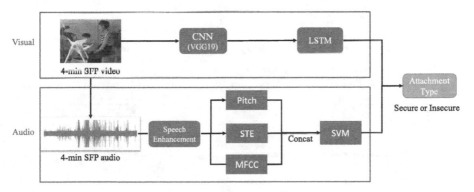

Fig. 2. The architecture of proposed method using audio and visual features. The pipeline of vision based method is shown in the red frame. And the method using audio features is shown in the green frame. (Color figure online)

4   Methods

In this section, we will describe the details of proposed methods including the video part and audio part. The pipeline is shown as Fig. 2.

4.1   Vision Based Method

Considering our problem, the length of the input video is up to 4 min. We need to extract the behavior patterns which requires the network is deep both spatially and temporally. In addition, since the appearance of the two classes is similar, multiple stream networks and 3D convolutional networks relying on spatial information do not perform well enough. The LSTM network shows excellent performance in extracting spatial and temporal information, especially long-term temporal features. Thus, we select a model based on the RNN network with Long Short-Term Memory (LSTM) units [27]. As for feature network, we choose a Convolutional Neural Network (CNN) model to extract features from the input video frame by frame. We extract the last convolutional layer of the VGG19 model pre-trained on the ImageNet dataset [28]. In particular, the image at each time step is resized to $224 \times 224 \times 3$ to adapt for the input of network. The shape of this last convolutional feature cube is $14 \times 14 \times 512$. Thus, for each frame, a 196×512 feature vector is extracted by the CNN. After that, we use a soft attention mechanism to take the expectation of the 14×14 regions through

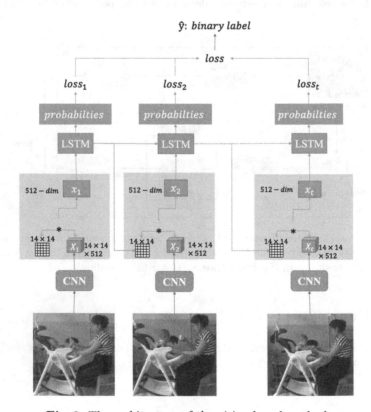

Fig. 3. The architecture of the vision-based method

a learned weight map at each time step. Finally, a feature vector of 512-dim is fed into the follow-up LSTM network. For each video we select T frames for the training of LSTM. We train an one-layer LSTM model where the dimension of the LSTM hidden layer is 1024. The cross-entropy loss function we used is defined as follows:

$$L = -\sum_{t=1}^{T}\sum_{i=1}^{C} y_{t,i} \log \hat{y}_{t,i} \qquad (1)$$

where T is the number of time-steps of a video ($T = 17$ in our experiments), C is the number of classes($C = 2$ in our experiments), $y_{t,i}$ is the ground truth label and $\hat{y}_{t,i}$ is the class probabilities at time-step t. The architecture is shown as Fig. 3.

4.2 Audio Channel Method

The attachment prediction is a complex affect computing problem which needs consideration of multichannel information, like action, emotion and voice. Voice is the most direct and effective way for infants to express their emotion and psychological needs, especially crying. Verbal expression of mothers is also an important aspect of maternal sensibility. Considering our dataset, there exists some regular patterns between voice and attachment type. Secure attachment infants may cry in still-face episode, but their mothers will comfort them and resume the interaction with infants effectively in reunion episode. However, insecure attachment infants may ignore mothers or wail in still-face episode and will respond negatively to mothers' comfort behaviors or scream in reunion episode. Based on these conclusions, we try to extract effective audio features for attachment classification.

Speech Enhancement. The main voices during Still-Face experiments are from infants, mother and some noises like the sound of wind and traffic noise. Mothers' audio interactions can be classified as: maternal vocalization (meaningful vocalizations, laugh, singing, animals' sound) or other noise (clap the hand, snap fingers or snap the tongue, voices from toys, mouth's noise). Similarly, infants' audio production can be defined as: infant vocalization (babbling vocalizations, laugh and cry) or other noise (voices from toys) [29].

To eliminate noise, we conduct speech enhancement using a fully convolutional neural network [30]. During the still-face episode, main source of voice we want to reserve is cry of infants. Therefore, we train the model on a dataset which consist of a public infant cry audio corpus and sound of wind, traffic noise. The cry dataset is designed for cry-cause factors analysis, including belly pain, burping, discomfort, hungry and tired. The noise is selected from freesound website. Then we evaluate on the audios of 2-min still-face episode with trained model to remove the noise and obtain the clean audios.

Feature Extraction. The first step in any automatic speech recognition system is to extract features i.e. identify the components of the audio signal that are good for identifying the linguistic content and discarding all the other stuff which carries information like background noise. We extract pitch frequency, short-time energy (STE) and Mel Frequency Cepstral Coefficient (MFCC) features because they are important audio features in emotion recognition according to the research [31–34]. The mentioned features can be summarized as follows:

Pitch Frequency. The fundamental frequency is important for classification purposes in this problem. Adult fundamental frequency ranges between 85–180 Hz. However, infant crying fundamental frequency is characterized by its high pitch (250–700 Hz). Pitch provides useful information about describing the cry envelope or the shape that is formed by the successive frames in the case of infant crying [24]. In this research study, the pitch detection algorithm is based on autocorrelation method [35].

Short-Time Energy. The original signal is $x(n)$. After the framing step, a Hamming window is applied on each frame.

$$y_i(n) = w(n) \times x((i-1) \times inc + n), \quad 1 \le n \le L, 1 \le i \le f_n \tag{2}$$

and $w(n)$ is the window function, L is the length of a frame, inc is the length of frame step ($L = 200$ and $inc = 80$ in our experiments), f_n is the number of frames. The short-time energy (STE) of a signal $y_i(n)$ is defined as:

$$E_i = \sum_{n=0}^{L-1} y_i(n)^2, 1 \le i \le f_n \tag{3}$$

Mel Frequency Cepstral Coefficient (MFCC). Mel Frequency Cepstral Coefficients is a feature widely used in automatic speech and speaker recognition. It provides a representation of the short-term power spectrum of a signal. In our experiments, we extract MFCC as follows:

step 1 Firstly, frame the signal into 32 ms frames. Frame step is 10 ms, which allows some overlap to the frames. The Next steps are applied to every single frame, one set of 12 MFCC coefficients is extracted for each frame.
step 2 Take the Discrete Fourier Transform of the frame, then we would generally perform a 512 point FFT and keep only the first 257 coefficients.
step 3 Compute the Mel-spaced filterbank. A set of 26 triangular filters is applied to the periodogram power spectral estimate from step 2.
step 4 Take the log of each of the 26 energies from step 3. This leaves us with 26 log filterbank energies.
step 5 Take the Discrete Cosine Transform (DCT) of the 26 log filterbank energies to give 26 cepstral coefficients. At last, only the lower 12 of the 26 coefficients are kept.

The resulting features (12 numbers for each frame) are Mel Frequency Cepstral Coefficients.

After the feature extraction, we concat all the features and use support vector machines (SVM) algorithm considering the class number and size of dataset. The complete architecture is shown as the green frame in Fig. 2.

5   Experiments and Discussion

5.1   Vision Part Experiments

Setup and Training Details. Three types of videos from two different viewpoints were used as our input. They were 2-min Still-Face episode, 2-min Reunion episode and the entire videos (including Still-face and Reunion episodes, 'SF + RU' for short) respectively. We selected usual hold-out validation and one-subject-out testing procedure, the latter can detect individual attachment and adapt to the real-world applications. For hold-out validation, the learning rate was 0.001 and the train/test ratio was 7/3. More specifically, we used 23 secure and 23 insecure subjects as training set, 9 secure and 9 insecure subjects as testing set. We trained each model 350 epochs and saved a checkpoint every 10 epochs during the training process. For one-subject-out testing, the number of epoch was 350. Hence, we would have 35 checkpoints for each subject. We selected the checkpoint which has the highest accuracy on all subjects as our final output. All models were trained by the gradient descending optimization algorithm.

Table 1. Hold-out validation accuracy for different inputs. We highlight in bold the highest accuracy. Inf-mot presents the viewpoint of infant-mother.

Viewpoint	Episode		
	Still-face	Reunion	SF + RU
Infant	0.38	0.60	0.61
Inf-mot	0.50	0.67	**0.78**

Table 2. Confusion matrix of hold-out validation using inf-mot SF + RU video.

		Predicted	
		Secure	Insecure
Actual	Secure	7	2
	Insecure	2	7

Table 3. One-subject-out testing accuracy using infant and inf-mot video.

	Infant	Inf-mot
Secure	0.50	0.72
Insecure	0.47	0.69
Average	0.48	0.70

Results and Discussion. In Table 1, we report the test accuracy of different inputs. Each line in Table 1 refers to a different viewpoint. Besides, we considered three types video clips in each viewpoint, including Still-Face episode, Reunion episode and 'SF + RU'. On the one hand, the results of Table 1 indicates that the accuracy using Reunion episode as input is apparently higher than

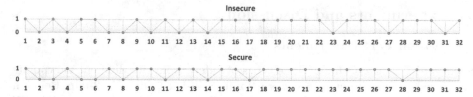

Fig. 4. One-subject-out testing result of inf-mot viewpoint. '1' presents the subject was classified correctly.

that of Still-Face episode. We obtained a random chance even lower accuracy using the Still-Face episode video. Through Still-Face episode, we can extract some patterns of infant personality. However, Still-Face video clips can't provide enough information to distinguish attachment types. In addition, 'SF + RU' performed a higher classification accuracy than only using Reunion episode as inputs. On the other hand, we can clearly note that for the same episodes, the classification accuracy of infant-mother viewpoint is obviously higher than that of infant. This probably indicates that mother's behaviors play an important role in the attachment prediction. It also proves that maternal sensitivity to infant's distress is of great importance to attachment security.

In Table 2, we compute the confusion matrix of entire procedure videos from of inf-mot viewpoint. We compared our method with the improved dense trajectories (iDT) on the infant-mother 'SF + RU' dataset [39]. We extracted Trajectory, HOG, HOF, MBH descriptors [36–38]. Each video was thus represented by a $2 \times 32 \times (15 + 48 + 54 + 96) = 13632$ dimensional Fisher vector for each descriptor type. Then, we normalized and concatenated the Fisher vectors. Finally, libsvm was used for classification [40]. However, the hold-out validation accuracy is only 0.60 which slightly exceeds random accuracy. It indicates that traditional hand-craft feature is not enough to solve this problem.

Figure 4 shows the results of one-subject-out testing procedure. We considered the SF+RU videos from inf-mot viewpoint. The top line is the result of insecure group, the bottom line is the result of secure group, '1' represents that the model can classify the subject correctly. Table 3 reports the accuracy of one-subject-out testing. We found that the result from inf-mot viewpoint is better than that from infant viewpoint for the insecure group, which is consistent with the result of hold-out validation.

The accuracy of the two testing are not as excellent as the results on public action recognition datasets which reach 92% or over. We believe that the complication of the problem itself leads to this result. We analyzed the videos of miss-detected subjects and found that there were some similar actions between the secure subjects and miss-detected insecure subjects. In SF episode, the infants did not make substantial physical movements and they were also calm during RU episode. In addition, we analyzed the videos of the insecure subjects that were correctly classified, and found that the infant movement range is very large in both SF and RU episode, including waving arms up and down, kicking the legs, reaching out to mother, asking for a hug from her and sucking fingers or toys. Hence, our classification mainly depends on the differences among infant movement ranges. It is easier to make a mistake if there is only tiny difference between two types. Besides, there are three different insecure attachment types, ambivalent attachment, avoidant attachment and disorganized attachment, so it may help improving the accuracy if making a further discrimination among these three types. Another possible way is to add infant emotion changes as a feature into our method. We will explore this idea in future work. As a whole, our model shows a good performance on classification, given that the total accuracy reached 0.78 on the infant-mother viewpoint dataset.

5.2 For Audio Part Experiments

Figure 5 shows the result of speech enhancement on a sample audio. We can see that speech enhancement can remove the noise in audio effectively and the waveform is more clean after enhancement. Thus we can extract more effective features.

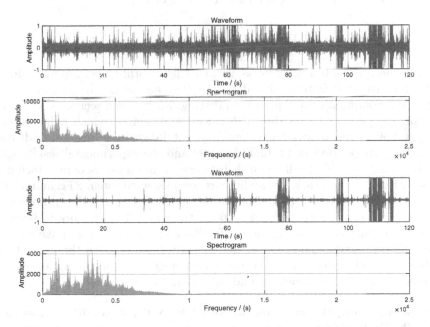

Fig. 5. Waveform and Spectrogram of a sample audio before (top) and after (bottom) enhancement

Table 4. The confusion matrix

		Predicted	
		Secure	Insecure
Actual	Secure	4	7
	Insecure	2	13

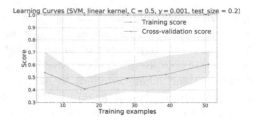

Fig. 6. The learning curve of 10 fold cross validation.

We conduct grid search to determine optimal values for the model. In Table 4, we compute the confusion matrix of test results on this dataset. The accuracy of classification is 65% which indicates the audio features we selected are effective for the attachment prediction problem. Figure 6 shows the relationship between scores of training, cross-validation and the number of training examples. The model inevitably has the overfitting problem. But we can see that the accuracy will probably increase if there are more training examples.

6   Conclusion and Future Work

In this paper, we built a novel dataset consisting of Still-Face Paradigm videos, audios and attachment type labeled by the Stranger Situation Assessment. We proposed a computational approach to distinguish between secure and insecure attachment types based on the SFP videos and audios, which helps parents and psychology researchers know better about infant attachment in advance. For vision part, we designed a LSTM model and the experimental results show that our model is able to distinguish between secure and insecure attachment types with a great accuracy. For audio part, we extracted pitch, STE and MFCC features which are verified to be feasible for attachment prediction problem.

Due to the complexity of problem itself, there is still much work to be done. We need to consider extracting more affective audio features and make a better fusion of visual and audio features. Besides, we should expand the dataset and further distinguish among the insecure attachment types, including ambivalent attachment, avoidant attachment and disorganized attachment.

Acknowledgment. This work was supported by National Key R&D Program of China (2017YFB1002503).

References

1. Bowlby, J.: Attachment theory and its therapeutic implications. Adolesc. Psychiatry **6**, 5–33 (1978)
2. Firestone, L.: Disorganized Attachment
3. Braungart-rieker, J., Garwood, M., Powers, B., Wang, X.: Parental sensitivity, infant affect, and affect regulation: predictors of later attachment. Child Dev. **72**, 252–270 (2001)
4. Braungart-rieker, J., Zentall, S., Lickenbrock, D., Ekas, N., Oshio, T., Planalp, E.: Attachment in the making: mother and father sensitivity and infants' responses during the still-face paradigm. J. Exp. Child Psychol. **125**, 63–84 (2014)
5. Tronick, E., Als, H., Adamson, L., Wise, S., Brazelton, T.: The infant's response to entrapment between contradictory messages in face-to-face interaction. Pediatrics **62**, 403–403 (1978)
6. Cohn, J.: Additional components of the still-face effect: commentary on Adamson and Frick. Infancy **4**, 493–497 (2003)
7. Ainsworth, M., Blehar, M., Waters, E., Wall, S.: Patterns of Attachment: A Psychological Study of the Strange Situation. Psychology Press, London (2015)
8. Ainsworth, M., Blehar, M., Waters, E., Wall, S.: Patterns of Attachment: Assessed in the Strange Situation and at Home. Erlbau, Hillsdale (1978)
9. Zeng, Z., et al.: Audio-visual affect recognition. IEEE Trans. Multimedia **9**, 424–428 (2007)
10. Qu, J., Leerkes, E.: Patterns of RSA and observed distress during the still-face paradigm predict later attachment, compliance and behavior problems: a person-centered approach Dev. Psychobiol. **00**, 707–721 (2018)
11. Hammal, Z., Cohn, J., Messinger, D.: Head movement dynamics during play and perturbed mother-infant interaction. IEEE Trans. Affect. Comput. **6**, 361–370 (2015)
12. Egmose, I., et al.: Relations between automatically extracted motion features and the quality of mother-infant interactions at 4 and 13 months. Front. Psychol. **8**, 2178 (2017)
13. Messinger, D., Mahoor, M., Chow, S., Cohn, J.: Automated measurement of facial expression in infant-mother interaction: a pilot study. Infancy **14**, 285–305 (2009)
14. Cohen, D., et al.: Do parentese prosody and fathers' involvement in interacting facilitate social interaction in infants who later develop autism? Plos One **8**, e61402 (2013)
15. Weisman, O., et al.: Dynamics of non-verbal vocalizations and hormones during father-infant interaction. IEEE Trans. Affect. Comput. **7**, 337–345 (2015)
16. Leclère, C., et al.: Interaction and behaviour imaging: a novel method to measure mother-infant interaction using video 3D reconstruction. Transl. Psychiatry **6**, e816 (2016)
17. Herath, S., Harandi, M., Porikli, F.: Going deeper into action recognition: a survey. Image Vis. Comput. **60**, 4–21 (2017)
18. Simonyan, K., Zisserman, A.: Two-stream convolutional networks for action recognition in videos
19. Tran, D., Bourdev, L., Fergus, R., Torresani, L., Paluri, M.: Learning spatiotemporal features with 3D convolutional networks
20. Donahue, J., et al.: Long-term recurrent convolutional networks for visual recognition and description
21. Zunino, A., et al.: Video gesture analysis for autism spectrum disorder detection

22. Elayadi, M., Kamel, M., Karray, F.: Survey on speech emotion recognition: features, classification schemes, and datasets. Pattern Recogn. **44**, 572–587 (2011)
23. Davis, S., Mermelstein, P.: Comparison of parametric representations for mono-syllabic word recognition in continuously spoken sentences. IEEE Trans. Acoust. Speech Sig. Process. **28**, 357–366 (1980)
24. Osmani, A., Hamidi, M., Chibani, A.: Machine learning approach for infant cry interpretation
25. Zhang, S., Zhang, S., Huang, T., Gao, W., Tian, Q.: Learning affective features with a hybrid deep model for audio-visual emotion recognition. IEEE Trans. Circuits Syst. Video Technol. **28**, 3030–3043 (2017)
26. Anbarjafari, G., Noroozi, F., Marjanovic, M., Njegus, A., Escalera, S.: Audio-visual emotion recognition in video clips
27. Sharma, S., Kiros, R., Salakhutdinov, R.: Action recognition using visual attention. Arxiv Preprint Arxiv:1511.04119 (2015)
28. Simonyan, K., Zisserman, A.: Very deep convolutional networks for large-scale image recognition. Arxiv Preprint Arxiv:1409.1556 (2014)
29. Noroozi, F., Sapiński, T., Kamińska, D., Anbarjafari, G.: Vocal-based emotion recognition using random forests and decision tree. Int. J. Speech Technol. **20**, 239–246 (2017)
30. Park, S., Lee, J.: A fully convolutional neural network for speech enhancement. Arxiv Preprint Arxiv:1609.07132 (2016)
31. Zeng, Z., Hu, Y., Fu, Y., Huang, T., Roisman, G., Wen, Z.: Audio-visual emotion recognition in adult attachment interview
32. Kamińska, D., Sapiński, T., Anbarjafari, G.: Efficiency of chosen speech descriptors in relation to emotion recognition. EURASIP J. Audio Speech Music Process. **2017**, 3 (2017)
33. Noroozi, F., Marjanovic, M., Njegus, A., Escalera, S., Anbarjafari, G.: Audio-visual emotion recognition in video clips. IEEE Trans. Affect. Comput. **10**, 60–75 (2017)
34. Haq, S., Jackson, P.: Multimodal Emotion Recognition. IGI Global, Hershey (2011)
35. Rabiner, L.: On the use of autocorrelation analysis for pitch detection. IEEE Trans. Acoust. Speech Sig. Process. **25**, 24–33 (1977)
36. Chaudhry, R., Ran, A., Hager, G., Vidal, R.: Histograms of oriented optical flow and binet-cauchy kernels on nonlinear dynamical systems for the recognition of human actions
37. Dalal, N., Triggs, B.: Histograms of oriented gradients for human detection
38. Dalal, N., Triggs, B., Schmid, C.: Human detection using oriented histograms of flow and appearance
39. Wang, H., Schmid, C.: Action recognition with improved trajectories
40. Chang, C., Lin, C.: LIBSVM: a library for support vector machines. ACM Trans. Intell. Syst. Technol. (TIST) **2**, 27 (2011)

Early Diagnosis of Alzheimer's Disease Based on Selective Kernel Network with Spatial Attention

Huanhuan Ji[1], Zhenbing Liu[1], Wei Qi Yan[2(✉)], and Reinhard Klette[2]

[1] Guilin University of Electronic Technology, Guilin, China
[2] Auckland University of Technology, Auckland, New Zealand
wyan@aut.ac.nz

Abstract. Alzheimer's disease (AD) is a neurodegenerative disorder which leads to memory and behaviour impairment. Early discovery and diagnosis can delay the progress of this disease. In this paper, we propose a new deep learning method called selective kernel network with attention for early diagnosis of AD using magnetic resonance imaging. Generally, deep learning methods for high-accuracy recognition are based on structure of deep neural networks by stacking a myriad of convolutional layers in the model. In this paper, the structure of SKANet is constructed similarly to that of ResNeXt by repeating residual blocks with the same topology and group convolution for saving computational costs. Different from ResNeXt, the primary convolution is replaced by using selective kernel convolution to adaptively adjust the receptive field based on imported information. Then, attention mechanism is added to the bottom of the block to emphasize on important features and suppress unnecessary ones for more accurate representation of the network. The block is termed as selective kernel with attention block that consists of a sequence of operations followed by the order: a convolution with kernel size 1×1, a selective kernel convolution, a convolution with kernel size 1×1, and spatial attention mechanism. The effectiveness of this proposed model is verified based on the Alzheimer's Disease Neuroimaging Initiative dataset. Our experimental results show superiority of the proposed model for the early diagnosis of AD. The classification accuracy of AD and mild cognitive impairment reaches up to 98.82%.

Keywords: Alzheimer's disease diagnosis · MRI · Spatial attention mechanism · Selective kernel

1 Introduction

Alzheimer's disease (AD) is a neurodegenerative disorder characterised by aphasia, amnesia, visual impairment, and behavioural changes such as dementia. This disease can be divided into three stages including *normal control* (NC), *mild cognitive impairment* (MCI), and AD. There are distinctions of the brain structure

© Springer Nature Switzerland AG 2020
S. Palaiahnakote et al. (Eds.): ACPR 2019, LNCS 12047, pp. 503–515, 2020.
https://doi.org/10.1007/978-3-030-41299-9_39

between the three stages. MCI, a prodromal stage of AD, is an important phase for clinical trials. In MCI, individuals have a mild change in behaviour that can be observed easily by anybody who is close to the patients. AD is an incurable disease that often occurs on the elderly. But we can diagnose it early and take a timely solution to delay the deterioration of this disease. The early diagnosis of AD has become very important and attracted numerous researchers to investigate this problem. It is reported [3] that there will be 1 out of 85 people troubled with this disease by 2050.

Magnetic resonance imaging (MRI) provides a clear structure of the brain in a 3D view noninvasively. Thus, we capture the mild changes due to the atrophic process [9]. A plethora of *computer-aided diagnosis* (CAD) approaches [1,24,31] have been proposed for the early diagnosis of AD based on MRI. Meanwhile, *white matter* (WM) and *grey matter* (GM) of brain play a pivotal role on the diagnosis of AD. A series of methods have been proposed based on the GM and WM from MRI for the diagnosis of AD [7,8,30]. In this paper, GM and WM from MRI are preprocessed and classified using our proposed method.

Multiple machine learning methods have been proposed to aid the diagnosis of AD. The diagnosis through MRI using machine learning methods is split into three independent phases [31]: (1) Predetermination of *region-of-interest* (ROI), (2) extraction or selection of features from ROIs, and (3) construction of a classifier for classifications.

T-test [26] was proposed to extract texture features, classify the images, and diagnose AD by using collaborative representation. The multi-atlas-based method [25] exploits useful information in feature representations to classify the AD. A *support vector machine* (SVM) was used as a classifier to classify the features extracted from each atlas space; the results of multiple atlases were combined by using majority voting methods for making a final decision. A new multimodal data fusion and classification method was set forth [33] based on a kernel combination on the diagnosis of AD which provides a unified way to combine heterogeneous data, particularly for the cases that different types of data cannot be directly concatenated. Generally, the classifiers based on those data from ROIs often suffer overfitting due to its high dimensionality compared with a small number of samples for model training. Other research work also has been adopted for the diagnosis of AD with a sub-optimal performance because of the heterogeneous nature between the features extraction and classifier training. Thus, various methods [2,5,23,35] have been proposed to improve the diagnosis accuracy of AD.

Deep learning is a revolutionary methodology compared with traditional machine learning approaches. Instead of separating feature extraction and classifier training, deep learning can train a model and learn its parameters from end to end without an engagement of human specialists. *Convolutional neural networks* (CNNs) have significantly pushed computer vision forward based on the rich representation of digital images [13].

Deep learning also shows its powerful capability on the early diagnosis of AD recently [22]. A sparse autoencoder [10] was proposed to learn a set of bases from natural images and applied convolution to extract features from the MRI so as to classify the AD. LeNet-5 and convolutional neural networks were employed [32] to classify Alzheimer's disease with an accuracy of 96.85%. The 16-layer VGGNet was modified [4] for the 3-way classification between AD, MCI, and NC. A hierarchical and fully convolutional network [21] was put forward to identify multiscale discriminative location for the diagnosis of AD using MRI. Suk et al. [36] embarked on a CNN that applied multiple sparse regression models to various parameters of regularisation for AD diagnosis. Ensemble learning [16,29] applies voting methods to the final results with the base classifiers for the early diagnosis of AD.

From our observations, we see that the structure of deep learning models is prone to be complex. From LeNet [19] to ResNet [11], the networks are becoming increasingly much deeper and richer for feature representations. ResNet is constructed by stacking residual blocks with the same topology along with a skip connection. Inception [37–39] shows its depth as a neural network and also takes prominent effect on solving problems in computer vision. Xception [6] and ResNeXt [41] empirically reveal that the cardinality can boost the representation of networks and save the time of computations.

Generally, *receptive fields* (RFs) share the corresponding size in every layer for neural networks. Spillmann et al. [34] stated that the RF sizes can be adjusted by using stimuli. Li et al. [20] proposed selective kernel networks to adaptively adjust the RF size by using selective kernel (SK) convolution. In this paper, SK convolution is employed for our proposed network. SK convolution consists of three operations: Splitting, fusing, and selecting. During splitting, two kernels of the sizes 3×3 and 5×5 are adapted to extract the features from the input. A fusing operation is used to aggregate the information from the two branches so as to produce the global information and generate weights for the two branches. The two weights from the fusing operator are multiplied with the corresponding feature maps from the splitting operation during the selecting operation. Thus, the exports from two branches are operated by using element-wise summation to generate the final output of SK convolution.

The residual block of ResNeXt is used as the base for our model. In our proposed model, the main convolution is replaced by using SK convolution. The spatial attention mechanism is added to the end of this block so as to learn the attention in the spatial domain. Our work, reported in this paper, mainly has the following contributions:

- ResNeXt is set as the base architecture which is constructed by repeating the block with the same topology to share and save the hyperparameters (width and size of filters). Thus, a group convolution is introduced to reduce the computational cost.
- SK convolution is employed for our network to select the appropriate RF size and to increase the power of representations.
- Spatial attention is used to emphasise on those important features and suppress unnecessary ones to improve the representation of networks.

The rest of this paper is organised as follows. In Sect. 2, we introduce the progress of convolutional neural networks. Our method based on the proposed deep neural network is introduced in Sect. 3. In Sect. 4, experimental results are demonstrated along with our proposed model. Section 5 concludes.

2   Related Work

Convolutional neural networks with a deep structure by stacking convolutional layers have made a series of breakthroughs for image classification recently. YOLOv3 has been used on the human behaviour recognition with a good performance compared with traditional machine learning methods [27].

As the increase of layers, gradient vanishing and model degradation problems turned up during the training of neural networks. *Batch normalization* (BN) is introduced to reduce the problem of gradient vanishing. He et al. [11] used a shortcut to reduce gradient vanishing and model degradation. A residual learning framework is offered to train the network. ResNet has shown that residual learning is easily to be optimised and can gain high accuracy from considerably increased depth. With the stack of residual blocks, ResNet has shown a higher accuracy and superior performance on the classification.

Inception networks [37–39] are with the successes of all multi-branch networks where every branch is designed elaborately. The appearance of inception made the model much wider. Inception v1 (GoogleNet) [37] was designed by using a sufficient size of filters to fit various RFs. Meanwhile, 1×1 convolution is introduced so as to reduce the dimension of the inputs and computational cost. Thus, the 5×5 convolution kernel is replaced by using two 3×3 convolution operators to reduce the complexity; correspondingly, $n \times n$ convolution is replaced by using $n \times 1$ and $1 \times n$ kernels; the 1-dimensional convolution is employed to accelerate the computations. Our model follows the idea of the Inception networks with filters for multiple branches.

The significant contribution of group convolution is to reduce the computational cost. AlexNet [17] is the neural network that firstly took advantage of group convolution successfully. Then, the group convolution was employed on ResNeXt [41] to reduce computations and improve the outcomes of classifications. In this paper, group convolution is introduced based on our designed neural network to reduce the computational cost.

An attention mechanism is an important part of our proposed model to capture useful information. Attention mechanisms have been employed on various fields such as image captioning [42]. According to an attention mechanism, the informative features are strengthened; simultaneously, the less useful ones are suppressed [14,15,18,28]. Wang et al. [40] created a residual attention network by stacking attention modules to make the model working. Furthermore, SENet [12] was proposed to use a gating mechanism on channels to improve the representation and performance of this network. Beyond channel attention, CBAM [43] also introduced spatial attention in the network to capture the valid information of inputs in the same way.

Unlike existing networks, our proposed neural network introduces the SK convolution to adaptively adjust the RF sizes and spatial attention to improve the representation of networks.

3 Methods

In this section, the details of our model are delineated which encapsulate the architecture of this neural network. First, our model is constructed based on ResNeXt with SK convolution. The spatial attention mechanism is added to the block for strengthening representation capability of the network.

In our model, SK convolution is employed to adaptively adjust the RF sizes. SK convolution is implemented by introducing two branches with kernel sizes of 3×3 and 5×5. The SK convolution is constructed by using the three operations of splitting, fusing, and selection.

The image slices from MRI are sent to the model for training the classifier from end to end. Then, the features of these slices, extracted from a convolutional layer with kernel size 1×1, are sent to the SK convolution for adaptively adjusting RF size. SK convolution is conducted by following two paths with kernel sizes 3×3 and 5×5. The splitting outputs are grouped into two branches.

During the splitting operation, a convolution operation based on a 3×3 kernel is the standard; 5×5 convolution is a dilated one with dilation size 2. The outputs from two branches are fused by using element-wise summation. Global average pooling is employed on the fusing operation to conduct channel-wise statistics. Full connection is applied to the channel-wise statistics by using adaptive selection so as to squeeze the network. Soft attention is added into the squeezed network, mixed with selected spatial scales and different weights which are applied to the two branches.

The weights are multiplied with the features extracted from splitting operations with different kernel sizes. We fuse outputs from the two branches via element-wise summation; the final outputs of SK convolution are conducted with kernel size 1×1; the spatial attention module is embedded to the bottom of the block.

In the spatial attention, the outputs of 1×1 convolution are operated by using max pooling and average pooling to generate two 2D maps on each channel. Then, the two maps are concatenated and convoluted by using a standard convolution of size 7×7 to produce a 2D spatial attention map which determines how to emphasise or suppress the feature maps.

Similar to ResNeXt, our proposed model is constructed by using the stack of bottleneck blocks called *selective kernel attention* (SKA). Each SKA block consists of a sequence of operations, following the order: a standard convolution with 1×1 kernel, a SK convolution, a standard convolution with 1×1 kernel, and spatial attention as shown in Fig. 1.

Different to ResNeXt, the SK convolution is used to replace the large kernel convolution in the bottleneck block of ResNeXt so as to select appropriate RF sizes in an adaptive manner, the spatial attention is added to the bottom of the block. The architecture of our proposed model is similar to the ResNeXt for two reasons: (1) Compared with standard convolution, it has a lower computational cost by introducing the group convolution; (2) the network is one of the popular networks constructed by repeating a block with the same topology so as to reduce the choices of hyperparameters and the complexity of the design.

In this section, we introduce our neural network having 50 layers, called SKA-50. Table 1 displays the structure of the 50-layer network with the number of stacked blocks on each stage. During each stage, path M is set as 2 which means that there are two different kernels that are selected and aggregated. The group number G is set as 32 to control the cardinality on each path; r is the reduction ratio to control the numbers of parameters in the fusing operation.

4   Experiments

The data for our experiments is taken from the public dataset *Alzheimer's Disease Neuroimaging Initiative* (ADNI) at adni.loni.usc.edu. ADNI was designed to develop clinical, imaging, genetic, and biochemical biomarkers for the early detection and tracking AD. In this paper, 490 MRI images were grouped into 143 AD (108 train samples, 35 test samples), 203 MCI (153 train samples, 50 test samples), and 144 NC (108 train samples, 36 test samples) in .nii format which were used to validate the effect of our model. The proportion between male and female samples in each category is roughly equal.

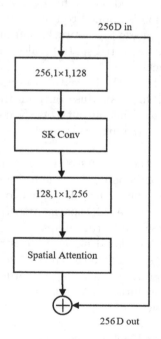

Fig. 1. A block of an SKANet.

There are minor head movements and noises in the MRI images downloaded directly from ADNI. We corrected the images to exclude the negative influence due to head movements by using *statistical parametric mapping* (SPM). Then, the images were resized into $192 \times 192 \times 160$. WM and GM are segmented from the resized images by using SPM. Then, the segmented GM and WM images were split into 192 slices in `.tif` format. According to prior knowledge, consecutive sets of 20 slices with significant brain structure from each MRI were selected from GM and WM. The slices from WM were added to the corresponding slices of GM and resized to 224×224 as the inputs of our model.

All experiments were implemented based on Python 3.5 and Pytorch framework. Our model was trained from scratch for 60 epochs on an Nvidia Tesla P100 GPU. The Adam optimiser was used for model training; the learning rate was set to be 10^{-3} at the beginning; the attenuation started from 10^{-5}; the size of mini batch was initialised as 20; the loss was measured by using the cross-entropy function. The data was preprocessed by using SPM based on MATLAB.

Our model was mainly validated on the early diagnosis of AD (i.e., AD *vs* MCI, and MCI *vs* NC). Classification *accuracy* (ACC), *specificity* (SPE),

Table 1. The 50-layer networks whose structures are similar to that of ResNeXt. The three columns show the structures of ResNeXt-50, SKNet-50, and SKANet-50 with $32 \times 4D$, respectively. The filter sizes and feature map dimensions are shown in brackets.d

Output	ResNeXt-50	SKNet-50	SKANet-50
112×112	7×7, 64, stride 2		
56×56	3×3 max pool, stride 2		
56×56	$\begin{bmatrix} 1 \times 1, 128 \\ 3 \times 3, 128, \; G=32 \\ 1 \times 1, 256 \end{bmatrix} \times 3$	$\begin{bmatrix} 1 \times 1, 128 \\ SK[M=2, G=32, r=16], 128 \\ 1 \times 1, 256 \end{bmatrix} \times 3$	$\begin{bmatrix} 1 \times 1, 128 \\ SK[M=2, G=32, r=16], 128 \\ 1 \times 1, 256 \\ \text{spatial attention} \end{bmatrix} \times 3$
28×28	$\begin{bmatrix} 1 \times 1, 256 \\ 3 \times 3, 256, \; G=32 \\ 1 \times 1, 512 \end{bmatrix} \times 4$	$\begin{bmatrix} 1 \times 1, 256 \\ SK[M=2, G=32, r=16], 256 \\ 1 \times 1, 512 \end{bmatrix} \times 4$	$\begin{bmatrix} 1 \times 1, 256 \\ SK[M=2, G=32, r=16], 256 \\ 1 \times 1, 512 \\ \text{spatial attention} \end{bmatrix} \times 4$
14×14	$\begin{bmatrix} 1 \times 1, 512 \\ 3 \times 3, 512, \; G=32 \\ 1 \times 1, 1024 \end{bmatrix} \times 6$	$\begin{bmatrix} 1 \times 1, 512 \\ SK[M=2, G=32, r=16], 512 \\ 1 \times 1, 1024 \end{bmatrix} \times 6$	$\begin{bmatrix} 1 \times 1, 512 \\ SK[M=2, G=32, r=16], 512 \\ 1 \times 1, 1024 \\ \text{spatial attention} \end{bmatrix} \times 6$
7×7	$\begin{bmatrix} 1 \times 1, 1024 \\ 3 \times 3, 1024, \; G=32 \\ 1 \times 1, 2048 \end{bmatrix} \times 3$	$\begin{bmatrix} 1 \times 1, 1024 \\ SK[M=2, G=32, r=16], 1024 \\ 1 \times 1, 2048 \end{bmatrix} \times 3$	$\begin{bmatrix} 1 \times 1, 1024 \\ SK[M=2, G=32, r=16], 1024 \\ 1 \times 1, 2048 \\ \text{spatial attention} \end{bmatrix} \times 3$
1×1	7×7 global average pool, 2-D f_c, softmax		

sensitivity (SEN), and the *area under receiver operating characteristic curve* (AUC) are used to evaluate the results of classification. The AUC is calculated based on all possible pairs of SEN and SPE obtained by changing the thresholds of the classification scores which are yielded by using the trained networks.

Let TP, TN, FP, and FN be the rates of true positive, true negative, false positive, and false negative, respectively. The common definitions are as follows:

$$ACC = \frac{TP+TN}{TP+TN+FP+FN} \tag{1}$$

$$SEN = \frac{TP}{TP+FN} \tag{2}$$

$$SPE = \frac{TN}{TN+FP} \tag{3}$$

Our model is constructed based on ResNeXt with SK convolution and a spatial attention mechanism. In this section, we will assert the effects of SK convolution and spatial attention by controlling variables to execute the experiments on the 50-layer networks. The results from our model are compared with those of the ResNeXt-50 network with SK convolution called SK-50, ResNeXt network with the spatial attention, and ResNeXt-50 to verify the effectiveness of our idea.

Table 2. Accuracy, sensitivity, specificity, and AUC of various classification methods for MCI and NC.

Models	ACC (%)	SEN (%)	SPE (%)	AUC
ResNeXt-50	84.88	72.22	94.00	0.96
ResNeXt-50+attention	86.05	75.00	94.00	0.96
SK-50	87.21	80.56	92.00	0.94
SKA-50	89.53	86.11	92.00	0.97

Table 2 shows our results based on MCI and NC from those models. From Table 2, we see that the accuracy of SK-50 is higher than ResNeXt-50 that reflects the model has become better after introduced the SK convolution. Furthermore, both accuracy and sensitivity of SK-50 are higher than ResNeXt-50. Meanwhile, the performance of SKA-50 is superior to the ResNeXt with spatial attention that also illustrates the effectiveness of SK convolution.

Compared to ResNeXt-50, the accuracy and sensitivity of ResNeXt with the attention block have been greatly improved that show the effectiveness of spatial attention. Compared to the SK-50, the accuracy and sensitivity of SKA-50 have increased by 2.32% and 5.55%, respectively. The AUC of SKA-50 also was raised, which shows the superior impact of the spatial attention.

Table 3. Accuracy, sensitivity, specificity, and AUC of various classification methods for AD and MCI.

Models	ACC (%)	SEN (%)	SPE (%)	AUC
ResNeXt-50	95.29	100.00	88.57	1.00
ResNeXt-50+attention	97.65	98.00	97.14	0.99
SK-50	97.65	96.00	100.00	1.00
SKA-50	98.82	98.00	100.00	1.00

We furthermore validate our model for classification based on AD and MCI to verify the effectiveness of SK convolution and spatial attention. Table 3 shows the classification results based on AD and MCI of those models. From Table 3, we see that our proposed model with SK convolution and spatial attention has taken a great step that also illustrates the effectiveness of SK convolution and spatial attention. In general, our model has significant improvement after introduced the SK convolution and spatial attention. The receiver operating characteristic (ROC) curves based on MCI and NC are shown in Fig. 2.

From Fig. 2, we see that our model has the best AUC on the early diagnosis of AD compared with other models. In general, the performance of our model with SK convolution is superior to the model without SK convolution. The model becomes better after adding the spatial attention block. From Tables 2 and 3, we see that our model SKA-50 with SK convolution and spatial attention has the best performances for the early diagnosis of AD.

We compared our results with previous work regarding the early diagnosis of AD. Table 4 shows the results of previous studies and our results. From Table 4, we see that our results have a better performance on the classifications of MCI. Our model has a higher accuracy of classification based on AD and MCI.

Table 4. Comparisons of the results of previous studies and our method (%)

References	AD vs NC			AD vs MCI			MCI vs NC		
	ACC	SEN	SPE	ACC	SEN	SPE	ACC	SEN	SPE
Sarraf et al. [32]	96.85	-	-	-	-	-	-	-	-
Ortiz et al. [29]	90.00	86.00	94.00	84.00	79.00	89.00	83.00	67.00	95.00
Lian et al. [21]	90.00	82.00	97.00	-	-	-	-	-	-
Suk et al. [36]	91.02	92.72	89.94	-	-	-	73.02	77.60	68.22
Ji et al. [16]	98.59	97.22	100.00	97.65	96.00	100.00	88.37	80.56	94.00
Our method	98.59	97.22	100.00	**98.82**	**98.00**	100.00	**89.53**	**86.11**	92.00

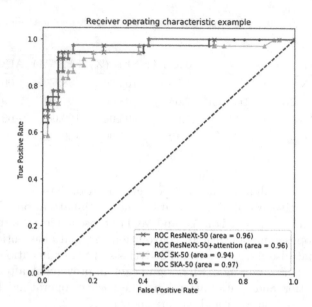

Fig. 2. The ROC curves for the comparisons between MCI and NC using our own methods.

5   Conclusion

In this paper, we constructed a model based on the ResNeXt with SK convolution and spatial attention to boost the representation of the network. Firstly, our model is constructed by stacking blocks with the same topology to reduce the hyperparameters and complexity of design. Then, selective kernel convolution was offered to adaptively adjust the RF sizes to capture more useful information through using dynamic selection mechanism. The spatial attention mechanism was attached to the module so as to improve the representation power of the network. In order to verify the effectiveness, we conducted the experiments for early diagnosis of AD. Our results show that the SKANet demonstrates superior performances for the early diagnosis of AD.

Acknowledgments. This project is supported by the study abroad program for graduate student of the Guilin University of Electronic Technology China and the National Natural Science Foundation of China under grants (61866009). The data used in this paper was downloaded from Alzheimer's Disease Neuroimaging Initiative (ADNI) (adni.loni.usc.edu). We are grateful to everyone who provided their support for this research project.

References

1. Arbabshirani, M.R., Plis, S., Sui, J., Calhoun, V.D.: Single subject prediction of brain disorders in neuroimaging: promises and pitfalls. NeuroImage **145**, 137–165 (2017)
2. Arribas, J., Calhoun, V., Adali, T.: A automatic Bayesian classification of healthy controls, bipolar disorder, and schizophrenia using intrinsic connectivity maps from fMRI data. IEEE Trans. Bio-med. Eng. **57**(12), 2850–2860 (2010)
3. Brookmeyer, R., Johnson, E., Ziegler-Graham, K.: Forecasting the global burden of Alzheimer's disease. J. Alzheimers Assoc. **3**(3), 186–191 (2007)
4. Billones, D., Demetria, D., Hostallero, D.: DemNet: a convolutional neural network for the detection of Alzheimer's disease and mild cognitive impairment. In: TENCON. IEEE, Singapore (2016)
5. Cheng, B., Zhang, D., Chen, S., Shen, D.: Predicting clinical scores using semi-supervised multimodal relevance vector regression. In: Suzuki, K., Wang, F., Shen, D., Yan, P. (eds.) MLMI 2011. LNCS, vol. 7009, pp. 241–248. Springer, Heidelberg (2011). https://doi.org/10.1007/978-3-642-24319-6_30
6. Chollet, F.: Xception: deep learning with depthwise separable convolutions. In: CVPR. IEEE, Piscataway (2016)
7. Chyzhykand, D., Grana, M., Savio, A., Maiora, J.: Hybrid dendritic computing with kernel-LICA applied to Alzheimer's disease detection in MRI. Neurocomputing **75**(1), 72–77 (2012)
8. Cuingnet, R., Gerardin, E., Tessieras, J.: Automatic classification of patients with Alzheimer's disease from structural MRI: a comparison of ten methods using the ADNI database. NeuroImage **56**(2), 766–781 (2011)
9. Frisoni, G.B., Fox, N.C., Jack, C.R., Scheltens, P., Thompson, P.M.: The clinical use of structural MRI in Alzheimer disease. Nat. Rev. Neurol. **6**(2), 67–77 (2010)
10. Gupta, A., Ayhan, M., Maida, A.: Natural image bases to represent neuroimaging data. In: ICML 2013, USA, pp. 987–994 (2013)
11. He, K., Zhang, X., Ren, S., Sun, J.: Deep residual learning for image recognition. In: CVPR, pp. 770–778. IEEE, Piscataway (2016)
12. Hu, J., Shen, L., Sun, G.: Squeeze-and-excitation networks. In: CVPR, pp. 7132–7141. IEEE, Piscataway (2018)
13. Huang, G., Liu, Z., Laurens, M.: Densely connected convolutional networks. In: CVPR, pp. 4700–4708. IEEE, Piscataway (2017)
14. Itti, L., Koch, C., Niebur, E.: A model of saliency-based visual attention for rapid scene analysis. IEEE TPAMI **20**(11), 1254–1259 (1998)
15. Itti, L., Koch, C.: Computational modelling of visual attention. Nat. Rev. Neurosci. **2**(3), 194–203 (2001)
16. Ji, H., Liu, Z., Yan, W., Klette, R.: Early diagnosis of Alzheimer's disease using deep learning. In: ICCCV, Korea (2019)
17. Krizhevsky, A., Sutskever, I., Hinton, G.: ImageNet classification with deep convolutional neural networks. In: NIPS (2012)
18. Larochelle, H., Hinton, G.: Learning to combine foveal glimpses with a third-order Boltzmann machine. In: NIPS (2010)
19. Lecun, Y., Bottou, L., Bengio, Y., et al.: Gradient-based learning applied to document recognition. Proc. IEEE **86**(11), 2278–2324 (1998)
20. Li, X., Wang, W., Hu, X., Yang, J.: Selective kernel networks. In: CVPR, pp. 510–519. IEEE, Piscataway (2019)

21. Lian, C., Liu, M., Zhang, J., Shen, D.: Hierarchical fully convolution network for joint atrophy localization and Alzheimer's disease diagnosis using structural MRI. IEEE Trans. PAMI **12**, 1–14 (2018)
22. Litjens, G., Kooi, T., Bejnordi, B., Setio, A., Ciompi, F., Ghafoorian, M.: A survey on deep learning in medical image analysis. Med. Image Anal. **42**, 60–88 (2017)
23. Liu, F., Wee, C., Chen, H., Shen, D.: Inter-modality relationship constrained multi-modality multi-task feature selection for Alzheimer's disease and mild cognitive impairment identification. NeuroImage **84**, 466–475 (2014)
24. Liu, M., Zhang, D., Chen, S., Xue, H.: Joint binary classifier learning for ECOC-based multi-class classification. IEEE Trans. Pattern Anal. Mach. Intell. **38**(11), 2335–2341 (2016)
25. Liu, M., Zhang, D., Shen, D.: View-centralized multi-atlas classification for Alzheimer's disease diagnosis. Hum. Brain Mapp. **36**(5), 1847–1865 (2015)
26. Liu, Z., Xu, T., Ma, C., Yang, H.: T-test based Alzheimer's disease diagnosis with multi-feature in MRIs. Multimedia Tools Appl. **77**(22), 29687–29703 (2018)
27. Lu, J., Yan, W., Nguyen, M.: Human behaviour recognition using deep learning. In: AVSS (2018)
28. Mnih, V., Heess, N., Graves, A.: Recurrent models of visual attention. In: NIPS (2014)
29. Ortiz, A., Munilla, J., Gorriz, M.: Ensembles of deep learning architectures for the early diagnosis of the Alzheimer's disease. Int. J. Neural Syst. **26**(7), 1650025 (2016)
30. Ortiz, A., Munilla, J., Martínez-Murcia, F.J., Górriz, J.M., Ramírez, J.: Learning longitudinal MRI patterns by SICE and deep learning: assessing the Alzheimer's disease progression. In: Valdés Hernández, M., González-Castro, V. (eds.) MIUA 2017. CCIS, vol. 723, pp. 413–424. Springer, Cham (2017). https://doi.org/10.1007/978-3-319-60964-5_36
31. Rathore, S., Habes, M., Iftikhar, M.A., Shacklett, A., Davatzikos, C.: A review on neuroimaging-based classification studies and associated feature extraction methods for Alzheimer's disease and its prodromal stages. NeuroImage **155**, 530–548 (2017)
32. Sarraf, S., Tofighi, G.: Deep learning-based pipeline to recognize Alzheimer's disease using fMRI data. In: Future Technologies Conference, pp. 816–820. IEEE, San Francisco (2016)
33. Shen, D., Davatzikos, C.: HAMMER: hierarchical attribute matching mechanism for elastic registration. IEEE Trans. Med. Imaging **21**(11), 1421–1439 (2002)
34. Spillmann, L., Dresp-Langley, B., Tseng, C.: Beyond the classical receptive field: the effect of contextual stimuli. J. Vis. **15**(9), 7 (2015)
35. Suk, H., Lee, S., Shen, D.: Deep sparse multi-task learning for feature selection in Alzheimer's disease diagnosis. Brain Struct. Funct. **221**(15), 2569–2587 (2016)
36. Suk, H., Lee, S., Shen, D.: Deep ensemble learning of sparse regression models for brain disease diagnosis. Med. Image Anal. **37**, 101–113 (2017)
37. Szegedy, C., et al.: Going deeper with convolutions. In: CVPR, pp. 1–9. IEEE, Piscataway (2015)
38. Szegedy, C., Vanhoucke, V., Ioffe, S., Shlens, J., Wojna, Z.: Rethinking the inception architecture for computer vision. In: CVPR, pp. 2818–2826. IEEE, Piscataway (2016)
39. Szegedy, C., Ioffe, S., Vanhoucke, V.: Inception-v4, Inception-ResNet and the impact of residual connections on learning. In: AAAI, San Francisco (2017)
40. Wang, F., et al.: Residual attention network for image classification. In: CVPR, pp. 3156–3164. IEEE, Piscataway (2017)

41. Xie, S., Girshick, R., Dollar, P., Tu, Z., He, K.: Aggregated residual transformations for deep neural networks. In: CVPR, pp. 1492–1500. IEEE, Piscataway (2017)
42. You, Q., Jin, H., Wang, Z., Fang, C., Luo, J.: Image captioning with semantic attention. In: CVPR, pp. 4651–4659. IEEE, USA (2016)
43. Woo, S., Park, J., Lee, J.-Y., Kweon, I.S.: CBAM: convolutional block attention module. In: Ferrari, V., Hebert, M., Sminchisescu, C., Weiss, Y. (eds.) ECCV 2018. LNCS, vol. 11211, pp. 3–19. Springer, Cham (2018). https://doi.org/10.1007/978-3-030-01234-2_1

Detection of One Dimensional Anomalies Using a Vector-Based Convolutional Autoencoder

Qien Yu[✉], Muthusubash Kavitha, and Takio Kurita[✉]

Department of Information Engineering, Hiroshima University,
Higashi-hiroshima, Hiroshima 739-8521, Japan
{d182819,tkurita}@hiroshima-u.ac.jp

Abstract. Anomaly detection is important to significant real life entities such as network intrusion and credit card fraud. Existing anomaly detection methods were partially learned the features, which is not appropriate for accurate detection of anomalies. In this study we proposed vector-based convolutional autoencoder (V-CAE) for one dimensional anomaly detection. The core of our model is a linear autoencoder, which is used to construct a low-dimensional manifold of feature vectors for normal data. At the same time, we used vector-based convolutional neural network (V-CNN) to extract the features from vector data before and after the linear autoencoder that makes the model learned deep features for efficient anomaly detection. This unsupervised learning method used only normal data in the training phase. We used the combined abnormal score calculated from two reconstruction errors: (i) error between the input and output of the whole architecture and (ii) error between the input and output of the linear encoder. Compared with the nine state-of-the-arts methods, our proposed V-CAE shows effective and stable results of AUC with 0.996 in estimating anomalies based on several benchmark datasets.

Keywords: Anomaly detection · Unsupervised learning · Autoencoder · Vector-based convolutional neural network

1 Introduction

Deep learning has achieved encouraging performance in many visual task applications, which were included with labels. The cost of labeling increases, as the amount of data increases. Generally unusual data appeared in real life entities cannot be effectively trained by a classification model because of less number of data. Hence anomaly detection algorithms is used to identify unusual/abnormal samples by training the model using normal samples [1]. For example, the practical application of anomaly intrusion detection demonstrates the anomaly detection task as shown in Fig. 1. The red color points in the plot represents the abnormal data and it took various positions and values due to different external factors.

© Springer Nature Switzerland AG 2020
S. Palaiahnakote et al. (Eds.): ACPR 2019, LNCS 12047, pp. 516–529, 2020.
https://doi.org/10.1007/978-3-030-41299-9_40

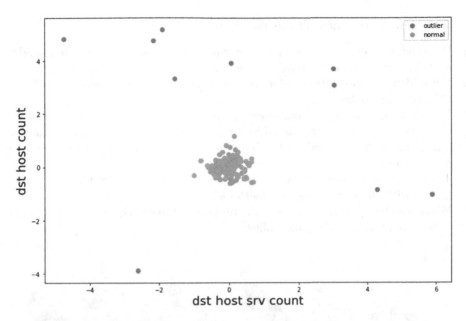

Fig. 1. Demonstration of intrusion detection on KDD99 dataset. The two features of network connections are represented as 'dst host count' and 'dst host srv count'. The green and red color points indicated the normal and abnormal network connections (Color figure online)

In general, anomaly detection tasks used large number of normal samples to train the model parameters Θ to generate the feature distribution $p(x)$ for normal samples. However, in training phase the number of abnormal samples are very small or sometimes not available to identify the abnormal samples in the test phase. In this case, only normal samples can be used to optimize the parameters of the model and hence the abnormal score $S(x)$ can be calculated using the test data for identifying the abnormal samples.

Varying number of neurons and layers has been observed to largely affect the performance of the anomaly detection models [2,3]. Several intrusion classification models have focused deep belief networks with stacked Restricted Boltzmann Machine and showed superior performance in identifying anomalies [4,5]. Inspired from the aforementioned studies, we proposed to develop a V-CAE model for anomaly detection. The core of the proposed architecture is a linear autoencoder, which is used to find the sub space of the normal data by using the feature vectors extracted by the vector-based convolutional neural network (V-CNN) [6]. The V-CNN is used to extract non-linear feature vector from the input vector by 2-D convolutional neural network. The proposed V-CAE framework for identifying anomalies is shown in Fig. 2. In this study, we used input data as a vector form and only the features extracted from the normal input data are used to train our proposed model.

The main contributions of this paper are as follows:

1. We used a autoencoder based on mutual information to enable the encoder and decoder to learn the most significant features of the input data.
2. We added a linear autoencoder to construct a low-dimensional manifold of the normal samples.
3. We used combined abnormal score computed from two different reconstruction errors: first one is calculated between the input and output of the model, and the second one is calculated between the input and output of the linear autoencoder.
4. The effectiveness of the proposed method is experimentally evaluated by comparing with the state-of-the-art methods.
5. We conducted ablation study on our proposed framework by removing linear autoencoder in detecting anomalies.

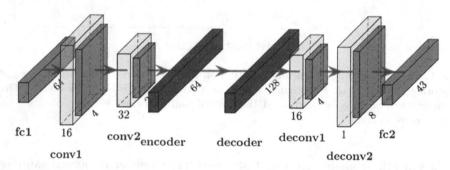

Fig. 2. Pipeline of the proposed approach for anomaly detection. The parameters of the model shown in this figure are selected according to the features of KDD dataset. In fc1, conv1, conv2 the activation functions are leaky relu; In deconv1, deconv2 the activation functions are relu; In fc2 the activation function is tanh. The parameters in this figure are the size of the output data of each layer. The input data is 43-dimensional vector data and first passes through fc1 layer. Then the data dimension becomes 64-dimensional. And the data is converted into matrix with $8 \times 8 \times 1$ format as the input of conv1 after fc1; After output from conv2 layer, the data is changed into vector form, as input of linear encoder. The dimension of output data of linear encoder is 64; The process of data output from decoder to the whole model is similar to the previous process.

2   Related Works

Anomaly detection has always been the focus of researchers, especially in the fields of finance, information security, video surveillance and medical imaging. The traditional methods are used to measure the similarity between data based on distance [7], density [8], angle [9], isolation and [10], clustering [11], etc. These algorithms are actually similar in lower dimension, because the core assumption is that "the representation of abnormal points is different from normal points and also it is a minority group". However, most similarity based algorithms will face the curse of dimensionality, that is, common similarity measures (such as Euclidean distance) will often fail on high-dimensional data [12,13].

In order to solve this problem, many methods have been proposed, including:

1. Dimension reduction or feature selection [14].
2. Subspace methods, such as detection and merging on multiple low-dimensional spaces, random projection (randomly generating multiple subspaces and modeling separately on each subspace, feature bagging) and random forest.
3. Graph based methods are used to represent the relationships and extracted features of data [15].
4. Intrinsic dimensionality based reverse nearest neighbors methods [16].

Furthermore, based on the availability of data labels, anomaly detection technology can be divided into the following two types:

Supervised anomaly detection: The supervised anomaly detection mode assumes that we have labeled normal data and abnormal data. The most typical method is to transform the problem into a special two-class problem and establish a predictive classification model. Many general machine learning classification algorithms can be applied to model training [17]. The predicted data can be used to determine whether it is normal or abnormal. The supervised anomaly detection mode mainly has two application difficulties. Firstly, in the training data, the amount of abnormal data is far less than the amount of normal data, which brings a common data imbalance problem in the field of machine learning and data mining. Secondly, it is very challenging to obtain accurate and representative anomaly class label data. Researchers have been proposed sampling, price sensitivity, active learning and other methods to solve the above two problems. However, in practical application, the supervised anomaly detection model is still very limited.

Unsupervised anomaly detection: Unsupervised anomaly detection does not need to label data sets, and only normal data in the training set, so it has the widest applicability. This technique contains an implicit assumption that normal samples occur more frequently and are easier to obtain than abnormal samples. This assumption is also based on the fact that the number of abnormal samples in the data set is far lesser than the number of normal samples. Khreich et al. [18] used one-class support vector machine (SVM) to map the data to high-dimensional space by kernel function, looking for hyperplanes to maximize the interval between the data and the origin of coordinates. Tax et al. [19] used support vector domain description (SVDD) method to map the data to high-dimensional space by using kernel function to find the hypersphere as small as possible to wrap the normal data. Yang et al. [20] modeled the normal data with Gaussian mixture model and estimated the parameters with maximum likelihood. When anomaly detection is carried out, the probability that it belongs to normal data can be obtained by bringing its features into the model. Liu et al. [21] used isolation forest method for anomaly detection. This method is suitable for the case where there are few abnormal points, and adopts the method of constructing multiple decision trees for anomaly detection. It is entirely based on the concept of isolation to detect anomalies without any distance or density measurement. He et al. [11] heuristically divided the data set into large and small clusters. If an example belongs to a large cluster, the abnormal score is calculated by using the example and the large cluster to which it belongs; if an example belongs to a small cluster, the abnormal score is calculated by using the example and the nearest large cluster.

3   Proposed Method

3.1   Overview

This paper proposes an anomaly detection method based on vector-based convolutional autoencoder. The flow chart of our proposed method is shown in Fig. 2. In this study, we consider the anomaly detection in one dimensional feature vectors. After the one dimensional feature vector is fed into the first fully connected layer, the vector data is converted to two dimensional matrix form. Then the deep features are extracted by the standard convolutional layers. The core of our model is a linear autoencoder, whose function is to reduce the dimension of data and finds the linear subspace of the normal samples. It is expected that this linear autoencoder in the middle of the convolutional autoencoder can help to find the tight boundary of the normal samples. The reconstructed vector by the linear autoencoder are used to reconstruct the output vector by using the deconvolutional layers. In the test phase, an abnormal sample is detected by using the scores defined by using two reconstruction errors of the convolutional autoencoder and the linear autoencoder.

3.2   Vector-Based Convolutional Autoencoder

In order to extract the non linear manifold of the normal data, we adopt the vector-based convolution autoencoder. As shown in Fig. 2, the vector-based autoencoder includes an input layer, fully connected (FC) layers, a linear autoencoder, convolution layers before and after the linear autoencoder and output layer.

Let $X = \{\mathbf{x}_1, \mathbf{x}_2, ..., \mathbf{x}_n\}$ be the set of one dimensional feature vectors in the normal data and $\mathbf{x}_i \in R^m$ where m is the dimension of each sample.

The input vector first passes through FC layer and is converted into the two dimensional array. Then convolution neural network is used to extract the features of the input vector. The extracted features are converted into the vector form and then it is used as a input into the linear autoencoder. This procedure is defined by a function $C(.)$ and the flattened feature vector $\widehat{\mathbf{x}_i} \in R^d$ is given as

$$\widehat{\mathbf{x}_i} = C(\mathbf{x}_i) \tag{1}$$

where $d = l \times h \times ch$ and h, l, and ch are the width, the height and the number of channels of the output of the conv2 layer, respectively.

The linear autoencoder extracts the dimension reduced feature vector $\mathbf{z}_i$ from the flattened feature vector $\widehat{\mathbf{x}}_i$ as

$$\mathbf{z_i} = \mathbf{W}\widehat{\mathbf{x}}_i + \mathbf{b} \tag{2}$$

where $\mathbf{W} \in R^{d \times k}$ and $\mathbf{b} \in R^k$ are the weights and the bias of the linear encoder. The dimension of the extracted feature vector $\mathbf{z}_i$ is shown as k. The approximation $\widehat{\mathbf{y}}_i$ of $\widehat{x}_i$ is calculated by

$$\widehat{\mathbf{y}}_i = \mathbf{W}'\mathbf{z_i} + \mathbf{b}' \tag{3}$$

where $\mathbf{W}' \in R^{k \times d}$ and $\mathbf{b}' \in R^d$ are the weights and the bias of the linear decoder, respectively.

The approximation $\widehat{\mathbf{y}}_i$ of $\widehat{\mathbf{x}}_i$ by the linear autoencoder is reshaped into the original tensor format. It is used as the input of the next deconvolution layers. Finally, the output vector $\mathbf{y_i}$ of the vector-based convolutional autoencoder is obtained through another FC layer. This procedure is defined by function $C''(.)$ as

$$\mathbf{y_i} = C'(\widehat{\mathbf{y}}_i). \tag{4}$$

The loss function is defined based on the mean squared errors (MSEs) of the convolutional autoencoder and the embedded linear autoencoder as

$$\ell = \alpha \left\{ \frac{1}{n} \sum_0^n (\mathbf{x}_i - \mathbf{y}_i)^2 \right\} + (1 - \alpha) \left\{ \frac{1}{n} \sum_0^n (\widehat{\mathbf{x}}_i - \widehat{\mathbf{y}}_i)^2 \right\}, \alpha \in [0,1] \tag{5}$$

where the first term is the mean squred errors (MSE) between the input vector $\mathbf{x_i}$ and its approximation $\mathbf{y_i}$ by the convolutional autoencoder and the second term is the mean squared errors (MSE) between the feature vector $\widehat{\mathbf{x}}_i$ and its approximation $\widehat{\mathbf{y}}_i$ by the linear autoencoder. The parameter α is used to adjust the degree of contribution of these two MSEs to the objective function ℓ.

3.3 Anomaly Scores

In the test phase, the model calculates the anomaly score of each test sample $\mathbf{x}$. Again the anomaly score is defined based on the reconstruction error $S_1(\mathbf{x})$ of the convolutional autoencoder and the reconstruction error $S_2(\mathbf{x})$ of the linear autoencoder as

$$S(\mathbf{x}) = \lambda S_1(\mathbf{x}) + (1 - \lambda)S_2(\mathbf{x}) \tag{6}$$

where λ is the tuning parameter that can be adjusted according to the tasks. The reconstruction error $S_1(\mathbf{x})$ between the input vector $\mathbf{x}$ and it approximation $\mathbf{y}$ by the convolutional autoencoder is defined as

$$S_1(\mathbf{x}) = ||\mathbf{x} - \mathbf{y}||^2 \tag{7}$$

Similarly, the reconstruction error $S_2(\mathbf{x})$ between the feature vector $\widehat{\mathbf{x}}$ and its approximation by the linear autoencoder is defined as

$$S_2(\mathbf{x}) = ||\widehat{\mathbf{x}} - \widehat{\mathbf{y}}||^2 \tag{8}$$

In order to evaluate the impact of the overall anomaly detection performance, the anomaly score are normalized. At first, the anomaly scores $S = \{S(\mathbf{x}_i)|\mathbf{x}_i \in X\}$ for all training samples X are calculated and the maximum $max(S)$ and the minimum $min(S)$ of the anomaly scores are obtained. Then the anomaly score $S(\mathbf{x})$ for the new samples is normalized as

$$p = \frac{S(\mathbf{x}) - min(S)}{max(S) - min(S)} \tag{9}$$

4   Experimental Setup

4.1   Data Set

In order to confirm the effectiveness and the efficiency of the proposed method, we have performed experiments using three benchmark data sets which are KDD99, Optdigits, default of credit card clients. We first carried out experiments on KDD99 abnormal intrusion data, treating the 'normal' class data in the training phase and defining other classes as abnormal data. The test set contains normal as well as abnormal data. Optdigits data is experimented by treating one class (class '3') being an anomaly, while another class (class '1') is considered as the normal data. Default of credit card clients data set is an open source data set of a foreign organization. The content of the data includes some attributes such as gender, education, marriage, age, etc. It also includes the credit card consumption and bill situation of the user over a period of time. 'Payment next month', which only includes 0 or 1, is one of the feartures from data indicates whether the user has repaid the credit card bill, '1' indicates repayment, and we classify this sample with 1 as category 1; Similarly, and '0' indicates no repayment. We classify the samples with features of 'Payment next month' which equal to '1' into class '1'; Similarly, We classify the samples with features of 'Payment next month' which equal to '0' into class '0'.

The data sets used in our experiments are converted into binary data sets, i.e. normal and abnormal data. The class which is considered as normal data is used to train the model. The labels of the data sets are converted into binary labels, which are used during testing. We calculated the abnormal scores of the test sets in each data set and selected an appropriate threshold to distinguish them. The original data set is randomly divided into training/testing with a ratio of 7/3. The details of the data set used in our experiments are shown in Table 1.

4.2   Parameter Settings and Evaluation

We used adam to optimize the network parameters. The proposed method is implemented in tensorflow. The parameter α is adjusted depending on the data sets. The training was done with 1,000, 2,000 and 400 epochs for KDD99, Optdigits and default of credit card clients, respectively. In the experiments, we compared the proposed method with nine state-of-the-art methods, including

Table 1. Details of the benchmark datasets used for performance evaluation.

Datasets	Features	Normal class	Abnormal class
KDD99	43	1	Others
Optdigits	47	1	3
Default of credit card clients	24	0	1

several traditional supervised methods and unsupervised methods. The proposed method is compared with the four most advanced supervised methods including active learning (AL) [22], feature packing (FB) [23], local outlier factor (LOF) [24] and pattern window (PW) [25]. The proposed method is compared with the four unsupervised methods including sparse coding (SC) [26], $L21 - SRC(L21)$ [27], reverse nearest neighbors (RNN) [28] and self-representation outlier detection (SRO) [29]. In addition to these eight methods, the proposed method is also compared with the sparse reconstruction (SR) method proposed by Hou et al. [30]. We computed area under the curve (AUC) value using receiver operating curve analysis as the main evaluation measure for performance evaluation. If the AUC score is large, then the performance of the anomaly detection algorithm is good. Furthermore, we used precision, recall and F1 score to evaluate the performance of the proposed system.

5 Results

5.1 Comparison with the State-of-the-art Methods

Tables 2 and 3 presented the results of our experiments. Based on this our method showed more robust performance, which is better than those of the nine state-of-the-arts methods. Among all the compared models, our model scored the highest AUC score on all the three data sets, especially it is much higher compared over the latest method [30]. Experiments on KDD99 and Optdigits selected the best α of 0.5 for detecting the anomaly data. Experiment with default of credit card clients, the best α is chosen as 0.4. By adjusting the values of λ, the detection results of the model will also change accordingly. On the KDD99 and Optdigits, the choice of λ has little influence on the model, because the distribution of S_1 and S_2 terms is enough to separate the abnormal data, so the choice of λ is 0.5. But on default of credit card clients dataset, the choice of λ has great influence on the results. We chose $\lambda = 0.6$ by assigning more weight on approximating S_1 which plays a leading role in the data set of default of credit card clients for detection.

Figure 3 shows the distribution of anormaly scores on the KDD99 dataset. Based on the distribution of anomaly scores of S_1 and S_2, it is suggested that these two terms are sufficient to distinguish normal from abnormal data.

The detection results of the proposed model on default of credit card clients are not as good as those of the other two datasets. It is because it showed

Table 2. Performance comparison of ours and the state-of-the-art methods in terms of AUC

Data sets	FB	AL	LOF	PW	SC
KDD99	0.140	0.2971	0.134	0.196	0.627
Optdigits	0.577	–	0.523	0.734	0.589
Default of credit card clients	0.535	0.484	0.524	0.643	0.496
Data sets	L21	RNN	SRO	SR	Ours
KDD99	0.799	0.798	0.368	0.819	**0.996**
Optdigits	0.833	0.767	0.515	0.722	**0.996**
Default of credit card clients	0.599	0.506	0.600	0.606	**0.657**

Table 3. Performance statistics of our proposed model in terms of precision, recall and F1-score

Data sets	Precision	Recall	F1-socre
KDD99	0.9932	0.9932	0.9932
Optdigits	0.9677	0.9928	0.9831
Default of credit card clients	0.6200	0.6201	0.6192

very poor relationship between the data and its features. Figure 4 shows the correlation of features and the data set. It clearly explained that there is no significant correlation between the characteristics of some of its features (sex, education, marriage and age) and the categories of the data sets. Therefore, the appearance of these irrelevant features increases the difficulty of finding the anomaly data.

5.2   Ablation Study on Proposed Framework

The core of our model is using linear autoencoder, whose function is to reduce the dimensions as well as compress the boundary of normal data. If the linear autoencoder is removed from the framework, the anomaly score can be calculated only from the term S_1. As shown in Fig. 3(c), removing the linear autoencoder has no effect on KDD99 dataset, and the same is true for Optdigits dataset. For default of credit card clients dataset, the detection performance will be drastically reduced. As can be seen from Table 4, the proposed framework without linear autoencoder showed less value of AUC compared over the framework with linear autoencoder. Furthermore, the precision, recall and F1-score of our method with linear autoencoder is significantly better than those values without linear autoencoder.

As can be seen from Fig. 5, the results of our proposed method are better than without autoencoder on KDD99 data set and Optdigits dataset. Meanwhile, the results on default of credit card clients data set, our proposed framewok with

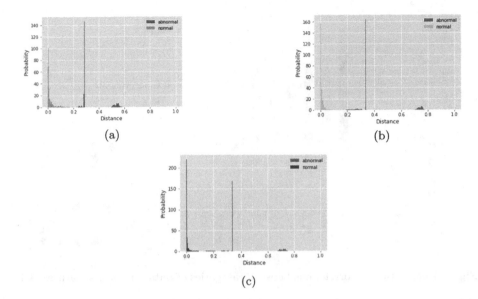

Fig. 3. Detection of distributions of abnormal scores using our proposed method on KDD99. (a) distribution of S_1 (b) distribution of S_2 and (c) without linear Autoencoder.

Table 4. Performance of our proposed framework with and without linear autoencoder interms of AUC

Datasets	With linear AE	Without linear AE
KDD99	**0.9973**	0.9953
Optdigits	**0.9986**	0.9967
Default of credit card clients	**0.6570**	0.5894

linear autoencoder is better than that without linear autoencoder. It proved that our proposed V-CAE structure has potential ability to detect abnormal samples. In addition, as shown in Tables 5 and 6, the precision, recall and F1-score results of our method with or without linear autoencoder are almost similar. According to Table 7, those measures on default of credit card clients data set, our method with linear autoencoder showed better performance than those values without linear autoencoder.

As can be seen from Table 8, We removed the vector-based convolutional neural network (V-CNN) before and after the linear Autoencoder, and only used the linear Autoencoder for experiments. We found our results were better than the results of structure without the V-CNN.

Overall, experimental results can clearly explained that our proposed system with linear autoencoder can perfectly separate the abnormally distributed data as shown in Fig. 3. The reconstruction error of the abnormal data is always larger than that of normal data and thus the results demonstrated that the

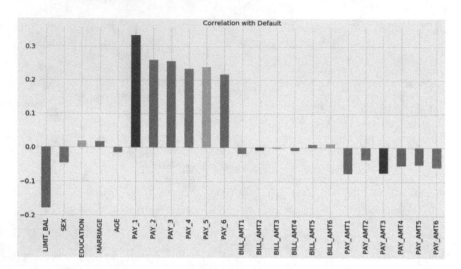

Fig. 4. Interpreting of correlations between categories of data set and its features [31]

Table 5. Performance of our proposed framework using with and without linear autoencoder on KDD99

Measures	With linear AE	Without linear AE
Precision	**0.9932**	0.9908
Recall	**0.9932**	0.9908
F1 socre	**0.9932**	0.9908

Table 6. Performance of our proposed framework using with and without linear autoencoder on Optdigits

Measures	With linear AE	Without linear AE
Precision	**0.9946**	0.9945
Recall	**0.9946**	0.9942
F1 socre	**0.9945**	0.9940

anomaly data can be well detected by our proposed V-CAE approach. However, the differences of reconstruction errors between the normal and abnormal data is not very high on the credit card data set. Hence it is difficult to distinguish the anomalies from the normal data set. But still our proposed model achieved the second highest score in detecting anomalies on credit card data set.

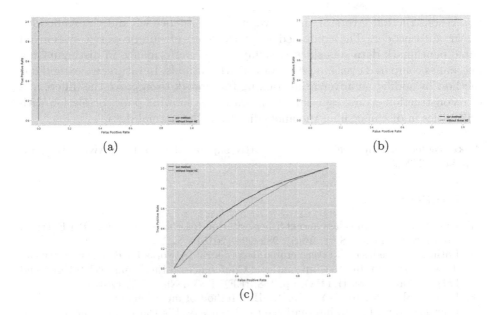

Fig. 5. Comparison of ROC curves for abnormal scores on different data sets. (a) KDD, (b) optdigits and (c) default of credit card clients.

Table 7. Performance of our proposed framework using with and without linear autoencoder on default of credit card clients

Measures	With linear AE	Without linear AE
Precision	**0.6200**	0.5613
Recall	**0.6201**	0.5698
F1 socre	**0.6192**	0.5649

Table 8. Performance of our proposed framework with and without V-CNN interms of AUC

Datasets	With V-CNN	Without V-CNN
KDD99	**0.9973**	0.9958
Optdigits	**0.9986**	0.9914
Default of credit card clients	**0.6570**	0.5268

6   Conclusions

We introduced a new vector-based convolutional autoencoder model for anomaly detection tasks. The proposed model transformed the vector-based data sets into graphs. In addition it is capable to detect the tight boundary of the normal data by reducing the dimensionality of the linear sub space of the extracted non-linear feature vectors. The anomalies are detected by using the scores defined by two reconstruction errors of the convolutional autoencoder and the embedded linear autoencoder. The combined anomaly score of our proposed system on three benchmark data sets showed highly robust performance in distinguishing anomalies compared over nine state-of-the-art methods. In future, we extend our method by adding an adversarial training framework to improve the differences of reconstruction errors, especially if the data sets showing poor correlation with difficulties in distinguishing the anomalies from the normal data.

Ackknowledgement. This work was partly supported by JSPS KAKENHI Grant Number 16K00239.

References

1. Dufrenois, F.: A one-class kernel fisher criterion for outlier detection. IEEE Trans. Neural Netw. Learn. Syst. **26**(5), 982–994 (2015)
2. Potluri, S., Diedrich, C.: Accelerated deep neural networks for enhanced intrusion detection system. In: 21st International Conference on Emerging Technologies and Factory Automation (ETFA), pp. 1–8. IEEE Press, New York (2016)
3. Kim, J., Shin, N., Jo, S.-Y., Kim, S.-H.: Method of intrusion detection using deep neural network. In: 4th International Conference on Big Data and Smart Computing, pp. 313–316. IEEE Press, New York (2017)
4. Alom, M.Z., Bontupalli, V., Taha, T.M.: Intrusion detection using deep belief networks. In: National Aerospace and Electronics Conference, pp. 339–344. IEEE Press, New York (2015)
5. Qu, F., Zhang, J.-T., Shao, Z.-T., Qi, S.-Z.: Intrusion detection model based on deep belief. In: the 2017 VI International Conference on Network, Communication and Computing, pp. 97–101. ACM Press, New York(2017)
6. Kavitha, M.S., Kurita, T., Park, S.-Y., Chien, S.-I., Bae, J.-S., Ahn, B.-C.: Deep vector-based convolutional neural network approach for automatic recognition of colonies of induced pluripotent stem cells. PLoS ONE **12**(12), 1–18 (2017)
7. Ramaswamy, S., Rastogi, R., Shim, K.: Efficient algorithms for miningoutliers from large data sets. ACM SIGMOD Rec. **29**(2), 427–438 (2000)
8. Breunig, M.M., Kriegel, H.P., Ng, R.T., Sander, J.: LOF: identifying density-based local outliers. ACM SIGMOD Rec. **29**(2), 93–104 (2000)
9. Kriegel, H.P., Zimek, A.: Angle-based outlier detection in high-dimensional data. In: 14th ACM SIGKDD International Conference on Knowledge Discovery and Data Mining, pp. 444–452. ACM Press, New York (2008)
10. Liu, F.T., Ting, K.M., Zhou, Z.H.: Isolation forest. In: International Conference on Data Mining, pp. 413–422. IEEE Press, New York (2008)
11. He, Z., Xu, X., Deng, S.: Discovering cluster-based local outliers. Pattern Recogn. Lett. **24**(9–10), 1641–1650 (2003)
12. Zimek, A., Schubert, E., Kriegel, H.P.: A survey on unsupervised outlier detection in high-dimensional numerical data. Stat. Anal. Data Min.: ASA Data Sci. J. **5**(5), 363–387 (2012)
13. Ro, K., Zou, C., Wang, Z., Yin, G.: Outlier detection for high-dimensionaldata. Biometrika **102**(3), 589–599 (2015)
14. Pang, G., Cao, L., Chen, L., Liu, H.: Learning homophily couplings from Non-IID data for joint feature selection and noise-resilient outlier detection. In: 26th International Joint Conference on Artificial Intelligence, pp. 2585–2591. Morgan Kaufmann Press, San Francisco (2017)

15. Akoglu, L., Tong, H., Koutra, D.: Graph based anomaly detection and description: a survey. Data Min. Knowl. Discov. **29**(3), 626–688 (2015)
16. Radovanović, M., Nanopoulos, A., Ivanović, M.: Reverse nearest neighbors in unsupervised distance-based outlier detection. IEEE Trans. Knowl. Data Eng. **27**(5), 1369–1382 (2015)
17. Fujimaki, R., Yairi, T., Machida, K.: An anomaly detection method for spacecraft using relevance vector learning. In: Ho, T.B., Cheung, D., Liu, H. (eds.) PAKDD 2005. LNCS (LNAI), vol. 3518, pp. 785–790. Springer, Heidelberg (2005). https://doi.org/10.1007/11430919_92
18. Khreich, W., Khosravifar, B., Hamou-Lhadj, A., Talhi, C.: An anomaly detection system based on variable N-gram features and one-class SVM. Inf. Softw. Technol. **91**, 186–197 (2017)
19. Tax, D.M.J., Duin, R.P.W.: Support vector domain description. Pattern Recogn. Lett. **20**(11–13), 1191–1199 (1999)
20. Yang, X., Latecki, L.J., Pokrajac, D.: Outlier detection with globally optimal exemplar-based GMM. In: SIAM International Conference on Data Mining, pp. 145–154. SIAM Press, Philadelphia (2009)
21. Liu, F.-T., Ting, K.-M., Zhou, Z.-H.: Isolation-based anomaly detection. ACM Trans. Knowl. Discov. Data (TKDD) **6**(1), 1–39 (2012)
22. Sun, G., Cong, Y., Xu, X.: Active lifelong learning with "watchdog". In: The 32th AAAI Conference on Artificial Intelligence, pp. 4107–4114. AAAI Press, Palo Alto (2018)
23. Lazarevic, A., Kumar, V.: Feature bagging for outlier detection. In: 11th ACM SIGKDD International Conference on Knowledge Discovery in Data Mining Table of Contents, pp. 157–166. ACM Press, New York (2005)
24. Breunig, M.M., Kriegel, H.-P., Ng, R.T., Sander, J.: LOF: identifying density-based local outliers. SIGMOD Rec. **29**(2), 93–104 (2000)
25. Yeung, D.-Y., Chow, C.: Parzen-window network intrusion detectors. In: Object Recognition Supported by User Interaction for Service Robots, vol. 4, no. 4, pp. 385–388 (2002)
26. Adler, A., Elad, M., Hel-Or, Y., Rivlin, E.: Sparse coding with anomaly detection. Signal Process. Syst. **79**(2), 179–188 (2015)
27. Cong, Y., Yuan, J., Liu, J.: Sparse reconstruction cost for abnormal event detection. In: 2011 IEEE Conference on Computer Vision and Pattern Recognition (CVPR 2011), pp. 3449–3456. IEEE Press, New York (2011)
28. Radovanovi, M., Nanopoulos, A., Ivanovi, M.: Reverse nearest neighbors in unsupervised distance-based outlier detection. IEEE Trans. Knowl. Data Eng. **27**(5), 1369–1382 (2015)
29. You, C., Robinson, D.P., Vidal, R.: Provable self-representation based outlier detection in a union of subspaces. In: 2017 IEEE Conference on Computer Vision and Pattern Recognition (CVPR 2017), pp. 4323–4332. IEEE Press, New York (2017)
30. Hou, D.-D., Cong, Y., Sun, G., Liu, J.: Anomaly detection via adaptive greedy model. Neurocomputing **330**, 369–379 (2019)
31. Analysis of credit card default dataset of Taiwan for machine learning. https://github.com/KaushikJais/Credit-Card-Default/blob/master/Credit%20Card%20Default%20(Final%20Submission)%20(1).ipynb. Accessed 19 Feb 2019

Detection of Pilot's Drowsiness Based on Multimodal Convolutional Bidirectional LSTM Network

Baek-Woon Yu, Ji-Hoon Jeong, Dae-Hyeok Lee, and Seong-Whan Lee[✉]

Korea University, Seoul 02841, Republic of Korea
sw.lee@korea.ac.kr

Abstract. The drowsiness of pilot causes the various aviation accidents such as an aircraft crash, breaking away airline, and passenger safety. Therefore, detecting the pilot's drowsiness is one of the critical issues to prevent huge aircraft accidents and to predict pilot's mental states. Conventional studies have been investigated using physiological signals such as brain signals, electrodermal activity (EDA), electrocardiogram (ECG), respiration (RESP) for detecting pilot's drowsiness. However, these studies have not sufficient performance to prevent sudden aviation accidents yet because it could detect the mental states after drowsiness occurred and only focus on whether drowsiness or not. To overcome the limitations, in this paper, we propose a multimodal convolutional bidirectional LSTM network (MCBLN) to detect drowsiness or not as well as drowsiness level using the fused physiological signals (electroencephalography (EEG), EDA, ECG, and RESP) for the pilot's environment. We acquired the physiological signals for the pilot's simulated aircraft environment across seven participants. The proposed MCBLN extracted the features considering the spatial-temporal correlation of between EEG signals and peripheral physiological measures (PPMs) (EDA, ECG, RESP) to detect the current pilot's drowsiness level. Our proposed method achieved the grand-averaged 45.16% (±1.01) classification accuracy for 9-level of drowsiness. Also, we obtained 84.41% (±1.34) classification accuracy for whether the drowsiness or not across all participants. Hence, we have demonstrated the possibility of the not only drowsiness detection but also 9-level of drowsiness for the pilot's aircraft environment.

Keywords: Drowsiness level detection · Pilot's mental state · Multimodal fusion · Multimodal convolutional bidirectional LSTM

1 Introduction

Recent artificial intelligence (AI) technologies have developed for improving quality of daily life with various research fields. Therefore, advanced pattern recognition (PR) and machine learning methods for brain-computer interface (BCI) or vision-based systems, are promising techniques for detecting the user's behavior [6, 24] or mental state [1–4, 35] and controlling external devices [5, 7–11, 36–38]. In previous studies for detecting mental state on the driving environment, some research groups have applied for the

© Springer Nature Switzerland AG 2020
S. Palaiahnakote et al. (Eds.): ACPR 2019, LNCS 12047, pp. 530–543, 2020.
https://doi.org/10.1007/978-3-030-41299-9_41

vision technique or unimodal BCI [12–15]. The detection accuracy of the camera-based vision approach has sufficient performance for simulated driving environment, however, there are some limitations; the detection of mental state could perform only after the drowsiness, the users cannot wear glasses, stare at the front only during a long driving time [3]. In addition, the unimodal BCI approach has inefficient detection performance yet due to signal's characteristics; high signal-to-noise ratio (SNR) or BCI illiteracy [16].

Recently, the importance of the pilot's mental state has increased due to the development of various aircraft technology. Flight tasks require a high degree of attention, and pilots are exposed to extreme noise and vibration for long periods during the tasks and gradually increase their fatigue. If the level of drowsiness caused by fatigue increases, the probability of occurring the accident will be increased. When the drowsiness level is increased, many changes occur in our body, for example, the behavior is slowed, and response time to external stimuli is also decrease. There are many works to handle at the cockpit and it is also important for pilots to be careful that passengers do not feel anxiety on the aircraft by reacting quickly to external changes. There are many reasons for human accidents in the airplane, but among the causes, the drowsiness of pilot occupies a large proportion [17]. To prevent this, various studies have been researched and multimodal BCI is one of these studies. The multimodal BCI have used for not only brain signals (electroencephalogram (EEG), magnetoencephalography (MEG)) but also peripheral physiological measures (PPMs) such as electrocardiogram (ECG), electrodermal activity (EDA), and respiration (RESP) [18–21].

In the previous studies for detecting user's drowsiness have proceeded to binary classification only, which classifies the normal state or drowsiness state. These conventional studies used a general machine learning methods such as power spectral density (PSD) as feature extraction [22, 23], linear discriminant analysis (LDA) [1, 22, 23], support vector machine (SVM) [12, 13, 25], random forest (RF) [26, 27], and k-nearest neighbors (KNN) [28] as classifiers. Recent studies have been studied for deep learning methods appeared and the studies of pilot's drowsiness proceed actively. Various deep learning methods from basic artificial neural networks (ANN) [29] to convolutional neural networks (CNN) [30, 31] have emerged. The deep learning method has developed continuously and it is likely that the study of user's drowsiness can be developed.

In this paper, we propose a framework, called as multimodal convolutional bidirectional LSTM network (MCBLN), for the pilot's drowsiness level detection. It attaches importance to modality fusion based on temporal and spatial information dependencies learning. The developed framework has an effective performance on fused signal classification tasks. First, we introduce the CNN model to extract spatial-temporal dependencies features according to each modality. Second, the features extracted at the spatial-temporal dimension are fused by considering the correlation between modalities through the dense layer. Finally, bidirectional LSTM is introduced for the analysis of sequential information contained in the fused features. The proposed framework has achieved the best performance (45.16% (±1.01)) to classify 9-level of drowsiness compared with four conventional methods and showed a high binary detection performance (84.41% (±1.34)).

To gain insight into drowsiness level detection, this paper is organized as follows. First, we introduce the drowsiness experiment and the acquisition and preprocessing of EEG signals. Second, we present the MCBLN framework with the learning process and implementation details. Third, we present the overall performance of the model and compare it with four competitive methods. Finally, we present conclusions with an outlook of future applications on drowsiness level detection systems.

2 Methods

2.1 Participants

Seven healthy participants (6 males and 1 female, aged 25.6 ($\pm$5)), who had the flight experience over 300 h in the Taean Flight Education Center as an affiliated education organization of Hanseo University, participated in our experiment. Before the start of the experiment, all participants were informed about the entire experimental protocols and consent according to the Helsinki declaration. All participants were asked to refrain from alcohol and coffee and take sleep (6–8 h) before the day. At the end of the experiment, all participants were instructed to fill out the questionnaires for recognition of their status and evaluation of the paradigm. The experimental environments and protocols were reviewed and approved by the Institutional Review Board at Korea University [1040548-KU-IRB-18-92-A-2].

2.2 Experimental Environment

The simulator system presented the flight environment of Cessna 172. The cockpit was consisted of the wide visual display, providing the 210 angles view of what the pilot would see outside the aircraft, a flight yoke, an electric control loading, and other control options for presentation of realistic flight environments. The wireless keypad had been attached to the yoke to get the pilot's response. The EEG signals were measured by the BrainAmp amplifier (Brain Products GmbH, Germany). The sampling frequency was 1000 Hz and a 60 Hz of notch filter was applied for DC noise reduction. 60 EEG channels were located on the scalp according to the standard international 10–20 system (Fp1-2, F1-6, Fz, FT7-10, FC1-6, T7-8, C1-6, Cz, CP1-6, CPz, TP7-10, P1-8, P7-8, Pz, PO3-4, PO7-10, POz, O1-2, Oz, and AF3-4). In addition, electrooculography (EOG) (horizontal with electrodes at the outer canthi (left: F7, right: F8) and vertical with electrodes above/below the left eye (above: AF7, below: AF8) were also acquired. The reference electrode was placed at the FCz channel, and the ground was placed at the AFz channel. Before the acquisition, the skin impedance of EEG electrodes was adjusted below 10 kΩ by injecting conductive gel. The PPMs sensors were measured by the Flexcomp system with biograph infinity software (Thought Technology Ltd., Canada). The sampling frequency was 256 Hz (Fig. 1).

2.3 Experimental Paradigm

We designed the experimental paradigm to induce the pilot's drowsiness level. The pilot's physiological signals such as EEG and PPMs were recorded based on this paradigm. The participants carried out a monotonous flight of night time (high: 3,000 feet,

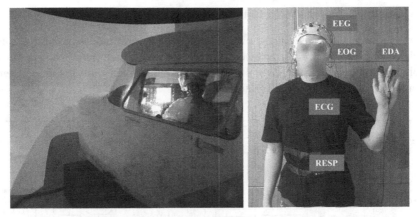

Fig. 1. Experimental environment for acquiring the pilot's physiological signals (EEG, ECG, EDA, and RESP).

heading: 0 degrees, speed: 100 km/h) for one hour to induce drowsiness. The beep sound appeared every 1 min. After appearing this sound, the participants should input one number using the keypad, depending on the subjective drowsiness level which is related to the Karolinska sleepiness scale (KSS). KSS consists of 9 levels and level 1 means clearly alert and level 9 means extremely drowsy. If the participants failed to input the KSS or fly an aircraft successfully, we regarded that part as drowsiness level (Fig. 2).

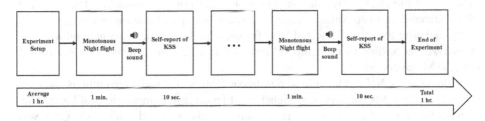

Fig. 2. Experimental paradigm for KSS-based pilot's drowsiness level acquisition.

2.4 Data Pre-processing

The EEG signals were down-sampled from 1000 Hz to 250 Hz and were applied bandpass filtering with 1–50 Hz using zero-phase, second-order, Butterworth filter. The PPMs were digitally filtered by a 30 Hz low pass filter. After that, EEG and PPMs data for 1 h were segmented into 1-s epochs without overlap. The label of each epoch applied to KSS index which was measured after 1 min equally. Therefore, we collected the 3,600 samples per each participant.

2.5 Multimodal Convolutional Bidirectional LSTM Network (MCBLN)

Figure 3 shows an overview of the proposed model for multimodal based drowsiness level detection. The proposed framework extracts the spatial-temporal features through convolution and fuses it with the multimodal signals, and then analyzes the sequential information included in the fused feature.

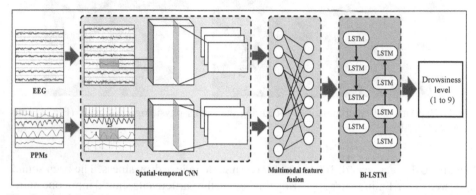

Fig. 3. Overview of the proposed framework for drowsiness level detection using multimodal signals.

The EEG input data comprises electrode by sampling point (60×250), and PPMs input data is the channel by sampling point (4×250) of each modality. The EEG and PPMs are entered to convolution layers for spatial-temporal feature extraction respectively, and after the fully connected layer for multimodal fusion, sequential information is analyzed according to the temporal order of features through bidirectional LSTM (Bi-LSTM) layers.

Commonly, CNN can learn local non-linear features through convolutions and non-linearity. Also, CNN can represent higher-level features as compositions of lower-level features through multiple layers of processing. In this proposed framework, CNN was divided up into two layers to deal with a large number of input channels better. In the first layer, each filter performs a convolution over time. The filter size is one-tenth (1×25) of the sampling rate of 250 Hz and is designed to consider the frequency features above 10 Hz [32]. In EEG data, alpha and beta frequency bands are included in more than 10 Hz, which is known as frequency bands associated with drowsiness [22, 23]. In the second layer, each filter performs spatial filtering with the weights for all possible pairs of electrodes with filters of the preceding temporal convolution. Since there is no activation function between the temporal and spatial convolution layers, the layers could be combined into one layer. We used the exponential linear units (ELUs) as activation functions (Table 1).

The spatial-temporal features of EEG and PPMs, which were extracted by the convolution layer, were fused considering the correlation of two modalities. After flattening and concatenating the spatial-temporal features of each modality, the two modalities with different features can be effectively fused by input the fully connected layer.

Table 1. Details of the MCBLN framework. The symbol [] contains the kernel size, the number of feature maps, and the type of layers, respectively

Layers	Output size	Construction
Input (EEG)	60 × 250	–
Input (PPMs)	4 × 250	
Temporal	60 × 226 × 50	[1 × 25, map 50, conv]
	4 × 226 × 50	
Spatial	1 × 226 × 50	[60 × 1, map 50, conv]
	1 × 226 × 50	[4 × 1, map 50, conv]
Flatten	100 × 226	–
Feature fusion	50 × 226	Fully connected
Sequence	200	Bi-LSTM (units 100, dropout 0.5)
Sequence	100	Bi-LSTM (units 50, dropout 0.5)
Classification	100	Fully connected
	9	Softmax

The LSTM network is one of the best methods to consider time sequences such as physiological signals. To know how information is processed in a single LSTM, consider a sequence of input that spans n time steps: $X = (x_1, x_2, x_3, \cdots, x_n)$. At time t, using input x_t, and previous hidden state h_{t-1}, the memory cell selects what to keep or forget from the previous states using the forget gate f_t (1). The memory cell computes the current state c_t in two steps. At first, the cell calculates a memory cell candidate state $\tilde{c}_t$ (4). Next, using previous cell state c_{t-1} and input gate i_t (2), the cell decides how much information to write in the current state c_t (5). The output gate decides how much information h_t (6) will be transferred into the next cell using output gate o_t (3).

$$f_t = \sigma\left(W_{jh}h_{t-1} + W_{fx}x_t + b_f\right) \qquad (1)$$

$$i_t = \sigma\left(W_{jh}h_{t-1} + W_{ix}x_t + b_i\right) \qquad (2)$$

$$o_t = \sigma(W_{oh}h_{t-1} + W_{ox}x_t + b_o) \qquad (3)$$

$$\tilde{c}_t = \tanh(W_{ch}h_{t-1} + W_{cx}x_t + b_c) \qquad (4)$$

$$c_t = f_t \circ c_{t-1} + i_t \circ \tilde{c}_t \qquad (5)$$

$$h_t = o_t \circ \tanh(c_t) \qquad (6)$$

W denotes weight matrices or weight vectors, b denotes biases. σ is the logistic function and $\circ$ is the Hadamard product operator.

After the two modalities were fused, we adopted the Bi-LSTM network. Bi-LSTM network is one kind of the recurrent neural networks that can process a sequence or a temporal series of data. It consists of two LSTM blocks that process sequential information in two opposite directions simultaneously. One block processes the sequence instances of fused data forwardly starting from its first-time instance to the end-time instance. The other block processes the same fused data in the reverse order. Each block generates its own output. The final output of the Bi-LSTM at each sequence instance is calculated by combining the two outputs of each LSTM block. We only consider the last output generated from the Bi-LSTM blocks after passing the whole input data. The proposed model consists of two Bi-LSTM layers. Finally, the output from each cell in the Bi-LSTM block was classified as 9-class using a fully connected layer and a softmax layer.

3   Experimental Results

In this section, to validate the performance of the proposed framework, we gave the overall performance evaluated with 4-fold cross-validation for EEG, PPMs, and fused multimodal data. The confusion matrix of each KSS index classification accuracy was evaluated about each physiological signal combination. Also, we compared with other common structures and competitive works to prove the fusion of physiological signals and the spatial-temporal superiority of our proposed structure. Finally, we analyzed the impact of the MCBLN framework on the multimodal fusion and the spatial-temporal information with some discussions.

3.1   Classification Performances for Drowsiness Level

The proposed MCBLN was trained to detect drowsiness levels for each participant. The performance was obtained by 4-fold cross-validation. 75% of the samples were randomly selected for training and the remaining 25% were reserved for validation. We showed the recognition effects with classification accuracy, standard deviation (STD), sensitivity, and specificity on the validation sets. Tables 2, 3 and Fig. 4 presented the performance of the MCBLN framework on the drowsiness dataset.

Table 2. Overall 9-class performances using MCBLN framework

Modality	Accuracy (%)	Standard deviation
All (EEG+PPMs)	45.16	±1.01
EEG	40.89	±1.77
PPMs	34.50	±0.07

The training dataset was used to train the MCBLN model, whereas the validation dataset was used to verify the accuracy and the effectiveness of the trained MCBLN

Table 3. Overall 2-class performance of the MCBLN framework

Modality	Accuracy (%)	STD	Sensitivity	Specificity
All (EEG+PPMs)	84.41	±1.34	0.95	0.89
EEG	81.98	±0.92	0.93	0.73
PPMs	74.01	±0.97	0.84	0.64

model for classification of the nine classes of each modality. Each MCBLN used 30 training epochs. At the end of 30 training epochs, the network loss convergence value of EEG, PPMs, and EEG+PPMs was 1.3295, 1.6144, and 1.2506 of each modality.

From Table 2, we confirmed that the MCBLN model was stably effective on the dataset including EEG. The average accuracy of 4-fold cross-validation for each modality was 45.16%, 40.89%, and 34.50% with STD 1.01, 1.77, and 0.07. The maximum classification accuracy showed using multimodality (EEG+PPMs). The maximum and minimum accuracy showed significant difference about 10%.

Previous studies researched 2-class classification, and we applied our method to 2-class classification. The performance was significant. Of them, the fusion of modality showed the best performance. When we used all modalities, the average accuracy was 84.41% with STD 1.34. Also, the sensitivity was 0.95 and the specificity was 0.89. And in the case of using PPMs only, it was lower than that of using all modalities. The average accuracy was 74.01% with STD 0.97. Also, the sensitivity was 0.84 and the specificity was 0.64. In the case of using EEG only, all indicators were the lowest. The average accuracy was 81.98% with STD 0.92. Also, the sensitivity was 0.93 and the specificity was 0.73.

The confusion matrices of each modality were shown in Fig. 4, which give details of the strength and weakness of each modality. Each column of the confusion matrix represents the target class and each row represents the predicted class. The element (i, j) is the percentage of samples in class j that is classified as class i. From confusion matrices, we can see that the extremely alert or drowsiness level (from 1 or 9 KSS index) was recognized with very high accuracy of over 80% in EEG and EEG+PPMs. And through the confusion matrix, the short temporal data (1 s) of PPMs could identify the potential for classification of alert and drowsiness (KSS $\geq$ 7).

3.2 Comparison Performances with Conventional Methods

The proposed method had effective performances from the overall performance because of the multimodality fusion. In order to explore the ability of the proposed method on recognition tasks deeply, we compared our proposed method with several commonly used structures and other conventional works.

Convolutional and recurrent layers are often used to extract the temporal dependency in deep neural networks [33, 34]. Although direct performance comparison using advanced methods was impossible due to different experimental protocol and paradigm, we compared the classification performances with commonly used methods in signal processing. Here, we introduce some of the details for these compared methods.

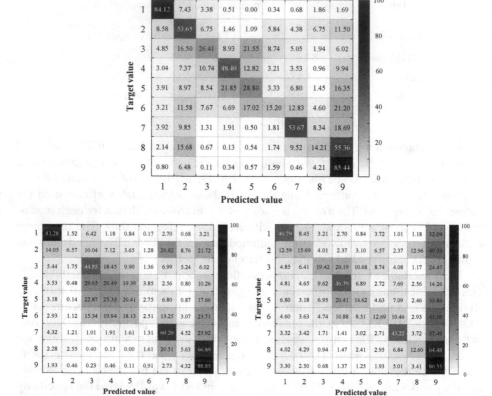

Fig. 4. Confusion matrices of classification accuracy for each modality. (up: EEG+PPMs, left-down: EEG, right-down: PPMs).

(1) PSD-SVM: It extracted the frequency PSD features and used SVM classifier to determine mental fatigue [13].
(2) CCA-SVM: It fed selected features from canonical correlation analysis (CCA) into SVM classifiers [1].
(3) CNN: It proposed a channel-wise CNN with raw EEG data on driver's cognitive performance prediction tasks in [31].
(4) Bi-LSTM: It used the Bi-LSTM architecture, which consisted of two Bi-LSTM layers.

We chose the four above-mentioned methods. There are some conventional methods for drowsiness classification such as PSD-SVM in [13]. It extracted PSD while neglecting the spatial information of physiological signals, and then fed into SVM classifier. CCA-SVM calculated the PSD from the frequency features of the physiological signals and chose the features considering the canonical correlation coefficient between PSD and drowsiness level. Deep learning methods made an effort to improve recognition performance taking full consideration for spatial or temporal features. We chose the four competitive works for further comparison, including PSD-SVM, CCA-SVM, CNN, and

Bi-LSTM. From the above, we chose the two feature-based methods and the two deep learning frameworks to test and evaluate our proposed framework.

According to the structure parameters introduced in the other papers, we reproduced these methods using OpenBMI Toolbox [16] to analyze our EEG or PPMs signals. We applied these methods with 4-fold cross-validation and received the mean accuracy of each method. Figure 5 showed the classification accuracies of the compared methods.

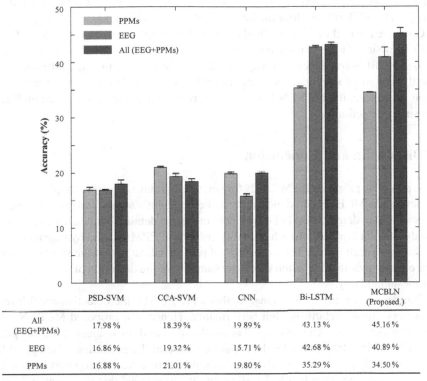

	PSD-SVM	CCA-SVM	CNN	Bi-LSTM	MCBLN (Proposed.)
All (EEG+PPMs)	17.98 %	18.39 %	19.89 %	43.13 %	45.16 %
EEG	16.86 %	19.32 %	15.71 %	42.68 %	40.89 %
PPMs	16.88 %	21.01 %	19.80 %	35.29 %	34.50 %

Fig. 5. Comparison of classification performances for drowsiness level with the conventional methods.

We took CNN and Bi-LSTM for comparison to investigate the importance of extracting spatial or temporal features. In CNN, using the channel-wise CNN as proposed in [31], we identified the classification performance of drowsiness using only spatial information. In Bi-LSTM, it was configured to omit spatial-temporal analysis through convolution layers in the proposed method and input each raw modality data (EEG, PPMs, EEG+PPMs) into Bi-LSTM, which consists of two layers. From Fig. 5, CNN and Bi-LSTM had significant performances and achieved 19.89% and 43.13% with EEG+PPMs, respectively. It means that recurrent layers are possible to extract the temporal information from EEG and PPMs signals.

Some non-deep learning methods had been conducted to explore different kinds of modality analysis tasks. We used the 4-fold cross-validation to estimate the classification performances among MCBLN and the other competitive methods. Frequency features were often used in the above-mentioned studies, including PSD-SVM and CCA-SVM. PSD-SVM got the mean accuracy of 17.98% with EEG+PPMs while CCA-SVM had a better performance of 21.01% with PPMs. However, these two methods were worse than other Bi-LSTM methods, which reflected the robust capacity of Bi-LSTM methods on learning representations. The classification accuracies of these deep learning methods were various, and the range was from 15% to 43%. Among the methods, CNN, Bi-LSTM, and MCBLN worked well in the whole dataset when using the multimodal data as input. The STD of these four methods varied from 0.07 to 1.77.

Compared with these four methods, MCBLN showed the best performance for the pilot's drowsiness level classification tasks with considerable advantages. This means that our MCBLN framework can capture effective information using multimodal data robustly, due to its significant abilities for spatial-temporal feature-based multi-modal fusion. Therefore, the MCBLN framework delivers noticeable performance on 9-level drowsiness detection.

4   Discussion and Conclusion

In this paper, we proposed the MCBLN framework for robust detecting the pilot's drowsiness level. The MCBLN could measure whether the pilot's mental state is drowsiness or not as well as to detect 9-level of drowsiness more in a detail. To do that, we applied the spatial-temporal features of each modality (EEG and PPMs) fused by considering the correlation of each signal. The experimental results indicated that the spatial-temporal feature-based multimodal fusion was more suitable for the detection of drowsiness level than for unimodality.

Compared with the conventional methods [1, 13, 31], the detection for 9-level of drowsiness showed relatively high performance. Hence, our proposed MCBLN could prove that our method can reliably classify the alert and drowsiness states for pilots. When classifying drowsiness levels, it can be seen that the extraction of temporal features through Bi-LSTM was significant. In unimodal results, the proposed method had a lower performance than Bi-LSTM. However, it was verified that maintaining the temporal characteristics and fusing all modality affects performance improvement. Also, improvement of CNN-based spatial-temporal feature extraction method could result in better performance. Although direct performance comparison using advanced methods was impossible due to different experimental protocol and paradigm, we compared the classification performances with commonly used methods in signal processing.

The unimodal signal (EEG) was a suitable modality for detecting drowsiness, but the short multimodal signal (EEG+PPMs) showed the possibility of performance enhancement. In the conventional studies, for most physiological signals, except for EEGs, long-time data (approximately from 30 s to 1 min) was required to analyze drowsiness [13]. However, our experimental results showed that drowsiness can be detected using only short temporal data of 1 s, given the spatial-temporal features between physiological signals.

MCBLN showed the stable performance of drowsiness detection and suggested the possibility of drowsiness level classification. Since the structure of MCBLN models is not limited to physiological signals, it is expected to be applicable to various multimodal fusion systems. But, it is not enough to prevent accidents. It needs to focus on the prediction of the drowsiness time through the participant's drowsiness level propensity through a fusion of multimodality for accident prevention. In addition, current multimodal based drowsiness detection is uncomfortable due to using a lot of signal acquisition devices. To solve this limitation, we will have developed the optimal combination of signals for real-world environment.

Acknowledgement. This work was supported by Defense Acquisition Program Administration (DAPA) and Agency for Defense Development (ADD) of Korea (06-201-305-001, A Study on Human-Computer Interaction Technology for the Pilot Status Recognition).

References

1. Han, S.-Y., Kim, J.-W., Lee, S.-W.: Recognition of pilot's cognitive states based on combination of physiological signals. In: Proceedings of the 7th International Winter Conference on Brain-Computer Interface (BCI), Korea, pp. 1–5 (2019)
2. Lee, M., et al.: Connectivity differences between consciousness and unconsciousness in non-rapid eye movement sleep: a TMS–EEG study. Sci. Rep. **9**, 1–9 (2019)
3. Balandong, R.P., Ahmad, R.F., Saad, M.N.M., Malik, A.S.: A review on EEG-based automatic sleepiness detection systems for driver. IEEE Access **6**, 22908–22919 (2018)
4. Won, D.O., Hwang, H.J., Dähne, S., Müller, K.R., Lee, S.-W.: Effect of higher frequency on the classification of steady-state visual evoked potentials. J. Neural Eng. **13**(1), 1–11 (2016)
5. Kim, I.-H., Kim, J.-W., Haufe, S., Lee, S.-W.: Detection of braking intention in diverse situations during simulated driving based on EEG feature combination. J. Neural Eng. **12**(1), 1–12 (2015)
6. Roh, M.C., Shin, H.K., Lee, S.-W.: View-independent human action recognition with volume motion template on single stereo camera. Pattern Recognit. Let. **31**(7), 639–647 (2010)
7. Jeong, J.-H., Shim, K.-H., Cho, J.-H., Lee, S.-W.: Trajectory decoding of arm reaching movement imageries for brain-controlled robot arm system. In: Conference Proceedings Engineering in Medicine and Biology Society (EMBC), Germany, pp. 1–4 (2019)
8. Choi, I.-H., Kim, Y.-G.: Head pose and gaze direction tracking for detecting a drowsy driver. Appl. Math. Inf. Sci. **9**, 505–512 (2015)
9. Kwak, N.-S., Muller, K.-R., Lee, S.-W.: A lower limb exoskeleton control system based on steady state visual evoked potentials. J. Neural Eng. **12**(5), 1–14 (2015)
10. Lee, M.-H., Williamson, J., Won, D.-O., Fazli, S., Lee, S.-W.: A high performance spelling system based on EEG-EOG signals with visual feedback. IEEE Trans. Neural Syst. Rehabil. Eng. **26**(7), 1443–1459 (2018)
11. Kim, J.-H., Bießmann, F., Lee, S.-W.: Decoding three-dimensional trajectory of executed and imagined arm movements from electroencephalogram signals. IEEE Trans. Neural Syst. Rehabil. Eng. **23**(5), 867–876 (2014)
12. Wei, C.-S., Wang, Y.-T., Lin, C.-T., Jung, T.-P.: Toward drowsiness detection using non-hair-bearing EEG-based BCI. IEEE Trans. Neural Syst. Rehabil. Eng. **26**(2), 400–406 (2018)
13. Zhang, X., et al.: Design of a fatigue detection system for high speed trains based on driver vigilance using a wireless wearable EEG. Sensors **17**, 1–21 (2017)
14. Kim, K.-T., Suk, H.-I., Lee, S.-W.: Commanding a brain-controlled wheelchair using steady-state somatosensory evoked potentials. IEEE Trans. Neural Syst. Rehabil. Eng. **26**(3), 654–665 (2016)

15. Rumagit, A.M., Akbar, I.A., Igasaki, T.: Gazing time analysis for drowsiness assessment using eye gaze tracker. Telkomnika **15**(2), 919–925 (2017)
16. Lee, M.-H., et al.: EEG dataset and OpenBMI toolbox for three BCI paradigms: an investigation into BCI illiteracy. Gigascience **8**(5), 1–16 (2019)
17. Yen, J.-R., Hsu, C.-C., Yang, H., Ho, H.: An investigation of fatigue issues on different flight operations. J. Air Transp. Manag. **15**, 236–240 (2009)
18. Lee, M.-H., Fazli, S., Mehnert, J., Lee, S.-W.: Subject-dependent classification for robust idle state detection using multi-modal neuroimaging and data-fusion techniques in BCI. Pattern Recognit. **48**(8), 2725–2737 (2015)
19. Lee, M.-H., Fazli, S., Mehnert, J., Lee, S.-W.: Hybrid brain-computer interface based on EEG and NIRS modalities. In: Proceedings of the 2nd International Winter Conference on Brain-Computer Interface (BCI), Korea, pp. 1–15 (2014)
20. Fazli, S., Lee, S.-W.: Brain computer interfacing: a multi-modal perspective. J. Comput. Sci. Eng. **7**(2), 132–138 (2013)
21. Yeom, S.-K., et al.: Spatio-temporal dynamics of multimodal EEG-fNIRS signals in the loss and recovery of consciousness under sedation using midazolam and propofol. PLoS ONE **12**(11), 1–22 (2017)
22. Nguyen, T., Ahn, S., Jang, H., Jun, S.C., Kim, J.G.: Utilization of a combined EEG/NIRS system to predict driver drowsiness. Sci. Rep. **7**, 1–10 (2017)
23. Ahn, S., Nguyen, T., Jang, H., Kim, J.G., Jun, S.C.: Exploring neuro-physiological correlates of drivers' mental fatigue using simultaneous EEG, ECG, and fNIRS. Front. Hum. Neurosci. **10**, 1–14 (2016)
24. Park, U., Choi, H.C., Jain, A.K., Lee, S.-W.: Face tracking and recognition at a distance: a coaxial and concentric PTZ camera system. IEEE Trans. Inf. Forensics Secur. **8**(10), 1665–1677 (2013)
25. Dimitrakopoulos, G.N., et al.: Functional connectivity analysis of fatigue reveals different network topological alterations. IEEE Trans. Neural Syst. Rehabil. Eng. **26**(4), 1–14 (2018)
26. Hong, S., Kwon, H., Choi, S.H., Park, K.S.: Intelligent system for drowsiness recognition based on ear canal EEG with PPG and ECG. Inf. Sci. **453**, 302–322 (2018)
27. Mårtensson, H., Keelan, O., Ahlström, C.: Driver sleepiness classification based on physiological data and driving performance from real road driving. IEEE Trans. Intell. Transp. Syst. **20**, 421–430 (2018)
28. Chen, J., Wang, H., Hua, C.: Assessment of driver drowsiness using electroencephalogram signals based on multiple functional brain networks. Int. J. Psychophysiol. **133**, 120–130 (2018)
29. De Narois, C.J., Bourdin, C., Stratulat, A., Diaz, E., Vercher, J.-L.: Detection and prediction of driver drowsiness using artificial neural network models. Accid. Anal. Prev. **126**, 95–104 (2019)
30. Wu, E.Q., Peng, X.Y., Zhang, C.Z., Lin, J.X., Sheng, R.S.F.: Pilot's fatigue status recognition using deep contractive autoencoder network. IEEE Trans. Instrum. Meas. **68**, 3907–3919 (2019)
31. Hajinoroozi, M., Mao, Z., Jung, T.-P., Lin, C.-T., Huang, Y.: EEG based prediction of driver's cognitive performance by deep convolutional neural network. Signal Process. Image Commun. **47**, 549–555 (2016)
32. Lawhern, V.J., Solon, A.J., Waytowich, N.R., Goedon, S.M., Hung, C.P., Lance, B.J.: EEGNet: a compact convolutional neural network for EEG-based brain-computer interfaces. arXiv, 1–30 (2018)
33. Sakhavi, S., Guan, C., Yan, S.: Learning temporal information for brain-computer interface using convolutional neural networks. IEEE Trans. Neural Netw. Learn. Syst. **29**(11), 5619–5629 (2018)

34. Hefron, R.G., Borghetti, B.J., Christensen, J.C., Kabban, C.M.S.: Deep long short-term memory structures model temporal dependencies improving cognitive workload estimation. Pattern Recognit. Lett. **94**, 96–104 (2017)
35. Lee, M., et al.: Network properties in transitions of consciousness during propofol-induced sedation. Sci. Rep. **1**(7), 16791 (2017)
36. Bulthoff, H.H., Lee, S.-W., Poggio, T.A., Wallraven, C.: Biologically motivated computer vision. Lecture Notes in Computer Science, vol. 2525. Springer, Heidelberg (2003). https://doi.org/10.1007/3-540-36181-2
37. Jeong, J.-H., Lee, M.-H., Kwak, N.-S., Lee, S.-W.: Single-trial analysis of readiness potentials for lower limb exoskeleton control. In: Proceedings of the 7th International Winter Conference on Brain-Computer Interface (BCI), Korea, pp. 50–52 (2017)
38. Lee, S.-H., Lee, M., Jeong, J.-H., Lee, S.-W.: Towards an EEG-based intuitive BCI communication system using imagined speech and visual imagery. In: Proceedings of the IEEE International Conference on Systems, Man and Cybernetics, pp. 4409–4414. IEEE, Bari (2019)

Deriving Perfect Reconstruction Filter Bank for Focal Stack Refocusing

Asami Ito[1], Akira Kubota[1(✉)], and Kazuya Kodama[2]

[1] School of Science and Engineering, Chuo University, Tokyo, Japan
kubota@elect.chuo-u.ac.jp
[2] National Institute of Informatics, Tokyo, Japan

Abstract. This paper presents a digital refocusing method that transforms the captured focal stack directly into a new focal stack under different focus settings. Assuming Lambertian scenes with no occlusions, this paper theoretically shows that there exist a set of filters that perfectly reconstructs focal stack under Gaussian aperture from that captured under Cauchy one. The perfect reconstruction filters are derived in linear and space-invariant using a layered scene representation. Numerical simulations using synthetic focal stacks showed that the root mean squared errors are quite small and less than 10^{-9}, indicating the derived filters allow perfect reconstruction.

Keywords: Digital refocusing · Focal stack · Aperture · Perfect reconstruction filters

1 Introduction

Digital refocusing is a post image processing that generates images with different focus settings such as focal length (or focus depth), depth-of-field and aperture shape. So called light field cameras [1] allow this post-capture refocusing easily; based on a desired lens model, refocused images can be generated by weighted averaging of the light intensities recorded in the captured light field [2]. However, the resolution of the refocused images becomes low. This is because the recorded light field is a two-dimensional array of sub-aperture images at every incident angles and the resolution of each image is much lower than that of the imaging plane.

To acquire light fields in higher resolution, image recovery approaches have been actively studied [3]. Some approaches attempt to recover the light field from its focal stack (a set of multiple images with different focus depths) captured with a single camera [4–11]. In these approaches, the spatial resolution of the sub-aperture image becomes equal to that of the camera.

Aizawa et al. presented a light field recovery method [4] for Lambertian scenes that consist of a few layers with no occlusion. Modeling both the focal stack and the sub-aperture image as linear combinations of the unknown layers at different

© Springer Nature Switzerland AG 2020
S. Palaiahnakote et al. (Eds.): ACPR 2019, LNCS 12047, pp. 544–553, 2020.
https://doi.org/10.1007/978-3-030-41299-9_42

depths, they derived an equation that holds between them and solved it iteratively. This method can appropriately shift the layers according to their depths without depth estimation. However, the equation cannot be completely solved when the number of layers is larger than two; in fact, at the lower frequencies, the solution (i.e., the sub-aperture image) tends to diverge. Alonso et al. recently extended this method for general scenes that consist of many layers [5]. They recovered the layers using the least-square optimization in the Fourier domain and reconstructed the sub-aperture image as a projection of the recovered layers.

Several similar approaches using a layered (or volumetric) scene representation have been presented [6–10]. They applied the techniques in computer tomography to light field recovery based on the fact that a focal stack can be modeled as a set of projections of the light field [12]. The light field was recovered via deconvolution of the back-projected focal stack. Although these methods also do not involve depth estimation, they work well only if the range of the focus depths is much wider than that of the actual scene depth.

Recently, our previous work [11] has designed an optimum aperture in capturing focal stacks, named Cauchy aperture [13], such that the light field is perfectly recovered from the focal stack in theory. In this paper, we present a focal stack refocusing method that reconstructs a desired focal stack under different aperture from a captured focal stack under Cauchy aperture. For non-occluded Lambertian scenes, we theoretically show that there exist filter bank that perfectly reconstructs the desired focal stack. The stable and perfect reconstruction filters are derived in linear and space-invariant based on a layered scene representation.

There are some focal stack refocusing methods that directly reconstruct a new focal stack from the captured focal stack. Based on the layered scene representation, these method also derived reconstruction filters in the Fourier domain [14–16]. However reconstructing new focal stack was still ill-conditioned in the lower frequencies and the filters are stable only for the focal stack captured with shallow depth-of-field. In contrast, our filtering method allows perfect refocusing without error in theory.

2   Problem Description

A focal stack is a set of multiple images that are focused at different depths. This paper deals with a problem of focal stack refocusing where a focal stack captured under Cauchy aperture is transformed into a new focal stack under a desired aperture.

Letting z_i ($i = 1, 2, ..., N$) be the focus depths, denote $\{f_i(\boldsymbol{x})\}$ and $\{g_i(\boldsymbol{x})\}$ be the captured and the desired focal stacks, respectively. Here the vector $\boldsymbol{x}$ represents the image coordinate (x, y). Our goal is to derive a filter bank $\{r_{ij}(\boldsymbol{x})\}$ that perfectly reconstructs the desired focal stack $\{g_i(\boldsymbol{x})\}$ directly from the captured focal stack $\{f_i(\boldsymbol{x})\}$ by the following filtering-and-adding formula:

$$g_i(\boldsymbol{x}) = \sum_{j=1}^{N} r_{ij}(\boldsymbol{x}) * f_j(\boldsymbol{x}), \qquad i = 1, 2, \ldots, N, \tag{1}$$

where $*$ denotes two-dimensional convolution. Note that all the derived filters are linear and spatial invariant; thus our refocusing method requires no depth estimation and is independent of the scene texture.

In this work, we assume that the focus depths are changed such that $z_1 < z_2 < \cdots < z_N$ holds with $1/z_i - 1/z_{i+1}$ constant and the magnification difference among the images due to different focus depths is already corrected on the basis of the image $f_1(\boldsymbol{x})$. We also assume a target scene consists of Lambertian objects with no occlusion.

3   Filter Bank for Focal Stack Refocusing

3.1   Focal Stack Model Using a Layered Scene Representation

In this subsection, image formation models for the captured and the desired focal stacks are described. These models are based on a layered (or volumetric) scene representation [4,5,15,17].

Let $\phi_i(\boldsymbol{x})$ be focus slices defined as in-focus regions in the captured image $f_i(\boldsymbol{x})$. Based on the thin lens model, the captured focal stack can be modeled by a combination of the focus slices as

$$f_i(\boldsymbol{x}) = \sum_{j=1}^{N} h_{ij}(\boldsymbol{x}) * \phi_j(\boldsymbol{x}), \quad i = 1, 2, \ldots, N, \tag{2}$$

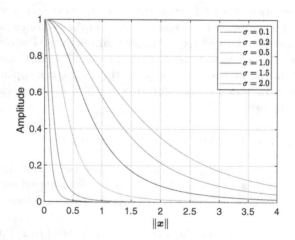

Fig. 1. Characteristics of PSF $h_{12}(\boldsymbol{x})$ (The maximum amplitude is normalized to 1).

where the function $h_{ij}(x)$ represents a point spread function (PSF). The PSFs are expressed by scaled versions of Cauchy function [19] as follows:

$$h_{ij}(x) = \begin{cases} \dfrac{1}{|i-j|^2\sigma^2}\left[\left(\dfrac{\|x\|}{|i-j|\sigma}\right)^2 + 1\right]^{-3/2} , & (i \neq j) \\ \delta(x) & (i = j) \end{cases} \qquad (3)$$

where σ is a parameter that determines the minimum amount of blur degree. Note that the PSFs $h_{ii}(x)$ becomes two-dimensional Dirac delta function $\delta(x)$. For example, the PSF $h_{12}(x)$ is plotted in Fig. 1.

In the same manner, the desired focal stack is modeled as a combination of the focus slices:

$$g_i(x) = \sum_{j=1}^{N} k_{ij}(x) * \phi_j(x), \quad i = 1, 2, \ldots, N. \qquad (4)$$

where the functions $k_{ij}(x)$ are PSFs. Let $a(x, y)$ be a desired aperture function, the PSFs become scaled versions of the function a as follows:

$$k_{ij}(x) = \begin{cases} \dfrac{1}{(i-j)^2\rho^2} a\left(\dfrac{x}{(i-j)\rho}, \dfrac{y}{(i-j)\rho}\right) & (i \neq j) \\ \delta(x) & (i = j) \end{cases} \qquad (5)$$

where ρ is a parameter that determines the minimum amount of blur degree.

3.2   Focal Stack Model in the Fourier Domain

The models in Eqs. (2) and (4) can be simply expressed in the Fourier domain. Taking Fourier transform of Eq. (2) gives the following matrix-vector formula:

$$f = H\phi. \qquad (6)$$

The two vectors are defined as

$$f = \begin{pmatrix} F_1(u) \\ F_2(u) \\ \vdots \\ F_N(u) \end{pmatrix} \quad \text{and} \quad \phi = \begin{pmatrix} \Phi_1(u) \\ \Phi_2(u) \\ \vdots \\ \Phi_N(u) \end{pmatrix},$$

where $F_i(u)$ and $\Phi_i(u)$ represent the Fourier transforms of $f_i(x)$ and $\phi_i(x)$, respectively, and the vector $u = (u, v)$ denotes the frequency counterpart to the image coordinate $x = (x, y)$. The matrix H is given by the following symmetric Toeplitz form (especially called Kac-Murdock-Szegö (KMS) matrix [18]):

$$H = \begin{pmatrix} 1 & C(u) & C^2(u) & \cdots & C^{N-1}(u) \\ C(u) & 1 & C(u) & \cdots & C^{N-2}(u) \\ C^2(u) & C(u) & 1 & \ddots & \vdots \\ \vdots & \ddots & \ddots & \ddots & C(u) \\ C^{N-1}(u) & \cdots & \cdots & C(u) & 1 \end{pmatrix}, \qquad (7)$$

where the function $C(\boldsymbol{u})$ is simply derived to be the exponential function of

$$C(\boldsymbol{u}) = \exp\left(-2\pi\sigma\|\boldsymbol{u}\|\right). \tag{8}$$

Similarly, taking the Fourier transform of Eq. (4) gives

$$\boldsymbol{g} = K\boldsymbol{\phi}. \tag{9}$$

The vector $\boldsymbol{g}$ is defined by

$$\boldsymbol{g} = \begin{pmatrix} G_1(\boldsymbol{u}) \\ G_2(\boldsymbol{u}) \\ \vdots \\ G_N(\boldsymbol{u}) \end{pmatrix},$$

where $G_i(\boldsymbol{u})$ is the Fourier transform of $g_i(\boldsymbol{x})$. The matrix K is given by form of Hermitian Toeplitz matrix such that the element $K_{ij}(\boldsymbol{u})$ is expressed by

$$K_{ij}(\boldsymbol{u}) = \begin{cases} A\left[(i-j)\rho u, (i-j)\rho v\right] & (i \neq j) \\ 1 & (i = j) \end{cases}, \tag{10}$$

where $A(u,v)$ is the Fourier transform of the desired aperture function $a(x,y)$.

3.3   Deriving Perfect Reconstruction Filters

In the frequencies where H^{-1} exists, eliminating the focus slice images $\boldsymbol{\phi}$ from Eqs. (6) and (9) yields

$$\boldsymbol{g} = \left(KH^{-1}\right)\boldsymbol{f}. \tag{11}$$

Each element of the matrix KH^{-1} specifies the frequency characteristic of the reconstruction filters $r_{ij}(\boldsymbol{x})$, noted by $R_{ij}(\boldsymbol{u})$.

In all the frequencies except the direct current (DC), since $C(\boldsymbol{u}) \neq \pm 1$, the matrix H is invertible and its inverse is found to be

$$H^{-1} = \frac{1}{1 - C^2(\boldsymbol{u})} \begin{pmatrix} 1 & -C(\boldsymbol{u}) & & & \\ -C(\boldsymbol{u}) & 1+C^2(\boldsymbol{u}) & -C(\boldsymbol{u}) & & \\ & \ddots & \ddots & \ddots & \\ & & -C(\boldsymbol{u}) & 1+C^2(\boldsymbol{u}) & -C(\boldsymbol{u}) \\ & & & -C(\boldsymbol{u}) & 1 \end{pmatrix}, \quad \text{for} \quad \|\boldsymbol{u}\| \neq 0. \tag{12}$$

Consequently, the filter bank is derived based on Eq. (11) as follows:

$$R_{ij}(u) =$$

$$\begin{cases} \dfrac{K_{i1}(u) - C(u)K_{i2}(u)}{1 - C^2(u)} & (j = 1) \\[2ex] \dfrac{-C(u)K_{i(j-1)}(u) + [1 + C^2(u)]K_{ij}(u) - C(u)K_{i(j+1)}(u)}{1 - C^2(u)} & \\ \hspace{4cm} (j = 2, \ldots, N-1) & \text{for } \|u\| \neq 0. \\[2ex] \dfrac{-C(u)K_{i(N-1)}(u) + K_{iN}(u)}{1 - C^2(u)} & (j = N) \end{cases}$$

$$(13)$$

At the DC where H^{-1} does not exist, Eq. (6) cannot be solved for ϕ. Despite of this, taking the limit to zero of the above obtained results in Eq. (13) identifies the DC component of the filters:

$$\lim_{u \to 0} R_{ij}(u) = \begin{cases} \dfrac{1}{2} - \dfrac{\rho}{4\pi\sigma}\left[\dfrac{\partial}{\partial u}A(0,0) + \dfrac{\partial}{\partial v}A(0,0)\right] & (j = 1) \\[2ex] 0 & (j = 2, \ldots, N-1) \\[2ex] \dfrac{1}{2} + \dfrac{\rho}{4\pi\sigma}\left[\dfrac{\partial}{\partial u}A(0,0) + \dfrac{\partial}{\partial v}A(0,0)\right] & (j = N) \end{cases} \qquad i = 1, 2, \ldots, N. \quad (14)$$

The obtained result shows that all the reconstruction filters do not diverge and are stable.

4 Simulation

4.1 Preparation of Focal Stacks

The test image "Lenna" (a 24-bit color image with 512×512 pixels) were used as the texture of the target scene. The focus slice images $\phi_i(x)$ $(i = 1, 2, \ldots, N)$ were created by equally dividing the test image into N thin rectangles from the left edge. Using them, the input focal stack $\{f_i(x)\}$ was synthetically generated based on the imaging model of Eq. (6). The generated focal stack for $N = 8$ is shown in Fig. 2.

The ground truth focal stacks were also generated based on the imaging model of Eq. (9) for performance evaluation. In this evaluation, to precisely evaluate whether the presented method achieves perfect reconstruction, root mean squared error (RMSE) of the reconstructed focal stack was measured in double precision floating-point number.

4.2 Result

We conducted simulations for the case when the desired aperture function $a(x)$ is a two-dimensional Gaussian function with standard deviation of ρ. In this case, the elements of the matrix K is given by

$$K_{ij}(u) = \exp\left(-2\pi^2|i-j|^2\rho^2\|u\|^2\right) \tag{15}$$

and the DC components of the reconstruction filters are computed as follows:

$$\lim_{u \to 0} R_{ij}(u) = \begin{cases} \dfrac{1}{2} & (j = 1, N) \\ 0 & (j = 2, \dots, N-1) \end{cases} \qquad i = 1, 2, \dots, N. \tag{16}$$

When the maximum blur amount in the input focal stack, $(N-1)\sigma$, was fixed to 3 pixels, we reconstructed focal stacks with Gaussian aperture of the maximum blur amount of $(N-1)\rho = 6$ pixels. The measured RMSEs for $N = 4, 8, 16, 32$ and 64 are shown in Table 1. All RMSEs are quite small and less than 10^{-9}, indicating that perfect reconstruction can be achieved by the presented method.

For example, the refocused images g_1 and g_8 of the reconstructed focal stack images for $N = 8$ are shown in Fig. 3. It can be seen that both images have shallower depth-of-field (DOF) than those in the input focal stack. Such shortened DOF effect can be perfectly produced through the presented linear filters without estimating depth map.

Table 1. RMSEs of reconstructed refocused focal stack with Gaussian aperture for different numbers of depths in double precision. The maximum blur amounts in the input and the reconstructed focal stacks are $(N-1)\sigma = 3$ and $(N-1)\rho = 6$ pixels, respectively.

Number of focus depths, N	4	8	16	32	64
RMSE $(\times 10^{-10})$	0.0131	0.0160	0.0773	0.198	1.76

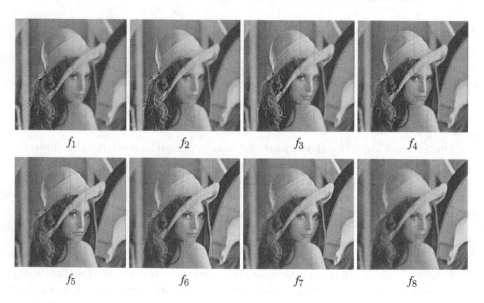

Fig. 2. Synthetically generated focal stack with Cauchy aperture of $7\sigma = 3$ for $N = 8$.

g_1 g_8

Fig. 3. Example images of the reconstructed focal stack with Gaussian aperture of $7\rho = 6$ for $N = 8$.

Extended DOF effect can also be produced by the presented method. When N is fixed to 8, we reconstructed focal stacks with Gaussian aperture of the maximum blur amount (7ρ) of 0, 1 and 2. As shown in Table 2, all measured RMSEs are below 10^{-11} and much smaller than those in the previous result. This result indicates that the derived filters allow perfect reconstruction in theory.

The reconstructed refocused images g_1 for $7\rho = 0, 1$ and 2 are shown in Fig. 4. All the refocused images have extended DOF compared with the image f_1 in the input focal stack. Notably, the image g_1 of $7\rho - 0$ was successfully reconstructed with all regions in-focus.

Table 2. RMSEs of reconstructed refocused focal stack with Gaussian aperture for different blur amounts for $N = 8$ in double precision.

The maximum blur amount, 7ρ	0	1	2
RMSE $(\times 10^{-12})$	1.09	0.763	1.19

$g_1, 7\rho = 0$ $g_1, 7\rho = 1$ $g_1, 7\rho = 2$

Fig. 4. Example g_1 images of the reconstructed focal stack with Gaussian aperture for $N = 8$.

5   Conclusion

This paper presented a focal stack refocusing method that reconstructs a desired focal stack under different aperture from a captured focal stack under Cauchy aperture. For non-occluded Lambertian scenes, this paper theoretically showed that there exists filter bank that perfectly reconstructs the desired focal stack. The stable and perfect reconstruction filters were derived in linear and space-invariant based on a layered scene representation. Results of numerical simulations showed that the root mean squared errors are quite small and less than 10^{-9}, indicating the derived filters allow perfect reconstruction.

In future, we will evaluate the performance of the presented method through experiments using real captured focal stack images. To do this, we design not only aperture but also a lens system to create Cauchy PSF as precisely as possible on the captured focal stacks.

References

1. Ng, R., Levoy, M., Bredif, M., Duval, G., Horowitz, M., Hanrahan, P.: Light field photography with a hand-held plenoptic camera. Stanford University Computer Science Technical Report CSTR 2(11), pp. 1–11 (2005)
2. Levoy, M., Chen, B., Vaish, V., Horowitz, M., Mcdowall, I., Bolas, M.: Synthetic aperture confocal imaging. ACM Trans. Graph. **23**(3), 822–831 (2004)
3. Wu, G., et al.: Light field image processing: an overview. IEEE J. Sel. Topics Signal Process. **11**(7), 926–954 (2017)
4. Aizawa, K., Kodama, K., Kubota, A.: Producing object based special effects by fusing multiple differently focused images. IEEE Trans. Circuits Syst. Video Technol. **10**(2), 323–330 (2000)
5. Alonso, J.R., Fernandez, A., Ferrari, J.A.: Reconstruction of perspective shifts and refocusing of a three-dimensional scene from a multi-focus image stack. Appl. Opt. **55**(9), 2380–2386 (2016)
6. Pendu, M.L., Guillemot, C., Smolic, A.: A Fourier disparity layer representation for light fields. IEEE Trans. Image Process. (2019). https://doi.org/10.1109/TIP.2019.2922099
7. Kodama, K., Kubota, A.: Efficient reconstruction of all-in-focus images through shifted pinholes from multi-focus images for dense light field synthesis and rendering. IEEE Trans. Image Process. **22**(11), 4407–4421 (2013)
8. Levin, A., Durand, F.: Linear view synthesis using a dimensionality gap light field prior. In: Proceedings of IEEE Conference on Computer Vision and Pattern Recognition, pp. 1831–1838 (2010)
9. Perez, F., Perez, A., Rodriguez, M., Magdaleno, E.: Lightfield recovery from its focal stack. Math. Imaging Vis. **56**, 573–590 (2016)
10. Trujillo-Sevilla, J.M., et al.: Restoring integral images from focal stacks using compressed sensing techniques. Disp. Technol. **12**, 701–706 (2016)
11. Kubota, A., Ito, A., Kodama, K.: Deriving synthetic filter bank for perfect reconstruction of light field from its focal stack. In: Proceedings of 2018 Asia-Pacific Signal and Information Processing Association Annual Summit and Conference (2018). https://doi.org/10.23919/APSIPA.2018.8659454
12. Ng, R.: Fourier slice photography. ACM Trans. Graph. **24**(3), 735–744 (2005)

13. Kubota, A.: Synthesis filter bank and pupil function for perfect reconstruction of all-in-focus image from focal stack. In: Proceedings of SPIE International Conference on Quality Control by Artificial Vision, p. 103380A (2017)
14. Kubota, A., Aizawa, K.: Arbitrary view and focus image generation: rendering object-based shifting and focussing effect by linear filtering. In: Proceedings of International Conference on Image Processing, pp. I-489–I-492 (2002)
15. Kubota, A., Aizawa, K.: Reconstructing arbitrarily focused images from two differently focused images using linear filters. IEEE Trans. Image Proces. **14**(11), 1848–1859 (2005)
16. Alonso, J.R.: Fourier domain post-acquisition aperture reshaping from a multi-focus stack. Appl. Opt. **56**, D60–D65 (2017)
17. Schechner, Y.Y., Kiryati, N., Basri, R.: Separation of transparent layers using focus. Int. J. Comput. Vis, **39**(1), 25–39 (2000)
18. Grenander, U., Szego, G.: Toeplitz Forms and Their Applications. University of California Press, Berkeley (1958)
19. Kotz, S., Nadarajah, S.: Multivariate t Distributions and Their Applications. Cambridge University Press, Cambridge (2004)

Skeleton-Based Labanotation Generation Using Multi-model Aggregation

Ningwei Xie[✉], Zhenjiang Miao, and Jiaji Wang

School of Computer and Information Technology, Institute of Information Science,
Beijing Jiaotong University, Haidian District, Beijing, China
{18120323,zjmiao,12112069}@bjtu.edu.cn

Abstract. Labanotation is a well-known notation system for effective dance recording and archiving. Using computer technology to generate Labanotation automatically is a challenging but meaningful task, while existing methods cannot fully utilize spatial characteristics of human motion and distinguish subtle differences between similar human movements. In this paper, we propose a method based on multi-model aggregation for Labanotation generation. Firstly, two types of feature are extracted, the joint feature and the Lie group feature, which reinforce the representation of human motion data. Secondly, a two-branch network architecture based on Long Short-Term Memory (LSTM) network and LieNet is introduced to conduct effective human movement recognition. LSTM is capable to model long-term dependencies in temporal domain, and LieNet is a powerful network for spatial analysis based on Lie group structure. Within the architecture, the joint feature and the Lie group feature are fed into LSTM model and LieNet model respectively for training. Furthermore, we utilize score fusion methods to fuse the output class scores of the two branches, which performs better than any of the single models, due to complementarity between LSTM and LieNet. In addition, skip connection is applied in the structure of LieNet, which simplifies the training procedure and improves the convergence behavior. Evaluations on standard motion capture dataset demonstrate the effectiveness of proposed method and its superiority compared with previous works.

Keywords: Labanotation · Motion capture data · LSTM · Lie group feature · Class score fusion

1 Introduction

Traditional dance is a precious intangible cultural asset, which however is gradually disappearing with the development of modern society. The protection of these performing arts has become an urgent task. Labanotation is a widely used symbolic system designed for effectively recording and archiving human movements [1]. Similar to the music scores for recording songs, Labanotation provides intuitive and convenient symbolic representation of human motion, that facilitates the process of movement recording, choreography and dance practicing [2]. Hence, Labanotation plays an important role in the protection of traditional dance. Manually drawing down the notation by observing

© Springer Nature Switzerland AG 2020
S. Palaiahnakote et al. (Eds.): ACPR 2019, LNCS 12047, pp. 554–565, 2020.
https://doi.org/10.1007/978-3-030-41299-9_43

human movements is a complicated and inefficient task. Even through some previous works, for example Laban Writer [3], LED&LINTEL [4] and Labanatory [5], simplified this process in aid of computer, the intervention of professional practitioners is still required. Thus, the research of automatic Labanotation generation is of great significance.

In this field, the common approach is first acquiring human motion data by motion capturing techniques, and then generating Labanotation automatically by computer algorithms [6]. Some attempts mainly focused on spatial analysis of human movements. Hachimura and Nakamura [7] proposed a spatial analysis method for generating upper limb movement notations. The features they utilized were merely extracted from static arm poses, while important lower limb movements are ignored. The methods put forward by [8, 9] mapped dancing motions to Labanotation according to a series of predefined rules, which lacks robustness and flexibility. However simply analyzing spatial characteristics of motion sequences is not sufficient to recognize complex human movements.

Other works introduced methods based on temporal modelling of motion sequences. A template matching method introduced by Zhou et al. [10] matched unclassified movement segments with movement templates prestored in the database using dynamic time warping (DTW), in order to find similar movement type. Owing to the diversity of height and weight between different dancers, the method failed to achieve promising accuracy. Li and Miao [11] presented a method using the hidden Markov models (HMMs). However, the time complexity of training and testing increases with the number of movement categories. Moreover, HMMs are not conducive for modeling long-term dependence in temporal domain. To addressed this problem, Zhang and Miao [12] firstly applied neural networks, Extreme-Learning Machine (ELM) and Long Short-Term Memory (LSTM) in the field of Labanotation generation, and made a great progress in accuracy. However, most temporal modelling methods [11, 12, 18] simply convert the motion capture data to body joint positions in the world coordinate system and extracted features in the Euclidean space. This feature extracting implementation causes a loss of information about joint rotation, so that these methods fail to capture subtle differences between similar human movements.

To solve these problems, we combine temporal modelling with more elaborate spatial analysis. In the field of skeleton-based action recognition, there are some studies based on manifold learning theories, providing a new direction for spatial analysis. [15] presented a novel neural network architecture LieNet, to learn better Lie group feature for action recognition. Inspired by [15], we propose a new modeling method using LSTM and LieNet to recognize and classify movement segments to their corresponding Laban symbols. We extract two types of features, the joint feature and the Lie group feature to fully capture spatial information underlying in raw motion capture data. The joint feature is obtained by computing the relative positions of joints, while the Lie group feature represents the rotation between joints and the rotation between bones in Riemannian space. Then the two types of features are fed into LSTM and LieNet respectively. Class score fusion technique is later applied on the SoftMax layers of LSTM and LieNet, which aggregates the two models in the output level. In this way, we integrate the long-term temporal dependence modeling of LSTM with the powerful spatial analysis of

LieNet, which leads to a better recognition performance. Besides, we add skip connection [16] in LieNet structure, so that the modified network (which we call residual LieNet) has a simpler training procedure and steadier convergence behavior. In summary, our contributions are listed as follows:

- A new multi-model aggregation architecture using class score fusion is built to combine temporal modelling of LSTM with elaborate spatial analysis of LieNet.
- We explore the method for Lie group representation of motion capture data, which captures the information of joint rotation. We apply LieNet as a branch of the proposed architecture, and implement the skip connection in LieNet structure to improve its convergence behavior in training while maintaining its recognition performance.
- The proposed multi-model method performs better than any of the single models, and achieve higher accuracy compared with previous methods in Labanotation generation.

2   Labanotation

Labanotation is a standardized notation system for reliable human motion analysis and recording. It was originally designed by Rudolf von Laban [2] for dance profession in the early 20th century, but now it has been applied to arbitrary human motion. An example of a Labanotation score with three pages is shown in the right side of Fig. 1. The left side of Fig. 1 shows the staff of Labanotation scores, where each of the 9 column represents a human body part. The vertical line in the middle is defined as the body center which divides the body into the left and right sides. The columns drawn in bold are used to record the most important part in Labanotation, the support (the transference of weights) and leg movements.

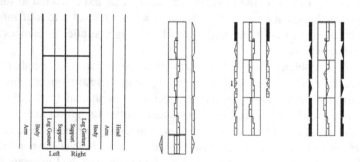

Fig. 1. The staff and example of a Labanotation score [11].

Another composition of the Labanotation score is the Laban symbols, consisting of horizontal direction symbols and vertical level symbols (see Fig. 2). There are nine directions defined in the horizontal plane: place (or centre), forward, left, right, backward, left-forward, left-backward, right-forward, right-backward, represented by different shapes. The three level symbols, low, middle and high, are expressed by three

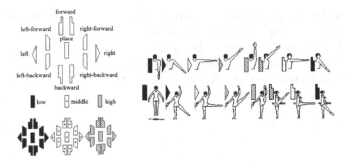

Fig. 2. Human gestures and Laban symbols [11].

kinds of shadings to indicate the vertical dimension. In total, there are 27 types of Laban symbols and their corresponding relationship with human gestures is shown in Fig. 2.

In this paper, we focus on the analysis and the recognition of lower limb movements and transference of body weight, which are essential for the generation of Labanotation.

3 Proposed Method

As shown in Fig. 3, the proposed architecture consists of three parts: the feature extraction, the movement recognition using two neural network branches (LSTM and residual LieNet), and the class score fusion which generates the final recognition results. The recognition results conduct the generation of Laban symbols.

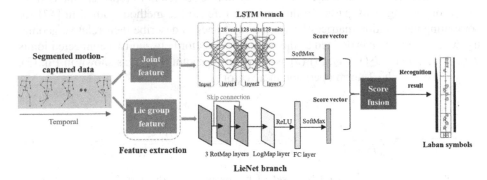

Fig. 3. Two-branch network architecture for Labanotation generation.

3.1 Feature Extraction

In the motion data which is captured by the OptiTrack device [19] and stored in BVH (Bio-vision hierarchical) files [20], the 3D human skeleton model is described as a tree structure with 26 joints. The BVH file records the rotation of each joint relative to its parent joint as Euler angles in each frame of the motion segment, as well as the initial position offset of each joint to its parent joint.

Joint Feature. Firstly, we utilize the relative joint position feature (the joint feature) proposed in [12] as the input of LSTM branch. Suppose that in a body part B, the set of world coordinates of the interesting joints is $J = \{J_1, J_2, \ldots, J_N\}$, where N is the number of joints. By computing the relative position vector $\vec{g}_i(t)$ between every two adjacent joints, we obtain the resulting joint feature vector $G(t) = \{\vec{g}_1(t), \vec{g}_2(t), \ldots, \vec{g}_{N-1}(t)\}$ at the time instance t, defined in Eq. (1).

$$\vec{g}_i(t) = J_i(t) - J_j(t) = [(x_{J_{i,t}} - x_{J_{j,t}}), (y_{J_{i,t}} - y_{J_{j,t}}), (z_{J_{i,t}} - z_{J_{j,t}})],$$
$$i = 1, \ldots, N, j = i - 1 \tag{1}$$

Lie Group Feature. Secondly, we explore a Lie group representation of the motion data, to obtain appropriate Lie group features for LieNet. The set of 3×3 rotation matrices in $\mathbb{R}^3$ forms the special orthogonal group SO_3 which is a matrix Lie group. Each sample can be considered as a curve on the Lie group $SO_3 \times \ldots \times SO_3$. Firstly, we compute the rotation matrices between connected joints in each frame. Suppose that the Euler angles of a joint J_0 are (r, p, y). The corresponding rotation matrix R_J is computed in Eq. (2).

$$R_J = R_z(-r)R_x(-p)R_y(-y)$$
$$= \begin{pmatrix} \cos r \cos y - \sin r \sin p \sin y & -\sin r \cos p & \cos r \sin y + \sin r \sin p \cos y \\ \sin r \cos y + \cos r \sin p \sin y & \cos r \cos p & \sin r \sin y - \cos r \sin p \cos y \\ -\cos p \sin y & \sin p & \cos p \cos y \end{pmatrix}$$
$$\tag{2}$$

The rotation matrices between joints may not be sufficient to represent the coordination of moving body parts. To this end, we refer to the method studied in [17] for computing the rotation matrices between body bones to describe their relative geometry, where the bone e is defined as the normalized vector between two connected joints $J_1(x_1, y_1, z_1)$ and $J_2(x_2, y_2, z_2)$ (J_1 is the precursor joint of J_2), as shown in Eq. (3). $\|\bullet\|_2$ represents the 2-norm operation of a vector.

$$e = \frac{J_2 - J_1}{\|J_2 - J_1\|_2} = \frac{[(x_2 - x_1), (y_2 - y_1), (z_2 - z)]}{\sqrt{(x_2 - x_1)^2 + (y_2 - y_1)^2 + (z_2 - z_1)^2}} \tag{3}$$

Let $E = \{e_1, e_2, \ldots, e_M\}$ be a set of bones in the body part B, where M is the number of bones. Considering any pair of bones e_i and e_j in E, their relative geometry is described by the rotation matrix $R_{i,j}$. We build the local coordinate system of e_j by defining its starting joint as origin and its direction as the positive direction of x-axis. Then e_i is translated into the local system with its starting joint coinciding with the origin. With the process (see Fig. 4), we consequently obtain the two transformed vectors $\hat{e}_i$ and $\hat{e}_j$. Specially, the axis-angle representation (ω, θ) for the rotation matrix $R_{i,j}$ is calculated in Eqs. (4) and (5).

$$\omega = \frac{\hat{e}_i \otimes \hat{e}_j}{\|\hat{e}_i \otimes \hat{e}_j\|} \tag{4}$$

$$\theta = \arccos(\hat{e}_i \cdot \hat{e}_j) \tag{5}$$

The axis-angle representation can be easily converted to $R_{i,j}$ using Rodrigues' rotation formula.

As a result, for a lower limb (a leg) $B = (J, E)$, the resulting Lie group feature vector at the time instance t, denoted as $R(t)$ is obtained by stacking the two kinds of rotation matrices, as shown in the Eq. (6).

$$R(t) = (R_1(t), R_2(t), \ldots, R_N(t), R_{1,2}(t), \ldots, R_{M-1,M}(t)) \tag{6}$$

The total number of rotation s is $K = N + C_M^2$ (C_M^2 is the combination formula).

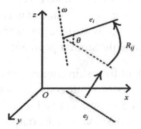

Fig. 4. Process of computing rotation matrices between bones.

3.2 Movement Recognition Using Neural Networks

LSTM and LieNet. LSTM is an advanced neural network structure for modeling long-term dependencies and is widely applied to time-series data.

In our architecture, the LSTM branch is a three-layer LSTM network followed by a SoftMax output layer, where each LSTM layer contain 128 neurons. The LSTM network can efficiently learn useful information from the input joint feature and the SoftMax layer outputs the classification score, that is the probability distribution of each movement class.

LieNet [15] is a novel neural network structure specially designed for 3D human action recognition based on Lie group features. Similar to convolutional networks (ConvNets), the LieNet is equipped with fully connected convolution-like layers named RotMap, and accordingly the pooling layers named RotPooling. The functions of the RotMap layer $f_r^{(k)}$ and the LogMap layer $f_l^{(k)}$ are expressed in Eqs. (7) and (8) respectively.

$$\begin{aligned} &f_r^{(k)}((R_1^{k-1}, R_2^{k-1}, \ldots, R_K^{k-1}); W_1^k, W_2^k, \ldots, W_K^k) \\ &= (W_1^k R_1^{k-1}, W_2^k R_2^{k-1}, \ldots, W_K^k R_K^{k-1}) = (R_1^k, R_2^k, \ldots, R_K^k) \end{aligned} \tag{7}$$

$$\begin{aligned} &f_l^{(k)}((R_1^{k-1}, R_2^{k-1}, \ldots, R_K^{k-1})) \\ &= (\log(R_1^{k-1}), \log(R_2^{k-1}), \ldots, \log(R_K^{k-1})) \end{aligned} \tag{8}$$

$(R_1^{k-1}, R_2^{k-1}, \ldots, R_K^{k-1})$ is the input Lie group feature in the k-th layer. $W_i^k \in \mathbb{R}^3$ is a transformation matrix. Both the feature and the weight spaces correspond to a Lie group $SO_3 \times \ldots \times SO_3$. The calculation of the logarithm map $\log(\cdot)$ is explicated in [15].

In our network structure, the LieNet branch is consist of three RotMap layers, a LogMap layer, and regular layers including a rectified linear unit (ReLU) layer, a fully-connected (FC) layer and a SoftMax output layer. Considering that the Lie group feature extracted from lower limb data is much less complex than that from the whole skeleton, we don't use the RotPooling layer. Specially, the dimensionality of the weight in the FC layer is $c \times d_{k-1}$, where c is the class number and d_{k-1} is the dimensionality of the LogMap layer's output.

Applying the Skip Connection in LieNet. The optimization procedure of LieNet is time-consuming and unstable when we add more layers or evaluate it on more complex tasks. The Residual network (ResNet) is an advanced network structure which improves the training procedure of very deep convolutional networks. Thus, we apply the idea of residual learning in LieNet to solve the similar problem.

In ResNets, an identity block utilizes skip connections and identity mappings allowing the activations or the gradients to be directly propagated between distant layers. Figure 5 shows the basic structure of an identity block, which use skip connection to perform element-wised addition of the input x to the output of the second layer $y = F(x)$. $F(x)$ is the function that the block previously needs to learn, shown in Eq. (9). Let W_1 and W_2 be the weight matrices of the two layers respectively, and $\sigma(\cdot)$ denote the activation function ReLU.

$$y = F(x) = W_2\sigma(W_1x) \tag{9}$$

After applying the skip connection, the output of the second layer becomes $H(x)$ in Eq. (10).

$$y = H(x) = F(x) + x = W_2\sigma(W_1x) + x \tag{10}$$

So, the objective function to learn becomes the residual $F(x) := H(x) - x$.

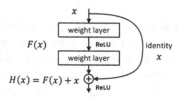

Fig. 5. The basic structure of an identity block.

These two expressions have the same learning effect; however, the residuals are much easier to be optimized by SDG algorithm. The residual learning can gain more computation capacity without sacrificing the accuracy, and solve the gradient vanishing problem in deep neural networks.

Similarly, we equip LieNet with skip connection (see Fig. 6), which starts from the input of the second RotMap layer to its output. Therefore, the new input of the third RotMap layer $\left(R_1^{2'}, R_2^{2'}, \ldots, R_K^{2'} \right)$ is shown in Eq. (11).

$$\left(R_1^{2'}, R_2^{2'}, \ldots, R_K^{2'} \right) = \left(R_1^1 R_1^2, R_2^1 R_2^2, \ldots, R_K^1 R_K^2 \right) \tag{11}$$

Different from regular identity blocks, we replace the addition operation with the matrix multiplication, which guarantees the validation of rotation matrices.

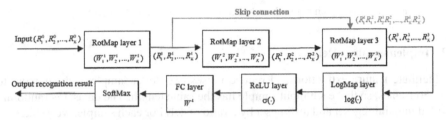

Fig. 6. The structure of residual LieNet with skip connection.

3.3 Class Score Fusion

To take full advantages of the two models (LSTM and LieNet) fed with different features and further improve the recognition accuracy, we evaluate two commonly-used fusion methods for the SoftMax scores, fully-connected fusion and multiply fusion. Fully-connected fusion method, illustrated in Fig. 7, trains a single fully-connected layer behind the two branches to learn the final class label y'.

Multiply fusion [13], expressed in Eq. (12), performs an element-wised multiplication of s_1, s_2 and then finds the index of the maximum score to be y'. Let $\circ$ denote the element-wised multiplication between vectors, $\arg\max(\cdot)$ denote the function to find the index of the maximum score.

$$y' = \underset{index}{\arg\max}(s_1 \circ s_2)) \tag{12}$$

4 Experiments

4.1 Datasets

We use a standard motion capture dataset [18] to evaluate our method. The dataset contains 1610 segmented lower limb movements of 80500 frames in total, stored separately in BVH files. The sampling rate of motion-capture data is 150 frames per second, each segment includes a single dance movement and lasts from 90 to 200 frames. According to their correspondent relationship with Laban horizontal symbols, the segments are classified into 8 categories, each category contains about 200 segments.

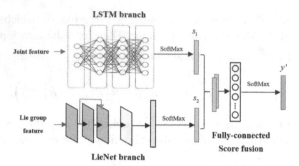

Fig. 7. Two-branch network architecture with fully-connected score fusion.

4.2 Implementation Details

To facilitate feature extraction and reduce noises, we down-sample each segment to $T = 50$ frames, so that each input sample has the same length. The dataset is randomly divided into training set and testing set by a ratio of 4:1. For each sample, we extract the joint feature $G(t)$ and the Lie group feature $R(t)(t \in T)$, corresponding to a label y.

The LSTM branch is fed with the joint feature and trained to minimize the cross-entropy loss function by Adam optimizer. The learning rate is initially set to 0.001, with a decay rate equal to 0.005 and the batch size is 30. To avoid over-fitting, the Dropout rate of the 3 LSTM layers is set to 0.2. As for the LieNet branch fed with the Lie group feature, the optimizing strategy is a modified Stochastic gradient descent (SGD) algorithm, which can compute Riemannian gradients in the procedure of backpropagation for the RotMap layers. The learning rate is fixed to 0.005, and the batch size is 16. The transformation (weight) matrices of the RotMap layers are randomly initialized, with a decay of 0.0005.

4.3 Results Analysis

In the standard dataset, each movement segment records a dance movement which is symbolized by a certain Laban symbol. In order to generate correct Laban symbols, we need to accurately recognize and classify the segments to their corresponding categories. Therefore, the criterion of method performance evaluation is the average recognition accuracy in different body parts (which are represented by supporting columns in Labanotation).

Compare to State-of-the-Arts. We compare the performance of proposed method with other existing methods, including two short-term temporal modelling methods using DTW [10], HMM [11], and a long-term temporal modelling method using LSTM [12]. Specially, [12] utilized the joint feature mentioned in above and trained a two-layer LSTM model with a SoftMax output layer to perform movement recognition. As shown in Table 1, the proposed method performs better than the state-of-the-arts, and contributes to a more accurate Labanotation generation. This result demonstrates the obvious advantage of multi-model aggregation method, which supplements the sequential model LSTM with more elaborate spatial analysis of LieNet and improves the overall performance.

Ablation Study. To evaluate the effectiveness of each part in the proposed architecture, we conduct an ablation study (see Table 2).

Firstly, we examine the effectiveness of the skip connection in the residual LieNet. On one hand, the convergence behavior of conventional LieNet and residual LieNet is compared. We train the two models for 150 epochs with same Lie group features and training hyper-parameters, and draw the curves of their objective value as a function of the iteration number, which are shown in Fig. 8. The curves of conventional LieNet on the train and validation sets are colored red and orange respectively, while the curves of residual LieNet on the same train and validation sets are colored blue and grey. According to the result, residual LieNet can converge smoothly to a stable solution, however the conventional LieNet needs to run more epochs to achieve the same status. On the other hand, the performance of the two is also compared in Table 2. Residual LieNet gains a higher accuracy than conventional LieNet. Therefore, residual LieNet can simplify the training procedure and also gives better recognizing results.

Secondly, we evaluate the performance of the complete architecture. The performance of two branches, the LSTM model and the residual LieNet model, are evaluated to prove their effectiveness as a part of the proposed architecture. Then we compare the complete architecture with two different score fusion methods, fully-connected fusion and multiple fusion. The multiply score fusion obtains higher accuracy than any of the single branch, which demonstrates that the segments which the LSTM model fails to recognize are different from those misclassified by residual LieNet. It means that to some

Table 1. Performance comparison of proposed method and previous methods.

Method	Accuracy (Left leg)	Accuracy (Right leg)	Average
DTW [10]	83.90	80.23	82.07
HMM [11]	92.07	92.55	92.31
LSTM [12]	96.26	97.81	97.04
Proposed method	**98.12**	**98.49**	**98.31**

Table 2. Ablation study of the proposed method.

Method	Accuracy (Left leg)	Accuracy (Right leg)	Average
Conventional LieNet	97.29	97.54	97.42
Residual LieNet	97.31	97.73	97.52
LSTM (3 layers)	96.26	97.81	97.04
Fully-connected fusion	97.31	97.81	97.56
Multiply fusion	**98.12**	**98.49**	**98.31**

extent, the modelling capacity of LSTM and residual LieNet is complementary. However, the fully-connected fusion has almost the same performance as the single branch. Fed with score vectors of two trained models which generate very low loss values, the single fully-connected layer is easy to train overfitted [14]. Thus, the multiple fusion is chosen to aggregate the two branches.

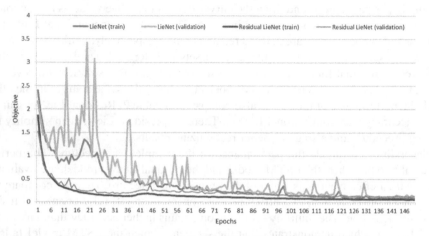

Fig. 8. The convergence behavior of LieNet and residual LieNet.

5   Conclusion

We introduce a method using multi-model aggregation for automatic Labanotation generation of motion capture data. The proposed method combines the long-term temporal modelling with more elaborate spatial analysis. We extract two types of feature which fully represent the spatial characteristics of human movement. Besides the joint feature, a Lie group representation of motion capture data is explored to capture the information about joint rotation. We introduce a two-branch network architecture using LSTM and LieNet with class score fusion technique to conduct movement recognition. The architecture combines the long-term temporal modeling of LSTM on joint feature with effective spatial analysis of LieNet on Lie group feature. To solve the time-consuming and unstable training procedure of LieNet, we implant skip connection in its structure, which improves the convergence behavior. Experiments are put forward to prove that our method outperforms other existing works, due to its efficient feature extraction and the well-designed multi-model architecture.

Acknowledgement. This work is supported by the NSFC 61672089, 61273274, 61572064, and National Key Technology R&D Program of China 2012BAH01F03.

References

1. Loke, L., Larssen, A.T., Robertson, T.: Labanotation for design of movement-based interaction. In: Australasian Conference on Interactive Entertainment (2005)
2. Guest, A.H.: Labanotation: The System of Analyzing and Recording Movement. Psychology Press, London (2014)
3. Venable, L.: Archives of the dance: Labanwriter: there had to be a better way. Dance Res. J. Soc. Dance Res. **9**(2), 76–88 (1991)
4. Edward, F.: LED&LINTEL: a windows mini-editor and interpreter for Labanotation [EB/OL]. http://donhe.topcities.com/pubs/led.heml. Accessed 16 Nov 2016
5. Misi, G.: Labanatory [EB/OL]. http://labanatory.com/eng/software.html. Accessed 16 Nov 2016
6. Hachimura, K.: Digital archiving of dance by using motion-capture technology. In: New Directions in Digital Humanities for Japanese Arts and Cultures, pp. 167–182. Nakanishiya Publishing, Kyoto (2008)
7. Hachimura, K., Nakamura, M.: Method of generating coded description of human body motion from motion-captured data. In: 10th IEEE International Workshop on Robot & Human Interactive Communication, pp. 122–127. IEEE (2001)
8. Guo, H., Miao, Z., Zhu, F., Zhang, G., Li, S.: Automatic labanotation generation based on human motion capture data. In: Li, S., Liu, C., Wang, Y. (eds.) CCPR 2014. CCIS, vol. 483, pp. 426–435. Springer, Heidelberg (2014). https://doi.org/10.1007/978-3-662-45646-0_44
9. Choensawat, W., Nakamura, M., Hachimura, K.: GenLaban: a tool for generating labanotation from motion capture data. Multimedia Tools Appl. **74**(23), 10823–10846 (2015)
10. Zhou, Z., Miao, Z., Wang, J.: A system for automatic generation of Labanotation from motion capture data. In: 13th IEEE International Conference of Signal Process, pp. 1031–1034. IEEE (2016)
11. Li, M., Miao, Z.: Automatic Labanotation generation from motion-captured data based on hidden Markov models. In: 4th Asia Conference of Pattern Recognition (2017)
12. Zhang, X., Miao, Z., Zhang, Q., Wang, J.: Skeleton-based automatic generation of Labanotation with neural networks. J. Electron. Imaging **28**(2), 23–26 (2019)
13. Li, C., Wang, P., Wang, S., Hou, Y., Li, W.: Skeleton-based action recognition using LSTM and CNN. In: IEEE International Conference on Multimedia & Expo Workshops (2017)
14. Zhang, S., et al.: Fusing geometric features for skeleton-based action recognition using multilayer LSTM networks. IEEE Trans. Multimedia **20**(9), 2330–2343 (2018)
15. Huang, Z., Wan, C., Probst, T., Van Gool, L.: Deep learning on lie groups for skeleton-based action recognition. In: Computer Vision & Pattern Recognition (2016)
16. He, K., Zhang, X., Ren, S., Sun, J.: Identity mappings in deep residual networks. In: Leibe, B., Matas, J., Sebe, N., Welling, M. (eds.) ECCV 2016, Part IV. LNCS, vol. 9908, pp. 630–645. Springer, Cham (2016). https://doi.org/10.1007/978-3-319-46493-0_38
17. Vemulapalli, R., Chellappa, R.: Rolling rotations for recognizing human actions from 3D skeletal data. In: The IEEE Conference on Computer Vision and Pattern Recognition (2016)
18. Wang, J., Miao, Z.: A method of automatically generating Labanotation from human motion capture data. In: International Conference on Pattern Recognition, pp. 845–859 (2018)
19. NaturalPoint Corporation: OptiTrack Documentation Center [EB/OL]. http://wiki.optitrack.com/. Accessed 16 Nov 2016
20. Meredith, M., Maddock, S.: Motion capture file formats explained. Department of Computer Science, University of Sheffield (2001)

Genetic Programming-Based Simultaneous Feature Selection and Imputation for Symbolic Regression with Incomplete Data

Baligh Al-Helali[✉], Qi Chen[✉], Bing Xue[✉], and Mengjie Zhang[✉]

School of Engineering and Computer Science, Victoria University of Wellington,
PO Box 600, Wellington 6400, New Zealand
{baligh.al-helali,Qi.Chen,Bing.Xue,Mengjie.Zhang}@ecs.vuw.ac.nz

Abstract. Symbolic regression via genetic programming has been used successfully for empirical modeling from given data sets. However, real-world data sets might contain missing values. Although there are different approaches to dealing with incomplete data sets for classification, symbolic regression with missing values has been rarely investigated. Similarly, only a few studies have been conducted on feature selection for symbolic regression, but none of them addresses the incompleteness issue. In this work, a genetic programming-based method for simultaneous imputation and feature selection is developed. This method selects the predictive features for the incomplete features whilst constructing their imputation models. Such models are designed to be suitable for data sets with mixed numerical and categorical features. The performance of the proposed method is compared with state-of-the-art widely used imputation methods from three aspects: the imputation accuracy, the feature selection effectiveness, and the symbolic regression performance.

Keywords: Symbolic regression · Genetic programming · Incomplete data · Imputation · Feature selection

1 Introduction

Genetic programming (GP) is one of a collection of techniques inspired by Darwinian evolution and known as evolutionary computation (EC). GP automatically generates computer programs for performing a user-defined task [11]. GP can be utilized as a classifier and as a regressor based on the aimed task. One of the typical applications of GP is symbolic regression (SR). SR aims to discover mathematical models that express a dependent feature/variable in terms of a number of independent features/variables from a given data set [11]. SR has many advantages over traditional regression methods such as getting regression models without prior assumptions [3]. Therefore, SR has a broad range of applications in many real-world areas [7].

S. Palaiahnakote et al. (Eds.): ACPR 2019, LNCS 12047, pp. 566–579, 2020.
https://doi.org/10.1007/978-3-030-41299-9_44

Unfortunately, real-world data sets might contain missing values that are difficult to be handled properly by many learning methods. The missing values are categorized into three main types: missing at random (MAR), missing completely at random (MCAR), and missing not at random (MNAR) [8]. There are several approaches to dealing with incomplete data sets. The first one is to delete incomplete instances/features and perform learning using only the complete data portion [10]. Another approach is to use methods that can learn directly from incomplete data such as decision trees [14]. The imputation approach is to estimate missing values and to learn using a complete data set with the imputed ones [8]. Imputation methods are classified as single imputation and multiple imputation [8]. The research on dealing with missing values has focused on the classification task. The investigation of performing traditional regression with missing values has been considered in several studies such as in [12], but SR with missing values has been rarely investigated [1].

GP-based imputation (GPI) has been adopted for classification with missing values in [15–17]. Each feature having missing values is considered as a target variable using other features are used as predictive variables. The presented studies are aimed to address the classification task. A method formed by combining GPI and KNN to handle the missing values for SR is proposed in [1]. This method is called GP-KNN and the main idea is that, instead of using all instances to build the GP imputation models for features having missing values, only k nearest instances are used to build such imputers for the missing values. These methods have the advantage of not requiring any presumptions. However, they build regression models regardless of the data type of the incomplete feature, which reduces their effectiveness on data sets with categorical features. Moreover, the reduction of the number of selected features is not considered.

Feature selection is the process of choosing a useful subset from a given set of features [21]. Feature selection has been shown to be useful for improving the performance of the learning process, but it is a challenging process [20]. EC techniques have been used successfully for feature selection to improve the performance of different learning tasks such as clustering and classification [21]. Due to its natural selection ability, GP is one of the EC techniques that has attained more investigations for feature selection in recently published studies [4,13,19]. For SR, a feature selection method for improving the generalisation ability of GP is proposed in [6]. In [2], artificial bee colony programming is proposed for SR with feature selection. In their work, the missing data are removed. However, to the best of our knowledge, no study has been conducted to investigate feature selection for SR on incomplete data.

In this work, we present a GP-based method for simultaneous feature selection and imputation. This method is applied for SR with incomplete data. The intended contributions of this study are:

1. A GP-based method that can construct imputation models for each incomplete feature and select features as predictors in the imputation model simultaneously.

2. GP imputation models that can deal with both numerical and categorical features.
3. A multiple imputation process that utilizes several GP models to impute the missing values.
4. An experimental analysis of the proposed method compared with state-of-the-art imputation methods on real-world data sets.

2   The Proposed Method

GP-based imputation (GPI) can be considered as a feature selection process by nature. The features involved in a constructed GP imputation model represent a set of selected features. However, the main goal in GPI is to provide better imputation performance. For each incomplete feature, the fitness function is designed to minimize the imputation error, which represents the accuracy of predicting the missing values in the incomplete features.

In this work, reducing the number of selected features is also explicitly considered when constructing GP imputation models. This is achieved by developing a GP-based imputation and feature selection method (GPIFS). In this method, the imputation error and the number of selected features are minimized simultaneously during the evolutionary process of GP models (Sect. 2.1). This process results in several GP imputation models for each incomplete feature. These models are then used to estimate the missing values in a GPIFS imputation process (Sect. 2.2) and to select a feature subset by a GPIFS feature selection process (Sect. 2.3).

2.1   The GPIFS Approach

Assuming that the same imputation performance could be achieved by a smaller feature subset, the selected feature set using GPI can still be shrunk as the basic fitness function of considering imputation performance only does not intend to minimise the number of used features. To further reduce the number of features, a modified fitness function is used in GPIFS as shown in Eq. (1).

$$Fitness = \alpha * Imp\_Error + (1 - \alpha) * \frac{\#Selected\_Features}{\#All\_Features} * Imp\_Error \quad (1)$$

where Imp_Err is the imputation error obtained by a GP imputation model, $\alpha \in [0,1]$ is a constant value, $\#Selected\_Features$ represents the number of features that appear in the constructed GP model, and $\#All\_Features$ is the number of all available features.

The fitness function considers both the number of features and the imputation error. When combining them into a single fitness function, α is used to balance their relative importance with respect to each other. The larger α, the more importance is enforced for the imputation performance over the reduction of the number of features. Since the imputation accuracy is supposed to be more important than the number of features, α is set to be 0.8.

The pseudo-code of the GPIFS process is shown in Algorithm 1. The input is an incomplete training data set, D. The output is a set of GP imputation models, G, that contains N GP imputation models for each incomplete feature. Denote $G_{F,g}$ as the GP imputer for the feature F obtained by taking the best GP model of the gth GP run. Having more than one model reduces the uncertainty when estimating the missing values using one model similar to using the feedback from several experts rather than depending only on one. In each GP model, the data set is reformed to make the incomplete feature as the prediction target variable and the other features are used as a terminal set in the modeling process. Although, any feature except F can be a terminal variable including the incomplete ones, only complete instances are used in the evaluation process.

Algorithm 1: GPIFS training

Input : D: an incomplete training data set.
Output: G: a set of GP-imputation models.
1 Let $G = \phi$;
2 **foreach** *Incomplete feature F* **do**
3   Create a copy of the data considering F as an imputation target variable and the other features as input variables (predictors);
4   Build N GP-based models for F, $\{G_{F,g}\}_{g=1}^{N}$, each minimizes the fitness function in Eq. (1);
5   Append the constructed GP models $\{G_{F,g}\}_{g=1}^{N}$ to the set G;
6 **end**

To make the imputation method more effective when imputing data sets with mixed types (categorical and numerical features), the GPIFS model can be a classifier or a regressor depending on the data type of the feature to be imputed. The imputation error part in the fitness function is then computed accordingly. In either case, the incomplete feature is the target of the modeling process. GPIFS regression models are constructed if the feature type is numerical while the GPIFS models are classifiers if the feature is categorical.

GPIFS Regression Models: For imputing a numerical feature, F, GP is used to construct regression models where the other features are involved in the terminal set as predictors with F as a prediction target. The imputation error of GPIFS models is the relative square error (RSE) shown in Eq. (2).

$$RSE = \frac{\sum_{i=1}^{m}(y_i - t_i)^2}{\sum_{i=1}^{m}(t_i - \bar{t})^2} \tag{2}$$

where m is the number of instances, y_i is the predicted value of the ith instance, t_i is the desired value of the ith instance, and $\bar{t}$ is the average of the desired values t_i, $i = 1, 2, 3..., m$.

GPIFS Classification Models: When the data type of the incomplete feature is categorical, GP classification is adopted to produce the imputation models. Although it has a similar evolutionary process, GP-based classifiers differ from GPIFS regressors in the imputation error. For the categorical features, the imputation error is the classification error rate shown in Eq. (3).

$$Error\_rate = \frac{\sum_{j=1}^{m}(I_{c_i=\hat{c}_i})}{m} \tag{3}$$

where m is the number of instances, $\hat{c}_i$ is the predicted class label of the ith instance, c_i is the desired class label of the ith instance, and $I_{c_i=\hat{c}_i}$ is a misclassification indicator defined as below (Eq.(4)).

$$I_{c_i=\hat{c}_i} = \begin{cases} 0, & \text{if } c_i = \hat{c}_i, \\ 1, & \text{otherwise.} \end{cases} \tag{4}$$

As the output value of GP individual is numerical, the GP classifier requires to determine the corresponding class label. For this purpose, a rule is used to translate the GP program to a class label. If there are C distinct categorical values in an incomplete feature, they are considered as C classes in a classification task. The classes are assigned C intervals along the range of the numerical output value of the GP model, val, using the class determination rule defined as in Eq. (5) [22]:

$$class(val) = \begin{cases} class_1 : & : val \leq T_1, \\ class_2 : & : T_1 < val \leq T_2, \\ \dots \\ class_j : & : T_{j-1} < val \leq T_j, \\ \dots \\ class_C : & \text{otherwise,} \end{cases} \tag{5}$$

where $T_1, T_2, ..., T_C$ are predefined class boundaries over the range of val.

2.2   GPIFS for Imputation

The models obtained by GPIFS construction process are used to impute incomplete instances using a GPIFS imputation process given in Algorithm 2. This process outputs a complete instance.

During the training process, if the feature F has missing values in the training data, then GPIFS models are constructed. In this case, for a given instance, I, the missing value at the feature F, $I[F]$, is imputed using the mean of the N GPIFS predictions if F is numerical or their mode is used if F is categorical. On the other hand, if there are no such constructed models, i.e. F has no missing values in the training set, the missing value $I[F]$ is simply estimated using the mean (mode) of the available training values of the corresponding numerical (categorical) feature.

Algorithm 2: GPIFS imputation

Input : I: an incomplete instance.

         G: a set of trained GPIFS imputation models.

Output: Complete instance

1 **foreach** *Missing value in I (I[F])* **do**

2   **if** *There are trained GPIFS models for F in G* **then**

3   **if** *F is categorical* **then**

4   Impute $I[F]$ using the mode of the predictions of $G_{F,g}$ classifiers, i.e. $I[F] \leftarrow mode\{G_{F,g}(I)\}_{g=1}^{N}$

5   **else**

6   Impute $I[F]$ using the mean of the predictions of $G_{F,g}$ regressors, i.e. $I[F] \leftarrow mean\{G_{F,g}(I)\}_{g=1}^{N}$

7   **end**

8   **else**

9   Use simple imputation for $I[F]$ using a simple method (the mean for numerical features and the mode for the categorical ones.)

10   **end**

11 **end**

Suppose we have a data set with four features where F_2 and F_3 have missing values. If we set $N = 3$, then we get three GPIFS models for each incomplete feature. Let I be an input instance with a missing value at the feature F_2, the constructed three models, $\{G_{F_2,1}(I), G_{F_2,2}(I), G_{F_2,3}(I)\}$ are then applied by taking the other features as predictors to obtain predictions for the missing value. If F_2 is numerical, the mean of the obtained predictions is used to fill in the missing value, otherwise, the mode is used.

2.3 GPIFS Feature Selection

For feature selection, the features selected by the constructed GPIFS models are combined. The proposed process of feature selection is given in Algorithm 3. For each feature F with its N GPIFS imputation models, $\{G_{F,g}\}_{g=1}^{N}$, let $SF_{F,g}$ refer to the set of features involved in the gth model, $G_{F,g}$. The selected features for imputing the feature F are then the ones that appear in every GPIFS model, i.e. the intersection of the features selected by all models (Eq. (6)).

$$SF_F = \bigcap_{g=1}^{N} SF_{F,g}. \qquad (6)$$

For example, given a data set with four features where F_2 has missing values, let the three constructed GPIFS imputation models for feature F_2 be as shown in Fig. 1. We have that $SF_{F_2,1} = \{F_1, F_3\}$, $SF_{F_2,2} = \{F_1, F_3\}$, and $SF_{F_2,3} = \{F_1, F_3, F_4\}$. The selected feature set by the GFIFS selection algorithm is the intersection of these sets, i.e. $SF_{F_2} = SF_{F_2,1} \bigcap SF_{F_2,2} \bigcap SF_{F_2,3} = \{F_1, F_3\}$.

Algorithm 3: GPIFS feature selection

    Input : G: a set of trained GPIFS imputation models.
    Output: SF: a set of selected features.
1 **foreach** *Feature F* **do**
2 | **if** *there are trained GPIFS models for F in G* **then**
3 | | $SF_F = \phi$;
4 | | **for** $g = 1$ *to* N **do**
5 | | | Let $SF_{F,g}$ be the set of features involved in the imputer $G_{F,g}$;
6 | | | $SF_F = SF_F \cap SF_{F,g}$;
7 | | **end**
8 | **else**
9 | | Set SF_F to be all other features but F.
10 | **end**
11 | Append the selected features for F, SF_F to the pool of selected features SF;
12 **end**

Combining the features obtained by several models gives more confidence that the selected predictors are those that necessarily contribute to the imputation of the incomplete feature. They are the survival predictors in every imputation model despite the feature reduction pressure. The selected predictors are associated with their corresponding incomplete features. This obtained association knowledge is fed to an imputation method (e.g. regression-based imputation) to be utilized to get better performance.

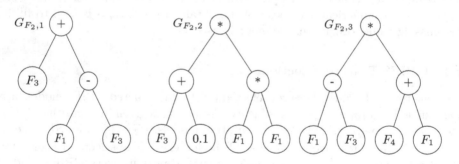

Fig. 1. Three GP-based imputation models for the feature F_2.

3   Experimental Setup

Five regression data sets are used for evaluating the proposed method. The number of features in these data sets varies from 33 to 252. Table 1 shows the information of the used data sets. More details can be found in the data repository OpenML [18]. For each data set, a train/test sets pair is produced by a random split of the 70:30 ratio. For each pair, different MAR missingness probabilities are imposed to generate incomplete data sets. The considered missingness probabilities are 10%, 20%, and 30% of the instances, on 10% of the features, each repeated 30 times resulting in 90 incomplete train/test pairs for each data set.

Table 1. Statistics of the used data sets

Data set	# Instances	# Features
Bank32nh (Bank)	8192	32
kc1-numeric (Kc1)	145	94
MIP-2016-regression (MIP)	1090	147
Mtp	4450	202
yprop_4_1 (Yprop)	8885	251

Table 2 shows the parameter settings in GP for both GPIFS models and SR models. These parameters are determined empirically. Since GP is a stochastic method, for each experiment, 30 independent GP runs are performed under the GP framework provided by distributed evolutionary algorithms in python (DEAP) [9].

For the purpose of comparisons, some popular regression-based imputation methods are considered as benchmark methods. These methods are linear regression (LR), polynomial regression (PR), random forest (RF), and classification and regression trees (CART). These methods are implemented using a powerful and flexible imputation approach which is multivariate imputation by chained equations (MICE) [5]. MICE is based on chained equations that iteratively generates an imputation model for each feature using other features as predictors. The comparisons are based on the statistical significance pairwise Wilcoxon test with the significance level of 0.05.

Table 2. The settings of GP parameters

Parameter	Value
Generations	100
Population size	1024
Crossover rate	0.9
Mutation rate	0.1
Elitism	Top-5 individuals
Selection method	Tournament
Tournament size	7
Maximum depth	17
Initialization	Ramped-half and half
Function set	$+, -, *$, protected %
Terminal set	Features and constants $\in U(-1, 1)$

4   Results and Analysis

4.1   Imputation Performance

To evaluate the proposed imputation process (Algorithm 2), it is compared with different imputation methods on different data sets as shown in Fig. 2. The imputation error is computed using RSE (Eq. (2)) as the difference between the original complete data and the imputed data. For each missingness probability, the plots are drawn using the mean imputation errors over 30 copies of synthetically generated incomplete (test) data sets.

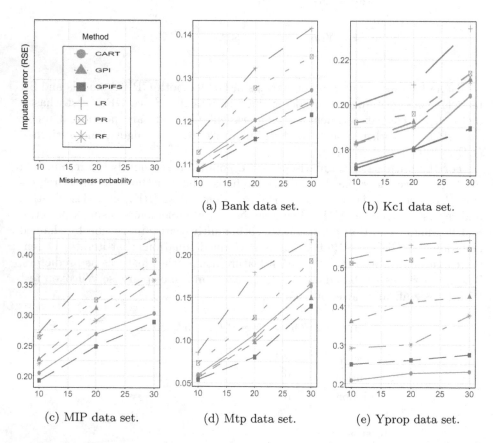

(a) Bank data set.   (b) Kc1 data set.

(c) MIP data set.   (d) Mtp data set.   (e) Yprop data set.

Fig. 2. The imputation results of using different methods.

The results show that GPIFS has the lowest imputation errors on the data sets Bank, Kc1, MIP, and Mtp. Such results might be own to the use of several imputation models to impute each missing value. Data types of the features seem to be related to the performance of different methods. Some methods work by regressing the incomplete features regardless of their data types. The limitation of this approach occurs when dealing with mixed data sets, i.e. data with both categorical and numerical features. We considered each feature with less than ten distinct values as a categorical feature and we got 4/32, 22/94, 41/147, 13/202, and 205/251 categorical features in the data sets Bank, Kc1, MIP, Mtp, and Yprop, respectively.

It can be noticed that LR, PR, and GPI have the poorest performance on data sets with a high number of categorical features such as Kc1, MIP, and Yprop. On the other hand, CART achieved the best performance on Yprop. This can be due to the high ratio of categorical features in this data set, where CART deals with categorical data effectively. The results on this data set have a high variance between the errors of using different methods. GPIFS is better than most benchmark methods on this data set.

Overall, LR is the worst on all data sets as it requires the presumption of a linear relationship between features. GPI has a good performance on data with more numerical features and CART is more suitable for categorical features. However, GPIFS shows a stable and consistently good performance with different levels of mixed data types on different data sets.

4.2   SR Performance

Figure 3 shows the comparisons between the proposed imputation method and the benchmark imputation methods with respect to the SR performance. Each imputation method is applied to the synthetic incomplete data sets to produce

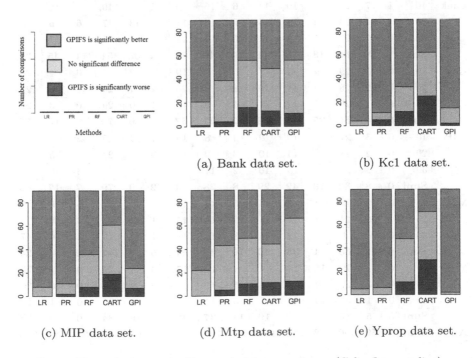

(a) Bank data set.   (b) Kc1 data set.

(c) MIP data set.   (d) Mtp data set.   (e) Yprop data set.

Fig. 3. SR performance significance total comparisons. (Color figure online)

imputed complete data sets. The imputed data sets are then used to perform 30 independent SR runs and the significant difference between the obtained results is measured using Wilcoxon test. In this figure, the green (red) color refers to the number of cases in which GPIFS is significantly better (worse) and the yellow color to the non-significant difference cases.

It can be seen that GPIFS leads to better SR results comparing to the other methods although it emphasizes reducing the number of used features. This can be noticed on all data sets. The only exception is when compared to CART on data sets with large numbers of categorical features such as Yprop. However, also on Yprop, GPIFS significantly outperforms all other methods. In total, the use of GPIFS won 1408, drew 638, and lost 204 out of the 2250 pairs of comparisons against the use of other methods. The detailed results are shown in Table 3. The column "+" ("−") refers to the number of cases in which GPIFS outperforms (is outperformed by) the use of the corresponding method, whereas "=" means there is no significant difference. Column "M" is for the missingness probabilities. The shown results indicate that by reducing the irrelevant features, SR has better predictability than the use of the full feature set.

Table 3. SR performance significance detailed test results of GPIFS vs. benchmark methods.

Data	M	LR			CART			RF			PR			GPI		
		−	=	+	−	=	+	−	=	+	−	=	+	−	=	+
Bank	10	1	7	22	6	10	14	6	15	9	1	10	19	2	12	16
	20	0	6	24	2	12	16	5	14	11	0	13	17	6	16	8
	30	0	7	23	5	14	11	5	11	14	3	12	15	3	17	10
	Sum	**1**	**20**	**69**	**13**	**36**	**41**	**16**	**40**	**34**	**4**	**35**	**51**	**11**	**45**	**34**
Kc1	10	0	1	29	9	14	7	5	6	19	2	2	26	0	4	26
	20	0	2	28	9	11	10	3	7	20	2	4	24	2	6	22
	30	0	1	29	7	12	11	4	8	18	1	0	29	0	3	27
	Sum	**0**	**4**	**86**	**25**	**37**	**28**	**12**	**21**	**57**	**5**	**6**	**79**	**2**	**13**	**75**
MIP	10	0	3	27	6	15	9	3	8	19	1	4	25	2	6	22
	20	0	2	28	6	14	10	3	9	18	0	3	27	2	5	23
	30	0	3	27	7	13	10	2	11	17	1	2	27	3	6	21
	Sum	**0**	**8**	**82**	**19**	**42**	**29**	**8**	**28**	**54**	**2**	**9**	**79**	**7**	**17**	**66**
Mtp	10	0	12	18	4	10	16	5	12	13	3	10	17	5	16	9
	20	0	6	24	4	11	15	2	13	15	1	14	15	4	18	8
	30	0	4	26	3	12	15	3	14	13	1	14	15	3	20	7
	Sum	**0**	**22**	**68**	**11**	**33**	**46**	**10**	**39**	**41**	**5**	**38**	**47**	**12**	**54**	**24**
Yprop	10	0	1	29	11	15	4	1	12	17	0	3	27	0	1	29
	20	0	3	27	9	13	8	5	11	14	0	2	28	0	0	30
	30	0	1	29	10	13	7	5	14	11	0	1	29	0	1	29
	Sum	**0**	**5**	**85**	**30**	**41**	**19**	**11**	**37**	**42**	**0**	**6**	**84**	**0**	**2**	**88**
All	**Sum**	**1**	**59**	**390**	**98**	**189**	**163**	**57**	**165**	**228**	**16**	**94**	**340**	**32**	**131**	**287**

Notably, GPIFS has an advance over the performance of GPI. Such improvement is mainly achieved by two modifications; multiple model imputation and classification-based imputation. Instead of using one GP model in GPI, GPIFS imputation combines N models for estimating the missing values. Moreover, GPI works by regressing the incomplete features regardless of their data types which limits its applicability when imputing categorical features.

4.3   Feature Selection Performance

To examine the effectiveness of the GPIFS selection process (Algorithm 3), the imputation errors of using different imputation methods with all features and with the selected features are compared. Experiments similar to those shown in Sect. 4.1 are conducted but this time GPIFS selected features for each incomplete feature are fed as predictors to the imputation methods.

The average imputation (test) errors over 30 copies of each data set per missingness probability are given in Table 4. The table presents the imputation performance with and without feature selection and the average number of selected features. Column "FS" refers to the feature subset selected by the proposed GPIFS method and "Full" is the use of all features (without feature selection). Column "T" refers to the pair-wise Wilcoxon significance test of the difference between the results of using the corresponding method with and without feature selection. The symbol "+" ("−") means that the use of the method with selected features outperforms (is outperformed by) the use of the same method with all features, whereas "=" refers to no significant difference. The number of the selected features is the average of the averages of the selected features for each feature rounded to an integer number.

Table 4. The imputation performance of different methods with all features and with GPIFS selected features.

Data	#Features			PR			LR			CART			RF		
	M%	Full	FS	Full	FS	T	Full	FS	T	Full	FS	T	Full	FS	T
Bank	10	32	6	0.1127	0.1122	=	0.1171	0.1103	=	0.1106	0.1004	=	0.1107	0.1066	=
	20	32	6	0.1276	0.1208	+	0.1322	0.1134	+	0.1202	0.1056	+	0.1182	0.1101	+
	30	32	4	0.1349	0.1339	=	0.1414	0.1278	+	0.1272	0.1194	+	0.1241	0.1194	+
Kc1	10	94	14	0.1922	0.1901	=	0.1998	0.1758	=	0.1732	0.1563	+	0.1826	0.1764	=
	20	94	9	0.1961	0.1906	=	0.2090	0.1867	+	0.1807	0.1730	−	0.1903	0.1834	+
	30	94	7	0.2141	0.2030	+	0.2340	0.2001	+	0.2042	0.1885	+	0.2106	0.2055	+
MIP	10	144	35	0.2635	0.2512	=	0.2704	0.2340	=	0.2044	0.1980	+	0.2204	0.2133	=
	20	144	26	0.3243	0.3012	+	0.3784	0.3458	+	0.2687	0.2591	=	0.2905	0.2705	+
	30	144	23	0.3904	0.3587	+	0.4256	0.3866	+	0.3029	0.2936	=	0.3578	0.3357	+
Mtp	10	202	32	0.0733	0.0698	=	0.0857	0.0675	=	0.0594	0.0532	+	0.0579	0.0418	+
	20	202	28	0.1266	0.1157	+	0.1797	0.1501	+	0.1067	0.0955	+	0.1009	0.0910	+
	30	202	22	0.1936	0.1567	+	0.2182	0.1842	+	0.1640	0.1435	+	0.1674	0.1434	+
Yprop	10	251	25	0.5103	0.4924	+	0.5226	0.5032	+	0.2077	0.2002	+	0.2912	0.2872	=
	20	251	19	0.5190	0.5025	=	0.5580	0.5439	+	0.2265	0.2012	+	0.3004	0.2890	+
	30	251	20	0.5474	0.5245	+	0.5706	0.5778	=	0.2303	0.2266	+	0.3757	0.3677	=

The results show that in addition to the reduction of the used features, GPIFS feature selection leads to a significant improvement in the imputation accuracy. The feature reduction ratios defer from data to another. The highest reduction ratios are on the Yprop data set with around 92% and the data with the lowest reduction is MIP. Another note is that the higher missingness probability the higher feature reduction ratio.

These results indicate the intra-correlation between the predictor features. Normally, the relationships between predictor features are supposed to be weak as a high correlation between different features implies high redundancy. High reduction of the used features with good imputation performance is not only considered as a success of the GPIFS method but also it can be used as a sign of the potential usability of feature selection for predicting the main task regression target.

5   Conclusions

In this work, a GP-based method for simultaneous feature selection and imputation is proposed. This method is applied to SR on data sets with missing values. The proposed method constructs imputation models for each incomplete feature while selecting its predictors in the imputation models simultaneously. These models are designed to be able to impute both numerical and categorical features. The obtained results show that the proposed feature selection approach is effective not only in reducing the number of the used features, but also in improving the imputation performance and the symbolic regression accuracy.

There are important directions that can be further investigated in the near future. For example, a method to construct efficiently GP-based imputation models along with feature selection in big data. Another one is the applicability of the proposed method on different machine learning tasks such as classification and clustering. These issues need extensive future investigations on high-dimensional and large-scale data sets considering comparisons with existing state-of-the-art methods.

References

1. Al-Helali, B., Chen, Q., Xue, B., Zhang, M.: A hybrid GP-KNN imputation for symbolic regression with missing values. In: Mitrovic, T., Xue, B., Li, X. (eds.) AI 2018. LNCS (LNAI), vol. 11320, pp. 345–357. Springer, Cham (2018). https://doi.org/10.1007/978-3-030-03991-2_33
2. Arslan, S., Ozturk, C.: Multi hive artificial bee colony programming for high dimensional symbolic regression with feature selection. Appl. Soft Comput. **78**, 515–527 (2019)
3. Austel, V., et al.: Globally optimal symbolic regression. arXiv preprint arXiv:1710.10720 (2017)
4. Bhardwaj, H., Sakalle, A., Bhardwaj, A., Tiwari, A., Verma, M.: Breast cancer diagnosis using simultaneous feature selection and classification: a genetic programming approach. In: 2018 IEEE Symposium Series on Computational Intelligence (SSCI), pp. 2186–2192. IEEE (2018)

5. Buuren, S.V., Groothuis-Oudshoorn, K.: MICE: multivariate imputation by chained equations in R. J. Stat. softw. **15**, 1–68 (2010)
6. Chen, Q., Zhang, M., Xue, B.: Feature selection to improve generalization of genetic programming for high-dimensional symbolic regression. IEEE Trans. Evol. Comput. **21**(5), 792–806 (2017)
7. Davidson, J.W., Savic, D.A., Walters, G.A.: Symbolic and numerical regression: experiments and applications. Inf. Sci. **150**(1–2), 95–117 (2003)
8. Donders, A.R.T., Van Der Heijden, G.J., Stijnen, T., Moons, K.G.: A gentle introduction to imputation of missing values. J. Clin. Epidemiol. **59**(10), 1087–1091 (2006)
9. Fortin, F.A., Rainville, F.M.D., Gardner, M.A., Parizeau, M., Gagné, C.: Deap: evolutionary algorithms made easy. J. Mach. Learn. Res. **13**(Jul), 2171–2175 (2012)
10. García-Laencina, P.J., Sancho-Gómez, J.L., Figueiras-Vidal, A.R.: Pattern classification with missing data: a review. Neural Comput. Appl. **19**(2), 263–282 (2010)
11. Koza, J.R.: Genetic Programming II, Automatic Discovery of Reusable Subprograms. MIT Press, Cambridge (1992)
12. Loh, P.L., Wainwright, M.J.: High-dimensional regression with noisy and missing data: provable guarantees with non-convexity. In: Advances in Neural Information Processing Systems, pp. 2726–2734 (2011)
13. Nag, K., Pal, N.R.: Genetic programming for classification and feature selection. In: Bansal, J.C., Singh, P.K., Pal, N.R. (eds.) Evolutionary and Swarm Intelligence Algorithms. SCI, vol. 779, pp. 119–141. Springer, Cham (2019). https://doi.org/10.1007/978-3-319-91341-4_7
14. Quinlan, J.R.: C4.5: Programs for Machine Learning. Elsevier, San Francisco (2014)
15. Tran, C.T., Zhang, M., Andreae, P.: Multiple imputation for missing data using genetic programming. In: Proceedings of the 2015 Annual Conference on Genetic and Evolutionary Computation, pp. 583–590. ACM (2015)
16. Tran, C.T., Zhang, M., Andreae, P.: A genetic programming-based imputation method for classification with missing data. In: Heywood, M.I., McDermott, J., Castelli, M., Costa, E., Sim, K. (eds.) EuroGP 2016. LNCS, vol. 9594, pp. 149–163. Springer, Cham (2016). https://doi.org/10.1007/978-3-319-30668-1_10
17. Tran, C.T., Zhang, M., Andreae, P., Xue, B.: Multiple imputation and genetic programming for classification with incomplete data. In: Proceedings of the Genetic and Evolutionary Computation Conference, pp. 521–528. ACM (2017)
18. Vanschoren, J., Van Rijn, J.N., Bischl, B., Torgo, L.: Openml: networked science in machine learning. ACM SIGKDD Explor. Newsl. **15**(2), 49–60 (2014)
19. Viegas, F., et al.: A genetic programming approach for feature selection in highly dimensional skewed data. Neurocomputing **273**, 554–569 (2018)
20. Xue, B., Zhang, M.: Evolutionary computation for feature manipulation: key challenges and future directions. In: 2016 IEEE Congress on Evolutionary Computation (CEC), pp. 3061–3067. IEEE (2016)
21. Xue, B., Zhang, M.: Evolutionary feature manipulation in data mining/big data. ACM SIGEVOlution **10**(1), 4–11 (2017)
22. Zhang, M., Ciesielski, V.: Genetic programming for multiple class object detection. In: Foo, N. (ed.) AI 1999. LNCS (LNAI), vol. 1747, pp. 180–192. Springer, Heidelberg (1999). https://doi.org/10.1007/3-540-46695-9_16

A Generative Adversarial Network Based Ensemble Technique for Automatic Evaluation of Machine Synthesized Speech

Jaynil Jaiswal(✉) , Ashutosh Chaubey , Sasi Kiran Reddy Bhimavarapu ,
Shashank Kashyap , Puneet Kumar , Balasubramanian Raman ,
and Partha Pratim Roy

Department of Computer Science and Engineering, Indian Institute of Technology
Roorkee, Roorkee 247667, India
{jjaynil,achaubey,breddy,skashyap,pkumar99}@cs.iitr.ac.in,
{balarfcs,proy.fcs}@iitr.ac.in

Abstract. In this paper, we propose a method to automatically compute a speech evaluation metric, Virtual Mean Opinion Score (vMOS) for the speech generated by Text-to-Speech (TTS) models to analyse its human-ness. In contrast to the currently used manual speech evaluation techniques, the proposed method uses an end-to-end neural network to calculate vMOS which is qualitatively similar to manually obtained Mean Opinion Score (MOS). The Generative Adversarial Network (GAN) and a binary classifier have been trained on real natural speech with known MOS. Further, the vMOS has been calculated by averaging the scores obtained by the two networks. In this work, the input to GAN's discriminator is conditioned with the speech generated by off-the-shelf TTS models so as to get closer to the natural speech. It has been shown that the proposed model can be trained with a minimum amount of data as its objective is to generate only the evaluation score and not speech. The proposed method has been tested to evaluate the speech synthesized by state-of-the-art TTS models and it has reported the vMOS of 0.6675, 0.4945 and 0.4890 for Wavenet2, Tacotron and Deepvoice3 respectively while the vMOS for natural speech is 0.6682 on a scale from 0 to 1. These vMOS scores correspond to and are qualitatively explained by their manually calculated MOS scores.

Keywords: Automatic speech evaluation · Text-to-speech · Conditional GAN · Binary classifier · Virtual Mean Opinion Score

---

J. Jaiswal and A. Chaubey—Denotes equal contribution.

Electronic supplementary material The online version of this chapter (https:// doi.org/10.1007/978-3-030-41299-9_45) contains supplementary material, which is available to authorized users.

S. Palaiahnakote et al. (Eds.): ACPR 2019, LNCS 12047, pp. 580–593, 2020.
https://doi.org/10.1007/978-3-030-41299-9_45

1 Introduction

The capability of generative models in generating realistic data examples has given birth to several text-to-speech (TTS) systems [1–3]. They find applications in areas such as - man and machine interaction, aid to visually impaired people, smart devices, personal digital assistants, security and authentication, vehicle control and automation, gaming and animation, etc. [4] However, the task of measuring the quality of the speech produced by TTS systems using traditional speech evaluation methods such as - Mean Opinion Score (MOS), paired-comparison test, etc. [5] is very challenging. It requires measuring the values of acoustic parameters such as - pitch, frequency, amplitude, etc. and getting their ranges corresponding to real natural speech so that the generated fake speech can be evaluated in comparison to these parameters. These methods suffer from the subjective variance caused by human intervention during the evaluation. Therefore, the need to build speech processing systems that are capable of automatically evaluating machine generated speech is increasing at a fast pace [6]. By automating the evaluation process of TTS models, model prototyping and testing can be accelerated significantly. The need for an objective metric builds upon this as it would be consistent, quick and cheaper than variants of MOS.

In this paper, a novel approach to evaluate the human-ness of a given speech sample has been proposed. The primary goal is to be able to automatically evaluate the speech generated by TTS models thereby excluding the need of having speech experts for calculating the MOS metric. The proposed model is an ensemble of a Generative Adversarial Network (GAN) and a binary classifier, resnet-v2-50 which outputs the scores about the quality of the generated speech. Then a 'Virtual MOS' (vMOS) is determined by ensembling the evaluation scores computed by the GAN and the binary classifier. The proposed system has been tested to evaluate the speech synthesized by state-of-the-art TTS models trained on benchmark speech datasets [7–10]. It has been shown that the scores outputted by the proposed model are in-line with the actual MOS scores already available for the TTS models. The correspondence of manual MOS and automatic vMOS scores proves the applicability of the proposed model to automatically evaluate the synthetic speech just as effectively as the manual approaches.

The rest of the paper is organized as follows. Existing techniques in context of speech evaluation have been reviewed in Sect. 2. Section 3 formulates the problem of developing an automatic speech evaluation method comparable to MOS approach in terms of its effectiveness. The proposed methodology has been detailed in Sect. 4. Section 5 presents the implementation details and results for various cases. The broader implications of the achieved results and the scope for future improvements have been concluded in Sect. 6.

2   Related Work

Speech evaluation is typically performed by human experts either manually by assigning opinion scores to speech samples or semi-automatically by formulating and optimizing relevant objective functions. The existing techniques in this field are briefly reviewed in this section.

2.1   Speech Evaluation Metrics

Evaluation of synthetic speech generated by TTS systems such as Deep Voice [1], Tacotron [2], Wavenet [4], etc. is done by natural speech experts. They are made to listen to the output speech and asked for their opinions on its human-ness. The scores assigned by various experts are recorded and the evaluation metrics such as - Mean Opinion Score (MOS), paired-comparison score, etc. are calculated by taking their weighted average [5,6].

2.2   Subjective Methods of Speech Recognition and Evaluation

The evaluation of speech has been carried out in terms of social cues such as - ethnicity, social class, speaker-age, etc. through pair-wise comparision of a speech sample in contrast to another one and in a questionnaire-based manner [11,12]. While these methods were able to provide a legitimate assessment of the output speech, they required human intervention to carry out the assessment process. In the context of speech recognition, Tyagi et. al. [13] used selectively-biased linear discriminant analysis and Chorowski et. al. [14] used Attention-based model. Researchers have been successful to recognize user-specific [15] and multi-lingual speech [16]. Various efforts to improve speech recognition and evaluation have been made such as - modelling and evaluation of the speech duration [17], instrumental measure-based speech enhancement [18], speech evaluation by analysing the correlation between pitch frequencies of input spectrum and processed spectrum [19], etc. The speech processing methods thus developed performed decently to recognize the natural speech and evaluate it manually [6]. However, they were not adequate to automatically evaluate the machine synthesized speech.

2.3   Objective Function Based Speech Evaluation

There are several issues with subjective speech evaluation methods , most prominent of them being the time and cost involved in the process. Another issue is that a separate analysis of speech needs to be done at each stage to check which properties or features are not properly modelled in the TTS system. This is not feasible to do with subjective evaluation methods. To overcome these problems, there have been some attempts at evaluating machine synthesized speech using objective methods. Objective evaluation of speech involves representing the speech evaluator as a function and using the output of this function as a score to rate a speech sample, thus eliminating the need of manual intervention. It helps in developing good quality TTS voice during its initial stages as well. Well-known objective measures, viz., Mel Cepstral Distortion and Dynamic Time Warping distance have been used as the objective measures during the optimization stage for TTS synthesis [20]. Their results concluded the need for further research on objective measures as human speech was too complicated to be evaluated on the basis of simple objective measures mentioned above.

Motivated by the successful use of neural networks as function approximators, we have attempted to model the evaluator as a function using deep neural network. Hereafter, this function is referred as the *scoring function* and its effectiveness for the task of automatic speech evaluation has been demonstrated as advocated by the results. In contrast to the currently used manual techniques of analysing the human-ness of synthetic speech, the proposed method uses an end-to-end neural network to automatically calculate a score qualitatively similar to manually obtained MOS.

3   Problem Formulation

Consider the input text $\hat{S}$ and the speech generated by a TTS system T conditioned on $\hat{S}$ denoted as $X = \hat{G}(\hat{S})$. An automatic speech evaluation system is supposed to produce the score $V(\hat{G}(\hat{S}))$ in a way to fulfil following constraints:

- The generated score $V(\hat{G}(\hat{S}))$ should correspond to the quality of the generated speech X in terms of human-ness and emotion.
- The score $V(\hat{G}(\hat{S}))$ should be comparable to till date used metric for generated speech evaluation, i.e. MOS.

We need to find some mapping from the generated speech space to the vMOS space conditioned on the prior of input text. It has been shown that the vMOS outputted by our method for the speech generated by well-known TTS Models for input sentence $\hat{S}$, corresponds well to the MOS already available for them.

Definition 1. (Anthropomorphic Score)
Anthropomorphic Score is defined to gauge the goodness of a TTS model in terms of synthesizing human-like speech. Normalization of vMOS in order to directly compare it with MOS is not valid as they vary in their absolute scales. MOS ranges from 0 to 5 with value 4.58 for natural speech while vMOS varies from 0 to 1 with value 0.6682 for the natural speech. vMOS value 1 for natural speech corresponds to 4.58 and not 5. In this case, Anthropomorphic Score provides a better comparative framework to evaluate the capability of a TTS model by comparing the values of same scale. Anthropomorphic Score A of a TTS model T is determined as per the following formula.

$$A = \frac{vMOS \ for \ T}{vMOS \ of \ natural \ speech} \tag{1}$$

4   Proposed Methodology

The proposed model uses an ensemble of Generative Adversarial Network (GAN) Discriminator along with a binary classifier, resnet-v2-50 for calculating vMOS. The details of various steps involved in this process are described as follows.

4.1   Pre-processing

The transcripts to generate the corresponding speech are available along-with the training speech data. The speech generated by off-the-shelf TTS models and human voice audio are interpreted in terms of intensity of the signal and then plotted along the log mel scale to obtain images of the spectrogram of size 64 × 64 × 3. These images are now used as the speech samples rather than the audio. The latent vector of dimension 100×1 has been used as the input to the generator (as random Gaussian noise) along with the pre-processed text sequence. The speech generated by the generator is also the image of the waveform with a dimension of 64 × 64 × 3. We use speech (waveforms as images) as input and their corresponding ground truth. Real and fake nature of the speech is denoted by 1 and 0 respectively as the labels to train the classifier.

4.2   Training

A schematic of the proposed model during training is summarized in Fig. 1. The architecture has a discriminator along with a vanilla binary classifier to classify the input speech as real or fake. The reason behind adding a classifier on top of the network is that training the classifier encourages it to automatically learn features that are inherent to human voice. The discriminator network is trained alternatively with real speech and the enhanced generated speech.

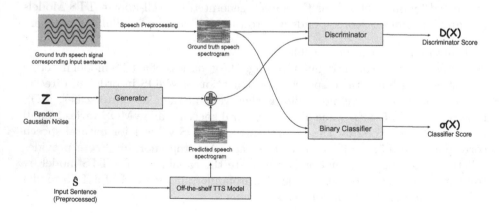

Fig. 1. Overview of the proposed model during training stage

Phase I: GAN Training. Generative Adversarial Networks have been well known for their effectiveness in image generation tasks since their introduction in 2014 [21]. GANs are composed of two deep networks namely the Generator and the Discriminator. Training of GANs takes place as a minmax game between Generator and Discriminator where the Generator tries to minimize the value function $V_f(D, G)$ and the Discriminator tries to maximize it.

$$V_f(D, G) = E_{x \sim P(x)}[log D(x)] + E_{z \sim P(z)}[log(1 - D(G(z)))] \tag{2}$$

where $D(x)$ is the score outputted by the discriminator, which is the probability that the image input to the Discriminator is real and $G(z)$ is the image generated from the Generator given input z from a Gaussian prior.

The proposed method uses a *Conditional* GAN [22] for the training. It uses a Generator-Discriminator model for training the GAN as is used traditionally. During training, the random Gaussian noise **z** input to the Generator is conditioned with the input sentence $\hat{S}$. Then the Generator of the network is removed and only the Discriminator is used to output the discriminator score. Here GANs have been used despite them not being very good at waveform generation because the task is not to generate speech but to evaluate its goodness on a relative scale. To provide a way to calculate a metric very similar in quality to MOS, known generated speeches of previous models and datasets are passed from the proposed Discriminator.

Since the discriminator, in a speech synthesizing GAN discriminates between the generated speech and the real speech, it will itself learn the features which are specific to real speech. So, after the discriminator has been trained, we use it as the scoring function, and the score $D(X)$ produced by it for an input speech X is used as a measure of the human-ness of the input speech. However, the problem with this approach is that the speech output by a Conditional GAN [22] on the input $\hat{S}$, is very weak in quality compared to the natural speech, due to which our discriminator can fool the generator easily which will disable it from learning different features in the spectrogram. To tackle this issue, instead of the generator directly producing the speech spectrogram, our generator outputs the error to the speech spectrogram output by some off-the-shelf TTS model corresponding to input text sentence, so that the corrected spectrogram becomes close to natural speech spectrogram.

Phase II: Binary Classifier Training. The second part of the proposed ensemble model is a binary classifier, resnet-v2-50 that is used to classify whether the input speech spectrogram is real or fake. Binary classifiers learn the features corresponding to the two different classes, i.e. real speech and generated speech in our case. Based on the learned features, it outputs the probability of the input speech belonging to one of the two classes. If we want to calculate the quality of some speech spectrogram, indirectly we want to know what is the likelihood of that speech being real. So, we can directly use this binary classifier as a model which outputs the quality score of some given input speech. To train the binary classifier, generated speech of some TTS models corresponding to some input

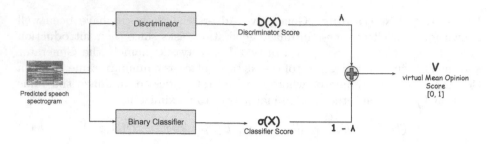

Fig. 2. Overview of the proposed model during inference stage

sentence is labelled as fake (zero) and the natural speech corresponding to same
sentence is considered as real (one). After training the model in such a way, it
will output the quality score corresponding to some speech spectrogram input
to it.

4.3 Inference Through Ensembling

Figure 2 illustrates the inferencing process to compute the vMOS by ensembling
the evaluation scores produced by GAN and the binary classifier. Here $D(\hat{G}(\hat{S}))$
is the score output by the Discriminator for a speech spectrogram $\hat{G}(\hat{S})$ generated by some TTS model corresponding to a sentence $\hat{S}$, and $\sigma(\hat{G}(\hat{S}))$ is the
classifier score outputted by the binary classifier. These two scores are combined as per the hyper-parameter λ to compute the speech quality score as per
Eq. (3). The final vMOS score is a weighted combination of the scores outputted
by the Discriminator and the binary classifier. The computed vMOS scores for
various evaluation-cases have been presented in Sect. 5.6 and the intermediate
calculations have been included in the Supplementary Material.

$$V(\hat{G}(\hat{S})) = \lambda * D(\hat{G}(\hat{S})) + (1 - \lambda) * \sigma(\hat{G}(\hat{S})) \tag{3}$$

5 Implementation and Results

This section discusses the experimental implementation and analyses the results.
The proposed model has been trained with LJSpeech dataset using PyTorch[1].
Four test-cases have been formulated to generate the vMOS evaluation scores
for synthetic speech samples containing single-speaker, multi-speaker, emotional
and gender-specific speech utterances.

5.1 Experimental Set-Up

The model training has been carried out on NVidia Tesla K80 GPU machine with
24 GB RAM and 4992 CUDA cores. Model evaluation has been performed on
Intel(R) Core(TM) i7-7700U, 3.60 GHz, 16 GB RAM CPU machine. The implementation results for all the use-cases are presented in the following sections.

---

[1] https://pytorch.org/.

5.2 TTS Models Evaluated

The following off-the-shelf TTS models have been considered for evaluation.

- **Tacotron** is a sequence-to-sequence architecture. It takes raw text as input and converts it into spectrogram. Then Griffin-Lim algorithm is used to synthesize speech by approximating the spectrogram into waveforms [2].
- **DeepVoice3** is a fully convolutional end-to-end architecture for speech generation from text. It uses alternate vocoders instead of Griffin Lim algorithms such as – WORLD, WaveNet, etc. It is capable of scaling up to 2000 different voices with good quality speech output [3].
- **WaveNet2** is a generative model for raw-audio. It takes acoustic and linguistic features as input and generates human-like voices. It models the waveforms directly using a network trained with real speech recordings [4].

5.3 Datasets

Following datasets have been considered for evaluation using proposed method.

- **LJSpeech dataset** [7] consists of 13,100 samples of sentences spoken by a single-speaker from 7 non-fiction books. It contains short audio clips of single speaker whose length vary from 1 to 10 s.
- **VCTK Dataset** [8] is a multi-speaker data released by CSTR (Centre for Speech Technology Voice Cloning Toolkit) containing speech utterances by native English speakers in various accents.
- **RAVDESS** [9] The Ryerson Audio-Visual Database of Emotional Speech and Song contains audio samples of 24 professional speakers of North American accent. It contains spoken and sung speech utterances including angry, calm, fear, happy, neutral and sad emotions.
- **IEMOCAP** [10] The Interactive Emotional Dyadic Motion Capture database is a multi-speaker database containing scripted and improvised speech samples annotated into emotion categories anger, excited, frustration, happiness, neutral and sadness.

The proposed model has been trained with LJSpeech data and evaluated with the samples of aforementioned datasets. The intention of considering these datasets is to capture variations in terms of spoken or sung utterances, scripted or improvised samples and the samples labelled with gender and emotion information. For LJSpeech and VCTK data, the evaluation is done for the synthetic speech. However, for the emotional datasets RAVDESS and IEMOCAP, the evaluation has been carried out for natural speech only. The aim is to check whether human-ness for the scripted emotional speech is maintained or not.

5.4 Parameter Settings

The generator-discriminator pair has been trained using Adam Optimizer [23] ($\beta_1 = 0.5$) with learning rate of LR = 0.0002. The number of iterations for

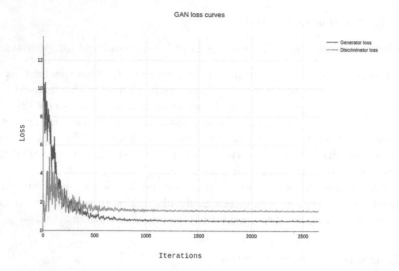

Fig. 3. GAN loss curve during training

the training was set to be 3000. This number was decided after observing the convergence of loss values during the training. As depicted in Fig. 3 the losses started converging after just 1000 iterations. Both the generator and discriminator were trained in a 1:1 ratio of optimization step. A single iteration consists of one discriminator optimization step followed by one generator optimization step. Due to the unavailability of sufficient GPU vRAM, the batch size was kept to 1.

5.5   Model Architecture

GAN. The complete architecture of the generator and the discriminator has been provided in Table 1. The Generator contains 5 transpose convolution layers which upsample the latent z dimension conditioned on the input speech $\hat{S}$, to the speech spectrogram dimension. The Discriminator contains 5 convolution layers which classify the speech spectrogram input to it as real or fake.

Binary Classifier. The binary classifier is a neural network with ResNetv2-50 [24] architecture and has been trained completely separately from the rest of the model. In a separate pipeline to train the binary classifier, we take a pre-trained ResNet model [25] with its weights frozen for first 7 layers. The weights of the rest of the layers are initialized from scratch and the output dimension of the last fully connected layer is kept to 128. The detailed architecture of the Binary classifier has been depicted in the Supplementary Material.

5.6   Results and Discussion

This section presents the results for the four evaluation-cases that are formulated as per the datasets mentioned in Sect. 5.3.

Table 1. Architecture details of the generator and discriminator

Layer (kernel-size)	Stride	No. of filters
Generator		
ConvTranspose2D (4×4)	1	512
BatchNorm	–	–
ReLU	–	–
ConvTranspose2D (4×4)	2, pad $= 1$	256
BatchNorm	–	–
ReLU	–	–
ConvTranspose2D (4×4)	2, pad $= 1$	128
BatchNorm	–	–
ReLU	–	–
ConvTranspose2D (4×4)	2, pad $= 1$	64
BatchNorm	–	–
ReLU	–	–
ConvTranspose2D (4×4)	2, pad $= 1$	128
Tanh	–	–
Discriminator		
Conv2D (4×4)	2, pad $= 1$	64
LeakyReLU $(\alpha = 0.2)$		–
Conv2D (4×4)	2, pad $= 1$	128
BatchNorm	–	–
LeakyReLU $(\alpha = 0.2)$	–	–
Conv2D (4×4)	2, pad $= 1$	256
BatchNorm	–	–
LeakyReLU $(\alpha = 0.2)$	–	–
Conv2D (4×4)	2, pad $= 1$	512
BatchNorm	–	–
LeakyReLU $(\alpha = 0.2)$	–	–
Conv2D (4×4)	2	1
Sigmoid	–	–

Where α the negative slope of LeakyReLU function

Case-1 Non-emotional Single and Multi-speaker Speech Data. The speech synthesized by various off-the-shelf TTS models for the transcripts of single-speaker LJSpeech data and multi-speaker VCTK data have been evaluated in this evaluation-case. The vMOS scores are presented in Table 2. As defined in Eq. (1), the Anthropomorphic scores, denoting the good-ness of TTS models in terms of producing human-like speech from text are presented in Fig. 4.

The vMOS scores are in correspondence with the MOS scores. Tacotron and DeepVoice3 have a MOS of 3.82 and 3.62 suggesting Tacotron to be a slightly better model than DeepVoice3 which is also depicted by the vMOS scores. The vMOS scores for *Natural Speech* reported in Table 2 correspond to the training samples from LJSpeech and VCTK datasets respectively. The difference between them is because of the variations in speech quality while recording human voice.

The Anthropomorphic Score, A for VCTK dataset with WaveNet2 came out to be 1.06. $A > 1$ suggests the quality of synthesized speech to be better than that of natural speech. We analysed the speech outputs of WaveNet2 and found them to be actually clearer and louder than the corresponding data samples.

Table 2. Evaluation of the speech synthesized by TTS models

TTS model	LJSpeech dataset		VCTK dataset	
	MOS	vMOS	MOS	vMOS
DeepVoice3	3.62	0.4863	3.62	0.4917
Tacotron	3.82	0.4966	3.82	0.4923
WaveNet2	4.53	0.6624	4.53	0.6725
Natural Speech	4.58	0.7009	4.58	0.6354

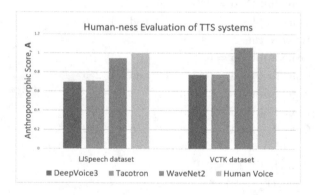

Fig. 4. Anthropomorphic Scores denoting the human-ness of TTS models

Case-2 Emotional Speech Data of Spoken and Sung Samples. Table 3 depicts the vMOS scores for the evaluation of emotion-labelled spoken and sung speech samples from RAVDESS dataset.

Case-3 IEMOCAP Data with Scripted and Improvised Samples. The MOS scores for scripted and improvised speech samples from IEMOCAP speech dataset are shown in Table 4.

Table 3. Evaluation of emotional speech from **RAVDESS** dataset

Emotion class	vMOS (speaking)	vMOS (singing)
Angry	0.6380	0.6372
Calm	0.6497	0.6227
Fear	0.6350	0.6390
Happy	0.6183	0.6352
Neutral	0.6457	0.6210
Sad	0.6290	0.6191

Table 4. Evaluation of emotional speech from **IEMOCAP** dataset

Emotion class	vMOS (scripted)	vMOS (improvised)
Anger	0.6716	0.6669
Excited	0.6685	0.6654
Frustration	0.6729	0.6839
Happiness	0.6771	0.6688
Neutral	0.6724	0.6629
Sadness	0.6733	0.6600

Table 5. Evaluation of gender-specific speech from RAVDESS & IEMOCAP datasets

Gender	vMOS (RAVDESS)	vMOS (IEMOCAP)
Male	0.6232	0.6607
Female	0.6352	0.6715

Case-4 Gender Specific Speech Samples. This evaluation-case evaluates the speech samples for gender classification. The resulting vMOS scores are shown in Table 5. Though the proposed model was trained on single female speaker dataset, it performed equivalently while evaluating the male speaker's voice.

Discussion. The evaluator assigns average vMOS of 0.6682 to an actual natural speech rather than a score of 1. This behaviour has been implicitly learned to be the same as how even MOS of real speech is not 5. We observe that the model producing more human-like speech for each word performs better on our evaluation standards. WaveNet2 performs significantly better than its competition and comes very close to how our evaluator judges natural speech.

The first evaluation-case demonstrates the effectiveness of the proposed model to evaluate the human-ness of the synthetic speech. TTS models capable of generating speech output corresponding to various emotional classes are not yet completely developed [26]. For that reason, we have evaluated natural emotional speech samples in case 2, 3 and 4. The vMOS scores computed for various emotion and gender classes ranged from 0.6183 to 0.6839 which is very close to the average vMOS score of natural speech i.e. 0.6682. Although these score showed negligible differences among themselves, the usability of the proposed model to detect the human-ness of the speech samples involving variations in terms of emotional and gender-specific information has been demonstrated.

6   Conclusion

In this paper, we present an automatic speech evaluation method capable of assessing the speech generated by TTS models as effectively as manual approaches such as - MOS, paired-comparison tests, etc. On evaluating the speech generated by benchmark TTS models, i.e. Wavenet2, Tacotron and Deepvoice3, the proposed method computed the vMOS as 0.6675, 0.4945 and 0.4890 respectively. It was also able to analyse the goodness of a TTS model in terms of average Anthropomorphic Scores of 0.7338, 0.7417 and 1.0017 for DeepVoice3, Tacotron and WaveNet2 respectively where a score of 1 corresponds to completely naturalistic speech synthesis. In spite of being trained on single-speaker speech dataset, the proposed model performed well in evaluating the speech including multi-speaker utterances and emotional variations. It can be improved further by training with multi-speaker and emotional speech datasets. In future, we will focus to improve the proposed model by explicitly training it with multi-speaker and emotional datasets. Another research dimension could be to focus on conditioning the existing TTS systems for emotional speech synthesis by incorporating a new evaluation score, along-with the already proposed score of human-ness, that could measure the correctness of desired emotion in the output speech.

References

1. Arik, S.Ö., et al.: Deep voice: real-time neural text-to-speech. In: Proceedings of the 34th International Conference on Machine Learning, vol. 70, pp. 195–204. JMLR.org (2017)
2. Wang, Y., et al.: Tacotron: towards end-to-end speech synthesis. arXiv preprint arXiv:1703.10135 (2017)
3. Ping, W.: Deep voice 3: scaling text-to-speech with convolutional sequence learning. arXiv preprint arXiv:1710.07654 (2017)
4. van den Oord, A., et al.: Wavenet: a generative model for raw audio. arXiv preprint arXiv:1609.03499 (2016)
5. Salza, P.L., Foti, E., Nebbia, L., Oreglia, M.: MOS and pair comparison combined methods for quality evaluation of text-to-speech systems. Acta Acust. United Acust. **82**(4), 650–656 (1996)
6. Ghate, P., Shirbahadurkar, S.D.: A survey on methods of TTS and various test for evaluating the quality of synthesized speech. Int. J. Dev. Res. **07**, 15236–15239 (2017)
7. Ito, K.: The LJ speech dataset (2017)
8. Veaux, C., Yamagishi, J., MacDonald, K., et al.: CSTR VCTK corpus: English multi-speaker corpus for CSTR voice cloning toolkit. University of Edinburgh, The Centre for Speech Technology Research (CSTR) (2017)

9. Livingstone, S.R., Russo, F.A.: The Ryerson audio-visual database of emotional speech and song (RAVDESS): a dynamic, multimodal set of facial and vocal expressions in North American English. PloS One **13**(5), e0196391 (2018)
10. Busso, C., et al.: IEMOCAP: interactive emotional dyadic motion capture database. Lang. Resour. Eval. **42**(4), 335 (2008)
11. Dwivedi, R.C., et al.: Acoustic parameters of speech: lack of correlation with perceptual and questionnaire-based speech evaluation in patients with oral and oropharyngeal cancer treated with primary surgery. Head Neck **38**(5), 670–676 (2016)
12. Sebastian, R.J., Ryan, E.B.: Speech cues and social evaluation: markers of ethnicity, social class, and age. In: Recent Advances in Language, Communication, and Social Psychology, pp. 112–143. Routledge (2018)
13. Tyagi, V., Ganapathiraju, A., Wyss, F.I.: Method and system for selectively biased linear discriminant analysis in automatic speech recognition systems. US Patent 9,679,556, 13 June 2017
14. Chorowski, J.K., Bahdanau, D., Serdyuk, D., Cho, K., Bengio, Y.: Attention-based models for speech recognition. In: Advances in Neural Information Processing Systems, pp. 577–585 (2015)
15. Gajic, B., Narayanan, S.S., Parthasarathy, S., Rose, R.C., Rosenberg, A.E.: System and method of performing user-specific automatic speech recognition. US Patent 9,058,810, 16 June 2015
16. Renals, S.: Automatic Speech Recognition-ASR Lecture: Multilingual Speech Recognition (2017)
17. Russell, M., Cook, A.: Experimental evaluation of duration modelling techniques for automatic speech recognition. In: ICASSP 1987, IEEE International Conference on Acoustics, Speech, and Signal Processing, vol. 12, pp. 2376–2379. IEEE (1987)
18. Moore, A.H., Parada, P.P., Naylor, P.A.: Speech enhancement for robust automatic speech recognition: evaluation using a baseline system and instrumental measures. Comput. Speech Lang. **46**, 574–584 (2017)
19. Otani, T., Togawa, T., Nakayama, S.: Speech evaluation apparatus and speech evaluation method. US Patent App. 15/703,249, 29 March 2018
20. Sailor, H.B., Patil, H.A.: Fusion of magnitude and phase-based features for objective evaluation of TTS voice. In: The 9th International Symposium on Chinese Spoken Language Processing, pp. 521–525. IEEE (2014)
21. Goodfellow, I., et al.: Generative adversarial nets. In: Advances in Neural Information Processing Systems, pp. 2672–2680 (2014)
22. Mirza, M., Osindero, S.: Conditional generative adversarial nets. arXiv preprint arXiv:1411.1784 (2014)
23. Kingma, D.P., Ba, J.: Adam: a method for stochastic optimization (2014). arxiv:1412.6980. Comment: Published as a Conference Paper at the 3rd International Conference for Learning Representations, San Diego (2015)
24. He, K., Zhang, X., Ren, S., Sun, J.: Deep residual learning for image recognition. CoRR, abs/1512.03385 (2015)
25. He, K., Zhang, X., Ren, S., Sun, J.: Deep residual learning for image recognition. In: Proceedings of the IEEE Conference on Computer Vision and Pattern Recognition, pp. 770–778 (2016)
26. An, S., Ling, Z., Dai, L.: Emotional statistical parametric speech synthesis using LSTM-RNNs. In: 2017 Asia-Pacific Signal and Information Processing Association Annual Summit and Conference (APSIPA ASC), pp. 1613–1616. IEEE (2017)

Large-Scale Font Identification
from Document Images

Subhankar Ghosh[1]([⊠]), Prasun Roy[1], Saumik Bhattacharya[2],
and Umapada Pal[1]

[1] CVPR Unit, Indian Statistical Institute, Kolkata, Kolkata, India
sgcs2005@gmail.com, prasunroy.pr@gmail.com, umapada@isical.ac.in
[2] Indian Institute of Technology, Kharagpur, Kharagpur, India
saumiksweb@gmail.com

Abstract. Identification of font in document images has many applications in modern character recognition systems. Visual font recognition task is a challenging yet popular problem in pattern recognition as many designers try to identify a font for their designs from available images. However, the identification problem is particularly difficult because of the large number of probable fonts that can be used for a particular design. Moreover, there can be multiple fonts in a database which contain similar visual features; therefore, it is difficult to identify the exact font class. In this paper, we explore the font recognition problem for a database of 10,000 fonts using convolutional neural network (CNN) architecture. To the best of our knowledge, no previous approach has explored the font identification problem with these many classes. We performed extensive experiments to quantify our results for synthetic document images as well as natural document images. We have achieved 63.45% top-1 accuracy and 70.76% top-3 accuracy in character level, in addition we also observed 57.18% top-1 accuracy and 62.11% top-3 accuracy in word level even in the presence of rotation and scaling, which demonstrate the effectiveness of the proposed method.

Keywords: Font classification · Document image · Visual font recognition

1 Introduction

Font plays an important role in digital designs. Often designers want to identify and use fonts for their applications from some reference sources. This identification of font from an image or digital design is known as visual font recognition problem (VFR) [10]. VFR has huge application in design editing, image editing, automatic scene text translation, text style transferring etc.. Though, VFR has huge commercial value. manual identification of fonts from image is highly inaccurate due to the similar features of different font types from same font family. Nowadays, several popular online platforms, like Fontspring, WhatTheFont

© Springer Nature Switzerland AG 2020
S. Palaiahnakote et al. (Eds.): ACPR 2019, LNCS 12047, pp. 594–600, 2020.
https://doi.org/10.1007/978-3-030-41299-9_46

THE QUICK BROWN FOX	*Air Atlas*
THE QUICK BROWN FOX	*AlteSchwabacher-DemiBold*
THE QUICK BROWN FOX	*AnAkronism*
THE QUICK BROWN FOX	*Androidi Pisa*
THE QUICK BROWN FOX	*ApolloASM*

THE QUICK BROWN FOX	*TEXTANR*
THE QUICK BROWN FOX	*LABRAT*
THE QUICK BROWN FOX	*KOMIKATR*
THE QUICK BROWN FOX	*AuntBerthaNF*
THE QUICK BROWN FOX	*Avond*

Fig. 1. Some of the sample font types that are used in our database

etc., try to predict the font from an uploaded image. However, these platforms require that the uploaded image resolution is high and it should have proper pre-processing. Even then, the suggested results of these platforms are unsatisfactory in most of the cases. The difficulties in VFR problem is explored by Chen et al. [2] to establish the importance of a robust VFR system. The problem becomes even more challenging for document images as the character in each document typically has small resolution, and the document might contain different degradations that make the identification task difficult. In [4], authors proposed typographical attributes to solve the multifont classification problem for 7 fonts in different sizes ranging from 7 to 18. Zhu et al. [11] considered the classification problem as a texture identification problem for 24 frequently used Chinese fonts and 32 frequently used English fonts. Gabor filter based approach is proposed in [6] to identify font types and font styles from scanned document images. In [9], author used representative stroke templates to identify English and Chinese fonts. In [7], the authors used global texture analysis and support vector machines to classify six common English fonts with different training instances. In [8], the author used eigenimages to find out a font from scanned characters within a database of 2763 English fonts. Presently, some researchers have used deep neural network architectures to identify fonts from images. Chen et al. [2] proposed a scalable solution based on nearest class mean classifier. In [10], authors used domain adaptation technique based on Stacked Convolutional Auto-Encoder to address VFR problem for 2383 English fonts. Though, the present CNN based techniques exhibit promising results, they are tested for limited number even though the number of font types that we use in different real world designs are much larger. Moreover, the algorithms do not consider stylish fonts with ornamentations, and focus more on commonly used fonts. In this paper, we consider a dataset of 10,000 fonts with stylish and common font types. We consider different scales and rotational angle on the dataset to propose a robust solution to the font recognition problem for document images. Rest of this paper is organized as follows. Details regarding our work is presented in Sect. 2. Experimental results are discussed in Sect. 3. Finally, a conclusive discussion and future directions of the work are mentioned in Sect. 4.

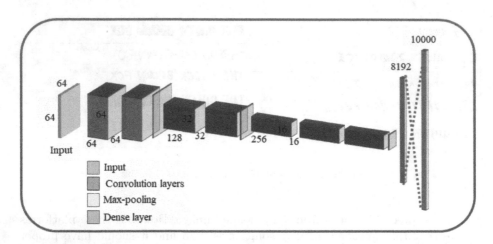

Fig. 2. CNN architecture used for character-level font identification.

2 Proposed Work

The main challenge in a font recognition task is to correctly identify different fonts from same font family as they often exhibit similar font features. To accommodate different fonts with large varieties of style features, we consider the MC-GAN dataset [1] for our font recognition task. The dataset contains 10,000 Latin fonts in gray scale. As shown in the Fig. 1, there are many fonts in the database which contain ornamentations. First we compute a rectangular bounding box around each character which is resized such that the longer dimension of the rectangle becomes 64 pixels. We pad the smaller dimension on both sides accordingly such that the size of each character becomes 64 × 64. We then generate words with character length between 3–8 selecting a font randomly from our font database. To make the database even more exhaustive and close to real world problems, we randomly apply scaling on each generated word with a scaling factor between 0.75 to 1.75. We also apply rotation on the scaled images with a random rotational angle between −10° to +10°. We also vary the inter-character spacing in a word to mimic the resizing effect that we typically observe in real world digital designs. With the data augmentation, our dataset contains 0.2 million training images and 40,000 test images (Fig. 2).

2.1 Character Segmentation

Unlike some of the font recognition algorithms, that process the entire text region as one image, we first try to predict the probable fonts from each character and then apply a decision rule to recognize the font in a document. This strategy is particularly helpful if we have words that are written using two or more fonts.

To extract the characters in a document, first we apply maximally stable extremal regions (MSER) [3] to detect text regions. However, MSER does not

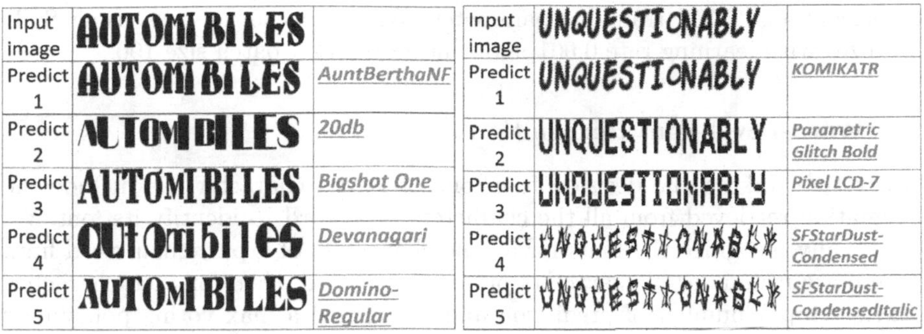

Input image	**AUTOMBILES**		Input image	*UNQUESTIONABLY*	
Predict 1	**AUTOMBILES**	*AuntBerthaNF*	Predict 1	*UNQUESTIONABLY*	*KOMIKATR*
Predict 2	**ALTOMBILES**	*20db*	Predict 2	UNQUESTIONABLY	*Parametric Glitch Bold*
Predict 3	**AUTOMIBILES**	*Bigshot One*	Predict 3	UNQUESTIONABLY	*Pixel LCD-7*
Predict 4	**OUtOmibiles**	*Devanagari*	Predict 4	UNQUESTIONABLY	*SFStarDust-Condensed*
Predict 5	**AUTOMBILES**	*Domino-Regular*	Predict 5	UNQUESTIONABLY	*SFStarDust-CondensedItalic*

Fig. 3. Prediction output of the algorithm. In the first row we have the input document images and in the following rows we have the top-5 predicted fonts.

segment a character completely, especially for the characters with a 'loop', e.g. 'B', 'P', 'R' etc. Thus, we also apply Otsu's thresholding on the image, and multiply the thresholded image with the MSER detected image to get sharper character regions. To detect each character individually, we estimate the slant angle and the inflection points in a word, and use the segmentation points [5] to detect the characters individually.

As most of the CNN architectures take square images as input, we have applied a pre-processing on the segmented characters to generate square input images. We first estimated the minimum rectangular bounding box for a segmented character. Let us assume that for a particular character the dimension of the bounding box is $h_c \times w_c$, where h_c and w_c are the height and width of the bounding rectangle respectively. We rescale the image such that M_c becomes 64 pixel, where $M_c = max(h_c, w_c)$. After that, we zero pad the other dimension on the both side such that we get a square image of dimension 64×64, and the character is placed at the center of the square image.

2.2 Model Architecture

As extraction of font features for such a large number of classes is an extremely challenging work, we use CNN for the classification task. Following the VGG architecture, our CNN has three convolution block. Except the last one, each convolution block contains two convolution layers followed by a max-pooling of 2. The last convolution block has three convolution layer followed by a max-pooling and flattening operation. The layers in the first convolution block contains 64 filters each, whereas the layers in second and third block contain 128 and 256 filters respectively. All the layers use filter size of 3×3 with ReLU activation. After the third convolution block, we flatten the output and add two dense layer with 8192 nodes and 10,000 nodes respectively. The first dense layer has ReLU activation whereas the second dense layer has softmax activation. We use 0.5 dropout in between two dense layers to reduce the chances of overfitting. The model contains 217M trainable parameter. We train the network using SGD optimizer with learning rate 0.001, momentum 0.5 and batch size 100.

2.3   Word-Level Font Recognition

Rather than taking the decision from one observed character, we include the information retrieved from all the characters of a word to identify its font. We individually recognize the fonts for all the characters of a particular word first. If there are n number of characters present in a word, then for each character, we select top k number of prediction and performed a max voting policy over nk number of detected fonts to decide our results. Usually, we fix $k = 3$ for our experiments. If we do not receive any winner for a particular word, we increase k until we have a font that achieves majority.

3   Result

To test the accuracy of the proposed font detection algorithm, we test the framework on 40000 test images for both character-level and font-level analyses. First, for individual character images, we randomly select character out of 0.2M training images for training, and we take characters randomly from the test images for validation. In this setting, we achieved 73.57% validation accuracy without introducing augmentation, and 63.45% validation accuracy with rotation and scaling. Next, we tried to evaluate the generalisation of the font features that the network is learning. So, in the next setting, we randomly select 22 uppercase character and use these characters for training and other 4 uppercase characters were used for validation. This way, the network was evaluated on the character sets that it has never seen. In this case, we achieved 43.23% validation accuracy without augmentation and 38.53% validation accuracy with data augmentation.

To understand the performance of the network with max-voting policy in character level, we test the network on the entire dataset. It turned out that for $k = 3$, the model achieved 51.68% top-1 accuracy and 65.38% top-3 accuracy without augmentation, and 46.59% top-1 accuracy and 62.11% top-3 accuracy with augmentation. It can be observed that the word level classification accuracy is lower than the character level classification accuracy. This is because of the maximum voting policy. Though, there are more samples in the word level task, the correct font should be predicted in most of the cases irrespective of the distortion or the characters present.

In Fig. 3, we have shown two cases where the proposed approach correctly found the document fonts. In this case, we use the character-level classification followed by max-voting policy to detect the final font. On a close observation, it can be realized that all the classified fonts with top-5 accuracy have many similar features with the input document. This proves that the network has learnt to extract important features like character style, ornamentations, stroke width, shadow etc. In Fig. 4, we have shown some of the cases where the proposed approach failed to find the correct font. However, the misclassified fonts still share

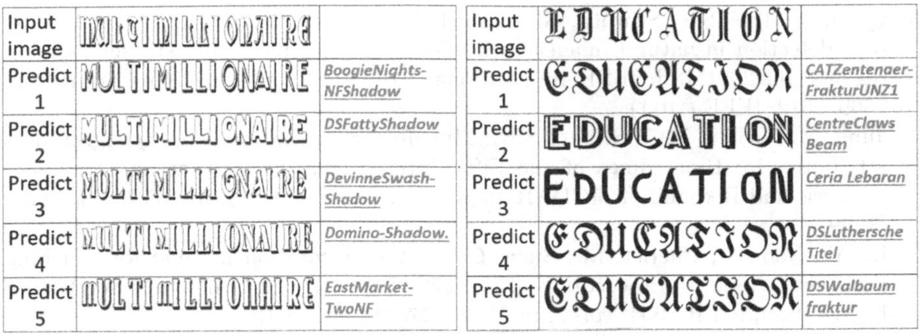

Input image	МULTIMILLIONAIRE		Input image	EDUCATION	
Predict 1	MULTIMILLIONAIRE	BoogieNights-NFShadow	Predict 1	EDUCATION	CATZentenaer-FrakturUNZ1
Predict 2	MULTIMILLIONAIRE	DSFattyShadow	Predict 2	EDUCATION	CentreClaws Beam
Predict 3	MULTIMILLIONAIRE	DevinneSwash-Shadow	Predict 3	EDUCATION	Ceria Lebaran
Predict 4	MULTIMILLIONAIRE	Domino-Shadow.	Predict 4	EDUCATION	DSLuthersche Titel
Predict 5	MULTIMILLIONAIRE	EastMarket-TwoNF	Predict 5	EDUCATION	DSWalbaum fraktur

Fig. 4. Transfer of poses from a particular condition image using proposed algorithm.

some font features with the input document font. We could not compare our algorithm with the existing font recognition algorithm as none of them are designed to handle such a big font database and recognition performances are extremely poor. In real world scenarios, to detect the font from an image, we need to apply text segmentation first. Segmentation is itself a challenging problem. Another problem is that there is no well recognized labelled dataset of document images with known fonts. To compute the performance of the proposed approach, we considered 200 document images with labeled font types, and apply MSER algorithm to extract the text regions [3]. With the MSER segmentation algorithm, the proposed method achieved 41.5% top-1 accuracy and 57% top-3 accuracy.

4   Discussion and Future Scope

Detection of font from images and documents is a challenging task due to several reasons. One of the main reasons is the large number of the font types that we typically use in different documents and design. To the best of our knowledge, our work is the first attempt to recognize font where the database has 10,000 unique font types. We have shown that the proposed network model is robust to data augmentations like scaling and rotations. In future, we are planning to extend this problem for damaged document so that the detected font can directly be used for the digital restoration of archival documents. We are also looking into CNN based sequential modelling to develop more robust decision algorithm to detect font from words.

References

1. Azadi, S., Fisher, M., Kim, V.G., Wang, Z., Shechtman, E., Darrell, T.: Multi-content GAN for few-shot font style transfer. In: Proceedings of the IEEE Conference on Computer Vision and Pattern Recognition, pp. 7564–7573 (2018)
2. Chen, G., et al.: Large-scale visual font recognition. In: Proceedings of the IEEE Conference on Computer Vision and Pattern Recognition, pp. 3598–3605 (2014)

3. Chen, H., Tsai, S.S., Schroth, G., Chen, D.M., Grzeszczuk, R., Girod, B.: Robust text detection in natural images with edge-enhanced maximally stable extremal regions. In: 2011 18th IEEE International Conference on Image Processing, pp. 2609–2612. IEEE (2011)

4. Jung, M.C., Shin, Y.C., Srihari, S.N.: Multifont classification using typographical attributes. In: Proceedings of the Fifth International Conference on Document Analysis and Recognition. ICDAR 1999 (Cat. No. PR00318), pp. 353–356. IEEE (1999)

5. Li, Y., Naoi, S., Cheriet, M., Suen, C.Y.: A segmentation method for touching italic characters. In: Proceedings of the 17th International Conference on Pattern Recognition, ICPR 2004, vol. 2, pp. 594–597. IEEE (2004)

6. Ma, H., Doermann, D.S.: Gabor filter based multi-class classifier for scanned document images. In: ICDAR, vol. 3, p. 968. Citeseer (2003)

7. Ramanathan, R., Soman, K., Thaneshwaran, L., Viknesh, V., Arunkumar, T., Yuvaraj, P.: A novel technique for English font recognition using support vector machines. In: 2009 International Conference on Advances in Recent Technologies in Communication and Computing, pp. 766–769. IEEE (2009)

8. Solli, M., Lenz, R.: FyFont: find-your-font in large font databases. In: Ersbøll, B.K., Pedersen, K.S. (eds.) SCIA 2007. LNCS, vol. 4522, pp. 432–441. Springer, Heidelberg (2007). https://doi.org/10.1007/978-3-540-73040-8_44

9. Sun, H.M.: Multi-linguistic optical font recognition using stroke templates. In: 18th International Conference on Pattern Recognition (ICPR 2006), vol. 2, pp. 889–892. IEEE (2006)

10. Wang, Z., et al.: Deepfont: identify your font from an image. In: Proceedings of the 23rd ACM International Conference on Multimedia, pp. 451–459. ACM (2015)

11. Zhu, Y., Tan, T., Wang, Y.: Font recognition based on global texture analysis. IEEE Trans. Pattern Anal. Mach. Intell. $23(10)$, 1192–1200 (2001)

Aggregating Motion and Attention for Video Object Detection

Ruyi Zhang[(✉)], Zhenjiang Miao, Cong Ma, and Shanshan Hao

Beijing Jiaotong University, No. 3 Shangyuancun, Haidian, Beijing 100044, China
{17120331,zjmiao,13112063,17120308}@bjtu.edu.cn

Abstract. Video object detection plays a vital role in a wide variety of computer vision applications. To deal with challenges such as motion blur, varying viewpoints/poses, and occlusions, we need to solve the temporal association across frames. One of the most typical solutions to maintain frame association is exploiting optical flow between consecutive frames. However, using optical flow alone may lead to poor alignment across frames due to the gap between optical flow and high-level features. In this paper, we propose an Attention-Based Temporal Context module (ABTC) for more accurate frame alignments. We first extract two kinds of features for each frame using the ABTC module and a Flow-Guided Temporal Coherence module (FGTC). Then, the features are integrated and fed to the detection network for the final result. The ABTC and FGTC are complementary to each other and can work together to obtain a higher detection quality. Experiments on the ImageNet VID dataset show that the proposed framework performs favorable against the state-of-the-art methods.

Keywords: Video object detection · Optical flow · Self-attention · End-to-end

1 Introduction

Object detection, aiming at locating and classifying particular objects in an image or throughout an entire video sequence, is a fundamental task in computer vision. In recent years, the development of deep neural networks has contributed a lot to the progress of this task. Yet, many existing object detection methods [1–5] are specially designed for images. Directly applying image-level detecting techniques to the video domain usually fails to get satisfactory performance, since frames tend to be deteriorated by issues such as motion blur, rare poses, and occlusions. Beyond the object detection method for images, temporal information in videos can be exploited to improve the detection performance.

So far, existing video object detection methods [6–12, 17] can be roughly divided into two categories. One category depends on manually-designed post-processing rules [6–9, 12]. These methods first detect each frame independently on a still image detector and then apply hand-crafted rules across the time dimension to refine the final detection results. Generally, the association rules are enforced independently of training. Methods of this type are neither end-to-end nor optimal. By contrast, methods such as FGFA [10]

© Springer Nature Switzerland AG 2020
S. Palaiahnakote et al. (Eds.): ACPR 2019, LNCS 12047, pp. 601–610, 2020.
https://doi.org/10.1007/978-3-030-41299-9_47

and MANet [11] learn to establish temporal consistency by multi-frame aggregation in both training and testing stages. Moreover, they can be trained in an end-to-end manner. In these methods, optical flow is applied to capture temporal information to enhance features in the current frame. However, the optical flow only predicts the displacement of pixels between the original images. Directly applying it to high-level feature may lead to inaccurate spatial correspondences.

To alleviate the issue mentioned above, in this work, we propose an Attention-Based Temporal Context module (ABTC) to enhance frame temporal consistency. This module models pixel-level consistency across frames. Specifically, given features F_t and $F_{t+\tau}$ (or $F_{t-\tau}$) of a reference frame and a neighboring frame, ABTC first computes corresponding weights based on the similarity between any two locations across the two frames. Then, it selectively extracts neighboring spatial information based on the temporal context information. Compared to optical flow-based methods, ABTC can obtain relevant information from high-level features of neighboring frames and bridge the gap between the original frames and high-level features. Finally, the output of the ABTC module is integrated with the output of an optical flow module to form a more effective representation for each frame. This representation is then fed to a detector to get the final detection result.

By incorporating rich temporal information, our model can deal with the challenging issues including appearance changes and occlusions. Extensive experiments conducted on the ImageNet VID dataset demonstrate that our method outperforms state-of-the-art methods in detection accuracy.

2 Related Work

2.1 Object Detection for Still Images

It has been over two decades since the academic community studied object detection. Recently, deep convolutional neural networks have achieved great success on the task of video object detection. Existing object detectors mainly fall into two streams, namely, one-stage methods and two-stage methods. One-stage methods such as YOLO [4] and SSD [5] directly utilize features produced from a feature extraction network to predict class labels and the corresponding locations of objects. On the other hand, two-stage methods such as Fast R-CNN [1], Faster R-CNN [2], and R-FCN [3] need to extract proposals in the first stage and then perform fine-grained object classification and regression based on the proposals. These methods are more flexible for integration and extension. Therefore, we take R-FCN as the basic framework and then extends it for video object detection.

2.2 Object Detection for Videos

Video object detection is increasingly popular in the literature since the introduction of the ImageNet VID dataset. Comparing with images, ample temporal information can be employed to assist object detection in videos. Building relationships in both space and time of objects properties across frames is key to accurate video object detection.

Researchers have designed several video object detectors [6–12, 17] that can be divided into two settings, namely, box-level methods [6–9, 12] and feature-level methods [10, 11, 17]. Seq-NMS [6] links boxes if the IOU of two boxes from consecutive frames is higher than a certain threshold. Then, boxes within the sequences constructed before are rescored and re-ranked through a method named "Seq-NMS". TPN [7] proposes a novel network to generate high-quality tubelet proposals efficiently and exploits LSTM to construct temporal coherence. TCNN [9] utilizes optical flow to propagate bounding boxes from neighbor frames and also adopts a different strategy for tubelet classification and rescoring.

In feature-level methods, FGFA [10] combines the warped features from adjacent frames to enhance the features of the current frame by using optical flow. MANet [11] employs optical flow information for both pixel-level aggregation and instance-level aggregation to incorporate temporal information in an end-to-end manner. However, the optical flow based feature propagation may fail to align the frames in some cases. To alleviate this issue, we introduce a novel module for dense frame matching in feature space.

2.3 Self-attention Mechanism

The self-attention mechanism has become a hot topic in academia and achieved remarkable success in various tasks since the introduction of [13]. The inspiration of attention mechanism comes from how human perception works to recognize objects across videos. Humans focus attention selectively on parts of the visual space to acquire information when and where they needed other than the whole scene. Specifically, a self-attention module computes the response at a position in a sequence (e.g., a sentence) by attending to all positions and taking their weighted mean values of all positions. [14] aligns the source and the target words by applying soft attention to the task of machine learning. The work [18] applies self-attention to the task of scene segmentation. It aims to capture rich context for powerful feature representation. The work [19] introduces a non-local operator based on self-attention mechanism. Experimental results in [19] show that the performance of both video classification and object detection can be improved via the non-local operator. Unlike previous works, we apply self-attention to the task of video object detection and design a module to select temporal context information for better cross-frame alignment.

3   Approach

3.1   Overview

The full architecture of the proposed approach is shown in Fig. 1. Specifically, given a video sequence $X = \{X_1, \ldots, X_i, \ldots\}$, we aim to use the proposed model to generate the detection results $S = \{S_1, \ldots, S_i, \ldots\}$, where S_i is the detection results corresponding to X_i. All frames are fed forward into a convolutional network N_{feat} to extract the intermediate features $F = \{F_i, \ldots, F_i, \ldots\}$. To find an effective representation of current time t, the intermediate features $F_t, F_{t-\tau}$ (or $F_{t+\tau}$) are taken as the input of two modules, i.e., the Flow-Guided Temporal Coherence module (FGTC) as well as the Attention-Base Temporal Context module (ABTC) to capture temporal information. FGTC exploits optical flow information to propagate features and maintain the temporal coherence across frames. Meanwhile, ABTC learns to selectively extract temporal context information based on self-attention mechanism. The combination of two kinds of features obtained from the two modules is taken as the input of the detection network described in MANet [11]. Both of the two modules allow us to better handle the challenging deteriorated frames commonly seen in videos. We will describe the two modules in detail in the following sections.

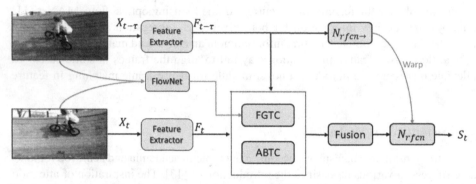

Fig. 1. The full architecture of the proposed approach. Only a neighboring frame $X_{t-\tau}$ and the reference frame X_t are shown for simplicity. Intermediate feature maps $F_{t-\tau}$ and F_t are extracted from a convolutional network and fed to both the FGTC and ABTC modules. Their outputs are then aggregated to obtain the representation of the current frame for the following detection module. The detector is standard but an extra module is added to enhance the features of the region of interests. The output of the Fusion module is fed into the specially designed detector to produce the final results.

3.2 Flow-Guided Temporal Coherence Module

We next explain how the Flow-Guided Temporal Coherence module (FGTC) establishes temporal consistency by using optical flow information. This design is motivated by FGFA [10]. Specifically, at each time step, FGTC takes current frame X_t and a neighbor frame $X_{t-\tau}$ (or $X_{t+\tau}$) as input and computes as follows:

$$F_t = N_{feat}(X_t) \tag{1}$$

$$F_{t-\tau} = N_{feat}(X_{t-\tau}) \tag{2}$$

$$F_{t-\tau \to t} = W\big(F_{t-\tau}, N_{flow}(X_{t-\tau}, X_t)\big) \tag{3}$$

$$F^e_{t-\tau \to t}, F^e_t = \varepsilon(F_{t-\tau \to t}, F_t) \tag{4}$$

$$c_{t-\tau \to t} = exp\left(\frac{F^e_{t-\tau \to t}(p) \cdot F^e_t(p)}{\left|F^e_{t-\tau \to t}(p)\right| \cdot \left|F^e_t(p)\right|}\right) \tag{5}$$

$$\sum_{j=i-\tau}^{i+\tau} c_{j \to t} = 1 \qquad (6)$$

$$F_i = \sum_{j=i-\tau}^{i+\tau} c_{j \to t} F_{j \to t} \qquad (7)$$

Here F_t and $F_{t-\tau}$ denote the intermediate features of the reference frame and a neighboring frame extracted from a convolutional network N_{feat}. While $N_{flow}(X_{t-\tau}, X_t)$ indicates a flow field from $X_{t-\tau}$ to X_t estimated by an optical flow network N_{flow}. $W(\cdot)$ is a bilinear warping function exploited on each location of $F_{t-\tau}$. The warped features $F_{t-\tau \to t}$ are computed based on the flow field obtained before as Eq. (3) shows. Next, we embed the two features F_t and $F_{t-\tau \to t}$ with a tiny neural network for similarity measurement. $\varepsilon(\cdot)$ in Eq. (4) denotes the embedding function. Then, as Eqs. (5) and (6) shows the cosine similarity metric is applied to measure the relationship between the warped features and the reference features and the measurement result is then normalized and exploited for adaptive feature aggregation. As Eq. (7) shows, F_i is the final flow-guided enhanced feature that incorporates temporal information from time $t - \tau$ to time $t + \tau$.

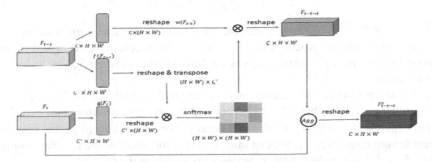

Fig. 2. The detail of Attention-Based Temporal Context module (Color figure online)

3.3 Attention-Based Temporal Context Module

The Flow-Guided Temporal Coherence module propagates temporal information by estimating flow filed between frames. However, only exploiting flow field for feature-level calibration may lead to unsatisfactory spatial correspondence. The reason is that optical flow predicts the displacement of raw pixels and directly using it for the alignment of high-level features may introduce interference. To alleviate this issue, in this section, we introduce a novel ABTC module for feature alignment and enhancement which covers all the space-time pixel locations of feature space. Next, we explain how ABTC incorporates temporal context information using the attention-based temporal context clues.

To capture context information in the time axis, the proposed module compares every space-time locations across frames. Then, the comparison result is utilized to generate a temporal context feature for aligning with the feature of the reference frame. Figure 2 shows the details of the proposed ABTC module. The operation can be summarized into three steps as follows.

The orange lines in the figure represent the first step. This is to generate weights based on feature similarities. Just as Fig. 2 shows, supposing F_t and $F_{t-\tau} \in R^{C \times H \times W}$ are the intermediate features of the reference frame and a neighbor frame, we first embed them into separate convolutional layers to get features with reduced dimensions. Thus we can get $g(F_t) \in R^{C' \times H \times W}$ and $f(F_{t-\tau}) \in R^{C' \times H \times W}$. Then, we perform reshape and transpose operations in turn on $f(F_{t-\tau})$ to $R^{(H \times W) \times C'}$. Meanwhile, we reshape $g(F_t)$ to $R^{C' \times (H \times W)}$, where (H × W) is the number of pixels. The multiplication of the two matrices is the similarity between two feature cells. The above process can be formulated as follows:

$$s_{ij} = [f(F_{t-\tau}^i), g(F_t^j)], \tag{8}$$

where i and j are the locations of the two intermediate feature maps. [·] represents the operation to get similarities s_{ij} between any two cells across features.

Then, we can get normalized correspondence weights by applying a softmax layer:

$$\hat{s}_{ij} = \frac{\exp(s_{ij})}{\sum_{i=1}^{H \times W} s_{ij}}, \tag{9}$$

where $\hat{s}_{ij} \in R^{(H \times W) \times (H \times W)}$ measures the ith position's impact on jth position. This self-attention calculation process simulates the attention mechanism. After that, spatial relations between the two feature maps is established. The resulting $\hat{s}_{ij}$ can be seen as attention maps.

The blue lines indicate the second step. A temporal context feature $F_{t-\tau \to t}$ is produced from step 2. We embed $F_{t-\tau}^i$ to a space that shares the same dimension with $F_{t-\tau}$ and perform reshape on it to get $w(F_{t-\tau}^i)$. Equation (9) shows how to get the temporal context feature maps $F_{t-\tau \to t}$. We can infer from the formula that the temporal context feature $F_{t-\tau \to t}$ is a weighted sum of across all positions of the map $w(F_{t-\tau}^i)$:

$$F_{t-\tau \to t}^j = \sum_i \hat{s}_{ij} w(F_{t-\tau}^i) \tag{10}$$

$F_{t-\tau \to t}$ is the feature that incorporates temporal context to align with the reference feature F_t.

Then, the procedure of aligning features is represented with green lines as follows:

$$F_{t-\tau \to t}^* = Agg(F_{t-\tau \to t}, F_t) \tag{11}$$

Here, Agg(·) is the function for feature aggregation between $F_{t-\tau \to t}^*$ and F_t. Specifically, the two features are taken as input to a tiny neural network to produce adaptive weights for feature fusion:

$$Agg(F_{t-\tau \to t}, F_t) = W_{t-\tau \to t} \cdot F_{t-\tau \to t} + W_t \cdot F_t \tag{12}$$

Similarly, the output of this module which absorbs temporal context clues from time t − τ to time t + τ can be formulated as follows:

$$F_i^* = \sum_{j=i-\tau}^{i+\tau} s_{j \to t} F_{j \to t} \tag{13}$$

The outputs of the FGTC and ABTC are summed on element-wise:

$$F_{final}^t = F_i + F_i^* \tag{14}$$

Finally, F_{final}^t is fed into the detection network like detector for the final result.

We name the detector used in the proposed method TR-FCN. R-FCN [3] is a fully convolutional detector. It achieves excellent performance both on speed and accuracy. Based on R-FCN, a tiny neural network is introduced to predict the movement between the proposals among nearby frames to the current frame. We use $N_{rfcn \rightarrow}$ represents the proposals of frame $X_{t-\tau}$. Similar to FGTC described in Sect. 3.2, TR-FCN aligns proposal features by optical flow propagation. The warp operation shows in Fig. 1 indicate the propagation procedure. Such instance-level aggregation further builds temporal information in the detection network, ablative study shows the effectiveness of this detector. Details of the two modules will be further clarified in Sect. 3.4.

3.4 Implementation Details

We take the pre-trained ResNet-101 model [20] as the feature extraction network and make some modifications to it. The specific changes follow the practice of [10]. FlowNet [15] is exploited as the network for optical flow, which is the pioneer of applying deep convolutional networks to optical flow estimation. In order to match the dimension of the intermediate feature, the output of the flow network needs to be downscaled to half.

We use 1×1 convolution layers with 256 filters to implement the embedding function $f(\cdot)$ and $g(\cdot)$. For the realization of Agg($\cdot$), we use 1×1 convolution with 256 filters followed by two 1×1 convolution layers with 16 and 2 filters, respectively. The design of the detection network follows MANet [11]. The main difference from a standard detector is that it adds an additional instance-level aggregation module by use of optical flow to make temporal consistency.

4 Experiments

4.1 Dataset and Setup

We evaluate the proposed approach against state-of-the-art video object detectors on the ImageNet VID [13] dataset. ImageNet VID is one of the most popular benchmark datasets for video object detection. It contains 3862 training videos, 555 validation videos and 937 test videos for 30 categories. The annotations for the test set is not released. We report all results on the validation set following the protocols of FGFA [10] and performance is measured in terms of the mean average precision (mAP).

In addition to the ImageNet VID training set, the ImageNet DET training set is also used for training. Note that we only use the 30 categories shared by both of the two datasets. Our model is trained in three stages. We set the frame sampling interval to 15 in the first two stages and 10 in the third stage. The method is implemented with MXNet and trained on a single NVIDIA P40 GPU.

4.2 Ablation Study

First of all, to analyze the effectiveness of various components in the proposed method, we conducted four modifications of our approach in our ablative experiments, the experimental results are shown in Table 1.

Version (a) is the baseline R-FCN with ResNet-101. It obtains a test mAP of 70.9%. For purer analysis, the models listed in Table 2 are all evolved from this strong baseline by applying corresponding temporal module.

Comparing with the baseline R-FCN, version (b) employs FGTC, this module effectively renders temporal information from neighbor frames. It achieves a result of 73.2%, which brings 2.3% improvement.

We then investigate the contribution of ABTC in version (c). When we add it to version (b), we can obtain a 3.5% improvement of test mAP comparing to the single baseline. This module compares every space-time locations across frames in feature space to capture temporal context information.

Table 1. Ablation studies on the ImageNet VID validation set. FGTC represents Flow-Guided Temporal Coherence module, ABTC represents Attention-Based Temporal Context module, and TR-FCN refers to the standard R-FCN with temporal information rendered by instance-level aggregation.

Feature extractor	ResNet-101				
Versions	(a)	(b)	(c)	(d)	(e)
FGTC		√	√	√	√
ABTC			√		√
TR-FCN				√	√
mAP (%)	70.9	73.2	74.1	76.2	77.8

Version (d) exploits TR-FCN instead of R-FCN as the detector. The main difference is that TR-FCN incorporates temporal information by use of optical flow. Specifically, it predicts movements of proposals among nearby frames and aligns them with the proposals obtained from the reference frame.

Version (e) is the proposed method. Comparing to version (d), it improves mAP by 1.6%, indicating that these components are complementary and they can work together to obtain a higher detection quality.

To sum up, the two features produced from both modules can represent useful spatial-temporal information from neighbor frames, and the combination of them is quite necessary for the detection performance. With all the above modules, the overall mAP is improved from 70.9% to 77.8%.

4.3 Detection Results

We compare the proposed method against several existing state-of–the-art methods on the task of video object detection, including Seq-NMS [6], TPN [7], TCN [8], TCNN

[9], FGFA [10], MANet [11] and D&T [12]. The results of Seq-NMS [6], TPN [7], TCN [8], TCNN [9] and D&T [12] are obtained from the original papers. The remaining methods are implemented using the code provided by the authors on a platform with NVIDIA P40 GPU.

As we can see from Table 2, compared with the existing methods for video object detection, the proposed method achieves the best performance. It surpasses the R-FCN based detector by a large margin of ~7 points, proving the effectiveness of our model. Comparing with box-level methods [6–9, 12], feature-level methods [10, 11, 17] exploit temporal information during both training and testing and can be trained end-to-end. Therefore, they usually perform better on accuracy. Comparing with FGFA [10] and MANet [11] which establish temporal information by exploiting optical flow, our method outperforms these methods via the adoption of multi-module collaboration strategy, which greatly improves the results.

Table 2. Quantitative results of our proposed method, comparing with state-of-the-art solutions on ImageNet VID validation set.

Methods	mAP (%)	Backbone
Seq-NMS [6]	52.2	VGGNet
TCN [8]	47.5	GoogLeNet
TPN [7]	68.4	GoogLeNet
R-FCN [3]	70.9	ResNet101
TCNN [9]	73.8	GoogLeNet
DFF [17]	69.9	ResNet101
D(&T loss) [12]	75.8	ResNet101
FGFA [10]	73.2	ResNet101
MANet [11]	76.2	ResNet101
Ours	**77.8**	ResNet101

5   Conclusions

In this paper, we present a unified, end-to-end trainable spatiotemporal CNN model for video object detection. The key components are two modules FGTC and ABTC that extracts two kinds of features for each frame respectively. Specifically, FGTC adaptively propagates features over time via optical flow. To align the frame features more precisely, we propose the ABTC module which aims to render the temporal context for spatial correspondence between features across frames. The two features are combined for the benefit of their complementarity. Experimental results show that the proposed framework achieves 77.8% mAP on ImageNet VID, which outperforms existing state-of-the-art methods. The ablative results show ABTC is complementary to flow-based feature propagation modules, demonstrating the generalization ability of our method.

Acknowledgements. This work is supported by the NSFC 61672089, 61703436, 61572064, 61273274 and CELFA.

References

1. Girshick, R.: Fast R-CNN. In: ICCV (2015)
2. Ren, S., He, K., Girshick, R., Sun, J.: Faster R-CNN: towards real-time object detection with region proposal networks. In: NIPS (2015)
3. Dai, J., Li, Y., He, K., Sun, J.: R-FCN: object detection via region-based fully convolutional networks. In: NIPS (2016)
4. Redmon, J., Divvala, S., Girshick, R., Farhadi, A.: You only look once: unified, real-time object detection. In: CVPR (2016)
5. Liu, W., et al.: SSD: single shot multibox detector. In: Leibe, B., Matas, J., Sebe, N., Welling, M. (eds.) ECCV 2016. LNCS, vol. 9905, pp. 21–37. Springer, Cham (2016). https://doi.org/10.1007/978-3-319-46448-0_2
6. Han, W., et al.: Seq-NMS for video object detection. arXiv preprint arXiv:1602.08465 (2016)
7. Kang, K., et al.: Object detection in videos with tubelet proposal networks. In: CVPR (2017)
8. Kang, K., Ouyang, W., Li, H., Wang, X.: Object detection from video tubelets with convolutional neural networks. In: CVPR (2016)
9. Kang, K., et al.: T-CNN: tubelets with convolutional neural networks for object detection from videos. In: T-CSVT (2017)
10. Zhu, X., Wang, Y., Dai, J., Yuan, L., Wei, Y.: Flow-guided feature aggregation for video object detection. In: ICCV (2017)
11. Wang, S., Zhou, Y., Yan, J., Deng, Z.: Fully motion-aware network for video object detection. In: Ferrari, V., Hebert, M., Sminchisescu, C., Weiss, Y. (eds.) ECCV 2018. LNCS, vol. 11217, pp. 557–573. Springer, Cham (2018). https://doi.org/10.1007/978-3-030-01261-8_33
12. Feichtenhofer, C., Pinz, A., Zisserman, A.: Detect to track and track to detect. In: ICCV (2017)
13. Vaswani, A., et al.: Attention is all you need. In: NIPS (2017)
14. Bahdanau, D., Cho, K., Bengio, Y.: Neural machine translation by jointly learning to align and translate arXiv preprint arXiv:1409.0473 (2014)
15. Dosovitskiy, A., et al.: FlowNet: learning optical flow with convolutional networks. In: ICCV (2015)
16. Russakovsky, O., et al.: ImageNet large scale visual recognition challenge. IJCV **115**, 211–252 (2015)
17. Zhu, X., Xiong, Y., Dai, J., Yuan, L., Wei, Y.: Deep feature flow for video recognition. In: CVPR (2017)
18. Fu, J., Liu, J., Tian, H., et al.: Dual Attention Network for Scene Segmentation. arXiv preprint arXiv:1809.02983v4 (2018)
19. Wang, X., Girshick, R., Gupta, A., He, K.: Non-local neural networks. arXiv preprint arXiv:1711.07971v3 (2018)
20. He, K., Zhang, X., Ren, S., Sun, J.: Deep residual learning for image recognition. In: CVPR (2016)

Convexity Preserving Contraction
of Digital Sets

Lama Tarsissi[1,2]([✉]), David Coeurjolly[3], Yukiko Kenmochi[1],
and Pascal Romon[2]

[1] LIGM, Université Gustave Eiffel, CNRS, ESIEE Paris, Marne-la-vallée, France
lama.tarsissi@gmail.com
[2] LAMA, Université Gustave Eiffel, CNRS, Marne-la-vallée, France
[3] Université de Lyon, CNRS, LIRIS, Lyon, France

Abstract. Convexity is one of the useful geometric properties of digital
sets in digital image processing. There are various applications which
require deforming digital convex sets while preserving their convexity. In
this article, we consider the contraction of such digital sets by removing
digital points one by one. For this aim, we use some tools of combinatorics on words to detect a set of removable points and to define such
convexity-preserving contraction of a digital set as an operation of re-
writing its boundary word. In order to chose one of removable points
for each contraction step, we present three geometrical strategies, which
are related to vertex angle and area changes. We also show experimental
results of applying the methods to repair some non-convex digital sets,
which are obtained by rotations of convex digital sets.

Keywords: Digital convexity · Digital set contraction · Christoffel
words · Lyndon words

1 Introduction

Convexity is one of the useful geometric properties of digital sets in digital image
processing. There are various applications which require deforming digital convex
sets while preserving their convexity. In this article, we consider the contraction
of such digital sets by removing digital points one by one. For this aim, we use
some tools of combinatorics on words to detect a set of removable points and to
define such convexity-preserving contraction of a digital set as an operation of
re-writing its boundary word.

A relation between combinatorics on words [16] and digital geometry [15] has
afforded many advantages to both areas and has led to interesting results. Indeed,
such a connection realizes through the so-called Freeman coding, introduced by
Freeman in 1961 [13], that allows to uniquely determine a 4- or 8-connected

---

This work was partly funded by the French National Research Agency, grant agree-
ments ANR-10-LABX-58 (Labex Bézout) and ANR-15-CE40-0006 (CoMeDiC).

S. Palaiahnakote et al. (Eds.): ACPR 2019, LNCS 12047, pp. 611–624, 2020.
https://doi.org/10.1007/978-3-030-41299-9_48

finite set of points in the discrete plane by means of its boundary word, i.e. a word over the alphabet of cardinal four $A = \{0, 1, 2, 3\}$. This code is the bridge between these two worlds. In this article, we deal with one of the main notions of convexity for polyominos presenting in literature: digital convexity. During the last decades, many definitions were given; it started with Minsky and Papert in 1969 [18] and then Hubler (further notions and related studies can be found in [4,5,9,10]). A subset C in a digital image is convex if the straight line segment joining any two pixels P and Q of C lies entirely in C. We also present another notion of digital convexity with a help of the convex hull of C. In this article, we rely on a recent result by Berleck and et al. [3], who have defined a set to be digitally convex if and only if the Lyndon factorization of its boundary word is made of Christoffel words. Based on this definition, we show how we can deflate a digitally convex set by choosing correct digital points to remove and preserving the convexity at the same time. We can find some previous studies on inflating digitally convex sets [11]. One of the main contributions of this article is to give three different strategies that allow us to determine an order to follow in removing digital points.

This article is decomposed into four parts. We first introduce digital convexity and give some related notions about two families of words. Then, we show the link between digital convexity and combinatorics on words. In the second part, we do modifications on the border of a digital convex set, and show all the perturbations obtained after removing a digital point. The third part is for presenting three different techniques in choosing a certain order for deflating digital convex sets. The last part is dedicated for showing experimental results, where we apply our methods to a rotated digital convex set, which is not convex anymore, in order to repair its convexity by making the convex hull and deflating it by removing some digital points.

2   Digital Convexity and Combinatorics on Words

In this section, we introduce the basic notions needed in the following sections. Precisely, we recall basic notions in combinatorics on words. Then we define digital convexity and two particular families in combinatorics on words, that are Christoffel and Lyndon words [1, 16].

2.1   Basic Notions of Combinatorics

An **alphabet** A is a finite set of symbols such that its elements are called **letters**. A **word** w over an alphabet A is a finite sequence of letters over A. In another way, a word w is obtained by *concatenating* letters of A, we write $w \in A^*$, where A^* represents the set of all the words formed by A, with the **empty word** ϵ being the identity element for the concatenation. We denote $w^n \in A^*$ the concatenation of the word w, n times, such that: $w^n = w.w.w \cdots w$ where w is repeated n times with the convention of $w^0 = \epsilon$. A word w is said **primitive** if it is not the power of a nonempty word.

Let $w \in A^*$. The **length** of w is the number of letters of w denoted by $|w|$. Note that $|\epsilon| = 0$. For all $\ell \in A$, $|w|_\ell$ denotes the number of occurrences of the letter ℓ in the word w, so that: $|w| = \sum_{\ell \in A} |w|_\ell$.

Two words w and w' are **conjugate** of order k, with $k \neq 0$, denoted by $w \equiv_k w'$, if and only if there exist $u, v \in A^*$ such that $|u| = k$, $w = u.v$ and $w' = v.u$. When the exact value of k is not relevant, we simply write $w \equiv w'$. We can set several order relations over A^*. Here, we choose the total **lexicographic order**, or simply the *dictionary order* that is a total order and denoted by $<$.

Definition 1. *Let $w_1, w_2 \in A^*$, we say that:*

$$w_1 < w_2 \quad if \quad \begin{cases} w_2 = w_1.w', \text{ where } w' \in A^*; \text{ or} \\ w_1 = u.a.v_1 \text{ and } w_2 = u.b.v_2 \text{ where } u, v_1, v_2 \in A^* \text{ and } a < b \in A. \end{cases}$$

2.2 Polyomino and Boundary Words

Let us consider the lattice $\mathbb{Z}^2$, which is the set of all the vectors having integer components. The canonical basis of the Euclidean vector space $\mathbb{R}^2$ is $\{e_1, e_2\}$. We call by a finite discrete set any finite subset of $\mathbb{Z}^2$. We define a path in $\mathbb{Z}^2$ from point X to point Y a sequence of points $(p_i)_{1 \leq i \leq n}$ with $p_i \in \mathbb{Z}^2$ where $p_1 = X$ and $p_n = Y$. A path in $\mathbb{Z}^2$ is 4-connected if for any $1 \leq i \leq n-1$, $p_{i+1} - p_i \in \{\pm e_1, \pm e_2\}$. A *polyomino* P is a simply 4-connected finite set of digital points, without hole, that can be visualized as set of a unit squares. The Freeman chain code allows us to represent any path over $\mathbb{Z}^2$ as a word over an alphabet of four letters $A = \{0, 1, 2, 3\}$ [13] such that:

- 0 denotes a right horizontal step;
- 2 denotes a left horizontal step;
- 1 denotes an upside vertical step;
- 3 denotes a downward vertical step;

The boundary word of a polyomino P, denoted by $Bd(P)$, is the word obtained by coding its perimeter using the alphabet A. For each polyomino, we can have an equivalent class of words that are the conjugates to each other. We choose the representative of these words to be the one that starts from the left lowermost side of the polyomino.

In the last decades, many researchers [3] have studied the digital convexity and worked on finding the link between it and combinatorics on words. In the following two sections, we introduce two families of words, that are essential for this work and we introduce some results.

2.3 Christoffel Words and Digital Line Segments

The first family of words is *Christoffel words*. From the geometric point of view, we can consider any Christoffel word as the discretization of a line segment of rational slope [1,2]. In this case, we restrict our work to the binary alphabet $A = \{0, 1\}$, and start by defining a Christoffel path which is a sequence of unitary steps joining two points of integer lattice. Let a and b be two coprime numbers, the *lower* Christoffel path of slope a/b is the path joining the origin $O(0,0)$ to the point (b,a) and respecting the following characteristics:

1. it is the nearest path below the line segment joining these two points;
2. there is no point of $\mathbb{Z}^2$ between the path and line segment.

Analogously, the *upper* Christoffel path is the path that lies above the line segment. By convention, the Christoffel path is exactly the *lower* Christoffel path. We remark that the *upper* Christoffel path of slope a/b is the image of the *lower* Christoffel path obtained by a rotation of angle $180°$ (see Fig. 1 for example).

Using the binary alphabet $A = \{0,1\}$, and by assigning to each horizontal (resp. vertical) step the letter 0 (resp. 1), we get the *Christoffel word* w of slope $\frac{a}{b}$, denoted by $C(\frac{a}{b})$, where the fraction $\frac{a}{b}$ is exactly $\frac{|w|_1}{|w|_0}$ (see Fig. 2). We define the morphism $\rho : A^* \longrightarrow \mathbb{Q} \cup \{\infty\}$ by:

$$\rho(\epsilon) = 1 \text{ and } \rho(w) = \frac{|w|_1}{|w|_0}, \ \forall w \neq \epsilon \in A^*;$$

where $\frac{1}{0} = \infty$. This morphism determines the slope for each given word in A^*. In particular:

1. for $a = 0$, we get the Christoffel word of slope 0, $C(\frac{0}{1}) = 0$.
2. for $b = 0$, we have the Christoffel word of slope ∞, $C(\frac{1}{0}) = 1$.

Christoffel words have several properties and characterizations [1,8,16]. Here, we only mention a few of them. Let $\widetilde{w} = w_n \cdots w_2.w_1$ be the **reversal** of the word $w = w_1 w_2 \ldots w_n$; we say that w is a **palindrome** if $\widetilde{w} = w$.

Property 1. The Christoffel word of slope $\frac{a}{b} \neq 0, \infty$, can be written as: $C\left(\frac{a}{b}\right) = 0p1$, where p is a palindrome.

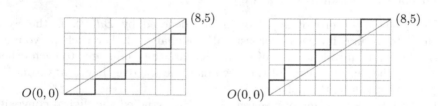

Fig. 1. The lower and upper Christoffel paths of slope 5/8 respectively.

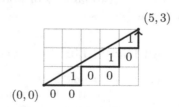

Fig. 2. The Christoffel path from $(0, 0)$ to $(5, 3)$ and the Christoffel word $C(\frac{3}{5}) = 00100101$.

In addition, any Christoffel word can be factorized into two Christoffel words, using the standard factorization [2] defined as follows.

Property 2. Let w be any Christoffel word of slope $\frac{a}{b}$, w can be written in a unique way as $w = (w_1, w_2)$, where both w_1 and w_2 are Christoffel words and of minimal length equal to 1.

Note that the slopes of the new Christoffel words, obtained by the standard factorization, bound the initial slope.

Property 3. Let $w = (w_1, w_2)$, such that $\rho(w_1) = \frac{a}{b}$ and $\rho(w_2) = \frac{c}{d}$. We have:

1. $\frac{a}{b} < \rho(w) < \frac{c}{d}$,

2. $\det(\rho(w_1), \rho(w_2)) = \det \begin{pmatrix} |w_1|_0 & |w_2|_0 \\ |w_1|_1 & |w_2|_1 \end{pmatrix} = cb - ad = 1$.

Using this factorization, we are able to define two functions ϕ_0 and ϕ_1 from $A^* \times A^*$ into itself as follows: $\phi_0(w_1, w_2) = (w_1, w_1w_2)$; $\phi_1(w_1, w_2) = (w_1w_2, w_2)$.

Property 4. Any Christoffel word is constructed by an iteration of these two functions ϕ_0 and ϕ_1 on $(0, 1)$.

Consequently, the standard factorization of any Christoffel word is unique, i.e, there exists a unique point of the Christoffel path, where we can do this factorization. This point is exactly the closest point of the Christoffel path with respect to the line segment, that divides the initial Christoffel word into two concatenated Christoffel words. For Fig. 2, we note that the standard factorization of $C(\frac{3}{5}) = (001, 00101)$.

2.4 Lyndon Factorization

The second family of words that is useful for our purpose is the family of *Lyndon words* [17].

Definition 2. *A word $w \in A^*$ is a Lyndon word if for all $u, v \in A^* \setminus \{\epsilon\}$ such that $w = u.v$, we have $w < v.u$.*

In other words, we can say that w is a Lyndon word if w is the smallest word among all its conjugates. Based on this definition, we can deduce one of the properties of this family of words. In fact, *Lyndon* words are primitive words; otherwise, we can have an equality between at least one of the conjugates of w. Therefore, if w is a Lyndon word, $w.w$ is not Lyndon. For example, let $w = 0110$, and X be the set of all the possible conjugates of w. The smallest element of X in lexicographic order is $w' = 0011$. Therefore, w is not a Lyndon word. A factorization was introduced in 1958 by Chen, Fox and Lyndon [7].

Definition 3. *Every non-empty word w admits a unique factorization as a lexicographically decreasing sequence of Lyndon words: $w = l_1^{n_1} l_2^{n_2} \cdots l_k^{n_k}$, s.t $\rho(l_1) > \rho(l_2) > \cdots > \rho(l_k)$ where $n_i \geq 1$ and l_i are Lyndon words.*

This factorization is called **Lyndon factorization**. In 1980 Duval proved that this factorization can be computed in a linear time [12]. For example, the Lyndon factorization of $w = 100101100101010$ is $(1)(001011)(0010101)(0)$.

2.5   Digital Convexity by Combinatorics

Now, we introduce the link between these two families of words and digital convexity for polyominos.

Definition 4. *A polyomino P is said digitally convex if the convex hull of P, $conv(P)$, satisfies $conv(P) \cap \mathbb{Z}^2 \subset P$.*

Each digital convex polyomino P can be bounded by a rectangle box. We denote the points of intersection of each side of the bounding box with P by (W)est, (N)orth, (E)ast and (S)outh, starting from the leftmost side and in a clockwise order as shown in Fig. 3.

The boundary word of P, $Bd(P)$, can be divided into 4 sub-words, w_1, w_2, w_3 and w_4, where each word codes respectively the WN, NE, ES and SW side of P. We note that w_1 starts with the letter 1, while w_2, w_3 and w_4 start with 0, 3 and 2 respectively. Each word uses only two letters among the four of the alphabet, and hence, we deal on each side with a binary word.

In this section, we consider the WN-word of the boundary word of the digital convex set, and introduce the result of Brleck et al. [3] about digital convexity: a characterization for this path using the two previous families of words.

Property 5. [3]   A word w is WN-convex if and only if its unique Lyndon factorization $w_1^{n_1} w_2^{n_2} \ldots w_k^{n_k}$ is such that all w_i are primitive Christoffel words.

See Fig. 4 for an example of Lyndon factorization: $(1)^1(011)^1(01)^2(0001)^1(0)^1$ where 1, 011, 01, 0001 and 0 are all Christoffel words. The Christoffel words are arranged in a weakly decreasing order of slopes: $\frac{1}{0} > \frac{2}{1} > \frac{1}{1} > \frac{1}{3} > \frac{0}{1}$. With this result, the WN convex path is decomposed into line segments with decreasing slopes using a linear algorithm [3]. As mentioned before, we know that line segments are represented by Christoffel words, and the decreasing lexicographic order that appears in the Lyndon factorization can explain the condition of decreasing order of slopes.

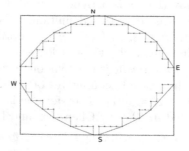

Fig. 3. A digital convex polyomino bounded in a rectangle, where W, N, E and S are the first points of intersection with the left, upper, right and lower sides of the bounding box.

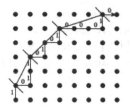

Fig. 4. Example of a WN-convex word 101101010001, formed of Christoffel words using the Lyndon factorization.

Due to this, we have the following property:

Property 6. [3] Each vertex of the convex hull of any digital convex set C is exactly at the end of each factor of the Lyndon factorization of its boundary word $Bd(C)$.

Such a vertex is exactly at the factor of either 10, 03, 32 or 21, depending on which side of C it is located, either in WN, NE, ES and SW respectively.

3   Contraction of Digitally Convex Set

Deflating a digital convex set in a geometric point of view can be obtained by removing, step by step, at most one integer point. In the digital sense, we can see that this deformation is investigated by removing one pixel at each step. In [11], the authors introduced the *split operator*, that determines the unique accepted position, for each line segment, to add a digital point and conserving the convexity. In this paper, we work on removing digital points under the same condition. Since the vertices, as mentioned before, are digitally represented by the factor 10 in the WN path, then removing a point means switching the factor 10 (resp. 03, 32 and 21) to 01, (resp. 30, 23 and 12).

3.1   Removable Points and Contraction

The points that we are allowed to remove on the boundary are exactly the vertices of the convex hull. In fact, we might face several possibilities depending on the position of the points and the convex shape. In this section, we will mention all the possible situations that we can encounter and the different algorithms that we can use in order to remove a single point from the digital convex set at each step. Given a digitally convex set C, on the boundary word $Bd(C)$, we can have the vertices of the convex hull of C, also called corners. These vertices can be grouped in two categories. The first ones, called the internal vertices, that belong to one of the boundary sub-words. The second ones are external or critical vertices that are at the intersection between two consecutive sub-words. Let w_i and w_{i+1} be two consecutive sub-words, the corner is exactly the last letter of w_i and the first letter of w_{i+1}.

We define such contraction operator on Christoffel words. Note that Christoffel words on WN, NE, ES and SW sides are binary words over the binary alphabets $\{0,1\}$, $\{0,3\}$, $\{2,3\}$ and $\{1,2\}$ respectively. Christoffel words on each side are defined as follows: $x = 0h1$, $y = 3k0$, $z = 2l3$ and $t = 1m2$ where h, k, l and m are palindroms.

3.2   Switch Operator

In this section, we consider the four boundary binary sub-words w_1, w_2, w_3 and w_4 that belong to $A^* = \{0,1,2,3\}^*$. On these sub-words, we set an operation at a certain position k, from A^* into A^* such that it switches the two consecutive letters of this word at position k.

Definition 5. *Let $w = l_1 l_2 \ldots l_k l_{k+1} \ldots l_n$ be a word in $\{0,1,2,3\}^*$,*

$$\text{switch}_k(w) = l_1 l_2 \ldots l_{k+1} l_k \ldots l_n,$$

where $02 = 20 = 13 = 31 = \epsilon$.

Given a boundary word $Bd(C)$, thanks to Property 6, we can apply this switch operator to $Bd(C)$ at the last letter of each Christoffel word obtained by the Lyndon factorization of $Bd(C)$. In this way, we remove a corner of the form 10, 03, 32 or 21. We have two cases, depending on the position of a corner. The first case is when the corner belongs to one of the four paths, either WN, NE, ES or SW. In this case, the operation is directly applied at position k of the Christoffel word w, where $k = |w|$, as seen in Fig. 5. In the second case, we deal with the corners that are at the intersection position between w_1, w_2, w_3 and w_4. In fact, only at these positions, it might occurs some particular situations. We might obtain, after a switch operation, one of the factors 02, 20, 13, and 31 that will be replaced by ϵ by definition. But in this case, a re-arrangement for the four boundary words has to be considered; see an example in Fig. 6.

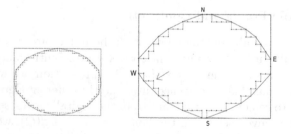

Fig. 5. The fourth boundary word of C is: $w_4 = (2)(12222)(1222)(122)^2(12)^2$. The switch operator is applied on the fifth factor, at position 2. We obtain the new boundary word $w_4' = (2)(12222)(1222)(122)(12122)(12)$.

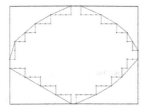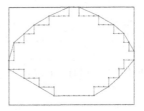

Fig. 6. Re-arrangement of the boundary words, w_4 and w_3, after removing the factor 31 from the boundary word $Bd(C)$ of the digital convex set C.

To keep the set 4-connected, we must avoid the existence of the factor $1^k 0^l 3^k 2^l$ and its conjugates on the boundary word. Therefore, if after a certain operation, one of these four factors appear in the boundary word, the switch operation must be canceled.

By applying the switch operator locally all around the boundary word and subwords, the possible words that can be obtained are of the following form:

Theorem 1. *Let u and v be two consecutive Christoffel words such that $|u| = k$, and $\rho(u) > \rho(v)$. By applying the switch operator one of these cases appears:*

1. $\text{switch}_k(uv) = w^\ell$ where $\rho(w) < \rho(u)$ and $\ell \geq 1$;
2. $\text{switch}_k(uv) = m'_1 \ldots m'_i$ where $\rho(m'_1) < \rho(u)$ and $\rho(m'_1) > \ldots \rho(m'_i)$.

In order to give the proof, we must recall Pick's theorem [19] that gives the area of a triangle using its interior and boundary integer points:

Theorem 2 *[18]. The area of a given polygon in Euclidean plane, whose vertices have integer coordinates, is equal to $A = i + \frac{b}{2} - 1$ where i is the number of interior integer points and b is the number of boundary integer points of the polygon.*

After this theorem, we can pass to a sketch of the proof of Theorem 1.

Proof From Properties 2 and 3, we write $m_1 = u.v$ and $m_2 = x.y$ with $\rho(u) < \rho(m_1) < \rho(v)$ and $\rho(x) < \rho(m_2) < \rho(y)$. Recall that the area of a triangle ABC is equal to $\frac{1}{2} \det(\overrightarrow{AB}, \overrightarrow{AC}) = \frac{1}{2} \det \begin{pmatrix} x_{\overrightarrow{AB}} & x_{\overrightarrow{AC}} \\ y_{\overrightarrow{AB}} & y_{\overrightarrow{AC}} \end{pmatrix} = i + \frac{b}{2} - 1$. Hence, we can conclude that: $i = \frac{\det(\overrightarrow{AB}, \overrightarrow{AC}) - b + 2}{2}$. With a simple vectorial calculation we can conclude that if $\det(\rho(m_1), \rho(m_2)) = 1$, we get $i = 0$. Hence we get the first case for $\ell = 1$ since in this case we get $b = 3$, that are exactly the points A, B and C. We can note that this result is the inverse of the split operator defined in [11]. Similarly, and by noticing that each Christoffel word is obtained by applying several morphisms of ϕ_0 and ϕ_1 defined in Property 4, we can prove the other cases.

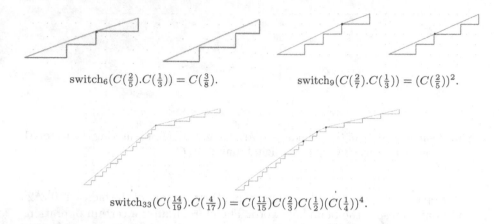

$$\text{switch}_6(C(\tfrac{2}{5}).C(\tfrac{1}{3})) = C(\tfrac{3}{8}).$$ $$\text{switch}_9(C(\tfrac{2}{7}).C(\tfrac{1}{3})) = (C(\tfrac{2}{5}))^2.$$

$$\text{switch}_{33}(C(\tfrac{14}{19}).C(\tfrac{4}{17})) = C(\tfrac{11}{15})C(\tfrac{2}{3})C(\tfrac{1}{2})(C(\tfrac{1}{4}))^4.$$

Fig. 7. Examples of three different cases of switch operator.

Some examples are illustrated in Fig. 7.

The number of possible new segments is related to the solutions of the a variant of the *Knapsack problem* in integer linear programming: given a triangle T with rational vertices in a $[0, N] \times [0, N]$ domain ($N \in \mathbb{Z}$), the number of vertices in the convex hull of $T \cap \mathbb{Z}^2$ is in $O(\log N)$ [14]. This result has been used to efficiently compute the convex hull of interior grid points of rational convex polygon [6]. In other words, the number of $m_i's$ and the value of k is bounded by $O(\log N)$.

4   Choosing the Removal Order

As mentioned above, we can find all points that can be removed in order to deflate a digital convex set C while preserving its convexity. These points are exactly the vertices of the convex hull of C. Here arises the natural question: is there any specific order for choosing which digital point to remove at each step? If so, is there a preferred order? In fact, we investigate and test three ordering techniques to deflate C.

Let V be a removable point, and V_p and V_n be its previous and next vertices of the convex hull of C. After removing V from C, let us assume that the set of new vertices **W** is obtained, instead of V.

4.1   Ordering by Determinant Obtained from Two Consecutive Line Segments

For each removable point V, we calculate

$$\triangle(V) = \frac{1}{2} \det(\overrightarrow{V_pV}, \overrightarrow{VV_n}),$$

which is the (signed) area of triangle V_pVV_n. Then, all removable points are listed in increasing order with respect to this area measure, and the point with minimum value is chosen for each elementary contraction step. The list is updated locally whenever a point is removed.

4.2   Ordering by Area Change

For each removable point V, we calculate

$$\triangle_c(V) = \triangle(V) - \frac{1}{2} \sum_{W \in \mathbf{W}} \det(\overrightarrow{V_p W}, \overrightarrow{W W_n})$$

which is the area difference between the convex hulls of C and $C \setminus \{V\}$. Similarly to the previous case, all removable points are listed in increasing order with respect to this area change, and the point with minimum value is chosen for each elementary contraction step. Once a point is removed, the list is updated locally.

4.3   Ordering by Angle-Sum Change

For each removable point V, we calculate the difference between the line-segment angle at V and the sum of such angles at all $W \in \mathbf{W}$, such that

$$\theta_c(V) = \theta(V) - \sum_{W \in \mathbf{W}} \theta(W)$$

where

$$\theta(V) = \cos^{-1} \left(\frac{\overrightarrow{V_p V} \cdot \overrightarrow{V V_n}}{\|\overrightarrow{V_p V}\| \|\overrightarrow{V V_n}\|} \right).$$

Similarly to the previous cases, all removable points are listed in increasing order with respect to this angle measure, and the point with minimum value is chosen from the list for each elementary contraction step. Once a point is removed, the list is updated locally.

5   Experimental Results

Deforming digitally convex sets with digital convexity preservation is crucial, and has different applications in digital image processing. One of them is rotating digitally convex sets without loss of their convexity, which is not trivial at all.

Figure 8 shows the experimental results of applying our methods to a rotated digitized ellipse, which was originally digitally convex before the rotation but is not anymore (see Fig. 8(b)). In order to repair the digital convexity, we made the convex hull and digitized it (see Fig. 8(c)). As the difference between the rotated digitized ellipse and the digitized convex hull are 45 pixels, which are colored in gray in Fig. 8(b), we decided to remove 45 pixels while preserving the digital convexity by using our method with the three different strategies. Note that, in Sect. 4, the point to be removed at each iteration has to be with a minimal value for the three strategies. We might encounter the case where several points can give us the same

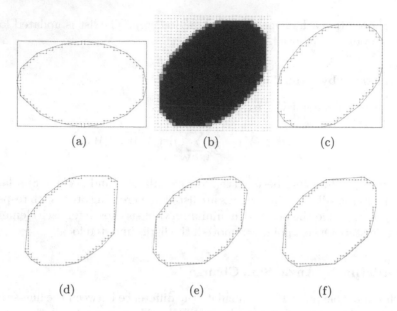

(a)   (b)   (c)

(d)   (e)   (f)

Fig. 8. (a) A digitized ellipse, which is digitally convex; (b) its rotated image with angle $\frac{\pi}{4}$ (in blue), which is not anymore digitally convex; (c) the convex hull of the rotated shape of (b) where the difference from the rotated image is visualized with 45 gray pixels in (b); the contraction results for the digitized convex hull of the rotated digitized ellipse of (b) after removing 45 points with the three ordering strategies: by (d) determinant, (e) area change, and (f) angle-sum change. (Color figure online)

minimal value. Our algorithm chooses in a random way a removal point among all the points having the minimal value at each iteration. Using this algorithm, the results after 45 iterations can be seen in Fig. 8(d, e, f), respectively. Table 1 also shows the results of the three different strategies after 22, 150 and 200 iterations.

While all of them preserve digital convexity, the deformation tendencies are slightly different. The first removal-point ordering strategy based on area tends to extend vertical and horizontal line segments while it keeps curved parts as well. On the other hand, it is seen that the second (resp. third) strategy based on area (resp. angle) change tend to straighten curved parts.

Table 1. The contraction results for the digitized convex shape illustrated in Fig. 8(c) after removing 22, 150 and 200 points with the three ordering strategies: determinant, area change, and angle-sum change.

Iter.	22	150	200
Det.			
Area			
Angle			

6   Conclusion

We presented the method for contraction of digitally convex sets by removing digital points one by one while preserving their digital convexity, with a help of combinatorics on words. Such tools enable us to detect a set of removable points and to define such convexity-preserving contraction of a digital set as an operation of re-writing its boundary word. In order to choose one of removable points for each contraction step, we present three geometrical strategies, which are related to vertex angle and area changes of the convex hull of a digitally convex set. We showed experimental results of applying the methods to repair some non-convex digital sets, which are obtained by a rotation of convex digital sets. Our further work will consist of deforming non-convex sets while preserving their convex and concave parts. Another idea can be to write the algorithm needed to add a point at each step instead of removing.

References

1. Berstel, J., Lauve, A., Reutenauer, C., Saliola, F.: Combinatorics on words: Christoffel words and repetition in words (2008)
2. Borel, J.-P., Laubie, F.: Quelques mots sur la droite projective réelle. J. de théorie des nombres de Bordeaux **5**(1), 23–51 (1993)

3. Brlek, S., Lachaud, J.-O., Provençal, X., Reutenauer, C.: Lyndon+ Christoffel=digitally convex. Pattern Recognit. **42**(10), 2239–2246 (2009)
4. Castiglione, G., Frosini, A., Munarini, E., Restivo, A., Rinaldi, S.: Combinatorial aspects of L-convex polyominoes. Eur. J. Comb. **28**(6), 1724–1741 (2007)
5. Castiglione, G., Frosini, A., Restivo, A., Rinaldi, S.: Enumeration of l-convex polyominoes by rows and columns. Theor. Comput. Sci. **347**(1–2), 336–352 (2005)
6. Charrier, E., Buzer, L.: Approximating a real number by a rational number with a limited denominator: a geometric approach. Discrete Appl. Math. **157**(16), 3473–3484 (2009)
7. Chen, K., Fox, R., Lyndon, R.: Free differential calculus IV. The quotient groups of the lower central series. Ann. Math. **68**, 81–95 (1958)
8. Christoffel, E.: Observatio arithmetica. Annali di Matematica Pura ed Applicata (1867–1897), **6**(1), 148–152 (1875)
9. Del Lungo, A., Duchi, E., Frosini, A., Rinaldi, S.: Enumeration of convex polyominoes using the ECO method. In: DMCS, pp. 103–116 (2003)
10. Del Lungo, A., Duchi, E., Frosini, A., Rinaldi, S.: On the generation and enumeration of some classes of convex polyominoes. Electron. J. Comb. **11**(1), 60 (2004)
11. Dulio, P., Frosini, A., Rinaldi, S., Tarsissi, L., Vuillon, L.: First steps in the algorithmic reconstruction of digital convex sets. In: Brlek, S., Dolce, F., Reutenauer, C., Vandomme, É. (eds.) WORDS 2017. LNCS, vol. 10432, pp. 164–176. Springer, Cham (2017). https://doi.org/10.1007/978-3-319-66396-8_16
12. Duval, J.: Mots de lyndon et périodicité. RAIRO, Informatique théorique **14**(2), 181–191 (1980)
13. Freeman, H.: On the encoding of arbitrary geometric configurations. IRE Trans. Electron. Comput. **2**, 260–268 (1961)
14. Hayes, A.C., Larman, D.G.: The vertices of the knapsack polytope. Discrete Appl. Math. **6**(2), 135–138 (1983)
15. Klette, R., Rosenfeld, A.: Digital Geometry: Geometric Methods for Digital Picture Analysis. Morgan Kaufmann Publishers Inc., San Francisco (2004)
16. Lothaire, M.: Algebraic Combinatorics on Words. Encyclopedia of Mathematics and Its Applications, vol. 90. Cambridge University Press, Cambridge (2002)
17. Lyndon, R.: Identities in finite algebras. Proc. Am. Math. Soc. **5**(1), 8–9 (1954)
18. Minsky, M., Papert, S.: Perceptrons. MIT Press, Cambridge (1969)
19. Pick, G.: Geometrisches zur zahlenlehre. Sitzungsberichte des Deutschen Naturwissenschaftlich-Medicinischen Vereines für Böhmen "Lotos" in Prag., vol. 47–48, pp. 1899–1900 (1906)

Parallax-Tolerant Video Stitching with Moving Foregrounds

Muhammad Umer Kakli[1], Yongju Cho[1,2($\boxtimes$)], and Jeongil Seo[2]

[1] Korea University of Science and Technology, Daejeon, South Korea
umar.isb@etri.re.kr
[2] Electronics and Telecommunications Research Institute, Daejeon, South Korea
{yongjucho,seoji}@etri.re.kr

Abstract. The parallax artifacts introduced due to movement of objects across different views in the overlapping area drastically degrade the video stitching quality. To alleviate such visual artifacts, this paper extend our earlier video stitching framework [1] by replacing a deep learning based object detection algorithm for parallax detection, and an optical flow estimation algorithm for parallax correction. Given a set of multi-view overlapping videos, the geometric look-up tables (G-LUT) are generated by stitching a reference frame from the multi-view input videos, which map the input video frames to the panorama domain. We propose to use a deep learning based approach to detect the moving objects in the overlapping area to identify the G-LUT control points which get affected by parallax. To compute the optimal locations of these parallax affected G-LUT control points we propose to use patch-match based optical flow (CPM-flow). The adjustment of G-LUT control points in the overlapping area may cause some unwanted geometric distortions in the non-overlapping area. Therefore, the G-LUT control points in close proximity of moving objects are also updated to ensure the smooth geometric transition between the overlapping and the non-overlapping area. Experimental results on challenging video sequences with very narrow overlapping areas ($\sim$3% to $\sim$10%) demonstrate that video stitching framework with the proposed parallax minimization scheme can significantly suppress the parallax artifacts occurring due to the moving objects. In comparison to our previous work, the computational time is reduce by $\sim$26% with the proposed scheme, while the stitching quality is also marginally improved.

Keywords: Video stitching · Parallax · Object detection · Optical flow

1 Introduction

Image stitching refers to a process of generating a seamless wide Field-of-View (FoV) high-resolution image from multiple partially overlapped low-resolution images captured from different viewpoints. It is an extensively studied topic

© Springer Nature Switzerland AG 2020
S. Palaiahnakote et al. (Eds.): ACPR 2019, LNCS 12047, pp. 625–639, 2020.
https://doi.org/10.1007/978-3-030-41299-9_49

Fig. 1. Parallax artifacts due to movement of objects in the overlapping area.

in computer vision with the number of commercially available solutions. We refer to [2] for a comprehensive survey of image stitching methods. A typical image stitching pipeline consists of an alignment step followed by a composition step. In the alignment step, the geometric relationship such as homography is estimated between the overlapping images. Photometric correction and blending is applied in the composition step to ensure the color consistency and smooth transition in the overlapping area of the resultant stitched image. The image stitching methods proposed in the literature mostly employ a 2D projective transformation for the alignment of overlapping input images. The underlying assumption of such methods is that the images were taken with a rotating camera or the scene appears to be planar [3]. These conditions can rarely be satisfied in any practical scenario. Often the scene is non-planar and captured from different viewpoints using a multi-camera rig. Due to different center of projections (CoP) of cameras, the captured images suffer from the parallax as the relative position of objects varies across the different viewpoints. As the 2D projective transforms cannot handle the parallax well [8]. Therefore, stitching such images with a global 2D projective transform leads to the visible artifacts such as blurring or ghosting. Few examples of parallax artifacts are shown in Fig. 1.

In video stitching, either the frames of multiple synchronized video streams are stitched independently or a single alignment model is used to stitch all video frames (single alignment video stitching framework). Apart from image stitching issues, handling of visible parallax artifacts when objects move across the multiple views is a difficult problem to solve [10]. The problem becomes even more challenging if the images/videos to be stitched having a narrow overlapping area [1]. The majority of existing schemes exploit the information about matched features to minimize the parallax artifacts. However, the existence of reliable and enough number of matched features cannot be guaranteed for the multi-view videos with narrow overlaps especially if the scene is mostly composed of homogeneous information.

In this paper, we propose a parallax minimization scheme based on our previous video stitching framework [1] to efficiently handle the parallax artifacts induced by the movement of objects across multiple views in narrow overlapping area. In video stitching framework, firstly one frame is stitched form the multi-view input videos and geometric look-up table (G-LUT) and a blending mask (BM) for every input video are generated. The G-LUTs are used to map the multi-view video frames to the panorama domain, and BMs are used to apply the linear blending in the overlapping area. For every frame of multi-view input videos, we propose to detect the presence and location of moving objects in the overlapping area with YOLO (You Only Look Once v3) [18]. Once the location of the moving object is determined, we robustly identify the G-LUT control points belongs to the moving object. Next, we propose to use patch-match based optical flow (CPM-flow) [20] to find the optimal locations of these G-LUT control points to minimize the parallax artifacts in the overlapping area. To assure the geometric consistency between overlapping and non-overlapping area, the non-overlapping G-LUT control points around the moving object are also updated with a weighted function, as proposed in [1]. We tested the video stitching framework with the proposed changes on videos having very narrow overlap and diverse range of object movements. The experimental results show that it can efficiently stitch the narrow overlapping multi-view videos in the presence of moving objects without any significant parallax artifacts.

An earlier version of this research work was presented in [1]. The main difference between [1] and this paper is: we propose to use YOLO object detection to identify the moving object in the overlapping area. In the previous work, we used Structural SIMilarity (SSIM) Index [16] to identify the parallax affected G-LUT control points, which requires at least one frame without any object in the overlapping area for comparison. Moreover, SSIM based approach cannot handle cases with varying exposure and color changes between the comparing images [17]. Therefore, detection with YOLO is a better choice for detection of moving objects than the SSIM. In our previous work, the motion vectors of parallax affected G-LUT control points are computed by generating an HLBP descriptor which is matched using L2 distance in a pre-defined search area. The performance of HLBP descriptor can be greatly affected if the area around G-LUT control points is not very distinctive. Furthermore, the computational complexity of descriptor based matching scheme increases directly with an increase amount of parallax, pre-defined search space, and the number of G-LUT control points to be adjusted in the overlapping area. In this work, we propose to use a random search based optical flow algorithm (CPM-flow) due to its robustness and better computational performance. The abovementioned changes in video stitching framework decreased the computational time by ~26% with slight improvement in the stitching quality as compared to our previous work in [1].

The rest of the paper is organized as follows. Section 2 reviews the earlier work on parallax reduction in image and video stitching. In Sect. 3, we describe the video stitching framework and the proposed parallax minimization scheme. Experimental details and results are discussed in Sect. 4. Finally, the conclusions are summarized in Sect. 5.

2   Related Work

2.1   Image Stitching

The early image stitching methods were based on global 2D projective transforms estimated from matching feature points. Such projective transformations cannot handle the parallax well [8]. To overcome the parallax artifacts, multi-band blending and seam-cut based solutions have been proposed in the literature. These methods can only produce satisfactory results, when the amount of parallax is very small. For the input images with large viewpoint changes or with narrow overlapping areas, the performance of these algorithm decreases drastically. Gao et al. [4] proposed to divide an image scene into a ground and a distant plane. A separate homography is computed for each plane, and the final output is obtained by the weighted average of both homographies. As-projective-as-possible (APAP) warping method proposed by Zaragoza et al. [5] divides the image into grids cells. The grid cells are updated with the spatially-variant homographies computed on the basis of matched features. APAP often introduce shape and perspective discontinuity from overlapping to the non-overlapping area. To avoid this issue, Chang et al. [6] proposed shape-preserving-half-perspective (SPHP) warping method, which assure the smooth transition between overlapping and non-overlapping area. Zhang et al. proposed a hybrid model for image alignment by combining homography warping and the content-preserving warping (CPW) [8]. These warping based methods are prone to the misalignments if the parallax is very large [3]. Furthermore, these methods are computationally intensive and very much dependent on the accuracy of matched features.

Fig. 2. Example of few failure cases of graph-cut methods for narrow overlaps.

2.2   Video Stitching

Video stitching has comparatively received less attention than that of image stitching. It is often regarded as the temporal extension of image stitching. While retaining the challenges of image stitching, it also suffers from additional issues, such as different camera optical centers, drift and jittering, dominant foregrounds movement, and non-planar scenes. A number of application specific video stitching solutions have been proposed over the years. Video stitching frameworks targeting surveillance cameras are proposed in [11] and [9]. An extension of [8] in the both spatial and temporal domain is presented in [10]. A video stitching solution of sporting events is described in [7]. Some work on independently moving mobile devices is reported in [10–15], which concentrate on jointly solving the problem of video stitching and stabilization.

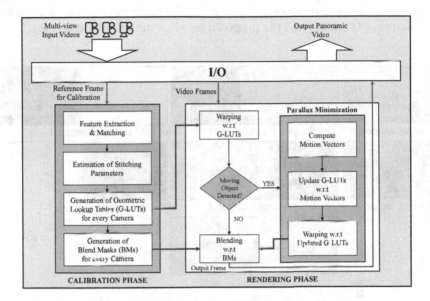

Fig. 3. Block diagram of parallax-tolerant video stitching framework of [1].

The majority of existing video stitching solutions utilize seam-cut methods as a post-processing step by either updating an existing seam or by estimating a new seam for every frame. Inclusion of seam-cut methods often produces visually unpleasant movement of seam position around the moving objects in a stitched video. Additionally, seam-cut methods cannot handle the case of narrow overlapping area. If the size of a moving object is fairly large and cannot fit completely in the overlapping area. The estimated seam will pass through the moving object and the misalignment will become prominent in the resultant stitched image as shown in Fig. 2. In contrast, the proposed scheme does not require any post-processing seam-cut algorithm and can robustly deal with the case of narrow overlapping area without any dependency on matched features.

3   Video Stitching Framework

The block diagram of the video stitching framework is shown in Fig. 3 [1]. The video stitching framework has a calibration phase and a rendering phase. The stitching parameters are calculated once during the calibration phase, and G-LUT and BMs are then generated for every input video. The G-LUTs and BMs are further used in the rendering phase for each frame. During the rendering phase, if a moving object is not detected in the overlapping area then G-LUTs and BMs from the calibration phase are used to render the output panoramic frame. If a moving object is detected in the overlapping area, G-LUTs control points affected by parallax are corrected, and the updated G-LUTs are used to render the output of that particular frame. Figure 4 shows an example of two input images, the warped images using G-LUTs, the BMs, and the output panoramic image. In the following text, we will explain the structure of G-LUTs, and the proposed parallax minimization scheme.

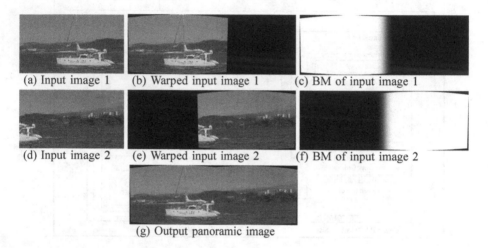

(a) Input image 1   (b) Warped input image 1   (c) BM of input image 1

(d) Input image 2   (e) Warped input image 2   (f) BM of input image 2

(g) Output panoramic image

Fig. 4. (a) and (d) Two input images, (b) and (e) input images after warping with G-LUTs, (c) and (f) blending masks of input images, (g) output panoramic image.

3.1 Overview of Geometric Look-Up Tables (G-LUTs)

Given N number of overlapping multi-view videos with F frames in each, the goal is to generate G-LUTs which maps the every video frame to the panorama domain with transformation function T, as proposed in [1]. Let $F_{(i,j)}$ denotes the jth frame from the ith video, and $\overline{F}_{(i,j)}$ is the respective warped frame in panorama domain, which is obtained by applying the transformation function T_i. In order to generate G-LUT, the reference input frame $F_{(i,j)}$ is divide into to a uniform grid of size $(X \times Y)$. The input grid control points are denoted by $CP_{(i,j)}$, and the corresponding control point locations in panorama domain are denoted by $\overline{CP}_{(i,j)}$. G-LUT stores the mapping information $CP_{(i,j)} \rightarrow \overline{CP}_{(i,j)}$ of all control points in quadrilateral fashion. The input video frames are warped to the panorama domain with OpenGL texture mapping. A graphical representation of G-LUTs for the two input images are shown in Fig. 5.

3.2   Proposed Parallax Minimization Algorithm

In order to detect a moving object in the overlapping area, we propose to use You Only Look Once (YOLO v3) [18], a deep learning based object detection algorithm. The YOLO v3 is trained with Common Objects in COntext (COCO) dataset [19] having 172 different object categories, thus, making it a suitable choice for our task. Figure 6 shows some object detection results in overlapping area with YOLO.

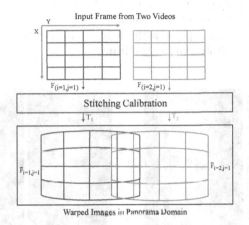

Fig. 5. Graphical representation of G-LUTs for two input images [1].

The G-LUT control points belong to moving objects are the potential candidates suffering from parallax. As soon as a moving object is detected in the overlapping area, we consider G-LUT control points of that moving object as the parallax affected control points, and denoted them as a subset M. The control points in subset M are denoted by $\overline{CP}_{(i,j)}^{P_k}$, where P_k represents the kth control point suffered from parallax.

The next step is to compute the optimal locations of the control points in subset M, such that the parallax artifacts are minimized in the overlapping area. To do so, we propose to use CPM-flow due to its robustness and computational efficacy. The classical optical flow estimation methods such as Horn-Schunck [21] and Lucas-Kanade [22] are well-known for their limitations in the case of large displacements [20]. On the other hand, CPM-flow incorporates a patch-match based random search strategy in a coarse-to-fine fashion which can efficiently handle the large as well as the very tiny motions. Therefore, we compute the motion vectors $v_{(i,j)}^{P_k}$ for each control point of subset M using CPM-flow (some computed motion vectors are shown in Fig. 7). The optimal locations $\widetilde{CP}_{(i,j)}^{P_k}$ of control points in subset M are then obtained as:

$$\widetilde{CP}_{(i,j)}^{P_k} = \overline{CP}_{(i,j)}^{P_k} + v_{(i,j)}^{P_k} \tag{1}$$

632   M. U. Kakli et al.

This adjustment of control points in the overlapping area often introduce the geometric distortion in the non-overlapping area closer to control points $\overline{CP}^{P_k}_{(i,j)}$. To preserve the geometric consistency between the overlapping and non-overlapping area, we use sigmoid function to assign weights to the control points in non-overlapping area based upon the nearest estimated motion vectors $(v^{P_k}_{(i,j)})$ of parallax affected control points [1]. A higher weightage is assign to the control points in close vicinity of the moving object, and decrease gradually to the neighboring control points to ensure the overall smoothness.

Once all G-LUT control points are updated, we use OpenGL texture mapping to render the input frames to the panorama domain. Warped images in the panoramic domain are multiplied with their respective BMs and finally added up to generate the resultant panoramic frame.

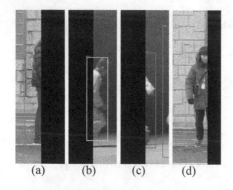

(a)   (b)   (c)   (d)

Fig. 6. Object detection with YOLO in overlapping area.

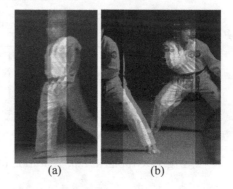

(a)   (b)

Fig. 7. Computed motion vectors in overlapping area.

(a) Taekwondo I   (b) Taekwondo II

(c) Playground I   (d) Playground II

Fig. 8. A frame from each sequence of test video data.

4   Experimentation Details and Results

4.1   Test Dataset

Due to unavailability of benchmark video stitching dataset, we collected our own dataset with two 4 K overlapping cameras placed on a rig. We kept the overlapping area very narrow roughly between ~3% to ~10%. The test dataset covers diverse challenging cases where objects horizontally crossing the overlapping area (Playground II and Taekwondo I), object moving towards camera with scale change (Playground I), multiple occluded objects in the overlapping area (Taekwondo II), and the movement of both object and cameras (Taekwondo II). Please note that the same test dataset was used in our previous work [1]. Figure 8 shows a single frame from each test video.

4.2   Evaluation of Stitching Quality

In order to evaluate the stitching quality, we first stitched all frames of the test videos without incorporating any parallax correction schemes, and then with the proposed parallax minimization scheme. Figure 9 shows the video stitching results on some selected frames with and without the application of parallax minimization scheme (the overlapping area is cropped and zoomed out for better visibility). In first, third, and fifth column, we show the results without utilization of the proposed parallax minimization scheme. The parallax artifacts are very prominent and marked in red. The second, fourth, and sixth column present the results with the proposed parallax minimization scheme. It is evident from the results that the video stitching with proposed scheme can suppress the parallax artifacts in order to improve the video stitching quality in the presence of moving foregrounds.

To evaluate the stitching quality in quantitative manner, we computed the Structural SIMilarity (SSIM) index [16] on high-frequency information of the overlapping area, as proposed in [17] for assessing the quality of stitched images. The SSIM value of zero corresponds to the worst alignment, while a value of one represents the perfect alignment (i.e. identical images). The stitching score is referred as the average SSIM value of all frames. Figure 10 compares the stitching quality score with and without application of the proposed parallax minimization scheme, and with our previous work in [1]. It can be seen that the stitching score with the proposed scheme is over 0.9 for all the test videos, while, the value remains below 0.7 for the case when parallax minimization scheme is not employed. The proposed scheme is marginally better in terms of stitching score than our previous work in [1]. The qualitative and quantitative results verifies that the utilization of parallax minimization schemes improved the structural alignment in the overlapping area by ~30%, which validates their effectiveness to suppress the parallax artifacts. The stitched video results with and without application of the proposed parallax minimization scheme are available at [23].

Before   After   Before   After   Before   After

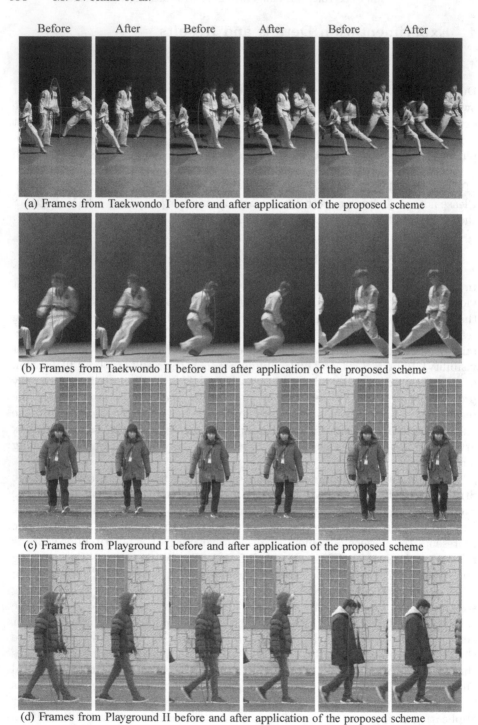

(a) Frames from Taekwondo I before and after application of the proposed scheme

(b) Frames from Taekwondo II before and after application of the proposed scheme

(c) Frames from Playground I before and after application of the proposed scheme

(d) Frames from Playground II before and after application of the proposed scheme

Fig. 9. Results of the proposed scheme on selected frames of test videos. (Color figure online)

4.3   Comparison with Conventional Approaches

We also compared the results of our proposed scheme with state-of-the-art parallax tolerant stitching schemes: APAP [5], SPHP [6], and [1]. APAP computes the multiple homographies for local regions which allows the flexible warping, while, SPHP imposes a similarity transform constraint to preserve the structural naturalness from overlapping to the non-overlapping area. However, we cannot directly compare our video stitching results with APAP and SPHP, as both methods are proposed mainly for the images. To avoid the temporal inconsistency by stitching each frame independently for both APAP and SPHP, we only compare the results on single frames for fairness of comparison. Figure 11 shows the results on selected frames from our test videos. The first row shows the stitched frames without any parallax minimization framework applied. The second, third, and fourth row presents the results of APAP, SPHP, and our previous work in [1], respectively. The results of the proposed scheme are shown fifth row (area around overlapping region is cropped and zoomed for the sake of better visibility). It can be clearly seen that the proposed scheme and our previous work in [1] outperformed in all cases. APAP performs poorly on all selected frames. The geometric misalignment issues in non-overlapping area, blockiness artifacts, and perspective distortions can be clearly seen in all results. SPHP also suffered severely from some perspective distortion and alignment issues. The accuracy of both APAP and SPHP is very much dependent on the accuracy of matched features. In case of narrow overlapping area, there are not enough number of matched features for APAP and SPHP to produce the acceptable results. On the other hand, the proposed scheme does not depend on matching or tracking of the features, hence, it can handle the case of narrow overlapping area effectively. In terms of computational time, APAP takes ~28 s, SPHP takes ~16 s,

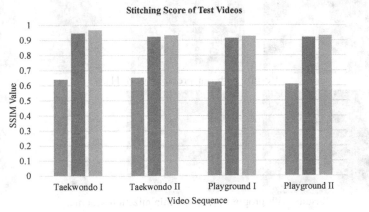

Fig. 10. Stitching quality score for test videos.

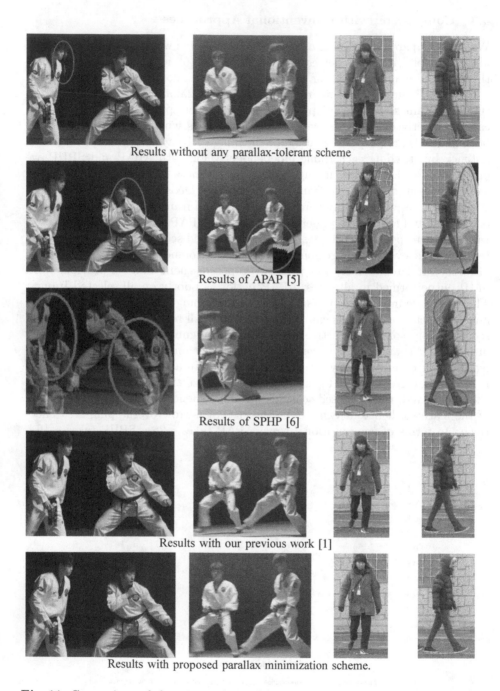

Results without any parallax-tolerant scheme

Results of APAP [5]

Results of SPHP [6]

Results with our previous work [1]

Results with proposed parallax minimization scheme.

Fig. 11. Comparison of the proposed parallax minimization scheme with APAP [5], SPHP [6], and [1].

and our previous work in [1] takes ~4.9 s to process a frame. In comparison, the proposed scheme only takes ~3.6 s.

As we showed in Fig. 2, the seam-cut methods cannot handle the case of the narrow overlapping area. Therefore, we have not compared the results with video stitching algorithms utilizing graph-cut based methods such as [3] and [9].

4.4   Timing Analysis

The proposed scheme is implemented in C++ with OpenCV library. All experiments are conducted on a machine with Intel Core i7-2600K 3.4 GHz GPU with 32 GB of RAM. OpenGL is used for the texture mapping of G-LUTs.

The computational time of the proposed scheme is measured at three instances: time to detect the moving object using YOLO (DT), time to estimate the CPM-flow and the correction of control points in overlapping and non-overlapping area (CT), and the time to render a frame with updated G-LUT (RT). For every component, we took the average value from all video frames and reported in Table 1. On average, the proposed scheme takes around ~3.64 to process a frame. DT takes the least time of ~0.077 s, followed by CT, which takes ~1.28 s. RT is most expensive component of the proposed algorithm with ~2.29 s.

Table 1. Comparison of computation time of the proposed scheme with [23].

Video sequence	Computational time (seconds)							
	Proposed scheme				Our previous work in [23]			
	DT	CT	RT	Total time	DT	CT	RT	Total time
Taekwondo I	0.086	1.36	2.25	3.69	0.014	2.84	2.29	5.09
Taekwondo II	0.082	1.25	2.27	3.60	0.013	2.33	2.24	4.62
Playground I	0.072	1.31	2.34	3.72	0.014	2.05	2.28	4.36
Playground II	0.070	1.20	2.32	3.59	0.014	3.17	2.30	5.88
Average	0.077	1.28	2.29	3.64	0.014	2.60	2.28	4.97

Table 1 also compare the computation time of the proposed scheme with our previous work in [1]. Overall, the proposed scheme is ~1.36 times faster than [1]. The main difference can be notice for CT. The CPM-flow is ~2 times faster than the HLBP descriptor based matching scheme. It can be seen that the CT in the case of Playground II (large parallax) is significantly higher than rest of the test video, as mentioned in Sect. 1. In contrast, CPM-flow takes almost same time for all cases. The SSIM based parallax detection scheme is faster than the YOLO based object detection scheme. The OpenGL based rendering almost takes the same time in both cases. We believe that the GPU based implementation of the proposed scheme could achieve the real-time performance.

5   Conclusions and Future Work

In this paper, we proposed a scheme to minimize the visual parallax artifacts, which occurs due to movement of objects in the overlapping area. The proposed scheme utilize state-of-the-art object detection and optical flow algorithms along with the G-LUTs. The results on challenging videos with very narrow overlapping area confirms its efficacy to remove the visual distortions caused by the moving foregrounds. Moreover, the proposed scheme is computationally efficient, and it does not require any feature matching or tracking between overlapping images or across the frames.

Future work includes the GPU implementation of the proposed scheme to achieve the real-time performance. Another interesting direction is the direct exploitation of deep learning based approaches for solving the alignment issues in image/video stitching.

Acknowledgments. This work was supported by Korea Government (MSIT) 19ZR1120 (Development of 3D Spatial Media Core Technology).

References

1. Kakli, M.U., Cho, Y., Seo, J.: Minimization of parallax artifacts in video stitching for moving foregrounds. IEEE Access **6**, 57763–57777 (2018)
2. Szeliski, R.: Image alignment and stitching: a tutorial. Found. Trends Comput. Graph. Vis. **2**(1), 1–104 (2006)
3. Lin, K., Jiang, N., Cheong, L.-F., Do, M., Lu, J.: SEAGULL: seam-guided local alignment for parallax-tolerant image stitching. In: Leibe, B., Matas, J., Sebe, N., Welling, M. (eds.) ECCV 2016, Part III. LNCS, vol. 9907, pp. 370–385. Springer, Cham (2016). https://doi.org/10.1007/978-3-319-46487-9_23
4. Gao, J, Kim, S. J., Brown, M. S.: Constructing image panoramas using dual-homography warping. In: Proceedings of the IEEE CVPR, pp. 49–56 (2011)
5. Zaragoza, J., Chin, T., Tran, Q., Brown, M.S., Suter, D.: As-projective-as-possible image stitching with moving DLT. IEEE TPAMI. **36**(7), 1285–1298 (2014)
6. Chang, C.-H., Sato, Y., Chuang, Y.-Y.: Shape-preserving half projective warps for image stitching. In: IEEE CVPR, pp. 3254–3261 (2014)
7. Shimizu, T., Yoneyama, A., Takishima, Y.: A fast video stitching method for motion-compensated frames in compressed video streams. In: ICCE, pp. 173–174 (2006)
8. Zhang, F., Liu, F.: Parallax-tolerant image stitching. In: IEEE CVPR, pp. 3262–3269 (2014)
9. He, B., Yu, S.: Parallax-robust surveillance video stitching. Sensors **16**(1), 7 (2015)
10. Jiang, W., Gu, J.: Video stitching with spatial-temporal content preserving warping. In: IEEE CVPRW, pp. 42–48 (2015)
11. He, B., Zhao, G., Liu, Q.: Panoramic video stitching in multi-camera surveillance system. In: IVCNZ, pp. 1–6 (2010)
12. Lin, K., Liu, S., Cheong, L.-F., Zeng, B.: Seamless video stitching from hand-held camera inputs. Comput. Graph. Forum **35**(5), 479–487 (2016)
13. Guo, H., Liu, S., He, T., Zhu, S., Zeng, B., Gabbouj, M.: Joint video stitching and stabilization from moving cameras. IEEE TIP **25**(11), 5491–5503 (2016)

14. Su, T., Nie, Y., Zhang, Z., Sun, H., Li, G.: Video stitching for handheld inputs via combined video stabilization. In: SIGGRAPH Asia Technical Briefs, pp. 25:1–25:4 (2016)

15. Nie, Y., Su, T., Zhang, Z., Sun, H., Li, G.: Dynamic video stitching via shakiness removing. IEEE TIP **27**(1), 164–178 (2018)

16. Wang, Z., Bovik, A.C., Sheikh, H.R., Simoncelli, E.P.: Image quality assessment: from error visibility to structural similarity. IEEE TIP **13**(4), 600–612 (2004)

17. Qureshi, H.S., Khan, M.M., Hafiz, R., Cho, Y., Cha, J.: Quantitative quality assessment of stitched panoramic images. IET Image Process. **6**(9), 1348–1358 (2012)

18. Redmon, J., Divvala, S., Girshick, R., Farhadi, A.: You only look once: Unified, real-time object detection. In: IEEE CVPR, pp. 779–788 (2016)

19. Lin, T.-Y., et al.: Microsoft COCO: common objects in context. In: Fleet, D., Pajdla, T., Schiele, B., Tuytelaars, T. (eds.) ECCV 2014, Part V. LNCS, vol. 8693, pp. 740–755. Springer, Cham (2014). https://doi.org/10.1007/978-3-319-10602-1_48

20. Hu, Y., Song, R., Li, Y.: Efficient coarse-to-fine patchmatch for large displacement optical flow. In: IEEE CVPR, pp. 5704–5712 (2016)

21. Horn, B.K.P., Schunck, B.G.: Determining optical flow. Artificial Intelligence (1981)

22. Lucas, B., Kanade, T.: An iterative image registration technique with an application to stereo vision. In: IJCAI, pp. 674–679 (1981)

23. Video results of proposed scheme. http://sites.google.com/site/parallaxminconf-2019. Accessed 29 Sept 2019

Prototype-Based Interpretation of Pathological Image Analysis by Convolutional Neural Networks

Kazuki Uehara[✉][iD], Masahiro Murakawa[iD], Hirokazu Nosato[iD], and Hidenori Sakanashi[iD]

National Institute of Advanced Industrial Science and Technology (AIST),
1-1-1 Umezono, Tsukuba, Ibaraki, Japan
{k-uehara,m.murakawa,h.nosato,h.sakanashi}@aist.go.jp
https://www.aist.go.jp

Abstract. The recent success of convolutional neural networks (CNNs) attracts much attention to applying a computer-aided diagnosis system for digital pathology. However, the basis of CNN's decision is incomprehensible for humans due to its complexity, and this will reduce the reliability of its decision. We improve the interpretability of the decision made using the CNN by presenting them as co-occurrences of interpretable components which typically appeared in parts of images. To this end, we propose a prototype-based interpretation method and define prototypes as the components. The method comprises the following three approaches: (1) presenting typical parts of images as multiple components, (2) allowing humans to interpret the components visually, and (3) making decisions based on the co-occurrence relation of the multiple components. Concretely, we first encode image patches using the encoder of a variational auto-encoder (VAE) and construct clusters for the encoded image patches to obtain prototypes. We then decode prototypes into images using the VAE's decoder to make the prototypes visually interpretable. Finally, we calculate the weighted combinations of the prototype occurrences for image-level classification. The weights enable us to ascertain the prototypes that contributed to decision-making. We verified both the interpretability and classification performance of our method through experiments using two types of datasets. The proposed method showed a significant advantage for interpretation by displaying the association between class discriminative components in an image and the prototypes.

Keywords: Interpretability · Explainable AI · Prototypes · Convolutional neural networks · Pathological images

---

The part of this work is based on results obtained from a project commissioned by the New Energy and Industrial Technology Development Organization (NEDO).

S. Palaiahnakote et al. (Eds.): ACPR 2019, LNCS 12047, pp. 640–652, 2020.
https://doi.org/10.1007/978-3-030-41299-9_50

1   Introduction

Developing accurate computer-aided diagnosis (CAD) algorithms is essential for rapid diagnosis. CAD is a foremost research topic in medical imaging, including digital pathology. The recent advance of digital scanners enables the applications of machine learning techniques to CAD algorithms.

Deep convolutional neural networks (CNNs) have achieved significant performance for visual tasks [10], and they attract considerable attention in digital pathology [12]. However, the interpretability (transparency) of the rationale behind their decision is an issue for reliability, and many researchers tackled the problems [1].

The objective of this study is to improve the interpretability of the basis behind decisions made using a CNN. How the network reached its decision can be interpretable by ascertain the essential elements for the decision from the learned information. To achieve this, we propose a prototype-based interpretation method. The method improves the interpretability by presenting the decisions as co-occurrence relation of interpretable components which typically appeared in parts of images. We define prototypes as the components and introduce the following three steps to decision-making: (1) presenting typical parts of images as multiple components, (2) allowing humans to interpret the components visually, and (3) making decisions based on the co-occurrence relation of the multiple components. For the first step, we encode image patches using an encoder of variational auto-encoder (VAE) to make them class discriminative and then construct clusters for the encoded image patches. The centroids of the clusters are used for prototypes which correspond to representative parts of images. For the second step, we decode prototypes into images using the VAE's decoder because the prototypes are encoded vectors that are incomprehensible for humans. For the last step, we calculate the weighted combinations of the prototype occurrence for image-level classification. The combinations of the weights enable us to obtain which prototypes were important in contributing to the decision.

The main contributions of our work are the following: (i) We propose a prototype-based interpretation method for a decision made using a CNN. (ii) We verify our method on two types of datasets, namely Kylberg texture dataset and PCAM dataset. (iii) The proposed method shows a significant advantage for interpretation by displaying the association between class-discriminative parts and the prototypes.

The remainder of this paper is organized as follows. Section 2 provides related studies. Section 3 explains our proposed method in detail. Section 3.1 presents experiments and discussion. Finally, Sect. 4 provides a summary of this study.

2   Related Work

In the recent years, many studies related to explainability or interpretability of neural networks have been conducted owing to their importance [6,13,14,16].

In this section, we refer solely to studies related to CNNs. Most of the studies adopted saliency based visualization methods [13,14,16]. However, these activation-based explaining methods cannot help understand the basis of classification because they only highlight a part of an image.

The studies closely related to our work were conducted by Li et al. [11] and Chen et al. [3]. Li et al. [11] proposed prototypes-based case base reasoning into neural network to improve interpretability, and they visualized prototypes by using an auto-encoder. Chen et al. [3] designed their network to minimize the distance between the prototypes and the local image patches of training samples in the latent feature space. However, their prototypes learn the complete images or parts of images that appear directly in training samples; thus, their methods may fail to predict classes that contain variation, which is unacceptable for predicting widely variable classes of histopathological images.

3   Prototype-Based Classification

In this section, we describe our prototype-based classification. The method comprises three phases as stated in Sect. 1. At first, we extract prototypes to obtain representative points that typically appear in training images. Moreover, we convert the prototypes to images to make them visually interpretable. We then calculate image-level features based on the prototypes for image-level classification because the prototypes only represent the local parts of images. Finally, we train a linear classifier to make a decision based on the combinations between class importance and prototypes.

At the last of this section, we describe two interpretation processes by using the proposed method.

3.1   Constructing Prototypes and Visualization Modules

The prototypes are representative points that typically appear in the local parts of training data. Moreover, they must be interpretable for humans. To obtain good representative points, we calculate cluster centers of local image patches in a latent space.

To this end, we construct prototypes $P = \{p_k\}_{k=1}^q$, as shown in Fig. 1. Let images be $X = \{x_i\}_{i=1}^n$ and corresponding labels be $Y = \{y_i\}_{i=1}^n$. First, we simultaneously train an encoder f and a decoder g by using CNN to map image patches $\hat{X} = \{\hat{x}_j\}_{j=1}^m$ into latent space and remap the encoded features to an original space. We then construct clusters for the features extracted from the patches. We remap the prototypes that are encoded in a latent space onto an original image space by using the decoder when we interpret the prototypes. In our method, the encoder is used as a feature extractor, while the decoder is used as a visualizer. Thus, the objective of this phase is to obtain three modules, which are prototypes, a feature extractor, and a feature visualizer.

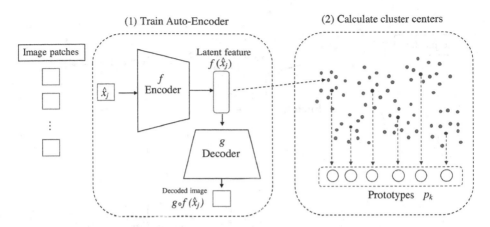

Fig. 1. Procedures to construct prototypes. (1) Train an auto-encoder to obtain both feature extractor and visualizer. (2) Calculate cluster centers to obtain representative points in the encoded image patches.

Training Encoder and Decoder. We adopt a variational auto-encoder (VAE) [8] to obtain the feature extractor and visualizer. VAE is an auto-encoder model comprising an encoder-decoder network; it has two objectives: to estimate both probability density distribution and examples obtained from the distributions. To estimate the distributions from observations, we calculate the difference between the estimated distributions and the prior distribution by using the following equation:

$$D_{KL}(f, \hat{X}) = \mathcal{N}(f(\hat{X}); \mu, diag(\sigma \odot \sigma)) || \mathcal{N}(f(\hat{X}); 0, I) \tag{1}$$

$$= \frac{1}{2} \sum_{j=1}^{m} (\sigma_j^2 + \mu_j^2 - 1 - log\sigma_j^2) \tag{2}$$

where $\odot$ presents a Hadamard production, and f denotes an encoder. μ_j is estimated mean vector from an image $\hat{x}_j$, while σ_j^2 is estimated variance.

To estimate images from given distributions, we calculate the difference between input images and estimated images by using following equation:

$$\mathcal{L}_r(f, g, \hat{X}) = \frac{1}{m} \sum_{j=1}^{m} ||g \circ f(\hat{x}_j) - \hat{x}_j||_2^2, \tag{3}$$

where g denotes decoding based on the CNN.

To make the distribution class-discriminative, we introduce an additional objective by adding another classifier h to the end of the encoder to train the VAE as shown in Fig. 2. The classifier is a single layer perceptron with a soft-max function. The objective, classification loss is calculated by following equation:

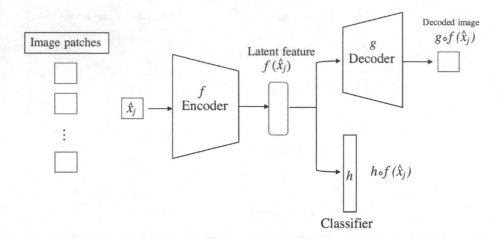

Fig. 2. Illustration for our class discriminative auto-encoder.

$$\mathcal{L}_c(h \circ f, \hat{X}, \hat{Y}) = -\sum_{j=1}^{m} \sum_{c=1}^{K} \hat{y}_{jc} * log(h \circ f(x_j)_c), \qquad (4)$$

where $\hat{y}_{jc}$, corresponding to image patch $\hat{x}_j$ means a binary indicator for class c.

By integrating all objectives together, the loss function to be minimized for the VAE is calculated as follows:

$$\mathcal{L}(f, g, h, \hat{X}, \hat{Y})_{VAE} = w_1 \mathcal{L}_r(f, g, \hat{X}) + w_2 \mathcal{L}_c(f, h, \hat{X}, \hat{Y}) + D_{KL}(f, \hat{X}), \quad (5)$$

where w_1 and w_2 present weights to balance between image reconstruction and class prediction.

Constructing Prototypes. The prototypes are representative data points of local features; thus, we use the k-means algorithm to obtain them. Ensuring the following image-level features to class-discriminative, we conduct the k-means for each class. After calculating the cluster centers for each class, we concatenate these cluster centers as a set of prototypes for training data.

3.2 Image-Level Feature

For image-level classification, we calculate image-level features by using the set of prototypes derived from encoded image patches.

To calculate image-level feature, we spatially aggregate a similarity between encoded image patches and each prototype (Fig. 3). This feature calculation is inspired by bags-of-features (BoF) scheme [4]. BoF calculates its feature based on a hard-assignment for local feature aggregation. To embed relative information between encoded image and all prototypes to image-level feature, we calculate the relative similarity s^i_{jk} [2] in soft-assignment manner by using following equation:

$$s^i_{jk} = \frac{e^{-\alpha||f(\hat{x}_j) - p_k||^2}}{\sum_{k=1}^{q} e^{-\alpha||f(\hat{x}_j) - p_k||^2}}, \tag{6}$$

where α is a parameter that controls relative values corresponding distance. Increasing α indicates the assignment close to hard assignment. After calculate the similarity, we accumulate the similarites to each prototype as image-level feature z_i for an image x_i by following equation:

$$z_{ik} = \sum_{j}^{m_i} s^i_{jk}, \tag{7}$$

where m_i is the number of image patches extracted from an image x_i. The kth element of the image feature is aggregation of similarities between all image patches and kth prototype.

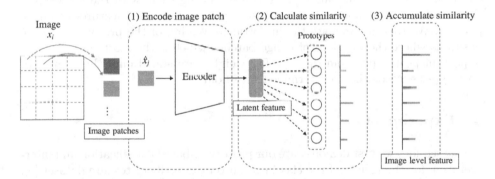

Fig. 3. Procedure to calculate an image-level feature. (1) Encode image patches by using the trained encoder. (2) Calculate similarity between encoded features and each prototypes. (3) Accumulate the similarities as an image-level feature.

3.3 Train a Linear Classifier

The objective of this step is to train a classifier for image-level classification using image-level features stated in the above section. In this step, we associate each prototype with a decision using a linear classifier. We adopt a single layer perceptron as the classifier. An optimization process is executed to minimize the cross-entropy loss represented in the following equation.

$$\mathcal{L}_c(Z, Y) = -\sum_{i=1}^{n} \sum_{c=1}^{K} y_{ic} * log(H(z_i)_c), \tag{8}$$

where $\{z_i \in Z\}_{i=1}^{n}$ presents image-level feature and $H(z_i)$ is class probability estimated by the classifier.

3.4 Process for Interpretation

This sub-section describes two processes, namely prototype-based analysis and heat-map visualization, for interpretation of decisions made using the proposed method.

Prototype-Based Analysis. The purpose of this prototype-based analysis is to show which prototypes have contributed to the classification of each class. In addition, we visualize the prototypes through decoder for visual interpretation. As stated in the above section, decisions are made using the linear combination of image features, which is calculated by accumulating the similarity between encoded local image patches and prototypes. Thus, we can treat the weights in the classifier are the importance of class discrimination.

Heat-Map Visualization. The second interpretation process is to highlight class-discriminative parts in images. Moreover, our method suggests prototypes concerning class-discriminative parts by associating them with the prototypes. We calculate the similarity between prototypes and encoded features of image patches. We then accumulate the importance weights of the prototypes to the positions where the locations of image patches which is the most similar to the important prototypes from the top. The weights are added until the number of prototypes exceeds the predetermined threshold.

4 Experiments

In this section, we first demonstrate our prototype-based classification and interpretation for the results of the classification on the Kylberg texture dataset [9]. Our method is then applied to pathological images, using the PCAM dataset [15]; further its performance is compared with that of [11], which is an interpretation method for CNN.

4.1 Network Architecture and Training

We adopted a CNN for an encoder and a decoder. Both the encoder and decoder had three layers for convolution. The kernel sizes for each layer of the encoder were 7, 5, and 3, respectively, and 2, 4, and 6 for the decoder. The kernels were applied in two strides with padding sizes 3, 2, and 1 for the encoder and 0, 1, and 2 for the decoder. The output of each layer was transformed by batch-normalization [7] and rectified linear unit (Relu), except for the last layer of the decoder. The output for the last layer was transformed by a sigmoid function to reconstruct images.

4.2   Experiment for Kylberg Texture Dataset

Dataset. We adopted the Kylberg texture dataset because some characteristics in texture images are identical to those of pathological images. For example, similar patterns appear in an arranged manner, and pathological and the textured images do not have any directions.

The Kylberg texture dataset comprises 28 texture classes. Each class comprises 160 texture images, having 576×576 pixels. We divided the complete images into 72×72 pixels of image patches to construct the prototypes. We set 10 to the number of prototypes for each class, thus the number of the total prototypes was 280. To use this texture dataset for the image classification task, we separated 90% of this dataset for training data and the remaining 10% for test data. For observing the training condition, we validated 10% of the training data.

Results. For the classification of the Kylberg texture dataset, our method yielded an accuracy of 96.0%.

Figure 4 shows the three examples of prototypes for each class. These prototypes were visualized by using the decoder. Although the prototypes were not generated from encoded images but the centers of the feature's distributions in the latent space, most prototypes can be clearly visualized. The prototypes illustrate representative parts for each class.

For interpreting the decision for the classification, we selected two types of images, namely "canvas" (Fig. 5) and "floor1" (Fig. 6). Images in the "canvas" have even textural patterns, while images in the "floor1" have locally structured patterns. In these figures, the left images show a heat-map analysis for input images. Redder parts indicates similar to the important prototypes, i.e. more class-discriminative parts in the image. The center images show original images with marks indicating the closest patches to the prototypes, and the right images show the prototypes with the largest importance weights for class prediction. The prototypes are arranged in descending order of the importance and only the top 10 are displayed.

In Fig. 5, the important prototypes well represent the canvas texture. Focusing on the importance weights, the values for the first seven prototypes exhibit higher importance, while the remaining three values are significantly low, because the lower prototypes are not clear and No. 9 is derived from the different class. The top seven prototypes can have similar information for humans because their appearance are similar each other.

In this experiment, we set the fixed number for the prototypes in each class. However, an appropriate number of prototypes can be vary according to class characteristics. Thus, pruning redundant prototypes would be necessary to improve the interpretability for human users.

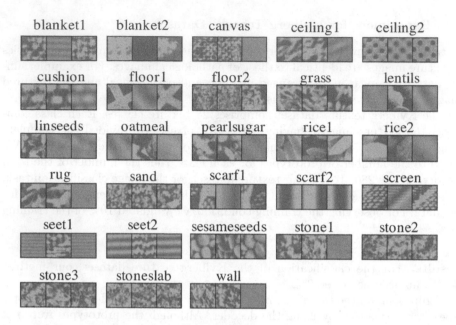

Fig. 4. Examples of three decoded prototypes obtained by our method in each class for Kylberg dataset.

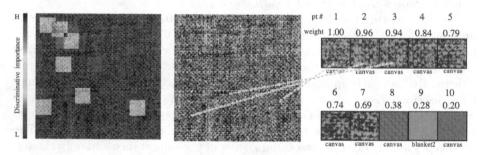

Fig. 5. Visual interpretation for classification result of "canvas" class in the Kylberg dataset.

Fig. 6. Visual interpretation for classification result of "floor1" class in the Kylberg dataset.

Figure 6 shows that the related patches exhibit a structure similar to the prototypes, particularly for the prototypes No. 8 and 10 obtained from "cushion" class and "scarf2" class, respectively. Although these prototypes are not class related, they complement the shape to characterize for the classification. Reason for the compensation is the number of the prototypes for the "floor1" class was insufficient to characterize the class. As stated in above, determining appropriate number of prototype is needed to provide deficient prototypes.

Through these analyses, we have shown that our method can be used for decision interpretation by utilizing weighted relation between prototypes and class discriminative importance. This is a further advantage compared with conventional saliency-based methods such as Grad-CAM [13].

4.3   Experiment for PCAM Dataset

Dataset. We used PCAM dataset [15], which is derived from the CAMELYON '16 dataset [5]. It comprises 327,680 images extracted from the pathologic scans of lymph node sections, and the images are divided into 262,144 of the training dataset, 32,768 of the validation dataset, and 32,768 of the test dataset. The class labels indicate binary class, either tumor or non-tumor. Each image has 96 × 96 pixels, and we cropped 32 × 32 pixels of image patches to construct the prototypes. We set 200 to the number of prototypes for each class due to the large variation of the pathological images.

Results. We compared our method with two methods, which are a state-of-the-art DenseNet-based method (non-interpretable) [15] and a prototype-based CNN method (interpretable) [11].

Table 1. Classification results for PCAM dataset.

Method	Accuracy
P4M-DenseNet [15]	89.8%
Li et al. [11]	76.5%
Proposed method	77.0%

Table 1 summarizes the detection results for the PCAM dataset. Although the interpretable methods could not attain the state-of-the-art performance for this dataset, both methods yielded a performance of approximately 77%. Comparing our method with that of Li et al. [11], our method exhibits slightly improved detection results.

Figure 7 illustrates a part of prototypes obtained from each class, namely tumor and non-tumor, respectively. Although some prototypes are unclear, most of them are well interpretable. In contrast, the method [11] failed to obtain interpretable prototypes as reported by Chen et al. [3].

tumor   non-tumor

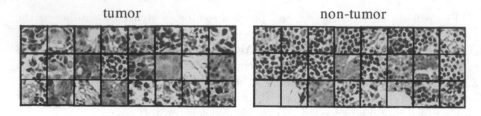

Fig. 7. A part of visualized prototypes obtained from tumor and non-tumor classes.

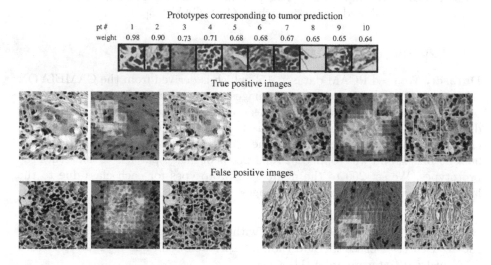

Fig. 8. Prototypes corresponding to decision for tumor class are shown, and true positive and false positive examples are displayed with heat-map and with marks to the important prototypes. The heat-map was created by accumulating weights of all prototypes whose importance exceed 0.1.

Figure 8 shows the prototypes obtained from the tumor class and interpretation for examples of images classified as tumor (true positives and false positives). As shown in the figure, the false positive images include cells similar to the prototypes.

Figure 9 shows the prototypes obtained from the non-tumor class and interpretation for examples of images classified as non-tumor (true negatives and false negatives). These true negative and false negative images include a part similar to the prototypes, especially a part highlighted by heat-map is more similar than other parts.

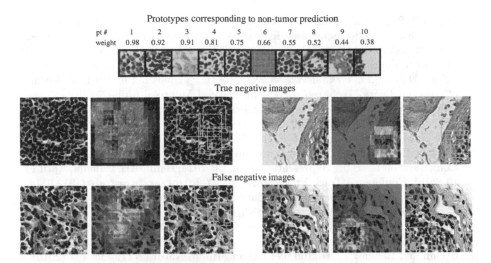

Fig. 9. Prototypes corresponding to decision for non-tumor class are shown, and true negative and false negative examples are displayed with heat-map and with marks to the important prototypes. The heat-map was created by accumulating weights of all prototypes whose importance exceed 0.1.

5   Conclusion

This paper has proposed a prototype-based classification method to improve the interpretability of decision made using the CNN for pathological images. We verified the method by using two textural datasets, namely the Kylberg and the PCAM datasets. Visualized prototypes and their associated weights make the rationale for the classification results interpretable. Furthermore, our heat-map based visualization has more advantages than conventional saliency-based methods owing to displaying association between class-discriminative parts with the prototypes.

References

1. Adadi, A., Berrada, M.: Peeking inside the black-box: a survey on explainable artificial intelligence (XAI). IEEE Access **6**, 52138–52160 (2018)
2. Arandjelović, R., Gronat, P., Torii, A., Pajdla, T., Sivic, J.: NetVLAD: CNN architecture for weakly supervised place recognition. In: IEEE Conference on Computer Vision and Pattern Recognition (2016)
3. Chen, C., Li, O., Barnett, A., Su, J., Rudin, C.: This looks like that: deep learning for interpretable image recognition. arXiv arXiv:1806.10574 (2018)
4. Csurka, G., Dance, C.R., Fan, L., Willamowski, J., Bray, C.: Visual categorization with bags of keypoints. In: Workshop on Statistical Learning in Computer Vision, ECCV, pp. 1–22 (2004)

5. Bejnordi, B.E., et al.: Diagnostic assessment of deep learning algorithms for detection of lymph node metastases in women with breast cancer. JAMA **318**(22), 2199–2210 (2017). https://doi.org/10.1001/jama.2017.14585. The CAMELYON16 Consortium machine learning detection of breast cancer lymph node metastases

6. Erhan, D., Bengio, Y., Courville, A., Vincent, P.: Visualizing higher-layer features of a deep network. Tech. rep. 1341, University of Montreal (June 2009)

7. Ioffe, S., Szegedy, C.: Batch normalization: accelerating deep network training by reducing internal covariate shift. In: Proceedings of the 32nd International Conference on Machine Learning, vol. PMLR37, pp. 448–456 (2015)

8. Kingma, D.P., Welling, M.: Auto-encoding variational bayes. In: 2nd International Conference on Learning Representations, ICLR 2014, Banff, AB, Canada, April 14–16, 2014, Conference Track Proceedings (2014). http://arxiv.org/abs/1312.6114

9. Kylberg, G.: The Kylberg texture dataset v. 1.0. External report (blue series) 35, Centre for Image Analysis, Swedish University of Agricultural Sciences and Uppsala University, Uppsala, Sweden (September 2011). http://www.cb.uu.se/~gustaf/texture/

10. LeCun, Y., Bengio, Y., Hinton, G.: Deep learning. Nature **521**, 436–444 (2015)

11. Li, O., Chen, H.L.C., Rudin, C.: Deep learning for case-based reasoning through prototypes: a neural network that explains its predictions. In: Proceedings of AAAI, pp. 123–456 (2018)

12. Liu, Y., et al.: Detecting cancer metastases on gigapixel pathology images. Tech. rep. arXiv arXiv:1703.02442 (2017)

13. Selvaraju, R.R., Cogswell, M., Das, A., Vedantam, R., Parikh, D., Batra, D.: Grad-cam: visual explanations from deep networks via gradient-based localization. In: International Conference on Computer Vision (ICCV). IEEE (2017)

14. Simonyan, K., Vedaldi, A., Zisserman, A.: Deep inside convolutional networks: visualising image classification models and saliency maps. CoRR arXiv:1312.6034 (2013)

15. Veeling, B.S., Linmans, J., Winkens, J., Cohen, T., Welling, M.: Rotation equivariant CNNs for digital pathology. In: Frangi, A.F., Schnabel, J.A., Davatzikos, C., Alberola-López, C., Fichtinger, G. (eds.) MICCAI 2018. LNCS, vol. 11071, pp. 210–218. Springer, Cham (2018). https://doi.org/10.1007/978-3-030-00934-2_24

16. Zeiler, M.D., Fergus, R.: Visualizing and understanding convolutional networks. In: Fleet, D., Pajdla, T., Schiele, B., Tuytelaars, T. (eds.) ECCV 2014. LNCS, vol. 8689, pp. 818–833. Springer, Cham (2014). https://doi.org/10.1007/978-3-319-10590-1_53

Design of an Optical Filter to Improve Green Pepper Segmentation Using a Deep Neural Network

Jun Yu[1], Xinzhi Liu[1,2], Pan Wang[1,3], and Toru Kurihara[1(✉)] (iD)

[1] Kochi University of Technology, Kochi, Japan
218005n@gs.kochi-tech.ac.jp, 234008d@ugs.kochi-tech.ac.jp,
kurihara.toru@kochi-tech.ac.jp
[2] Hefei University of Technology, Hefei, China
xinzhiliu@mail.hfut.edu.cn
[3] Taiyuan University of Technology, Taiyuan, China

Abstract. Image segmentation is a challenging task in computer vision fields. In this paper, we aim to distinguish green peppers from large amounts of green leaves by using hyperspectral information. Our key aim is to design a novel optical filter to identify the bands where peppers differ substantially from green leaves. We design an optical filter as a learnable weight in front of an RGB filter with a fixed weight, and classify green peppers in an end-to-end manner. Our work consists of two stages. In the first stage, we obtain the optical filter parameters by training an optical filter and a small neural network simultaneously at the pixel level of hyperspectral data. In the second stage, we apply the learned optical filter and an RGB filter in a successive manner to a hyperspectral image to obtain an RGB image. Then we use a SegNet-based network to obtain better segmentation results at the image level. Our experimental results demonstrate that this two-stage method performs well for a small dataset and the optical filter helps to improve segmentation accuracy.

Keywords: Hyperspectral image · Segmentation · DNN for design · Optical filter · SegNet · Agriculture

1 Introduction

The automatic picking of green peppers is a challenging task in agricultural applications because the colors of green peppers and leaves are similar. An automatic harvesting system for vegetables and fruits is attracting widespread interest in research fields, such as computer vision, robots [7], and the Internet of Things. There are several methods for distinguishing a green pepper from its leaves in the research fields of image processing and computer vision. Recently, some research has focused on hue, saturation, and the value of the color space to classify a

---

X. Liu and P. Wang—This work was done during an internship at Kochi University of Technology.

S. Palaiahnakote et al. (Eds.): ACPR 2019, LNCS 12047, pp. 653–666, 2020.
https://doi.org/10.1007/978-3-030-41299-9_51

green pepper and its leaves [3]. An alternative approach was developed using a three-dimensional (3D) sensor for the detection and segmentation of fruits and vegetables in 3D space [5]. Despite this, there remains a need for an algorithm to identify crops for which leaves and the remaining backgrounds are similar colors. The final goal of this research is to obtain a good segmentation results for automatic picking or prediction of green pepper growth. For such a purpose, the object detection is not suitable, since the detection results cannot be used for cutting stem of fruit or estimation of fruit size although it can be used for counting.

One approach to improve the system is to incorporate hyperspectral imaging, which collects and transfers various information at different wavelengths; hence, it can provide rich information for classifying vegetables and leaves, even if the colors are similar.

Hyperspectral image segmentation is most commonly used in remote sensing [2,8]. Its research purpose is to achieve land cover classification, including vegetation mapping. In the agricultural field, hyperspectral imaging has been used to detect green citrus using pixel-wise linear discriminant analysis followed by spatial image processing [9]. Its purpose is to improve segmentation results using all the information in high-dimensional space. Therefore, unlike RGB, it does not need to degenerate to three dimensions from a higher spectral dimension.

Image segmentation is a major topic in image processing and computer vision, which can build a solid foundation for understanding images and solving another computer vision tasks. Traditional image segmentation is based on threshold methods [10], edge detection methods [11], and clustering methods. Much more research in recent years has used deep learning to improve the performance of image segmentation.

A breakthrough was achieved by fully convolutional neural networks (FCN), which convert a fully connected layer to a convolutional layer [6,12]. Despite the successful results of the FCN model for dense pixel-wise segmentation, it still has some problems regarding losing the spatial resolution of feature maps and some boundary information of objects. To solve these problems, the SegNet [1] structure was proposed, which adopts encoder and decoder architecture to improve performance. The encoder part of SegNet is the same as that of the first 13 convolutional layers of the VGG16 network [13], and the corresponding decoder part uses pooling indices to upsample the feature map. Because of the pooling indices and fully convolutional layer, SegNet is smaller and quicker than other architectures in the training and evaluation stages. However, most image segmentation studies have focused only on RGB space. Few studies have addressed the continuous wavelength in the real world.

In this paper, we present a novel approach to use a hyperspectral image and the power of a deep neural network to design an optical filter that improves the segmentation of green peppers. Although a hyperspectral image carries a large amount of information, a hyperspectral camera is expensive and hyperspectral data processing requires a large amount of memory; hence, there are some difficulties in introducing hyperspectral imaging to many farmers. We propose a method that uses hyperspectral information as much as possible in spite of using a widely available, cheap RGB camera. A conceptual diagram is shown in Fig. 1.

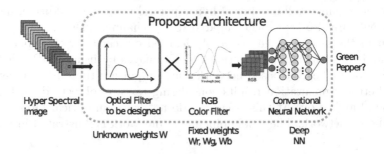

Fig. 1. The conceptual diagram of the proposed method. The conventional deep neural networks focus on RGB color image. In this research we focus on spectral data, and we take into account of spectrum in deep neural networks with RGB filter. Hereby, this architecture uses hyperspectral information as much as possible in spite of using a widely available, cheap RGB camera.

The proposed method has three steps: (1) pixel-wise learning to design optical filter transmittance for a small number of hyperspectral images; (2) SegNet-based two-dimensional (2D) learning to improve the segmentation results; and (3) the implementation of an optical filter as a physical device and the estimation of a segmentation map using the captured RGB images. We report the first and second steps in this paper.

The contributions of this paper are as follows:

- We present a two-stage architecture for small hyperspectral datasets.
- We present the end-to-end design of an optical filter to improve segmentation results.
- We use hyperspectral information as much as possible using a widely available, cheap RGB camera.

Our paper is organized as follows: In the Sect. 2, we describe the entire pipeline of our method and the details of the architecture of our neural network. In the Sect. 3, we explain the explicit results of our experiment, such as the training process and segmentation results. In the Sect. 4, we summarize the entire paper and report our future work in these fields.

2   Method

2.1   Overview

Our work is divided into two stages. In the first stage, an optical filter is designed to find the effective bands where peppers and green leaves have the biggest difference. We make a model including optical filter with learnable weights, RGB filter with fixed weights, and a small neural network. We train the optical filter and the small neural network simultaneously with pixel-wise samples from hyperspectral images, which means the end-to-end design of an optical filter. Even one hyperspectral image has 480×640 samples so that it's enough to learn by using only small number of hyperspectral images. In addition, memory requirements are also kept low compared to 121 channels 2DCNN.

Then in the second stage, we apply learned optical filter to hyperspectral images and then calculate conventional RGB images by using RGB color filter response. We use SegNet-based network whose input are these RGB images for getting better segmentation results considering texture, object edges, lighting, background and so on. At this stage, we do not learn an optical filter but just SegNet-based network. The whole pipeline of our work is illustrated in Fig. 2.

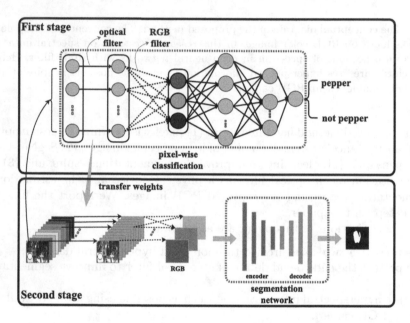

Fig. 2. The pipeline of our method. The learned optical filter at pixel level in the first stage is transferred to the second stage for further refined segmentation at image level. RGB filter is fixed represented by black dashed line, and the learnable parts are enclosed by red dashed line. (Color figure online)

2.2   Spectral Analysis

We take hyperspectral image as input for the purpose of using hyperspectral information to get a better accuracy. Hyperspectral image has C channels, where C is the number of spectral bands and is of size $H \times W$. In our task, we used 121 bands from 400 nm to 1000 nm with an interval of 5 nm. The size of input image is 480×640 and the value of hyperspectral image ranges from 0 to 4095, which is different from that of RGB image. For a better understanding of the design of an optical filter, we made a simple display for the spectral curve of peppers and green leaves, which is illustrated in Fig. 3(b), and the spectral curve is drawed by taking average of some regions on a sample image which is showed in Fig. 3(a), where we use green and blue rectangle to pick pepper and leaf samples respectively. It seems that peppers differ a lot from green leaves mainly in infrared band (700–800 nm) and green band (500–600 nm), and this fact motivates us to design an optical filter to select useful bands for classification and segmentation.

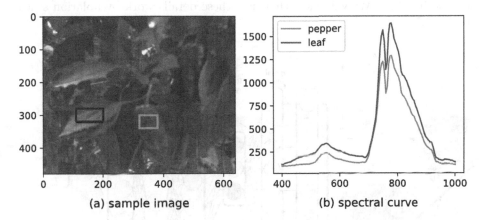

| (a) sample image | (b) spectral curve |

Fig. 3. Spectral analysis on the sample image. All the curves in (b) are drawed by taking average of some regions in the sample image which illustrated in (a). We pick pepper samples within green rectangle and leaf samples within blue rectangle. (Color figure online)

2.3 Pixel-Wise Segmentation

Due to the fact that peppers are well distinguished from green leaves in some specific bands, we introduce an optical filter to pass useful bands and suppress relatively useless bands. However, hyperspectral image is expensive to obtain and we have only 10 images for training out of 20 images. Thus, we trained the optical filter at pixel level to make the most of resources.

There lies two merits. First, it solves the problem of datasets for training a neural network, for we have $10 \times 480 \times 640$ available samples to select for training datasets. Second, pixel-wise data take less memory than a whole image, thus leading to more efficient for learning an optical filter. We design a neural network for pixel-wise segmentation. Initially, we take pixel-wise input with size 1×121 and let it get through an optical filter, operated by element-wise multiplication. Considering physical meaning, we limit the weights in optical filter between 0 and 1. Secondly, an RGB filter follows the optical filter to simulate the process of obtaining RGB image in optical instruments, such as cameras and smart phones. Thirdly, we use a small neural network containing 3 dense layers for binary classification and get first segmentation results based on pixel-wise classification.

We add Batch Normalization (BN) [4] layer after RGB filter and apply L1 regularization to the first and second layer in the small neural network, with weight decay 0.1 and 0.01, respectively. Moreover, inspired by [14], we expand features after RGB filter by adding 4 other components, $G/R, B/R, G/(G + R), B/(B + R)$, to achieve better classification results. The whole neural network is trained in an end-to-end manner, which means optical filter is trained together with the later small neural network. We use binary cross-entropy loss function as the objective function in the training stage. We present the whole pixel-wise pipeline in Fig. 4. We will show that how these details work by ablation study in Sect. 3.5.

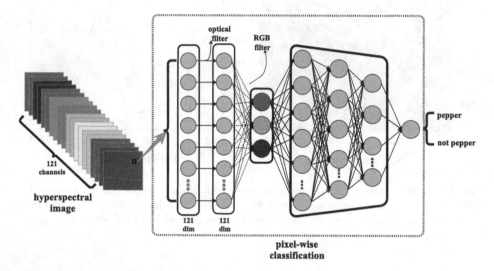

Fig. 4. The pipeline of pixel-wise classification. A pixel data gets through optical filter, RGB filter and a small neural network sequentially, to be predicted a pepper or not. Different from dense layer, there is element-wise connection within optical filter.

2.4   SegNet-Based Network

In this part, we design our segmentation network inspired by SegNet. It means we discard the fully connected layers in favor of retaining higher resolution feature maps. Overall, our network has an encoder and a corresponding decoder, followed by a pixelwise classification layer, which is illustrated in Fig. 5. The encoder network consists of 8 convolutional layers. Each encoder layer has a corresponding decoder layer. Similar to SegNet, batch normalization is used behind each convolutional layer, and the deepest encoder output is upsampled step by step through 4 blocks. However, compared to SegNet for semantic segmentation including many classes, our task is a binary classification task essentially. Therefore, we use bilinear interpolation in upsampling layer adopting simple manner, and this also reduces the memory for remembering the pooling indices proposed in SegNet. Ultimately, the final decoder output is fed to a sigmoid classifier to produce class probabilities for each pixel independently. In the second stage, binary cross-entropy loss is used for training SegNet-based network.

All the input images are resized to 480×480 (square). We use 3×3 kernel with stride 1 in all convolutional layers and 2×2 kernel with stride 2 in all max-pooling layers.

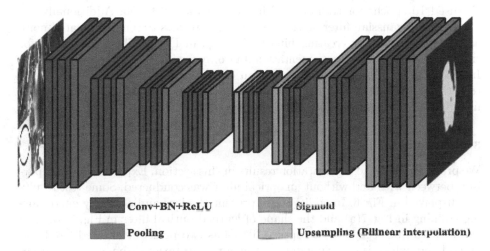

Conv+BN+ReLU   **Sigmoid**

Pooling   **Upsampling (Bilinear interpolation)**

Fig. 5. The structure of our SegNet-based network. Inspired by SegNet, we utilize convolutional encoder-decoder network for refined segmentation. There are only convolutional layer, pooling layer and upsampling within, and the extracted feature is upsampled through bilinear interpolation step by step, which is similar to the SegNet.

3 Experiments and Results

3.1 Datasets

Since there are no available hyperspectral pepper datasets on the internet, hyperspectral images of green peppers were taken by our own hyperspectral camera (NH-2 by EBA JAPAN CO., LTD.). In the first stage, we selected 10 images to pick samples fort training and the rest 10 images were used for testing. In order to increase the diversity of training data, we picked leaf samples both near and far from green peppers, both bright and dark in appearance. In addition, each sample was divided by white reflectance calculated by taking average of the white reflectance standards in hyperspectral image, to compensate spectral reflectance. We selected 120445 samples including 56,967 positive samples (green peppers) and 63478 negative samples (green leaves). In the second stage, we used 18 images for training and 2 images for testing. Furthermore, data augmentation was applied because 18 images were far from enough to train such a deep convolutional neural network. Specifically, we conducted data augmentation to each RGB image generated from hyperspectral images by horizontal flips, shift, scale transformation and rotation, with a result that the whole dataset is enlarged by a factor of 200. We used Photoshop to make segmentation label for whole 20 images.

3.2   Implementation Detail

In first stage, we adopted Nadam with a learning rate 0.001 for optimization. We used 10 epochs for training and batch size was set to 50. Additionally, we applied 5×5 gaussian filter on the segmentation results to remove isolated noise as post-processing for creating binarized image in Fig. 6. This is not used for training step in pixel-wise segmentation, but only for identifying resulting image. In second stage, Adagrad was used as an optimizer and the batch size was set to 4 through 7-epoch training. We implemented our models and trained them using a NVIDIA GeForce1080Ti GPU.

3.3   Results of Pixel-Wise Network

We present pixel-wise segmentation results in this section. Experimental comparison between with and without an optical filter was conducted. Some test results are displayed in Fig. 6. In addition, we present the loss and accuracy curve during training in Fig. 7(a) and the shape of learned optical filter in Fig. 7(b). As a result, it can be seen that the optical filter has two peaks in infrared band and green band, which is corresponding to the our expectations from spectral curves of peppers and green leaves.

3.4   Results by SegNet-Based Network

Our SegNet-based network used the optical filter parameters learned in the first stage, and then performed image segmentation on RGB images. Experimental comparison between with and without an optical filter was conducted to demonstrate that an optical filter really improved segmentation of green peppers. We also compared our network with FCN, which indicates that our network is more effect on the ad-hoc task. All the experimental results are depicted in Fig. 8.

In Fig. 8, it can be observed that the results with an optical filter are more fine-grained than those without an optical filter. For instance, some backgrounds are mistaken for green peppers in column (e) while they are correctly classified in column (f). This is because there is much background information similar to green pepper in the image and the optical filter can help to find where green peppers differ from backgrounds, so that it is easier for classifier to judge whether one pixel belongs to a pepper or not. Furthermore, we observe that our network outperforms FCN (with optical filter) in this ad-hoc task from column (d) and column (e).

Following FCN, we report three kinds of metrics to measure the segmentation accuracy of our network. Let p_{ij} represents the number of pixels that belong to class i but are predicted to belong to class j. Since this is a binary classification task, we set $k = 1$. The quantitative results are shown in Table 1.

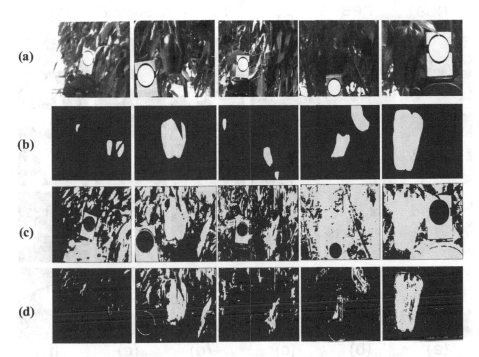

Fig. 6. The results by pixel-wise classification Row (a) is the RGB images generated from hyperspectral images. Row (b) is the corresponding label made by Photoshop. Row (c) is the output result without an optical filter. Row (d) is the output result with an optical filter.

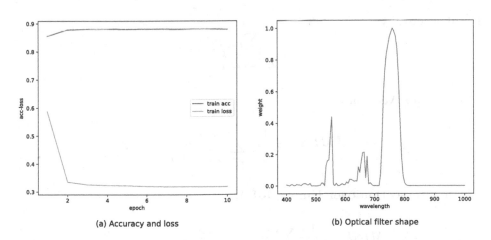

(a) Accuracy and loss

(b) Optical filter shape

Fig. 7. The results by pixel-wise learning: (a) shows the training accuracy and loss as the epoch increases, (b) shows the shape of the learned optical filter. (Color figure online)

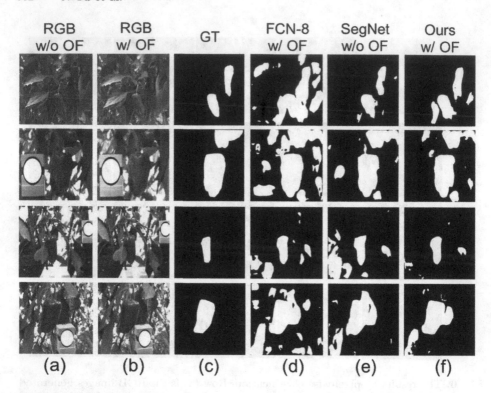

Fig. 8. The results by SegNet-based network. From left column to right column, original images, RGB images through optical filter, ground truth. results by FCN with optical filter, results without optical filter and results with optical filter, respectively.

Pixel Accuracy:

$$PA = \frac{\sum_{i=0}^{k} p_{ii}}{\sum_{i=0}^{k} \sum_{j=0}^{k} p_{ij}}$$

Mean Pixel Accuracy:

$$MPA = \frac{1}{k+1} \sum_{i=0}^{k} \frac{p_{ii}}{\sum_{j=0}^{k} p_{ij}}$$

Mean Intersection over Union (IoU):

$$MIoU = \frac{1}{k+1} \sum_{i=0}^{k} \frac{p_{ii}}{\sum_{j=0}^{k} p_{ij} + \sum_{j=0}^{k} p_{ji} - p_{ii}}$$

Table 1. Comparison results between 3 models measured by PA, MPA, and MIoU. It can be observed that an optical filter does work and our network outperforms FCN-8s.

	FCN-8s	Ours without OF	Ours with OF
PA	81.2%	90.9%	94.0%
MPA	57.0%	73.9%	79.4%
MIoU	56.8%	71.2%	78.2%

3.5 Ablation Study

In order to study the effects of each component in pixel-wise segmentation model, we gradually modified this model and compared their differences. The overall visual comparison is illustrated in Fig. 9. Each column represents a model with its configurations shown in the top. The red sign indicates the main improvement compared with the previous model. A detailed discussion is provided as follows.

BN and L1 Regularization. Batch normalization (BN) is added after RGB filter, and L1 regularization is adopted in the first and second layers of the small neural network. This is because we observed a high accuracy on the training set and a poor performance on the samples which had not been seen before. As a result, the introduction of BN and L1 regularization can indeed improve qualitative performance visually and alleviate overfitting problems. As can be seen from the 2nd and 3rd columns in Fig. 9, there is a significant improvement on the quality of segmentation results.

Feature Expansion. We expand features for binary classification by introducing 4 other features, $G/R, B/R, G/(G + R), B/(B + R)$, derived from R, G, B components. Intuitively, more features can help to make more accurate classification, which can be proved by the results. Accordingly, a slight improvement can be observed from the 4th and 5th columns in Fig. 9.

Optical Filter. The key insight of this paper is the design of the optical filter. It not only improves the accuracy in pixel-wise segmentation, but also refines the results in further segmentation. For example, some backgrounds like reflectance standard holder were mistaken for green peppers in the 3rd column while they were correctly classified in the 4th column, as shown in Fig. 9.

We also show some quantitative results for each model above in Table 2. By the ablation study, we can clearly observe that how each component contributes to this model. Thanks to the improvements mentioned above, we can get better optical filter parameters and employ it in SegNet-based network as a strengthening equipment for fine-grained segmentation.

	1st	2nd	3rd	4th	5th
BN+L1?	×	✓	✓	✓	
Optical filter?	×	×	✓	✓	
Feature expansion?	×	×	×	✓	

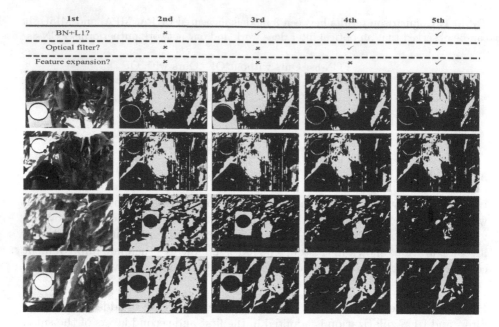

Fig. 9. Overall visual comparisons for showing the effects of each component in pixel-wise model. Each column represents a model with its configurations in the top. The red sign indicates the main improvement compared with the previous model. (Color figure online)

Table 2. Quantitative comparison results in pixel-wise model on accuracy between 4 models with slight differences, where OF represents an optical filter, L1 represents L1 regularization and FE represents feature expansion.

	Training	Validation	Test
w/o OF	90.48%	90.92%	70.02%
w/o OF+BN+L1	83.86%	82.73%	76.97%
w/ OF+BN+L1	85.72%	85.92%	81.19%
w/ OF+BN+L1+FE	87.58%	86.64%	87.94%

4   Conclusion

We have presented a two-stage model that achieves great performance on the green pepper segmentation problem, which can be widely applied in agricultural fields. The aim of this research is not object detection for green pepper counting, but segmentation for automatic picking or growth prediction of green pepper.

Our method consists of two parts. Initially, we obtain an optical filter by training it together with a small neural network in an end-to-end manner at pixel level. Then, in order to eliminate the pixel-wise limitation and improve segmentation results, we apply learned optical filter to the hyperspectral image to get filtered image in RGB channel for further segmentation by SegNet-based network. As discussed in the paper and presented by the experimental results, the optical filter plays an important role throughout the whole model, with its significant improvements on the accuracy and quality of segmentation results.

However, there still remains some drawbacks. Since there was no enough memory for us to train an optical filter at image level, we had to divide this work into two parts and trained the optical filter separately. In the future, we aim to build upon this framework to overcome these drawbacks and investigate the possibility of training an optical filter in an integrated end-to-end manner.

Acknowledgments. This work was supported by Cabinet Office grant in aid, the Advanced Next-Generation Greenhouse Horticulture by IoP (Internet of Plants), Japan.

References

1. Badrinarayanan, V., Kendall, A., Cipolla, R.: SegNet: a deep convolutional encoder-decoder architecture for image segmentation. IEEE Trans. Pattern Anal. Mach. Intell. **39**(12), 2481–2495 (2017)
2. Chen, Y., Jiang, H., Li, C., Jia, X., Ghamisi, P.: Deep feature extraction and classification of hyperspectral images based on convolutional neural networks. IEEE Trans. Geosci. Remote Sens. **54**(10), 6232–6251 (2016). https://doi.org/10.1109/TGRS.2016.2584107
3. Eizentals, P., Oka, K.: 3D pose estimation of green pepper fruit for automated harvesting. Comput. Electron. Agric. **128**, 127–140 (2016)
4. Ioffe, S., Szegedy, C.: Batch normalization: accelerating deep network training by reducing internal covariate shift. arXiv preprint arXiv:1502.03167 (2015)
5. Lehnert, C., English, A., McCool, C., Tow, A.W., Perez, T.: Autonomous sweet pepper harvesting for protected cropping systems. IEEE Robot. Autom. Lett. **2**(2), 872–879 (2017)
6. Long, J., Shelhamer, E., Darrell, T.: Fully convolutional networks for semantic segmentation. In: Proceedings of the IEEE Conference on Computer Vision and Pattern Recognition, pp. 3431–3440 (2015)
7. McCool, C., Sa, I., Dayoub, F., Lehnert, C., Perez, T., Upcroft, B.: Visual detection of occluded crop: for automated harvesting. In: 2016 IEEE International Conference on Robotics and Automation (ICRA), pp. 2506–2512. IEEE (2016)
8. Melgani, F., Bruzzone, L.: Classification of hyperspectral remote sensing images with support vector machines. IEEE Trans. Geosci. Remote Sens. **42**(8), 1778–1790 (2004). https://doi.org/10.1109/TGRS.2004.831865
9. Okamoto, H., Lee, W.S.: Green citrus detection using hyperspectral imaging. Comput. Electron. Agric. **66**(2), 201–208 (2009). https://doi.org/10.1016/j.compag.2009.02.004. http://www.sciencedirect.com/science/article/pii/S0168169909000258

10. Otsu, N.: A threshold selection method from gray-level histograms. IEEE Trans. Syst. Man Cybern. **9**(1), 62–66 (1979)
11. Senthilkumaran, N., Rajesh, R.: Edge detection techniques for image segmentation-a survey of soft computing approaches. Int. J. Recent Trends Eng. **1**(2), 250 (2009)
12. Shelhamer, E., Long, J., Darrell, T.: Fully convolutional networks for semantic segmentation. IEEE Trans. Pattern Anal. Mach. Intell. **39**(4), 640–651 (2017). https://doi.org/10.1109/TPAMI.2016.2572683
13. Simonyan, K., Zisserman, A.: Very deep convolutional networks for large-scale image recognition. arXiv preprint arXiv:1409.1556 (2014)
14. Stigell, P., Miyata, K., Hauta-Kasari, M.: Wiener estimation method in estimating of spectral reflectance from RGB images. Pattern Recogn. Image Anal. **17**(2), 233–242 (2007)

Eye Contact Detection from Third Person Video

Yuki Ohshima$^{(\boxtimes)}$ and Atsushi Nakazawa

Department of Informatics, Kyoto University, Kyoto, Japan
ohshima@ii.ist.i.kyoto-u.ac.jp,
nakazawa.atsushi@i.kyoto-u.ac.jp

Abstract. Eye contact is fundamental for human communication and social interactions; therefore much effort has been made to develop automated eye-contact detection using image recognition techniques. However, existing methods use first-person-videos (FPV) that need participants to equip wearable cameras. In this work, we develop an novel eye contact detection algorithm taken from normal viewpoint (third person video) assuming the scenes of conversations or social interactions. Our system have high affordability since it does not require special hardware or recording setups, moreover, can use pre-recorded videos such as Youtube and home videos. In designing algorithm, we first develop DNN-based one-sided gaze estimation algorithms which output the states whether the one subject looks at another. Afterwards, eye contact is found at the frame when the pair of one-sided gaze happens. To verify the proposed algorithm, we generate third-person eye contact video dataset using publicly available videos from Youtube. As the result, proposed algorithms performed 0.775 in precision and 0.671 in recall, while the existing method performed 0.484 in precision and 0.061 in recall, respectively.

Keywords: Eye contact · Deep neural nets · Image recognition · Third person video · Human communication

1 Introduction

Eye contact (mutual gaze) is a fundamental part of human communication and social interaction. In psychology, the 'eye-contact effect' is the phenomenon in which perceived eye contact with another human face affects certain aspects of the concurrent and/or immediately following cognitive processing [16]. Thus, eye contact greatly affects human behaviour in areas such as affective perceptions [1] and social interactions [4]. Eye contact is also used in medicine, such as in the diagnosis of autism spectrum disorders (ASDs) [7]. In dementia nursing, making appropriate eye contact is an important skill for communicating with patients [6]. Our research mainly focuses on the nursing scenario, particularly on evaluating tender-care nursing skills for dementia by examining facial communication behaviours between caregivers and patients, such as the number of eye contact

© Springer Nature Switzerland AG 2020
S. Palaiahnakote et al. (Eds.): ACPR 2019, LNCS 12047, pp. 667–677, 2020.
https://doi.org/10.1007/978-3-030-41299-9_52

Fig. 1. Eye contact detection from third person video. Our algorithm detects the one-sided gaze from one subject to another. When one-sided gaze occurs at each other, the eye contact (mutual gaze) is engaged.

events and the relative facial positions and distances between caregiver and care receiver [12]. However, existing methods are mostly using first-person-camera that need participants to equip wearable cameras, therefore application scenario has been limited.

There are several work that tackled this problem. One representative examples is the GazeFollow algorithm that obtains the point of regard (PoR) from third person video [15]. This method combines eye appearance and scene visual saliency, however, as shown in Fig. 2, such 'general' PoR estimation algorithms do not work for the conversational scenes where people are looking at each other.

In this paper, we develop a novel eye contact detection algorithm taken from normal viewpoint (third person video) such as the scenes of conversations or

Fig. 2. Failure cases of Gazefollow's gaze target detection [15]. Since hands are salient in the scene, Gazefollow wrongly infers the gaze directions to the hands.

social interactions. Our system have high affordability and varieties of applications including communication analysis for Youtube and home videos. In designing algorithm, we assume the eye contact consists of the pair of *one-sided gaze* – situations when a subject looks at another subject – as illustrated by Fig. 1. Namely, we first develop a detection algorithm of one-sided gaze from an input frame. When one-sided gaze occurs each other, e.g. from subject A to B and from subject B to A, at the same time, we assume an eye contact (mutual eye contact) is engaged.

We developed three DNN-based one-sided gaze estimation algorithms that take eye locations, gaze directions and eye images, and compared the results. To evaluate the algorithms, we develop a third person eye contact video dataset that includes five videos taken from Youtube, in which people converse in front of a camera. In every frame in the video, facial positions and gaze states are annotated. Experimental result shows that proposed algorithm s that use both gaze directions and eye images outperformed the approaches that only use gaze directions and a state-of-the-art.

The remainder of the paper is organized as follows. First we describe the related work regarding eye contact detection and its applications. Next, we show our one-sided gaze detection algorithms and the third person eye contact video dataset, followed by experimental result and conclusion.

2 Related Work

Detecting eye contact events is important for understanding social communication and designing communication robots. Therefore, several studies in this area have been conducted. Smith et al. [17] proposed an algorithm to detect *gaze-locking* (looking at a camera) faces using eye appearances and PCA plus multiple discriminate analysis. Ye et al. developed a pioneering algorithm that detects mutual eye gaze using wearable glasses [18]. In recent years, deep-learning-based approaches are being implemented for eye-contact detection. Mitsuzumi et al. developed the DNN-based eye contact detection algorithm (DeepEC) [11] that uses only cropped eye regions for eye-contact detection and performed better than existing methods. Eunji et al. develop the DNN-based PiCNN detector that accepts the facial region and output both facial postures and eye contact states [3]. Zhang et al. presented an eye-contact detection algorithm based on their deep neural network (DNN) based gaze estimations [19].

In robotics, Petric et al. developed an eye contact detection algorithm that uses facial images taken with a camera embedded in a robot's eyes [13] to develop robot-assisted ASD-diagnosis systems. These eye contact detection algorithms depend on facial landmark detection libraries or gaze estimation algorithms with which it is assumed that subject faces are not occluded.

Image-based gaze estimation algorithms have also been recently studied. The current trend in this area is deep learning-based approaches, namely, learning and predicting gaze directions according to datasets that describe the relation between facial images, facial landmarks, and gaze points. For example, Lu et al. developed a head pose-free gaze estimation method by synthesizing eye images from small samples. However, their method requires personal-dependent eye image samples taken under experimental setups [10]. Zhang et al. proposed a DNN algorithm that inputs eye images and 3D head poses obtained from facial landmark points [20]. They also developed a DNN-based algorithm using full facial images without occlusions [21]. Krafka et al. developed a DNN-based eye gaze estimation algorithm that inputs full facial images as well as eye images [8]. However, these methods use first person videos for eye contact detection. On the contrary, Recasens et al. proposes a method to estimate where a person is looking at from a third person image using the facial image and image saliency [14]. In our experiment, we compare their methods and proposed algorithms. Lian et al. proposed a network that used multi-scale field of gaze direction estimation instead of image saliency and recorded higher accuracy than Gazefollow [9].

3   Algorithms

As described above, since eye contact is a pair of one-sided gazes – the state the one person looking at another and vise versa – we first develop the algorithm to detect one-sided gaze then apply it to all combinations between the individuals in the scene. Finally, we find the engagement of mutual gazes by finding the pair of one-sided gaze at the same time. In the followings, we describe our four DNN-based one-sided gaze detection algorithms, then show a mutual-gaze detection algorithm.

3.1   One-Sided Gaze Detection Algorithms

We implemented four one-sided gaze detection algorithms and compared the performances. Specifically, we first apply OpenFace [2] to find facial boundaries, facial parts and gaze directions of all subjects in images, and use them for one-sided gaze detection as illustrated in Fig. 3. The details of the algorithms are as follows.

(a) **Baseline Algorithm (Baseline)** takes a gaze direction of subject A and eye positions of subjects A and B as inputs of a neural network which consists of two fully connected layers. Each fully connected layer has six units for person A's gaze estimation and four units for eye positions of both persons and uses Leaky ReLU as activation function. After passing through the output

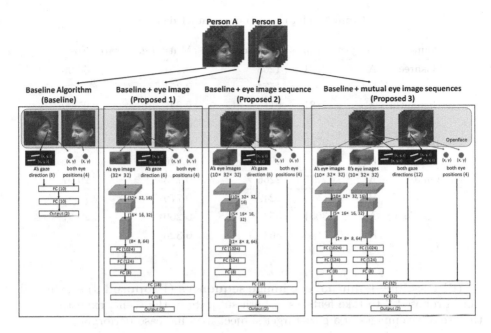

Fig. 3. The structures of the one-sided eye contact detection algorithms. Our model first detects gaze directions, eye positions, and eye images using OpenFace and uses these information as the inputs of the CNN.

layer, we use softmax function to obtain the probability. The loss function is cross entropy between the output probability of our system and correct value 0 or 1.

(b) **Baseline + eye image (Proposed 1)** takes eye images of subject A in addition to the inputs of the baseline algorithm. We obtain eye images, using facial landmark points of OpenFace. Eye images are converted to grayscale, resized to 32×32 and normalized by global contrast normalization (GCN). We obtain features of eye image by six 2D-CNN layers and merge with other information (gaze direction and eye positions) with two fully connected layers. The number of convolutional filters is 16 for first two layers, 32 for next two layers, and 64 for last two layers. In all convolutional layers, filter size is 3×3 and zero padding is conducted. We set 2D-max pooling layers every two convolutional layers. Pool size is 2×2 and stride is $(2, 2)$. We use three fully connected layers to lower the dimension of eye image feature vector before merging. Each layer has 1024, 124, and 8 units.

(c) **Baseline + eye image sequence (Proposed 2)** takes the image sequences for inference. Namely, we extend the architecture of Proposed 1 to take image sequence by using 3D-convolutional layers instead of 2D-convolutional layers and using 3D-max pooling layers instead of 2D-max pooling layers. Pool size is $2 \times 2 \times 2$ and stride is $(2, 2, 2)$. Each convolutional layer uses LeakyReLU as activation function. The number of image sequence is set to ten.

Table 1. Third person eye contact dataset.

Name	Subject	Frames	One-sided gaze	Mutual gaze rate	Fps
Sushree	A	3001	1208	0.457	30.00
	B		1574	0.351	
Samantha	A	2881	911	0.347	25.00
	B		785	0.403	
HCA	A	2789	572	0.330	30.00
	B		591	0.320	
Ollie	A	2775	2266	0.773	29.97
	B		2249	0.779	
Nishikawa	A	3001	554	0.209	29.97
	B		363	0.320	

(d) **Baseline + mutual eye image sequences (Proposed 3)** is extended from Proposed 2 to take features of both subjects in the scene, expecting that the use of eye images of a gaze-target subject may increase performance.

3.2 Mutual Eye Contact Detection

Finally, we detect the engagement of mutual eye contact by finding the frame when one-sided gaze occurs each other, e.g. from subject A to B and subject B to A. We apply this step for all combinations of subjects in a frame.

4 Third-Person Eye-Contact Video Dataset

For learning and evaluating the algorithms, we prepared the third-person eye-video dataset using publicly available videos from YouTube where two or more people are seating and talking in front of a camera. The details and example frames are shown in Table 1 and Fig. 4. At each frame in videos, an annotator provides gaze states between subjects, namely whether subject A looks at subject B (1 or 0) and vice versa. Subject names (A or B), facial regions and eye contact states are included in the dataset.

5 Result

We compared four proposed algorithms and existing Gazefollow [14]. The learning and testing are performed by leave-one-out manner. In all experiments, we used a Nadam optimizer [5] with the learning rate set to 0.001, the decay as 0.004 per epoch and β_1 and β_2 as 0.9 and 0.999, respectively.

The experimental result is shown in Table 2. Overall, proposed algorithms outperformed Baseline and Gazefollow. Specifically, Proposed 2 performed best

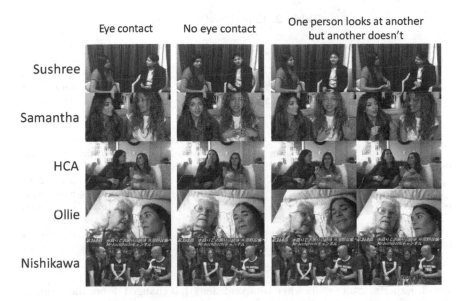

Fig. 4. Example frames of our dataset.

on average in F_1 measure, followed by Proposed 1 and 3. This shows the eye image provides key information for eye contact detection although the image-based gaze direction, provided by OpenFace, are used in Baseline. In addition, Proposed 2 outperformed Proposed 1 due to the use of temporal inference.

The Gazefollow could not perform the same performance even as Baseline. This is because the algorithm uses the scene saliency for estimating gaze target that mostly emphasizes the human hands as shown in Fig. 2 even when the subject looks at the other's face. Thus, we expect the combination of the proposed algorithms and Gazefollow-like saliency information may realize the gaze target estimation for general third person videos.

Proposed 3 used eye images of both subject A and B for inference, but could not outperform Proposed 2 on average. However, in several cases such as Sushree A, Samantha B and Ollie A, Proposed 3 performed considerably better performance than Proposed 2 as shown in Fig. 7. These data have relatively larger occurrence rate of mutual gaze (eye contact) than others as shown in Table 2. Therefore, Proposed 3, that use the eye information of both side, worked better than Proposed 1 and 2 that only use the information of the gaze source subject.

The performance of the 'Nishikawa' is fair to other footage. In the video, subjects A and B are the left-most and right-most individuals. As shown in Fig. 5, when the subject B looks at the subject A, the head direction often goes to the backward of the screen therefore the facial parts are occluded. For this reason, the gaze direction cannot be estimated correctly. In addition, the eye images of the subject B are smaller than others as shown in Fig. 6.

Table 2. Experimental results.

Video		Sushree		Samantha		HCA		Ollie		Nishikawa		
Subject		A	B	A	B	A	B	A	B	A	B	total
Baseline	Precision	0.464	0.515	0.251	0.145	0.206	0.212	0.400	0.793	0.095	0.000	0.335
	Recall	0.530	0.933	0.642	0.236	1.000	1.000	0.003	0.406	0.063	0.000	0.451
	F_1	0.495	0.664	0.361	0.179	0.341	0.350	0.005	0.537	0.076	0.000	0.384
Proposed1	Precision	0.956	0.625	0.580	0.969	0.892	0.849	0.970	0.951	0.887	0.345	0.769
	Recall	0.341	0.999	0.993	0.603	0.691	0.936	0.395	0.566	0.953	0.289	0.642
	F_1	0.503	0.769	0.733	0.743	**0.778**	**0.891**	0.562	0.709	**0.919**	0.315	0.700
Proposed2	Precision	0.757	0.677	0.957	0.995	0.644	0.675	0.843	0.818	0.853	0.553	**0.775**
	Recall	0.262	0.960	0.812	0.752	0.802	0.992	0.151	1.000	0.879	0.399	**0.671**
	F_1	0.390	**0.794**	**0.879**	0.856	0.714	0.803	0.257	**0.900**	0.866	**0.464**	**0.719**
Proposed3	Precision	0.840	0.768	0.799	0.922	0.627	0.839	0.976	0.958	0.390	0.093	0.641
	Recall	0.466	0.662	0.940	0.869	0.535	0.548	0.486	0.624	0.982	0.576	0.635
	F_1	**0.600**	0.711	0.863	**0.894**	0.577	0.663	**0.649**	0.756	0.559	0.160	0.638
Gazefollow [14]	Precision	0.400	**1.000**	0.202	0.000	0.000	0.000	0.447	0.808	0.000	0.000	0.484
	Recall	0.005	0.014	0.052	0.000	0.000	0.000	0.094	0.174	0.000	0.000	0.061
	F_1	0.010	0.483	0.082	0.000	0.000	0.000	0.156	0.286	0.000	0.000	0.109

Table 3. Experimental results when time scale length is changed. From the comparison of F_1 values, we concluded that the time scale of 10 frames is optimal for eye contact detection.

Video		Sushree		Samantha		HCA		Ollie		Nishikawa		
Subject		A	B	A	B	A	B	A	B	A	B	total
1frame Proposed1	Precision	0.956	0.625	0.580	0.969	0.892	0.849	0.970	0.951	0.887	0.345	0.769
	Recall	0.341	0.999	0.993	0.603	0.691	0.936	0.395	0.566	0.953	0.289	0.642
	F_1	**0.503**	0.769	0.733	0.743	**0.778**	**0.891**	0.562	0.709	**0.919**	0.315	0.700
5frames	Precision	0.693	0.891	0.957	0.950	0.686	0.883	0.940	0.954	0.894	0.000	**0.899**
	Recall	0.294	0.474	0.876	0.948	0.748	0.868	0.543	0.381	0.886	0.000	0.556
	F_1	0.413	0.619	**0.915**	**0.949**	0.716	0.875	0.688	0.544	0.890	0.000	0.687
10frames Proposed2	Precision	0.757	0.677	0.957	0.995	0.644	0.675	0.843	0.818	0.853	0.553	0.775
	Recall	0.262	0.960	0.812	0.752	0.802	0.992	0.151	1.000	0.879	0.399	**0.671**
	F_1	0.390	0.794	0.879	0.856	0.714	0.803	0.257	**0.900**	0.866	**0.464**	**0.719**
20frames	Precision	0.810	0.623	0.772	0.997	0.314	0.597	0.955	0.980	0.968	0.000	0.777
	Recall	0.223	0.992	0.798	0.749	0.582	0.961	0.583	0.110	0.709	0.000	0.542
	F_1	0.349	0.765	0.785	0.855	0.408	0.737	**0.724**	0.198	0.819	0.000	0.638
30frames	Precision	0.840	0.857	0.907	0.739	0.832	0.894	0.874	0.889	0.903	0.286	0.857
	Recall	0.262	0.835	0.289	0.964	0.537	0.543	0.442	0.742	0.874	0.066	0.583
	F_1	0.399	**0.846**	0.438	0.836	0.653	0.676	0.587	0.809	0.888	0.107	0.694

Fig. 5. Erroneous examples of Nishikawa B's gaze detection.

| Sushree-B | Samantha-B | HCA-B | Ollie-B | Nishikawa-B |

Fig. 6. Difference in resized eye images when looking at the other. Nishikawa B's eye looks different from other eyes.

6 Conclusion

	frm1	frm2	frm3	frm4	frm5	frm6
Ground Truth	0 1	0 1	0 1	0 1	0 1	0 1
baseline	1 1	1 1	1 1	1 1	1 1	1 1
Proposed 1	0 1	0 1	0 1	0 1	0 1	0 1
Proposed 2	0 1	0 1	0 1	0 1	0 1	0 1
Proposed 3	0 0	0 0	0 0	0 0	0 0	0 0
gazefollow	0 0	0 0	0 0	0 0	0 0	0 0

	frm1	trm2	frm3	frm4	frm5	frm6
Ground Truth	1 1	1 1	1 1	1 1	1 1	1 1
baseline	0 0	0 0	0 0	0 0	0 0	0 0
Proposed 1	1 0	1 0	1 0	1 0	1 1	1 1
Proposed 2	0 1	0 1	0 1	1 1	1 1	0 1
Proposed 3	1 1	1 1	1 1	1 1	1 1	1 1
gazefollow	0 0	0 0	0 0	0 0	0 0	0 0

Fig. 7. Examples difference between the algorithms. Red text indicates erroneous detection.

In this study, we show an eye contact detection algorithm from third person video. Leveraged by CNN-based algorithms that input facial positions, gaze directions and eye images, proposed algorithms achieved about 80% in accuracy and outperformed existing approaches. For future, we want to deal with the case when estimation of facial landmark points does not work and extend this method for general gaze target estimation tasks.

Acknowledgements. This work was supported by JST CREST Grant Number JPMJCR17A5 and JSPS KAKENHI 17H01779, Japan.

References

1. Adams Jr., R.B., Kleck, R.E.: Effects of direct and averted gaze on the perception of facially communicated emotion. Emotion **5**(1), 3 (2005)
2. Baltrušaitis, T., Robinson, P., Morency, L.P.: OpenFace: an open source facial behavior analysis toolkit. In: 2016 IEEE Winter Conference on Applications of Computer Vision (WACV), pp. 1–10. IEEE (2016)
3. Chong, E., et al.: Detecting gaze towards eyes in natural social interactions and its use in child assessment. Proc. ACM Interact. Mobile Wearable Ubiquit. Technol. **1**(3), 43 (2017)
4. Csibra, G., Gergely, G.: Social learning and social cognition: the case for pedagogy. Process. Change Brain Cogn. Dev. Atten. Perform. XXI **21**, 249–274 (2006)
5. Dozat, T.: Incorporating nesterov momentum into adam (2016). http://cs229. stanford.edu/proj2015/054_report.pdf. Accessed 25 Aug 2018
6. Gineste, Y., Pellissier, J.: Humanitude: comprendre la vieillesse, prendre soin des hommes vieux(2007). A. Colin
7. Joseph, R.M., Ehrman, K., McNally, R., Keehn, B.: Affective response to eye contact and face recognition ability in children with ASD. J. Int. Neuropsychol. Soc. **14**(06), 947–955 (2008)
8. Krafka, K., et al.: Eye tracking for everyone. In: The IEEE Conference on Computer Vision and Pattern Recognition (CVPR) (June 2016)
9. Lian, D., Yu, Z., Gao, S.: Believe it or not, we know what you are looking at!. In: Jawahar, C.V., Li, H., Mori, G., Schindler, K. (eds.) ACCV 2018. LNCS, vol. 11363, pp. 35–50. Springer, Cham (2019). https://doi.org/10.1007/978-3-030-20893-6_3
10. Lu, F., Sugano, Y., Okabe, T., Sato, Y.: Gaze estimation from eye appearance: a head pose-free method via eye image synthesis. IEEE Trans. Image Process. **24**(11), 3680–3693 (2015)
11. Mitsuzumi, Y., Nakazawa, A., Nishida, T.: Deep eye contact detector: robust eye contact bid detection using convolutional neural network. In: Proceedings of the British Machine Vision Conference (BMVC) (2017)
12. Nakazawa, A., Okino, Y., Honda, M.: Evaluation of face-to-face communication skills for people with dementia using a head-mounted system (2016)
13. Petric, F., Miklić, D., Kovačić, Z.: Probabilistic eye contact detection for the robot-assisted ASD diagnostic protocol. In: Lončarić, S., Cupec, R. (eds.) Proceedings of the Croatian Computer Vision Workshop, Year 4, pp. 3–8. Center of Excellence for Computer Vision, University of Zagreb, Osijek (October 2016)
14. Recasens, A., Khosla, A., Vondrick, C., Torralba, A.: Where are they looking? In: Advances in Neural Information Processing Systems (NIPS) (2015)
15. Recasens, A., Vondrick, C., Khosla, A., Torralba, A.: Following gaze in video. In: Proceedings of the IEEE Conference on Computer Vision and Pattern Recognition, pp. 1435–1443 (2017)
16. Senju, A., Johnson, M.H.: The eye contact effect: mechanisms and development. Trends Cogn. Sci. **13**(3), 127–134 (2009). https://doi.org/10.1016/j.tics.2008.11. 009. http://linkinghub.elsevier.com/retrieve/pii/S1364661309000199
17. Smith, B.A., Yin, Q., Feiner, S.K., Nayar, S.K.: Gaze locking: passive eye contact detection for human-object interaction. In: Proceedings of the 26th Annual ACM Symposium on User Interface Software and Technology, pp. 271–280. ACM (2013)
18. Ye, Z., Li, Y., Liu, Y., Bridges, C., Rozga, A., Rehg, J.M.: Detecting bids for eye contact using a wearable camera. In: 2015 11th IEEE International Conference and Workshops on Automatic Face and Gesture Recognition (FG), vol. 1, pp. 1–8. IEEE (2015)

19. Zhang, X., Sugano, Y., Bulling, A.: Everyday eye contact detection using unsupervised gaze target discovery. In: Proceedings of the 30th Annual ACM Symposium on User Interface Software and Technology, pp. 193–203. ACM (2017)

20. Zhang, X., Sugano, Y., Fritz, M., Bulling, A.: Appearance-based gaze estimation in the wild. In: Proceedings of the IEEE Conference on Computer Vision and Pattern Recognition, pp. 4511–4520 (2015)

21. Zhang, X., Sugano, Y., Fritz, M., Bulling, A.: It's written all over your face: full-face appearance-based gaze estimation (2016). http://arxiv.org/abs/1611.08860.   https://perceptual.mpi-inf.mpg.de/wp-content/blogs.dir/12/files/2016/11/zhang16_arxiv.pdf

Semi-supervised Early Event Detection

Liping Xie[1,2(✉)], Chen Gong[2], Jinxia Zhang[1], Shuo Shan[1], and Haikun Wei[1]

[1] Key Laboratory of Measurement and Control of CSE, Ministry of Education,
School of Automation, Southeast University, Nanjing 210096, China
{lpxie,jinxiazhang,230189536,hkwei}@seu.edu.cn
[2] Jiangsu Key Laboratory of Image and Video Understanding for Social Safety,
School of Computer Science and Engineering, Nanjing University of Science
and Technology, Nanjing 210094, China
chen.gong@njust.edu.cn

Abstract. Early event detection is among the keys in the field of event detection due to its timeliness in widespread applications. The objective of early detection (ED) is to identify the specified event of the video sequence as early as possible before its ending. This paper introduces semi-supervised learning to ED, which is the first attempt to utilize the domain knowledge in the field of ED. In this setting, some domain knowledge in the form of pairwise constraints is available. Particularly, we treat the segments of complete events as must-link constraints. Furthermore, some segments do not overlap with the event are put together with the complete events as cannot-link constraints. Thus, a new algorithm termed semi-supervised ED (SemiED) is proposed, which could make better early detection for videos. The SemiED algorithm is a convex quadratic programming problem, which could be resolved efficiently. We also discuss the computational complexity of SemiED to evaluate its effectiveness. The superiority of the proposed method is validated on two video-based datasets.

Keywords: Semi-supervised learning · Domain knowledge · Early event detection

1 Introduction

While event detection has received rapid accumulation over the past decades [15, 19], little attention has been given to early detection (ED) [10,23,25]. ED aims to make a reliable early identification for video sequences, where the temporal information needs to be well exploited. Different with event detection where the result is given after the video ends, ED gives sequential results, i.e., each frame contained in the video corresponds to an independent detection score. An efficient algorithm can obtain an early identification as soon as possible based on these frame-level outputs, and the accuracy is also guaranteed. The potential applications of ED can range form healthcare [4], environmental science [21], artificial intelligence [14,24], etc. For example, the timeliness in human-computer

© Springer Nature Switzerland AG 2020
S. Palaiahnakote et al. (Eds.): ACPR 2019, LNCS 12047, pp. 678–690, 2020.
https://doi.org/10.1007/978-3-030-41299-9_53

interaction is the key issue to provide an efficient and comfortable communication especially in this age of artificial intelligence. Therefore, ED will become more and more important in the future.

Over the past few years, several works have been proposed for ED. The max-margin early event detector (MMED) [10] is the first and also the most representative model for ED. Subsequently, [23] applies multi-instance learning to early detection based on MMED and obtain better performance in both of accuracy and timeliness. In addition, [23] further extends the model in an online manner to deal with streaming sequences and make it available in large-scale applications. Some frameworks with several assumptions are also presented in recent years. [7] tries to give a probability ration test based on probabilistic reliable-inference, where the model is still trained on the whole video sequences. [16] designs a RankBoost-based approach where the length of the testing video needs to be provided before testing. [20] suffers the same problem with [16], which is also unpractical to be utilized in real-world applications.

The application of domain knowledge to various data mining fields attracts increasing attention due to its effectiveness [26]. Normally, the forms of domain knowledge could be expressed in several ways: class labels, pairwise constraints and other prior information [6]. In most cases, the domain knowledge in the form of pairwise constraints is utilized since the cost of labelling is expensive in real-world applications. In contrast, whether the pairs of samples belong to the same type of class or not is relatively practical to obtain. In general, the pairs from the same class are treated as the must-link constraints, and the pairs form different classes are for cannot-link constraints. Moreover, the pairwise constraints can be derived from labeled data but not vice versa. Pairwise constraints have been applied in many applications such as facial expression recognition [11], image retrieval [2,5], dimension reduction [6,26] and image segmentation [12,13].

Although studied extensively of domain knowledge in various applications, few attentions have been given to ED. Introducing domain knowledge [22] to ED raises a natural and new problem: how to combine the detection function with the provided domain knowledge to obtain a more accurate score for each frame. There are two key issues existed in semi-supervised early event detection. The most critical one is to make full use of the domain knowledge to help ED model learning. The objective of ED is to improve the performance in terms of both timeliness and accuracy. In addition, how to define the pairwise constraints is another problem since the label is given for video sequence and not for segments. Therefore, the segments utilized for model training have no label information.

In this paper, we study early detection where the domain knowledge is available in the form of pairwise constraints. A novel algorithm, termed semi-supervised early detection (SemiED), is proposed. Inspired by the success of MMED, we follow the framework of structured output SVM (SOSVM), and the monotonicity of the detection function is learnt by the pairwise constraints. Based on MMED, we introduce semi-supervised learning to SOSVM. Particularly, the pairwise constraints for domain knowledge in SemiED consists of two parts: (1) each video sequence from the datasets contains a complete event.

We thus treat these complete events as the must-link constraints since they all correspond to the greatest value of the detection function; (2) we extract some segments before the event fires or after it ends, i.e., these segments have no overlap with the complete events. We put these segments together with the complete events as the cannot-link constraints. We develop an efficient algorithm for problem optimization. The comparison of computational complexity of MMED and the proposed SemiED is also provided. The superiority of SemiED is validated on two popular video-based datasets with various complexities.

2   SemiED: Semi-supervised Early Event Detection

Notations: In this paper, the video sequences for training and their associated ground truth annotations of interest are denoted as (X^1, y^1), $(X^2, y^2), ..., (X^n, y^n)$. Here, two elements are contained in $y^i = [s^i, e^i]$ to indicate the beginning and ending of the event for the i-th training sample X^i. n denotes the total number of training videos. The length of the i-th training video X^i is represented as l^i. For every frame time $t = 1, ..., l^i$, we adopt y_t^i to denote the partial event of y^i that has occurred, i.e., $y_t^i = y^i \cap [1, t]$. y_t^i may be empty if no event is contained. In addition, $\mathcal{Y}(t)$ is utilized to denote all the possible segments from the 1-st to the t-th frames: $\mathcal{Y}(t) = \{y \in \mathbb{N}^2 | y \subset [1, t]\} \cup \{\emptyset\}$. The segment $y = \emptyset$, indicates y is empty and there is no event occurs. If a video sequence X has the number of frames, which is denoted as l, then $\mathcal{Y}(l)$ indicates the set that all possible segments are contained. Note that, for an arbitrary segment $y = [s, e] \in \mathcal{Y}(l)$, X_y is the subset of X from the s-th to the e-th frames.

2.1   Formulation

Following [10], we employ structured output SVM for early event detection. The monotonicity of the detection function is achieved by extracting various pairwise constraints and ranking them based on the information contained. Before the formulation, we first give the detection function as follows:

$$f(X_y, w, b) = \begin{cases} w^T \varphi(X_y) + b & \text{if } \mathcal{Y} \neq \emptyset, \\ 0 & \text{otherwise.} \end{cases}$$

where $\varphi(X_y)$ is the feature representation of segment X_y. In this paper, we utilize $f(X_y)$ to denote $f(X_y, w, b)$ for brevity. The fundamental principle of ED is that the score of a positive training sample $X_{y^i}^i$ is greater than the that of any other segment from the same video sequence, i.e., $f(X_{y^i}^i) > f(X_y^i), \forall y \neq y^i$. The early event detection based on SOSVM can be written as follows:

$$\min_{\{w,b,\xi^i \geq 0\}} \frac{1}{2} \|w\|_F^2 + \gamma \sum_{i=1}^{n} \xi^i,$$

$$\text{s.t. } f(X_{y_t^i}^i) \geq f(X_y^i) + \Delta(y_t^i, y) - \frac{\xi^i}{\mu(\frac{|y_t^i|}{|y^i|})}, \tag{1}$$

$$\forall i, \forall t = 1, ..., l^i, \forall y \in \mathcal{Y}(t).$$

where $|\cdot|$ represents the length of the segments, and $\mu(\cdot)$ is a rescaling factor of slack variable. Following [10], the piece-wise function with linearity is employed as follows:

$$\mu(x) = \begin{cases} 2x & 0 < x \leq 0.5, \\ 1 & 0.5 < x \leq 1 \ \ and \ \ x = 0. \end{cases}$$

$\triangle(\cdot)$ is an adaptive margin of the pairwise segments and denoted as the loss of the detector for outputting y when the desired output is y^i, i.e, $\triangle(y_t^i, y) = 1 - overlap(y^i, y)$.

In this paper, we use M and C to denote the number of must-link and cannot-link constraints respectively. We first denote $\Phi = [\varphi(X_y^1), \varphi(X_{y^1}^1), \varphi(X_y^2), \varphi(X_{y^2}^2)...\varphi(X_y^n), \varphi(X_{y^n}^n)] \in \mathbb{R}^{d \times 2n}$, $\varphi(X_y^i)$ is the segment from the 1-st frame to the event beginning. $\varphi(X_{y^i}^i)$ represents the feature vector of the complete event in video sample X^i. For brevity, we use $\Phi = [\varphi(y_1), \varphi(y_2)...\varphi(y_{2n})] \in \mathbb{R}^{d \times 2n}$. Then, the objective function of the domain knowledge can be written as minimizing $J(w)$:

$$
\begin{aligned}
& J(w) \\
&= \frac{\beta}{2n_M} \sum_{(y_i, y_j) \in M} (f(y_i) - f(y_j))^2 - \frac{\alpha}{2n_C} \sum_{(y_i, y_j) \in C} (f(y_i) - f(y_j))^2 \\
&= \frac{\beta}{2n_M} \sum_{(y_i, y_j) \in M} (w^T \varphi(y_i) - w^T \varphi(y_j))^2 - \frac{\alpha}{2n_C} \sum_{(y_i, y_j) \in C} (w^T \varphi(y_i) - w^T \varphi(y_j))^2
\end{aligned}
\tag{2}
$$

The concise form of $J(w)$ can be further denoted as follows:

$$J(w) = \frac{1}{2} \sum_{i,j} (w^T \varphi(y_i) - w^T \varphi(y_j))^2 S_{ij} \tag{3}$$

where

$$
S_{ij} = \begin{cases} -\dfrac{\alpha}{n_C}, & if (y_i, y_j) \in C, \\ \dfrac{\beta}{n_M}, & if (y_i, y_j) \in M, \\ 0, & otherwise. \end{cases}
\tag{4}
$$

Then we reformulate (3) as:

$$
\begin{aligned}
J(w) &= \frac{1}{2}\sum_{i,j}(w^T\varphi(y_i) - w^T\varphi(y_j))^2 S_{ij}\\
&= \frac{1}{2}\sum_{i,j}(w^T\varphi(y_i)\varphi(y_i)^T w + w^T\varphi(y_j)\varphi(y_j)^T w - 2w^T\varphi(y_i)\varphi(y_j)^T w)S_{ij}\\
&= \sum_{i,j}(w^T\varphi(y_i)\varphi(y_i)^T w - \sum_{i,j}w^T\varphi(y_i)\varphi(y_j)^T w)S_{ij}\\
&= \sum_{i,j}w^T\varphi(y_i)S_{ij}\varphi(y_i)^T w - \sum_{i,j}w^T\varphi(y_i)S_{ij}\varphi(y_j)^T w\\
&= \sum_{i}w^T\varphi(y_i)D_{ii}\varphi(y_i)^T w - w^T\Phi S\Phi^T w\\
&= w^T\Phi(D - S)\Phi^T w\\
&= w^T\Phi L\Phi^T w
\end{aligned}
$$

$$(5)$$

where D denotes a diagonal matrix where the entries are rows sums of S, i.e., $D_{ij} = \sum_j S_{ij}$. $L = D - S$ is called the Laplacian matrix. Therefore, the regularizer (5) can further be simplified as minimizing $J(w)$, where

$$
J(w) = w^T\Phi L\Phi^T w \tag{6}
$$

where L could be computed by the domain knowledge. Therefore, the semi-supervised early event detection (SemiED) can be formulated as follows:

$$
\min_{\{w,b,\xi^i\geq 0\}} \frac{1}{2}\|w\|_F^2 + \gamma\sum_{i=1}^n \xi^i + w^T\Phi L\Phi^T w,
$$

$$
\text{s.t. } f(X_{y_t^i}^i) \geq f(X_y^i) + \Delta(y_t^i, y) - \frac{\xi^i}{\mu(\frac{|y_t^i|}{|y^i|})}, \tag{7}
$$

$$
\forall i, \forall t = 1, ..., l^i, \forall y \in \mathcal{Y}(t).
$$

During the experiments, we set $\alpha = 1$ and $\beta = 20$ in SemiED, which are the empirical values. The detailed information of optimization and analysis of computational complexity can be seen in the following.

2.2   Optimization

Before the detailed optimization procedure, we first reformulate the problem (7) as follows:

$$
\min_{\{w,b,\xi^i\geq 0\}} \frac{1}{2}w^T(I_d + \Phi L\Phi^T)w + \gamma\sum_{i=1}^n \xi^i,
$$

$$
\text{s.t. } f(X_{y_t^i}^i) \geq f(X_y^i) + \Delta(y_t^i, y) - \frac{\xi^i}{\mu(\frac{|y_t^i|}{|y^i|})}, \tag{8}
$$

$$
\forall i, \forall t = 1, ..., l^i, \forall y \in \mathcal{Y}(t).
$$

Here, I_d represents the identity matrix with size of $d \times d$. Then, we set $u = [w^T \ b \ \xi^1 \ \xi^2 ... \xi^n]^T \in \mathbb{R}^N$, $N = d + 1 + n$. d represents the length of the feature vector, n is the total number of sequences. Problem (8) can be rewritten as follows:

$$\min_u \frac{1}{2} u^T H u + h u,$$

$$\text{s.t.} \ \ f(X_{y_t^i}^i) \geq f(X_y^i) + \Delta(y_t^i, y) - \frac{\xi^i}{\mu(\frac{|y_t^i|}{|y^i|})}, \tag{9}$$

$$\forall i, \forall t = 1, ..., l^i, \forall y \in \mathcal{Y}(t),$$

$$u_i \geq 0, i \in (d + 2, d + 1 + n).$$

Here, H is a block diagonal matrix with the 1-st block as $I_d + \Phi L \Phi^T$ and 0 for others. $h = [0...0 \ 0 \ \gamma...\gamma]$ with n nonzero elements. We can see that this is a convex quadratic programming problem which has massive pairwise constraints. In this paper, the constraint generation strategy utilized in [18] is employed, which reduces the memory consumption by several iterations. It has been validated that the convergence to the global minimum could be guaranteed.

2.3   Computational Complexity

This section presents the discussion of the computational cost for the proposed algorithm SemiED with the comparison to MMED, which is the most representative framework for ED and the proposed SemiED is designed based on MMED.

(1) **MMED**: The optimization of MMED is a standard quadratic programming problem. According to the computational complexity theory stated in [17], the complexity of MMED is $O((n + d + 1)^3 S)$ and roughly estimated by $O((n + d)^3 S)$. Here, n is the number of training samples, d is the dimension of the feature and S denotes the size of the problem encoding in binary.

(2) **SemiED**: Similar to MMED, the solution of SemiED is a quadratic programming problem, and the complexity of optimization procedure is also $O((n + d)^3 S)$. In addition, SemiED needs to compute the regularizer and the most expensive operation contained is the computation of $(I_d + \Phi L \Phi^T)$, which needs the complexity of $O(n^2 d + n d^2)$. Therefore, the computational complexity of SemiED is $O((n + d)^3 S)$. We can see that SemiED shares the same complexity with MMED although domain knowledge is adopted in SemiED.

3   Experiments

This section validates the performance in terms of accuracy, timeliness and training time cost on two video-based datasets: Weizmann dataset and UvA-NEMO dataset. There are 10 different actions and two smiles in Weizmann and UvA-NEMO respectively. During our experiments, the specified event "Bend" on Weizmann dataset and the event "spontaneous smile" on UvA-NEMO dataset are chosen for detection. In the following, we first introduce the datasets, experiment setup, as well as evaluation criteria before the analysis of experimental results.

3.1   Datasets

Weizmann Dataset. It [3] is created from 9 subjects and 90 video sequences are contained, i.e, every subject performs 10 actions. The specified actions are: Bend, Jack, Jump, Pjump, Run, Side, Skip, Walk, Wave1 and Wave2. Following [10], we concatenate all actions of the same subject to construct a longer video sequence. Particularly, we put the targeted event "Bend" at the end of each longer sequence. In this paper, the AlexNet architecture [9] is utilized for frame-level feature extraction, in which the features have the dimensionality of 4096. Then, we apply PCA for feature dimension reduction, and 1000-dimensional features are obtained. Note that we utilize the leave-one-out cross validation for Weizmann dataset due to the limitation of video samples.

(a) Bend/Run/Side on Weizmann dataset

(b) Spontaneous/Deliberate smile on UvA-NEMO dataset

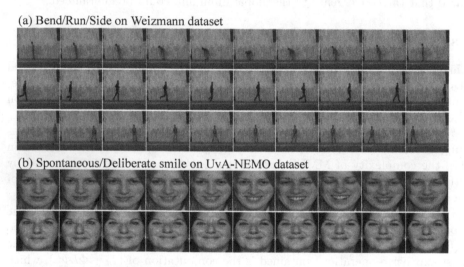

Fig. 1. Examples from two video-based datasets. Note that the length of each sequence is set to the same as "10" for better illustration, which does not represent the actual length. (a) Bend/Run/Side on Weizmann dataset; (b) Spontaneous/Deliberate smile on UvA-NEMO dataset.

Table 1. Statistics of benchmark datasets.

	#Sample					#Feature
	Training				Test	
	Video	Pairwise constrnts	M constrnts	C constrnts		
Weizmann	8	120	16	8	1	1000
UvA-NEMO	800	1200	1600	800	440	944

UvA-NEMO Dataset. It is built for the analysis of the dynamic difference from spontaneous/deliberate smiles. It consists of 1240 smile videos, where 597 are spontaneous. All the videos contained begin with a neutral or near-neutral frame, and the same situation with the ending frame. During the experiments, the Local Binary Patterns (LBP) [1] is adopted for frame-level feature extraction. Particularly, we crop each frame into 4×4 blocks and the number of neighboring points is set to 2^3. Thus, the dimensionality of the extracted feature is $4 \times 4 \times 59 = 944$.

We present some examples for illustration from each dataset in Fig. 1, and summarize the statistics in Table 1.

3.2 Experiment Setup

Compared Approaches. The results of SemiED are compared with two baseline methods (FrmPeak, FrmAll) [23] and MMED [10]. FrmPeak and FrmAll are frame-based SVMs, where the detection result is obtained by classifying each frame contained in the video. The difference between FrmPeak and FrmAll is the samples used for training. FrmPeak utilizes only the peak frames for training while in FrmAll, all the frames are employed. MMED and the proposed SemiED are segment-based methods.

Experiment Setting. In this paper, we tune the parameter γ contained in the proposed SemiED model and the parameter in SVM on set $\{10^i | i = -5, -4, ..., 3, 4, 5\}$, and the best performance is reported. Note that the SVMs utilized for FrmPeak and FrmAll are linear. During the experiments, the overlap of the extracted pairwise constraints is set to be lower than 0.7. This is to guarantee the discrimination for the training set. We conduct all the compared experiments on the computer with the following configurations: Intel(R) Xeon(R) Core-20, CPU E5-2650 v3-2.3 GHz, Memory 48 GB, LINUX operating system with Matlab 2015a.

3.3 Evaluation Criteria

This section presents the introduction of the evaluation criteria employed during our experiments: F-score [9], AUC [27], AMOC [8] and the training time curve. F-score is a measurement for detection accuracy, which is usually adopted for binary classification in statistical analysis. The two main variables in F-score are the precision p and the recall r on the testing set, which is computed as follows: $p = \frac{|y \cap y^*|}{|y|}$, $r = \frac{|y \cap y^*|}{|y^*|}$, where y^* is the segment contains the event and y is the output event that detected. AUC is short for the area under the Receiver Operating Characteristic (ROC) curve, which is also used for the evaluation of accuracy. In AUC, TPR denotes that the model fires during the event of interest, and FPR is that the model fires before the beginning or after the ending. AMOC represents the Activity Monitoring Operating Curve, which is used to evaluate the timeliness of detection. In AMOC, Normalized Time to Detection

(NTtoD) is computed, and the value is the lower the better. NTtoD is denoted as NTtoD= $\frac{t-s+1}{e-s+1}$, where t, s, and e denote the current-, the starting- and the ending-frame, respectively.

3.4 Results on Weizmann Dataset

The compared performance with different methods can be seen in Figs. 2, 3 and Table 2. F-score is adopted on this dataset instead of AUC and AMOC since there is no negative training sequences contained and the return type of FPR is void. Similarly, only the results of segment-based methods are reported since SVM could not be used for FrmAll and FrmPeak due to the lack of negative samples. From the results, we can see that: (1) The F-score of SemiED is obviously better than that of MMED on the whole event fraction. This demonstrates the effectiveness of SemiED and is consistent with the analysis that the utilization of domain knowledge could help model learning for better detection function. (2) The training time cost of SemiED is comparable to that of MMED. This demonstrates that the additional computation of regularizer has no increasing for the complexity. This is also consistent with the analysis of computational complexity in Sect. 2.3. Note that the training cost of SemiED does not strictly increase with the training size. This is because the number of iterations needed for each quadratic optimization is not fixed, and it is varied from the number of training sequences. (3) The comparison of timeliness is shown in Fig. 3. The results further demonstrate the effectiveness of the proposed SemiED. The NTtoD of SemiED is obviously lower than that of MMED, which is used to denote the normalized detection time.

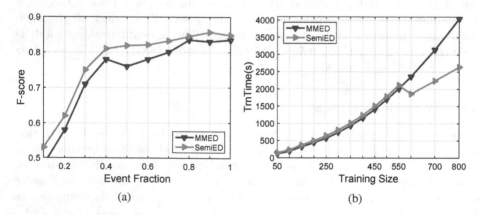

(a)   (b)

Fig. 2. Experimental results on Weizmann dataset. (a) F-score comparison for the performance of accuracy. The score of F-score is the higher the better. (b) Comparison of training time cost for the performance of efficiency. The value of time cost is the lower the better.

Table 2. Comparisons of F-score (mean and deviation) value and the training time cost (s) of segment-based methods on Weizmann dataset. The F-score results are provided with the best event fraction seen. The training time cost is reported by setting the number of training sequences as 450.

	MMED		SemiED	
	F-score	Time	F-score	Time
Weizmann	0.835 ± 0.020	1415	0.857 ± 0.121	1505

3.5   Results on UvA-NEMO Dataset

AUC, AMOC and training cost are utilized for comparison on UvA-NEMO dataset. From the results in Figs. 3, 4, and Table 3, some conclusions could be obtained: (1) Segment-based methods obviously outperform baselines in terms of both accuracy and timeliness. The time cost of FrmAll is obviously much greater than other methods. Although the cost of FrmPeak is much lower than the segment-based methods, the accuracy and timeliness of FrmPeak could not be accepted. (2) Compared with MMED, SemiED achieves much better performance of the timeliness, especially when the value of FPR is less than 0.3. Note that we usually give attention to the low FPR since the value is meaningless when greater. (3) The time cost of SemiED is approximately the same as that of MMED, in which both MMED and SemiED are much lower compared with FrmAll. This demonstrates that the utilization of domain knowledge brings no additional complexity. (4) Figure 3 illustrates the NTtoD comparison of MMED and the proposed SemiED on UvA-NEMO dataset. The lower value NTtoD is, the better timeliness the detection is. The results demonstrate that the

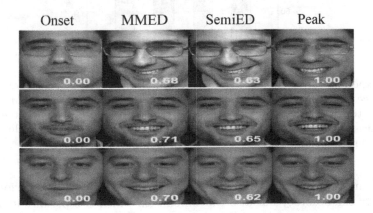

Fig. 3. NTtoD of some examples from UvA-NEMO dataset. The four frames in each row from left to right are: the onset, the ones MMED and SemiED make a detection and the peak one. The number on each frame denotes the value of NTtoD. For a testing sequence, the length from the onset to the peak frame is normalized as "1". Therefore, the value of NTtoD is the lower the better.

Table 3. Comparisons of AUC (mean and standard deviation) value and the training time cost (s) of the different methods on UvA-NEMO dataset. The training time cost is reported by setting the number of training sequences as 200.

	FrmPeak		FrmAll		MMED		SemiED	
	AUC	Time	AUC	Time	AUC	Time	AUC	Time
UvA-NEMO	0.62 ± 0.02	0.09	0.65 ± 0.01	4705	0.78 ± 0.01	713	0.81 ± 0.01	759

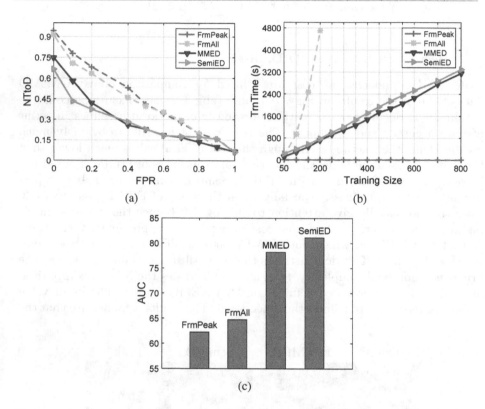

Fig. 4. Experimental results on UvA-NEMO dataset. (a) AUC comparison for the performance of accuracy. The value of AUC is the higher the better. (b) AMOC comparison for the performance of timeliness. The value of AMOC is the lower the better. (c) Comparison of training time cost for the comparison of efficiency. The value of time cost is the lower the better.

exploitation of domain knowledge makes the detector achieve better performance of not only the accuracy but also the timeliness, which is especially important for ED.

4   Conclusions

Early event detection is a relatively new and challenging problem, which receives few attentions in the past years. Motivated by the semi-supervised learning and the success of representative MMED, we propose an efficient semi-supervised early event detection algorithm called SemiED in this paper. SemiED exploits both the must-link and cannot-link constraints for the early detection function. The domain knowledge is thus well utilized to obtain better detection performance. We also provide the theoretical analysis to demonstrate that the domain knowledge utilized has no additional computation of the complexity compared with MMED. Extensive experiments and comparisons on two video-based datasets illustrate that the proposed SemiED enjoys much better performance in terms of both accuracy and timeliness than that of MMED while the training time cost is comparable with each other.

Acknowledgments. This work was supported in part by the National NSF of China under Grant 61802059, in part by the NSF of Jiangsu under Grant BK20180365, in part by the Innovation Fund of Key Laboratory of Measurement and Control of CSE through Southeast University under Grant MCCSE2018B01 and in part by the Innovation Fund of Jiangsu Key Laboratory of Image and Video Understanding for Social Safety through Nanjing University of Science and Technology under Grant 30918014107.

References

1. Ahonen, T., Hadid, A., Pietikainen, M.: Face description with local binary patterns: application to face recognition. IEEE Trans. Pattern Anal. Mach. Intell. **28**(12), 2037–2041 (2006)
2. Alemu, L.T., Pelillo, M.: Multi-feature fusion for image retrieval using constrained dominant sets. arXiv preprint arXiv:1808.05075 (2018)
3. Blank, M., Gorelick, L., Shechtman, E., Irani, M., Basri, R.: Actions as space-time shapes. In: Tenth IEEE International Conference on Computer Vision (ICCV 2005) Volume 1, vol. 2, pp. 1395–1402. IEEE (2005)
4. Bonifait, L., et al.: Detection and quantification of airborne norovirus during outbreaks in healthcare facilities. Clin. Infect. Dis. **61**(3), 299–304 (2015)
5. Chung, Y.A., Weng, W.H.: Learning deep representations of medical images using siamese cnns with application to content-based image retrieval. arXiv preprint arXiv:1711.08490 (2017)
6. Dai, A.M., Le, Q.V.: Semi-supervised sequence learning. In: Advances in Neural Information Processing Systems, pp. 3079–3087 (2015)
7. Davis, J.W., Tyagi, A.: Minimal-latency human action recognition using reliable-inference. Image Vis. Comput. **24**(5), 455–472 (2006)
8. Fawcett, T., Provost, F.: Activity monitoring: Noticing interesting changes in behavior. In: ACM SIGKDD International Conference on Knowledge Discovery and Data Mining, pp. 53–62 (1999)
9. Hammerla, N.Y., Halloran, S., Plötz, T.: Deep, convolutional, and recurrent models for human activity recognition using wearables. arXiv preprint arXiv:1604.08880 (2016)

10. Hoai, M., De la Torre, F.: Max-margin early event detectors. Int. J. Comput. Vis. **107**(2), 191–202 (2014)
11. Meng, Z., Liu, P., Cai, J., Han, S., Tong, Y.: Identity-aware convolutional neural network for facial expression recognition. In: 2017 12th IEEE International Conference on Automatic Face & Gesture Recognition (FG 2017), pp. 558–565. IEEE (2017)
12. Murez, Z., Kolouri, S., Kriegman, D., Ramamoorthi, R., Kim, K.: Image to image translation for domain adaptation. In: Proceedings of the IEEE Conference on Computer Vision and Pattern Recognition, pp. 4500–4509 (2018)
13. Pape, C., Matskevych, A., Hennies, J., Kreshuk, A.: Leveraging domain knowledge to improve em image segmentation with lifted multicuts. arXiv preprint arXiv:1905.10535 (2019)
14. Rautaray, S.S., Agrawal, A.: Vision based hand gesture recognition for human computer interaction: a survey. Artif. Intell. Rev. **43**(1), 1–54 (2015)
15. Satkin, S., Hebert, M.: Modeling the temporal extent of actions. In: Daniilidis, K., Maragos, P., Paragios, N. (eds.) ECCV 2010. LNCS, vol. 6311, pp. 536–548. Springer, Heidelberg (2010). https://doi.org/10.1007/978-3-642-15549-9_39
16. Su, L., Sato, Y.: Early facial expression recognition using early rankboost. In: IEEE International Conference and Workshops on Automatic Face and Gesture Recognition (FG), pp. 1–7 (2013)
17. Tseng, P., et al.: A simple polynomial-time algorithm for convex quadratic programming (1988)
18. Tsochantaridis, I., Joachims, T., Hofmann, T., Altun, Y.: Large margin methods for structured and interdependent output variables. J. Mach. Learn. Res. **6**(Sep), 1453–1484 (2005)
19. Venugopalan, S., Rohrbach, M., Donahue, J., Mooney, R., Darrell, T., Saenko, K.: Sequence to sequence-video to text. In: Proceedings of the IEEE International Conference on Computer Vision, pp. 4534–4542 (2015)
20. Wang, J., Wang, S., Ji, Q.: Early facial expression recognition using hidden markov models. In: International Conference on Pattern Recognition (ICPR), pp. 4594–4599 (2014)
21. Wilcox, T.M., et al.: Understanding environmental dna detection probabilities: a case study using a stream-dwelling char salvelinus fontinalis. Biol. Conserv. **194**, 209–216 (2016)
22. Xie, L., Tao, D., Wei, H.: Joint structured sparsity regularized multiview dimension reduction for video-based facial expression recognition. ACM Trans. Intell. Syst. Technol. (TIST) **8**(2), 28 (2017)
23. Xie, L., Tao, D., Wei, H.: Early expression detection via online multi-instance learning with nonlinear extension. IEEE Trans. Neural Netw. Learn. Syst. **30**(5), 1486–1496 (2019)
24. Xie, L., Wei, H., Zhao, J., Zhang, K.: Automatic feature extraction based structure decomposition method for multi-classification. Neurocomputing **173**, 744–750 (2016)
25. Xie, L., Zhao, J., Wei, H., Zhang, K., Pang, G.: Online kernel-based structured output svm for early expression detection. IEEE Sign. Process. Lett. **26**(9), 1305–1309 (2019)
26. Zhang, D., Zhou, Z.H., Chen, S.: Semi-supervised dimensionality reduction. In: Proceedings of the 2007 SIAM International Conference on Data Mining, pp. 629–634. SIAM (2007)
27. Zweig, M.H., Campbell, G.: Receiver-operating characteristic (roc) plots: a fundamental evaluation tool in clinical medicine. Clin. Chem. **39**(4), 561–577 (1993)

A Web-Based Augmented Reality Approach to Instantly View and Display 4D Medical Images

Huy Le[(✉)], Minh Nguyen, and Wei Qi Yan

School of Engineering, Computer and Mathematical Sciences, WZ Building,
Saint Paul Street, Auckland, New Zealand
huy.le@aut.ac.nz

Abstract. In recent years, development in non-invasive and painless medical imaging such as Computed Tomography (CT) or MRI (magnetic resonance imaging), has improved the process of diseases diagnosis and clarification, including tumours, cysts, injuries, and cancers. The full-body scanner with superior spatial resolution provides essential details of complicated anatomical structures for effective diagnostics. However, it is challenging for a physician in glance over a large data-set of over hundreds or even thousands of images (2D "slices" of the body). Consider a case when a doctor wants to view a patient's CT or MRI scans for analysing, he needs to review and compare among many layers of 2D image stacks (many 2D slices make a 3D stack). If the patient is scanned multiple time (three consecutive months, for instance) to confirm the growth of the tumours, the dataset is turned to be 4D (time-stamp added). The manual analysing process is time-consuming, troublesome and labour-intensive. The innovation of Augmented Reality (AR) in the last few decades allows us to illuminate this problem. In this paper, we propose an AR technique which assists the doctor in instantly accessing and viewing a patient's set of medical images quickly and easily. The doctor can use an optical head-mounted display such as the Google Glass, or a VR Headset such as the Samsung Gear VR, or a general smartphone such as the Apple iPhone X. He looks at one palm-sized AR Tag with patient's document embedded with a QR code, and the smart device could detect and download the patient's data using the decrypted QR code and display layers of CT or MRI images right on top of the AR tag. Looking in and out from the tag allows the doctor to see the above or below of the current viewing layer. Moreover, shifting the looking orientation left or right allows the doctor to see the same layer of images but in different timestamp (e.g. previous or next monthly scans). Our obtained results demonstrated that this technique enhances the diagnosing process, save cost and time for medical practice.

Keywords: Augmented Reality · Medical imaging · Computer vision · Magnetic Resonance Imaging (MRI) · Computed Tomography (CT)

© Springer Nature Switzerland AG 2020
S. Palaiahnakote et al. (Eds.): ACPR 2019, LNCS 12047, pp. 691–704, 2020.
https://doi.org/10.1007/978-3-030-41299-9_54

1 Introduction

Computed Tomography (CT) or MRI (magnetic resonance imaging) uses X-rays or large magnet and radio waves to look at tissues, organs and structures inside the patient body. They are the standard procedures used in hospitals worldwide[1]. Health care professionals use these scans to help diagnose a variety of conditions, from tumours, cysts, injuries, to cancers. However, visually scanning and examining a sequence of the CT Scan or MRI is a real challenge to radiologist when diagnosing disease. After each scanning process, a sequence of images of the inner organs of a human would be produced. The image slides in a sequence have some small distance (inter-slice gap of a millimetre or less is common) from the adjacent ones. They can be stored as files in one computer folder or printed as a stack of photos. Multiple folders or physical stacks are needed if there are more than one scanning times. Traditionally, the doctor needs to examine each picture and compare with adjacent ones (the above and below slides) to identify spots of possible tumours. Things could get a lot more problematical when there are many scans of the same organ over different timestamps, e.g. a patient's head is being scanned every week or month to confirm the growing of a particular brain tumour. This process is relatively time-consuming, and possible human errors may occur [11]. The aids of an automatic intelligent vision system could be a future solution; however, the accuracy of such artificial detection is not yet robust and reliable at this stage.

Fig. 1. A medical doctor examines patient's scanned results.

---

[1] CAT scans and PET scans are also common.

In practice, a patient goes for a non-invasive medical scanning, either CT, MRI, CAT, or PET scans, depends on the doctor's suggestions or requirements. After attaining the files or printed folder of scan films; the physician or specialist starts to examine one particular projected internal organ image (2D information: x, y). Figure 1 demonstrates one particular doctor visit described in this paper. Frequently, the doctor needs to compare it to the above or below scanned layers (3D information: x, y, z) to detect a possible tumour. Once found, he may look at the same image slides from previous scan dates (4D information: x, y, z, t) to confirm the growing of it. The problem is that the short-term memory (STM) of human is limited, the memory storage is very fragile, and with distraction, information can be lost quickly. It causes many issues to the practitioners to identify and confirm the growth of a tumour. Fortunately, the rising of Portable Smart Devices and Augmented Reality (AR) recently can help us to overcome this problem. In fact, AR is not a new technology, Ivan Sutherland and his colleagues introduced AR technology in 1960s [4]. It only started to gain more popularity due to the recent advances in display and processing technologies. Today, AR has been widely utilised in many areas, such as education [15], entertainment, manufacturing, product design, health, military training, tele-robotics [13].

In this paper, we propose an application that can detect the patient's data on the go (via a pre-registered QR code); and instantly display their CT/MRI images to help the doctor in achieving and viewing the organ's data during a diagnostic or even surgery. This specifically aims to enhance the observation of complex 4D medical images, save time and labour; thus, it could potentially help improve the medical and health-care services in general. The proposed application presents the following features:

- The capability to detect the embed QR code within the ID photo of patients.
- It could efficiently display consecutive CT/MRI images of the patient and navigate between them by zooming and out of the patient facial image/tag. Moving the camera side-way will enable the display of images in adjacent time-frames (scanning data from previous months).
- It can run smoothly on different Internet-enable mobile platforms such as iOS, Android, and Microsoft Windows.

The structure is as followed. It includes a brief cover of CT/MRI and AR-related works in Sect. 2. It is trailed by Sect. 3 – system design and implementation. Section 4 demonstrates several trials and testing to verify the system performances of the proposed solution for which to work on different mobile devices and platforms. Finally, conclusion and future work are covered in Sect. 5.

2 Background and Related Works

2.1 Computed Tomography (CT) and MRI (Magnetic Resonance Imaging)

CT and MRI have revolutionised the observation and study of the internal organs by allowing doctors to look into them non-invasively. In the course of a CT examination, the patient lies horizontally in a flat-bed; he is slowly moving across the

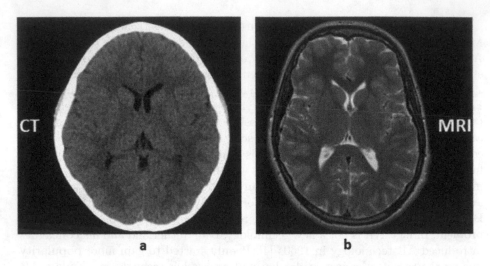

Fig. 2. Difference between a MRI scan and a CT scan.

portal while rotating the x-ray tube. The assembly of the x-ray detector rotates around a patient to achieve more projection images. This information is utilised by processing techniques such as filtered back-projection for reconstructing axial plane (x–y plane). The images from the coronal planes (x–z planes) and sagittal planes (y–z planes) then are rebuilt from the images of the axial level. The modern CT scans will be acquired quickly. A full-body scan, with a spatial resolution of less than 1 mm, can be carried out for about 15 s. The potential of this approach to evaluate disease and to evaluate trauma is incredible. The trauma centres have high efficiency in the inspection of inner organs with these CT scans, so the trauma surgeon can act rapidly and save a patient's life in many cases [6].

In an MRI scan, the patients are also placed on a flat table containing a large tube scanner with powerful magnets. The MRI operates on the long axis of the pipe through a significant magnetic field. To produce an image, the transmit tendrils send a radio frequency pulse (RF) to the patient and receive the body's electromagnetic signal. This similar matrix is used for the MRI images. In each pixel of an image with a symbolic value corresponding to its shade of grey, the MRI stores images in a similar model. Much more acquisition than CT scans is required for modern MRI scans. For example, a brain MRI often takes about 60 min. However, MRI has the remarkable resolution of contrast, which makes it possible to distinguish between different body tissues such as grey matter and white matter (Fig. 2b), and this is an advantage of an MRI scans over a CT scan (Fig. 2a) [6].

2.2 Challenges of Diagnostic on Modern CT and MRI Data

Radiologists with information overload were challenged by the rapid improvements in spatial resolution in both the CT and MRI (generally less than 1 mm).

For example, a chest CT consists of approximately 500 axial plane images. A 500 slice of the torso is a matrix of 500 (Z - direction) layers of image, and each is 262×144 pixels in one slice. The data set would include 131 million pixels with 500 axial slices (Z - direction). Each point has a unique grey level that correlates with the tissue density in the body at that point. The 2D display monitors or flat printed papers are used for displaying these points. The radiologist must also examine coronal plane and sagittal plane images in addition to axial plane images, which could be a total of over $1,000$ images for a single CT chest examination. The radiologist should pass through each 2D section and construct a 3D volume structure mentally; this process can be very challenging depending on his experience and the complexity of the anatomy [7].

Consequently, innovative viewing methods are desirable due to the large and complex data set. Today, Augmented Reality (AR) refers to a live view of a real-world physical environment whose elements are fused with computer-generated images that create a mixed reality. The augmentation is generally done with environmental elements in real-time and in a semantic context. The information about the environment becomes interactive and digitally usable by using the latest AR techniques and technologies.

Fig. 3. Virtual Human Brain is integrated in real environment using AR

2.3 Augmented Reality and Its Applications

Augmented reality (AR) allows users to see virtual objects. It is different from virtual reality (VR) that the users are not immersed in a virtual world, and so they apprehend the hybrid virtual environment and reality (see Fig. 3). A robust AR system also needs to capture real and virtual 3-D components and be interactive in real-time. The AR is useful for displaying the virtual elements on top of the physical environment. AR application has appeared in several domains around us, such as manufacture. In just more than ten years, AR technology has grown and demonstrated to be an innovative and efficient solution to contribute in solving some of the critical concerns of simulating, assisting and improving manufacturing processes before they are implemented. This would determine that phases, such as design, planning, machining, are carried out right-of-the-first without the need for further rework and adjustments [12]. The display of information and the overlay of images are context-sensitive, meaning they depend on the objects observed [1,2,4]. Although AR is generally considered a visualising device, our perceived reality can be enhanced through a strategy for all the senses. Headphone and microphone can be used, for instance, for the integration of virtual sound from the room [3]. The augmented reality is used to improve our feeling. When a virtual interface is found, haptic AR systems monitor virtual objects and provide haptic feedback by tactile effectors.

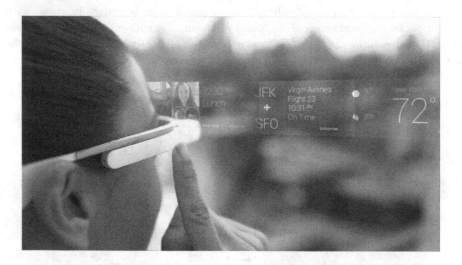

Fig. 4. Head mounted projective display

AR systems with screenings are designed to project virtual data directly to the real environment, which will be easier if the planar surfaced projector forms the real environment, but complex areas can efficiently be increased with multiple calibrations. Scalable Resolution, better ergonomics and an unlimited theoretical field of view are examples of Spatial Appreciated Realities (SAR). They may also reduce the strain of the eye, given the real depth of the virtual object [3]. In the phase of product design, AR becomes a significant part of the prototyping process. AR systems support the visualization of information through the extension of virtual objects to the real world. For instance, in the automotive industry, AR was used to evaluate interior design by overlaying several car interior designs, which in the initial development phase are usually available only as 3D models, for real car body mock-up [8]. However, only a few systems support product development and modification using 2D or 3D interaction tools in an AR-based environment. Constructions of primitives can be made and displayed in a 3-Dimensional space in Construct 3-Dimensional, developed for mathematics and geometry education. However, it is impossible to create complex components that combine these primitives because this system does not support modelling functions [10].

Head-mounted projective displays (as seen in Fig. 4) images of the project from headgear to background reflective surfaces, which reflect the user's light. These structures provide an enormous vision subject and prevent parallax distortions caused by a discrepancy in the distance between the pupil. This head-mounted display would be the ideal device for this task, as when it is mounted to the head, the device is relatively more stable compared to holding the devices by hand. Moreover, it is needed during surgery as the doctor use their hands for the operation. Therefore, hand-held devices are not suitable in this case; hygiene is also another problem for the use of hand-held devices during surgery. However, at this stage, the head mounting device is still cumbersome, expensive, and its resolution is relatively limited. As a result, we still utilise hand-held devices for testing the result of this project.

3 Our Design and Implementation Framework

In this proposed AR application, we develop a process to simulate the medical image slides viewing and navigating between slides by letting the camera to look at a real physical object (an AR marker in this case). Position and orientation of the physical object will identify what to be displayed. This is a relatively simple AR approach; it was chosen to make realism by mapping pictures on the virtual model mixed with the reality [17]. Figure 5 demonstrate the idea in a glance. There is an invisible stack of images place on top of the marker. At one time, only one slide of the stack is visible; the user can move the marker in and out to make the upper or lower slide appeared.

In slightly more detail, assume that we have a set of scanned images $I = F(d,t)$ with d is the layer number of the image stack, and t is the timestamp when the image is scanned. I could contain hundreds or thousands of images, and $F(d,t)$ represents a particular image slide (at layer d and time t) to be displayed. The image could either be RGB or grey-scaled, as what is shown in Fig. 2. Using AR, to display the virtual image quickly and precisely on the physical world coordinate, we need one AR marker. We start with a simple thick border Hiro square marker, as seen in Fig. 5. First, the black square border is detected to estimate the geometry of the marker. From that, we could find its position (x, y, z) and orientation (α, β, γ). With this information, we could estimate the relative distance d_M and the angle ϕ_M of the marker corresponding to the camera. These two parameters will be used to identify (d, t), to find which image $I = F(d,t)$ in the image data set to be displayed on top of the marker.

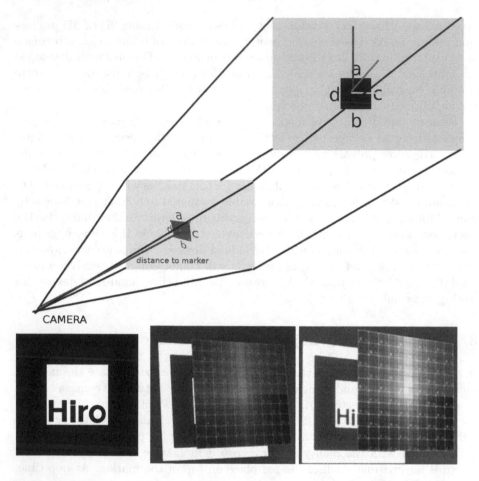

Fig. 5. Principle Idea of the project: Top is marker position, bottom-left is a template marker, and bottom middle and right are two viewing examples

3.1   Locating of the Virtual Image Slide

The first problem to be solved is: in the stack (N images), which slide $n \in [1..N]$ are made visible and which are made invisible at a particular time. The distance between the marker and the camera is calculated to be d. We specify d_{min}, d_{max}, which are the minimum and maximum allowed distance of the marker. The scanning slide I_n will only appear within this distance. The layer number n^{th} is calculated as:

$$n = N \times \frac{(d - d_{min})}{(d_{max} - d_{min})} \tag{1}$$

After identifying the image slide I_n, we could display it right on top of the marker. The problem is if the marker is faraway from the camera, the mark appeared small, and it makes the image slide also appeared small. To enable the best visibility, its appearance should be large enough, and its displaying size should also be relatively the same (compared to other layers). To achieve it, we translate the position of the virtual image slide disproportional to the distance of the marker. The translate distance of T is calculated as:

$$T = f - d \tag{2}$$

with f is the focal length and d is the marker distance. The effect of that is to push the rendered image slide towards the viewers when the marker gets faraway (hence, appeared smaller on the live video stream). This is to make sure that the image slide always appears at the same distance f to the observer.

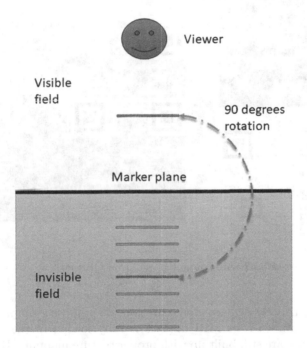

Fig. 6. The rotation to make image slide visible to the viewer (right).

3.2   Virtual Image Layer Arrangement for a Efficient Displaying

Previously, Sect. 3.1 demonstrates how our proposed system loads and displays a selected image slide I_n and overlays it on top of an AR tag at a suitable distance to the viewer. If there is only one image, it is easy and straight forward. However, one CT or MRI dataset $I = F(d, t)$ might contain hundreds or thousands of images. For instance, each CT and MRI dataset provided at http://www.pcir.org/researchers/, contains a few folders, each folder has 150+ images which represent one image slide of the scanned body part. The next problem is how to effectively load, arrange, and display them efficiently on a web-based application/website. Loading a single image and applying texture mapping every time the distance d changed is inefficient and time-consuming. However, displaying the 200+ layer of 3D planes will likely exhaust lower-end systems. For an instant, we set up an AR.js web page to display the 200+ layer of images and change the opacity of each layer to display the selected image. Testing on a Windows PC, with Core i5-6599 CPU, 16.0 GB Ram, one Google Chrome Version 75.0; we can only achieve a maximum of 2.8 fps. This speed is not acceptable for proper use in practice.

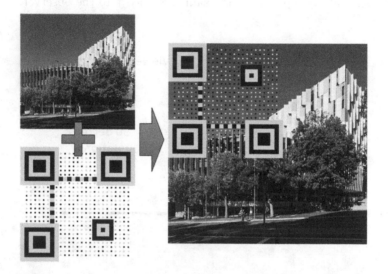

Fig. 7. An example of pictorial QR marker

To speed up this 3D rendering presentation, we propose a simple but efficient fix. First, the entire CT or MRI dataset is fully loaded in the browser, and those 200+ planes are still built up with proper texture mapping. However, they are initially located behind the marker plane, e.g. marker plane is covering the image stack. As a result, they are not being rendered by the WebGL engine. The displaying scan-slide is rotated by 90° and appeared above the marker plane. As it is now not covered by that plane, it is being rendered by the WebGL engine. The idea is demonstrated in Fig. 6. For each frame, only one image slide is rotated. Thus, only one is needed to be rendered. This approach makes the system much more light-weight. Thanks to it, we now can effectively achieve 45–50 fps with this 200+ image dataset. Moreover, this is found to be almost independent of the number of image slides, e.g. 200+, 300+, or 400+ images, will not severely affect the rendering speed. It is because, end of the day, only one image slide is being rendered at a time.

3.3   Pictorial QR as an AR Marker

Assume that the loading and displaying of the data set is not a problem now. The next issue to be solved is the flexibility of accessing and loading different datasets. A static Hiro marker is not useful because it could not be used to specify the scanning datasets, e.g. of different patients. We have to make a marker which can encode patient information, and a different marker is used to load and display different patient's dataset. Our proposed solution is an AR marker with QR Code embedded in an RGB photo ID of the patient.

QR Code is widely known for its excellent storage capability and error correction [16]. The hidden information of QR code is easily decoded by various built-in smartphones QR Code Scanner tools. However, document space occupied and uninteresting presentation are the drawbacks of the QR Code that could lead to losing the audience interest. In this section, we would like to introduce a transparent QR code which could encrypt any information without occupying too many spaces of the document as shown in Fig. 7. The four square boxes are preserved to reliable detection of the marker target. The black and white binary data/errors correction keys are minimised into 20% of their original sizes to generate more transparent area. For instance, it does not fully cover the original figure and it could be applied to any regions of the figure. Moreover, the detection performance of this marker is also equivalent to the pure QR code (approx. 90% of rate). The AR-tag will be set up to contain the patient photo ID and a unique URL of the website which display the patient's scanning dataset. The proposed marker could be easily generated at https://cv.aut.ac.nz/qrCreation.

3.4   The Overall Loading and Displaying Process

Systematically, the application work-flow as what is shown in Fig. 8. The viewer uses their smart device or webcam to scan the QR Code of the AR Tag. Once detected, the website will automatically be loaded on the viewer's Internet browser. The web application starts and tries to connect with the device camera. Then the tag image detected and the QR code captured will be confronted with the defined marker pattern stored in the server. If the images were matched with defined markers, the system renders the corresponding CT/MRI images on the top of the marker. User can see the CT/MRI image in the sequence by moving the device in or out, left or right.

4   Initial Results

A web system at cerv.aut.ac.nz/ar/mri presents a simple demonstration for this application, here, the Messi tag could be used as an AR marker (Fig. 9). Reader can use this tag and access the website to visualise our proposed system.

From our initial experiments, the visible results of the application applied to an image set were relatively satisfactory. Some screen-shots of the applications are shown in Fig. 10. We have tried the application on various devices including

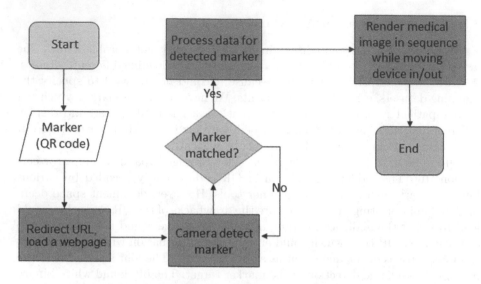

Fig. 8. Our proposed AR system

Fig. 9. Messi tag to be used as a demonstration on cerv.aut.ac.nz/ar/mri

mobile phones and tablets; some can be named are the iPhone 8, the Samsung Note 8, the iPad Pro and the Microsoft Surface Pro 4. All the devices can run and display the image slides relatively well. However, from the observation point of evaluation, the Microsoft Surface Pro 4 performs much better with smoother rendering and display. It is likely due to the dedicated GPU Intel I3 2.4 GHz,

Geforce NVIDIA 940 of the machine. The project is currently at its initial stage; many improvements and test cases are needed to fully evaluate the system. However, we believe it could be a useful tool in practice at some near future due to its flexibility, capability, portability and lightweight. In particular, during the process of disease diagnosing, surgery doctors can access the data set of scanned images quickly, and comfortably view the brain CT/MRI images in sequence to find and diagnose the potential health issues. A large number of printed image stacks can be forgotten.

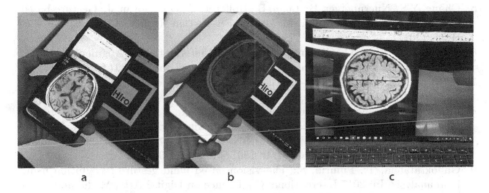

Fig. 10. Experiment on iPhone 8 (a), Samsung Note 8 (b) and Microsoft Surface Pro 4 (c)

5   Conclusion and Future Work

AR technology permeates our daily lives and literally transforms our worldview. It is evident that the application of augmented reality is far-reaching. The purpose of this paper was to save time and unnecessary efforts during the patient operation and surgery, the practitioner, can virtually view all CT/MRI images in 4D sequence. Even though, this system currently works with smart devices like smart-phones and tablet; the aiming devices are optical head-mounted displays such as the Google Glass [14]. These devices are mounted to the doctor's head, allows more stable controls of the image navigation (heads do not shake as much as hands). By that, the doctor can easily view the image in the surgery or diagnosing process.

The proposed technique could be used to integrate other image processing techniques. The help of augmented reality would provide efficient results and would help diagnose other cysts with accurate results in other parts of the body at an early stage. In addition, a gesture-based feature [5,9] that enables users to use their hand to touch on the rendered CT/MRI image then zoom in or zoom out would be implemented. These would assist the doctor to have more controls and improve the accuracy in the disease diagnosing considerably.

References

1. Azuma, R., Baillot, Y., Behringer, R., Feiner, S., Julier, S., MacIntyre, B.: Recent advances in augmented reality. IEEE Comput. Graph. Appl. **21**(6), 34–47 (2001)
2. Azuma, R.T.: A survey of augmented reality. Presence Teleoperators Virtual Environ. **6**(4), 355–385 (1997)
3. Bartlett, J.: The use of augmented reality in the operating room: a review
4. Carmigniani, J., Furht, B.: Augmented reality: an overview. In: Furht, B. (ed.) Handbook of Augmented Reality, pp. 3–46. Springer, New York (2011). https://doi.org/10.1007/978-1-4614-0064-6_1
5. Chang, Y.S., Nuernberger, B., Luan, B., Höllerer, T., O'Donovan, J.: Gesture-based augmented reality annotation. In: 2017 IEEE Virtual Reality (VR), pp. 469–470. IEEE (2017)
6. Douglas, D., Wilke, C., Gibson, J., Boone, J., Wintermark, M.: Augmented reality: advances in diagnostic imaging. Multimodal Technol. Interact. **1**(4), 29 (2017)
7. Ferroli, P., et al.: Advanced 3-dimensional planning in neurosurgery. Neurosurgery **72**(suppl_1), A54–A62 (2013)
8. Fründ, J., Gausemeier, J., Matysczok, C., Radkowski, R.: Using augmented reality technology to support the automobile development. In: Shen, W., Lin, Z., Barthès, J.-P.A., Li, T. (eds.) CSCWD 2004. LNCS, vol. 3168, pp. 289–298. Springer, Heidelberg (2005). https://doi.org/10.1007/11568421_29
9. Ganokratanaa, T., Pumrin, S.: The vision-based hand gesture recognition using blob analysis. In: 2017 International Conference on Digital Arts, Media and Technology (ICDAMT), pp. 336–341. IEEE (2017)
10. Kaufmann, H., Schmalstieg, D.: Mathematics and geometry education with collaborative augmented reality. ACM (2002)
11. Li, Y., Wang, B., Gao, Y., Zhou, J.: Affine invariant point-set matching using convex hull bisection. In: 2016 International Conference on Digital Image Computing: Techniques and Applications (DICTA), pp. 1–8. IEEE (2016)
12. Nee, A.Y., Ong, S., Chryssolouris, G., Mourtzis, D.: Augmented reality applications in design and manufacturing. CIRP Ann. **61**(2), 657–679 (2012)
13. Ong, S., Yuan, M., Nee, A.: Augmented reality applications in manufacturing: a survey. Int. J. Prod. Res. **46**(10), 2707–2742 (2008)
14. Rehman, U., Cao, S.: Augmented-reality-based indoor navigation: a comparative analysis of handheld devices versus Google glass. IEEE Trans. Hum.-Mach. Syst. **47**(1), 140–151 (2017)
15. Siegle, D.: Seeing is believing: using virtual and augmented reality to enhance student learning. Gift. Child Today **42**(1), 46–52 (2019)
16. Soon, T.J.: QR code. Synth. J. **2008**, 59–78 (2008)
17. Tawara, T., Ono, K.: A framework for volume segmentation and visualization using augmented reality. In: 2010 IEEE Symposium on 3D User Interfaces (3DUI), pp. 121–122. IEEE (2010)

Group Activity Recognition via Computing Human Pose Motion History and Collective Map from Video

Hsing-Yu Chen and Shang-Hong Lai[(⊠)]

Department of Computer Science, National Tsing Hua University, Hsinchu, Taiwan
andy19933@gapp.nthu.edu.tw, lai@cs.nthu.edu.tw

Abstract. In this paper, we propose a deep learning based approach that exploits multi-person pose estimation from an image sequence to predict individual actions as well as the collective activity for a group scene. We first apply multi-person pose estimation to extract pose information from the image sequence. Then we propose a novel representation called pose motion history (PMH), that aggregates spatio-temporal dynamics of multi-person human joints in the whole scene into a single stack of feature maps. Then, individual pose motion history stacks (Indi-PMH) are cropped from the whole scene stack and sent into a CNN model to obtain individual action predictions. Based on these individual predictions, we construct a collective map that encodes both the positions and actions of all individuals in the group scene into a feature map stack. The final group activity prediction is determined by fusing results of two classification CNNs. One takes the whole scene pose motion history stack as input, and the other takes the collective map stack as input. We evaluate the proposed approach on a challenging Volleyball dataset, and it provides very competitive performance compared to the state-of-the-art methods.

Keywords: Activity recognition · Action recognition · Human pose estimation · Deep learning

1 Introduction

In a scene consisting of a group of people, the collective activity can be seen as integration of actions for all individuals. To recognize individual human action, human pose, which is the configuration of all the main joints, is an important cue [19]. In fact, human actions, especially sport actions, are directly related to the spatio-temporal dynamics of human body parts or joints. For instance, the process of a volleyball player performing "setting" comprises representative evolution of his or her joints, which is different from the one of another player simply standing in a same place.

Previous works on this problem are basically appearance based [1,2,10,16,28]. Given a sequence of images, these appearance based approaches first used ground-truth tracking information or human detection plus human identity association

© Springer Nature Switzerland AG 2020
S. Palaiahnakote et al. (Eds.): ACPR 2019, LNCS 12047, pp. 705–718, 2020.
https://doi.org/10.1007/978-3-030-41299-9_55

to localize the bounding box of each individual in the group, then used CNNs to extract visual features from the corresponding region of each individual in each frame of the sequence, and constructed the rest of their RNN-based models upon these visual features. With recent impressive achievement of bottom-up multi-person pose estimation [5,25], we believe that it is sufficient now to use 2D human pose from pose estimation as the input for DNNs to learn the dynamics of both individual actions and collective activities in a group scene.

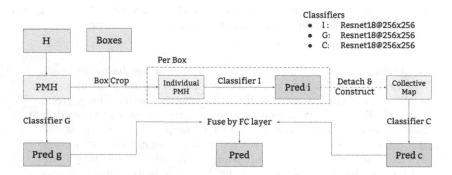

Fig. 1. Overview of the proposed system. **H** denotes joint heat maps. **Boxes** denotes bounding boxes given by annotation, or converted from pose estimation results. **Pred i** denotes the individual action prediction scores classified from Indi-PMH. **Pred g** and **Pred c** denote the collective activity prediction scores classified from PMH and collective map respectively.

In recent years, 2D pose information has been exploited in video-based action or activity recognition tasks to focus on human body parts or joints in the input sequence images. Some previous works proposed to use human joints as guidance to aggregate or attend to partial appearance or motion features from the whole RGB images or optical flow [4,6,12]. Lately, some works started to utilize joint confidence maps as input. For example, [24] proposed to apply spatial rank pooling on joint confidence maps and use body guided sampling on estimated human pose to obtain two kinds of complementary description images. The recognition task is then performed on these two descriptions. [9] proposed to use color coding to aggregate joint confidence maps from different time steps into a single stack of feature maps which is called PoTion. The activity of the whole scene is classified from PoTion.

In this paper we propose a novel approach utilizing multi-person pose information and fusion of individual actions through two novel representations, pose motion history and collective map, to recognize the action of each individual and the group activity. We use intensity retaining mechanism instead of color coding to perform temporal aggregation. The aggregated pose motion history feature map takes less memory than PoTion [9] and thus the recognition task can be done more efficiently. Also, the collective map that encodes both positions and actions of all individuals enhances the group activity recognition.

In this work, we adopt OpenPose [5] to retrieve the positions where people are in the sequence frames, and extract multi-person pose features without the need of cumbersome combination of human detection plus single-person pose estimation. Moreover, with the aid of the two novel representations that represents the spatio-temporal dynamics of multi-person human joints, and the integration of individual actions as well as individual positions in sequences, it is unnecessary to use RNN or its variants to learn the mapping from the input to individual action or group activity prediction. Instead, we use a simple CNN model, such as Resnet-18, for the classification tasks, while still achieving very competitive performance.

In summary, the contributions of this work are three-fold. Firstly, we propose two novel representations, pose motion history and collective map, for representing individual actions and the group activity of a multi-person scene. To the best of our knowledge, we are the first one to utilize multi-person pose estimation to classify both individual actions and collective activities at the same time. Secondly, we design a simple CNN model without the help of human tracking on these two novel representations for the classification task. Finally, we evaluate the proposed system on the Volleyball dataset, and achieve competitive performance even we just use simple CNNs without human identity association compared to the previous RNN-based works.

2   Related Works

Fig. 2. Some examples of applying OpenPose [5] on sequences of Volleyball dataset.

Action Recognition in Videos. Action recognition plays an essential role in various domain such as surveillance, robotics, health care, video searching, and human-computer interaction [36]. With recent revival of deep learning, given video sequences, many works successfully exploit the power of DNNs to learn spatio-temporal evolution of human actions, and report impressive results on several popular benchmarks [13,29,35]. Since deep learning based approaches outperform previous hand-crafted feature based methods, we only review deep learning based ones here. [29] used two-stream CNNs which consist of one spatial stream learning appearance features and one temporal stream learning motion features to recognize actions in video sequences. Several works proposed further enhancement based on this kind of multi-stream architectures [13,35,37]. [18] introduced an 3D CNN which extended traditional 2D CNNs to a 3D one to convolve spatio-temporal information. From then on, 3D CNNs have been used and improved in many works [14,32,33]. RNNs are also popular for action recognition in videos since they naturally extract temporal features from sequence input [11,26,34].

2D Pose-Based Action Recognition from Video. Since the target of action recognition is mainly human, pose is a natural input cue for classifying human actions in videos. Many works have proposed to use pose information in videos to learn spatial and temporal evolution of human actions [4,6,9,12,17,24]. Some used joint positions to further aggregate or pool appearance or motion features [4,6,12]. Some directly used estimated pose or joint confidence maps as the input for their models [9,17,24]. Our method is most similar to PoTion [9] in which color coding is applied on joint confidence maps to aggregate human joints information from different sequence images into single compact stack of feature maps. We use intensity retaining mechanism instead of color coding to construct our proposed pose motion history feature map stacks which consume less memory and thus results in higher learning efficiency. Our method is different from all the works above, since all these methods only deal with single person, double people settings, or predict only the activity of the whole sequence, while we not only predict the group activity, but also the action of each individual in the input sequence.

Group Activity Recognition from Video. Group activity recognition has attracted a lot of work in past years. Many former methods used hand-crafted features as input to structured models [7,8,20–23]. While with recent revival of deep learning, more and more papers started to take advantage of the superior classification performance of Deep Neural Networks [1,2,15,16,27,28,30,31,38]. In [1,16], hierarchical models consisting of two LSTMs are used, one for representing individual action dynamics of each person in a video sequence, and the other for aggregating these individual action dynamics. [1] combined human detection module into their hierarchical model through reusing appearance features for detection and recognition. [2,15,27] combined graph structures with DNNs to model the actions of individuals, their interactions, and the group activities. [27,28,30,38] utilized attention mechanisms to focus on more relevant individuals or temporal segments in video sequences.

3   Proposed System

Our goal is to recognize the individual action of each individual from a video by utilizing their pose information, and also recognize the collective activity of the whole group based on both collective pose information and also individual predictions. To achieve this goal, we propose a two-stream framework for group activity classification based on the pose motion history and collective map.

The overview of our framework is given in Fig. 1. For a given input sequence, first we apply a multi-person pose estimation algorithm, OpenPose [5], to estimate the joint positions of each individual and also the confidence maps (heatmaps) of these joints for each frame. OpenPose [5] would produce a joint heatmap stack of 18 channels where each channel is the heatmap corresponding to a certain joint. The value of each pixel in a heatmap indicating the probability a joint locating there. We depict some pose estimation results on the Volleyball dataset in Fig. 2. Second, to construct the whole scene pose motion history stack of each frame, we first multiply a intensity retaining weight w to the joint heatmap stack of the first sequence frame, and sum it to the second sequence frame. Then the pose motion history stack of the second sequence frame is multiplied by the same retaining weight w and summed to the joint heatmap stack of the third sequence frame. We repeat this process until we derive the pose motion history stack of the last sequence frame. This very pose motion history stack $\mathbf{P}$ is the input for the following two streams: individual stream and collective stream.

For the individual stream, we first obtain the bounding box of each individual by simply finding the minimal rectangle containing all joints of the individual based on the joint positions output by OpenPose [5]. Second, these individual bounding boxes are used to crop their corresponding individual pose motion history stacks $p_b, b = 1 \ldots B$, which are then input into our individual PMH CNN to classify individual actions. We then construct a collective map stack $\mathbf{M}$ of I channels where each channel representing one individual action, and for each individual, we fill its individual softmax scores for all actions to its bounding box area in the corresponding action channel. This collective map stack thus encodes individual positions and their actions in a simple feature map stack. See the subsequent sections for more details.

For the collective stream, we first simply input the pose motion history stack $\mathbf{P}$ into our collective PMH CNN to obtain initial collective activity predictions. Second we input the collective map stack $\mathbf{M}$ into another collective map CNN to obtain auxiliary collective activity predictions. Next these two parts of collective activity predictions are fused by a fusion FC layer, where the fused result is the final collective activity predictions.

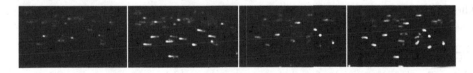

Fig. 3. Some examples of PMH maps computed from Volleyball dataset using retaining weight value $w = 0.95$.

3.1   Pose Motion History

Pose motion history(PMH) can be seen as an idea extended from MHI (Motion History Image) [3]. [3] proposed to form the temporal history of pixel points into a motion history image. In this image, more recently moving pixels are brighter. We observe that human actions are naturally highly related to the spatial-temporal dynamics of human joints. To form pose motion history, given an input multi-person sequence of T frames, we consider human joints in the sequence as the interest points. To represent their motion history from the past to the current frame, we apply the recursive overlaying mechanism on frame 2 to frame T as given by the following equation:

$$P_t = P_{t-1} * w + H_t, \tag{1}$$

where H_t is the whole scene 18-channel joint heatmap stack of frame t generated by OpenPose [5], w is the intensity retaining weight, and P_t denotes the 18-channel pose motion history stack of frame t. In this paper, the intensity retaining weight w is a fixed chosen value in $[0, 1]$ so that joint positions in latter frames would be more obvious than those in the earlier frames; however it could also be learned through training. Effect of different w values would be discussed in Sect. ?. We clip pixel values in P_t larger than 255 to 255. See Fig. 3 for pose motion history examples. We use OpenPose [5] to retrieve whole scene joint heatmap stacks and joint positions of the input sequence. For a frame image, OpenPose [5] would generate an 18-channel stack of whole scene joint heatmaps. The value of each pixel of a joint heatmap indicating the probability or the confidence that a joint locates at that position. The joint positions output by OpenPose [5] are grouped by individuals. We use the grouped joint positions in the individual stream to crop individual pose motion history stacks (Indi-PMH).

Given the grouped joint positions output by OpenPose [5], we calculate the bounding box of each individual through finding the minimal rectangle able to contain all joints of a person. We enlarge bounding boxes of the last frame with a scale s to crop the corresponding individual pose motion history stack for each individual from P_T. We crop from P_T since it contains all the joint motion history of the whole sequence from the first to the last frame. We enlarge the bounding boxes before cropping so that the cropped individual pose motion history stacks could contain more complete joint motion dynamics. We input these cropped individual pose motion history stacks (Indi-PMH) into a Resnet-18 to classify their individual actions.

3.2   Collective Map

The collective map stack of an input sequence could be computed through summing all the collective map stacks of each of its frames, but in this paper we simply use the one of the last sequence frame instead. We illustrate the construction of a collective map of a single frame in Fig. 4. We denote the number of individuals in an input sequence frame by N. To construct the collective map stack M_{H,W,A_I} of the sequence frame we first fill the stack with all zeros, where A_I denotes the number of individual action classes, H, W the height and the width of the frame. For each individual prediction $R_n, n \in [1, N]$ of this frame, which is a vector with A_I elements, we first apply softmax on it to restrict all the values of its elements in the range $[0, 1]$, so that each element represents the probability of a individual action class. Next we fill each softmax score to

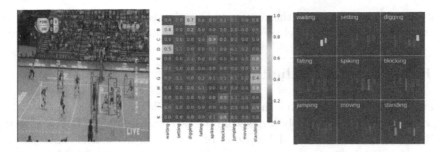

Fig. 4. An illustration for constructing the collective map of a sequence frame. In the leftmost picture, there are 11 players in the scene, denoted as **A** to **K**. First we construct a zero-filled stack. The matrix in the middle picture shows the softmax scores of individual action classes of each player. We fill each softmax score of each player to the area of his/her bounding box of the corresponding individual action channel in the collective map stack. The rightmost picture shows the result collective map stack.

the bounding box area of each individual to the corresponding individual action channel of M_{H,W,A_I}. The constructed collective map thus encodes both the positions and the actions of all the individuals.

We use two CNNs to obtain the collective activity prediction of the input sequence. The first CNN takes PMH of the last frame P_T as input, and the second CNN takes previously constructed collective map stack M_{H,W,A_I} as input. For both collective CNNs, we use Resnet-18 for the classification network. The final collective activity prediction is fused from the output of these two CNNs by an FC layer. We do not need hierarchical RNNs or similar recurrent networks here modeling the integration among individuals for collective activity recognition task like those in [16].

3.3 Training and Loss Function

We use Resnet-18 for each CNN for its learning efficiency. Separate softmax layers are applied on the outputs of individual PMH CNN, collective map CNN, collective PMH CNN, and collective fusion FC layer to obtain the predictions $p_{I,n}$, p_{C_1}, p_{C_2}, and p_{C_3}, respectively, where n is in $[1, N]$. For each CNN, we compute the loss between predictions and targets using the cross entropy, optimized by an Adam optimizer during training. The loss for a training sample of the individual part (individual PMH CNN) and the group part (collective map CNN, collective PMH CNN, and collective fusion FC layer) is defined, respectively, as follows:

$$L_I = -\sum_{i=1}^{A_I} \hat{\mathbf{p}}_i^I log\mathbf{p}_i^I, \; L_C = -\sum_{c=1}^{A_C} \hat{\mathbf{p}}_c^C log\mathbf{p}_c^C, \tag{2}$$

where $\hat{\mathbf{p}}_*^I$ and $\hat{\mathbf{p}}_*^C$ denote the one-hot-encoded ground truth probabilities for individual action classes and group activity classes, respectively, and $\mathbf{p}_*^I$ and $\mathbf{p}_*^C$

denote the softmax scores of the corresponding classes. Here, A_I and A_C are the total numbers of individual action classes and group activity classes, respectively.

4 Experimental Evaluation

We evaluate our approach on the challenging Volleyball dataset collected in [16], as it is the only relatively large-scale dataset with individual action, group activity labels, and individual locations of multi-person scenes. This dataset contains totally 4830 sequences trimmed from volleyball match videos, where 3493 for training plus validation, and 1337 for testing. Each sequence consists of 41 frames, and only the center frame is annotated with the ground truth bounding box of each player in the scene, the individual action of each player, and the group activity of the scene. We follow [16] to obtain the ground truth bounding boxes of people for those unannotated frames. There are totally 9 classes of individual actions, and 8 classes of group activities in this dataset.

4.1 Implementation Details

In previous works [1, 16] on the same Volleyball dataset, each sequence is trimmed to a temporal window of length $T = 10$, corresponding to 4 and 5 frames before the annotated frame, and 5 and 4 after the annotated frames respectively. We find out that there are quite a large amount of sequences contain obviously different camera views when trimmed by these configurations. We manually find all sequences with different camera views among the temporal window corresponding to 9 frames before the annotated frames, and 5 frames after the annotated frames and remove these sequences from training. During training, if temporal sampling is applied, we trim a temporal window of length $T = 10$ from the range of 9 frames before the annotated frame, and 5 frames after it for each sequence. Otherwise we use the same temporal window as in the testing stage by fixing the temporal window corresponding to 9 frames before the annotated frame plus the annotated frame itself for each sequence. We use Resnet-18 for all CNNs in our proposed approach for its efficiency, and Adam optimizer for optimizing the model parameters.

We try two training strategies: sequence-by-sequence like in typical RNN training procedure, and batch-by-batch like in typical CNN training procedure. With sequence-by-sequence training strategy, choosing different data preprocessing related hyper parameters is more convenient, such as trying different values of enlarging scale s for constructing Indi-PMH, and retaining weight w for constructing PMH. With batch-by-batch training strategy, we can accelerate the training process of our all-CNN based approach. We first store PMH and Indi-PMH to disk, and then train the individual and collective CNNs separately by random sampling large batches of the corresponding PMH/Indi-PMH data. For the sequence-by-sequence training strategy, due to GPU memory constraint, we random retrieve one sequence at once and accumulate parameter gradients of several forwards before per backward. We first use batch-by-batch strategy to train our individual PMH CNN and collective PMH CNN. Next, our collective map CNN is trained by loading and freezing the pretrained weights of the individual PMH CNN. Finally, the collective fusion FC layer is trained by loading and freezing the pretrained weights of these three CNNs. Sequence-by-sequence strategy is used to train both the collective map CNN and the collective fusion FC layer. With pretraining and freezing, the time spent for each epoch when training collective map CNN and the final collective fusion FC layer could thus be greatly reduced.

4.2 Pose Estimation Quality

Since we do not have ground truth pose annotation of Volleyball dataset, we evaluate the quality of pose estimation generated by OpenPose [5] by calculating the recall rate of ground truth individual bounding boxes given by the annotation of Volleyball dataset. The estimated bounding boxes are converted from the grouped joint positions output by OpenPose [5]. As joints are center points of human body parts, we increase each side of a converted bounding box by 5 pixels. In Table 1, we report the recall rates on the Volleyball dataset with different IoU threshold values: 0.5, 0.4, 0.3. We find the gap between different IoU threshold values is resulted from cases where OpenPose [5] cannot generate very complete pose estimation for some individuals with occlusion, or sometimes generating mixed pose for occluded people. Some examples are shown in Fig. 5. For the rest of our experiments, we set the IoU threshold value to 0.3 when finding the matched bounding box for each of the ground truth ones, to make use of those imperfect but still partially informative pose.

Table 1. Recall rates of ground truth bounding boxes given by the annotation of Volleyball dataset with different IoU threshold values.

Threshold	Recall (train)	Recall (test)
0.5	84.7	86.4
0.4	91.4	92.7
0.3	94.8	95.8

Fig. 5. Examples for bounding boxes converted from incomplete or mixed pose of individuals with occlusion generated by OpenPose [5]. The red boxes are ground truth bounding boxes, and the yellow ones are converted from the pose estimation results. (Color figure online)

LSTM vs. CNN for Individual Part. We try an LSTM architecture taking pose features extracted by a CNN (Resnet-18) from individual joint heatmaps cropped from the whole scene version. Since we do not know human identities association across sequence frames, we try two matching mechanisms: bounding box IoU based and long term pose feature similarity based. In the first mechanism, human identities across sequence frames are associated through matching pairs with highest bounding box IoUs between adjacent frames. In the second mechanism, identities are associated through matching instances in different frames with highest long-term pose feature similarities. We compare the LSTM based architecture with these two matching mechanisms to a CNN based architecture taking Indi-PMH as input, which does not need to know human identities association. For an input sequence, we can simply use the bounding boxes converted from pose estimation result of the last frame, enlarged with certain scale (to contain more history information), to crop Indi-PMH from the whole scene version. In this way, the matching stage can be totally removed and thus reduce the execution time. We compare the testing individual action accuracy and recall rate of these methods in Table 2. We use Resnet-18 as the CNN model. We can see that using CNN for the individual part not only results in better accuracy, but also better recall rate since it can avoid potential human identity miss associated by the matching mechanisms. Using a CNN architecture with our proposed PMH representation on the individual action recognition task saves us from the need of any matching mechanisms, which would help in a real-time sequence-to-sequence scenario as it takes less time for the inference.

Table 2. The performance of using LSTM based and CNN based architectures. We report the testing individual action accuracy, and testing recall rate in the second and third columns, respectively. "ID Assoc." at the fourth column stands for human identity association needed or not when forming representation for each individual across frames.

Method	Individual Acc.	Recall	ID Assoc.
LSTM-matching-1	67.9	85.9	Yes
LSTM-matching-2	72.3	**95.4**	Yes
Resnet-18 w/Indi-PMH	**74.3**	**95.4**	No

4.3 PMH vs. PoTion

The biggest difference between our PMH representation and PoTion [9] is the temporal aggregation mechanisms. In PoTion [9], temporal relationship between joint heat maps of different sequence frames is represented through color coding with at least 2 color channels, while in our proposed PMH it is represented through intensity retaining with only 1 channel needed. Since both PoTion [9] and our proposed PMH can be simply classified by CNNs, the number of parameters of the classification networks would be nearly the same. However our only-1-channel needed PMH would be naturally more efficient than PoTion [9], which takes at least 2 channels. Because the authors [9] did not release their code, we evaluate with our PoTion implementation with 3 color channels here. In Table 3, we report testing individual action and collective activity accuracy resulted from using PoTion and our proposed PMH on Volleyball dataset. We can see that although PMH only uses one color channel, our intensity retaining mechanism is still as effective as the color coding mechanism in PoTion [9].

Table 3. Accuracies of using PoTion and our PMH for individual action and group activity classification on Volleyball dataset. C denotes the number of color channels.

Representation	C	Individual Acc.	Collective Acc.
PoTion [9]	3	75.8	80.5
(Indi-)PMH	1	75.3	81.0

4.4 Collective Stream

The performance of our proposed collective PMH CNN, collective map CNN, and their fusion is reported at the bottom of Table 4. We use $w = 0.95$ and $s = 1.75$ in this experiment. We first pretrain the individual PMH CNN with learning rate $1e - 4$. Then we train the collective PMH CNN from the scratch, and the collective map CNN using pretrained individual PMH CNN for generating individual action predictions (only the last frame of each sequence), with learning rates $1e - 3$ and $1e - 4$, respectively. The collective fusion FC layer is finally trained using these pretrained CNNs with their parameters frozen with learning rate set to $1e - 2$. The individual and collective PMH CNNs are trained with batch-by-batch strategy, while the collective map CNN and collective fusion FC layer trained with sequence-by-sequence strategy. When training all these modules, random horizontal flipping is applied. Since in Volleyball dataset ground truth player positions and their individual action labels are provided, we conduct feasibility assessment of collective map representation first to see whether it is really discriminative for group activity recognition (using learning rate $1e - 2$). The results show that collective maps built from ground truth information generate high testing accuracy, thus proving its effectiveness. Also, fusing both collective CNNs results in about 3% higher accuracy than the best of the two, suggesting that PMH and collective map representations are both effective and complementary.

We compare the performance of our system with the state-of-the-art methods on Volleyball dataset in Table 4. Our approaches denoted with "EST" use representations constructed from pose estimation only; ours with "GT" use representations constructed from ground truth human positions and individual action labels, noting that they directly access the ground truth individual action labels, so they should not be compared to other previous methods but just for reference. All our models are trained with training settings described in Sect. 4.4. From Table 4, it is evident that our fusion approach provides the best result among all the state-of-the-art methods without using the ground truth information [1,27].

Table 4. Comparison of group activity accuracy by using the proposed system with the state-of-the-art methods on the Volleyball dataset. Accuracy results generated by [1, 27] without using ground truth information are denoted by MRF and PRO, respectively.

Method	Collective Acc.
HDTM [16] (GT)	81.9
SSU-temporal [1] (MRF/GT)	87.1/89.9
SRNN [2] (GT)	83.5
RCRG [15] (GT)	89.5
stagNet [27] (PRO/GT)	85.7/87.9
stagNet-attention [27] (PRO/GT)	87.6/89.3
PC-TDM [38] (GT)	87.7
SPA [30] (GT)	90.7
ours (collective PMH, Resnet-18) (EST)	84.6
ours (collective map, Resnet-18) (EST/GT)	77.3/92.7
ours (fusion) (EST/GT)	**87.7**/95.4

5   Conclusion

In this paper, we proposed two novel representations: pose motion history and collective map to represent the spatio-temporal dynamics of multi-person joints and the integration among individuals in a group scene. Based on these two representations, we developed a CNN based architecture without the need of any human identity association mechanisms, achieving superior performance on the challenging Volleyball dataset for the group activity recognition task. Future work would be to fuse pose estimation networks into an end-to-end model, and also to compensate the camera motion when constructing the pose motion history.

References

1. Bagautdinov, T., Alahi, A., Fleuret, F., Fua, P., Savarese, S.: Social scene understanding: end-to-end multi-person action localization and collective activity recognition. In: CVPR (2017)
2. Biswas, S., Gall, J.: Structural recurrent neural network (SRNN) for group activity analysis. In: WACV (2018)
3. Bobick, A.F., Davis, J.W.: The recognition of human movement using temporal templates. TPAMI **23**(3), 257–267 (2001)
4. Cao, C., Zhang, Y., Zhang, C., Lu, H.: Action recognition with joints-pooled 3D deep convolutional descriptors. In: IJCAI (2016)
5. Cao, Z., Simon, T., Wei, S.E., Sheikh, Y.: Realtime multi-person 2D pose estimation using part affinity fields. In: CVPR (2017)
6. Chéron, G., Laptev, I.: P-CNN: pose-based CNN features for action recognition. In: ICCV (2015)

7. Choi, W., Savarese, S.: A unified framework for multi-target tracking and collective activity recognition. In: Fitzgibbon, A., Lazebnik, S., Perona, P., Sato, Y., Schmid, C. (eds.) ECCV 2012. LNCS, vol. 7575, pp. 215–230. Springer, Heidelberg (2012). https://doi.org/10.1007/978-3-642-33765-9_16

8. Choi, W., Shahid, K., Savarese, S.: Learning context for collective activity recognition. In: CVPR (2011)

9. Choutas, V., Weinzaepfel, P., Revaud, J., Schmid, C.: PoTion: pose MoTion representation for action recognition. In: CVPR (2018)

10. Deng, Z., Vahdat, A., Hu, H., Mori, G.: Structure inference machines: recurrent neural networks for analyzing relations in group activity recognition. In: CVPR (2016)

11. Donahue, J., et al.: Long-term recurrent convolutional networks for visual recognition and description. In: CVPR (2015)

12. Du, W., Wang, Y., Qiao, Y.: RPAN: an end-to-end recurrent pose-attention network for action recognition in videos. In: ICCV (2017)

13. Feichtenhofer, C., Pinz, A., Zisserman, A.: Convolutional two-stream network fusion for video action recognition. In: CVPR (2016)

14. Hara, K., Kataoka, H., Satoh, Y.: Can spatiotemporal 3D CNNs retrace the history of 2D CNNs and ImageNet? In: CVPR (2018)

15. Ibrahim, M.S., Mori, G.: Hierarchical relational networks for group activity recognition and retrieval. In: Ferrari, V., Hebert, M., Sminchisescu, C., Weiss, Y. (eds.) ECCV 2018. LNCS, vol. 11207, pp. 742–758. Springer, Cham (2018). https://doi.org/10.1007/978-3-030-01219-9_44

16. Ibrahim, M.S., Muralidharan, S., Deng, Z., Vahdat, A., Mori, G.: Hierarchical deep temporal models for group activity recognition. TPAMI (2016)

17. Iqbal, U., Garbade, M., Gall, J.: Pose for action – action for pose. In: FG (2017)

18. Ji, S., Xu, W., Yang, M., Yu, K.: 3D convolutional neural networks for human action recognition. TPAMI **35**(1), 221–231 (2013)

19. Johansson, G.: Visual perception of biological motion and a model for its analysis. Percept. Psychophys. **14**(2), 201–211 (1973)

20. Khamis, S., Morariu, V.I., Davis, L.S.: A flow model for joint action recognition and identity maintenance. In: CVPR (2012)

21. Khamis, S., Morariu, V.I., Davis, L.S.: Combining per-frame and per-track cues for multi-person action recognition. In: Fitzgibbon, A., Lazebnik, S., Perona, P., Sato, Y., Schmid, C. (eds.) ECCV 2012. LNCS, vol. 7572, pp. 116–129. Springer, Heidelberg (2012). https://doi.org/10.1007/978-3-642-33718-5_9

22. Lan, T., Sigal, L., Mori, G.: Social roles in hierarchical models for human activity recognition. In: CVPR (2012)

23. Lan, T., Wang, Y., Yang, W., Robinovitch, S.N., Mori, G.: Discriminative latent models for recognizing contextual group activities. TPAMI **34**(8), 1549–1562 (2012)

24. Liu, M., Yuan, J.: Recognizing human actions as the evolution of pose estimation maps. In: CVPR (2018)

25. Newell, A., Huang, Z., Deng, J.: Associative embedding: end-to-end learning for joint detection and grouping. In: NIPS (2017)

26. Yue-Hei Ng, J., Hausknecht, M., Vijayanarasimhan, S., Vinyals, O., Monga, R., Toderici, G.: Beyond short snippets: deep networks for video classification. In: CVPR (2015)

27. Qi, M., Qin, J., Li, A., Wang, Y., Luo, J., Van Gool, L.: stagNet: an attentive semantic RNN for group activity recognition. In: Ferrari, V., Hebert, M., Sminchisescu, C., Weiss, Y. (eds.) ECCV 2018. LNCS, vol. 11214, pp. 104–120. Springer, Cham (2018). https://doi.org/10.1007/978-3-030-01249-6_7
28. Ramanathan, V., Huang, J., Abu-El-Haija, S., Gorban, A., Murphy, K., Fei-Fei, L.: Detecting events and key actors in multi-person videos. In: CVPR (2016)
29. Simonyan, K., Zisserman, A.: Two-stream convolutional networks for action recognition in videos. In: NIPS (2014)
30. Tang, Y., Wang, Z., Li, P., Lu, J., Yang, M., Zhou, J.: Mining semantics-preserving attention for group activity recognition. In: ACM MM (2018)
31. Tora, M.R., Chen, J., Little, J.J.: Classification of puck possession events in ice hockey. In: CVPR Workshop (2017)
32. Tran, D., Bourdev, L., Fergus, R., Torresani, L., Paluri, M.: Learning spatiotemporal features with 3D convolutional networks. In: ICCV (2015)
33. Tran, D., Wang, H., Torresani, L., Ray, J., Lecun, Y., Paluri, M.: A closer look at spatiotemporal convolutions for action recognition. In: CVPR (2018)
34. Veeriah, V., Zhuang, N., Qi, G.J.: Differential recurrent neural networks for action recognition. In: ICCV (2015)
35. Wang, L., et al.: Temporal segment networks: towards good practices for deep action recognition. In: Leibe, B., Matas, J., Sebe, N., Welling, M. (eds.) ECCV 2016. LNCS, vol. 9912, pp. 20–36. Springer, Cham (2016). https://doi.org/10.1007/978-3-319-46484-8_2
36. Wu, D., Sharma, N., Blumenstein, M.: Recent advances in video-based human action recognition using deep learning: a review. In: IJCNN (2017)
37. Wu, Z., Jiang, Y.G., Wang, X., Ye, H., Xue, X.: Multi-stream multi-class fusion of deep networks for video classification. In: ACM MM (2016)
38. Yan, R., Tang, J., Shu, X., Li, Z., Tian, Q.: Participation-contributed temporal dynamic model for group activity recognition. In: ACM MM (2018)

DeepRoom: 3D Room Layout and Pose Estimation from a Single Image

Hung Jin Lin and Shang-Hong Lai[✉]

Department of Computer Science, National Tsing Hua University, Hsinchu, Taiwan
vtsh.jn@gmail.com, lai@cs.nthu.edu.tw

Abstract. Though many deep learning approaches have significantly boosted the accuracy for room layout estimation, the existing methods follow the long-established traditional pipeline. They replace the front-end model with CNN and still rely heavily on post-processing for layout reasoning. In this paper, we propose a geometry-aware framework with pure deep networks to estimate the 2D as well as 3D layout in a row. We decouple the task of layout estimation into two stages, first estimating the 2D layout representation and then the parameters for 3D cuboid layout. Moreover, with such a two-stage formulation, the outputs of deep networks are explainable and also extensible to other training signals jointly and separately. Our experiments demonstrate that the proposed framework can provide not only competitive 2D layout estimation but also 3D room layout estimation in real time without post processing.

Keywords: Room layout estimation · Deep learning · Pose estimation

1 Introduction

The research on 3D scene understanding dates back to 1960s' simple Block World assumption [20], with the vision of reconstructing the global scene with local evidences, and nowadays it has become one of the most pivotal research area in the era of artificial intelligence and deep learning. The goal for scene understanding is to know the semantic meaning of each single object and also the environments constructed the scene. For the case of indoor scene, it is often referred to the topics like object detection and semantic segmentation at the object-level, and structure-level information such as spatial layout estimation. The effectiveness of the indoor layout estimation can be applied to applications such as the indoor navigation, localization, and the virtual object arrangement in the rooms.

The layout estimation for an interior room from an image can be represented in several levels of structure with different parameterization; for example, pixel-wise classification labeling, crossing rays originating from vanishing points, and the projection of 3D solid geometry models. Many layout estimation works are based on the underlying assumption of "Manhattan World" proposed by Coughlan [3] in 1999. It means scenes are composed of three dominant orthogonal orientations, and the walls are perpendicular to each other as well as the

© Springer Nature Switzerland AG 2020
S. Palaiahnakote et al. (Eds.): ACPR 2019, LNCS 12047, pp. 719–733, 2020.
https://doi.org/10.1007/978-3-030-41299-9_56

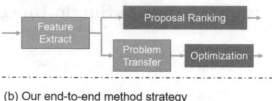

(a) Methods with post-processing

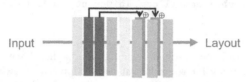

(b) Our end-to-end method strategy

Fig. 1. Difference between our proposed framework versus the previous methods in terms of system pipeline: (a) previous systems usually require post-processing, and (b) our end-to-end approach.

ceiling and floor. The cuboid model is applied to represent the room in most of the cases, in which the room enclosed by four walls, floor and ceiling. The earlier researches with machine learning approach tailored optimization and post-processing for the geometry reasoning on the hand-crafted features from the single-view image. In the succeeding methods with deep learning, existed works still built the framework as a two-tier pipeline with the deep neural networks for feature discription and the optimization step for the final estimate. These extra procedures, refining the layout estimation or extracting the 3D representation for the modeling of layout, usually take considerable amounts of computation and make them far from real-time applications. On the other hand, the widely used room layout datasets do not provide appropriate 3D annotated information, and it makes the problem of 3D modeling more challenging. Furthermore, very few deep learning models have been proposed for such 3D geometry recovery problems, and we simplify this with a deep neural network solution (Fig. 1).

This is not to deny the astonishing results achieved by the aforementioned methods. However, the time consumption of post-process rendered these methods unsuitable for time-efficient applications. And, many deep learning methods make the layout as the pixel-wise dense representation which is not enough to describe the 3D structure of layouts. To address these issues, in this paper, we propose a novel framework for predicting the 2D as well as 3D room layout estimation with efficient deep neural networks in a row. Under this framework, we will estimate the 2D layout as the intermediate representation, and then predict the 3D cuboid modeling parameters from the 2D representation. As the result, our method can estimate the layout in 2D and also 3D space completely through deep networks and provide the state-of-the-art results in real-time without post-processing.

The main contributions of this paper are threefold (Fig. 2),

- We decouple the layout estimation into two neural networks with the explainable intermediates separately in conjunction with effective training strategies;
- We believe that we are the first to model the 3D layout estimation task with two efficient end-to-end networks, and thus achieve real-time estimation;
- We demonstrate how to make good use of the existing datasets with the limitation of only the 2D layout annotations available to achieve the capability of 3D layout estimation from a single image.

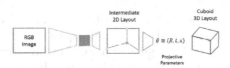

Fig. 2. An overview of our framework composed of two-stage networks.

2 Related Work

2.1 Room Layout Estimation

With the Manhattan assumption, Hoiem [8] proposed to estimate the outdoor scene geometry through learning appearance-based models of surfaces at different orientations and geometric context [7]. On the other hand, Hoiem [9] made the concept of geometric context into indoor scenes, and modified the labels into six classes for the indoor case: left-wall, right-wall, front-wall, ceiling, floor and objects, and which is also the most common classification modeling for the indoor layout estimation that inspired many later researches. In [9], they took features from color, texture, edges and vanishing point cues computed over each superpixel-segment, and applied a boosted decision tree classifier to estimate the likelihood of each possible label. Lee [14] alternatively used the orientation maps for the feature description which takes the layout and the objects oriented with three orthogonal orientations. In the traditional layout estimation approaches, the researchers extract several meaningful evidence from images, such as line segments [20,24], orthogonal orientations of line segment [14], superpixel [21], and contextual segments [7], or the volume form like geons [2]. However, these evidences fail in the cases of highly cluttered and occluded scenes containing less meaningful local features for the structure, and thus, some researched on inferring estimations through hyper volumetric reasoning [5,6,22].

With the rise of machine learning, structured learning [18] has been developed for the task, and its goal is to model the environment structure by generating hypotheses with incomplete low-level local features [5,6,17].

2.2   Room Layout Estimation with Deep Learning

With the successful modeling in the previous works, many have resorted to the deep learning approach due to its superior performance in several computer vision tasks. Some adopted the end-to-end supervised FCN (fully convolutional network) model [16] to the perspective of room layout estimation as a task of critical line detection, for instance the estimation of informative edges in [17] and coarse and fine layout joint prediction in [19]. Dasgupta [4], the winner of LSUN Room Layout Challenge 2015 [23], tackled the task with a two-tier framework: segment the planes and walls of the input image with a deep neural network first, and then optimize the output with the vanishing point estimation. The promising result of Dasgupta *et al.* inspired several subsequent works [19,25,26] to follow their two-tier pipeline, which consists of a FCN-like network for semantic segmentation and a layout optimization technique (e.g., layout hypotheses proposal-ranking pipeline in [4,17,19,25], and special optimization modules in [26]) for post-processing. For real-world application, however, the post-processing may be impractical if it is time-consuming.

2.3   Camera Pose and Geometry Learning

In early works, researchers viewed the locating 6-DoF camera pose task as a Perspective-N-Point (PnP) problem. For example, Arth [1] estimated pose of the query camera by solving a modified 3 point perspective (3PP) pose estimation problem after extracting the epipolar geometry of a query images and its nearby images. In recent works, researchers tend to estimate the 6-DoF camera pose via deep learning. PoseNet [12] is the first approach to regress the 6-DoF camera pose from a single RGB image through an end-to-end CNN. After that, a sequential works extended the PoseNet. To deal with the uncertainty of output predictions, Kendall [10] changed the network into a Bayesian model. And in [11], they notably improved the performance of PoseNet by introducing noval loss functions according to the geometry and scene reprojection error.

3   Layout Estimation in 2D Space

Our layout estimation framework can be decoupled into two stages, the 2D layout estimation in Sect. 3, and 3D cuboid model representation through projective parameters estimation in Sect. 4 and these two stage can be applied either jointly or separately.

3.1   Multi-purpose Layout Network

Under the Manhattan world assumption, we can consider each indoor scene as being composed of multiple orthogonal planes and the layout of regular room can be further simplified into the cuboid model. From this perspective, the layout estimation can also be regarded as a region segmentation problem on each surface of cuboid. To describe the segments of these regions, it can be parameterized by the densely segmentation, or the borders or the points of these polygons. In the previous deep learning methods, researchers proposed several representations, such as planar segment with semantic labels [4,15,19], scoring heatmap on the layout edges [17,26], and corner heatmaps [13]. From these works, we found that each representation has its pros and cons for the later usage, i.e. post-process methods. Consequently, we take the success of Lin [15] for fully convolutional network adopted ResNet 101 as the back-bone, which is the state-of-the-art layout estimation without post-process. We further advance the estimation for 2D layout in multiple representations through the multi-stream decoders, including the forms of *corners*, *edges* and the *semantic planes* simultaneously (Fig. 3).

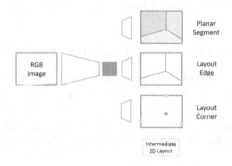

Fig. 3. The 2D layout estimation in multi-task network. The basic component is the ResNet101 back-bone network for feature extraction, and we make an independent decoders (up-sampling module) for each distinct targets, for planar segmentation, layout edge, and also layout corners.

Layout Segment. We refer the dense region segment to semantic planar described in [4] for five classes: front-wall, right-wall, left-wall, ceiling, and floor. It can then be formed as one semantic segmentation like problem with the labels on larger structure scale rather than object-level segments.

Layout Corner. RoomNet [13] estimates the corners for each possible layout structure in nearly fifties-channel heatmaps which is computationally inefficient. However, the corners labeled in captured scenes are given by two kinds of points, one is the room corner (*inner corner*), the other is the intersected points on the borders of image (*outer corner*), as illustrated in Fig. 4. In other words, we can categorize them into two classes rather than abundant channels nor as single one.

Layout Edge. The layout edge can be represented by the borders of the polygons. The detection of borders is to determine whether the pixel is the edges for the room layout. And it can be viewed as a binary classification problem.

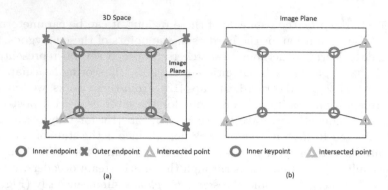

Fig. 4. (a) Showing each endpoint of the room cuboid and the projective image plane. (b) Indicating the projected points of inner and intersected ones.

3.2   Layout-Specific Objective Criterion

As proposed in [15], they find that if directly apply the vanilla semantic segmentation criterion on planar layout estimation, the result often suffers from distortion or tears apart from the center of planes and also "wavy curves" (rather than straight lines) mentioned in DeLay [4]. Hence, imposing extra smoothness criterion is necessary to alleviate the artifacts. The proposed loss function is given by,

$$\mathcal{L}_{seg}(x, target) = CE(x, target). \tag{1}$$

where x is the output of network for each single estimate, a five classes representation for semantic planar segmentation. And the smoothness term is given by,

$$\mathcal{L}_{smooth} = \ell(x, target) = |x - target|_1. \tag{2}$$

Loss of Corner and Edge Detection. The tasks for corner and edge detection can be viewed as binary classification on pixel-level, and thus the loss function can be given by the binary cross-entropy to determine whether one pixel belongs to a layout structure edge,

$$\mathcal{L}_{edge}(x, target) = BCE(x, target). \tag{3}$$

$$\mathcal{L}_{corner}(x, target) = \sum_i BCE(x_i, target_i). \tag{4}$$

For the corners, we categorize them into the inner and outer ones, the criterion would be the summed loss across the two-category corner maps.

Overall Loss. The criterion for the planar layout task is,

$$\mathcal{L}_{plane} = \mathcal{L}_{seg} + \lambda_s \mathcal{L}_{smooth}, \tag{5}$$

And the overall objective loss criterion for our network is the summation for these three branches. The overall loss function for model training is given by,

$$Loss_{Net2D} = \mathcal{L}_{plane} + \mathcal{L}_{edge} + \mathcal{L}_{corner}. \tag{6}$$

4   Layout Beyond Pixels

Most important of all, we further propose the second stage for 3D layout estimation in neural network, the novel approach compared to the existed works, by making use of those 2D intermediates from previous stage. The common rooms in the daily scenes are formed by the cuboid models. In the traditional works [6,14], they considered to model the 3D layout of various room scenes to be composed by boxes, and generated layout proposals based on 2D hand-craft cues and also optimization-based pipeline. We found that, however, in computational geometry, the 2D corners can be considered as the projection of 3D layout when depth information is reduced to the 2D space. Consequently, the task can then be converted to reconstruct the layout structure by estimating the projection of a cuboid, and we can then formulate the parameters for the transformation and projection. We thus parameterize the 3D layout in Sect. 4.1 with transformations and corresponding camera pose in the canonical 3D coordinate.

We make one neural network to predict the cuboid representation for the 3D layout in Sect. 4.2. However, we have no annotated 3D information for the supervised network, thus we resort to make use of the synthesized data with the strategy *abstract layout generation*, and then deliver the knowledge to the real case through transfer learning detailed in Sect. 4.3. With such formulation, we can estimates the 3D room layout with the representation of *projective parameters* from the 2D intermediate representation of layout estimated from stage one. And we now can make the 3D layout estimation framework end-to-end via deep networks.

4.1   Cuboid Model Parameterization

There are two components for the representation in our cuboid model, the scale of cuboid and camera pose. The parameters for the camera pose are decomposed into translation vector $\mathcal{T}$ and rotation matrix $\mathcal{R}$, in which we need three parameters for the position of camera and three parameters for the rotation angles along three coordinate axes, represented in quaternion. And three more parameters for the scaling along three axes of the unit box, template cube placing at the origin of the canonical space. Let $\mathcal{X}_{3D} \in \mathbb{R}^{3 \times N}$ denote the 3D coordinates of eight keypoints ($N = 8$) belonging to the unit box, and the locations of box keypoints viewed by a specific camera pose, and $\mathcal{X}_{2D} \in \mathbb{R}^{2 \times N}$ denotes the corresponding 2D coordinates in the image space. Thus the relationship between two coordinates is given by (Fig. 5)

$$\mathcal{X}_{2D} \equiv \pi \left(\mathcal{X}_{3D} | \mathcal{K}, P \right). \tag{7}$$

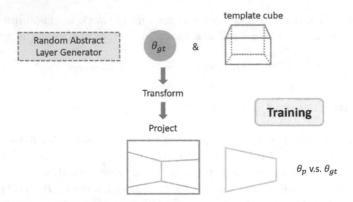

Fig. 5. The training of the regression model with the strategy of random generation of abstract layout. We can synthesize the paired samples, confident layout edge and the corresponding ground truth parameters θ_{gt}.

where $\mathcal{K}$ is the camera intrinsic matrix assumed to be given in the camera calibration procedure and $\mathcal{P}$ is the projection matrix given by

$$\mathcal{K} = \begin{bmatrix} f_x & 0 & c_x \\ 0 & f_y & c_y \\ 0 & 0 & 1 \end{bmatrix}, \tag{8}$$

$$P = [\mathcal{R}|\mathcal{T}] \in \mathbb{R}^{3 \times 4}. \tag{9}$$

Note that the rotation matrix $\mathcal{R} \in \mathbb{R}^{3 \times 3}$ and the translation vector $\mathcal{T} \in \mathbb{R}^3$ contain the extrinsic parameters for camera pose, and the rotation matrix is represented by a quaternion vector $\hat{q}$ as follows:

$$R = quat2mat(\hat{q}) \in \mathcal{SO}\,(3)\,, \tag{10}$$

Hence, we can extract the 3D cuboid layout from 2D space by estimating the *projective parameters* for the cuboid model.

4.2 Regression Forwarding Network

We formulate the task as one regression task by applying the CNN to learn these projective transformation parameters θ_t. Nevertheless, it is a challenge to train such a regression model since most of the datasets for the layout estimation do not provide any 3D annotations. The datasets for spatial layout are often annotated with 2D information such as the shapes of polygon of layout and the coordinates of the corners. And thus there is no easy way to retrieve 3D signals for supervised learning as one regression task.

So, we reformulate the problem as follows. The original task is to regress the model for the target parameters $\theta_t \in \mathbb{R}^9$ from the input $\in \mathbb{R}^{H \times W}$. Under this configuration, we can resort to the intermediate 2D layout representation

$\mathcal{E} \in \mathbb{R}^{H \times W}$, the estimate of 2D layout network in Sect. 3.1. The key value for the task decoupling is that the intermediate layout representation is easy to be synthesized by projecting the deformed cuboid onto the image plane, and we call it *abstract layout generation*. As a result we can acquire lots of reasonable samples through random generating target parameters θ_g as well as the corresponding 2D layout representation input $\equiv \mathcal{E}_g$ for the regression task, by using the transformation and projection modules described in Eq. 7.

With such a strategy, we can reform the ill-posed regression task and overcome the challenge of lacking 3D annotations in the existing datasets. The design of our regression network is composed of nine compounded layers of strided-convolutional layer with ReLU non-linearity activation and 1×1 convolutional layer acted as fully connected layers at the end of the network for the target projective parameters $\Theta \in \mathbb{R}^9$ of the cuboid layout representation.

4.3 End-to-End Learning Network

Besides the synthesized data, we need to make the trained regression model work on the real signals. under the configuration of our framework, the input of regression model is generic to the intermediate of previous stage, in the same space—2D layout representation. Though the estimated layout edge from 2D layout network is not as perfect as the one generated from ground truth, we can still make use of transfer learning strategy as Fig. 6 to make an extra network to learn and fit as close as to the model training on synthesized samples. Finally, we make an end-to-end framework for the 3D layout estimation via pure deep networks instead of any optimization or post-processing.

5 Experimental Results

We utilize LSUN Room Layout dataset [23], containing 4,000 training images, 394 validation, and 1,000 testing images, for evaluating 2D semantic planar segmentation and corner estimation results. Since there are no public labels for the testing set, we evaluate our method on the validation set with LSUN Room Layout official toolkit like the previous works. In addition, we evaluate the generalization capability of our model on the Hedau dataset [6], which is a challenging dataset due to its strict labeling. We can not evaluate any 3D accuracy metrics for our 3D layout estimation, for these two commonly used datasets do not contain any 3D annotations for the layout estimation. Instead, we also evaluate the 3D layout estimation results with 2D metrics on the re-projection of 3D layout.

Note that we only train our model on the training split of LSUN Room Layout and directly test on the testing split of Hedau dataset without fine-tuning on its training data. During the training, we apply random color jittering for slightly changing in the lightness and contrast of color images to increase the diversity of scenes. In addition, the time efficiency of our approach and other methods is also reported in our experiment.

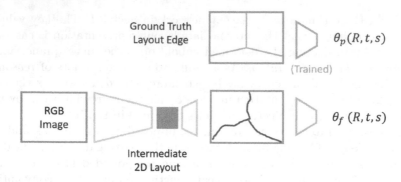

Fig. 6. The transfer learning pipeline for the 3D cuboid parameters estimate on real outputs of network.

5.1   Quantitative Results

We measure the performance of the proposed approach in 2D and 3D layout estimation through the following experimental evaluations: 2D pixel-wise accuracy for semantic planar segmentation in the single task and multi-task networks, 2D corner prediction accuracy for the keypoint corner detection, re-projected accuracy on 2D metrics on the estimated 3D projective parameters, and the visualization for the 3D cuboid rooms of the estimated parameters.

Pixel-Wise Accuracy of Layout Estimation. The performance of our layout estimation results are shown in Table 1. First, we take *DeepRoom 2D* for planar segmentation without any training strategies as our baseline model, and it can already achieve 9.75% error rate. And, the extended model *DeepRoom 2D multi-task* can reduce the error rate to 7.04%, which is 2.71% better than the baseline. Moreover, the performance of the ones trained with *Layout Degeneration* are comparable to the state-of-the-art method in the LSUN Challenge and we achieve 6.73% and 6.25% pixel-wise error rate for the single and multi-task networks, respectively. Furthermore, if we compare under a more fair condition, our proposed model can even beat the best performing method ST-PIO [26] (ST-PIO (2017) w/o optim.) without the extremely high-cost physical-inspired optimization but with the post-processing for proposal ranking.

We list the results from the direct 2D estimation networks and also the re-projected performance from the 3D parameter estimation network in Table 1. For the 3D projective parameters, *DeepRoom 3D*, which takes the ground truth generated edge map as input, can achieve similar performance as the 2D network in the metric of pixel-wise accuracy. Furthermore, the end-to-end approach of our *DeepRoom 2D/3D* achieves about 10% error rate, which is about similar level of the other state-of-the-art methods, LayoutNet [27], without post-processing.

Table 1. The pixel-wise accuracy performance benchmarking on LSUN Room Layout dataset for different approaches. Note that the data in the table is extracted from their papers.

Method	Pixel error (%)
Hedau [6]	24.23
DeLay [4]	10.63
CFILE [19]	7.95
RoomNet [13]	9.86
Zhang [25]	12.49
ST-PIO [26]	**5.48**
ST-PIO w/o optim. [26]	11.28
LayoutNet [27]	11.96
Ours 2D baseline, planar seg	9.75
Ours 2D multi-task	**6.73**
Ours 3D re-projected	6.89

5.2 Qualitative Results

First, we want to demonstrate the effects of our proposed layout objective criteria for layout segmentation. We show the visual outputs for our multi-stream networks with the full training strategies in Fig. 7. They mostly contain sharp but straight edges and strong consistencies in each predicted planar surface; and the inner-outer corner representation can successfully give the detection for the two kinds of keypoints in the layout. The detected layout edges are as impressive as the planar segmentation as they all produced by the same multi-task network.

For the evaluation on the estimated 3D room layout, we visualize the transformed cuboids along with the re-projected results in Fig. 7, in which we can see that the 3D layout estimation results are quite good only from a single image.

Table 2. The performance benchmarking on Hedau testing set.

Method	Pixel error (%)
Hedau [6]	21.20
Mallya [17]	12.83
DeLay [4]	9.73
CFILE [19]	8.67
RoomNet [13] recurrent 3-tier	8.36
Zhang [25]	12.70
ST-PIO [26]	**6.60**
DeepRoom (*ours*) 2D	**7.41**
DeepRoom (*ours*) 3D re-projected	9.97

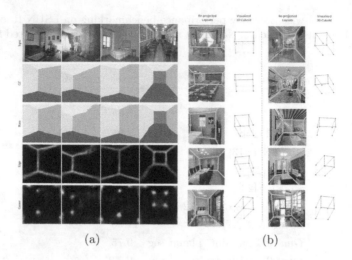

(a)   (b)

Fig. 7. (a) Some layout estimation results of the proposed multi-task network. (b) The representations for 3D cuboid and re-projected layout on LSUN Room.

In addition, we can observe that our model can be applied to different indoor datasets even without re-training. Table 2 shows that the accuracy of our model can almost achieve the state-of-the-art result.

5.3   Time Efficiency

Though our result is not overally the best for the 2D layout estimation metrics in the aforementioned two datasets, however, the most competitive advantage of our work is its computational efficiency since it is an end-to-end system without any optimization process or post-processing.

We implement our approach with PyTorch and perform all the experiments on the machine with single NVIDIA GeForce 1080 GPU and Intel i7-7700K 4.20 GHz CPU. For the analysis of time efficiency, Table 3 shows the consuming time for both network forwarding and post-processing time of the layout estimation methods. Although we cannot find fully released implementations of these papers, the listed entries in the column of the post-processing come from the official papers and cited ones, or the information from their released demo video. For the time consuming in the network forwarding column, several methods released their network configuration file for Caffe, and thus we can measure the time with official Caffe profiling tool and evaluate on our own machine under a fair competition.

Table 3. Comparison of time efficiency of the layout estimation methods in forwarding time and post-processing time (unit: seconds).

Method	Forward	Post-process	FPS
DeLay [4]	0.125	About 30	0.01
CFILE [19]	0.060	–	–
RoomNet [13]	0.168	–	5.96
Zhang [25]	–	About 180	–
ST-PIO [26]	0.067	About 10	0.1
LayoutNet [27]	0.039	0	25.64
DeepRoom (2D)	0.027	0	**36.95**
DeepRoom (2D/3D)	0.032	0	**31.25**

6   Conclusions

We proposed an end-to-end framework that is composed of two explainable networks for decoupling the 3D layout estimation task into two sub-tasks. They can also be jointly used to estimate the 3D cuboid representation of the spatial layout for the indoor scene. To the best of our knowledge, this is the first work that models the layout estimation as a two-stage deep learning forwarding pipeline instead of the conventional systems with an additional post-processing or optimization step. Furthermore, the combination of the two networks relies on the intermediate representation and it makes our framework pipeline open to the extensibility with using extra datasets for training and fine-tuning to achieve better outcomes.

References

1. Arth, C., Reitmayr, G., Schmalstieg, D.: Full 6DOF pose estimation from geo-located images. In: Lee, K.M., Matsushita, Y., Rehg, J.M., Hu, Z. (eds.) ACCV 2012. LNCS, vol. 7726, pp. 705–717. Springer, Heidelberg (2013). https://doi.org/10.1007/978-3-642-37431-9_54
2. Biederman, I.: Recognition-by-components: a theory of human image understanding. Psychol. Rev. **94**(2), 115 (1987)
3. Coughlan, J., Yuille, A.: Manhattan world: compass direction from a single image by bayesian inference. In: Proceedings of the Seventh IEEE International Conference on Computer Vision (1999). https://doi.org/10.1109/ICCV.1999.790349
4. Dasgupta, S., Fang, K., Chen, K., Savarese, S.: DeLay: robust spatial layout estimation for cluttered indoor scenes. In: Proceedings of the IEEE Conference on Computer Vision and Pattern Recognition, pp. 616–624 (2016)
5. Gupta, A., Hebert, M., Kanade, T., Blei, D.M.: Estimating spatial layout of rooms using volumetric reasoning about objects and surfaces. In: Advances in Neural Information Processing Systems, pp. 1288–1296 (2010)
6. Hedau, V., Hoiem, D., Forsyth, D.: Recovering the spatial layout of cluttered rooms. In: 2009 IEEE 12th International Conference on Computer vision, pp. 1849–1856. IEEE (2009)

7. Hoiem, D., Efros, A.A., Hebert, M.: Geometric context from a single image. In: 2005 Tenth IEEE International Conference on Computer Vision, ICCV 2005, vol. 1, pp. 654–661. IEEE (2005)
8. Hoiem, D., Efros, A.A., Hebert, M.: Recovering surface layout from an image. Int. J. Comput. Vis. **75**(1), 151–172 (2007)
9. Hoiem, D., Efros, A.A., Kanade, T.: Seeing the world behind the image: spatial layout for 3D scene understanding (2007)
10. Kendall, A., Cipolla, R.: Modelling uncertainty in deep learning for camera relocalization. In: 2016 IEEE International Conference on Robotics and Automation (ICRA), pp. 4762–4769. IEEE (2016)
11. Kendall, A., Cipolla, R.: Geometric loss functions for camera pose regression with deep learning. In: Proceedings of the CVPR, vol. 3, p. 8 (2017)
12. Kendall, A., Grimes, M., Cipolla, R.: PoseNet: a convolutional network for real-time 6-DOF camera relocalization. In: 2015 IEEE International Conference on Computer Vision (ICCV), pp. 2938–2946. IEEE (2015)
13. Lee, C.Y., Badrinarayanan, V., Malisiewicz, T., Rabinovich, A.: RoomNet: end-to-end room layout estimation. In: 2017 IEEE International Conference on Computer Vision (ICCV), pp. 4875–4884. IEEE (2017)
14. Lee, D.C., Hebert, M., Kanade, T.: Geometric reasoning for single image structure recovery. In: 2009 IEEE Conference on Computer Vision and Pattern Recognition, CVPR 2009, pp. 2136–2143. IEEE (2009)
15. Lin, H.J., Huang, S.W., Lai, S.H., Chiang, C.K.: Indoor scene layout estimation from a single image. In: 2018 24th International Conference on Pattern Recognition (ICPR) (2018)
16. Long, J., Shelhamer, E., Darrell, T.: Fully convolutional networks for semantic segmentation. In: Proceedings of the IEEE Conference on Computer Vision and Pattern Recognition, pp. 3431–3440 (2015)
17. Mallya, A., Lazebnik, S.: Learning informative edge maps for indoor scene layout prediction. In: Proceedings of the IEEE International Conference on Computer Vision, pp. 936–944 (2015)
18. Nowozin, S., Lampert, C.H., et al.: Structured learning and prediction in computer vision. Found. Trends® Comput. Graph. Vis. **6**(3–4), 185–365 (2011)
19. Ren, Y., Li, S., Chen, C., Kuo, C.-C.J.: A coarse-to-fine indoor layout estimation (CFILE) method. In: Lai, S.-H., Lepetit, V., Nishino, K., Sato, Y. (eds.) ACCV 2016. LNCS, vol. 10115, pp. 36–51. Springer, Cham (2017). https://doi.org/10.1007/978-3-319-54193-8_3
20. Roberts, L.G.: Machine perception of three-dimensional solids. Ph.D. thesis, Massachusetts Institute of Technology (1963)
21. Saxena, A., Chung, S.H., Ng, A.Y.: Learning depth from single monocular images. In: Advances in Neural Information Processing Systems, pp. 1161–1168 (2006)
22. Schwing, A.G., Urtasun, R.: Efficient exact inference for 3D indoor scene understanding. In: Fitzgibbon, A., Lazebnik, S., Perona, P., Sato, Y., Schmid, C. (eds.) ECCV 2012. LNCS, vol. 7577, pp. 299–313. Springer, Heidelberg (2012). https://doi.org/10.1007/978-3-642-33783-3_22
23. Princeton University: LSUN room layout estimation dataset (2015). http://lsun.cs.princeton.edu/. Accessed 30 Nov 2017
24. Waltz, D.: Understanding line drawings of scenes with shadows. In: Winston, P.H. (ed.) The Psychology of Computer Vision (1975)
25. Zhang, W., Zhang, W., Liu, K., Gu, J.: Learning to predict high-quality edge maps for room layout estimation. IEEE Trans. Multimed. **19**(5), 935–943 (2017)

26. Zhao, H., Lu, M., Yao, A., Guo, Y., Chen, Y., Zhang, L.: Physics inspired optimization on semantic transfer features: an alternative method for room layout estimation. In: Proceedings of the IEEE Conference on Computer Vision and Pattern Recognition (2017)
27. Zou, C., Colburn, A., Shan, Q., Hoiem, D.: LayoutNet: reconstructing the 3D room layout from a single RGB image. In: Proceedings of the IEEE Conference on Computer Vision and Pattern Recognition, pp. 2051–2059 (2018)

Multi-task Learning for Fine-Grained Eye Disease Prediction

Sahil Chelaramani[1(✉)], Manish Gupta[1(✉)] (iD), Vipul Agarwal[1],
Prashant Gupta[1], and Ranya Habash[2]

[1] Microsoft, Hyderabad, India
{sachelar,gmanish,vagarw,prgup}@microsoft.com
[2] Bascom Palmer Eye Institute, Miami, FL, USA
ranya@habash.net

Abstract. Recently, deep learning techniques have been widely used for medical image analysis. While there exists some work on deep learning for ophthalmology, there is little work on multi-disease predictions from retinal fundus images. Also, most of the work is based on small datasets. In this work, given a fundus image, we focus on three tasks related to eye disease prediction: (1) predicting one of the four broad disease categories – diabetic retinopathy, age-related macular degeneration, glaucoma, and melanoma, (2) predicting one of the 320 fine disease sub-categories, (3) generating a textual diagnosis. We model these three tasks under a multi-task learning setup using ResNet, a popular deep convolutional neural network architecture. Our experiments on a large dataset of 40658 images across 3502 patients provides ∼86% accuracy for task 1, ∼67% top-5 accuracy for task 2, and ∼32 BLEU for the diagnosis captioning task.

Keywords: Retinal imaging · Deep learning · Multi-task learning · Convolutional Neural Networks · Ophthalmology · Diagnosis caption generation

1 Introduction

Eye diseases significantly impact the quality of life for patients. Four of such diseases are glaucoma, diabetic retinopathy (DR), age-related macular degeneration (AMD), and uveal melanoma. About 6 to 67 million people have glaucoma globally. The disease affects about 2 million people in the United States[1]. In 2015, ∼415 million people were living with diabetes, of which ∼145 million have some of diabetic retinopathy[2]. Macular degeneration affected 6.2 million people globally[3]. Although it is a relatively rare disease, primarily found in the

---

[1] https://doi.org/10.1016/j.pop.2015.05.008, https://www.ncbi.nlm.nih.gov/pubmed/20711029.
[2] http://dx.doi.org/10.1016/S2214-109X(17)30393-5.
[3] https://doi.org/10.1016/S0140-6736(16)31678-6.

S. Palaiahnakote et al. (Eds.): ACPR 2019, LNCS 12047, pp. 734–749, 2020.
https://doi.org/10.1007/978-3-030-41299-9_57

Caucasian population, uveal melanoma is the most common primary intraocular tumor in adults with a mean age-adjusted incidence of 5.1 cases per million per year[4].

Motivation. Early diagnosis of these eye diseases can help in effective treatment or at least in avoiding further progression of these diseases. Limited availability of ophthalmologists, lack of awareness and consultation expenses restrict early diagnosis, especially in under-developed and developing countries. Hence, automated screening is critical. Deep learning techniques are very effective across text, speech and vision tasks. Recently, there has been a significant focus on applying deep learning techniques for medical imaging. In spite of the large variations and noise in medical imaging datasets, deep learning techniques have been successfully applied to multiple medical tasks. This motivated us to model the problems of disease classification and diagnosis generation for retinal fundus images.

Problem Definition

In this paper, we explore three critical tasks namely coarse-grained disease classification, fine-grained disease classification, and detailed disease diagnosis generation.

Given: A fundus image.

Predict: (1) One of the following classes: diabetic retinopathy, age-related macular degeneration, glaucoma, melanoma; (2) One of the 320 fine-grained disease classes; and (3) A detailed textual diagnosis.

Challenges

Disease prediction given fundus images is challenging because of the following:

- Gathering large amount of labeled data is difficult.
- Images have a lot of heterogeneity because of use of different lighting conditions, different devices, and different kinds of fundus images (like single field of view versus montage images).
- Sometimes, very minor changes in fundus images are indicative of particular diseases.
- There is no large corpus to learn robust embeddings for vocabulary of rather rare diagnosis keywords.
- Images in our dataset had variation in terms of field of view captured.
- Presence of artifacts like specular reflection, periphery haze, dust particle marks, fingerprints, blurred images, incorrectly stitched montage makes modeling complicated.
- Finally, images vary in terms of macula position – center, nasal, inferior, superior. Some images even have disc or the macula cut out.

Brief Overview of the Proposed Approach. The three proposed tasks are very related; hence we model them using a deep learning based multi-task learning setup. We jointly extract image description relevant to the three tasks using

---

[4] https://www.ncbi.nlm.nih.gov/pmc/articles/PMC5306463/.

deep convolutional neural networks (CNN). Using a fully connected hidden layer and a softmax layer appended to the output of the CNN, we make predictions for the first two tasks. For the diagnosis generation task, we attach a LSTM (Long Short-Term Memory network) decoder to the output of the CNN leading to a CNN-LSTM architecture. We use teacher forcing for the LSTM decoder and transfer learning to initialize CNN weights using ImageNet. The entire network is trained end-to-end using backpropagation with a linear combination of cross entropy loss across the three tasks. We experiment with a popular deep CNN model (ResNet [19]). We obtain high accuracies of 86.14 for task 1, 28.33 (top-1)/66.78 (top-5) for task 2 and 32.19 BLEU for task 3, making the proposed system practically usable as assistive AI. Using various case studies we show the effectiveness of the proposed approach. Figure 1 shows the architecture of our proposed disease prediction system.

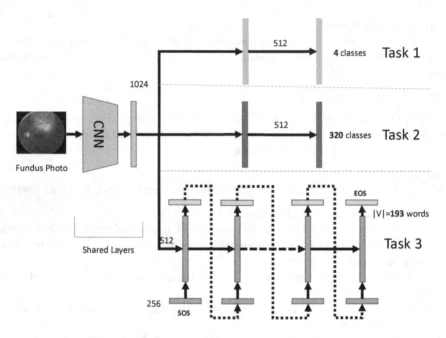

Fig. 1. Architecture for the proposed disease prediction system. Depicted are the shared layers of a CNN from which features are extracted and fed into the corresponding tasks.

Main Contributions

In summary, we make the following contributions in this paper.

- We propose three related critical tasks of (1) coarse-grained disease classification, (2) fine-grained disease classification and (3) detailed disease diagnosis generation.

- We model the first two tasks as multi-class classification problems, and the third one as image caption generation. Under a multi-task learning setup, we explore the effectiveness of ResNet, a popular deep convolutional network architecture for the three tasks, and an LSTM decoder for the third task. To the best of our knowledge, this is the first work to perform joint disease classification and diagnosis text generation for medical images.
- Using a dataset of 40658 fundus images for 3502 patients from 2004 to 2017, we show the effectiveness of the proposed methods. The code is made publicly available[5].

Paper Organization

The paper is organized as follows. We discuss related work in the areas of machine learning for eyecare, deep learning for medical imaging, and multi-task learning in Sect. 2. In Sect. 3, for each of the four eye diseases of focus, we discuss why diagnosing them early is critical, and the challenges in applying deep learning to fundus images corresponding to these diseases. In Sect. 4, we discuss various kinds of deep convolutional neural network methods that can be used for the task. In Sect. 5, we present dataset details, and also insights from analysis of results. Finally, we conclude with a summary in Sect. 6.

2   Related Work

In this section, we will first discuss interesting prediction problems on which researchers have worked in the field of ophthalmology. Further, we will discuss applications of deep learning specifically for medical imaging and also briefly discuss other popular papers that have shown the importance of multi-task learning with deep learning.

Machine Learning for Eyecare. Work on applying predictive analytics for eyecare includes the following: prediction of post-operative UCVA after LASIK surgeries [16], prediction of long term outcomes for glaucoma [33], discrimination between eyes with known glaucomatous progression and stable eyes [2], prediction of patients vision related quality of life from visual field and visual acuity [20], automated diagnosis of AMD [10], prediction of development of abnormal fields at follow-up in ocular hypertensive (OHT) eyes that had normal visual fields in baseline examination [35], retinal blood vessel segmentation [11], early diagnosis of diabetic retinopathy (DR) [15], automatic retina exudates segmentation, detecting microaneurysms in retina photographs, detecting exudates and cotton-wool spots in digital color fundus photographs and differentiating them from drusen for early diagnosis of diabetic retinopathy, classification of interferometry images pertaining to the dry eye syndrome, predicting whether a tumor in the eye is cancerous or dry, predicting if the eye tumor will spread to the liver [18], predicting posterior segment complications of cataract surgery or any

---

[5] https://github.com/SahilC/multitask-eye-disease-recognition.

other eye surgeries, predicting progression of epithelial ingrowth in patients following LASIK, and predicting clinical outcomes after intrastromal corneal ring segments implantation [39]. In this paper, we focus on the critical problem of screening patients concerning the four most important eye diseases using fundus images. Besides broad-level classification, we also attempt fine categorization, and automatic diagnosis caption generation.

Deep Learning for Medical Imaging. Motivated by immense success of deep learning techniques in general vision, speech as well as text problems, there has been a lot of focus on applying deep learning for medical imaging recently [14, 26]. Specifically, deep learning techniques have been applied to medical image data for neuro [31], retinal [15], pulmonary [23], digital pathology [21], breast [40], cardiac [1], abdominal [4], musculoskeletal [27] areas. Specifically, problem areas include image quality analysis [24], image segmentation [34], image/exam classification [41], object/lesion classification [9], and registration (i.e. spatial alignment) of medical images [6], image enhancement, and image reconstruction [17]. We focus on image classification and caption generation for retinal fundus images.

Multi-task Learning. In multi-task learning, multiple tasks are solved jointly, sharing inductive bias between them. Multi-task learning is inherently a multi-objective problem because different tasks may conflict, necessitating a trade-off. Multi-task learning has been used successfully across all applications of machine learning, from natural language processing [5] and speech recognition [7] to computer vision [13]. Multi-task learning can be done with soft [8] or hard parameter sharing [3]. To deal with specific problems in multi-task learning, several deep learning architectures have been proposed recently including Deep Relationship Networks [28], Fully-Adaptive Feature Sharing [29], and Cross-stitch Networks [30]. In this paper, we perform hard parameter sharing based multi-task learning across three eye diagnosis tasks.

3   Primer on Retinal Diseases

Diabetic retinopathy (DR) is a medical condition in which damage occurs to the retina due to diabetes mellitus. At least 90% of new cases could be reduced with proper treatment and monitoring of the eyes. If detected early, treatments like laser surgery, injection of corticosteroids or anti-VEGF agents into the eye can help avoid further progression. The only way to detect non-proliferative DR is by fundus photography, in which microaneurysms (microscopic blood-filled bulges in the artery walls) can be seen. Various kinds of anomalies in fundus images like microaneurysms, exudates, cotton wool spots, flame hemorrhages, dot-blot hemorrhages are indicative of the disease. Gulshan et al. [15] showed that deep learning techniques can predict DR with high accuracies.

Age-related macular degeneration (AMD) is a retinal condition which deteriorates the macula, the central portion of the visual field. Patients develop distortion and loss of central vision, which impacts their ability to recognize faces, drive, read, and perform normal activities of daily life. Since anti-VEGF medications and supplements may correct some damage and slow progression over time, early detection is critical. In AMD, fundus photos show progressive accumulation of characteristic yellow deposits, called drusen (buildup of extra-cellular proteins and lipids), in the macula (a part of the retina). Other image clues include geographic atrophy, increased pigment, and depigmentation. Lee et al. [25] show that deep learning is effective for classifying normal versus age-related macular degeneration OCT images.

Uveal melanoma is a cancer of the eye involving the iris, ciliary body, or choroid (collectively referred to as the uvea). When eye melanoma is spread to distant parts of the body, the five-year survival rate is about 15%. Melanoma tumors (with thickness around 2.5 mm) look like pigmented dome shaped mass that extends from the ciliary body or choroid. Fundus images of these lesions may have an orange lipofuscin pigmentation or subretinal fluid.

Glaucoma is a group of eye diseases which result in damage to the optic nerve and cause vision loss. Since this disease first affects the peripheral visual field and is otherwise asymptomatic, early detection and treatment are vital to saving nerve tissue. If treated early it is possible to slow or stop the progression of disease with medication, laser treatment, or surgery. Popular methods of detecting glaucoma from fundus images include analyzing cup to disc ratio, Inferior Superior Nasal Temporal (ISNT) features of cup and disc, and Optic Nerve Head atrophy. Fu et al. [12] adapt the U-Net [34] architecture specifically for optic disk segmentation, and hence glaucoma detection.

Overall, early detection is critical for each of these four diseases. For each disease, fundus images show clear distinguishable symptoms, making it plausible to diagnose them using deep learning methods on fundus images.

4   Approach

In this section, we first discuss how the three tasks can be modeled independently using popular deep learning architectures. Next, we discuss our proposed multi-task learning architecture.

4.1   Independent Task Learning

We pose the coarse grained classification (task 1) as a multi-class classification of four broad classes namely DR, AMD, melanoma and glaucoma. The fine-grained classification (task 2) is posed as a multi-class classification of 320 categories, where the coarse-grained classes have been divided further based on disease sub-types, severity of the diseases, regions of the eye involved, and specific visual symptoms. We model both tasks 1 and 2 using very deep CNN architectures. For these CNNs, we perform transfer learning, i.e., we initialize model weights using ImageNet pre-training to get a 1024-dimensional representation of a fundus image. This is then followed by two fully-connected layers with ReLU activation for the first layer and softmax activation for the second layer (of size 4 for task 1 and size 320 for task 2). We use cross entropy loss for both the tasks.

We pose the task of diagnosis generation (task 3) as an image captioning problem, where given a fundus image, we want the model to produce the textual description, describing the disease diagnosis in a detailed manner. The diagnosis generation task is parameterized by a single layered LSTM, which takes the features generated by the CNN encoder, and trains a language model conditioned on it. The features from the CNN encoder are projected to a suitable latent representation using a dense layer. This latent representation is then used to initialize the hidden state, along with passing the "start-of-sentence (SOS)" token as an input to our LSTM language model. The hidden state of each timestep is then projected to a vocabulary sized vector, on which softmax activation is applied. The softmax-normalized vector is then sampled to generate the corresponding word from the vocabulary. The generated word is then fed into the next timestep as an input. The LSTM language model is trained using teacher forcing to prevent training biases by setting the teacher forcing ratio to 0.5. We use the cross entropy loss averaged across all generated words for task 3.

For the CNN network, few architectures are popular like: AlexNet [22], Inception v3 [38], and VGGNet-19 [37]. Since ResNet has been to outperform other architectures for multiple vision tasks, we choose to use ResNet for our retinal disease prediction tasks.

4.2   Multi-task Learning

When multiple related prediction tasks need to be performed, multi-task learning (MTL) has been found to be very effective. Hard parameter sharing is the most commonly used approach to MTL in neural networks. It is generally applied by sharing the hidden layers between all tasks, while keeping several task-specific output layers. As shown in Fig. 1, we share the CNN encoder weights across all tasks. Task specific layers for each task are conditioned on the shared CNN encoder. For task 1 and 2, we use a fully-connected layer with ReLU, and then a softmax output layer. For task 3, we feed output of the CNN encoder to an LSTM as in the independent learning case. The final loss is computed as the weighted combination of the individual losses as shown in Eq. 1.

$$L = \lambda_1 L_1 + \lambda_2 L_2 + \lambda_3 L_3 \qquad (1)$$

where L_1, L_2 and L_3 are losses for tasks 1, 2 and 3 respectively, and λ_1, λ_2, and λ_3 are tunable hyper-parameters.

5   Experiments

5.1   Dataset

Our dataset consists of 40658 labeled fundus images corresponding to 3502 patients who visited the Bascom Palmer Eye Institute from 2004 to 2017, labeled by ~80 ophthalmologists. Labels are available at patient level, across four disease categories: DR, AMD, glaucoma or melanoma. Note that the original dataset consisted of normal eye images as well but those were filtered out so that we could focus more on images with anomalies. Of the 40658 images, by propagating patient level labels to individual images, 16418 are labeled as DR, 12878 as AMD, 6867 as glaucoma, and 4495 as melanoma. The fundus images have been captured using either OPTOS TX200 or OISCapture devices. The images vary in resolution and field of view. We resized all the images to a standard size of 224*224*3. We also normalized the pixel values to be between −1 and 1. While some images capture a small field of view, some others are montage images. Number of pictures in a montage also differ across images.

To obtain clean test data, we got 2436 (of the original 40658) images labeled manually. Thus, overall, we have the following dataset split: train (32489), validation (5733) and test (2436).

Each image is also labeled with a diagnosis description with an average length of 7.69 words. We filtered out diagnosis that occur less than 10 times. Our cleaned dataset contains 320 unique descriptions. These diagnosis contain a vocabulary of 237 words. Disease specific diagnosis vocabulary sizes are as follows: DR 130, AMD 59, Glaucoma 71, and Melanoma 15. Figure 2 shows the diagnosis caption length distribution in terms of words. Average length is 7.69 and the mode is 4 words. Of these 237 words, we retained the most frequent 193 words. Also, we set the maximum caption length to 14 and truncated longer captions. Finally, Table 1 shows top five most popular unigrams, bigrams and trigrams per disease.

5.2 Experimental Settings

Our multitask model is trained with stochastic gradient descent with Nesterov momentum. Our early experiments showed that Adam optimizer also performed similar to SGD. The model is trained using early stopping, which on average completes in 30 epochs. We set the learning rate for our model to 0.01, with momentum value of 0.9. Each model is trained on an Intel Xeon CPU machine with 32 cores and 2 Titan X GPUs (each with 12 GB RAM) with a batch size of 128. We split our dataset into sets of train, validation and test to 32489, 5733 and 2436 images respectively. We set λ_1, λ_2, λ_3 to 1 in Eq. 1. We additionally use an $L2$ weight regularization, whose coefficient we set to 10^{-6}. The teacher forcing ratio is set to 0.5. All experiments are conducted in PyTorch. For the LSTM decoder, we use embeddings of size 256, and hidden layer with 512 units. We learn embeddings dynamically and initialize with random embeddings. For details regarding the sizes of the weight matrices used in any model component refer to Fig. 1.

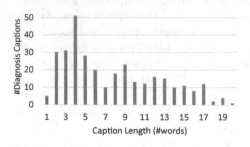

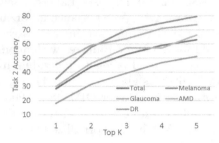

Fig. 2. Length distribution of captions (in words) in the original dataset

Fig. 3. Top-K accuracy values for fine-grained disease classification task across diseases for our best classifier.

Table 1. n-grams present in the disease diagnosis. The table highlights the most popular unigrams, bigrams and trigrams per disease.

	Diabetic retinopathy	AMD	Glaucoma	Melanoma
1-grams	retinopathy, diabetic, edema, macular, proliferative	macular, degeneration, age-related, nonexudative, retina	glaucoma, open-angle, stage, primary, severe	melanoma, malignant, choroidal, uvea, ciliary
2-grams	macular edema, diabetic retinopathy, diabetes mellitus, proliferative diabetic, type 2	macular degeneration, age-related macular, of retina, senile macular, degeneration of	glaucoma suspect, open-angle glaucoma, primary open-angle, severe stage, glaucoma severe	malignant melanoma, choroidal malignant, melanoma of, of uvea, ciliary body
3-grams	proliferative diabetic retinopathy, type 2 diabetes, 2 diabetes mellitus, associated with type, with type 2	age-related macular degeneration, senile macular degeneration, macular degeneration of, degeneration of retina, nonexudative senile macular	primary open-angle glaucoma, glaucoma severe stage, glaucoma moderate stage, open-angle glaucoma severe, glaucoma indeterminate stage	choroidal malignant melanoma, malignant melanoma of, melanoma of choroid, melanoma of uvea, melanoma of ciliary

5.3   Results

Broadly, our multi-task model mainly consists of two components – a shared representation across tasks, which is learned by a CNN and second, an independent task specific model. It is critical that since each task conditions on the shared representation that we select a suitable shared representation. We use Resnet-50 for this purpose since it has been shown to outperform several other CNN architectures across multiple visual recognition tasks. We leave exploration of other very deep CNN architectures as part of future work.

Table 2. Accuracy for MTL on different combinations of tasks using ResNet. N/A represents cases where evaluation does not make sense. T1 represents coarse grained classification, T2 represents fine-grained classification and T3 represents diagnosis generation.

MTL Train↓ Test→	T1 (Acc)	T2 (Acc)	T3 (BLEU)
T1	84.05	N/A	N/A
T2	N/A	26.11	N/A
T3	N/A	N/A	30.86
T1T2	84.29	26.97	N/A
T1T3	84.88	N/A	31.20
T2T3	N/A	26.55	31.22
T1T2T3	**86.14**	**28.33**	**32.19**

Table 2 highlights the performance on various tasks after training on a combination of them. This is the main result of our paper. Performance is measured using multi-class classification accuracy for task T1 and T2, and BLEU [32] score for task 3. Note that if we do not include a task in the MTL combination for training, we cannot use it for evaluation; hence some cells appear as "N/A" in the figure. From the results, it is clear that MTL with all the three tasks (T1, T2 and T3) leads to improved accuracies compared to independently learning for either of the three tasks. The combination leads to consistent gains across all tasks: 2.1, 2.2 and 1.33% points improvement for tasks 1, 2, and 3 respectively. MTL combinations of even two tasks usually leads to better accuracies compared to independent training.

5.4   Analysis

In Fig. 4, we present a qualitative analysis of our results. A set of 16 images are chosen at random from our test set. We generate predictions for these chosen images. These same images are submitted to a team of two ophthalmologists, who are asked to diagnose them. The fundus photo from the training set, the coarse grained model prediction, along with the generated diagnosis is presented

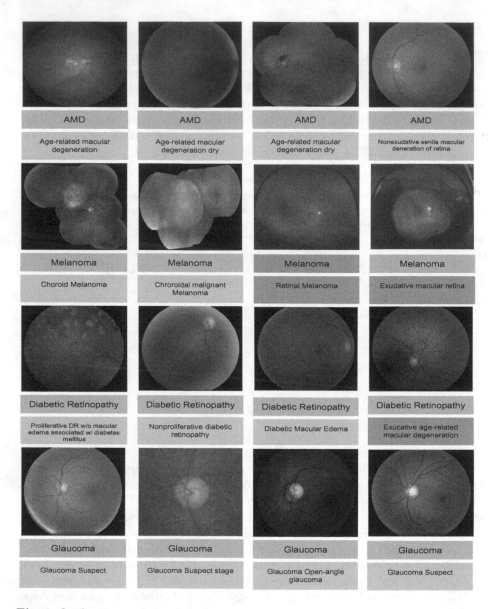

Fig. 4. Qualitative analysis of predictions made by proposed method. Errors made in coarse-grained and diagnosis generated are highlighted in red. (Color figure online)

in the figure. The disagreement between predictions and ophthalmologist's diagnoses are highlighted in red. The image in row 2, column 3 (2, 3) is a glaucoma patient and has been incorrectly labeled and captioned. For image (2, 4) the correct diagnosis is "choroidal melanoma", which has been incorrectly captioned. Image (3, 3) should have been diagnosed with "diabetic macular edema" but has

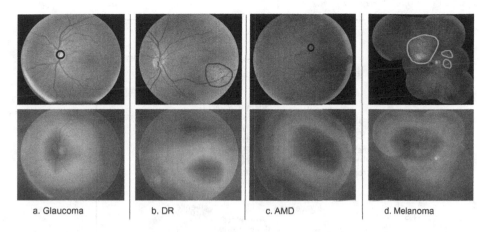

Fig. 5. Class activation mapping visualizations using Grad-CAM for the four different disease classes.

also been incorrectly captioned. We would like to highlight that while the model gets some diagnoses wrong, it does also manage to correctly capture a very fine-grained detailed description in the case of some of the other images like (3, 1).

Table 3 shows the confusion matrix for the best result for task T1. Interestingly the precision and recall across all classes is between 0.78 and 0.90, except for melanoma where the recall is as high as 0.94.

Figure 3 shows top-K accuracy values for the fine-grained disease classification task across diseases for our best classifier. As can be seen, the top-5 accuracy values for task-2 are ∼80% for melanoma but on average top-5 accuracy across all diseases is ∼67%. Further, we observed that for task 3, BLEU scores are good across all diseases: Melanoma (28.25), Glaucoma (43.60), AMD (31.87), and DR (28.62). BLEU scores remain similar across diseases even for larger sized captions.

Further, we used Grad-CAM [36] to visualize the regions of fundus image that are "important" for disease predictions from our best model from Table 2. Gradient-weighted Class Activation Mapping (Grad-CAM), uses the class-specific gradient information flowing into the final convolutional layer of a CNN to produce a coarse localization map of the important regions in the image. It captures how intensely the input image activates different channels by how important each channel is with regard to the class. Figure 5 shows class activation mapping visualizations for four images across the four diseases. The first row shows images with anomaly annotations by an ophthalmologist, pertaining to the corresponding disease. The second row shows class activation mappings obtained using Grad-CAM. We observe that the Grad-CAM activations highly correlate with expert annotations across all the four images.

Finally, in Fig. 6, we visualize the 2D t-SNE of the learned word embeddings by our output language model for diagnosis generation. We visualize a subset of

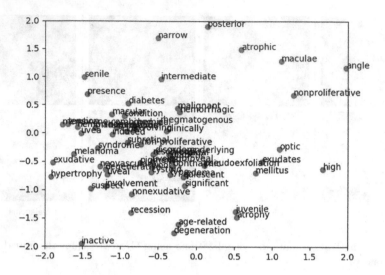

Fig. 6. t-SNE visualization of the learned word embeddings

Table 3. Confusion matrix for broad disease classification task

		Predicted			
		Melanoma	Glaucoma	AMD	DR
Actual	Melanoma	163	4	3	3
	Glaucoma	2	383	39	35
	AMD	11	29	737	50
	DR	31	31	76	837

the words after removing common language stop-words. As can be clearly seen in the plot, words such as "age-related", "degeneration", "atrophy", "juvenile" and "nonexudative" occur close together and are words generally associated with AMD. Words such as "diabetes", "proliferative", and "macular" are associated with Diabetic Retinopathy. Words such as "uvea", "melanoma", "malignant" occur close together. The cluster for Glaucoma is harder to discern, probably due to glaucoma having lesser training samples as compared to the others. Overall, we can infer that our model correctly learns to generate language by broadly grouping words associated with specific diseases.

6   Conclusion

In this paper we proposed the use of multi-task learning to improve fine-grained recognition of eye diseases using fundus images. We demonstrate that in such a setting, all three tasks benefit from such a training scheme, resulting in massive boosts in performance on all corresponding metrics. Such high performance models will contribute to early detection of diseases, and hence, be beneficial to providing early treatment to patients. In the future, we aim to explore additional auxiliary tasks to further improve our presented results. Also, we plan to experiment with images corresponding to more diseases.

References

1. Avendi, M.R., Kheradvar, A., Jafarkhani, H.: A combined deep-learning and deformable-model approach to fully automatic segmentation of the left ventricle in cardiac MRI. Med. Image Anal. **30**, 108–119 (2016)
2. Bowd, C., et al.: Glaucomatous patterns in frequency doubling technology (FDT) perimetry data identified by unsupervised machine learning classifiers. PLoS ONE **9**(1), e85941 (2014)
3. Caruana, R.: Multitask learning: a knowledge-based source of inductive bias. In: ICML, pp. 41–48 (1993)
4. Cheng, P.M., Malhi, H.S.: Transfer learning with convolutional neural networks for classification of abdominal ultrasound images. J. Digit. Imaging **30**(2), 234–243 (2017)
5. Collobert, R., Weston, J.: A unified architecture for natural language processing: deep neural networks with multitask learning. In: ICML, pp. 160–167 (2008)
6. de Vos, B.D., Berendsen, F.F., Viergever, M.A., Staring, M., Išgum, I.: End-to-end unsupervised deformable image registration with a convolutional neural network. In: Cardoso, M.J., et al. (eds.) DLMIA/ML-CDS -2017. LNCS, vol. 10553, pp. 204–212. Springer, Cham (2017). https://doi.org/10.1007/978-3-319-67558-9_24
7. Deng, L., Hinton, G., Kingsbury, B.: New types of deep neural network learning for speech recognition and related applications: an overview. In: ICASSP, pp. 8599–8603 (2013)
8. Duong, L., Cohn, T., Bird, S., Cook, P.: Low resource dependency parsing: cross-lingual parameter sharing in a neural network parser. In: IJCNLP, pp. 845–850 (2015)
9. Esteva, A., et al.: Dermatologist-level classification of skin cancer with deep neural networks. Nature **542**(7639), 115 (2017)
10. Fraccaro, P., et al.: Combining macula clinical signs and patient characteristics for age-related macular degeneration diagnosis: a machine learning approach. BMC Ophthalmol. **15**(1) (2015). Article number: 10
11. Fraz, M.M., et al.: An ensemble classification-based approach applied to retinal blood vessel segmentation. Biomed. Eng. **59**(9), 2538–2548 (2012)
12. Fu, H., et al.: Disc-aware ensemble network for glaucoma screening from fundus image. TMI **37**(11), 2493–2501 (2018)
13. Girshick, R.: Fast R-CNN. In: ICCV, pp. 1440–1448 (2015)
14. Greenspan, H., Van Ginneken, B., Summers, R.M.: Guest editorial deep learning in medical imaging: overview and future promise of an exciting new technique. TMI **35**(5), 1153–1159 (2016)
15. Gulshan, V., et al.: Development and validation of a deep learning algorithm for detection of diabetic retinopathy in retinal fundus photographs. JAMA **316**(22), 2402–2410 (2016)
16. Gupta, M., Gupta, P., Vaddavalli, P.K., Fatima, A.: Predicting post-operative visual acuity for LASIK surgeries. In: Bailey, J., Khan, L., Washio, T., Dobbie, G., Huang, J.Z., Wang, R. (eds.) PAKDD 2016. LNCS (LNAI), vol. 9651, pp. 489–501. Springer, Cham (2016). https://doi.org/10.1007/978-3-319-31753-3_39

17. Hammernik, K., et al.: Learning a variational network for reconstruction of accelerated MRI data. Magn. Reson. Med. **79**(6), 3055–3071 (2018)
18. Harbour, J.W.: Molecular prediction of time to metastasis from ocular melanoma fine needle aspirates. Clin. Cancer Res. **12**(19 Supplement), A77 (2006)
19. He, K., Zhang, X., Ren, S., Sun, J.: Deep residual learning for image recognition. In: Proceedings of the IEEE Conference on Computer Vision and Pattern Recognition, pp. 770–778 (2016)
20. Hirasawa, H., Murata, H., Mayama, C., Araie, M., Asaoka, R.: Evaluation of various machine learning methods to predict vision-related quality of life from visual field data and visual acuity in patients with glaucoma. Br. J. Ophthalmol. **98**(9), 1230–1235 (2014)
21. Janowczyk, A., Madabhushi, A.: Deep learning for digital pathology image analysis: a comprehensive tutorial with selected use cases. J. Pathol. Inform. **7**, 29 (2016)
22. Krizhevsky, A., Sutskever, I., Hinton, G.E.: ImageNet classification with deep convolutional neural networks. In: Advances in Neural Information Processing Systems, pp. 1097–1105 (2012)
23. Lakhani, P., Sundaram, B.: Deep learning at chest radiography: automated classification of pulmonary tuberculosis by using convolutional neural networks. Radiology **284**(2), 574–582 (2017)
24. Lalonde, M., Gagnon, L., Boucher, M.-C., et al.: Automatic visual quality assessment in optical fundus images. In: Vision Interface, vol. 32, pp. 259–264 (2001)
25. Lee, C.S., Baughman, D.M., Lee, A.Y.: Deep learning is effective for classifying normal versus age-related macular degeneration OCT images. Ophthalmol. Retin. **1**(4), 322–327 (2017)
26. Litjens, G., et al.: A survey on deep learning in medical image analysis. Med. Image Anal. **42**, 60–88 (2017)
27. Liu, F., Zhou, Z., Jang, H., Samsonov, A., Zhao, G., Kijowski, R.: Deep convolutional neural network and 3D deformable approach for tissue segmentation in musculoskeletal magnetic resonance imaging. Magn. Reson. Med. **79**(4), 2379–2391 (2018)
28. Long, M., Wang, J.: Learning multiple tasks with deep relationship networks. arXiv, 2 (2015)
29. Lu, Y., Kumar, A., Zhai, S., Cheng, Y., Javidi, T., Feris, R.: Fully-adaptive feature sharing in multi-task networks with applications in person attribute classification. In: CVPR, pp. 5334–5343 (2017)
30. Misra, I., Shrivastava, A., Gupta, A., Hebert, M.: Cross-stitch networks for multi-task learning. In: CVPR, pp. 3994–4003 (2016)
31. Nie, D., Zhang, H., Adeli, E., Liu, L., Shen, D.: 3D deep learning for multi-modal imaging-guided survival time prediction of brain tumor patients. In: Ourselin, S., Joskowicz, L., Sabuncu, M.R., Unal, G., Wells, W. (eds.) MICCAI 2016. LNCS, vol. 9901, pp. 212–220. Springer, Cham (2016). https://doi.org/10.1007/978-3-319-46723-8_25
32. Papineni, K., Roukos, S., Ward, T., Zhu, W.-J.: BLEU: a method for automatic evaluation of machine translation. In: ACL, pp. 311–318 (2002)
33. Rao, H.L., et al.: Accuracy of ordinary least squares and empirical bayes estimates of short term visual field progression rates to predict long term outcomes in glaucoma. Investig. Ophthalmol. Vis. Sci. **53**(14), 182 (2012)
34. Ronneberger, O., Fischer, P., Brox, T.: U-Net: convolutional networks for biomedical image segmentation. In: Navab, N., Hornegger, J., Wells, W.M., Frangi, A.F. (eds.) MICCAI 2015. LNCS, vol. 9351, pp. 234–241. Springer, Cham (2015). https://doi.org/10.1007/978-3-319-24574-4_28

35. Sample, P.A., et al.: Using machine learning classifiers to identify glaucomatous change earlier in standard visual fields. Investig. Ophthalmol. Vis. Sci. **43**(8), 2660–2665 (2002)
36. Selvaraju, R.R., Cogswell, M., Das, A., Vedantam, R., Parikh, D., Batra, D.: Grad-CAM: visual explanations from deep networks via gradient-based localization. In: ICCV, pp. 618–626 (2017)
37. Simonyan, K., Zisserman, A.: Very deep convolutional networks for large-scale image recognition. arXiv (2014)
38. Szegedy, C., Vanhoucke, V., Ioffe, S., Shlens, J., Wojna, Z.: Rethinking the inception architecture for computer vision. In: CVPR, pp. 2818–2826 (2016)
39. Torquetti, L., Ferrara, G., Ferrara, P.: Predictors of clinical outcomes after intrastromal corneal ring segments implantation. Int. J. Keratoconus Ectatic Corneal Dis. **1**, 26–30 (2012)
40. Jun, X., et al.: Stacked sparse autoencoder (SSAE) for nuclei detection on breast cancer histopathology images. TMI **35**(1), 119–130 (2015)
41. Xu, Y., et al.: Deep learning of feature representation with multiple instance learning for medical image analysis. In: ICASSP, pp. 1626–1630 (2014)

EEG-Based User Identification Using Channel-Wise Features

Longbin Jin, Jaeyoung Chang, and Eunyi Kim[(⊠)]

Department of Software at the Konkuk University, 120, Neungdong-ro, Gwangjin-gu, Seoul, Republic of Korea
{jinlongbin,martin4542,eykim}@konkuk.ac.kr

Abstract. During the last decades, the biometric signal such as the face, fingerprints, and iris has been widely employed to identify the individual. Recently, electroencephalograms (EEG)-based user identification has received much attention. Up to now, many types of research have focused on deep learning-based approaches, which involves high storage, power, and computing resource. In this paper, a novel EEG-based user identification method is presented that can provide real-time and accurate recognition with a low computing resource. The main novelty is to describe the unique EEG pattern of an individual by fusing the temporal domain of single-channel features and channel-wise information. The channel-wise features are defined by symmetric matrices, the element of which is calculated by the Pearson correlation coefficient between two-pair channels. Channel-wise features are input to the multi-layer perceptron (MLP) for classification. To assess the verity of the proposed identification method, two well-known datasets were chosen, where the proposed method shows the best average accuracies of 98.55% and 99.84% on the EEGMMIDB and DEAP dataset, respectively. The experimental results demonstrate the superiority of the proposed method in modeling the unique pattern of an individual's brainwave.

Keywords: Electroencephalograms (EEG) · User identification · Feature extraction · Channel-wise feature

1 Introduction

1.1 Background

The user identification has always been a challenging issue in the security area. It has many application areas such as access control systems, digital multimedia access, and transaction authentication. During the last decades, many studies have been investigated to identify a user from his unique biometrics such as the face, fingerprints, and iris. So far, many digital devices that were embedded with a biometric-based identification system have been developed. However, with the success of deep learning techniques in image synthesis and generation and pattern recognition, it is possible to imitate someone's biological signals.

To solve this problem, electroencephalograms (EEG) has been considered as an alternative for which can present an excellent potential for highly secure biometric-based

© Springer Nature Switzerland AG 2020
S. Palaiahnakote et al. (Eds.): ACPR 2019, LNCS 12047, pp. 750–762, 2020.
https://doi.org/10.1007/978-3-030-41299-9_58

user identification. As a typical kind of central nervous signal, the EEG signal directly reflects the strength and position of the brain activity with high temporal resolution. Besides, it has many advantages in universality, uniqueness, and robustness to spoofing attacks [1]. Benefit from many non-invasive and easy-to-wear brainwave measuring devices, it is easy to monitor electrical brain activity. Due to these advantages, EEG-based research has been carried out more actively.

1.2 Related Works

For EEG-based user identification, various signal processing and classification methods have been investigated. They have attempted to capture the unique features within a user and to identify the user based on them. Then, some methods have tried to extract the crucial features by human experts, and others have tried to extract from raw signals using deep learning techniques directly. So, the EEG-based user identification methods can mainly be divided into two approaches: deep features-based approaches and human crafted features-based ones. Table 1 summarizes the representative identification methods for the respective approaches.

Table 1. Representative works of EEG-based user identification

Approaches	Study	Features	Subjects
Deep features	Sun *et al.* [2]	CNN LSTM based features	109
	Mao *et al.* [3]	CNN	100
Human crafted features	Montri *et al.* [4]	Short-time Fourier transform	5
	Pinki *et al.* [5]	Daubechies wavelet transform	6
	Toshiaki *et al.* [6]	PCA and PLS	25
	Zaid *et al.* [7]	Wavelet Decomposition	7

Deep Features. Recently, feature learning has been applied to extract meaningful features for the raw data automatically to replace hand crafted features. These approaches were based on deep learning, especially the convolutional neural networks (CNN) and recurrent neural networks (RNN). Sun et al. [2] proposed a 1D convolutional RNN. The algorithm has been tested on a database with EEG data of 109 subjects and achieved an identification accuracy of 99.58%. Mao et al. [3] used a training procedure without the need for identifying or designing the EEG features manually. They have proposed a user identification method based on a CNN and achieved 97% accuracy in identifying different individuals. Although they can be universally used, the deep learning-based method requires a massive amount of training datasets and needs high computing resource in storage, power and computational complexity for training. To overcome the limitation, some paper uses human crafted features approaches.

Human Crafted Features. This approach focuses on finding distinctive features for user identification based on experience and knowledge. Toshiaki et al. [6] applied the principal component analysis (PCA) and partial least squares (PLS) for feature extraction of each EEG channel, which relies on classification algorithm latent dirichlet allocation (LDA) and quadratic discriminant analysis (QDA) to achieve accuracy of 72%. Montri et al. [4] processed EEG signals with short-time Fourier transform for each channel to modeling the unique pattern of a user. However, when compared to the deep features, the method showed low accuracy. Pinki et al. [5] utilized the Daubechies wavelet transform, whereas Zaid et al. [7] used wavelet decomposition to extract the feature of each EEG channel. Although their accuracy reached 97.5%, the number of subjects who participated in the experiments was too small to verify the universality of the system. As illustrated above, most of the recent works concentrated on modeling the unique pattern of the individual's EEG signal in the temporal resolution of a single channel. Relatively little attention has been employing the dependency between channels and analyzing their spatial interaction varying over time.

For practical application to the user identification, the system should satisfy some requirements: real-time, portable, and low computing resource. In spite of high accuracy, the system using deep features has limitations. Accordingly, a novel method of feature extraction for user identification is required.

1.3 Proposal

The purpose of this paper is to identify the individual quickly and accurately from their EEG signal. As mentioned above, most approaches in the literature have concentrated on analyzing the temporal information of the single channel. Although it is important information in modeling the uniqueness of the individual's EEG pattern, the interaction among different channels should also be considered. And the deep features from CNN take account for about 80% of computational costs during training stages. Therefore, we proposed a novel method to model an individual's EEG pattern that uses both time-series analyses and spatial correlation. To represent the spatial interactions between channels, a channel-wise feature is identified. It is a symmetric matrix, each element of which represents a correlation coefficient of two-pairwise channels. Accordingly, the proposed method is composed of two modules: feature extraction and classification which is shown in Fig. 1.

Two steps perform the feature extraction: (1) the single-channel features are first extracted, (2) then the channel-wise features are extracted. The single-channel features can extract the characteristic of the temporal domain and channel-wise features can explicitly model interdependencies between channels, and it shows the unique pattern of each user's EEG signal. The features are given to a classifier that discriminates each individual from their EEG features. Then a multi-layer perceptron (MLP) is adopted as classifier, because of its simplicity and efficiency. The proposed method was implemented on two public datasets, which are EEG Motor Movement/Imagery Dataset (EEG-MMIBD) [8] and Database for Emotion Analysis using Physiological Signals (DEAP) [9]. Then, it showed 98.55% and 99.84% accuracy on the respective datasets.

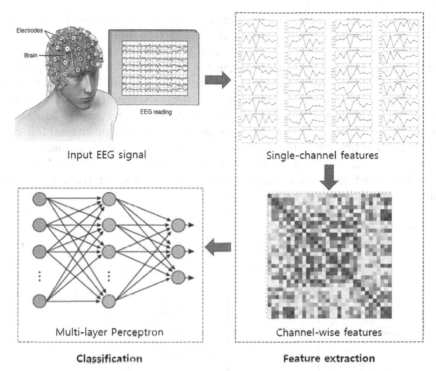

Classification   Feature extraction

Fig. 1. Overview of the proposed EEG-based user identification method

2 Feature Extraction

In this section, the methods of feature extraction are presented in detail. First, the feature extraction of the single-channel is introduced. Then the channel-wise feature is presented for modeling the interdependencies among channels.

2.1 Single-Channel Features

As illustrated above, the EEG signal directly reflects the strength and position of the brain activity with high temporal resolution. Accordingly, the features of the temporal domain can be easily extracted.

Let N denote the number of channels, T be the optimal times to be taken to obtain the EEG data of an individual, and R represents the sampling rate of the EEG signal. The EEGMMIDB contained 64 EEG channels ($N = 64$), then each signal was sampled at the rate of 160 Hz ($R = 160$). Accordingly, if we obtain the one minute ($T = 60$) EEG signal of an individual, the total amount of the raw data is $64 \times 160 \times 60$. In the case of the DEAP dataset, the 32 EEG channels ($N = 32$) were recorded by 128 Hz ($R = 128$). So the one minute ($T = 60$) EEG signal of an individual has a size of $32 \times 128 \times 60$ data. This raw data is likely to contain noisy data and too huge to apply for user identification.

To handle these problems and extract the temporal domain feature more effectively, the dimensionality reduction is processed. The EEG signal is first separated into several windows and then calculates the mean value of signals within a window. An overly wide window can lead to information overload, which causes the feature to be mixed up with other information. In the work of Candra et al. [10], they investigated the impact of various window size and then stated that the 3 s–10 s seconds' window size produced the better performance. We tested the window size of 3 s, 6 s, and 10 s. When using the window size of 10 s, it missed several peaks and valleys and resulted in the loss of some features. For the window size of 3 s and 6 s, the trend of data changes can be observed. However, when using the window size of 3 s, the data size of single-channel features is twice as much as which uses the window size of 6 s. So we used six seconds' window size in our method, and the data of each channel was split into 10 batches. Then the mean value of each batch was calculated. Finally, the EEG data was reduced to $N \times 10$ samples for each trial.

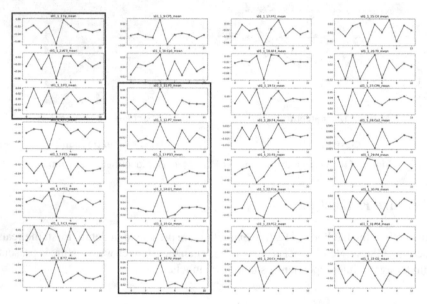

Fig. 2. Single-channel feature from 1st subject in the DEAP dataset

Figure 2 shows the extracted single-channel features from 1st subject in the DEAP dataset. Each subplot represents the changing trend of the single-channel features of one channel. The left half of the figure shows the EEG channels of the left human brain, and the right half of the figure shows the EEG channels of the right human brain. All the channel positions according to the 10–20 system. The horizontal axis of each subfigure represents time; the vertical axis represents the value of single-channel features, and they all change with time dynamically. As you can observe from Fig. 2, some channels have similar trends in data movement over time. For example, Fp1, AF3, F3 changes in similar patterns, P3, P7, PO3, O1, Oz, Pz are in similar patterns. On the contrary, some channels move conversely. For example, P3 and Fz, P7 and F4 are moves in reverse exactly. It means that some of the channels have a high correlation with each other, some of the channels have low interdependency, and other channels have no such relationship. It can prove that single-channel features can adequately describe the feature of the temporal domain. From such observations, we obtain the conclusion that the spatial interaction among channels should be considered to model the EEG pattern of an individual.

2.2 Channel-Wise Features

From Fig. 2, we can observe some interdependencies between channels which suggest that using the channel-wise feature is necessary to identify the individuals from their EEG more accurately. Here, *the channel-wise feature represents the interdependence between two channels, which is identified by the correlation of two pair channels.*

Then, we use the Pearson correlation coefficient to calculate the channel-wise feature. The Pearson correlation coefficient is a measure of the linear correlation between two variables X and Y. According to the Cauchy–Schwarz inequality, it has a value between $+1$ and -1, where 1 is a total positive linear correlation, 0 is no linear correlation, and -1 is total negative linear correlation [11].

These properties of the Pearson correlation coefficient can be used to quantify how similar the two channels change in patterns. Based on these properties, we compared every two pairwise channels of all N channels. Then the channel-wise features are computed as follows:

$$C_{i,j} = \frac{cov(S_i, S_j)}{\sigma_{S_i} \sigma_{S_j}}$$

where i, j refers to the channel number, and S_i and S_j are the single-channel features at the channel i and j. Accordingly, the channel-wise feature is described by $N \times N$ symmetric matrix.

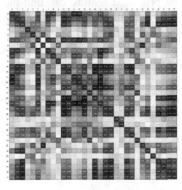

Fig. 3. Channel-wise feature of subject 1 in the DEAP dataset (Color figure online)

An example of the channel-wise feature is shown in Fig. 3, which is extracted from subject 1 in the DEAP dataset. As seen in the figure, the correlation is defined on two-pair channels, and it has different colors according to its strength of the correlation. If they have a strong positive correlation, the corresponding cell has a green color. If they have

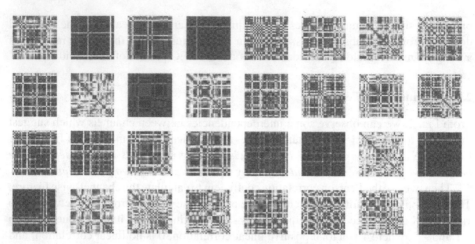

Fig. 4. Channel-wise features of 32 subjects in the DEAP dataset

a strong negative correlation, it is marked as a red color. Otherwise, if they have a week correlation, the cell has a white color.

Generally, a good feature should have a similar pattern for one class and different properties for different classes. To further examine the effectiveness of the proposed channel-wise features, we perform some preliminary experiments.

To observe its discriminative power, the channel-wise features were extracted from 32 subjects in the DEAP dataset. As shown in Fig. 4, although people may feel the same stimuli from watching the same video, the channel-wise feature varies from person to person. This result demonstrated that the proposed channel-wise feature could adequately describe the uniqueness of the respective individuals' EEG signals.

Besides, to examine if the proposed feature can find the common property within a subject, another experiment was performed. In the DEAP dataset, each subject was ordered to watch 40 music videos. Some examples of the channel-wise features for two different subjects are shown in Fig. 5. Figure 5(a) shows the features extracted from subject 1 and Fig. 5(b) shows the features for subject 22. Figures 5(a) and (b) consist of 40 channel-wise features, each of which is a channel-feature of EEG signal generated when each subject watches one music video. We can find that all channel-wise features of an individual from different stimuli are in similar patterns. It means each user has a unique pattern of EEG and demonstrates the channel-wise feature can correctly describe the uniqueness of an individual's brainwave.

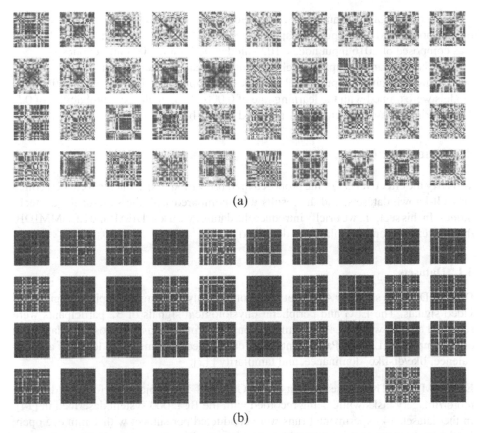

(a)

(b)

Fig. 5. (a) All the channel-wise features of subject 1 in the DEAP dataset. (b) All the channel-wise features of subject 22 in the DEAP dataset.

3   Classification

In this study, our goal is to develop a user identification system that can guarantee high accuracy with a low computing resource. For this, we present a novel EEG feature that has a high discriminative power in identifying individuals, which is called as channel-wise features. Based on these features, the classification was performed. Among many classification methods, the multi-layer perceptron was adopted in our system due to its simple architecture and low computational cost.

The channel-wise features were symmetric, so only the upper triangle of channel-wise features was flattened and input to the MLP and the others were discarded to reduce the input size and improve the computational time. The size of channel-wise features for 64 channels was 64×64, so the 2016 dimension vector was given to the MLP, and for 32 channels EEG signal, the size of channel-wise features are 32×32, so the input dimension was 496. The performance of the classifier was affected by some parameters such as the number of hidden layers and the number of their nodes at the respective layer. In this study, those parameters were chosen by experiments: for the DEAP data set that consists of 32 channels, the classifier was composed of 496 input nodes, 256 hidden nodes, and 32 output nodes; for the EMMIDB with 64 channels, the classifier was composed of 2016 input nodes, 1024 hidden nodes and 100 output nodes.

For the 64 channels input model, we found that stochastic gradient descent worked best, while 32 channels input model used the adam optimizer. Both of our identification models were trained with the learning rate of 0.0001 and the dropout probability of 0.5. Then during training, we randomly retrieved a mini-batch of size 64.

4   Experiments

To verify the effectiveness of the proposed method, various experiments were performed on well-known datasets, and the results were compared with the state-of-the-art techniques. In this section, we briefly introduce the dataset we used, DEAP and EEGMMIDB, then the experimental results were described.

4.1   Datasets

DEAP. DEAP is short for A dataset for emotion analysis using EEG, physiological and video signals. The EEG and peripheral physiological signals of 32 participants were recorded as each watched 40 one-minute-long excerpts of music videos. The data were downsampled to 128 Hz. Participants rated each video in terms of the levels of arousal, valence, like/dislike, dominance, and familiarity [12].

EEGMMIDB. The dataset consists of EEG data of 109 subjects performing different motor/imagery tasks while being recorded with the BCI2000 system described in [14]. In the dataset, 14 experimental runs were conducted per subject with 1-min eye open, 1-min eye close, and three sets of four tasks, including opening and closing fists and feet both physically and imaginarily. BCI2000 consists of 64 channels and the sampling frequencies were set to 160 for all channels. Table 2 illustrated two arrays for each of the 32 participants.

Table 2. Representation of EEGMMIDB and DEAP datasets

Datasets	Channels	Sampling Rate	Subjects	Trial
EEGMMIDB	64	160 Hz	109	12
DEAP	32	128 Hz	32	40

4.2   Experimental Results

Our goal is to develop an accurate EEG-based user identification system that is portable and flexible to various computing environments. For this, we presented channel-wise features that can efficiently describe the distinctive characteristics of an individual's EEG signal.

In this section, various experimental results were introduced. The performance of the proposed system was affected by several parameters: number of trials and number of channels. We changed these factors and analyzed their influence on performance. To verify the performance of our system, we used 10-fold validation.

First, the experiment was performed to prove the effectiveness of the proposed channel-wise features in user identification. For this, two methods were designed, one was to use only single-channel features, and the other was to use channel-wise features. Table 3 summarizes the accuracy of two methods for the respective dataset. The proposed method achieved the accuracy of 98.54% for EEGMMIDB and 99.84% for the DEAP dataset. When using single-channel features as input, the accuracies were much lower than the proposed model. For the EEG-based user identification, the most important thing is to find out the relationship between different channels. This is because the interdependency of different channels of each user is unique, and the channel-wise feature can well describe this interdependency. So the result states that using channel-wise features can significantly improve the accuracy of user identification systems.

Table 3. Performance comparison according to the features

Datasets	Channel-wise feature	Single-channel features
EEGMMIDB	98.55%	66.52%
DEAP	99.84%	87.10%

One of the greatest challenges when developing a machine learning-based classifier is to collect sufficient training data for the respective user. Accordingly, the proposed system should be working well on the fewer training dataset. To evaluate the robustness of the training data size, we performed the second experiment. We reduced the trial number in the train set and calculated the accuracy using the method of MLP. Figure 6 presents the accuracy of user identification when reducing the number of train trail.

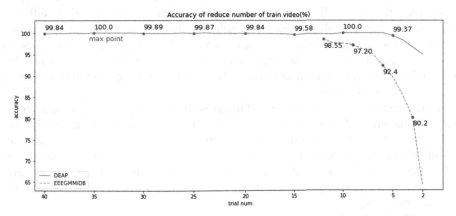

Fig. 6. Accuracy of user identification when reducing the number of train trials

As shown in Fig. 6, the proposed method can achieve high accuracy with relatively small training data. For the DEAP, when the training trial is less than 5, it still showed a good identification accuracy of 99.37%. Also, for the EEGMMIDB, the accuracy of using 9 trials was similar to the accuracy of using 12 trials. The reason for EEGMMIDB requires more trials than the DEAP dataset is there are 109 subjects in EEGMMIDB when there are only 32 subjects in DEAP datasets. To distinguish more users, the system needs more data to analyze the unique pattern of each user's EEG signal. However, the size of the data is much smaller than deep learning requires. So it can significantly reduce the training and testing time.

To measure the EEG signal of individuals more accurately, the device using more channels is required. For example, the Biosemi device with 32 channels and the device used in EEGMMIDB with 64 channels. It is natural for the more expensive device to guarantee higher performance. For real applications, the identification system should satisfy some requirements: high accuracy, real-time, and low price. Thus, the performance of the proposed method was also evaluated for various channel sizes. Here, 14 and 8 channel sizes were selected.

Table 4 shows the performance of the proposed method for various channel sizes.

Table 4. Accuracy of using the single-channel feature and channel-wise feature

Device	Channel	Accuracy on EEGMMIDB	Accuracy on DEAP
BCI2000	64	98.55%	–
Biosemi	32	94.52%	99.84%
Emotiv EPOC+	14	56.02%	99.52%
EEG IMEC	8	26.10%	97.11%

For DEAP, using the 8 channels in EEG IMEC still shows good accuracy with 97.11%. Also, for the EEGMMIDB, when using a 32 channels device, the accuracy of user identification reaches 94.52%. The reason why the two datasets show different results still has to do with the number of subjects. Just as longer indices are needed to identify more users, it needs more channels to identify a large number of people in a single system. The result shows that we can use the inexpensive device in real life to identify the appropriate number of users. But to identify a large number of users, it requires an expensive device.

Various experiments were performed to assess the validity of the proposed feature extraction method and the identification system. Through the whole of the experiments, we obtain the following conclusions: (1) the EEG channel-wise feature is adequate for user identification, (2) the proposed system showed robust results even with fewer data and channels, (3) due to the advantage of providing real-time and accurate recognition with a low computing resource, the proposed method has potential for practical applications.

4.3 Comparison

The results of the proposed identification system were compared with some of the EEG-based identification systems in Table 5. Sun et al. [2] proposed a 1D convolutional RNN. The algorithm has been tested on a database with EEG data of 109 subjects and achieved an identification accuracy of 99.58%. However, the use of deep learning would increase the initial cost of the user identification systems. Zaid et al. [7] used wavelet decomposition to extract the feature of each EEG channel. Although their accuracy reached 97.5%, the number of subjects who participated in the experiments was too small to verify the universality of the system. Fraschini et al. [13] proposed the use of eigenvector centrality of EEG signals as features to distinguish different subjects, and the authors also adopted the EEGMMIDB for evaluating the performance of the proposed biometric system. The system achieved an accuracy of 92.60% which is lower than the others. The proposed method showed acceptable accuracy, and it requires low computational cost than other deep learning models. This is mainly due to the fact that the proposed channel-wise feature can explicitly model an individual's distinctive brainwave characteristics.

Table 5. Comparison with some EEG-based identification systems

Study	Datasets	Subjects	Accuracy
Sun *et al.* [2]	EEGMMIDB	109	99.58%
Zaid *et al.* [7]	EEGMMIDB	7	97.50%
Fraschini *et al.* [13]	EEGMMIDB	109	92.60%
Proposed method	EEGMMIDB	109	98.55%
Proposed method	DEAP	32	99.84%

5 Conclusion

In this paper, single-channel features and channel-wise features of the spatiotemporal domain were extracted to model the uniqueness of the individual's EEG signal. Therefore, the low-cost classification method MLP has been used to identify the user. For experiments, two public datasets EEGMMIDB and DEAP datasets were used to evaluate the proposed method. The experimental results showed that the proposed method has a high average accuracy of 98.55% and 99.84% for the EEGMMIDB and DEAP dataset, respectively. Also, the proposed system showed robust results even with fewer data and channels. Moreover, this study offers some important insights into EEG-based user identification in practical application since it can provide real-time and accurate recognition with a low computing resource.

Acknowledgment. This research was supported by the MSIT (Ministry of Science, ICT), Korea, under the ITRC (Information Technology Research Center) support program (IITP-2019-2016-0-00465) supervised by the IITP (Institute for Information & communications Technology Planning

& Evaluation). And this research was supported by the MISP (Ministry of Science, ICT & Future Planning), Korea, under the National Program for Excellence in SW (No. 2018-0-00213, Konkuk University) supervised by the IITP (Institute of Information & communications Technology Planing & Evaluation)" (No.2018-0-00213, Konkuk University). Finally the authors would like to express our deepest gratitude to Ms. Jayoung Yang for her helpful comments and discussions.

References

1. Beijsterveldt, C., Boomsma, D.: Genetics of the human electroencephalogram (EEG) and event-related brain potentials (ERPs): a review. Hum. Genet. **94**(4), 319–330 (1994)
2. Sun, Y., Lo, F.P.-W., Lo, B.: EEG-based user identification system using 1D-convolutional long short-term memory neural networks. Expert Syst. Appl. **125**, 259–267 (2019)
3. Mao, Z., Yao, W.X., Huang, Y.: EEG-based biometric identification with deep learning. In: 2017 8th International IEEE/EMBS Conference on Neural Engineering, pp. 609–612 (2017)
4. Phothisonothai, M.: An investigation of using SSVEP for EEG-based user authentication system. In: Proceedings of the IEEE Asia-Pacific Signal and Information Processing Association Annual Summit and Conference, pp. 923–926 (2015)
5. Kumari, P., Vaish, A.: Brainwave based user identification system: a pilot study in robotics environment. Robot. Auton. Syst. **65**, 15–23 (2015)
6. Koike-Akino, T., et al.: High-accuracy user identification using EEG biometrics. In: 2016 IEEE 38th Annual International Conference of the Engineering in Medicine and Biology Society (EMBC), pp. 854–858 (2016)
7. Alyasseri, Z.A.A., Khader, A.T., Al-Betar, M.A., Papa, J.P., Alomari, O.A.: EEG feature extraction for person identification using wavelet decomposition and multi-objective flower pollination algorithm. IEEE Access **6**, 76007–76024 (2018)
8. Goldberger, A., et al.: Physiobank, physiotoolkit, and physionet: components of a new research resource for complex physiologic signals. Circulation **101**(23), e215–e220 (2000)
9. Koelstra, S., et al.: DEAP: a database for emotion analysis using physiological signals. IEEE Trans. Affect. Comput. **3**(1), 18–31 (2012)
10. Candra, H., et al.: Investigation of window size in classification of EEG-emotion signal with wavelet entropy and support vector machine. In: Proceedings of the 37th IEEE/EMBC, pp. 7250–7253 (2015)
11. Wikipedia page of Pearson correlation coefficient. https://en.wikipedia.org/wiki/Pearson_correlation_coefficient
12. DEAP dataset. https://www.eecs.qmul.ac.uk/mmv/datasets/deap/readme.html
13. Fraschini, M., Hillebrand, A., Demuru, M., Didaci, L., Marcialis, G.: An EEG-based biometric system using eigenvector centrality in resting state brain networks. IEEE Signal Process. Lett. **22**, 666–670 (2015)
14. EEG Motor Movement/Imagery Dataset. https://physionet.org/content/eegmmidb/1.0.0/

Author Index